BURTON'S

HISTORY OF MATHEMATICS:

An Introduction

Medieval and Renaissance Periods (Sixth Century to Sixteenth Century)

ca. 475–524	Boethius	636	Death of the Prophet Mohammed
ca. 476–550	Aryabhata	641	Library at Alexandria burned
ca. 625	Brahmagupta	732	Arabs defeated at Tours
ca. 826–901	Thabit-ibn-Korra	787	Start of the Viking Invasions
735–804	Alcuin of York	800	Charlemagne crowned Emperor
ca. 850–930	Abû Kâmil	ca. 1000	Leif Eriksson sails to America
ca. 1050–1130	Omar Khayyam	1066	Battle of Hastings
1114–1185	Bhaskara	1096–1099	The First Crusade
1114–1187	Gerard of Cremona	1200	University of Paris chartered
ca. 1120	Adelhard of Bath	1209	Cambridge University founded
ca. 1175–1250	Leonardo of Pisa	1215	Magna Carta signed
1436–1476	Regiomontanus	1258	Oxford University founded
1445–1514	Luca Pacioli	1271–1295	Travels of Marco Polo
1473–1543	Nicolas Copernicus	1293	Paper produced in Bologna
1494–1575	Francesco Maurolico	1337–1453	Hundred Years' War
1495–1552	Peter Apian	1348	Black Death in Europe
1500–1557	Niccolo Tartaglia	1440	Invention of Printing
1501–1576	Girolamo Cardan	1447	Founding of Vatican Library
1510–1558	Robert Recorde	1451	Death of Joan of Arc
1522–1565	Ludovico Ferrari	1453	Turks capture Constantinople
1526–1572	Raphael Bombelli	1472–1514	Leonardo da Vinci
1540–1603	François Vièta	1478	*Treviso Arithmetic*
1546–1601	Tycho Brache	1492	Columbus reaches West Indies
1548–1626	Pietro Cataldi	1513	Balboa discovers the Pacific
1550–1617	John Napier	1517	Luther's 95 Theses
1564–1642	Galileo Galilei	1564–1616	William Shakespeare
1571–1630	Johannes Kepler	1588	Defeat of Spanish Armada

Third Edition

BURTON'S

HISTORY OF
MATHEMATICS:
An Introduction

David M. Burton

University of New Hampshire

WCB **Wm. C. Brown Publishers**

Dubuque, IA Bogota Boston Buenos Aires Caracas Chicago
Guilford, CT London Madrid Mexico City Sydney Toronto

Book Team

Editor *Paula-Christy Heighton*
Developmental Editor *Daryl Bruflodt*
Production Editor *Carol M. Besler*
Designer *Kristyn A. Kalnes*
Art Editor *Jodi K. Banowetz*
Photo Editor *John C. Leland*

Wm. C. Brown Publishers
A Division of Wm. C. Brown Communications, Inc.

Vice President and General Manager *Beverly Kolz*
Vice President, Publisher *Earl McPeek*
Vice President, Director of Sales and Marketing *Virginia S. Moffat*
Vice President, Director of Production *Colleen A. Yonda*
National Sales Manager *Douglas J. DiNardo*
Marketing Manager *Julie Joyce Keck*
Advertising Manager *Janelle Keeffer*
Production Editorial Manager *Renée Menne*
Publishing Services Manager *Karen J. Slaght*
Royalty/Permissions Manager *Connie Allendorf*

Wm. C. Brown Communications, Inc.

President and Chief Executive Officer *G. Franklin Lewis*
Senior Vice President, Operations *James H. Higby*
Corporate Senior Vice President, President of WCB Manufacturing *Roger Meyer*
Corporate Senior Vice President and Chief Financial Officer *Robert Chesterman*

Copyedited by Lynn Brown

Cover Photo: The Bettmann Archive

Library of Congress Catalog Card Number: 94–70130

ISBN 0–697–16089–0

Printed in the United States of America by Wm. C. Brown Communications, Inc.,
2460 Kerper Boulevard, Dubuque, IA 52001

10 9 8 7 6 5 4 3 2 1

All these were honored in their generations, and were the glory of their times.

There be of them, that have left a name behind them, that their praises might be reported.

And some there be, which have no memorial; who are perished, as though they had never been; and are become as though they had never been born; and their children after them.

ECCLESIASTICUS 44:7–9

Contents

Preface

Since many excellent treatises on the history of mathematics are available, there may seem little reason for writing still another. But most current works are severely technical, written by mathematicians for other mathematicians or for historians of science. Despite the admirable scholarship and often clear presentation of these works, they are not especially well adapted to the undergraduate classroom. (Perhaps the most notable exception is Howard Eves's popular account, *An Introduction to the History of Mathematics*.) There seems to be room at this time for a textbook of tolerable length and balance addressed to the undergraduate student, which at the same time is accessible to the general reader interested in the history of mathematics.

In the following pages, I have tried to give a reasonably full account of how mathematics has developed over the past 5000 years. Because mathematics is one of the oldest intellectual instruments, it has a long story, interwoven with striking personalities and outstanding achievements. This narrative is basically chronological, beginning with the origin of mathematics in the great civilizations of antiquity and progressing through the first few decades of this century. The presentation necessarily becomes less complete for modern times, when the pace of discovery has been rapid and the subject matter more technical.

Considerable prominence has been assigned to the lives of the people responsible for progress in the mathematical enterprise. In emphasizing the biographical element, I can say only that there is no sphere in which individuals count for more than the intellectual life, and that most of the mathematicians cited here really did tower over their contemporaries. So that they will stand out as living figures and representatives of their day, it is necessary to pause from time to time to consider the social and cultural framework that animated their labors. I have especially tried to define why mathematical activity waxed and waned in different periods and in different countries.

Writers on the history of mathematics tend to be trapped between the desire to interject some genuine mathematics into a work and the desire to make the reading as painless and pleasant as possible. Believing that any mathematics textbook should concern itself primarily with teaching mathematical content, I have favored stressing the mathematics. Thus, assorted problems of varying degrees of difficulty have been interspersed throughout. Usually these problems typify a particular historical period, requiring the procedures of that time. They are an integral part of the text, and you will, in working them, learn some interesting mathematics as well as history. The level of maturity needed for this work is approximately the mathematical background of a college junior or senior. Readers with more extensive training in the subject must forgive certain explanations that seem unnecessary.

The title indicates that this book is in no way an encyclopedic enterprise. Neither does it pretend to present all the important mathematical ideas that arose during the vast sweep of time it covers. The inevitable limitations of space necessitate illuminating some outstanding landmarks instead of casting light of equal brilliance over the whole landscape. In keeping with this outlook, a certain amount of judgment and self-denial had to be exercised, both in choosing mathematicians and in treating their contributions. Nor was material selected exclusively on objective factors; some personal tastes and prejudices held sway. It stands to reason that not everyone will be satisfied with the choices. Some readers will raise an eyebrow at the omission of some household names of mathematics that have been either passed over in complete silence or shown no great hospitality; others will regard the scant treatment of their favorite topic as an unpardonable omission. Nevertheless, the path that I have pieced together should provide an adequate explanation of how mathematics came to occupy its position as a primary cultural force in Western civilization. The book is published in the modest hope that it may stimulate the reader to pursue the more elaborate works on the subject.

Anyone who ranges over such a well-cultivated field as the history of mathematics becomes so much in debt to the scholarship of others as to be virtually pauperized. The chapter bibliographies represent a partial listing of works, recent and not so recent, that in one way or another have helped my command of the facts. To the writers and to many others of whom no record was kept, I am enormously grateful.

New to This Edition

Readers familiar with the previous editions of *The History of Mathematics* will find that this edition maintains the same general organization and content. Nevertheless, the preparation of a third edition has provided the occasion for a variety of small improvements as well as several more significant ones. The largest change is a new chapter 13, ''Extensions and Generalizations: Hardy, Hausdorff, and Noether,'' which is devoted to some of the important aspects of twentieth-century mathematics. It focuses on the lives and works of such prominent mathematicians as G. H. Hardy, John Littlewood, Srinivasa Ramanujan, Maurice Fréchet, Felix Hausdorff, and Emmy Noether. Discussions of the Riemann Hypothesis and the Four-Color Problem are also included.

Another notable difference in this edition is the inclusion of several figures who were neglected in earlier editions. Specifically, material has been added concerning Zeno of Elea, Jordanus de Nemore, Simon Stevin, and George Boole. The present edition also pays increased attention to the influence of women on the development of mathematical thought by expanding the coverage of Mary Fairfax Somerville and Grace Chisholm Young.

Beyond these specific textual modifications, there are a number of relatively minor changes: further exercises have been introduced, bibliographies brought up-to-date, and certain numerical information kept current. Needless to say, an attempt has been made to correct all errors, typographical and historical, which crept into the original text.

Acknowledgments

I am indebted to a number of people in deciding which alterations would be desirable in this edition. Extensive and constructive comments were provided by:

Clinton M. Petty, University of Missouri–Columbia

Frank Kelly, University of New Mexico

Kenneth Shaw, Florida State University

Herb Wills, Florida State University

Many other friends, colleagues, and readers—too numerous to mention individually—were also kind enough to offer their suggestions. Although not all recommendations were adopted, they were all seriously considered in the revision process. Lastly, I am deeply grateful to my wife, Martha Beck Burton, for her constant encouragement and assistance in the project from beginning to end.

D. M. B.

CHAPTER 1

Early Number Systems and Symbols

To think the thinkable—that is the mathematician's aim.

C. J. KEYSER

1.1 Primitive Counting

The root of the term *mathematics* is in the Greek word *mathemata,* which was used quite generally in early writings to indicate any subject of instruction or study. As learning advanced, it was found convenient to restrict the scope of this term to particular fields of knowledge. The Pythagoreans are said to have used it to describe arithmetic and geometry; previously, each of these subjects had been called by its separate name, with no designation common to both. The Pythagoreans' use of the name would perhaps be a basis for the notion that mathematics began in Classical Greece during the years from 600 to 300 B.C. But its history can be followed much further back. Three or four thousand years ago, in ancient Egypt and Babylonia, there already existed a significant body of knowledge that we should describe as mathematics. If we take the broad view that mathematics involves the study of issues of a quantitative or spatial nature—number, size, order, and form—it is an activity that has been present from the earliest days of human experience. In every time and culture, there have been people with a compelling desire to comprehend and master the form of the natural world around them. To use Alexander Pope's words, ''This mighty maze is not without a plan.''

It is commonly accepted that mathematics originated with the practical problems of counting and recording numbers. The birth of the idea of number is so hidden behind the veil of countless ages that it is tantalizing to speculate on the remaining evidences of early humans' sense of number. Our remote ancestors of some 20,000 years ago— who were quite as clever as we are—must have felt the need to enumerate their livestock, tally objects for barter, or mark the passage of days. But the evolution of counting, with its spoken number words and written number symbols, was gradual and does not allow any determination of precise dates for its stages.

Anthropologists tell us that there has hardly been a culture, however primitive, that has not had some awareness of number, though it might have been as rudimentary as the distinction between one and two. Certain Australian aboriginal tribes, for instance, counted to two only, with any number larger than two called simply much or

many. South American Indians along the tributaries of the Amazon were equally des-
titute of number words. Although they ventured further than the aborigines in being
able to count to six, they had no independent number names for groups of three, four,
five, or six. In their counting vocabulary, three was called two-one, four was two-two,
and so on. A similar system has been reported for the Bushmen of South Africa, who
counted to ten ($10 = 2 + 2 + 2 + 2 + 2$) with just two words; beyond ten, the
descriptive phrases became too long. It is notable that such tribal groups would not
willingly trade, say, two cows for four pigs, yet had no hesitation in exchanging one
cow for two pigs and a second cow for another two pigs.

The earliest and most immediate technique for visibly expressing the idea of
number is tallying. The idea in tallying is to match the collection to be counted with
some easily employed set of objects—in the case of our early forebears, these were
fingers, shells, or stones. Sheep, for instance, could be counted by driving them one
by one through a narrow passage while dropping a pebble for each. As the flock was
gathered in for the night, the pebbles were moved from one pile to another until all the
sheep had been accounted for. On the occasion of a victory, a treaty, or the founding
of a village, frequently a cairn, or pillar of stones, was erected with one stone for each
person present.

The term *tally* comes from the French verb *tailler,* ''to cut,'' like the English word
tailor; the root is seen in the Latin *taliare,* meaning ''to cut.'' It is also interesting to
note that the English word *write* can be traced to the Anglo-Saxon *writan,* ''to scratch,''
or ''to notch.''

Neither the spoken numbers nor finger tallying have any permanence, although
finger counting shares the visual quality of written numerals. To preserve the record
of any count, it was necessary to have other representations. We should recognize as
human intellectual progress the idea of making a correspondence between the events
or objects recorded and a series of marks on some suitably permanent material, with
one mark representing each individual item. The change from counting by assembling
collections of physical objects to counting by making collections of marks on one
object is a long step, not only toward abstract number concept, but also toward written
communication.

Counts were maintained by making scratches on stones, by cutting notches in
wooden sticks or pieces of bone, or by tying knots in strings of different colors or
lengths. When the numbers of tally marks became too unwieldy to visualize, primitive
people arranged them in easily recognizable groups such as groups of five, for the
fingers of a hand. It is likely that grouping by pairs came first, soon abandoned in favor
of groups of five, ten, or twenty. The organization of counting by groups was a noteworthy improvement on counting by ones. The practice of counting by fives, say, shows
a tentative sort of progress towards reaching an abstract concept of ''five'' as contrasted
with the descriptive ideas ''five fingers'' or ''five days.'' To be sure, it was a timid
step in the long journey toward detaching the number sequence from the objects
being counted.

Bone artifacts bearing incised markings seem to indicate that the people of the Old
Stone Age had devised a system of tallying by groups as early as 30,000 B.C. The most
impressive example is a shinbone from a young wolf, found in Czechoslovakia in 1937;
about 7 inches long, the bone is engraved with 55 deeply cut notches, more or less
equal in length, arranged in groups of 5. (Similar recording notations are still used,

with the strokes bundled in fives, like 卌. Voting results in small towns are still counted in the manner devised by our remote ancestors.) For many years such notched bones were interpreted as hunting tallies and the incisions were thought to represent kills. A more recent theory, however, is that the first recordings of ancient people were concerned with reckoning time. The markings on bones discovered in French cave sites in the late 1880s are grouped in sequences of recurring numbers that agree with the numbers of days included in successive phases of the moon. One might argue that these incised bones represent lunar calendars.

Another arresting example of an incised bone was unearthed at Ishango along the shores of Lake Edward, one of the headwater sources of the Nile. The best archeological and geological evidence dates the site to 8500 B.C., or some 3000 years before the first settled agrarian communities appeared in the Nile valley. This fossil fragment was probably the handle of a tool used for engraving, or tattooing, or even writing in some way. It contains groups of notches arranged in three definite columns; the odd, unbalanced composition does not seem to be decorative. In one of the columns, the groups are composed of 11, 21, 19, and 9 notches. The underlying pattern may be $10 + 1$, $20 + 1$, $20 - 1$, and $10 - 1$. The notches in another column occur in 8 groups, in the following order: 3, 6, 4, 8, 10, 5, 5, 7. This arrangement seems to suggest an appreciation of the concept of duplication, or multiplying by two. The last column has 4 groups consisting of 11, 13, 17, and 19 individual notches. The pattern here may be fortuitous and does not necessarily indicate—as some authorities are wont to infer—a familiarity with prime numbers. Because $11 + 13 + 17 + 19 = 60$ and $11 + 21 + 19 + 9 = 60$, it might be argued that markings on the prehistoric Ishango bone are related to a lunar count, with the first and third columns indicating two lunar months.

A method of tallying that has been used in many different times and places involves the notched stick. Although this device provided one of the earliest forms of keeping records, its use was by no means limited to "primitive peoples," or for that matter, to the remote past. The acceptance of tally sticks as promissory notes or bills of exchange reached its highest level of development in the British Exchequer tallies, which formed an essential part of the government records from the twelfth century onward. In this instance, the tallies were flat pieces of hazelwood about 6–9 inches long and up to an inch thick. Notches of varying sizes and types were cut in the tallies, each notch representing a fixed amount of money. The width of the cut decided its value. For example, the notch of £1000 was as large as the width of a hand; for £100, as large as the thickness of a thumb; and for £20, the width of the little finger. When a loan was made, the appropriate notches were cut and the stick split into two pieces so that the notches appeared in each section. The debtor kept one piece and the Exchequer kept the other, so the transaction could easily be verified by fitting the two halves together and noticing whether the notches coincided (whence the expression "our accounts tallied"). Presumably, when the two halves had been matched, the Exchequer destroyed its section—either by burning it or by making it smooth again by cutting off the notches—but retained the debtor's section for future record. Obstinate adherence to custom kept this wooden accounting system in official use long after the rise of banking institutions and modern numeration had made its practice quaintly obsolete. It took an act of Parliament, which went into effect in 1826, to abolish the practice. In 1834, when the

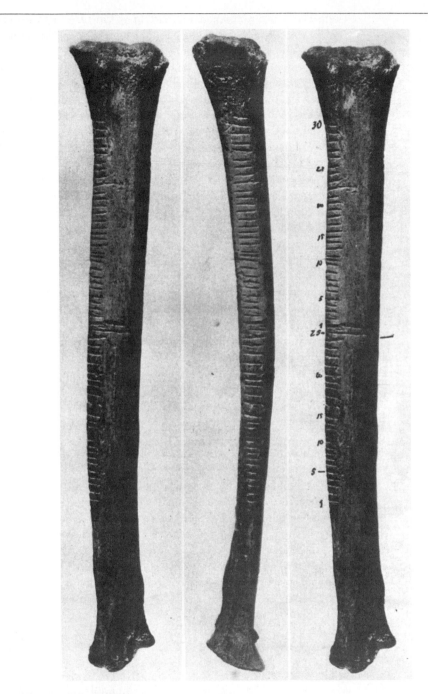

Three views of a Paleolithic wolfbone used for tallying. (The Illustrated London News Picture Library.)

long-accumulated tallies were burned in the furnaces that heated the House of Lords, the fire got out of hand, starting a more general conflagration that destroyed the old Houses of Parliament.

The English language has taken note of the peculiar quality of the double tally stick. Formerly, if someone lent money to the Bank of England, the amount was cut on a tally stick, which was then split. The piece retained by the bank was known as the foil, whereas the other half, known as the stock, was given the lender as a receipt for the sum of money paid in. Thus, he became a "stockholder" and owned "bank stock" having the same worth as paper money issued by the government. When the holder would return, the stock was carefully checked and compared against the foil in the bank's possession; if they agreed, the owner's piece would be redeemed in currency. Hence, a written certificate that was presented for remittance and checked against its security later came to be called a "check."

Using wooden tallies for records of obligations was common in most European countries and continued there until fairly recently. Early in this century, for instance, in some remote valleys of Switzerland, "milk sticks" provided evidence of transactions among farmers who owned cows in a common herd. Each day the chief herdsman would carve a six- or seven-sided rod of ashwood, coloring it with red chalk so that incised lines would stand out vividly. Below the personal symbol of each farmer, the herdsman marked off the amounts of milk, butter, and cheese yielded by a farmer's cows. Every Sunday after church, all parties would meet and settle the accounts. Tally sticks—in particular, double tallies—were recognized as legally valid documents until well into the 1800s. France's first modern code of law, the Code Civil, promulgated by Napoleon in 1804, contained the provision:

> The tally sticks which match their stocks have the force of contracts between persons who are accustomed to declare in this manner the deliveries they have made or received.

The variety in practical methods of tallying is so great that giving any detailed account would be impossible here. But the procedure of counting both days and objects by means of knots tied in cords has such a long tradition that it is worth mentioning. The device was frequently used in ancient Greece, and we find reference to it in the work of Herodotus (fifth century B.C.). Commenting in his *History,* he informs us that the Persian king Darius handed the Ionians a knotted cord to serve as a calendar:

> The King took a leather thong and tying sixty knots in it called together the Ionian tyrants and spoke thus to them: "Untie every day one of the knots; if I do not return before the last day to which the knots will hold out, then leave your station and return to your several homes."

In the New World, the number string is best illustrated by the knotted cords, called *quipus,* of the Incas of Peru. When the Spanish conquerors arrived in the sixteenth century, they observed that each city in Peru had an "official of the knots," who maintained complex accounts by means of knots and loops in strands of various colors. Performing duties not unlike those of the city treasurer of today, the quipu keepers recorded all official transactions concerning the land and subjects of the city and submitted the strings to the central government in Cuzco. The quipus were important in the Inca Empire, because apart from these knots no system of writing was ever developed there. The quipu was made of a thick main cord or crossbar to which were attached

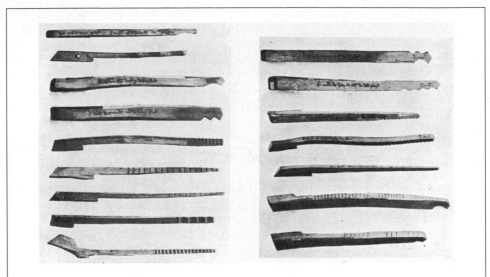

Thirteenth-century British Exchequer tallies. (By courtesy of the Society of Antiquaries of London.)

finer cords of different lengths and colors; ordinarily the cords hung down like the strands of a mop. Each of the pendent strings represented a certain item to be tallied; one might be used to show the number of sheep, for instance, another for goats, and a third for lambs. The knots themselves indicated numbers, the values of which varied according to the type of knot used and its specific position on the strand. A decimal system was used, with the knot representing units placed nearest the bottom, the tens appearing immediately above, then the hundreds, and so on; absence of a knot denoted zero. Bunches of cords were tied off by a single main thread, a summation cord, whose knots gave the total count for each bunch. The range of possibilities for numerical representation in the quipus allowed the Incas to keep incredibly detailed administrative records, despite their ignorance of the written word. More recent (1872) evidence of knots as a counting device occurs in India; some of the Santal headsmen, being illiterate, made knots in strings of four different colors to maintain an up-to-date census.

Over the long sweep of history, it seems clear that progress in devising efficient ways of retaining and conveying numerical information did not take place until primitive people abandoned the nomadic life. Incised markings on bone or stone may have been adequate for keeping records when human beings were hunters and gatherers, but the food producer required entirely new forms of numerical representation. Besides, as a means for storing information, groups of markings on a bone would have been intelligible only to the person making them, or perhaps to close friends or relatives; thus, the record was probably not intended to be used by people separated by great distances.

Deliberate cultivation of crops, particularly cereal grains, and the domestication of animals began, so far as can be judged from present evidence, in the Near East some 10,000 years ago. Later experiments in agriculture occurred in China and in the New

World. A widely held theory is that a climatic change at the end of the last ice age provided the essential stimulus for the introduction of food production and a settled village existence. As the polar ice cap began to retreat, the rain belt moved northward, causing the desiccation of much of the Near East. The increasing scarcity of wild food plants and the game on which people had lived forced them, as a condition of survival, to change to an agricultural life. It became necessary to count one's harvest and herd, to measure land, and to devise a calendar that would indicate the proper time to plant crops. Even at this stage, the need for means of counting was modest; and tallying techniques, although slow and cumbersome, were still adequate for ordinary dealings. But with a more secure food supply came the possibility of a considerable increase in population, which meant that larger collections of objects had to be enumerated. Repetition of some fundamental mark to record a tally led to inconvenient numeral representations, tedious to compose and difficult to interpret. The desire of village, temple, and palace officials to maintain meticulous records (if only for the purposes of systematic taxation) gave further impetus to finding new and more refined means of "fixing" a count in a permanent or semipermanent form.

Thus, it was in the more elaborate life of those societies that rose to power some 6000 years ago in the broad river valleys of the Nile, the Tigris-Euphrates, the Indus, and the Yangtze that special symbols for numbers first appeared. From these, some of our most elementary branches of mathematics arose, because a symbolism that would allow expressing large numbers in written numerals was an essential prerequisite for computation and measurement. Through a welter of practical experience with number symbols, people gradually recognized certain abstract principles; for instance, it was discovered that in the fundamental operation of addition, the sum did not depend on the order of the summands. Such discoveries were hardly the work of a single individual, or even a single culture, but more a slow process of awareness moving toward an increasingly abstract way of thinking.

We shall begin by considering the numeration systems of the important Near Eastern civilizations—the Egyptian and the Babylonian—from which sprang the main line of our own mathematical development. Number words are found among the word forms of the earliest extant writings of these people. Indeed, their use of symbols for numbers, detached from an association with the objects to be counted, was a big turning point in the history of civilization. It is more than likely to have been a first step in the evolution of humans' supreme intellectual achievement, the art of writing. Because the recording of quantities came more easily than the visual symbolization of speech, there is unmistakable evidence that the written languages of these ancient cultures grew out of their previously written number systems.

1.2 Number Recording of the Egyptians and Greeks

The writing of history, as we understand it, is a Greek invention; and foremost among the early Greek historians was Herodotus. Herodotus (circa 485–430 B.C.) was born at Halicarnassus, a largely Greek settlement on the southwest coast of Asia Minor. In early life he was involved in political troubles in his home city and forced to flee in exile to the island of Samos, and thence to Athens. From there Herodotus set out on travels whose leisurely character and broad extent indicate that they occupied many years. It is assumed that he made three principal journeys, perhaps as a merchant, collecting material and recording his impressions. In the Black Sea, he sailed all the way up the west coast to the Greek communities at the mouth of the Dnieper River, in what is now Russia, and then along the south coast to the foot of the Caucasus. In Asia Minor, he traversed modern Syria and Iraq, and traveled down the Euphrates, possibly as far as Babylon. In Egypt, he ascended the Nile River from its delta to somewhere near Aswan, exploring the pyramids along the way. Around 443 B.C., Herodotus became a citizen of Thurium in southern Italy, a new colony planted under Athenian auspices. In Thurium, he seems to have passed the last years of his life involved almost entirely in finishing the *History of Herodotus,* a book larger than any Greek prose work before it. The reputation of Herodotus as a historian stood high even in his own day. In the absence of numerous copies of books, it is natural that a history, like other literary compositions, should have been read aloud at public and private gatherings. In Athens, some twenty years before his death, Herodotus recited completed portions of his *History* to admiring audiences and, we are told, was voted an unprecedentedly large sum of public money in recognition of the merit of his work.

Although the story of the Persian Wars provides the connecting link in the *History of Herodotus,* the work is no mere chronicle of carefully recorded events. Almost anything that concerned people interested Herodotus, and his *History* is a vast store of information on all manner of details of daily life. He contrived to set before his compatriots a general picture of the known world, of its various peoples, of their lands and cities, and of what they did and above all why they did it. (A modern historian would probably describe the *History* as a guidebook containing useful sociological and anthropological data, instead of a work of history.) The object of his *History,* as Herodotus conceived it, required him to tell all he had heard but not necessarily to accept it all as fact. He flatly stated, ''My job is to report what people say, not to believe it all, and this principle is meant to apply to my whole work.'' We find him, accordingly, giving the traditional account of an occurrence and then offering his own interpretation or a contradictory one from a different source, leaving the reader to choose between versions. One point must be clear: Herodotus interpreted the state of the world at his time as a result of change in the past, and felt that the change could be described. It is this attempt that earned for him, and not any of the earlier writers of prose, the honorable title ''Father of History.''

Herodotus took the trouble to describe Egypt at great length, for he seems to have been more enthusiastic about the Egyptians than about almost any other people that he met. Like most visitors to Egypt, he was distinctly aware of the exceptional nature of the climate and the topography along the Nile: ''For anyone who sees Egypt, without having heard a word about it before, must perceive that Egypt is an acquired country, the gift of the river.'' This famous passage—often paraphrased to read ''Egypt is the gift of the Nile''—aptly sums up the great geographical fact about the country. In that sun-soaked, rainless climate, the river in overflowing its banks each year regularly deposited the rich silt washed down from the East African highlands. To the extreme limits of the river's waters there were fertile fields for crops and the pasturage of animals; and beyond that the barren desert frontiers stretched in all directions. This was the setting in which that literate, complex society known as Egyptian civilization developed.

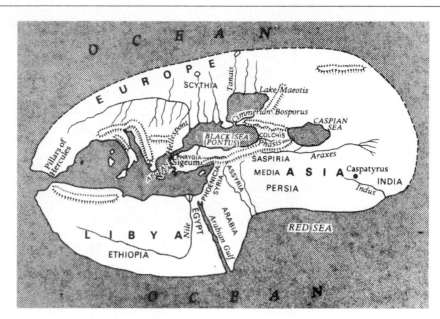

The habitable world according to Herodotus. (From Stories from Herodotus *by B. Wilson and D. Miller. Reproduced by permission of Oxford University Press.)*

The emergence of one of the world's earliest cultures was essentially a political act. Between 3500 and 3100 B.C., the self-sufficient agricultural communities that clung to the strip of land bordering the Nile had gradually coalesced into larger units until there were only the two kingdoms of Upper Egypt and Lower Egypt. Then, about 3100 B.C., these regions were united by military conquest from the south by a ruler named Menes, an elusive figure who stepped forth into history to head the long line of pharaohs. Protected from external invasion by the same deserts that isolated her, Egypt was able to develop the most stable and longest-lasting of the ancient civilizations. Whereas Greece and Rome counted their supremacies by the century, Egypt counted hers by the millennium; a well-ordered succession of thirty-two dynasties stretched from the unification of the Upper and Lower Kingdoms by Menes to Cleopatra's encounter with the asp in 31 B.C. Long after the apogee of Ancient Egypt, Napoleon was able to exhort his weary veterans with the glory of its past. Standing in the shadow of the Great Pyramid of Gizeh, he cried, ''Soldiers, forty centuries are looking down upon you!''

As soon as the unification of Egypt under a single leader became an accomplished fact, a powerful and extensive administrative system began to evolve. The census had to be taken, taxes imposed, an army maintained, and so forth, all of which required reckoning with relatively large numbers. (One of the years of the Second Dynasty was named Year of the Occurrence of the Numbering of all Large and Small Cattle of the North and South.) As early as 3500 B.C., the Egyptians had a fully developed number

The scene above is taken from the great stone macehead of Narmer, which J. E. Quibell discovered at Hierakonpolis in 1898. There is a summary of the spoil taken by Narmer during his wars, namely "cows, 400,000 🐂 𝔇𝔇𝔇𝔇 goats, 1,422,000, 🐐 𝔜 𝔇𝔇𝔇𝔇 ||⚲⚲|| and captives, 120,000, 𝔇 𝔇||."

Scene reproduced from the stone macehead of Narmer, giving a summary of the spoil taken by him during his wars. (From The Dwellers on the Nile *by E. W. Budge, 1977, Dover Publications, N.Y.)*

system that would allow counting to continue indefinitely with only the introduction from time to time of a new symbol. This is borne out by the macehead of King Narmer, one of the most remarkable relics of the ancient world, now in a museum at Oxford University. Near the beginning of the dynastic age, Narmer (who, some authorities suppose, may have been the legendary Menes, the first ruler of the united Egyptian nation) was obliged to punish the rebellious Libyans in the western Delta. He left in the temple at Hierakonpolis a magnificent slate palette—the famous Narmer Palette— and a ceremonial macehead, both of which bear scenes testifying to his victory. The macehead preserves forever the official record of the king's accomplishment, for the inscription boasts of the taking of 120,000 prisoners and a register of captive animals, 400,000 oxen and 1,422,000 goats.

Another example of the recording of very large numbers at an early stage occurs in the *Book of the Dead,* a collection of religious and magical texts whose principal aim was to secure for the deceased a satisfactory afterlife. In one section, which is believed to date from the First Dynasty, we read (the Egyptian god Nu is speaking): "I work for you, o ye spirits, we are in number four millions, six hundred and one thousand, and two hundred, ..."

The spectacular emergence of the Egyptian government and administration under the pharaohs of the first two dynasties could not have taken place without a method of writing; and we find such a method both in the elaborate "sacred signs," or hieroglyphics, and in the rapid cursive hand of the accounting scribe. The hieroglyphic system of writing is a picture script, in which each character represents a concrete object, the significance of which may still be recognizable in many cases. In one of the tombs near the Pyramid of Gizeh there have been found hieroglyphic number symbols in which the number one is represented by a single vertical stroke, or a picture of a staff, and a kind of horseshoe, or heelbone sign ∩ is used as a collective symbol to replace ten separate strokes. In other words, the Egyptian system was a decimal one (from the Latin *decem*, "ten") which used counting by powers of ten. That ten is so often found among ancient peoples as a base for their number systems is undoubtedly attributable to humans' ten fingers and to our habit of counting on them. For the same reason, a symbol much like our numeral 1 was almost everywhere used to express the number one.

Special pictographs were used for each new power of 10 up to 10,000,000: 100 by a curved rope, 1000 by a lotus flower, 10,000 by an upright bent finger, 100,000 by a tadpole, 1,000,000 by a person holding up his hands as if in great astonishment, and 10,000,000 by a symbol sometimes conjectured to be a rising sun.

1	10	100	1000	10,000	100,000	1,000,000	10,000,000

Other numbers could be expressed by using these symbols additively (that is, the number represented by a set of symbols is the sum of the numbers represented by the individual symbols), with each character repeated up to nine times. Usually, the direction of writing was from right to left, with the larger units listed first, then the others in order of importance. Thus, the scribe would write

to indicate our number

$$1 \cdot 100{,}000 + 4 \cdot 10{,}000 + 2 \cdot 1000 + 1 \cdot 100 + 3 \cdot 10 + 6 \cdot 1 = 142{,}136.$$

Occasionally, the larger units were written on the left, in which case the symbols were turned around to face the direction from which the writing began. Lateral space was saved by placing the symbols in two or three rows, one above the other. Because there was a different symbol for each power of ten, the value of the number represented was not affected by the order of the hieroglyphs within a grouping. For example,

$$|| \cap\cap\cap \; 9\,9 \;\text{🦋} \qquad\qquad 9\,9 \cap\cap\cap\,\text{🦋} \atop || \qquad\qquad \cap\cap\cap\; 9\,9\,\text{🦋}\; ||$$

all stood for the number 1232. Thus the Egyptian method of writing numbers was not a ''positional system''—a system in which one and the same symbol has a different significance depending on its position in the numerical representation.

Addition and subtraction caused little difficulty in the Egyptian number system. For addition, it was necessary only to collect symbols and exchange ten like symbols for the next higher symbol. This is how the Egyptians would have added, say, 345 and 678.

345
678
———
1023

| | | | ∩∩∩   999
| | ∩

| | | | ∩∩∩∩   999
| | | | ∩∩∩   999

———————————————

| | | | ∩∩∩∩   9999
| | | | ∩∩∩∩   9999
| | | | ∩∩∩   9
|

This converted would be

∩ | | | 9∩   99999
                              9999

and converted again,

| | | ∩∩   🦋

Subtraction was performed by the same process in reverse. Sometimes ''borrowing'' was used, wherein a symbol for the large number was exchanged for ten lower-order symbols to provide enough for the smaller number to be subtracted, as in the case

123
− 45
———
78

| | | ∩ ∩   9

| | | ∩∩∩
| | ∩

which, converted, would be

```
||||     nnnn
||||     nnnn
||||     nnn
|
|||       n nn
||          n
─────────────────
||  ||    nnnn
||  ||     nnn
```

Although the Egyptians had symbols for numbers, they had no generally uniform notation for arithmetical operations. In the case of the famous Rhind Papyrus (dating about 1650 B.C.), the scribe did represent addition and subtraction by the hieroglyphs $\wedge$ and $\wedge$, which resemble the legs of a person coming and going.

As long as writing was restricted to inscriptions carved on stone or metal, its scope was limited to short records deemed to be outstandingly important. What was needed was an easily available, inexpensive material to write on. The Egyptians solved this problem with the invention of papyrus. Papyrus was made by cutting thin lengthwise strips of the stem of the reedlike papyrus plant, which was abundant in the Nile Delta marshes. The sections were placed side by side on a board so as to form a sheet, and another layer was added at right angles to the first. When these were all soaked in water, pounded with a mallet, and allowed to dry in the sun, the natural gum of the plant glued the sections together. The writing surface was then scraped smooth with a shell until a finished sheet (usually 10 to 18 inches wide) resembled coarse brown paper; by pasting these sheets together along overlapping edges, the Egyptians could produce strips up to 100 feet long, which were rolled up when not in use. They wrote with a brushlike pen, and ink made of colored earth or charcoal that was mixed with gum or water. Thanks not so much to the durability of papyrus as to the exceedingly dry climate of Egypt, which prevented mold and mildew, a sizable body of scrolls has been preserved for us in a condition otherwise impossible.

With the introduction of papyrus, further steps in simplifying writing were almost inevitable. The first steps were made largely by the Egyptian priests who developed a more rapid, less pictorial style that was better adapted to pen and ink. In this so-called ''hieratic'' (sacred) script, the symbols were written in a cursive, or free-running, hand so that at first sight their forms bore little resemblance to the old hieroglyphs. It can be said to correspond to our handwriting as hieroglyphics corresponds to our print. As time passed and writing came into general use, even the hieratic proved to be too slow and a kind of shorthand known as ''demotic'' (popular) script arose. Hieratic writing is child's play compared with demotic, which at its worst consists of row upon row of agitated commas, each representing a totally different sign.

In both of these writing forms, numerical representation was still additive, based on powers of ten; but the repetitive principle of hieroglyphics was replaced by the device of using a single mark to represent a collection of like symbols. This type of notation may be called ''cipherization.'' Five, for instance, was assigned the distinctive mark $\succ$ instead of being indicated by a group of five vertical strokes.

1	2	3	4	5	6	7	8	9	10

20	30	40	50	60	70	80	90	100	1000

The hieratic system used to represent numbers is as shown in the table. Note that the signs for 1, 10, 100, and 1000 are essentially abbreviations for the pictographs used earlier. In hieroglyphics, the number 37 had appeared as

but in hieratic script it is replaced by the less cumbersome

The larger number of symbols called for in this notation imposed an annoying tax on the memory, but the Egyptian scribes no doubt regarded this as justified by its speed and conciseness. The idea of ciphering is one of the decisive steps in the development of numeration, comparable in significance to the Babylonian adoption of the positional principle.

Around the fifth century B.C., the Greeks of Ionia also developed a ciphered numeral system, but with a more extensive set of symbols to be memorized. They ciphered their numbers by means of the 24 letters of the ordinary Greek alphabet, augmented by 3 obsolete Phoenician letters (the digamma ς for 6, the koppa P for 90, and the sampi λ for 900). The resulting 27 letters were used as follows. The initial nine letters were associated with the numbers from 1 to 9; the next nine letters represented the first nine integral multiples of 10; the final nine letters were used for the first nine integral multiples of 100. The accompanying table shows how the letters of the alphabet (including the special forms) were arranged for use as numerals.

1 α	10 ι	100 ρ
2 β	20 κ	200 σ
3 γ	30 λ	300 τ
4 δ	40 μ	400 υ
5 ε	50 ν	500 φ
6 ϛ	60 ξ	600 χ
7 ζ	70 ο	700 ψ
8 η	80 π	800 ω
9 θ	90 ϙ	900 ϡ

Because the Ionic system was still a system of additive type, all numbers between 1 and 999 could be represented by at most three symbols. The principle is shown by

$$\psi\pi\delta = 700 + 80 + 4 = 784.$$

For larger numbers, the following scheme was used. An accent mark placed to the left and below the appropriate unit letter multiplied the corresponding number by 1000; thus ,β represents not 2 but 2000. Tens of thousands were indicated by using a new letter M, from the word *myriad* (meaning "ten thousand"). The letter M placed either next to or below the symbols for a number from 1 to 9999 caused the number to be multiplied by 10,000, as with

$$\delta M, \text{ or } \overset{\delta}{M} = 40,000$$

$$\rho\nu M, \text{ or } \overset{\rho\nu}{M} = 1,500,000.$$

With these conventions, the Greeks wrote

$$\tau\mu\epsilon M \, ,\beta\rho\mu\delta = 3,452,144$$

To express still larger numbers, powers of 10,000 were used, the double myriad MM denoting $(10,000)^2$, and so on.

The symbols were always arranged in the same order, from the highest multiple of 10 on the left to the lowest on the right, so accent marks sometimes could be omitted when the context was clear. The use of the same letter for thousands and units, as in

$$\delta\sigma\lambda\delta = 4234$$

gave the left-hand letter a local place value. To distinguish the numerical meaning of letters from their ordinary use in language, the Greeks added an accent at the end or a bar extended over them; thus, the number 1085 might appear as

$$,\alpha\pi\epsilon' \quad \text{or} \quad ,\overline{\alpha\pi\epsilon}.$$

The system as a whole afforded much economy of writing (whereas the Greek alphabetic numeral for 900 is a single letter, the Egyptians had to use the symbol ϡ nine times), but it required the mastery of numerous signs.

Multiplication in Greek alphabetic numerals was performed by beginning with the highest order in each factor and forming a sum of partial products. Let us calculate, for example, 24 × 53:

κ δ	24	
ν γ	× 53	
,α ξ	1000	60
σ ιβ	200	12
,ασ οβ	1200	72 = 1272

The idea in multiplying numbers consisting of more than one letter was to write each number as a sum of numbers represented by a single letter. Thus, the Greeks began by calculating 20 × 50 (κ by ν), then proceeded to 20 × 3 (κ by γ), then 4 × 50 (δ by ν), and finally 4 × 3 (δ by γ). This method, called Greek multiplication, corresponds to the modern computation

$$24 \times 53 = (20 + 4)(50 + 3)$$
$$= 20 \cdot 50 + 20 \cdot 3 + 4 \cdot 50 + 4 \cdot 3$$
$$= 1272.$$

The numerical connection in these products is not evident in the letter products, which necessitated elaborate multiplication tables. The Greeks had 27 symbols to multiply by each other, so they were obliged to keep track of 729 entirely separate answers. The same multiplicity of symbols tended to hide simple relations among numbers; where we recognize an even number by its ending in 0, 2, 4, 6, and 8, any one of the 27 Greek letters (possibly modified by an accent mark) could represent an even number.

An incidental objection raised against the alphabetic notation is that the juxtaposition of words and number expressions using the same symbols led to a form of number mysticism known as "gematria." In gematria, a number is assigned to each letter of the alphabet in some way and the value of a word is the sum of the numbers represented by its letters. Two words are then considered somehow related if they add up to the same number. This gave rise to the practice of giving names cryptically by citing their individual numbers. The most famous number was 666, the "number of the Beast," mentioned in the Bible in the Book of Revelations. (It is probable that it referred to Nero Caesar, whose name has this value when written in Hebrew.) A favorite pastime among Catholic theologians during the Reformation was devising alphabet schemes in which 666 was shown to stand for the name of Martin Luther, thereby supporting their contention that he was the Antichrist. Luther replied in kind; he concocted a system in which 666 forecast the duration of the papal reign and rejoiced that it was nearing an end. Readers of Tolstoy's *War and Peace* may recall that "L'Empereur Napoleon" can also be made equivalent to the number of the Beast.

Another number replacement that occurs in early theological writings concerns the word *amen,* which is αμην in Greek. These letters have the numerical values

$$A(\alpha) = 1, \quad M(\mu) = 40, \quad E(\eta) = 8, \quad N(\nu) = 50,$$

totaling 99. Thus, in many old editions of the Bible, the number 99 appears at the end of a prayer as a substitute for *amen.* An interesting illustration of gematria is also found in the graffiti of Pompeii: "I love her whose number is 545."

1.2   Problems

1. Express each of the given numbers in Egyptian hieroglyphics.

   (a) 1492   (d) 70,807
   (b) 1999   (e) 123,456
   (c) 12,321   (f) 3,040,279

2. Write each of the following Egyptian numbers in our system.

(a), (b), (c), (d) Egyptian hieroglyphic and hieratic numeral figures.

3. Perform the indicated operations and express the answers in hieroglyphics.

(a) Add

|| 99 ♉♉

and

||| ♉♉♉ ♉ |

(b) Add

|||| ♉ (((
||| (((

and

||| ∩ 99 (((
|| ((

(c) Subtract

||| ∩∩ 9
|||

from

|| ∩∩∩ 99
        ∩

(d) Subtract

||| ∩∩∩ 999
| ∩

from

|| ∩∩∩ ♉

4. Multiply the following number by ∩ (ten), expressing the result in hieroglyphics.

||| ∩∩∩ 999 ♉♉
     ∩∩∩ 99   ♉♉

Describe a simple rule for multiplying any Egyptian number by ten.

5. Write the Ionian Greek numerals corresponding to

(a) 396
(b) 1492
(c) 1999
(d) 24,789
(e) 123,456
(f) 1,234,567

6. Convert each of the following from Ionian Greek numerals to our system:

(a) ͺασλδ

(b) ͺβα

(c) $\overset{\varepsilon}{\text{M}}$,εφνε

(d) θͺͰͰΤΜ,βχμδ

7. Perform the indicated operations.

(a) Add νζ and φογ.
(b) Add σλβ and ͺλωπα.
(c) Subtract χμθ from ͺγφιβ.
(d) Multiply σπε by δ.

8. Another system of number symbols the Greeks used from about 450 to 95 B.C. is known as the "Attic" or "Herodianic" (after Herodian, a Byzantine grammarian of the second century, who described it). In this system, the initial letters of the words for 5 and the powers of 10 are used to represent the corresponding numbers; these are

Γ   the initial letter of *penta,* meaning "five."

△   the initial letter of *deka,* meaning "ten."

H   the initial letter of *hekaton,* meaning "hundred."

✕   the initial letter of *kilo,* meaning "thousand."

M   the initial letter of *myriad,* meaning "ten thousand."

The letter denoting 5 was combined with other letters to get intermediate symbols for 50, 500, 5000, and 50,000:

1	5	10	50	100	500

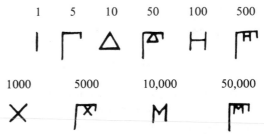

1000	5000	10,000	50,000

Other numbers were made up on an additive basis, with higher units coming before lower. Thus each symbol was repeated not more than four times. An example in this numeration system is

M Γ̄ ✕ Γ̄ △△ |||

= 10,000 + 5000 + 1000 + 50 + 20 + 3

= 16,073.

Write the Attic Greek numerals corresponding to

(a) 396
(b) 1492
(c) 1999
(d) 24,789
(e) 74,802
(f) 123,456

9. Convert the following from Greek Attic numerals to our system.

    (a) ΧΧ⌐ΗΗ⌐ΊΓΊΙ

    (b) Μ⌐Χ⌐ΊΔΔΔΓΊΙΙΙ

    (c) ΧΧ⌐ΗΗΗΗ⌐ΊΔΔΓΊΙΙ

    (d) ⌐ΊΜΜΜΓΊΔΙΙ

10. Perform the indicated operations and express the answers in Attic numerals.

    (a) Add   Χ⌐ΊΔΓΊΙ

       and   ΗΗ⌐ΊΔΔΔΔΙΙΙΙ

    (b) Add   ⌐ΊΧΧΧΧΗΗΗΔΔΙ

       and   ⌐ΊΗΗΓΊΔΔ

    (c) Subtract   ⌐ΊΗ⌐ΊΔΔΔΙΙ

       from   ΧΧΙ

    (d) Multiply ΗΔΙΙ

       by   ΔΔΓΊ

11. The Roman numerals, still used for such decorative purposes as clock faces and monuments, are patterned on the Greek Attic system in having letters as symbols for certain multiples of 5 as well as for numbers that are powers of 10. The primary symbols with their values are:

I	V	X	L	C	D	M
1	5	10	50	100	500	1000

The Roman numeration system is essentially additive, with certain subtractive and multiplicative features. If the symbols decrease in value from left to right, their values are added, as in the example

$$MDCCCXXVIII = 1000 + 500 + 300$$
$$+ 20 + 5 + 3 = 1828.$$

The representation of numbers that involve 4s and 9s is shortened by using a subtractive principle whereby a letter for a small unit placed before a unit of higher value indicates that the smaller is to be subtracted from the larger. For instance,

$$CDXCV = (500 - 100) + (100 - 10) + 5$$
$$= 495.$$

(This scheme incorporates features of a positional system, because IV = 4, whereas VI = 6.) However, there were definite rules:

    I could precede only V or X.
    X could precede only L or C.
    C could precede only D or M.

In place of new symbols for large numbers, a multiplicative device was introduced; a bar drawn over the entire symbol multiplied the corresponding number by 1000, whereas a double bar meant multiplication by 1000². Thus,

$$\overline{XV} = 15,000 \text{ and } \overline{\overline{XV}} = 15,000,000.$$

Write the Roman numerals corresponding to

    (a) 1492   (d) 74,802
    (b) 1066   (e) 123,456
    (c) 1999   (f) 3,040,279

12. Convert each of the following from Roman numerals into our system.

    (a) CXXIV   (d) DCCLXXXVII
    (b) MDLXI   (e) $\overline{XIX}$
    (c) MDCCXLVIII   (f) $\overline{XCXXV}$

13. Perform the indicated operations and express the answers in Roman numerals.

    (a) Add CM and XIX.
    (b) Add MMCLXI and MDCXX.
    (c) Add XXIV and XLVI.
    (d) Subtract XXIII from XXX.
    (e) Subtract CLXI from CCLII.
    (f) Multiply XXXIV by XVI.

1.3   Number Recording of the Babylonians

[handwritten margin note: Mesopotamia]

[handwritten margin note: Pictograph. Reed, Then stylus. restricted pics. → △ "cuneiform". Marked in CLAY = durable.]

Besides the Egyptian, another culture of antiquity that exerted a marked influence on the development of mathematics was the Babylonian. Here the term ''Babylonian'' is used without chronological restrictions to refer to those peoples who, many thousands of years ago, occupied the alluvial plain between the twin rivers, the Tigris and the Euphrates. The Greeks called this land ''Mesopotamia,'' meaning ''the land between the rivers.'' Most of it today is part of the modern state of Iraq, although both the Tigris and the Euphrates rise in Turkey. Humans stepped over the threshold of civilization in this region—and more especially in the lowland marshes near the Persian Gulf—about the same time that humans did in Egypt, that is, about 3500 B.C. or possibly a little earlier. Although the deserts surrounding Egypt successfully protected it against invasions, the open plains of the Tigris-Euphrates valley made it less defensible. The early history of Mesopotamia is largely the story of incessant invaders who, attracted by the richness of the land, conquered their deccadent predecessors, absorbed their culture, and then settled into a placid enjoyment of wealth until they were themselves overcome by the next wave of intruders.

Shortly after 3000 B.C., the Babylonians developed a system of writing from ''pictographs''—a kind of picture writing much like hieroglyphics. But the materials chosen for writing imposed special limitations of their own, which soon robbed the pictographs of any resemblance to the objects they stood for. Whereas the Egyptians used pen and ink to keep their records, the Babylonians used first a reed, later a stylus with a triangular end. With this they made impressions (rather than scratches) in moist clay. Clay dries quickly, so documents had to be relatively short and written all at one time, but they were virtually indestructible when baked hard in an oven or by the heat of the sun. (Contrast this with the Chinese method, which involved more perishable writing material such as bark or bamboo and didn't allow keeping permanent evidence of the culture's early attainments.) The sharp edge of a stylus made a vertical stroke (|) and the base made a more or less deep impression (▲), so that the combined effect was a head-and-tail figure resembling a wedge, or nail (⟙). Because the Latin word for ''wedge'' is *cuneus,* the resulting style of writing has become known as ''cuneiform.''

Cuneiform script was a natural consequence of the choice of clay as a writing medium. The stylus did not allow for drawing curved lines, so all pictographic symbols had to be composed of wedges oriented in different ways: vertical (⟙), horizontal (►), and oblique (◄ or ➚). Another wedge was later added to these three types; looked something like an angle bracket opening to the right (◄) and was made by holding the stylus so that its sides were inclined to the clay tablet. These four types of wedges had to serve for all drawings, because executing others was considered too tiresome for the hand or too time-consuming. Unlike hieroglyphics, which remained a picture writing until near the end of Egyptian civilization, cuneiform characters were gradually simplified until the pictographic originals were no longer apparent. The nearest the Babylonians could get to the old circle ◠ representing the sun was ✦, which was later condensed still further to ◄⟙. Similarly, the symbol for a fish, which began as ▷ ended up as ⟙⟙◄. The net effect of cuneiform script seems, to the uninitiated, ''like bird tracks in wet sand.'' Only within the last two centuries has anyone known what the many extant cuneiform writings meant, and indeed whether they were writing or simply decoration.

Because there were no colossal temples or monuments to capture the archeological imagination (the land is practically devoid of building stone), excavation came later to this part of the ancient world than to Egypt. It is estimated that today there are at least 400,000 Babylonian clay tablets, generally the size of a hand, scattered among the museums of various countries. Of these, some 400 tablets or tablet fragments have been identified as having mathematical content. Their decipherment and interpretation have gone slowly, owing to the variety of dialects and natural modifications in the language over the intervening several thousand years.

The initial step was taken by an obscure German schoolteacher, Georg Friedrich Grotefend (1775–1853), of Göttingen, who although well versed in classical Greek, was absolutely ignorant of Oriental languages. While drinking with friends, Grotefend wagered that he could decipher a certain cuneiform inscription from Persepolis provided that they would supply him with the previously published literature on the subject. By an inspired guess he found the key to reading Persian cuneiform. The prevailing arrangement of the characters was such that the points of the wedges headed either downward or to the right, and the angles formed by the broad wedges consistently opened to the right. He assumed that the language's characters were alphabetic; he then began picking out those characters that occurred with the greatest frequency and postulated that these were vowels. The most recurrent sign group was assumed to represent the word for ''king.'' These suppositions allowed Grotefend to decipher the title ''King of Kings'' and the names Darius, Xerxes, and Hystapes. Thereafter, he was able to isolate a great many individual characters and to read twelve of them correctly. Grotefend thus produced a translation that, although it contained numerous errors, gave an adequate idea of the contents. In 1802, when Grotefend was only 27 years old, he had his investigations presented to the Academy of Science in Göttingen (Grotefend was not allowed to read his own paper). But the overstated achievements of this little-known scholar, who neither belonged to the faculty of the university nor was even an Orientalist by profession, only evoked ridicule from the learned body. Buried in an obscure publication, Grotefend's brilliant discovery fell into oblivion, and decades later cuneiform script had to be deciphered anew. It is one of the whims of history that Champollion, the original translator of hieroglyphics, won an international reputation, while Georg Grotefend is almost entirely ignored.

Few chapters in the discovery of the ancient world can rival for interest the copying of the monumental rock inscriptions at Behistun by Henry Creswicke Rawlinson (1810–1895). Rawlinson, who was an officer in the Indian Army, became interested in cuneiform inscriptions when posted to Persia in 1835 as an advisor to the shah's troops. He learned the language and toured the country extensively, exploring its many antiquities. Rawlinson's attention was soon turned to Behistun, where a towering rock cliff, the ''Mountain of the Gods,'' rises dramatically above an ancient caravan road to Babylon. There, in 516 B.C., Darius the Great caused a lasting monument to his accomplishments to be engraved on a specially prepared surface measuring 150 feet by 100 feet. The inscription is written in thirteen panels in three languages—Old Persian, Elamite, and Akkadian (the language of the Babylonians)—all using a cuneiform script. Above the five panels of Persian writing, the artists chiseled a life-size figure in relief of Darius receiving the submission of ten rebel leaders who had disputed his right to the throne.

Although the Behistun Rock has been called by some the Mesopotamian Rosetta Stone, the designation is not entirely apt. The Greek text on the Rosetta Stone allowed Champollion to proceed from the known to the unknown, whereas all three passages of the Behistun trilingual were written in the same unknown cuneiform script. However, Old Persian, with its mainly alphabetic script limited to 43 signs, had been the subject of serious investigation since the beginning of the nineteenth century. This version of the text was ultimately to provide the key of admission into the whole cuneiform world.

The first difficulty lay in copying the long inscription. It is cut 400 feet above the ground on the face of a rock mass that itself rises 1700 feet above the plain. Since the stone steps were destroyed after the sculptors finished their work, there was no means of ascent. Rawlinson had to construct enormous ladders to get to the inscription and at times had to be suspended by block and tackle in front of the almost precipitous rock face. By the end of 1837, he had copied approximately half the 414 lines of Persian text; and using methods akin to those Grotefend worked out for himself 35 years earlier, he had translated the first two paragraphs. Rawlinson's goal was to transcribe every bit of the inscription on the Behistun Rock, but unfortunately war broke out between Great Britain and Afghanistan in 1839. Rawlinson was transferred to active duty in Afghanistan, where he was cut off by siege for the better part of the next two years. The year 1843 again found him back in Baghdad, this time as British consul, eager to continue to copy, decipher, and interpret the remainder of the Behistun inscription. His complete translation of the Old Persian part of the text, along with a copy of all the 263 lines of the Elamite, was published in 1846. Next he tackled the third class of cuneiform writing on the monument, the Babylonian, which was cut on two sides of a ponderous boulder overhanging the Elamite panels. Despite great danger to life and limb, Rawlinson obtained paper squeezes (casts) of 112 lines. With the help of the already translated Persian text, which contained numerous proper names, he assigned correct values to a total of 246 characters. During this work, he discovered an important feature of Babylonian writing, the principle of ''polyphony''; that is, the same sign could stand for different consonantal sounds, depending on the vowel that followed. Thanks to Rawlinson's remarkable efforts, the cuneiform enigma was penetrated, and the vast records of Mesopotamian civilization were now an open book.

From the exhaustive studies of the last half-century, it is apparent that Babylonian mathematics was far more highly developed than had hitherto been imagined. The Babylonians were the only pre-Grecian people who made even a partial use of a positional number system. Such systems are based on the notion of place value, in which the value of a symbol depends on the position it occupies in the numerical representation. Their immense advantage over other systems is that a limited set of symbols suffices to express numbers, no matter how large or small. The Babylonian scale of enumeration was not decimal, but sexagesimal (60 as a base), so that every place a ''digit'' is moved to the left increases its value by a factor of 60. When whole numbers are represented in the sexagesimal system, the last space is reserved for the numbers from 1 to 59, the next-to-last space for the multiples of 60, preceded by multiples of 60^2, and so on. For example, the Babylonian 3 25 4 might stand for the number

$$3 \cdot 60^2 + 25 \cdot 60 + 4 = 12,304$$

and not

$$3 \cdot 10^3 + 25 \cdot 10 + 4 = 3254$$

as in our decimal (base 10) system.

The Babylonian use of the sexagesimal place-value notation was confirmed by two tablets found in 1854 at Senkerah on the Euphrates by the English geologist W. K. Loftus. These tablets, which probably date from the period of Hammurabi (2000 B.C.), give the squares of all integers from 1 to 59 and their cubes as far as that of 32. The tablet of squares reads easily up to 7^2, or 49. Where we should expect to find 64, the tablet gives 1 4; the only thing that makes sense is to let 1 stand for 60. Following 8^2, the value of 9^2 is listed as 1 21, implying again that the left digit must represent 60. The same scheme is followed throughout the table until we come to the last entry, which is 58 1; this cannot but mean

$$58 \ 1 = 58 \cdot 60 + 1 = 3481 = 59^2.$$

The disadvantages of Egyptian hieroglyphic numeration are obvious. Representing even small numbers might necessitate relatively many symbols (to represent 999, no less than 27 hieroglyphs were required); and with each new power of 10, a new symbol had to be invented. By contrast, the numerical notation of the Babylonians emphasized two-wedge characters. The simple upright wedge ⊤ had the value 1 and could be used nine times, while the broad sideways wedge ◄ stood for 10 and could be used up to five times. The Babylonians, proceeding along the same lines as the Egyptians, made up all other numbers of combinations of these symbols, each represented as often as it was needed. When both symbols were used, those indicating tens appeared to the left of those for ones, as in

Appropriate spacing between tight groups of symbols corresponded to descending powers of 60, read from left to right. As an illustration, we have

which could be interpreted as $1 \cdot 60^3 + 28 \cdot 60^2 + 52 \cdot 60 + 20 = 319{,}940$. The Babylonians occasionally relieved the awkwardness of their system by using a subtractive sign ⊤ ►— . It permitted writing such numbers as 19 in the form $20 - 1$,

instead of using a tens symbol followed by nine units:

Babylonian positional notation in its earliest development lent itself to conflicting interpretations because there was no symbol for zero. There was no way to distinguish between the numbers

$$1 \cdot 60 + 24 = 84 \qquad \text{and} \qquad 1 \cdot 60^2 + 0 \cdot 60 + 24 = 3624,$$

since each was represented in cuneiform by

One could only rely on the context to relieve the ambiguity. A gap was often used to indicate that a whole sexagesimal place was missing, but this rule was not strictly applied and confusion could result. Someone recopying the tablet might not notice the empty space, and would put the figures closer together, thereby altering the value of the number. (Only in a positional system must the existence of an empty space be specified, so the Egyptians did not encounter this problem.) From 300 B.C. on, a separate symbol

called a divider, was introduced to serve as a placeholder, thus indicating an empty space between two digits inside a number. With this, the number 84 was readily distinguishable from 3624, the latter being represented by

The confusion was not ended, since the Babylonian divider was used only medially and there still existed no symbol to indicate the absence of a digit at the end of a number. About A.D. 150, the Alexandrian astronomer Ptolemy began using the *omicron* (o, the first letter of the Greek ουδεν, ''nothing''), in the manner of our zero, not only in a medial but also in a terminal position. There is no evidence that Ptolemy regarded o as a number by itself that could enter into computation with other numbers.

The absence of zero signs at the ends of numbers meant that there was no way of telling whether the lowest place was a unit, a multiple of 60 or 60^2, or even a multiple of $\frac{1}{60}$. The value of the symbol 2 24 (in cuneiform,) could be

$$2 \cdot 60 + 24 = 144.$$

But other interpretations are possible, for instance,

$$2 \cdot 60^2 + 24 \cdot 60 = 8640,$$

or if intended as a fraction,

$$2 + \frac{24}{60} = 2\tfrac{12}{30}.$$

Thus, the Babylonians of antiquity never achieved an absolute positional system. Their numerical representation expressed the relative order of the digits, and context alone decided the magnitude of a sexagesimally written number; since the base was so large, it was usually evident what value was intended. To remedy this shortcoming, let us agree to use a semicolon to separate integers from fractions, while all other sexagesimal places will be separated from one another by commas. With this convention, 25,0,3;30 and 25,0;3,30 will mean respectively,

$$25 \cdot 60^2 + 0 \cdot 60 + 3 + \frac{30}{60} = 90{,}003\tfrac{1}{2}$$

and

$$25 \cdot 60 + 0 + \frac{3}{60} + \frac{30}{60^2} = 1500\tfrac{7}{120} \, .$$

Note that neither the semicolon nor the comma had any counterpart in the original cuneiform texts.

The question how the sexagesimal system originated was posed long ago and has received different answers over time. According to Theon of Alexandria, a commentator of the fourth century, 60 was among all the numbers the most convenient since it was the smallest among all those that had the most divisors, and hence the most easily handled. Theon's point seemed to be that because 60 had a large number of proper divisors, namely 2, 3, 4, 5, 6, 10, 12, 15, 20, and 30, certain useful fractions could be represented conveniently; $\tfrac{1}{2}$, $\tfrac{1}{3}$ and $\tfrac{1}{4}$ by the integers 30, 20, and 15:

$$\tfrac{1}{2} = \tfrac{30}{60} = 0;30,$$

$$\tfrac{1}{3} = \tfrac{20}{60} = 0;20,$$

$$\tfrac{1}{4} = \tfrac{15}{60} = 0;15.$$

Fractions that had nonterminating sexagesimal expansions were approximated by finite ones, so that every number presented the form of an integer. The result was a simplicity of calculation that eluded the Egyptians, who reduced all their fractions to sums of fractions with numerator 1.

Others attached a ''natural'' origin to the sexagesimal system; their theory was that the early Babylonians reckoned the year at 360 days, and a higher base of 360 was chosen first, then lowered to 60. Perhaps the most satisfactory explanation is that it evolved from the merger between two peoples of whom one had adopted the decimal system, whereas the other brought with them a 6-system, affording the advantage of being divisible by 2 and by 3. (The origin of the decimal system is not logical but anatomical; humans have been provided with a natural abacus—their fingers and toes.)

The advantages of the Babylonian place-value system over the Egyptian additive computation with unit fractions were so apparent that this method became the principal instrument of calculation among astronomers. We see this numerical notation in full use in Ptolemy's outstanding work, the *Megale Syntaxis* (The Great Collection). The Arabs later passed this on to the West under the curious name *Almagest* (The Greatest). The *Almagest* so overshadowed its predecessors that until the time of Copernicus, it

was the fundamental textbook on astronomy. In one of the early chapters, Ptolemy
announced that he would be carrying out all his calculations in the sexagesimal system
to avoid "the embarrassment of [Egyptian] fractions."

Our study of early mathematics is limited mostly to the peoples of Mediterranean
antiquity, chiefly the Greeks, and their debt to the Egyptians and the inhabitants of the
Fertile Crescent. Nevertheless, some general comment is called for about the civilizations of the Far East, and especially about its oldest and most central civilization,
that of China. Although Chinese society was no older than the other river valley civilizations of the ancient world, it flourished long before those of Greece and Rome. In
the middle of the second millennium B.C., the Chinese were already keeping records
of astronomical events on bone fragments, some of which are extant. Indeed, by 1400
B.C., the Chinese had a positional numeration system that used nine signs.

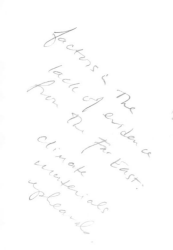

The scarcity of reliable sources of information almost completely seals from us
the history of the ancient Orient. In India, no mathematical text exists that can be
ascribed with any certainty to the pre-Christian era; and the first firm date that can be
connected with a Chinese work, namely the *Nine Chapters on the Mathematical Arts,*
is 150 B.C.

Much of the difference in availability of sources of information is to be ascribed
to differences in climate between the Near East and the Far East. The dry climate and
soil of Egypt and Babylonia preserved materials that would long since have perished
in more moist climates, materials that make it possible for us to trace the progress of
these cultures from the barbarism of the remote past to the full flower of civilization.
No other countries provide so rich a harvest of information about the origin and transmission of mathematics. "The Egyptians who lived in the cultivated part of the
country," wrote Herodotus in his *History,* "by their practice of keeping records of the
past, have made themselves much the best historians of any nation that I have
experienced."

If China had had Egypt's climate, there is no question that many a record would
have survived from antiquity, each with its story to tell of the intellectual life of earlier
generations. But the ancient Orient was a "bamboo civilization," and among the manifold uses of this plant was making books. The small bamboo slips used were prepared
by splitting the smooth section between two knots into thin strips, which were then
dried over a fire and scraped off. The narrowness of the bamboo strips made it necessary
to arrange the written characters in vertical lines running from top to bottom, a practice
that continues to this day. The opened, dried, and scraped strips of bamboo were laid
side by side, joined and kept in proper place by four crosswise cords. Naturally enough
the joining cords often rotted and broke, with the result that the order of the slips was
lost and could be reestablished only by a careful reading of the text. (Another material
used about that time for writing was silk, which presumably came into use because
bamboo books or wooden tablets were too heavy and cumbersome.) The great majority
of these ancient books were irretrievably lost to the ravages of time and nature. Those
few available today are known only as brief fragments.

Another factor making chronological accounts less trustworthy for China than for
Egypt and Babylonia is that books tended to accumulate in palace or government
libraries, where they disappeared in the great interdynastic upheavals. There is a story

that in 221 B.C., when China was united under the despotic emperor Shih Huang-ti, he tried to destroy all books of learning and nearly succeeded. Fortunately, many books were preserved in secret hiding places or in the memory of scholars, who feverishly reproduced them in the following dynasty. But such events make the dating of mathematical discoveries far from easy.

Modern science and technology, as all the world knows, grew up in western Europe, with the life of Galileo marking the great turning point. Yet between the first and fifteenth centuries, the Chinese, who experienced nothing comparable to Europe's Dark Ages, were generally much in advance of the West. Not until the scientific revolution of the later stages of the Renaissance did Europe rapidly draw ahead. Before China's isolation and inhibition, she transmitted to Europe a veritable abundance of inventions and technological discoveries, which were often received by the West with no clear idea of where they originated. No doubt the three greatest discoveries of the Chinese—ones that changed Western civilization, and indeed the civilization of the whole world—were gunpowder, the magnetic compass, and paper and printing. The subject of paper is of great interest; and we know almost to the day when the discovery was first made. A popular account of the time tells that Tshai Lun, the director of imperial workshops in A.D. 105, went to the emperor and said, ''Bamboo tablets are so heavy and silk so expensive that I sought for a way of mixing together the fragments of bark, bamboo, and fishnets and I have made a very thin material that is suitable for writing on.'' It took more than a thousand years for paper to make its way from China to Europe, first appearing in Egypt about 900 and then in Spain about 1150.

All the while mathematics was overwhelmingly concerned with practical matters that were important to a bureaucratic government: land measurement and surveying, taxation, the making of canals and dikes, granary dimensions, and so on. The misconception that the Chinese made considerable progress in theoretical mathematics is due to the Jesuit missionaries who arrived in Peking in the early 1600s. Finding that one of the most important governmental departments was known as the office of mathematics, they assumed that its function was to promote mathematical studies throughout the empire. Actually it consisted of minor officials trained in preparing the calendar. Throughout Chinese history the main importance of mathematics was in making the calendar, for its promulgation was considered a right of the emperor, corresponding to the issue of minted coins. In an agricultural economy so dependent on artificial irrigation, it was necessary to be forewarned of the beginning and end of the rainy monsoon season, as well as of the melting of the snows and the consequent rise of the rivers. The person who could give an accurate calendar to the people could thereby claim great importance.

Because the establishment of the calendar was a jealously guarded prerogative, it is not surprising that the emperor was likely to view any independent investigations with alarm. ''In China,'' wrote the Italian Jesuit Matteo Ricci (died 1610), ''it is forbidden under pain of death to study mathematics, without the Emperor's authorization.'' Regarded as a servant of the more important science astronomy, mathematics acquired a practical orientation that precluded the consideration of abstract ideas. Little mathematics was undertaken for its own sake in China.

1.3 Problems

1. Express each of the given numbers in Babylonian cuneiform notation.

   (a) 1000   (d) 1234

   (b) 10,000   (e) 12,345

   (c) 100,000   (f) 123,456

2. Translate each of the following into a number in our system.

   (a)

   (b)

   (c)

3. Express the fractions $\frac{1}{6}, \frac{1}{9}, \frac{1}{5}, \frac{1}{24}, \frac{1}{40}$, and $\frac{5}{12}$ in sexagesimal notation.

4. Convert the following numbers from sexagesimal notation to our system.

   (a) 1,23,45   (c) 0;12,3,45

   (b) 12;3,45   (d) 1,23;45

5. Multiply the number 12,3;45,6 by 60. Describe a simple rule for multiplying any sexagesimal number by 60; by 60^2.

6. Chinese bamboo or counting-rod numerals, which may go back to 1000 B.C., originated from bamboo sticks laid out on flat boards. The system is essentially positional, based on a ten scale, with blanks where we should put zeros. There are two sets of symbols for the digits 1, 2, 3, . . . , 9, which are used in alternate positions, the first set for units, hundreds, . . . , and the second set for tens, thousands, ...

Units, Hundreds, Ten thousands

Tens, Thousands, Hundred thousands

Thus, for example, the number 36,278 would be written

The circular symbol ◯ for zero was introduced relatively late, first appearing in print in the 1200s.

   Write the Chinese counting-rod numerals corresponding to

   (a) 1492   (d) 57,942

   (b) 1999   (e) 123,456

   (c) 1066   (f) 3,040,279

7. Convert the following into our numerals.

   (a)

   (b)

   (c)

   (d)

8. Multiply   by 10 and express the result in Chinese rod numerals. Describe a simple rule for multiplying any Chinese rod numerals by 10; by 10^2.

9. Perform the indicated operations.

   (a)

   (b)

   (c)

10. The traditional Chinese (brush form) numeral system shares some of the best features of both Egyptian hieroglyphic and Greek alphabetic numerals. It is an example of a vertically written multiplicative grouping system based on powers of 10. The digits 1, 2, 3, . . . , 9 are ciphered in this system, thus avoiding the repetition of symbols, and special characters exist for 100, 1000, 10,000, and 100,000.

Numerals are written from the top downward, so that

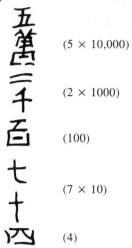

(5 × 10,000)

(2 × 1000)

(100)

(7 × 10)

(4)

represents

$$5 \cdot 10{,}000 + 2 \cdot 1000 + 100 + 7 \cdot 10 + 4$$
$$= 52{,}174.$$

Notice that if only one of a certain power of 10 is intended, then the multiplier 1 is omitted.

Express each of the given numbers in traditional Chinese numerals.

(a) 236   (d) 1066
(b) 1492   (e) 57,942
(c) 1999   (f) 123,456

11. Translate each of the following numerals from the Chinese system to our numerals.

(a)   (b)   (c)   (d)

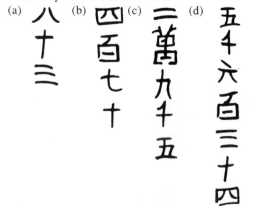

12. Multiply the given number by 10, expressing the result in Chinese numerals.

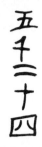

13. The Mayan Indians of Central America developed a positional number system with 20 as the primary base, along with an additive grouping technique (based on 5) for the numbers in the 20 block. The symbols for 1 to 19 were represented by combinations of dots and horizontal bars, each dot standing for 1 and each bar for 5.

1	•	6	⊟	11	⊟	16	⊟
2	••	7	⊟	12	⊟	17	⊟
3	•••	8	⊟	13	⊟	18	⊟
4	••••	9	⊟	14	⊟	19	⊟
5	—	10	⊟	15	⊟		

The Mayan year was divided into 18 months of 20 days each, with 5 extra holidays added to fill the difference between this and the solar year. Because the system the Mayans developed was designed mainly for calendar computations, they used $18 \cdot 20 = 360$ instead of 20^2 for the third position; successive positions after the third had a multiplicative value 20, so that the place values turned out to be

$$1, 20, 360, 7200, 144,000, \ldots.$$

Numerals were written vertically with the larger units above, and missing positions were indicated by a sign , which looked something like a small shell or a half-closed eye.

Thus,

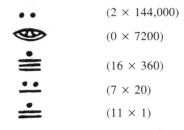

(2 × 144,000)

(0 × 7200)

(16 × 360)

(7 × 20)

(11 × 1)

represents

$$2 \cdot 144{,}000 + 0 \cdot 7200 + 16 \cdot 360 + 7 \cdot 20 + 11 = 290{,}311.$$

Write the Mayan numerals corresponding to

(a) 1492   (d) 57,942
(b) 1999   (e) 123,456
(c) 1066   (f) 3,040,279

14. Convert the following numerals from the Mayan system into ours.

(a)   (b)   (c)

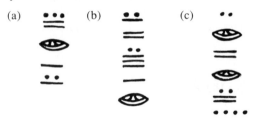

15. Perform the indicated operations shown here.

(a) Add

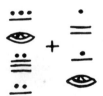

(b) Subtract

PLACE VALUES

$2{,}880{,}000 = 20^4 \times 18$

$144\,000 = 20^3 \times 18$

$7200 = 20^2 \times 18$

$360 = 20^1 \times 18$

$20 = 20^1$

$1 = 20^0$

(c) Multiply

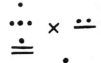

16. Multiply the given number by  (20), expressing the result in the Mayan system. Describe a simple rule for multiplying any Mayan number by 20; by 20².

17. How many different symbols are required to write the number 999,999 in (a) Egyptian hieroglyphics; (b) Babylonian cuneiform; (c) Ionian Greek numerals; (d) Roman numerals; (e) Chinese rod numerals; (f) traditional Chinese numerals; and (g) Mayan numerals?

Bibliography

Ascher, Marcia. *Ethnomathematics, A Multicultural View of Mathematical Ideas.* Pacific Grove, Calif.: Brooks Cole, 1991.

———. ''Before the Conquest.'' *Mathematics Magazine* 65 (1992): 211–218.

Ascher, Marcia, and Ascher, Robert. *Code of the Quipu.* Ann Arbor, Mich.: University of Michigan Press, 1981.

———. ''Ethnomathematics.'' *History of Science* 24 (1986): 125–144.

Boyer, Carl. ''Fundamental Steps in the Development of Numeration.'' *Isis* 35(1944): 153–158.

———. ''Note on Egyptian Numeration.'' *Mathematics Teacher* 52(1959): 127–129.

Chiera, E. *They Wrote on Clay: The Babylonian Tablets Speak Today.* Chicago: University of Chicago Press, 1938.

Cordrey, William. ''Ancient Mathematics and the Development of Primitive Culture.'' *Mathematics Teacher* 32(1939): 51–60.

Dantzig, Tobias. *Number: The Language of Science.* New York: Macmillan, 1939.

Grundlach, Bernard. ''A History of Numbers and Numerals.'' In *Historical Topics for the Mathematics Classroom.* Washington: National Council of Teachers of Mathematics, 1969.

Ifrah, Georges. *From One to Zero: A Universal History of Numbers.* Translated by Lowell Bair. New York: Viking, 1985.

Karpinski, Louis. *The History of Arithmetic.* Chicago: Rand McNally, 1925.

Menniger, Karl. *Number Words and Number Symbols: A Cultural History of Numbers.* Cambridge, Mass.: M.I.T. Press, 1969. (Dover reprint, 1992.)

Needham, Joseph. *Science and Civilization in China.* Vol. 3, *Mathematics and the Sciences of the Heavens and the Earth.* Cambridge: Cambridge University Press, 1959.

Ore, Oystein. *Number Theory and Its History.* New York: McGraw-Hill, 1948. (Dover reprint, 1988.)

Schmandt-Besserat, Denise. ''Reckoning Before Writing.'' *Archaeology* 32(May–June 1979): 23–31.

Scriba, Christopher. *The Concept of Number.* Mannheim: Bibliographisches Institut, 1968.

Seidenberg, A. ''The Ritual Origin of Counting.'' *Archive for History of Exact Sciences* 2(1962): 1–40.

———. ''The Origin of Mathematics.'' *Archive for History of Exact Sciences* 18(1978): 301–342.

Smeltzer, Donald. *Man and Number.* New York: Emerson Books, 1958.

Smith, David, and Ginsburg, Jekuthiel. *Numbers and Numerals.* Washington: National Council of Teachers of Mathematics, 1958.

Struik, Dirk. ''Stone Age Mathematics.'' *Scientific American* 179(Dec. 1948): 44–49.

———. ''On Chinese Mathematics.'' *Mathematics Teacher* 56(1963): 424–432.

Swetz, Frank. ''The Evolution of Mathematics in Ancient China.'' *Mathematics Magazine* 52(1979): 10–19.

Thureau-Dangin, F. ''Sketch of the History of the Sexagesimal System.'' *Osiris* 7(1939): 95–141.

Wilder, Raymond. ''The Origin and Growth of Mathematical Concepts.'' *Bulletin of the American Mathematical Society* 59(1953): 423–448.

———. *The Evolution of Mathematical Concepts: An Elementary Study.* New York: Wiley, 1968.

Zaslavsky, Claudia. *Africa Counts: Number Patterns in African Culture.* Boston: Prindle, Weber & Schmidt, 1973.

CHAPTER 2

Mathematics in Early Civilizations

In most sciences one generation tears down what another has built and what one has established another undoes. In Mathematics alone each generation builds a new story to an old structure.

HERMANN HANKEL

2.1 The Rhind Papyrus

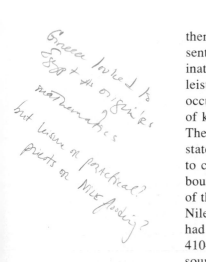

Greece looked to Egypt as the origin of mathematics but were or practical? priests or Nile flooding?

With the possible exception of astronomy, mathematics is the oldest and most continuously pursued of the exact sciences. Its origins lie shrouded in the mists of antiquity. We are often told that in mathematics all roads lead back to Greece. But the Greeks themselves had other ideas about where mathematics began. A favored one is represented by Aristotle, who in his *Metaphysics* wrote: "The mathematical sciences originated in the neighborhood of Egypt, because there the priestly class was allowed leisure." This is partly true, for the most spectacular advances in mathematics have occurred contemporaneously with the existence of a leisure class devoted to the pursuit of knowledge. A more prosaic view is that mathematics arose from practical needs. The Egyptians required ordinary arithmetic in the daily transactions of commerce and state government to fix taxes, to calculate the interest on loans, to compute wages, and to construct a workable calendar. Simple geometric rules were applied to determine boundaries of fields and the contents of granaries. As Herodotus called Egypt the gift of the Nile, we could call geometry a second gift. For with the annual flooding of the Nile Valley, it became necessary for purposes of taxation to determine how much land had been gained or lost. This was the view of the Greek commentator Proclus (A.D. 410–485), whose *Commentary on the First Book of Euclid's Elements* is our invaluable source of information on pre-Euclidean geometry:

> According to most accounts geometry was first discovered among the Egyptians and originated in the measuring of their lands. This was necessary for them because the Nile overflows and obliterates the boundaries between their properties.

Although the initial emphasis was on utilitarian mathematics, the subject began eventually to be studied for its own sake. Algebra evolved ultimately from the techniques of calculation, and theoretical geometry began with land measurement.

Most historians date the beginning of the recovery of the ancient past in Egypt from Napoleon Bonaparte's ill-fated invasion of 1798. In April of that year, Napoleon set sail from Toulon with an army of 38,000 soldiers crammed into 328 ships. He was intent on seizing Egypt and thereby threatening the land routes to the rich British possessions in India. Although England's Admiral Nelson destroyed much of the French fleet a month after the army debarked near Alexandria, the campaign dragged on another twelve months before Napoleon abandoned the cause and hurried back to France. Yet what had been a French military disaster was a scientific triumph. Napoleon had carried with his expeditionary force a commission on the sciences and arts, a carefully chosen body of 167 scholars—including the mathematicians Gaspard Monge and Jean-Baptiste Fourier—charged with making a comprehensive inquiry into every aspect of the life of Egypt in ancient and modern times. The grand plan had been to enrich the world's store of knowledge while softening the impact of France's military adventures by calling attention to the superiority of her culture.

The savants of the commission were captured by the British but generously allowed to return to France with their notes and drawings. In due course, they produced a truly monumental work with the title *Déscription de l'Egypte.* This work ran to 9 folio volumes of text and 12 volumes of plates, published over 25 years. The text itself was divided into four parts concerned respectively with ancient Egyptian civilization, monuments, modern Egypt, and natural history. Never before or since has an account of a foreign land been made so completely, so accurately, so fast, and under such difficult conditions.

The *Déscription de l'Egypte,* with its sumptuous and magnificently illustrated folios, thrust the riches of ancient Egypt on a society accustomed to the antiquities of Greece and Rome. The sudden revelation of a flourishing civilization, older than any known so far, aroused immense interest in European cultural and scholarly circles. What made the fascination even greater was that the historical records of this early society were in a script that no one had been able to translate into a modern language. The same military campaign of Napoleon provided the literary clue to the Egyptian past, for one of his engineers uncovered the Rosetta Stone and realized its possible importance for deciphering hieroglyphics.

Most of our knowledge of the order of mathematics in Egypt is derived from two sizable papyri, each named after its former owner—the Rhind Papyrus and the Golenischev Papyrus. The latter is sometimes called the Moscow Papyrus, since it reposes in the Museum of Fine Arts in Moscow. The Rhind Papyrus was purchased in Luxor, Egypt, in 1858 by the Scotsman A. Henry Rhind and was subsequently willed to the British Museum. When the health of this young lawyer broke down, he visited the milder climate of Egypt and became an archeologist, specializing in the excavation of Theban tombs. It was in Thebes, in the ruins of a small building near the Ramesseum, that the papyrus was said to have been found.

The Rhind Papyrus was written in hieratic script (a cursive form of hieroglyphics better adapted to the use of pen and ink) about 1650 B.C. by a scribe named Ahmes, who assured us that it was the likeness of an earlier work dating to the Twelfth Dynasty, 1849–1801 B.C. Although the papyrus was originally a single scroll nearly 18 feet long and 13 inches high, it came to the British Museum in two pieces, with a central portion missing. Perhaps the papyrus had been broken apart while being unrolled by someone

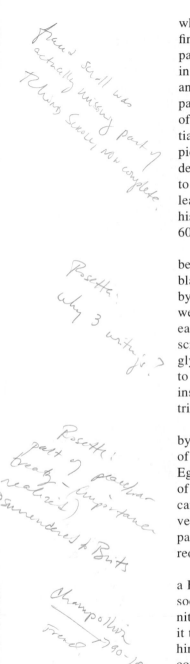

who lacked the skill for handling such delicate documents, or perhaps there were two finders and each claimed a portion. In any case, it appeared that a key section of the papyrus was forever lost to us, until one of those chance events that sometimes occur in archeology took place. About four years after Rhind had made his famous purchase, an American Egyptologist, Edwin Smith, was sold what he thought was a medical papyrus. This papyrus proved to be a deception, for it was made by pasting fragments of other papyri on a dummy scroll. At Smith's death (in 1906), his collection of Egyptian antiquaries was presented to the New York Historical Society, and in 1922, the pieces in the fraudulent scroll were identified as belonging to the Rhind Papyrus. The decipherment of the papyrus was completed when the missing fragments were brought to the British Museum and put in their appropriate places. Rhind also purchased a short leather manuscript, the Egyptian Mathematical Leather Scroll, at the same time as his papyrus; but owing to its very brittle condition, it remained unrolled for more than 60 years.

It was possible to begin the translation of the Rhind Papyrus almost immediately because of the knowledge gained from the Rosetta Stone. Finding this slab of polished black basalt was the most significant event of Napoleon's expedition. It was uncovered by officers of Napoleon's army near the Rosetta branch of the Nile in 1799, when they were digging the foundations of a fort. The Rosetta Stone is made up of three panels, each inscribed in a different type of writing: Greek down the bottom third, demotic script of Egyptian (a form developed from hieratic) in the middle, and ancient hieroglyphic in the broken upper third. The way to read Greek had never been lost; the way to read hieroglyphics and demotic had never been found. It was inferred from the Greek inscription that the other two panels carried the same message, so that here was a trilingual text from which the hieroglyphic alphabet could be deciphered.

The importance of the Rosetta Stone was realized at once by the French, especially by Napoleon, who ordered ink rubbings of it taken and distributed among the scholars of Europe. Public interest was so intense that when Napoleon was forced to relinquish Egypt in 1801, one of the articles of the treaty of capitulation required the surrender of the stone to the British. Like all the rest of the captured artifacts, the Rosetta Stone came to rest in the British Museum, where four plaster casts were made for the universities of Oxford, Cambridge, Edinburgh, and Dublin, and its decipherment by comparative analysis began. The problem turned out to be more difficult than imagined, requiring 23 years and the intensive study of many scholars for its solution.

The final chapter of the mystery of the Rosetta Stone, like the first, was written by a Frenchman, Jean François Champollion (1790–1832). The greatest of all names associated with the study of Egypt, Champollion had had from his childhood a premonition of the part he would play in the revival of ancient Egyptian culture. Story has it that at the age of 11, he met the mathematician Jean-Baptiste Fourier, who showed him some papyri and stone tablets bearing hieroglyphics. Although assured that no one could read them, the boy made the determined reply, ''I will do it when I am older.'' From then on, almost everything Champollion did was related to Egyptology; at the age of 13 he was reading three Eastern languages, and when he was 17, he was appointed to the faculty of the University of Grenoble. By 1822 he had compiled a hieroglyphic vocabulary and given a complete reading of the upper panel of the Rosetta Stone.

The Rosetta Stone, bearing the same inscription in hieroglyphics, demotic script, and Greek. (Copyright British Museum.)

Through many years hieroglyphics had evolved from a system of pictures of complete words to one that included both alphabetic signs and phonetic symbols. In the hieroglyphic inscription of the Rosetta Stone, oval frames called ''cartouches'' (the French word for ''cartridge'') were drawn around certain characters. Because these were the only signs showing special emphasis, Champollion reasoned that symbols enclosed by the cartouches represented the name of the ruler Ptolemy, mentioned in the Greek text. Champollion also secured a copy of inscriptions on an obelisk, and its base pedestal, from Philae. The base had a Greek dedication honoring Ptolemy and his wife Cleopatra (not the famous but ill-fated Cleopatra). On the obelisk itself, which was carved in hieroglyphics, are two cartouches close together, so it seemed probable that these outlined the Egyptian equivalents of their proper names. Moreover, one of them contained the same hieroglyphic characters that filled the cartouches found on the Rosetta Stone. This cross-check was enough to allow Champollion to make a preliminary decipherment. From the royal names he established a correlation between individual hieroglyphics and Greek letters. In that instant in which hieroglyphics dropped its shroud of insoluble mystery, Champollion, worn by the years of ceaseless effort, was rumored to cry, ''I've got it!'' and fall into a dead faint.

As a fitting climax to a life's study, Champollion wrote his *Grammaire Egyptienne en Encriture Hieroglyphique,* published posthumously in 1843. In it, he formulated a system of grammar and general decipherment that is the foundation on which all later Egyptologists have worked. The Rosetta Stone had provided the key to understanding one of the great civilizations of the past.

2.2   Egyptian Arithmetic

The Rhind Papyrus starts with a bold premise. Its content has to do with ''a thorough study of all things, insight into all that exists, knowledge of all obscure secrets.'' It soon becomes apparent that we are dealing with a practical handbook of mathematical exercises, and the only ''secrets'' are how to multiply and divide. Nonetheless, the 85 problems contained therein give us a pretty clear idea of the character of Egyptian mathematics. The Egyptian arithmetic was essentially ''additive,'' meaning that its tendency was to reduce multiplication and division to repeated additions. Multiplication of two numbers was accomplished by successively doubling one of the numbers and then adding the appropriate duplications to form the product. To find the product of 19 and 71, for instance, assume the multiplicand to be 71, doubling thus:

1	71
2	142
4	284
8	568
16	1136

Here we stop doubling, for a further step would give a multiplier of 71 that is larger than 19. Because 19 = 1 + 2 + 16, let us put checks alongside these multipliers to indicate that they should be added. The problem 19 times 71 would then look like this:

✔	1	71
✔	2	142
	4	284
	8	568
✔	16	1136
totals	19	1349

Adding those numbers in the right-hand column opposite the checks, the Egyptian mathematician would get the required answer, 1349; that is,

$$1349 = 71 + 142 + 1136 = (1 + 2 + 16)71 = 19 \cdot 71.$$

Had the number 19 been chosen as the multiplicand and 71 as the multiplier, the work would have been arranged as follows:

✓	1	19
✓	2	38
✓	4	76
	8	152
	16	304
	32	608
✓	64	1216
totals	71	1349

Because $71 = 1 + 2 + 4 + 64$, one has merely to add these multiples of 19 to get, again, 1349.

The method of multiplying by doubling and summing is workable because every integer (positive) can be expressed as a sum of distinct powers of 2; that is, as a sum of terms from the sequence, 1, 2, 4, 8, 16, 32, ... It is not likely that the ancient Egyptians actually proved this fact, but their confidence therein was probably established by numerous examples. The scheme of doubling and halving is sometimes called Russian multiplication because of its use among the Russian peasants. The obvious advantage is that it makes memorizing tables unnecessary.

Egyptian division might be described as doing multiplication in reverse—where the divisor is repeatedly doubled to give the dividend. To divide 91 by 7, for example, a number x is sought such that $7x = 91$. This is found by redoubling 7 until a total of 91 is reached; the procedure is shown herewith.

	1	7 ✓
	2	14
	4	28 ✓
	8	56 ✓
totals	13	91

Finding that $7 + 28 + 56 = 91$, one adds the powers of two corresponding to the checked numbers, namely, $1 + 4 + 8 = 13$, which gives the desired quotient. The Egyptian division procedure has the pedagogical advantage of not appearing to be a new operation.

Division was not always as simple as in the example just given, and fractions would often have to be introduced. To divide, say, 35 by 8, the scribe would begin by doubling the divisor, 8, to the point at which the next duplication would exceed the dividend, 35. Then he would start halving the divisor in order to complete the remainder. The calculations might appear thus:

	1	8
	2	16
	4	32 ✓
	$\frac{1}{2}$	4
	$\frac{1}{4}$	2 ✓
	$\frac{1}{8}$	1 ✓
totals	$4 + \frac{1}{4} + \frac{1}{8}$	35

Doubling 16 gives 32, so that what is missing is $35 - 32 = 3$. One first takes half of 8 to get 4, then half of 4 to get 2, and finally half of this to arrive at 1; when the fourth and the eighth are added, the needed 3 is obtained. Thus, the required quotient is $4 + \frac{1}{4} + \frac{1}{8}$.

In another example, division of 16 by 3 might be effected as follows:

1	3	✔
2	6	
4	12	✔
$\frac{2}{3}$	2	✔
$\frac{1}{3}$	1	✔
totals	$5 + \frac{1}{3}$	16

The sum of the entries in the left-hand column corresponding to the checks gives the quotient $5 + \frac{1}{3}$. It is extraordinary that to get one-third of a number, the Egyptian first found two-thirds of the number and then took one-half of the result. This is illustrated in more than a dozen problems of the Rhind Papyrus.

When the Egyptian mathematician needed to compute with fractions, he was confronted with many difficulties arising from his refusal to conceive of a fraction like $\frac{2}{5}$. His computational practice allowed him only to admit the so-called unit fractions; that is, fractions of the form $1/n$, where n is a natural number. The Egyptians indicated a unit fraction by placing an elongated oval over the hieroglyphic for the integer that was to appear in the denominator, so that $\frac{1}{4}$ was written as ⊖ or $\frac{1}{100}$ as ⊖. With the exception of $\frac{2}{3}$, for which there was a special symbol ⊖ all other fractions had to be decomposed into sums of unit fractions, each having a different denominator. Thus, for instance, $\frac{6}{7}$ would be represented as

$$\frac{6}{7} = \frac{1}{2} + \frac{1}{4} + \frac{1}{14} + \frac{1}{28}.$$

Although it is true that $\frac{6}{7}$ can be written in the form

$$\frac{6}{7} = \frac{1}{7} + \frac{1}{7} + \frac{1}{7} + \frac{1}{7} + \frac{1}{7} + \frac{1}{7},$$

the Egyptians would have thought it both absurd and contradictory to allow such repetitions. In their eyes there was one and one part only that could be the seventh of anything. The ancient scribe would probably have found the unit fraction equivalent of $\frac{6}{7}$ by the following conventional division of 6 by 7:

1	7	
$\frac{1}{2}$	$3 + \frac{1}{2}$	✔
$\frac{1}{4}$	$1 + \frac{1}{2} + \frac{1}{4}$	✔
$\frac{1}{7}$	1	
$\frac{1}{14}$	$\frac{1}{2}$	✔
$\frac{1}{28}$	$\frac{1}{4}$	✔
totals	$\frac{1}{2} + \frac{1}{4} + \frac{1}{14} + \frac{1}{28}$	6

To facilitate such decomposition into unit fractions, many reference tables must have existed, the simplest of which were no doubt committed to memory. At the beginning of the Rhind Papyrus, there is such a table giving the breakdown for fractions with numerator 2 and denominator an odd number between 5 and 101. This table, which occupies about one-third of the whole of the 18-foot roll, is the most extensive of the arithmetic tables to be found among the ancient Egyptian papyri that have come down to us. The scribe first stated what decomposition of $2/n$ he had selected; then, by ordinary multiplication, he proved that his choice of values was correct. That is, he multiplied the selected expression by the odd integer n to produce 2. Nowhere is there any inkling of the technique used to arrive at the decomposition.

Fractions $2/n$ whose denominators are divisible by 3 all follow the general rule

$$\frac{2}{3k} = \frac{1}{2k} + \frac{1}{6k}.$$

Typical of these entries is $\frac{2}{15}$ (the case $k = 5$), which is given as

$$\tfrac{2}{15} = \tfrac{1}{10} + \tfrac{1}{30}.$$

If we ignore the representations for fractions of the form $2/(3k)$, then the remainder of the $2/n$ table reads as shown herewith.

$\frac{2}{5} = \frac{1}{3} + \frac{1}{15}$	$\frac{2}{53} = \frac{1}{30} + \frac{1}{318} + \frac{1}{795}$
$\frac{2}{7} = \frac{1}{4} + \frac{1}{28}$	$\frac{2}{55} = \frac{1}{30} + \frac{1}{330}$
$\frac{2}{11} = \frac{1}{6} + \frac{1}{66}$	$\frac{2}{59} = \frac{1}{36} + \frac{1}{236} + \frac{1}{531}$
$\frac{2}{13} = \frac{1}{8} + \frac{1}{52} + \frac{1}{104}$	$\frac{2}{61} = \frac{1}{40} + \frac{1}{244} + \frac{1}{488} + \frac{1}{610}$
$\frac{2}{17} = \frac{1}{12} + \frac{1}{51} + \frac{1}{68}$	$\frac{2}{65} = \frac{1}{39} + \frac{1}{195}$
$\frac{2}{19} = \frac{1}{12} + \frac{1}{76} + \frac{1}{114}$	$\frac{2}{67} = \frac{1}{40} + \frac{1}{335} + \frac{1}{536}$
$\frac{2}{23} = \frac{1}{12} + \frac{1}{276}$	$\frac{2}{71} = \frac{1}{40} + \frac{1}{568} + \frac{1}{710}$
$\frac{2}{25} = \frac{1}{15} + \frac{1}{75}$	$\frac{2}{73} = \frac{1}{60} + \frac{1}{219} + \frac{1}{292} + \frac{1}{365}$
$\frac{2}{29} = \frac{1}{24} + \frac{1}{58} + \frac{1}{174} + \frac{1}{232}$	$\frac{2}{77} = \frac{1}{44} + \frac{1}{308}$
$\frac{2}{31} = \frac{1}{20} + \frac{1}{124} + \frac{1}{155}$	$\frac{2}{79} = \frac{1}{60} + \frac{1}{237} + \frac{1}{316} + \frac{1}{790}$
$\frac{2}{35} = \frac{1}{30} + \frac{1}{42}$	$\frac{2}{83} = \frac{1}{60} + \frac{1}{332} + \frac{1}{415} + \frac{1}{498}$
$\frac{2}{37} = \frac{1}{24} + \frac{1}{111} + \frac{1}{296}$	$\frac{2}{85} = \frac{1}{51} + \frac{1}{255}$
$\frac{2}{41} = \frac{1}{24} + \frac{1}{246} + \frac{1}{328}$	$\frac{2}{89} = \frac{1}{60} + \frac{1}{356} + \frac{1}{534} + \frac{1}{890}$
$\frac{2}{43} = \frac{1}{42} + \frac{1}{86} + \frac{1}{129} + \frac{1}{301}$	$\frac{2}{91} = \frac{1}{70} + \frac{1}{130}$
$\frac{2}{47} = \frac{1}{30} + \frac{1}{141} + \frac{1}{470}$	$\frac{2}{95} = \frac{1}{60} + \frac{1}{380} + \frac{1}{570}$
$\frac{2}{49} = \frac{1}{28} + \frac{1}{196}$	$\frac{2}{97} = \frac{1}{56} + \frac{1}{679} + \frac{1}{776}$
$\frac{2}{51} = \frac{1}{34} + \frac{1}{102}$	$\frac{2}{101} = \frac{1}{101} + \frac{1}{202} + \frac{1}{303} + \frac{1}{606}$

Ever since the first translation of the papyrus appeared, mathematicians have tried to explain what the scribe's method may have been in preparing this table. Of the many possible reductions to unit fractions, why is

$$\tfrac{2}{19} = \tfrac{1}{12} + \tfrac{1}{76} + \tfrac{1}{114}$$

chosen for $n = 19$ instead of, say,

$$\tfrac{2}{19} = \tfrac{1}{12} + \tfrac{1}{57} + \tfrac{1}{228}?$$

No definite rule has been discovered that will give all the results of the table.

The very last entry in the table, which is 2 divided by 101, is presented as

$$\frac{2}{101} = \frac{1}{101} + \frac{1}{202} + \frac{1}{303} + \frac{1}{606} .$$

This is the only possible decomposition of $\frac{2}{101}$ into no more than four different unit fractions with all the denominators less than 1000; and is a particular case of the general formula

$$\frac{2}{n} = \frac{1}{n} + \frac{1}{2n} + \frac{1}{3n} + \frac{1}{6n} .$$

By the indicated formula, it is possible to produce a whole new $2/n$ table consisting entirely of four-term expressions:

$$\frac{2}{3} = \frac{1}{3} + \frac{1}{6} + \frac{1}{9} + \frac{1}{18}$$

$$\frac{2}{5} = \frac{1}{5} + \frac{1}{10} + \frac{1}{15} + \frac{1}{30}$$

$$\frac{2}{7} = \frac{1}{7} + \frac{1}{14} + \frac{1}{21} + \frac{1}{42}$$

$$\frac{2}{9} = \frac{1}{9} + \frac{1}{18} + \frac{1}{27} + \frac{1}{54} .$$

Although the scribe was presumably aware of this, nowhere did he accept these values for his table (except in the last case, $\frac{2}{101}$), because there were so many other, ''simpler'' representations available. To the modern mind it even seems that the scribe followed certain principles in assembling his lists. We note that:

1. Small denominators were preferred, with none greater than 1000.

2. The fewer the unit fractions, the better; and there were never more than 4.

3. Denominators that were even were more desirable than odd ones, especially for the initial term.

4. The smaller denominators came first, and no two were the same.

5. A small first denominator might be increased if the size of the others was thereby reduced (for example, $\frac{2}{31} = \frac{1}{20} + \frac{1}{124} + \frac{1}{155}$ was preferred to $\frac{2}{31} = \frac{1}{18} + \frac{1}{186} + \frac{1}{279}$).

Why—or even whether—these precepts were chosen, we cannot determine.

Example. In an illustration of multiplying with fractions, let us find the product of $2 + \frac{1}{4}$ and $1 + \frac{1}{2} + \frac{1}{7}$. Notice that doubling $1 + \frac{1}{2} + \frac{1}{7}$ gives $3 + \frac{2}{7}$, which the Egyptian mathematicians would have written $3 + \frac{1}{4} + \frac{1}{28}$. The work may be arranged as follows:

1	$1 + \frac{1}{2} + \frac{1}{7}$
✔ 2	$3 + \frac{1}{4} + \frac{1}{28}$
$\frac{1}{2}$	$\frac{1}{2} + \frac{1}{4} + \frac{1}{14}$
✔ $\frac{1}{4}$	$\frac{1}{4} + \frac{1}{8} + \frac{1}{28}$
totals $\quad 2 + \frac{1}{4}$	$3 + \frac{1}{2} + \frac{1}{8} + \frac{1}{14}$

The mathematicians knew that twice the unit fraction $1/(2n)$ is the unit fraction $1/n$, so the answer would appear as $3 + \frac{1}{2} + \frac{1}{8} + \frac{1}{14}$.

Example. For a more difficult division involving fractions, let us look at a calculation that occurs in Problem 33 of the Rhind Papyrus. One is required here to divide 37 by $1 + \frac{2}{3} + \frac{1}{2} + \frac{1}{7}$. In the standard form for an Egyptian division, the computation begins:

$$
\begin{array}{cl}
1 & 1 + \frac{2}{3} + \frac{1}{2} + \frac{1}{7} \\[4pt]
2 & 4 + \frac{1}{3} + \frac{1}{4} + \frac{1}{28} \\[4pt]
4 & 8 + \frac{2}{3} + \frac{1}{2} + \frac{1}{14} \\[4pt]
8 & 18 + \frac{1}{3} + \frac{1}{7} \\[4pt]
16 & 36 + \frac{2}{3} + \frac{1}{4} + \frac{1}{28}
\end{array}
$$

with the value for $\frac{2}{7}$ recorded as $\frac{1}{4} + \frac{1}{28}$. Now the sum $36 + \frac{2}{3} + \frac{1}{4} + \frac{1}{28}$ is close to 37. By how much are we short? Or as the scribe would say, "What completes $\frac{2}{3} + \frac{1}{4} + \frac{1}{28}$ up to 1?" In modern notation, it is necessary to get a fraction x for which

$$\frac{2}{3} + \frac{1}{4} + \frac{1}{28} + x = 1;$$

or with the problem stated another way, a numerator y is sought that will satisfy

$$\frac{2}{3} + \frac{1}{4} + \frac{1}{28} + \frac{y}{84} = 1,$$

where the denominator 84 is simply the least common multiple of the denominators, 3, 4, and 28. Multiplying both sides of this last equation by 84 gives $56 + 21 + 3 + y = 84$, and so $y = 4$. Therefore, the remainder that must be added to $\frac{2}{3} + \frac{1}{4} + \frac{1}{28}$ to make 1 is $\frac{4}{84}$, or $\frac{1}{21}$. The next step is to determine by what amount we should multiply $1 + \frac{2}{3} + \frac{1}{2} + \frac{1}{7}$ to get the required $\frac{1}{21}$. This means solving for z in the equation

$$z\left(1 + \frac{2}{3} + \frac{1}{2} + \frac{1}{7}\right) = \frac{1}{21}.$$

Multiplying through by 42 leads to $97z = 2$ or $z = \frac{2}{97}$, which the Egyptian scribe found to be equal to $\frac{1}{56} + \frac{1}{679} + \frac{1}{776}$. Thus, the whole calculation would proceed as follows:

$$
\begin{array}{cl}
1 & 1 + \frac{2}{3} + \frac{1}{2} + \frac{1}{7} \\[4pt]
2 & 4 + \frac{1}{3} + \frac{1}{4} + \frac{1}{28} \\[4pt]
4 & 8 + \frac{2}{3} + \frac{1}{2} + \frac{1}{14} \\[4pt]
8 & 18 + \frac{1}{3} + \frac{1}{7} \\[4pt]
16 & 36 + \frac{2}{3} + \frac{1}{4} + \frac{1}{28} \quad \swarrow \\[4pt]
\frac{1}{56} + \frac{1}{679} + \frac{1}{776} & \frac{1}{21} \quad \swarrow \\[2pt]
\hline
\text{totals} \qquad 16 + \frac{1}{56} + \frac{1}{679} + \frac{1}{776} & 37
\end{array}
$$

The result of dividing 37 by $1 + \frac{2}{3} + \frac{1}{2} + \frac{1}{7}$ is $16 + \frac{1}{56} + \frac{1}{679} + \frac{1}{776}$.

There are several modern ways of expanding a fraction with numerator other than 2 as a sum of unit fractions. Suppose that $\frac{9}{13}$ is required to be expanded. Because $9 = 1 + 4 \cdot 2$, one procedure might be to convert $\frac{9}{13}$ to

$$\frac{9}{13} = \frac{1}{13} + 4\left(\frac{2}{13}\right).$$

The fraction $\frac{2}{13}$ could be reduced by means of the $2/n$ table and the results collected to give a sum of unit fractions without repetitions:

$$\frac{9}{13} = \frac{1}{13} + 4(\frac{1}{8} + \frac{1}{52} + \frac{1}{104})$$

$$= \frac{1}{13} + \frac{1}{2} + \frac{1}{13} + \frac{1}{26}$$

$$= \frac{2}{13} + \frac{1}{2} + \frac{1}{26}$$

$$= (\frac{1}{8} + \frac{1}{52} + \frac{1}{104}) + \frac{1}{2} + \frac{1}{26} .$$

The final answer would then be

$$\frac{9}{13} = \frac{1}{2} + \frac{1}{8} + \frac{1}{26} + \frac{1}{52} + \frac{1}{104} .$$

What makes this example work is that the denominators 8, 52, and 104 are all divisible by 4. We might not always be so fortunate.

Although we shall not do so, it can be proved that every positive rational number is expressible as a sum of a finite number of distinct unit fractions. Two systematic procedures will accomplish this decomposition; for the lack of better names let us call these the splitting method and Fibonacci's method. The splitting method is based on the so-called splitting identity

$$\frac{1}{n} = \frac{1}{n + 1} + \frac{1}{n(n + 1)} ,$$

which allows us to replace one unit fraction by a sum of two others. For instance, to handle $\frac{2}{19}$ we first write

$$\frac{2}{19} = \frac{1}{19} + \frac{1}{19}$$

and then split one of the fractions $\frac{1}{19}$ into $1/20 + 1/19 \cdot 20$, so that

$$\frac{2}{19} = \frac{1}{19} + \frac{1}{20} + \frac{1}{380} .$$

Again, in the case of $\frac{3}{5}$, this method begins with

$$\frac{3}{5} = \frac{1}{5} + \frac{1}{5} + \frac{1}{5}$$

and splits each of the last two unit fractions into $1/6 + 1/5 \cdot 6$; thus,

$$\frac{3}{5} = \frac{1}{5} + (\frac{1}{6} + \frac{1}{30}) + (\frac{1}{6} + \frac{1}{30}).$$

There are several avenues open to us at this point. Ignoring the obvious simplifications $\frac{2}{6} = \frac{1}{3}$ and $\frac{2}{30} = \frac{1}{15}$, let us instead split $\frac{1}{6}$ and $\frac{1}{30}$ into the sums $1/7 + 1/6 \cdot 7$ and $1/31 + 1/30 \cdot 31$, respectively, to arrive at the decomposition

$$\cdot\frac{3}{5} = \frac{1}{5} + \frac{1}{6} + \frac{1}{30} + \frac{1}{7} + \frac{1}{42} + \frac{1}{31} + \frac{1}{930} .$$

In general, the method is as follows. Starting with a fraction m/n, first write

$$\frac{m}{n} = \frac{1}{n} + \left(\underbrace{\frac{1}{n} + \cdots + \frac{1}{n}}_{m - 1 \text{ summands}}\right).$$

Now use the splitting identity to replace $m - 1$ instances of the unit fraction $1/n$ by

$$\frac{1}{n + 1} + \frac{1}{n(n + 1)},$$

thereby getting

$$\frac{m}{n} = \frac{1}{n} + \frac{1}{n + 1} + \frac{1}{n(n + 1)} + \left[\left(\frac{1}{n + 1} + \frac{1}{n(n + 1)}\right) + \cdots + \left(\frac{1}{n + 1} + \frac{1}{n(n + 1)}\right)\right].$$

$$m - 2 \text{ summands}$$

Continue in this manner. At the next stage, the splitting identity, as applied to

$$\frac{1}{n + 1} \quad \text{and} \quad \frac{1}{n(n + 1)},$$

yields

$$\frac{m}{n} = \frac{1}{n} + \frac{1}{n + 1} + \frac{1}{n(n + 1)} + \frac{1}{n + 2} + \frac{1}{(n + 1)(n + 2)} + \frac{1}{n(n + 1) + 1}$$

$$+ \frac{1}{n(n + 1)[n(n + 1) + 1]} + \cdots.$$

Although the number of unit fractions (and hence the likelihood of repetition) is increasing at each stage, it can be shown that this process eventually terminates.

The second technique we want to consider is credited to the thirteenth century Italian mathematician Leonardo of Pisa, better known by his patronymic, Fibonacci. In 1202, Fibonacci published an algorithm for expressing any rational number between 0 and 1 as a sum of distinct unit fractions; this was rediscovered and more deeply investigated by J. J. Sylvester in 1880. The idea is this. Suppose that the fraction a/b is given, where $0 < a/b < 1$. First find the integer n_1 satisfying

$$\frac{1}{n_1} \leq \frac{a}{b} < \frac{1}{n_1 - 1},$$

—or what amounts to the same thing, determine n_1 in such a way that $n_1 - 1 < b/a \leq n_1$. These inequalities imply that $n_1 a - a < b \leq n_1 a$, whence $n_1 a - b < a$. Subtract $1/n_1$ from a/b and express the difference as a fraction, calling it a_1/b_1:

$$\frac{a}{b} - \frac{1}{n_1} = \frac{n_1 a - b}{bn_1} = \frac{a_1}{b_1}.$$

This enables us to write a/b as

$$\frac{a}{b} = \frac{1}{n_1} + \frac{a_1}{b_1}.$$

The important point is that $a_1 = n_1 a - b < a$. In other words, the numerator a_1 of this new fraction is smaller than the numerator a of the original fraction.

If $a_1 = 1$, there is nothing more to do. Otherwise, repeat the process with a_1/b_1 now playing the role of a/b, to get

$$\frac{a}{b} = \frac{1}{n_1} + \frac{1}{n_2} + \frac{a_2}{b_2}, \qquad \text{where } a_2 < a_1.$$

At each successive stage, the numerator of the remainder fraction decreases. We must eventually come to a fraction a_k/b_k in which $a_k = 1$; for the strictly decreasing sequence $1 \leq a_k < a_{k-1} < \cdots < a_1 < a$ cannot continue indefinitely. Thus, the desired representation of a/b is reached, with

$$\frac{a}{b} = \frac{1}{n_1} + \frac{1}{n_2} + \cdots + \frac{1}{n_k} + \frac{1}{b_k},$$

a sum of unit fractions.

Let us examine several examples illustrating Fibonacci's method.

Example. Take $a/b = \frac{2}{19}$. To find n_1, note that $9 < \frac{19}{2} < 10$, and so $\frac{1}{10} < \frac{2}{19} < \frac{1}{9}$; hence, $n_1 = 10$. Subtraction gives

$$\frac{2}{19} - \frac{1}{10} = \frac{20 - 19}{19 \cdot 10} = \frac{1}{190}.$$

We may therefore represent $\frac{2}{19}$ as $\frac{2}{19} = \frac{1}{10} + \frac{1}{190}$.

Example. For a more penetrating illustration, we turn to the fraction $a/b = \frac{9}{13}$ once again. Dividing 9 into 13, one gets $1 < \frac{13}{9} < 2$, leading to $\frac{1}{2} < \frac{9}{13} < 1$; hence, $n_1 = 2$. This means that the first unit fraction in the decomposition of $\frac{9}{13}$ is $\frac{1}{2}$. Now

$$\frac{9}{13} - \frac{1}{2} = \frac{18 - 13}{13 \cdot 2} = \frac{5}{26},$$

which implies that

$$\tfrac{9}{13} = \tfrac{1}{2} + \tfrac{5}{26}.$$

As expected, the numerator of the remainder fraction is less than the numerator of the given fraction; that is, $5 < 9$. Now repeat the process with the fraction $\frac{5}{26}$. Because $5 < \frac{26}{5} < 6$, we get $\frac{1}{6} < \frac{5}{26} < \frac{1}{5}$ and $n_2 = 6$. Carrying out the arithmetic gives

$$\frac{5}{26} - \frac{1}{6} = \frac{30 - 26}{26 \cdot 6} = \frac{4}{156} = \frac{1}{39},$$

in consequence of which

$$\tfrac{5}{26} = \tfrac{1}{6} + \tfrac{1}{39}.$$

Putting the pieces together, we get our expansion for $\frac{9}{13}$:

$$\tfrac{9}{13} = \tfrac{1}{2} + \tfrac{1}{6} + \tfrac{1}{39}.$$

2.3   Three Problems from the Rhind Papyrus

Much space is taken up in the Rhind Papyrus by practical problems concerning the equitable division of loaves among a certain number of men or determining the amount of grain needed for making beer. These problems were simple and did not go beyond a linear equation in one unknown. Problem 24, for example, reads: "A quantity and its $\frac{1}{7}$ added become 19. What is the quantity?" Today with our algebraic symbolism, we should let x stand for the quantity sought and the equation to be solved would be

$$x + \frac{x}{7} = 19 \qquad \text{or} \qquad \frac{8x}{7} = 19.$$

Ahmes reasoned that because his notation did not admit the fraction $\frac{8}{7}$, "As many times as 8 must be multiplied to give 19, just as many times must 7 be multiplied to give the correct number." The scribe was using the oldest and most universal procedure for treating linear equations, the method of false position, or false assumption. Briefly, this method is to assume any convenient value for the desired quantity, and by performing the operations of the problem at hand, to calculate a number that can then be compared with a given number. The true answer has the same relation to the assumed answer that the given number has to the number thus calculated.

For instance, in solving the equation $x + x/7 = 19$, one assumes falsely that $x = 7$ (the choice is convenient because $x/7$ is easy to calculate). The left-hand side of the equation would then become $7 + \frac{7}{7} = 8$, instead of the required answer 19. Because 8 must be multiplied by $\frac{19}{8} = 2 + \frac{1}{4} + \frac{1}{8}$ to give the desired 19, the correct value of x is obtained by multiplying the false assumption, namely 7, by $2 + \frac{1}{4} + \frac{1}{8}$. The result is

$$x = (2 + \tfrac{1}{4} + \tfrac{1}{8})7 = 16 + \tfrac{1}{2} + \tfrac{1}{8}.$$

Actually, we could pose any convenient value for the unknown quantity, say $x = a$. If $a + a/7 = b$ and $bc = 19$, then $x = ac$ satisfies the equation $x + x/7 = 19$; for it is easily seen that

$$ac + \frac{1}{7}ac = \left(a + \frac{a}{7}\right)c = bc = 19.$$

We have seen that the Egyptians anticipated, at least in an elementary form, a favorite method of the Middle Ages, the false position. Once the method was learned from the Arabs, it became a prominent feature of European mathematics texts from the *Liber Abaci* (1202) of Fibonacci to the arithmetics of the sixteenth century. As algebraic symbolism developed, the rule disappeared from the more advanced works. Following is an example taken from *Liber Abaci*. A certain man buys eggs at the rate of 7 for 1 denarius and sells them at a rate of 5 for 1 denarius, and thus makes a profit of 19 denarii. The question is: How much money did he invest? Algebraically, this problem would be expressed by the equation

$$\frac{7x}{5} - x = 19.$$

The procedure of false position consists here in assuming 5 for the unknown; then $\frac{7}{5} \cdot 5 - 5 = 2$. This 2, in the expressive language of Fibonacci, "would be like 19" (it is related to 19 as 5 is to the sought number). Because $2(\frac{19}{2}) = 19$, the correct answer is

$$x = 5(\frac{19}{2}) = 47\frac{1}{2} .$$

Notice that the number posed by Fibonacci for the unknown was not arbitrarily chosen—when the coefficient of an unknown is a fraction, the number assumed for the unknown is the denominator of the fraction.

Thus far we have considered the rule of false position in which a single guess was made; but there was a variant that necessitated making two trials and noting the error due to each. This cumbersome rule of double false position, as it was sometimes called, can be explained as follows. To solve the equation $ax + b = 0$, let g_1 and g_2 be two guesses about the value of x, and let f_1 and f_2 be the corresponding failures, that is, the values $ag_1 + b$ and $ag_2 + b$, which would equal zero if the guesses were right. Then

(1) $\qquad ag_1 + b = f_1 \qquad$ and $\qquad$ (2) $\qquad ag_2 + b = f_2.$

On subtracting, it is clear that

(3) $$a(g_1 - g_2) = f_1 - f_2.$$

Multiplying equation (1) by g_2 and equation (2) by g_1 gives

$$ag_1g_2 + bg_2 = f_1g_2 \qquad \text{and} \qquad ag_2g_1 + bg_1 = f_2g_1.$$

When these last two equations are subtracted, the result is

(4) $$b(g_2 - g_1) = f_1g_2 - f_2g_1.$$

To finish the argument, divide (4) by (3) to get

$$-\frac{b}{a} = \frac{f_1g_2 - f_2g_1}{f_1 - f_2} .$$

But because $x = -b/a$, the value of x is found to be

$$x = \frac{f_1g_2 - f_2g_1}{f_1 - f_2} .$$

In summary, we have placed two false values for x in the expression $ax + b$, and from these trials we have been able to get the correct solution to the equation $ax + b = 0$.

To make things more specific, let us look at an actual example, for instance, the equation

$$x + \frac{x}{7} = 19, \qquad \text{or equivalently,} \qquad x + \frac{x}{7} - 19 = 0.$$

We take two guesses about the value of x, say $g_1 = 7$ and $g_2 = 14$. Then

$$7 + \tfrac{7}{7} - 19 = -11 = f_1 \qquad \text{and} \qquad 14 + \tfrac{14}{7} - 19 = -3 = f_2.$$

It follows that the true value of x is

$$x = \frac{f_1 g_2 - f_2 g_1}{f_1 - f_2} = \frac{(-11)14 - (-3)7}{(-11) - (-3)} = \frac{133}{8} = 16 + \frac{1}{2} + \frac{1}{8} .$$

Awkward as it seems, there is a certain element of simplicity in this primitive rule, and no wonder it was used even in the late 1800s. In his *Grounde of Artes*, Robert Recorde (1510–1558) wrote that he astonished his friends by proposing difficult questions and then, with the rule of falsehood, finding the true result from the chance answers of "such children or idiots that happened to be in the place."

Getting back to the Rhind Papyrus, we can consider Problem 28 the earliest example of a "think of a number" problem. Let us state this problem and Ahmes' solution in modern terms, adding a few clarifying details.

Example. Think of a number, and add $\frac{2}{3}$ of this number to itself. From this sum subtract $\frac{1}{3}$ its value and say what your answer is. Suppose the answer was 10. Then take away $\frac{1}{10}$ of this 10, giving 9. Then this was the number first thought of.

Proof. If the original number was 9, then $\frac{2}{3}$ is 6, which added makes 15. Then $\frac{1}{3}$ of 15 is 5, which on subtraction leaves 10. That is how you do it.

Here the scribe was really illustrating the algebraic identity

$$\left(n + \frac{2n}{3} \right) - \frac{1}{3}\left(n + \frac{2n}{3} \right) - \frac{1}{10}\left[\left(n + \frac{2n}{3} \right) - \frac{1}{3}\left(n + \frac{2n}{3} \right) \right] = n$$

by a simple example, in this case using the number $n = 9$. Having disclosed his "obscure secret," he added a traditional concluding phrase, "And that is how you do it."

Problem 79 is extremely concise and contains a curious set of data—which seems to indicate an acquaintance with the sum of a geometric series:

		Houses		7
		Cats		49
1	2801	Mice		343
2	5602	Sheaves		2401
4	11,204	Hekats (measures of grain)		16,807
total	19,607		total	19,607

This catalog of miscellany has suggested some fanciful ideas. Certain authorities regard these words as symbolic terminology given to the first five powers of 7. For at the right, we have the summation of 7, 7^2, 7^3, 7^4, and 7^5 by actual addition. At the left, the sum of the same series is given as $7 \cdot 2801$, with the multiplication carried out by the usual method of duplication. Because $2801 = (7^5 - 1)/(7 - 1)$, the result

$$7 \cdot 2801 = 7\left(\frac{7^5 - 1}{7 - 1} \right) = 7 + 7^2 + 7^3 + 7^4 + 7^5$$

is exactly what would be obtained by substitution in the modern formula for the sum S_n of n terms of a geometric series:

$$S_n = a + ar + ar^2 + \cdots + ar^{n-1} = a\frac{r^n - 1}{r - 1}.$$

(We note in the problem before us that $a = r = 7$ and $n = 5$.) Was such a formula, even for simpler cases, known to the Egyptians? There is no concrete evidence that it was. A more plausible interpretation of what is intended is something of the sort: "In each of seven houses there are seven cats; each cat kills seven mice; each mouse would have eaten seven sheaves of wheat; and each sheaf of wheat was capable of yielding seven hekat measures of grain. How much grain was thereby saved?" Or one may prefer the question, "Houses, cats, mice, sheaves, and hekats of grain—how many of these were there in all?"

Some 3000 years after Ahmes, Fibonacci included in his *Liber Abaci* the same series of powers of seven with one further term:

Seven old women were on the road to Rome;
Each woman had seven donkeys;
Each donkey carried seven sacks;
Each sack contained seven loaves of bread;
With each loaf were seven knives;
Each knife was in seven sheaths.
What is the total?

This rendering, coupled with the number seven, reminds us of an Old English children's rhyme, one version of which appears below:

As I was going to Saint Ives,
I met a man with seven wives.
Each wife had seven sacks;
Every sack had seven cats;
Every cat had seven kits;
Kits, cats, sacks, and wives,
How many were going to Saint Ives?

Here also, it is suggested that the sum total of a geometric progression be calculated, but there is a joker in the actual wording of the first and last lines. While the surprise twist is in all likelihood an Anglo-Saxon contribution, one can see how the same kind of problem perpetuated itself throughout centuries.

The Rhind Papyrus closes with the following prayer, expressing the principal worries of an agricultural community: "Catch the vermin and the mice, extinguish the noxious weeds; pray to the God Ra for heat, wind, and high water."

Looking at the extant Egyptian mathematical manuscripts as a whole, we find that they are nothing but collections of practical problems of the kind that are associated with business and administrative transactions. The teaching of the art of calculation appears to be the chief element in the problems. Everything is stated in terms of specific

Part of the Rhind Papyrus. (Copyright British Museum.)

numbers, and nowhere does one find a trace of what might properly be called a theorem or a general rule of procedure. If the criterion for scientific mathematics is the existence of the concept of proof, the ancient Egyptians confined themselves to ''applied arithmetic.'' Perhaps the best explanation of why the Egyptians never got beyond this relatively primitive level is that they had a natural, but unfortunate, idea of admitting only fractions with numerator one; thus even the simplest calculations became slow and laborious. It is hard to say whether the symbolism prevented the use of fractions with other numerators or whether the exclusive use of unit numerators was the reason for the symbolism they used to express fractions. The handling of fractions always remained a special art in Egyptian mathematics and can best be described as a retarding force on numerical procedures.

As evidenced by the Akhmin Papyrus (named after the city on the upper Nile where it was discovered), it appears that the methods of the scribe Ahmes were still in vogue centuries later. This document, written in Greek at some point between A.D. 500 and 800, closely resembles the Rhind Papyrus. Its author, like his ancient predecessor Ahmes, gave tables of fractions decomposed into unit fractions. Why did Egyptian mathematics remain so remarkably the same for more than 2000 years? Perhaps the Egyptians entered their discoveries in sacred books, and in later ages, it was considered heresy to modify the method or result. Whatever the explanation, the mathematical attainments of Ahmes were those of his ancestors and of his descendants.

2.3 Problems

1. Use the Egyptian method of doubling to find the following products:

   (a) $18 \cdot 25$;   (b) $26 \cdot 33$;
   (c) $85 \cdot 21$;   (d) $105 \cdot 59$.

2. Find, in the Egyptian fashion, the following quotients:

   (a) $184 \div 8$;
   (b) $19 \div 8$;
   (c) $47 \div 9$;
   (d) $1060 \div 12$;
   (e) $61 \div 8$.

3. Use the Egyptian method of multiplication to calculate the following products:

   (a) $(11 + \frac{1}{2} + \frac{1}{8})37$
   (b) $(1 + \frac{1}{2} + \frac{1}{4})(9 + \frac{1}{2} + \frac{1}{4})$
   (c) $(2 + \frac{1}{4})(1 + \frac{1}{2} + \frac{1}{4})$

4. (a) Show that the product of $\frac{1}{14}$ by $1 + \frac{1}{2} + \frac{1}{4}$ is equal to $\frac{1}{8}$ (Problem 12 of the Rhind Papyrus).

   (b) Show that the product of $\frac{1}{32} + \frac{1}{224}$ by $1 + \frac{1}{2} + \frac{1}{4}$ is equal to $\frac{1}{16}$ (Problem 15 of the Rhind Papyrus).

5. Problem 30 of the Rhind Papyrus asks the reader to find a quantity such that $\frac{2}{3} + \frac{1}{10}$ of it will make 10. Do this as the Egyptians would have done, first by confirming that

   $$13(\tfrac{2}{3} + \tfrac{1}{10}) = 9 + \tfrac{29}{30}$$

   and then determining by what amount $\frac{2}{3} + \frac{1}{10}$ should be multiplied to give $\frac{1}{30}$.

6. (a) Show that

   $$\frac{2}{n} = \frac{1}{3}\frac{1}{n} + \frac{5}{3}\frac{1}{n},$$

   hence that $2/n$ can be expressed as a sum of unit fractions whenever n is divisible by 5.

   (b) Use part (a) to obtain the unit fraction decompositions of $\frac{2}{25}$, $\frac{2}{65}$, and $\frac{2}{85}$ as given in the Rhind Papyrus.

7. (a) Show that

   $$\frac{2}{n} = \frac{1}{2}\frac{1}{n} + \frac{3}{2}\frac{1}{n},$$

   hence that $2/n$ can be expressed as a sum of unit fractions whenever n is divisible by 3.

   (b) Use part (a) to obtain the unit fraction decompositions of $\frac{2}{21}$, $\frac{2}{75}$, and $\frac{2}{99}$.

8. Show that

   $$\frac{2}{mn} = \frac{1}{m}\frac{1}{k} + \frac{1}{n}\frac{1}{k},$$

   where the number $k = (m + n)/2$. Use this to get the unit fraction decompositions of $\frac{2}{7}$, $\frac{2}{35}$, and $\frac{2}{91}$ as given in the Rhind Papyrus.

9. Verify that

   $$\frac{2}{n} = \frac{1}{n} + \frac{1}{2n} + \frac{1}{3n} + \frac{1}{6n}$$

   and use this fact to obtain the unit fraction decomposition of $\frac{2}{101}$ as given in the Rhind Papyrus.

10. Suppose that n is divisible by 7. Find a formula similar to that of Problem 6(a) that will represent $2/n$ as a sum of unit fractions.

11. Using the $2/n$ table, write $\frac{13}{15}$, $\frac{9}{49}$, and $\frac{19}{35}$ as sums of unit fractions without repetitions.

12. Represent $\frac{3}{7}$, $\frac{4}{15}$, and $\frac{7}{29}$ as sums of distinct unit fractions using (a) the splitting identity and (b) Fibonacci's method.

13. Show that if n divides $m + 1$, the fraction n/m can always be written as a sum of two unit fractions. Illustrate this with a specific example.

14. Expand $\frac{9}{13}$ as a sum of distinct unit fractions using the splitting identity.

15. Find a unit fraction representation of $\frac{2}{5}$ that involves at least six terms. Do the same for $\frac{2}{3}$.

16. Represent $\frac{2}{11}$ and $\frac{2}{17}$ as sums of distinct unit fractions using Fibonacci's method.

17. A method for writing $2/n$, where n is an odd number, as a sum of unit fractions, proceeds as follows. Given an integer m, put $2/n = 2m/(nm)$. If from among the divisors of nm a set can be chosen whose sum equals $2m$, take these divisors as the numerators of fractions whose denominators are nm. The result is a unit fraction decomposition of $2/n$. For $\frac{2}{19}$, we might let $m = 12$, so that $\frac{2}{19} = \frac{24}{228}$. From the divisors 1, 2, 3, 4, 6, 12, 19, . . . of 228, it is possible to find four sets of integers whose sums are each 24; specifically,

$$24 = 1 + 4 + 19 = 2 + 3 + 19$$
$$= 2 + 4 + 6 + 12$$
$$= 1 + 2 + 3 + 6 + 12.$$

Using these, one gets

$$\frac{2}{19} = \frac{1}{228} + \frac{4}{228} + \frac{19}{228} = \frac{1}{228} + \frac{1}{57} + \frac{1}{12};$$

$$\frac{2}{19} = \frac{2}{228} + \frac{3}{228} + \frac{19}{228} = \frac{1}{114} + \frac{1}{76} + \frac{1}{12};$$

$$\frac{2}{19} = \frac{2}{228} + \frac{4}{228} + \frac{6}{228} + \frac{12}{228} = \frac{1}{114} + \frac{1}{57} + \frac{1}{38} + \frac{1}{19};$$

$$\frac{2}{19} = \frac{1}{228} + \frac{2}{228} + \frac{3}{228} + \frac{6}{228} + \frac{12}{228} = \frac{1}{228} + \frac{1}{114} + \frac{1}{76}$$
$$+ \frac{1}{38} + \frac{1}{19}.$$

Applying this technique, obtain unit fraction expansions of $\frac{2}{15}$ and $\frac{2}{43}$. [*Hint:* Take $m = 4$ and $m = 12$, respectively.]

18. Consider the following variation of Problem 17 for writing $2/n$ as a sum of unit fractions. Choose an integer N having a set of divisors whose sum is $2N - n$, say,

$$N = d_1 M_1 = d_2 M_2 = d_3 M_3 \qquad \text{and}$$
$$2N = n + d_1 + d_2 + d_3.$$

Then

$$2 = \frac{n}{N} + \frac{d_1}{N} + \frac{d_2}{N} + \frac{d_3}{N}$$
$$= \frac{n}{N} + \frac{1}{M_1} + \frac{1}{M_2} + \frac{1}{M_3},$$

whence $2/n$ may be decomposed as

$$\frac{2}{n} = \frac{1}{N} + \frac{1}{M_1 n} + \frac{1}{M_2 n} + \frac{1}{M_3 n}.$$

In the case of $\frac{2}{19}$, we might take $N = 18$, so that

$$18 = 2 \cdot 9 = 6 \cdot 3 = 9 \cdot 2$$

and

$$2 \cdot 18 = 19 + 2 + 6 + 9.$$

Using these relations, one gets

$$\frac{2}{19} = \frac{1}{18} + \frac{1}{171} + \frac{1}{57} + \frac{1}{38}.$$

Apply this technique to obtain unit fraction expansions for $\frac{2}{15}$ and $\frac{2}{43}$. [*Hint:* Take $N = 12$ and $N = 36$.]

19. Problems 3–6 of the Rhind Papyrus describe four practical problems: the division of 6, 7, 8, and 9 loaves equally among 10 men. Solve each of these problems by false position, expressing the answers in unit fractions.

20. Problems 25, 26, and 27 of the Rhind Papyrus are as follows:

Problem 25. A quantity and its $\frac{1}{2}$ added together become 16. What is the quantity?

Problem 26. A quantity and its $\frac{1}{4}$ added together become 15. What is the quantity?

Problem 27. A quantity and its $\frac{1}{5}$ added together become 21. What is the quantity?

Solve each of these problems by false position, expressing the answers in unit fractions.

21. Problem 31 of the Rhind Papyrus states: A quantity and its $\frac{2}{3}$, its $\frac{1}{2}$, and its $\frac{1}{7}$ added together become 33. What is the quantity? Using modern notation, this calls for solving the equation

$$x + \frac{2x}{3} + \frac{x}{2} + \frac{x}{7} = 33.$$

Show that the scribe's answer

$$x = 14 + \frac{1}{4} + \frac{1}{56} + \frac{1}{97} + \frac{1}{194} + \frac{1}{388} + \frac{1}{679} + \frac{1}{776}$$

is correct.

$$\left[Hint: x = \frac{42 \cdot 33}{97} = 14 + \frac{28}{97}. \right]$$

22. Solve Problem 32 of the Rhind Papyrus, which states: A quantity, its $\frac{1}{3}$, its $\frac{1}{4}$, added together become 2. What is the quantity? Express the answer in the Egyptian fashion.

23. In Problem 70 of the Rhind Papyrus, one is asked to find the quotient when 100 is divided by $7 + \frac{1}{2} + \frac{1}{4} + \frac{1}{8}$; do this. [*Hint:* At some point in the calculation take $\frac{2}{3}$ of $7 + \frac{1}{2} + \frac{1}{4} + \frac{1}{8}$. Also note that the relation $8(7 + \frac{1}{2} + \frac{1}{4} + \frac{1}{8}) = 63$ implies that $\frac{2}{63}(7 + \frac{1}{2} + \frac{1}{4} + \frac{1}{8}) = \frac{1}{4}$.]

24. Problem 40 of the Rhind Papyrus concerns an arithmetic progression of five terms. It states:

Divide 100 loaves among 5 men so that the sum of the three largest shares is 7 times the sum of the two smallest.

(a) Solve this problem by modern techniques.

(b) Using the method of false position, the scribe assumed a common difference of $5 + \frac{1}{2}$ and the smallest share of 1 (hence, the five shares are 1, $6 + \frac{1}{2}$, 12, $17 + \frac{1}{2}$, 23). Obtain the correct answer from these assumptions.

2.4 Egyptian Geometry

The generally accepted account of the origin of geometry is that it came into being in ancient Egypt, where the yearly inundations of the Nile demanded that the size of landed property be resurveyed for tax purposes. Indeed, the name "geometry," a compound of two Greek words meaning "earth" and "measure," seems to indicate that the subject arose from the necessity of land surveying. The Greek historian Herodotus, who visited the Nile about 460–455 B.C., described how the first systematic geometric observations were made.

> They said also that this king [Sesostris] divided the land among all Egyptians so as to give each one a quadrangle of equal size and to draw from each his revenues, by imposing a tax to be levied yearly. But every one from whose part the river tore away anything, had to go to him and notify what had happened. He then sent the overseers, who had to measure out by how much the land had become smaller, in order that the owner might pay on what was left, in proportion to the entire tax imposed. In this way, it appears to me, geometry originated.

Whatever opinion is ultimately adopted regarding the first steps in geometry, it does seem safe to assume that in a country where cultivating even the smallest portion of fertile soil was a matter of concern, land measurement became increasingly important. To this end must be ascribed some of the remarkable results the Egyptians obtained in mathematics.

The task of surveying was performed by specialists whom the later Greeks called rope-stretchers, or rope-fasteners, because their main tool apparently was a rope with knots or marks at equal intervals. In a passage written about 420 B.C., the Greek philosopher Democritus (460–370 B.C.) testifies that in his time the Egyptian surveyors still ranked high among the great geometers, possessing a skill almost equal to his own. He proudly boasted, "No one can surpass me in the construction of plane figures with proof, not even the so-called rope-stretchers among the Egyptians."

What occupied the Egyptian geometers of some 4000 years ago? The mathematical papyri that have come down to us contain numerous concrete examples, without any theoretical motivation, of prescription-like rules for determining areas and volumes of the most familiar plane and solid figures. Such rules of calculation must be recognized as strictly empirical results, the accretion of ages of trial-and-error experience and observation. The Egyptians sought useful facts relating to measurement, without having to demonstrate such facts by any process of deductive reasoning. Some of their formulas were only approximately correct, but they gave results of sufficient acceptability for the practical needs of everyday life.

In the great dedicatory inscription, of about 100 B.C., in the Temple of Horus at Edfu, there are references to numerous four-sided fields that were gifts to the temple. For each of these, the areas were obtained by taking the product of the averages of the two pairs of opposite sides, that is, by using the formula

$$A = \tfrac{1}{4}(a + c)(b + d),$$

where a, b, c, d are the lengths of the consecutive sides. The formula is obviously incorrect. It gives a fairly accurate answer only when the field is approximately rectangular. What is interesting is that this same erroneous formula for the area of a quadrilateral had appeared 3000 years earlier in ancient Babylonia.

The geometrical problems of the Rhind Papyrus are those numbered 41–60, and are largely concerned with the amounts of grain stored in rectangular and cylindrical granaries. Perhaps the best achievement of the Egyptians in two-dimensional geometry was their method for finding the area of a circle, which appears in Problem 50:

> Example of a round field of a diameter 9 khet. What is its area? Take away $\frac{1}{9}$ of the diameter, namely 1; the remainder is 8. Multiply 8 times 8; it makes 64. Therefore it contains 64 setat of land.

The scribes' process for finding the area of a circle can thus be simply stated: Subtract from the diameter its $\frac{1}{9}$ part and square the remainder. In modern symbols, this amounts to the formula

$$A = \left(d - \frac{d}{9}\right)^2 = \left(\frac{8d}{9}\right)^2,$$

where d denotes the length of the diameter of the circle. If we compare this with the actual formula for the area of the circle, namely $\pi d^2/4$, then

$$\frac{\pi d^2}{4} = \left(\frac{8d}{9}\right)^2,$$

so that we get

$$\pi = 4(\tfrac{8}{9})^2 = 3.1605 \ldots$$

for the Egyptian value of the ratio of the circle's circumference to its diameter. This is a close approximation to $3\frac{1}{7}$, which many students find good enough for practical purposes.

In the Old Babylonian period (roughly 1800–1600 B.C.), the circumference of a circle was found by taking three times its diameter. Putting this equal to πd, we see that their calculation is equivalent to using 3 for the value of π. The Hebrews used the same value in the Old Testament, for example, in I Kings 7:23, wherein the dimensions of the bath in the temple of Solomon are described. The verse was written about 650 B.C., and may have been taken from temple records dating back to 900 B.C. It reads, "And he made a molten sea, 10 cubits from one brim to the other: it was round all about . . . : and a line of 30 cubits did compass it round about." A cuneiform tablet discovered at Susa by a French archeological expedition in 1936 (the interpretation of which was published in 1950) seems to indicate that the Babylonian writer adopted 3;7,30 or $3\frac{1}{8}$ as the value of π. This is at least as good as the approximation found by the Egyptians.

We have no direct knowledge about how the formula $A = (8d/9)^2$ for the area of a circle was arrived at, but it is possible that Problem 48 of the papyrus provides a hint. In this problem, the usual statement of what the author proposed to do was replaced by a figure that, although drawn quite roughly, most strongly suggests a square with four triangles at the vertices. In the middle of the figure is the demotic sign for 9. Thus it appears that the scribe formed an octagon from a square of side 9 units by

trisecting the sides and cutting off the four corner isosceles triangles (each triangle having an area of $\frac{9}{2}$ square units). The scribe may have concluded that the octagon was approximately equal in area to the circle inscribed in the square, because some portions of the inscribed circle lie outside the octagon and some portions lie inside, and these appear to be roughly equal.

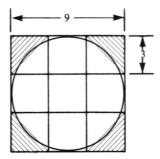

Now the area of the octagon equals the area of the original square less the areas of the four isosceles triangles made up of the four cut-off corners; that is,

$$A = 9^2 - 4(\tfrac{9}{2}) = 63.$$

This is nearly the value that is obtained by taking $d = 9$ in the expression $(8d/9)^2$. Thus a possible explanation of the area formula $A = (8d/9)^2$ is that it arose from considering the octagon as a first approximation to the circle inscribed in a square.

Problem 52 of the Rhind Papyrus calls for finding the area of a trapezoid (described as a truncated triangle) with apparently equal slanting sides; the lengths 6 and 4 of the parallel sides and the length 20 of an oblique side are given.

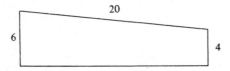

The calculation is carried out by means of the formula

$$A = \tfrac{1}{2}(b + b')h.$$

Did the author of the papyrus think that the area of a trapezoid was half the sum of the length of the parallel sides times the slant height, or was one oblique side intended to be perpendicular to the parallel sides? In the latter case, he would have been correct. It is not at all unlikely that the diagram, which is little more than a rough sketch, is badly drawn and that one of the seemingly equal sides is really meant to be perpendicular to the parallel sides.

There are only 25 problems in the Moscow Papyrus, but one of them contains the masterpiece of ancient geometry. Problem 14 shows that the Egyptians of about 1850 B.C. were familiar with the correct formula for the volume of a truncated square pyramid (or frustum). In our notation, this is

$$V = \frac{h}{3}(a^2 + ab + b^2),$$

where h is the altitude and a and b are the lengths of the sides of the square base and square top, respectively.

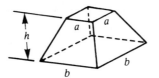

The figure associated with Problem 14 looks like an isosceles trapezoid,

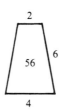

but the calculations indicate that the frustum of a square pyramid is intended. The exact text in this connection may be given:

Example of calculating a truncated pyramid. If you are told: a truncated pyramid of 6 for the vertical height by 4 on the base by 2 on the top: You are to square this 4; result 16. You are to double 4; result 8. You are to square this 2; result 4. You are to add the 16 and the 8 and the 4; result 28. You are to take $\frac{1}{3}$ of 6; result 2. You are to take 28 twice; result 56. See, it is of 56. You will find it right.

Although this solution deals with a particular problem and not with a general theorem, it is still breathtaking; some historians of mathematics have praised this achievement as the greatest of the Egyptian pyramids.

It is generally accepted that the Egyptians were acquainted with a formula for the volume of the complete square pyramid, and that it probably was the correct one,

$$V = \frac{h \cdot}{3}a^2.$$

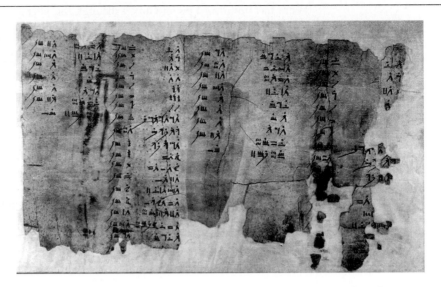

Extract from the Mathematical Leather Scroll, containing simple relations between fractions such as $\frac{1}{9} + \frac{1}{18} = \frac{1}{6}$. (Courtesy British Museum.)

In analogy with the formula $A = \frac{1}{2}bh$ for the area of a triangle, the Egyptians may have guessed that the volume of a pyramid was a constant times ha^2. We may suppose even that they guessed the constant to be 1/3. But the formula

$$V = \frac{h}{3}(a^2 + ab + b^2)$$

could not very well be a guess. It could have been obtained only by some sort of geometric analysis or by algebra from $V = (h/3)a^2$. It is not, however, an easy task to reconstruct a method by which they could have deduced the formula for the truncated pyramid with the materials available to them.

Any survey of the mathematics of the Egyptians ought to include a brief reference to the Great Pyramid at Gizeh, erected about 2600 B.C. by Khufu, whom the Greeks called Cheops. It provides monumental evidence of an appreciation of geometric form and a relatively high development of engineering construction, not to mention a very remarkable social and governmental organization. According to Herodotus, 400,000 workmen labored annually on the Great Pyramid for 30 years—four separate groups of 100,000, each group employed for three months. (Calculations indicate that no more than 36,000 men could have worked on the pyramid at one time without hampering one another's movements.) Ten years were spent constructing a road to a limestone quarry some miles distant, and over this road were dragged 2,300,000 blocks of stone averaging $2\frac{1}{2}$ tons and measuring 3 feet in each direction. These blocks were fitted together so perfectly that a knife blade cannot be inserted in the joints.

What has impressed people down through the years is not the aesthetic quality of the Great Pyramid but its size; it was the largest building of ancient times and one of the largest ever erected. When it was intact, it rose 481.2 feet (the top 31 feet are now missing), its four sides inclined at an angle of about 51° 51′ with the ground, and the base occupied 13 acres—an area equal to the combined base areas of the cathedrals of Florence and Milan, St. Peter's in Rome, and St. Paul's Cathedral and Westminster Abbey in London. Even more amazing was the accuracy with which it was put together. The base was almost a perfect square, no one of the four sides differing from the mean length of 755.78 feet by more than $4\frac{1}{2}$ inches. By using one of the celestial bodies, the Cheops builders were able to orient the sides of the pyramid almost exactly with the four cardinal points of the compass, the error being only fractions of one degree.

The Great Pyramid has down to the present fired adventurous minds to the wildest speculations. These pyramid mystics (or as they are sometimes uncharitably called, pyramidiots) have ascribed to the ancient builders all sorts of metaphysical intentions and esoteric knowledge. Among the extraordinary things claimed is that the pyramid was built so that half the perimeter of the base divided by the height should be exactly equal to π. While the difference between the two values

$$\pi = 3.1415926\ldots \qquad \text{and} \qquad \frac{2(755.78)}{481.2} = 3.14123\ldots$$

is only 0.00036 . . . , their closeness is merely accidental and has no basis in any mathematical law.

The Egyptian priests, according to a fiction that has crept into the recent literature, told Herodotus that the dimensions of the Great Pyramid were so chosen that the area of each face would be the same as the area of a square with sides equal to the Pyramid's height. Writing $2b$ for the length of a side of the base, a for the altitude of a face triangle, and h for the height of the pyramid, we find that Herodotus's relation is expressed by the equation

$$h^2 = \frac{1}{2}(2b \cdot a) = ab.$$

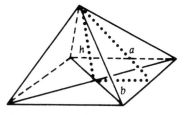

The Pythagorean theorem tells us that because a is the hypotenuse of a right triangle with legs b and h, then $h^2 + b^2 = a^2$, or $h^2 = a^2 - b^2$. Equating the two expressions for h^2, we get

$$a^2 - b^2 = ab.$$

When both sides are divided by a^2, this last equation becomes

$$1 - \left(\frac{b}{a}\right)^2 = \frac{b}{a} \text{ , or equivalently, } \left(\frac{b}{a}\right)^2 + \left(\frac{b}{a}\right) = 1.$$

Now the value of the positive root of the quadratic equation $x^2 + x = 1$ is $x = \frac{1}{2}(\sqrt{5} - 1)$. Then we get the ratio

$$\frac{b}{a} = \tfrac{1}{2}(\sqrt{5} - 1) = 0.6180339 \ldots \text{ ,}$$

the reciprocal of the "golden ratio," a value which has proved significant many times in mathematics and its applications.

How successful were the pyramid builders in achieving the golden ratio (if that, indeed, was their aim)? Checking with the actual measurements taken at the Great Pyramid, we see that

$$a = \sqrt{h^2 + b^2} = \sqrt{(481.2)^2 + (377.89)^2} = 611.85$$

leading to the value

$$\frac{b}{a} = 0.61762 \ldots \text{ .}$$

The theory that the Egyptians intended to use the golden ratio as a theoretical basis for building the Great Pyramid seems to have been first set down by a certain John Taylor, who in 1859 published *The Great Pyramid, Why Was It Built and Who Built It?* An amateur mathematician, Taylor spent 30 years collecting and comparing measurements reported by successive visitors to the Pyramid. Although he made a number of scale models of the Pyramid, he never set eyes on it himself. Because the only passage in Herodotus's *History* concerning its size reads, "Its base is square, each side is 800 feet long and its height is the same," a leap of faith would be required to justify Taylor's claim. Moreover, the dimensions Herodotus recorded are themselves way off the mark.

Another theory that is often taken as gospel is that the total area of the pyramid can be expressed in a way that leads to the golden ratio; that is, the area of the base is to the sum of the areas of the triangular faces as this sum is to the sum of the areas of the faces and base. Because the sum of the areas of the four face triangles is $4 \cdot \frac{1}{2}(2ba)$ and the area of the base is $(2b)^2$, the claim reduces to the assertion that

$$\frac{4b^2}{4ab} = \frac{4ab}{4ab + 4b^2} \text{ ,}$$

or in an equivalent form,

$$\frac{b}{a} = \frac{a}{a + b} \text{ .}$$

Using the previously calculated value for a, one finds that

$$\frac{a}{a + b} = \frac{611.85}{989.74} = 0.61819 \ldots \text{ ,}$$

whence the quotients b/a and $a/a + b$ are nearly the same. Whether this is a matter of accident or design is open to speculation.

There are some wilder theories. Some people maintained, for instance, that the Egyptians had erected the pyramids as dikes to keep the sands of the desert from moving and covering the cultivated area along the Nile. A popular belief during the Middle Ages was that they were granaries the captive Hebrews were forced to build for storing corn during the years of plenty. This legend has been preserved in the mosaics, done about A.D. 1250, of the Church of Saint Mark in Venice. Part of the pictorial narrative of the story of Joseph shows his brother being sent to fetch sheaves of grain from the pyramids. Speculation began to assume a more scientific appearance in 1864, when one highly regarded professor of astronomy (Charles Piazzi Smyth, the astronomer royal of Scotland) worked out to his own satisfaction a unit of measurement for the Great Pyramid, which he called the pyramid inch, equal to 1.001 of our inches. Using this mystical "pyramid inch" to measure the bumps and cracks along the walls of the interior passages and chambers, he concluded that the Great Pyramid was designed by God as an instrument of prophecy, a so-called Bible in stone. (The British Egyptologist Flinders Petrie wrote that he once caught one of the pyramid cultists surreptitiously filing down a stone protuberance in order to make its measurements conform to his theories.) If one knew how to read its message, there would be found in the pyramid all sorts of significant information about the history and future of humanity: the Great Flood, the birth of Christ, the beginning and end of World War I, and so on. When Smyth dated the start of the First World War as 1913, his believers jubilantly pointed out that he had erred "by only one year." Smyth and his followers posed fanciful, extravagant theories about the "secrets" locked in the measurements of the Great Pyramid. Their near miss in foretelling the date of the great war notwithstanding, these many enthusiastic speculations must be dismissed as stuff and nonsense.

Although we can be certain that the pyramid builders had already a fair knowledge of geometry, singularly little mathematics of this period has come down to us. Our two chief mathematical papyri, although different in age, may be said to represent the state of the subject at the time 2000–1750 B.C. Reviewing everything, we are forced to conclude that Egyptian geometry never advanced beyond an intuitive stage, in which the measurement of tangible objects was the chief consideration. The geometry of that period lacked deductive structure—there were no theoretical results, nor any general rules of procedure. It supplied only calculations, and these sometimes approximate, for problems that had a practical bearing in construction and surveying.

2.4   Problems

1. Solve the following geometrical problems from the Rhind Papyrus.

    (a) *Problem 41*. A cylindrical granary of diameter 9 cubits and height 10 cubits. What is the amount of grain that goes into it? [*Hint:* Use the Egyptian value for π, namely $4(\frac{8}{9})^2$, to get the scribe's answer.]

    (b) *Problem 51*. Example of a triangle of land. Suppose it is said to thee, what is the area of a triangle of side 10 khet and of base 4 khet? [*Hint:* The accompanying figure is apparently intended to be a right-angled triangle.]

    (c) *Problem 58*. If a pyramid (square) is $93\frac{1}{3}$ cubits high and the side of its base is 140 cubits long, what is its seked? [*Hint:* Given an isosceles triangle of base b and height h, we know that its seked equals $b/2h$, or the cotangent ratio of trigonometry. The seked in this problem has been associated with the slope of the faces of the Second Pyramid at Gizeh.]

2. (a) The Babylonians generally determined the area of a circle by taking it as equal to $\frac{1}{12}$ the square of the circle's circumference. Show that this is equivalent to letting $\pi = 3$.

(b) A Babylonian tablet excavated in 1936 asserts that when a more accurate determination of area is needed, the $\frac{1}{12}$ should be multiplied by 0;57,36, that is, by $\frac{24}{25}$. What value for π does this correction factor yield?

3. Archimedes (about 287–212 B.C.) in his book *Measurement of a Circle* stated: The area of a circle is to the square on its diameter as 11 to 14. Show that this geometric rule leads to $\frac{22}{7}$ for the value of π.

4. The sixth-century Hindu mathematician Aryabhata had the following procedure for finding the area of a circle: Half the circumference multiplied by half the diameter is the area of a circle. How accurate is this rule?

5. The Babylonians also knew a formula for the volume of a truncated square pyramid, namely,

$$V = h\left[\left(\frac{a+b}{2}\right)^2 + \frac{1}{3}\left(\frac{a-b}{2}\right)^2\right],$$

where h is the altitude and a and b are the lengths of the sides of the square (upper and lower) bases. Show that this reduces to the formula of the Moscow Papyrus.

6. A Babylonian table has been discovered in which the volume of the frustum of a cone is determined by using the (erroneous) formula

$$V = \frac{3}{2}h(r^2 + R^2),$$

where h is the height and r and R are the radii of the bases. Take $h = 6, r = 4, R = 2$ and compare the Babylonian result with the result from the correct formula

$$V = \frac{1}{3}\pi h(R^2 + rR + r^2).$$

7. Heron of Alexandria (circa A.D. 75?) found the volume of the frustum of a cone by calculations equivalent to using the formula

$$V = \frac{1}{4}\pi h(r + R)^2,$$

where h is the height and r and R are the radii of the bases. If $\frac{22}{7}$ is taken for the value of π, what answer would Heron have gotten for $h = 6, r = 4, R = 2$?

8. The text of Problem 10 of the Moscow Papyrus is illegible at some points, but a calculation is performed, using the equivalent of the formula

$$S = (1 - \tfrac{1}{9})^2(2d)d,$$

which seems designed to give the surface area of a hemispherical basket of diameter $d = 4\frac{1}{2}$. Show that if $4(\frac{8}{9})^2$ is set equal to π, this yields the right formula for the area of a hemisphere, namely $\pi d^2/2$.

9. (a) Starting with the area formula $A = \frac{1}{2}xy\sin\theta$ for a triangle in terms of two sides and the angle between them, use the accompanying figure to derive the formula

$$A = \tfrac{1}{4}(ad\sin A + ab\sin B + bc\sin C + cd\sin D)$$

for the area of a quadrilateral.

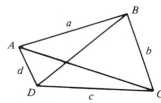

(b) Show that if A represents the area of the quadrilateral in part (a), then

$$A \le \frac{(a+c)(b+d)}{4},$$

with equality holding if and only if the quadrilateral is a rectangle. Thus, the ancient rule for the area of a quadrilateral overestimates the areas of all quadrilaterals that are not rectangles, so that tax assessors could well have continued to use this convenient formula long after they came to suspect that it never led them to underestimate the areas of quadrilateral fields.

10. (a) Prove that of all triangles having two given sides of lengths a and b, the one whose sides form a right angle encompasses the maximum possible area.

(b) Another formula giving the area of an arbitrary quadrilateral is

$$A^2 = (S - a)(S - b)(S - c)$$
$$(S - d) - T.$$

Here $S = \frac{1}{2}(a + b + c + d)$ is the semiperimeter and

$$T = abcd \cos^2 (A + C)/2,$$

where A and C are a pair of opposite vertex angles of the quadrilateral. Show that the maximum possible area corresponding to the given values a, b, c, d occurs when the angles A and C (and hence B and D also) are supplementary.

11. While measuring the Great Pyramid, Charles Piazzi Smyth (named for his godfather, Giuseppe Piazzi,

discoverer of the first known asteroid, Ceres) found a niche in the Queen's Chamber that was 185 pyramid inches long. Use this dimension to verify the accuracy of the following assertions Smyth made in his book *Our Inheritance in the Great Pyramid:*

(a) The length of the Grand Niche multiplied by 10π is equal to the height of the Great Pyramid.

(b) The square root of ten times the height of the north wall in the Queen's Chamber (182.4 pyramid inches) divided by the length of the Grand Niche is equal to π; that is,

$$\pi = \sqrt{\frac{10(182.4)}{185}}.$$

2.5   Babylonian Mathematics

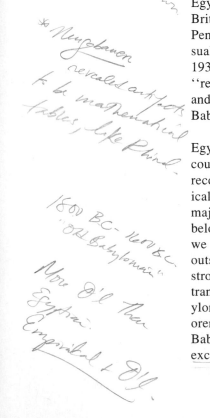

Most of what we know about the mathematics developed in Mesopotamia, first by the Sumerians and then later by the Akkadians and other people, is relatively new. This subject is called Babylonian mathematics, as if a single people had created it. Hitherto the great emphasis had been placed on the achievements of the Egyptians. For some time, it was known that the large Babylonian collections at the British Museum and the Louvre abroad and at Yale, Columbia, and the University of Pennsylvania in this country contained many undeciphered cuneiform tablets of unusual types. The exhaustive studies of Otto Neugebauer, which reached fruition in the 1930s, revealed these to be mathematical tables and texts, and thus a key to the "reading" of their contents was found. Chiefly through the decipherment, translation, and interpretation of this scholar, an entirely new light has been thrown on what the Babylonians contributed to the development of ancient mathematics.

In investigating Babylonian mathematics, we are much less fortunate than with Egyptian mathematics. Because the Babylonian mode of writing on clay tablets discouraged the compilation of long treatises, there is nothing among the Babylonian records comparable with the Rhind Papyrus. Nonetheless, several hundred mathematical tablets have been recovered, many in an excellent state of preservation. The great majority of these (about two-thirds) are "Old Babylonian," which is to say that they belong roughly to the period 1800–1600 B.C. Through this rich mine of source material we now know that except possibly for certain geometric rules, the Babylonians far outstripped the Egyptians in mathematics. Although Babylonian mathematics too had strong empirical roots that are clearly present in most of the tablets that have been translated, it seems to have tended towards a more theoretical expression. (The Babylonians can claim priority in several discoveries, most notably the Pythagorean theorem, usually ascribed to later mathematical schools.) The key to the advances the Babylonians made appears to have been their remarkably facile number system. The excellent sexagesimal notation enabled them to calculate with fractions as readily as with integers and led to a highly developed algebra. This was impossible for the Egyptians, for whom every operation with fractions involved a multitude of unit fractions, thereby making a difficult problem out of each division.

The Babylonians, freed by their remarkable system of numeration from the drudgery of calculation, became indefatigable compilers of arithmetic tables, some of them extraordinary in complexity and extent. Numerous tables give the squares of numbers 1 to 50, and also the cubes, square roots, and cube roots of these numbers. A tablet now in the Berlin Museum gives lists of not only n^2 and n^3 for $n = 1, 2, \ldots,$ 20, 30, 40, 50 but also the sum $n^2 + n^3$. It is surmised that this was used in solving cubic equations that had been reduced to the form $x^3 + x^2 = a$. Another large group of tables deals with the reciprocals of numbers. The standard format of such a table usually involves two columns of figures, such as

4	15
5	12
6	10
8	7;30
9	6;40
10	6
12	5
15	4
16	3;45
18	3;20

where the product of each pair of numbers is always 60. That is, each pair consists of a number on the left-hand side and its sexagesimal reciprocal on the right-hand side. These tables have certain gaps in them; missing are such numbers as 7, 11, 13, and 14, and some others. The reason is that only finite sexagesimal fractions were comprehensible to the Babylonians, and the reciprocals of these "irregular" numbers are nonterminating sexagesimals. For instance, in the sexagesimal expansion for $\frac{1}{7}$ the block 8,34,17 repeats itself infinitely often:

$$\tfrac{1}{7} = 0;8,34,17,8,34,17, \ldots .$$

(The analogous situation occurs in our own system, in which the reciprocal of, say, $\frac{1}{11} = 0.090909 \ldots$ is infinite when expanded decimally.) When an irregular number like 7 does appear in the first column, the statement is made that 7 does not divide, and an approximation is given. A Sumerian tablet of 2500 B.C. calls for dividing the number 5,20,0,0, by 7; the calculation is presented as

$$(5,20,0,0)(0;8,34,17,8) = 45,42,51;22,40,$$

where 5,20,0,0 has been multiplied by the reciprocal of 7 approximated to the fourth place. A later table seems to give upper and lower bounds on the size of $\frac{1}{7}$, namely,

$$8,34,16,59 < \tfrac{1}{7} < 8,34,18.$$

We can picture the scope of some tables of reciprocals from a tablet in the Louvre—dating from 350 B.C.—that comprises 252 entries of one-place to seventeen-place divisors, and one-place to fourteen-place reciprocals. This table is a list of numbers n and n' for which the product nn' equals 1 or some other power of 60. As a specific example, one line contains the values

2,59,21,40,48,54   20,4,16,22,28,44,14,57,40,4,56,17,46,40

which may be thought of as representing the product

$$(2 \cdot 60^5 + 59 \cdot 60^4 + \cdots + 48 \cdot 60 + 54) \times$$
$$(20 \cdot 60^{13} + 4 \cdot 60^{12} + \cdots + 46 \cdot 60 + 40) = 60^{19}.$$

It appears that calculations of this magnitude were necessary in the work of the astronomers of the time.

As suggested previously, the Babylonians did not carry out division by the clumsy duplication method of the Egyptians. Instead, they interpreted a divided by b to mean that a is multiplied by the reciprocal of b; that is, $a/b = a(1/b)$. After having found, either in a table or by calculation, the reciprocal of the divisor, they needed only to multiply it by the dividend. For this purpose, the Babylonian scribes had at their disposal multiplication tables, almost always giving the products of a certain number multiplied successively by 1, 2, 3, . . . 18, 19, 20 and then by 30, 40, and 50. On one tablet of 1500 B.C. are tables of 7, 10, $12\frac{1}{2}$, 16, 24, each multiplied by the foregoing series of values. Thus, the procedure for, say, 7 divided by 2 would be to multiply the reciprocal of 2 by 7:

$$7(0;30) = 0;210 = 3;30$$

which is of course the sexagesimal notation for $3\frac{1}{2}$.

Distinct from the table tablets are tablets that deal with algebraic and geometric problems. These generally present a sequence of closely related numerical problems, together with the relevant calculations and answers; the text often terminates with the words ''Such is the procedure.'' Although none of them gives general rules, the consistency with which the problems were treated suggests that the Babylonians (unlike the Egyptians) had some sort of theoretical approach to mathematics. The problems often seem to be intellectual exercises, instead of treatises on surveying or bookkeeping, and they indicate an abstract interest in numerical relations.

There are scores of clay tablets that indicate that the Babylonians of 2000 B.C. were familiar with our formula for solving the quadratic equation. This is well illustrated by an Old Babylonian text that contains the following problem:

I have added the area and two-thirds of the side of my square and it is 0;35. What is the side of my square?

It is often possible to translate such problems directly into our symbolism by replacing words like *length* (or *side*) and *width* by the letters x and y. In modern notation we would express the content of this problem as

$$x^2 + \tfrac{2}{3}x = \tfrac{35}{60} .$$

The details of solution are described by verbal instructions in the text as follows:

You take 1, the coefficient [of x]. Two-thirds of 1 is 0;40. Half of this, 0;20, you multiply by 0;20 and it [the result] 0;6,40 you add to 0;35 and [the result] 0;41,40 has 0;50 as its square root. The 0;20, which you have multiplied by itself, you subtract from 0;50, and 0;30 is [the side of] the square.

Old Babylonian cuneiform text containing 16 problems with solutions.
(Copyright British Museum.)

Converted to modern algebraic notation, these steps tell us that

$$x = \sqrt{\left(\frac{0;40}{2}\right)^2 + 0;35} - \frac{0;40}{2}$$

$$= \sqrt{0;6,40 + 0;35} - 0;20$$

$$= \sqrt{0;41,40} - 0;20$$

$$= 0;50 - 0;20 = 0;30.$$

Thus, the Babylonian instructions amount to using a formula equivalent to the familiar rule

$$x = \sqrt{\left(\frac{a}{2}\right)^2 + b} - \frac{a}{2}$$

for solving the quadratic equation $x^2 + ax = b$. Although the Babylonian mathematician had no "quadratic formula" that would solve all quadratic equations, the instructions in these concrete examples are so systematic that we can be pretty certain they were intended to illustrate a general procedure.

Historically, it is perhaps more appropriate to speak of rectangular instead of quadratic equations, for it was the problem of rectangles that gave rise to these equations. In the ancient world, the error was widespread that the area of a plane figure depended entirely on its perimeter; people believed that the same perimeter always confined the same area. Army commanders estimated the number of enemy soldiers according to the perimeter of their camp, and sailors the size of an island according to the time for its circumnavigation. The Greek historian Polybius tells us that in his time unscrupulous members of communal societies cheated their fellow members by giving them land of greater perimeter (but less area) than what they chose for themselves; in this way they earned reputations for unselfishness and generosity, while they really made excessive profits.

Evidently the problem of how the perimeter of a rectangle related to its area was systematically investigated in antiquity. A typical problem in early Babylonian mathematics was the following. Given the semiperimeter $x + y = a$ and the area $xy = b$ of a rectangle, find the length x and width y. How did they go about the solution? We can only speculate, for there is no explicit indication anywhere in the mathematical texts of this period of how one arrived at the result. The Babylonian mathematicians were empiricists and observers who worked with tables that presented the facts in an orderly fashion. In all likelihood, they must have constructed the tables for the different values the area might assume, the perimeter being kept constant. Thus, for a rectangle whose semiperimeter $x + y = a = 20$, the resulting areas might have been tabulated for the variations of

$$x = \frac{a}{2} + z \qquad \text{and} \qquad y = \frac{a}{2} - z$$

where z is one of the numbers 0 through 9.

	$x = \dfrac{a}{2} + z$	$y = \dfrac{a}{2} - z$	$b = xy$	$\left(\dfrac{a}{2}\right)^2 - b$
$z = 0$	10	10	100	0
$z = 1$	11	9	99	1^2
$z = 2$	12	8	96	2^2
$z = 3$	13	7	91	3^2
$z = 4$	14	6	84	4^2
$z = 5$	15	5	75	5^2
$z = 6$	16	4	64	6^2
$z = 7$	17	3	51	7^2
$z = 8$	18	2	36	8^2
$z = 9$	19	1	19	9^2

The lesson demonstrated by the numbers in the table is that the areas decrease with the growth of z, and that the difference $(a/2)^2 - b$ always equals the square of z; that is,

$$\left(\frac{a}{2}\right)^2 - b = z^2.$$

At some point, it surely dawned on the Babylonians that they could invert the procedure and ascertain z from the value of $(a/2)^2 - b$. This would give

$$z = \sqrt{\left(\frac{a}{2}\right)^2 - b}$$

and, as a result, the unknowns

$$x = \frac{a}{2} + \sqrt{\left(\frac{a}{2}\right)^2 - b} \quad \text{and} \quad y = \frac{a}{2} - \sqrt{\left(\frac{a}{2}\right)^2 - b}.$$

In the beginning, these conclusions were established empirically through observation of concrete facts; there was no logical speculation nor any deductive reasoning from proven theorems. The best that can be said for the ancient approach is that it substituted patience for brilliance. Later Babylonians would undoubtedly have realized that if the sum $x + y = a$ were given, then the larger quantity, say x, would exceed $a/2$ by a certain amount z. It is evident that since the sum $x + y$ is fixed, x can gain only what y loses. Thus,

$$x = \frac{a}{2} + z \quad \text{and} \quad y = \frac{a}{2} - z,$$

the sum of which is a. Substituting these values in the equation $xy = b$ leads to

$$\left(\frac{a}{2} + z\right)\left(\frac{a}{2} - z\right) = b,$$

whence

$$\left(\frac{a}{2}\right)^2 - z^2 = b.$$

The implication is that

$$z^2 = \left(\frac{a}{2}\right)^2 - b$$

and so

$$z = \sqrt{\left(\frac{a}{2}\right)^2 - b}.$$

The negative root was neglected, and this was usual until modern times. With the value of z known, x and y can now be obtained:

$$x = \frac{a}{2} + \sqrt{\left(\frac{a}{2}\right)^2 - b}, \qquad y = \frac{a}{2} - \sqrt{\left(\frac{a}{2}\right)^2 - b}.$$

This approach can be illustrated with a typical example. A cuneiform tablet in the Yale Babylonian collection asks (in specific numbers) for the solution of the two algebraic equations,

$$x + y = \tfrac{13}{2}, \qquad xy = \tfrac{15}{2}.$$

The Babylonian method just described calls for setting x and y equal to $\tfrac{13}{4}$, plus or minus a correction z; that is,

$$x = \tfrac{13}{4} + z, \qquad y = \tfrac{13}{4} - z.$$

Then the first equation is satisfied, because

$$x + y = (\tfrac{13}{4} + z) + (\tfrac{13}{4} - z) = 2(\tfrac{13}{4}) = \tfrac{13}{2},$$

and the second equation $xy = \tfrac{15}{2}$ becomes

$$(\tfrac{13}{4} + z)(\tfrac{13}{4} - z) = \tfrac{15}{2}.$$

This reduces to

$$\tfrac{169}{16} - z^2 = \tfrac{15}{2},$$

or

$$z^2 = \tfrac{169}{16} - \tfrac{15}{2} = \tfrac{49}{16}.$$

Thus $z = \tfrac{7}{4}$, and one finds immediately that

$$x = \tfrac{13}{4} + \tfrac{7}{4} = 5, \qquad y = \tfrac{13}{4} - \tfrac{7}{4} = \tfrac{3}{2}.$$

The same idea can be used if the difference $x - y$ is initially given, instead of $x + y$. Proceeding analogously, the Babylonians would have solved the system

$$x - y = a, \qquad xy = b$$

by putting

$$x = z + \frac{a}{2} \qquad \text{and} \qquad y = z - \frac{a}{2}$$

from which the solution then follows:

$$x = \sqrt{\left(\frac{a}{2}\right)^2 + b} + \frac{a}{2}, \qquad y = \sqrt{\left(\frac{a}{2}\right)^2 + b} - \frac{a}{2}.$$

More complicated algebraic problems were reduced, by various devices, to the fundamental systems

$$x \pm y = a, \qquad xy = b,$$

which we shall call normal form. For instance, one tablet contains the numerical equivalent of the problem.

$$x + y = \tfrac{35}{6}, \qquad x + y + xy = 14.$$

The values of x and y are given as $\frac{7}{2}$ and $\frac{7}{3}$, respectively, but the manner in which the solution was found is not given. It was probably effected by subtracting the first equation from the second, to get

$$xy = x + y + xy - (x + y) = 14 - \tfrac{35}{6} = \tfrac{49}{6}.$$

The problem then amounts to solving the system

$$x + y = \tfrac{35}{6}, \qquad xy = \tfrac{49}{6},$$

and by the procedure discussed earlier,

$$x = \tfrac{35}{12} + \tfrac{7}{12} = \tfrac{7}{2}, \qquad y = \tfrac{35}{12} - \tfrac{7}{12} = \tfrac{7}{3}.$$

A standard type of Babylonian problem consisted in keeping the condition $xy = b$ fixed but varying the second equation to arrive at more elaborate expressions in x and y. This is evidenced by another tablet, in which one is required to solve, in our notation,

$$xy = 600, \qquad (x + y)^2 + 120(x - y) = 3700.$$

Apparently the Babylonians were aware of the algebraic identity $(x + y)^2 = (x - y)^2 + 4xy$, which allowed them to convert $(x + y)^2$ to $(x - y)^2 + 2400$. When this substitution is made, the second equation becomes

$$(x - y)^2 + 120(x - y) = 1300,$$

a quadratic in $x - y$. An application of their quadratic formula gives the value of $x - y$:

$$x - y = \sqrt{(\tfrac{120}{2})^2 + 1300} - \tfrac{120}{2} = \sqrt{4900} - 60 = 70 - 60 = 10.$$

The Babylonian mathematician would then need to solve the system of equations

$$x - y = 10, \qquad xy = 600,$$

which would give no difficulty. In fact, the usual method of setting $x = z + 5$ and $y = z - 5$ gives rise to the solution $x = 30$, $y = 20$.

The Babylonians knew about quadratic equations of the form $x^2 + ax = b$ and $x^2 = ax + b$; and their respective solutions

$$x = \sqrt{\left(\frac{a}{2}\right)^2 + b} - \frac{a}{2}, \qquad x = \sqrt{\left(\frac{a}{2}\right)^2 + b} + \frac{a}{2},$$

were clearly and expressly taught through numerous examples. The negative square root, which would have led to a negative value of the solution x, was always neglected; nowhere in Babylonian mathematics were negative solutions to quadratics recognized. The type of quadratic $x^2 + b = ax$ seems also to have been well known but transformed by all sorts of ingenious devices to the system $x + y = a$, $xy = b$. The Babylonians' experience showed that $x^2 + b = ax$ led to two distinct solutions, namely, $x = a/2 + \sqrt{(a/2)^2 - b}$ and $x = a/2 - \sqrt{(a/2)^2 - b}$. Yet the idea of two values for one and the same quantity must have seemed a logical absurdity to the Babylonians, something to be circumvented at all costs.

Tablets at Yale University contain hundreds of similar problems (200 on one tablet alone), without solutions, arranged in systematic order. Only a few tablets have been preserved, so there must have been thousands of problems in the original series. In one case, the simultaneous equations for solution are

$$xy = 600, \qquad \tfrac{1}{2}(x + y)^2 - 60(x - y) = -100,$$

an extraordinary example of a negative number in the right-hand member. The concept of a negative number standing by itself—as distinguished from an indicated subtraction—was not current even in Europe 2500 years later.

In a final illustration of the algebraic character of Babylonian mathematics, let us consider a problem in which a reed, the usual measuring rod, of unknown size is used to measure the length and width of a rectangular field. The translated tablet reads:

> I have a reed. I know not its dimension. I broke off from it one cubit and walked 60 times along the length. I restored to it what I have broken off, then walked 30 times along the width. The area is 6,15. What is the original length of the reed?

The common unit for linear measures of land at the time was the ninda, or 12 cubits; thus, $\frac{1}{12}$ of a ninda was broken off from the rod of unknown dimension. If the original rod is assumed to have size x, then the length of the field was $60(x - \frac{1}{12})$, because the field was 60 times as long as the shortened rod. When the cubit was returned to the rod, the width of the field was 30 times the restored length of the complete rod, or $30x$. Because the area of the field is 375, it is found that

$$30x \cdot 60(x - \tfrac{1}{12}) = 375,$$

which leads to the quadratic equation

$$1800x^2 = 150x + 375.$$

On multiplying both sides of this equation by 1800, the author of the tablet would have gotten

$$(1800x)^2 = 150(1800x) + 1800 \cdot 375,$$

a quadratic in $1800x$. And setting $y = 1800x$ would give

$$y^2 = 150y + 1800 \cdot 375.$$

The instructions given in the cuneiform text are equivalent to substituting in the familiar formula

$$y = \sqrt{\left(\frac{a}{2}\right)^2 + b} + \frac{a}{2}$$

for a root of $y^2 = ay + b$. Adapting the rule to the numbers of the present problem, one gets

$$y = \sqrt{75^2 + (1800)(375)} + 75 = 825 + 75 = 900,$$

from which $x = \frac{1}{2}$ ninda is determined.

2.5 Problems

1. Write the fractions $\frac{19}{15}$, $\frac{5}{3}$, and $\frac{10}{9}$ in sexagesimal notation by

   (a) using the Babylonian method of finding the reciprocal of the denominator and then multiplying by the numerator, and
   (b) multiplying numerator and denominator by 60 and simplifying.

2. A tablet in the Yale Babylonian collection reads as follows:

   > I found a stone but did not weigh it; after I added $\frac{1}{7}$ and added $\frac{1}{11}$, I weighed it: [result] 1 mina. What was the original weight of the stone? The original weight was $\frac{2}{3}$ mina, 8 sheqels and $22\frac{1}{2}$ se.

   Use the equivalences 1 mina = 60 sheqels and a sheqel = 180 se to verify the indicated solution. [*Hint:* Call the original weight of the stone x, so that $(x + x/7) + \frac{1}{11}(x + x/7) = 60$ sheqels.]

3. Find the solution to the following ancient Babylonian problem:

   > There are two silver rings; $\frac{1}{7}$ of the first ring and $\frac{1}{11}$ of the second ring are broken off, so that what is broken off weighs 1 sheqel. The first diminished by its $\frac{1}{7}$ weighs as much as the second diminished by its $\frac{1}{11}$. What did the silver rings originally weigh?

   [*Hint:* Consider the system of equations

   $$\frac{x}{7} + \frac{y}{11} = 1, \qquad \frac{6x}{7} = \frac{10y}{11},$$

   where x and y are the weights of the two rings.]

4. A typical Babylonian problem of 1700 B.C. calls for finding the sides of a rectangle given its semiperimeter and area; that is, solve systems of equations of the type $x + y = a$, $xy = b$. Find the solution of the particular system $x + y = 10$, $xy = 16$. [*Hint:* The Babylonians might have used the identity $(x - y)^2 = (x + y)^2 - 4xy$ to find $x - y$.]

5. Another Babylonian problem is:

   > To the area of a rectangle, the excess of the length over the width is added, giving 120; moreover, the sum of the length and width is 24. Find the dimensions of the rectangle.

[*Hint:* The problem can be put in the form of two equations $xy + x - y = 120$, $x + y = 24$; if the substitution $y = z - 2$ is made, the system becomes $x + z = 26$, $xz = 144$.]

6. Using the Babylonian procedures, solve each of the systems below.

   (a) $x - y = 6$, $xy = 16$.
   (b) $x - y = 4$, $xy = 21$.
   (c) $x + y = 8$, $xy = 15$.

7. On a Babylonian tablet, the following problem is solved:

$$x + y = 27, \qquad xy + (x - y) = 183.$$

Solve by first letting $z = y + 2$ to get the system $x + z = 29$, $xz = 210$.

8. On a tablet in the British Museum, the following problem is solved.

   What the length is, the depth is also [except for a coefficient of 12]. A box is hollow. If I add its volume to its cross section, and get 1;10, and if the length measures 0;30, what is the width?

   In solving this problem, let x, y, z be the length, width, and depth of the box, respectively, so that

$$z = 12x, \qquad xyz + xy = 1;10, \qquad x = 0;30.$$

9. A classical example of the quadratic equation in Babylonian mathematics is found on a tablet in the British Museum, which states:

   I have added 7 times the side of my square to 11 times its surface to obtain 6;15. Reckon with 7 and 11.

   Solve for the scribe's answer of 0;30 for the side of the square. [*Hint:* The injunction to "reckon with 7 and 11" means simply that $11x^2 + 7x = 6;15$.

Multiply both sides of this equation by 11, thereby turning it into a quadratic in $11x$.]

10. Heron of Alexandria solved the quadratic equation $\frac{11}{14}x^2 + \frac{22}{7}x + x = 212$ by multiplying both sides by $11 \cdot 14$. Carry out Heron's calculations to obtain x.

11. An old Babylonian text reads as follows:

   I have added the areas of my two squares: [result] 25,25. [The side of] the second square is $\frac{2}{3}$ the side of the first plus 5.

   Find the lengths of the sides of the two squares.

12. Tabulate the values of $n^3 + n^2$ for $n = 1$ to $n = 10$, and use this table to solve the cubic equation $144x^3 + 12x^2 = 48$. [*Hint:* Multiply the given equation by 12.]

13. Using the procedure indicated in the hint, solve each of the following Babylonian problems:

   (a) $x = 30$, $xy - (x - y)^2 = 500$. [*Hint:* Subtract the second equation from the square of the first to get a quadratic in $x - y$. Solving this quadratic leads to a system of the form $x - y = a$, $xy = b$.]
   (b) $x + y = 50$, $x^2 + y^2 + (x - y)^2 = 1400$. [*Hint:* Subtract the square of the first equation from twice the second equation to get a quadratic in $x - y$.]
   (c) $xy = 600$, $(x + y)^2 + 60(x - y) = 3100$. [*Hint:* The formula $(x + y)^2 = (x - y)^2 + 4xy$ leads to a quadratic in $x - y$.]
   (d) $xy = 600$, $20(x + y) - (x - y)^2 = 900$. [*Hint:* The formula $(x - y)^2 = (x + y)^2 - 4xy$ leads to a quadratic in $x + y$.]
   (e) $xy = 600$, $x^2 + y^2 = 1300$. [*Hint:* Square the first equation to produce a system in $u = x^2$ and $v = y^2$.]
   (f) $x - y = 10$, $x^2 + y^2 = 1300$. [*Hint:* Subtract the square of the first equation from the second equation.]

2.6   Plimpton 322

Another oddity in the history of mathematics was brought to light when the Babylonian clay tablet Plimpton 322 (catalog number 322 in the G. A. Plimpton collection at Columbia University) was deciphered by Neugebauer and Sachs in 1945. This tablet is written in Old Babylonian script, which dates it somewhere between 1900 B.C. and 1600 B.C. The analysis of this extraordinary group of figures establishes beyond any doubt that the so-called Pythagorean theorem was known to Babylonian mathematicians more than a thousand years before Pythagoras was born. We recall that Pythagoras's result, which gives the relation between the lengths of the sides of a right triangle, is expressed succinctly in the formula $x^2 + y^2 = z^2$.

The text in question, Plimpton 322, is the right-hand part of a larger tablet with several columns. As is evident from the break at the left-hand side, this tablet was originally larger. The existence of modern glue on the break implies that the other part was lost after the tablet was excavated. The tablet is further marred by a deep chip near the middle of the right-hand edge and a flaked area in the top left corner. The list below conveys its contents.

119	169	1
3367	4825 (11521)	2
4601	6649	3
12709	18541	4
65	97	5
319	481	6
2291	3541	7
799	1249	8
481 (541)	769	9
4961	8161	10
45	75	11
1679	2929	12
161 (25921)	289	13
1771	3229	14
56	106 (53)	15

Three columns of numbers are preserved, each with a heading. The last column contains nothing but the numbers 1,2, . . . ,15, indicating that it enumerates the lines. The preceding two columns are more interesting and are headed by words that might be translated as "width" and "diagonal." It is not difficult to verify that they form the hypotenuse and leg of an integral-sided right triangle. In other words, if the numbers in the middle column are squared and one subtracts from each of them the square of the corresponding number in the first column, a perfect square results. For instance, the first row contains the equation

$$(169)^2 - (119)^2 = (120)^2.$$

The text involves several errors, and in the list, the original readings on the tablet appear in parentheses to the right of the corrected figures. In line 9, the occurrence of 541 instead of 481 is undoubtedly a scribal error, because in sexagesimal notation, 541 is written 9,1 and 481 is written 8,1. In line 13, the scribe wrote the square of 161 in place of the number itself, and the number in the last line is half the correct value. There remains an unexplained error in the second line.

The question naturally arises about how the Babylonians derived the numbers x, y, z satisfying the equation $x^2 + y^2 = z^2$. The values involved in Plimpton 322 are so large that they could not have been obtained simply by guesswork; using trial-and-error methods, one would have run across many simpler solutions before these. If the

Babylonians possessed a clearly discernible method for solving the Pythagorean equation, what was it? A clue is found in a fourth, but incomplete, column along the broken left-hand edge of the Plimpton tablet. It contains a list of the values z^2/x^2, which suggests that the relation $x^2 + y^2 = z^2$ was reduced to

$$\left(\frac{z}{x}\right)^2 - \left(\frac{y}{x}\right)^2 = 1.$$

If $\alpha = z/x$ and $\beta = y/x$, this becomes

$$\alpha^2 - \beta^2 = 1.$$

The problem would then be to construct right triangles whose sides have the rational lengths, 1, α, β, where $\alpha^2 - \beta^2 = 1$. Now, the critical step is recognizing that this last equation can be expressed as

$$(\alpha + \beta)(\alpha - \beta) = 1.$$

All the numbers concerned are rational, so if the product of two numbers is 1, they are reciprocals. That is, one number must be m/n and the other n/m, where m and n are integers. Setting

$$\alpha + \beta = \frac{m}{n} \qquad \text{and} \qquad \alpha - \beta = \frac{n}{m},$$

we find by addition that

$$\alpha = \frac{1}{2}\left(\frac{m}{n} + \frac{n}{m}\right)$$

and by subtraction that

$$\beta = \frac{1}{2}\left(\frac{m}{n} - \frac{n}{m}\right).$$

Consequently,

(1) $$\alpha = \frac{m^2 + n^2}{2mn}, \qquad \beta = \frac{m^2 - n^2}{2mn}.$$

But $y = \beta x$ and $z = \alpha x$; if we now put $x = 2mn$, so as to get a solution in integers, it follows that

$$x = 2mn, \qquad y = m^2 - n^2, \qquad z = m^2 + n^2.$$

These are well-known formulas for finding right triangles with sides of integral length and were used in Hellenistic times by Diophantus (circa 150), the most original mathematician of late antiquity.

To arrive at these formulas, apart from the ability to add and subtract fractions, one needs as the key result the algebraic formula

$$\alpha^2 - \beta^2 = (\alpha + \beta)(\alpha - \beta).$$

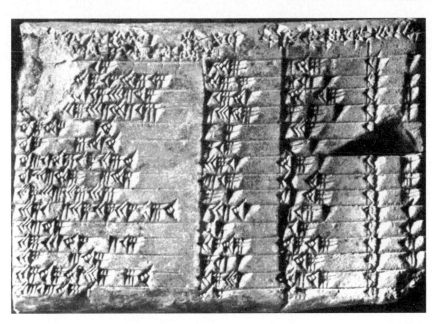

Babylonian tablet, Plimpton 322. (By courtesy of Columbia University.)

This may have been discovered by consideration of a figure like the one herewith. The shaded area $\alpha^2 - \beta^2$ can be dissected as shown and then rearranged as shown, that is, as a rectangle with sides of length $\alpha + \beta$ and $\alpha - \beta$. Hence, we have $\alpha^2 - \beta^2 = (\alpha + \beta)(\alpha - \beta)$.

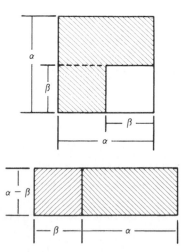

The accompanying table exhibits the values of m and n that give rise to the solutions in Plimpton 322. For example, taking $m = 12$ and $n = 5$ in formula (1), we arrive at

$$x = 120, \qquad y = 119, \qquad z = 169.$$

The latter two numbers are entries in the first line of the tablet. The only exception is in line 11. Here the choice $m = 2$ and $n = 1$ leads to $x = 4$, $y = 3$, $z = 5$, and each of these must be multiplied by 15 to produce the values listed. One interesting point that emerges from examining the table is that m and n always factor into products of powers of 2, 3, and 5.

m	n	$x = 2mn$	$y = m^2 - n^2$	$z = m^2 + n^2$
$2^2 \cdot 3$	5	120	119	169
2^6	3^3	3456	3367	4825
$3 \cdot 5^2$	2^5	4800	4601	6649
5^3	$2 \cdot 3^3$	13500	12709	18541
3^2	2^2	72	65	97
$2^2 \cdot 5$	3^2	360	319	481
$2 \cdot 3^3$	5^2	2700	2291	3541
2^5	$3 \cdot 5$	960	799	1249
5^2	$2^2 \cdot 3$	600	481	769
3^4	$2^3 \cdot 5$	6480	4961	8161
Exception		60	45	75
$2^4 \cdot 3$	5^2	2400	1679	2929
$3 \cdot 5$	2^3	240	161	289
$2 \cdot 5^2$	3^3	2700	1771	3229
3^2	5	90	56	106

The lists of m and n that make up the first two columns of the table are such that their reciprocals all have terminating sexagesimal expansions. If $1/N$ has the finite expansion

$$\frac{1}{N} = \frac{a_1}{60} + \frac{a_2}{60^2} + \cdots + \frac{a_k}{60^k},$$

then it can be written as $1/N = a/60^k$, whence $60^k = aN$. The implication is that N contains only the prime factors that appear in 60^k and therefore in 60. But because 60 has the factorization $60 = 2^2 \cdot 3 \cdot 5$, the permissible factors of N are 2, 3, and 5; that is to say, with suitable exponents α, β, γ, we must have $N = 2^\alpha \cdot 3^\beta \cdot 5^\gamma$.

It has been suggested that the values of z in the Plimpton tablet were not computed directly from $z = m^2 + n^2$ but from the equivalent formula

$$z = (m + n)^2 - 2mn.$$

This proposal furnishes an intriguing explanation of the scribal error in line 2 of the tablet (the case in which $m = 2^6 = 64$, $n = 3^3 = 27$). In using the displayed formula, the writer may have made two mistakes. First, he or she may have added the term $2mn$, when it should have been subtracted; and then, in calculating the term itself, may have written $2 \neq 60 \neq 27$, where $2 \neq 64 \neq 27$ was called for. This would produce the incorrect value

$$z = (64 + 27)^2 + 3240 = 8281 + 3240 = 11{,}521,$$

instead of

$$z = (64 + 27)^2 - 3456 = 8281 - 3456 = 4825.$$

Some of the most impressive treasures of the Babylonian past have been unearthed at Susa, capital of ancient Elam, a country bordering on Babylonia and often hostile to it. Susa has been more or less continuously excavated and for a longer time than any other site in southern Mesopotamia. Its shapeless mounds were identified by the British archeologist William Kennett Loftus, who directed his workmen in sinking the first trenches in 1854. But large-scale excavations did not really begin until the French archeological mission took over the diggings in 1884. In 1902, the mission discovered in the acropolis of Susa one of the outstanding landmarks in the history of humanity: the code of laws of King Hammurabi I (circa 1750 B.C.). The code is carved on a well-polished column of black diorite, which was carried back to Susa from Babylon as a trophy of war. Judged by present-day standards, the 285 articles are a strange mixture of the most enlightened adjudication with the most barbarous punishment. They stressed the principle of ''equivalent retaliation,'' according to which a punishment would be the equivalent of the wrong done: ''If a man has destroyed the eye of an aristocrat, they shall destroy his eye. . . .'' Although Hammurabi used to be called the first lawgiver, recent discoveries have shown that there were several earlier collections of Sumerian legal decrees.

Of more mathematical interest is a group of tablets uncovered by the French at Susa in 1936. These provide some of the oldest Babylonian examples of the use of the theorem of Pythagoras. One table computes the radius r of a circle that circumscribes an isosceles triangle of sides, 50, 50, and 60.

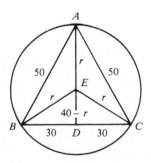

The solution goes as follows. The Pythagorean theorem is used first with regard to triangle ADB to get $AD = 40$. Because $r = AE$, we then have $ED = 40 - r$. A second application of the Pythagorean theorem, this time to the triangle EDB, leads to the equation

$$r^2 = 30^2 + (40 - r)^2,$$

which can be solved to give $r = \frac{2500}{80}$, or $r = 31;15$.

Another Old Babylonian tablet contains the problem,

> A beam of length 0;30 [stands in an upright position against a wall]. The upper end has slipped down a distance 0;6. How far did the lower end move [from the wall]?

The answer is correctly found with the aid of the Pythagorean theorem.

Until recently, scholarly opinion had differed about whether the ancient Egyptians were aware of even a single instance of the Pythagorean theorem, let alone acquainted with the general validity of the proposition. It is well known that as early as 4000 years ago the Egyptians had trained surveyors, the harpedonaptae, whose principal measuring instrument was a stretched rope. The precise orientation of the foundations of the immense structures of ancient Egypt with the four cardinal points of the compass led some historians to surmise that these "rope stretchers" were able to construct right angles using ropes divided by two knots into sections that were in the proportion 3:4:5. When the two ends of the rope were tied and the sections drawn taut around pegs laid out at the three knots, the rope would take the shape of a right triangle. However simple this approach may appear today, no surviving document from antiquity confirms that it actually took place.

The so-called Cairo Mathematical Papyrus, unearthed in 1938 and first examined in 1962, establishes conclusively that the Egyptians of 300 B.C. not only knew that the $(3, 4, 5)$ triangle was right-angled, but also that the $(5, 12, 13)$ and $(20, 21, 29)$ triangles had this property. Dating from the early Ptolemeic dynasties, this papyrus contains 40 problems of a mathematical nature, of which 9 deal exclusively with the Pythagorean theorem. One, for instance, translates as, "A ladder of 10 cubits has its foot 6 cubits from a wall; to what height will it reach?"

Two problems are particularly interesting, because they demonstrate the advance in Egyptian mathematical technique from the time of the Rhind Papyrus. These concern rectangles having areas of 60 square cubits and diagonals of 13 and 15 cubits. One is required to find the lengths of their sides. Writing, say, the first of the problems in modern notation, we have the system of equations

$$x^2 + y^2 = 169, \qquad xy = 60.$$

The scribe's method of solution amounts to adding and subtracting $2xy = 120$ from $x^2 + y^2 = 169$, to get

$$(x + y)^2 = 289, \qquad (x - y)^2 = 49;$$

or equivalently,

$$x + y = 17, \qquad x - y = 7.$$

Tablet in the Yale Babylonian Collection, showing a square with its diagonals.
(Yale Babylonian Collection, Yale University.)

From this it is found that $2y = 10$, or $y = 5$, and as a result $x = 17 - 5 = 12$.

The second problem,

$$x^2 + y^2 = 225, \qquad xy = 60,$$

is similar, except that the square roots of 345 and 105 are to be found. There were several methods for approximating the square root of a number that was not a perfect square. In this case, the scribe used a formula generally attributed to Archimedes (287–212 B.C.), which is also found in Babylonian texts,

$$\sqrt{a^2 + b} \approx a + \frac{b}{2a}.$$

The approximations arrived at are

$$\sqrt{345} \approx \sqrt{18^2 + 21} = 18 + \tfrac{21}{36} = 18 + \tfrac{1}{2} + \tfrac{1}{12}.$$

and

$$\sqrt{105} \approx \sqrt{10^2 + 5} = 10 + \tfrac{5}{20} = 10 + \tfrac{1}{4}.$$

2.6 Problems

1. In a Babylonian tablet, the following problem is found. Given that the circumference of a circle is 60 units and the length of a perpendicular from the center of a chord of the circle to the circumference is 2 units, find the length of the chord. In solving this problem, take $\pi = 3$.

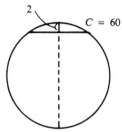

2. An Old Babylonian tablet calls for finding the area of an isosceles trapezoid whose sides are 30 units long and whose bases are 14 and 50. Solve this problem.

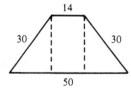

3. In another tablet, one side of a right triangle is 50 units long. Parallel to the other side and 20 units from this side, a line is drawn that cuts off a right trapezoid of area $5{,}20 = 320$ units. Find the lengths of the bases of the trapezoid. [*Hint:* If A is the area of the original triangle, then $320 + 15y = A = 25x$, and $\frac{1}{2}(x + y)20 = 320$.]

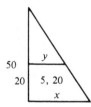

4. In a similar problem, a right triangle whose base is 30 units is divided into two parts by a line drawn parallel to the base. It is given that the resulting right trapezoid has an area larger by $7{,}0 = 420$ than the upper triangle, and that the difference between the height y of the upper triangle and the height z of the trapezoid is 20. If x is the length of the upper base of the trapezoid, these statements lead to the relations $\frac{1}{2}z(x + 30) = \frac{1}{2}xy + 420$, $y - z = 20$.

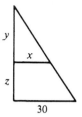

The problem calls for finding the values of the unknown quantities x, y, and z. [*Hint:* By properties of similar triangles, $y/(y + z) = x/30$.]

5. Another Old Babylonian problem calls for finding the length of the sides of an isosceles trapezoid, given that its area is 150, that the difference of its bases is 5 (that is, $b_1 - b_2 = 5$), and that its equal sides are 10 greater than two-thirds of the sum of its bases (that is, $s_1 = s_2 = \frac{2}{3}(b_1 + b_2) + 10$).

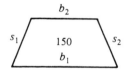

Solve this problem using an incorrect formula for the area of a trapezoid, namely,

$$A = \frac{b_1 + b_2}{2} \frac{s_1 + s_2}{2}.$$

6. A Babylonian tablet of 2000 B.C. gives two methods for calculating the diagonal d of a rectangle with sides of length 40 and 10 units. The first leads (in specific numbers) to the approximation

$$d \approx a + \frac{2ab^2}{3600},$$

where a is the larger side and b is the shorter side, and the second method to the approximation

$$d \approx a + \frac{b^2}{2a}.$$

Check the accuracy of these approximations to $\sqrt{1700}$ by squaring the respective answers.

7. Because a is smaller than $\sqrt{a^2 + b}$ when $b > 0$, whereas $a + b/a$ is larger, the Babylonian mathematician often approximated $\sqrt{a^2 + b}$ by taking the average of these two values, that is,

$$\sqrt{a^2 + b} \approx \frac{1}{2}\left(a + \left(a + \frac{b}{a}\right)\right)$$
$$= a + \frac{b}{2a}, \qquad 0 < b < a^2.$$

Use this formula to get rational approximations to $\sqrt{2}$, $\sqrt{5}$, and $\sqrt{17}$. [*Hint:* In the first case put $a = \frac{4}{3}$, $b = \frac{2}{9}$; in the second case, put $a = 2$, $b = 1$.]

8. An iterative procedure for closer approximations to the square root of a number that was not a square was obtained by Heron of Alexandria. In his work *Metrica* (discovered as recently as 1896 in Constantinople in a manuscript dating from the eleventh or twelfth century), he merely stated a rule that amounts to the following. If A is a nonsquare number and a^2 is the nearest perfect square to it, so that $A = a^2 \pm b$, a first approximation to $\sqrt{A}$ is the average of the values a and A/a;

$$x_1 = \frac{1}{2}\left(a + \frac{A}{a}\right).$$

This number can be used to get a more accurate approximation,

$$x_2 = \frac{1}{2}\left(x_1 + \frac{A}{x_1}\right),$$

and the process is repeated as often as desired.

(a) Find approximate square roots, through two approximations, of the numbers Heron used to explain his method, namely, 720 and 63.
(b) Show that Heron's first approximation was equivalent to a formula of Archimedes—which in turn was a generalization of the Babylonian method—namely,

$$\sqrt{a^2 \pm b} \approx a \pm \frac{b}{2a}. \qquad \left(a + \frac{b}{2a}\right)$$

Bibliography

Aaboe, Asgar. *Episodes from the Early History of Mathematics* (New Mathematical Library, No. 13). New York: Random House, 1964.

Archibald, Raymond C. "Babylonian Mathematics." *Science* 70(1929): 66–67.

———. "Mathematics Before the Greeks." *Science* 71(1930): 109–121.

———. "Babylonian Mathematics." *Isis* 26(1936): 63–81.

———. "Babylonian Mathematics with Special Reference to Recent Discoveries." *Mathematics Teacher* 29(1936): 209–219.

Botts, Truman. "Problem Solving in Mathematics: Every Positive Number is a Sum of Unit Fractions." *Mathematics Teacher* 58(1965): 596–600.

Bratton, Fred. *A History of Egyptian Archaeology.* New York: Thomas Y. Crowell, 1968.

Bruckheimer, M., and Salomon, Y. "Some Comments on R. J. Gillings' Analysis of the 2/n-Table in the Rhind Papyrus." *Historia Mathematica* 4(1977): 445–452.

Buck, R. Creighton. "Sherlock Holmes in Babylon." *American Mathematical Monthly* 87(1980): 335–345.

Budge, E. A. Wallis. *The Rosetta Stone.* Rev. ed. London: Harrison and Sons, 1950.

Chace, A. B. *The Rhind Mathematical Papyrus.* Reston, Va.: National Council of Teachers of Mathematics, 1979.

Gandz, Solomon. "The Babylonian Table of Reciprocals." *Isis* 25(1936): 426–431.

———. "Studies in Babylonian Mathematics I: Indeterminate Analysis in Babylonian Mathematics." *Osiris* 8(1938): 12–40.

———. "Studies in Babylonian Mathematics II: Conflicting Interpretations of Babylonian Mathematics." *Isis* 31(1939–40): 405–425.

———. "Studies in Babylonian Mathematics III: Isoperimetric Problems and the Origin of the Quadratic Equation." *Isis* 32(1940): 103–115.

Gillings, Richard J. "The Recto of the Rhind Mathematical Papyrus: How Did the Ancient Egyptian Scribe Prepare It?" *Archive for History of Exact Sciences* 12(1974): 291–298.

———. "Think of a Number Problems 28, 29 of the Rhind Mathematical Papyrus." *Mathematics Teacher* 54(1961): 97–100.

———. "Problems 1–6 of the Rhind Mathematical Papyrus." *Mathematics Teacher* 55(1962): 61–69.

———. "The Volume of a Truncated Pyramid in Ancient Egyptian Papyri." *Mathematics Teacher* 57(1964): 552–555.

———. "The Remarkable Mental Arithmetic of the Egyptian Scribes." *Mathematics Teacher* 59(1966): 372–381, 476–484.

———. *Mathematics in the Time of the Pharaohs.* Cambridge, Mass.: M.I.T. Press, 1972. (Dover reprint, 1982.)

———. "The Mathematics of Ancient Egypt." In *The Dictionary of Scientific Biography.* Vol. 15. New York: Charles Scribner (1978): 681–705.

Herodotus. *The History of Herodotus.* Translated by George Rawlinson. London: Everymans Library, 1927.

Hirsch, Harriet. "The Rhind Papyrus." In *Historical Topics for the Mathematical Classroom.* Washington: National Council of Teachers of Mathematics, 1969.

Jones, Phillip. "Recent Discoveries in Babylonian Mathematics I: Zero, Pi and Polygons." *Mathematics Teacher* 50(1957): 162–165.

———. "Recent Discoveries in Babylonian Mathematics II: The Oldest Known Problem Text." *Mathematics Teacher* 50(1957): 442–444.

———. "Recent Discoveries in Babylonian Mathematics III: Trapezoids and Quadratics." *Mathematics Teacher* 50(1957): 570–571.

Karpinski, Louis. "Algebraical Discoveries Among the Egyptians and Babylonians." *American Mathematical Monthly* 24(1917): 257–265.

Miller, G. A. "A Few Theorems Relating to the Rhind Mathematical Papyrus." *American Mathematical Monthly* 38(1931): 194–197.

Neugebauer, Otto. *The Exact Sciences in Antiquity.* 2nd ed. Providence: Brown University Press, 1957. (Dover reprint, 1969.)

Newman, James. "The Rhind Papyrus." *Scientific American* 187(August 1952): 24–27.

Robins, Gary, and Shute, Charles. *The Rhind Mathematical Papyrus: An Ancient Egyptian Text.* New York: Dover, 1990.

Seidenberg, A. "The Ritual Origin of Geometry." *Archive for History of Exact Sciences* 1(1962): 488–527.

———. "The Area of a Semi-Circle." *Archive for History of Exact Sciences* 9(1972): 171–211.

Smeur, A. J. E. M. "On the Value Equivalent to π in Ancient Mathematical Texts: A New Interpretation." *Archive for History of Exact Sciences* 6(1970): 249–270.

Tompkins, Peter. *Secrets of the Great Pyramid.* New York: Harper & Row, 1971.

Van der Waerden, B. L. *Science Awakening.* Translated by Arnold Dresden. Groningen: Noordhoff, 1954.

Wohlgemuth, Bernhart. "Egyptian Fractions." *Journal of Recreational Mathematics* 5(1972): 55–58.

CHAPTER 3

The Beginnings of Greek Mathematics

He is unworthy of the name of man who does not know that the diagonal of a square is incommensurable with its side.

PLATO

3.1 The Geometrical Discoveries of Thales

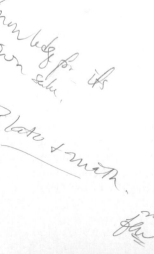

The Greeks made mathematics into one discipline, transforming a varied collection of empirical rules of calculation into an orderly and systematic unity. Although they were plainly heirs to an accumulation of Eastern knowledge, the Greeks fashioned through their own efforts a mathematics more profound, more abstract (in the sense of being more remote from the uses of everyday life), and more rational than any that preceded it. In ancient Babylonia and Egypt, mathematics had been cultivated chiefly as a tool, either for immediate practical application or as part of the special knowledge befitting a privileged class of scribes. Greek mathematics, on the other hand, seems to have been a detached intellectual subject for the connoisseur. The Greeks' habits of abstract thought distinguished them from previous thinkers; their concern was not with, say, triangular fields of grain but with "triangles" and the characteristics that must accompany "triangularity." This preference for the abstract concept can be seen in the attitude of the different cultures toward the number $\sqrt{2}$; the Babylonians had computed its approximations to a high accuracy, but the Greeks proved that it was irrational. The notion of seeking after knowledge for its own sake was almost completely alien to the older Eastern civilizations, so that in the application of reasoning to mathematics, the Greeks completely changed the nature of the subject. Plato's inscription over the door of his Academy, "Let no man ignorant of geometry enter here," was not the admonition of an eccentric but rather a tribute to the Greek conviction that through the spirit of inquiry and strict logic one could understand a person's place in an orderly universe.

All history is based on written documents. Although documentation concerning Egyptian and Babylonian mathematics is often very precise, the primary sources that can give us a clear picture of the early development of Greek mathematics are meager. In Greece, there was no papyrus such as was available in Egypt, no clays as in

Babylonia. Such "books" as were written must have been very few; and with the passing of time and ravages of the elements, little original material has survived. Consequently, early Greek history is a morass of myths, legends, and dubious anecdotes, preserved by writers who lived centuries later than the events under consideration. We depend on fragments and copies of copies many times removed from the original document. However scrupulous the copyist may have been in filling in obscure passages in an earlier text, we can never be sure how much the copyist had to call on one's own imagination or indeed, how well the copyist understood the original.

The Greeks were not always confined to the southeastern corner of Europe, their location in modern times. Although the Egyptians had kept to themselves, the Greeks were great travelers. Their colonization of the coasts and the offshore islands of Asia Minor from the eleventh to the ninth century B.C. was a prologue to later large-scale movements from mainland Greece. About the middle of the eighth century B.C., a network of Greek cities was founded on the coastal reaches of the Mediterranean, with scattered settlements as far afield as the eastern end of the Black Sea. Down to 650 B.C., the main vent for Greek expansion was lower Italy and Sicily; the many flourishing colonies there caused the whole area to be given the name Greece-in-the-West. Although the earlier migration to Asia Minor was probably the result of the Dorian conquest of large portions of the Balkan Peninsula, economic distress and political unrest in the homeland was the new incentive to spread overseas. An increase in population caused a crisis in land ownership, as well as a serious shortage of food. All these migrations not only provided an outlet for dissatisfied elements of the population at home but also served to establish foreign markets and to lay the material foundations of art, literature, and science. Although Hellenic culture had its beginnings in an expanded Greece, in due course peninsular Greece became only a part of "Greater Greece." By 800 B.C., there was, broadly speaking, a unity of language and custom throughout the ancient Mediterranean world.

The wave of colonization that took place outside of the Aegean from the eighth to the sixth century B.C. paved the way for an extraordinary breakthrough of reason and the attendant cultural advancement. Historians have called this phenomenon the Greek miracle. The miracle of Greece was not single but twofold—first the unrivaled rapidity and variety and quality of its achievement; then its success in permeating and imposing its values on alien civilizations. For this, the colonies were like conduits through which Greek culture flowed to the world of the "Barbarians," and the older Egyptian and Babylonian cultures streamed to the Greeks. It is remarkable that all the early Greek mathematics came from the outposts in Asia Minor, southern Italy, and Africa, and not from mainland Greece. It is as if the scanty Greek populations living next-door to the more developed societies had their wits sharpened by this contact, as well as having access to the knowledge gathered by them. The most decisive of all Greek borrowings was the art of writing with the convenient Phoenician alphabet. Each of the symbols of the Phoenician alphabet stood for a consonant; there were no signs for vowels. The Phoenician alphabet had more consonant symbols than the Greek language required, so the Greeks set out by selecting and adapting the consonant symbols they needed. Thereafter, they assigned vowel values to the remaining symbols, adding only such new signs as they needed (for instance α, which had a consonantal value in the Phoenician alphabet, became the symbol for the vowel A in the Greek alphabet). As in other matters, the Greek city-states vied with each other in the elaboration of the alphabet, with as many as ten different versions getting under way. Gradually, one of these local alphabets, the Ionian, gained the ascendancy; and after its official adoption by Athens in 403 B.C., it spread rapidly through the rest of Greece. Although the acceptance of alphabetic writing did not initiate anything like popular education, the ease with which it could be learned made possible a wider distribution of learning than had prevailed in the older cultures, where reading and writing were the property of a priestly class. (Although the Phoenician traders eventually spread the new device throughout the Mediterranean world, the intelligentsia of Egypt and Babylonia disdained the alphabet—possibly because they had invested lifetimes in learning the elaborate ideograms that were the mysterious delight of specialists.)

Coinage in precious metals was invented in the Greek cities of Asia Minor about 700 B.C., stimulating trade and giving rise to a money economy based not only on agriculture but also on movable goods. In rendering possible the accumulation of wealth, this new money economy permitted the formation of a leisure class from which an intellectual aristocracy could emerge. Aristotle recognized how important nonpractical activity is in the advancement of knowledge when he wrote in his *Metaphysics:*

> When all the inventions had been discovered, the sciences which are not concerned with the pleasures and necessities of life were developed first in the lands where men began to have leisure. This is the reason why mathematics originated in Egypt, for there the priestly class was able to enjoy leisure.

In most ancient civilized societies, an educated elite, usually priests, directed the activities of the community. Whether the priests were themselves the government (as in early Babylonia) or merely its servants (as in Egypt), proficiency in writing and mathematics was considered part of their special skills. In the structure of Egyptian bureaucracy, the man of learning held a position of great privilege and potential power. The Greek historian Polybius remarked that "the Egyptian priests obtained positions of leadership and respect because they surpassed their fellows in knowledge." Eastern learning was a mystery shared only by the specialists and not destined for the citizenry. Although an able and ambitious man had some opportunity to improve his lot through education, these hopes were seldom realized—just as very few of Napoleon's soldiers ever became field marshals. By contrast, Greek education was far more broadly based and designed to produce gentleman amateurs. Perhaps the difference was that the Greeks had no powerful priesthood that could monopolize learning as its own preserve; no sacred writings or rigid dogmas that required the mind's subservience. In any event, the first Greek intellectuals came not from the class of governmental managers but from people of affairs, for whom business was a profession and learning a pastime.

Geography shaped the pattern of Greek political life. In Egypt and Babylonia, it was easy to subject a large population to a single ruler, but in Greece, where every district was separated from the next by mountains or the sea, central control by an absolute monarch was impossible. Mountainous barriers were not enough to prevent invasion, but they were enough to prevent one state from being merged with another. Patriotic loyalty was to the native city—Athens, Corinth, Thebes, or Sparta—and not to Greece as a whole. In great emergencies the Greek states acted collectively, seeing that they must unite or be destroyed. During the Persian invasions of the later sixth and early fifth centuries B.C., they pooled their fighting forces to defeat Darius at Marathon (490 B.C.) and Xerxes at Salamis (480 B.C.), after a rearguard action by 300 Spartans at Thermopylae. On none of these occasions was the union successful or long-lasting, because with each victory the city-states would promptly fall out and exhaust themselves in long local wars. The lack of political unity made the outcome inevitable. The end came when Philip II of Macedonia overpowered the mixed Greek forces at the battle of Charonea in 338 B.C. and established himself as the head of all the Greek states except Sparta. Philip died two years later, and the power passed into the hands of his son, Alexander the Great, who achieved what no leader before had done. He unified Greece and carried Greek civilization to the limits of the known world. In 323 B.C., when Alexander died at age 32, he ruled over conquests of more than 2 million square miles. But neither Greeks nor Greek culture vanished with the change of masters. The years that followed—from the time of Alexander the Great into the first century B.C.—formed a brilliant period of history known to scholars as the Hellenistic Age.

The rise of Greek mathematics coincides in time with the general flowering of Greek civilization in the sixth century B.C. (By "Greek civilization" is usually meant the civilization that began in the Iron Age and reached its greatest brilliance, though not extent, in the fifth and fourth centuries B.C.) From the modest beginnings with the Pythagoreans, number theory and geometry developed rapidly, so that early Greek mathematics reached its zenith in the work of the great geometers of antiquity—Euclid, Archimedes, and Apollonius. Thereafter, the discoveries were less striking, although great names such as Ptolemy, Pappus, and Diophantus testify to memorable accomplishments from time to time. These pioneering contributors exhausted the possibilities of elementary mathematics to the extent that little significant progress was made, beyond what we call Greek mathematics, until the sixteenth century. What is more striking still is that almost all the really productive work was done in the relatively short interval from 350 to 200 B.C., and not in the old Aegean world but by Greek immigrants in Alexandria under the Ptolemies.

The first individuals with whom specific mathematical discoveries are traditionally associated are Thales of Miletus (circa 625–545 B.C.) and Pythagoras of Samos (circa 580–500 B.C.). Thales was of Phoenician descent, born in Miletus, a city of Ionia, at a time when a Greek colony flourished on the coast of Asia Minor. He seems to have spent his early years engaged in commercial ventures, and it is said that in his travels he learned geometry from the Egyptians and astronomy from the Babylonians. To his admiring countrymen of later generations, Thales was known as the first of the Seven Sages of Greece, the only mathematician so honored. In general, these men earned the title not so much as scholars as through statesmanship and philosophical and ethical wisdom. Thales is supposed to have coined the maxim "Know thyself," and when asked what was the strangest thing he had ever seen, he answered "An aged tyrant."

Ancient opinion is unanimous in regarding Thales as unusually shrewd in politics and commerce no less than in science, and many interesting anecdotes, some serious and some fanciful, are told about his cleverness. On one occasion, according to Aristotle, after several years in which the olive trees failed to produce, Thales used his skill in astronomy to calculate that favorable weather conditions were due the next season. Anticipating an unexpectedly abundant crop, he brought up all of the olive presses around Miletus. When the season came, having secured control of the presses, he was able to make his own terms for renting them out and thus realized a large sum. Others say that Thales, having proved the point that it was easy for philosophers to become rich if they wished, sold his olive oil at a reasonable price.

Another favorite story is related by Aesop. It appears that once one of Thales' mules, loaded with salt for trade, accidentally discovered that if it rolled over in a stream, the contents of its load would dissolve; on every trip thereafter, the beast deliberately repeated the same stunt. Thales discouraged this habit by the expedient of filling the mule's saddlebags with sponges instead of salt. This, if not true, is certainly well invented and more in character than the amusing tale Plato tells. One night, according to Plato, Thales was out walking and looking at the stars. He looked so intently at the stars that he fell into a ditch, whereupon an old woman attending him exclaimed, "How can you tell what is going on in the sky when you can't see what is lying at your feet?" This anecdote was often quoted in antiquity to illustrate the impractical nature of scholars.

As we have seen, the mathematics of the Egyptians was fundamentally a tool, crudely shaped to meet practical needs. The Greek intellect seized on this rich body of raw material and refined from it the common principles, thereby making the knowledge more general and more comprehensible and simultaneously discovering much that was new. Thales is generally hailed as the first to introduce using logical proof based on deductive reasoning rather than on experiment and intuition to support an argument. Proclus (about 450), in his *Commentary on the First Book of Euclid's Elements*, declared:

> Thales was the first to go into Egypt and bring back this learning [geometry] into Greece. He discovered many propositions himself and he disclosed to his successors the underlying principles of many others, in some cases his methods being more general, in others more empirical.

Modern reservations notwithstanding, if the mathematical attainments attributed to Thales by such Greek historians as Herodotus and Proclus are accepted, he must be credited with the following geometric propositions.

- Every angle inscribed in a semicircle is a right angle.

- A circle is bisected by its diameter.

- The base angles of an isosceles triangle are equal.

- If two straight lines intersect, the opposite angles are equal.

- The sides of similar triangles are proportional.

- Two triangles are congruent if they have one side and two adjacent angles respectively equal.

Because there is a continuous line from Egyptian to Greek mathematics, all of the listed facts may well have been known to the Egyptians. For them, the statements would remain unrelated, but for the Greeks they were the beginning of an extraordinary development in geometry. Conventional history inclines in such instances to look for some individual to whom the "miracle" can be ascribed. Thus, Thales is traditionally designated the father of geometry, or the first mathematician. Although we are not certain which propositions are directly attributable to him, it seems clear that Thales contributed something to the rational organization of geometry—perhaps the deductive method. For the orderly development of theorems by rigorous proof was entirely new and was thereafter a characteristic feature of Greek mathematics.

Several stories purport to illustrate Thales' interest in Egypt. According to legend, his most spectacular accomplishment while there was the indirect measurement of the height of the Great Pyramid by means of shadows. There are two versions of the story, one describing a very simple method of measurement and the other a more complex method. The earliest version is that Thales observed the length of the shadow of the pyramid at that hour of the day when a man's shadow is the same length as himself. Plutarch improved on this when he wrote in the *Convivium:*

> Although he [the king of Egypt] admired you [Thales] for other things, yet he particularly liked the manner by which you measured the height of the pyramid without any trouble or instrument; for, by merely placing your staff at the extremity of the shadow which the pyramid casts, you formed, by the impact of the sun's rays, two triangles and so showed that the height of the pyramid was to the length of the staff in the same ratio as their respective shadows.

Both versions of the story depend on the same geometric proposition, namely, that the sides of equiangular triangles are proportional. Thales, having thus conceived of two similar triangles, argued that the height h of the pyramid was to the length h' of the staff as the length s of the pyramid's shadow was to the length s' of the shadow cast by the staff when it was held vertically:

$$h/h' = s/s'.$$

Thales knew already that the distance along each side of the base of the Great Pyramid was 756 feet and that his staff was 6 feet long. It was necessary only to measure the shadow of the pyramid (the distance from the tip of the shadow to the center of the base of the pyramid) and the shadow of the staff. It was 342 feet from the tip of the pyramid's shadow to the edge of the base, and the shadow of the staff measured 9 feet.

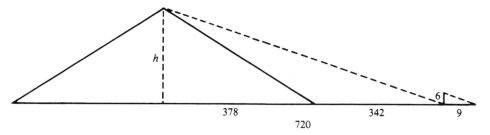

Now Thales had all the required dimensions, for three items of the proportion would give him the missing fourth item. The height of the Great Pyramid was

$$h = \frac{sh'}{s'} = \frac{(378 + 342)6}{9} = \frac{2}{3}720 = 480 \text{ feet.}$$

Another practical application of geometry attributed to Thales is determining how far a ship at sea is from the shore. How he used his knowledge of geometry for this purpose can only be conjectured. According to Proclus, Thales used the congruence theorem, which asserts that a triangle is completely determined if one side and two adjacent angles are known. The most probable assumption is that Thales, observing the ship from the top of a lookout tower (say of height h) used the proportionality of the sides of two similar right triangles. All he needed was a simple instrument with two legs forming a right angle, so that he could mark off the point E where the line of sight with the ship cut the leg parallel to the ground. This would produce similar triangles ACB and DCE.

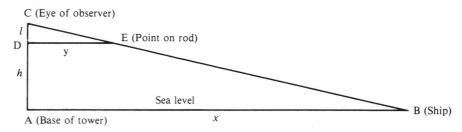

If x is the unknown distance of the ship, then by properties of similar triangles, one has

$$\frac{x}{y} = \frac{l + h}{l}$$

or

$$x = \frac{y(h + l)}{l}.$$

The only objection to this approach is that it does not depend directly on the theorem concerning two triangles in which corresponding sides and their adjacent angles are equal, as Proclus implied.

Another possible approach is that to find the distance x from the shore A to the ship B, one measures from A along a straight line perpendicular to AB an arbitrary length AC and determines its midpoint D. From C, construct a line CE perpendicular to AC (in a direction opposite to AB) and let E be the point on it which is in a straight line with B and D. Clearly, CE has the same length as AB, and CE can be measured, so that AB is known. This supposition is open to a different objection. It hardly seems credible that to ascertain the distance of the ship, the observer should have had to reproduce and measure on land an enormous triangle. Such an undertaking would have been so inconvenient as to deprive Thales' discovery of any practical value.

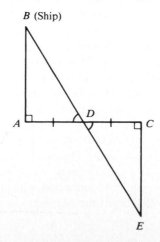

Among his contemporaries, Thales was more famous as an astronomer than as a mathematician. A legend that reappears from time to time is that he amazed his fellow Greeks by predicting a solar eclipse in 585 B.C. Herodotus records that the event took place during a battle between the Lydians and the Medes, and when day turned into night, the fighting ceased. The warring kings were so awed that they concluded a lasting peace. Although Thales' fame as a scientist rests mainly on this achievement, it is almost impossible to give credence to the tale. The astronomical records of that time were not accurate enough to allow anything like a precise forecast. Thales may well have had some knowledge of cycles of lunar eclipses—these were already known to the Babylonians—but his alleged prediction of the year of a solar eclipse would only have been a fortunate guess. A likely explanation is that he happened to be the "wise man" known to the people who saw this striking phenomenon, and so they assumed that Thales must have been able to foresee it.

Although Thales left no written record of any book or document behind him, he ranks high among mathematicians for his pioneering contribution to the logical development of geometry. He may even have been Pythagoras's teacher; some sources tell that Thales recognized the genius of the young Pythagoras, to whom he taught all that he knew.

3.2   Pythagorean Mathematics

The study of numbers in the abstract begins in sixth century B.C. Greece with Pythagoras and the Pythagoreans. Our knowledge of the life of Pythagoras is scanty, and little can be said with any certainty. Those scraps of information that have filtered down to us come from early writers who vied with each other in inventing fables concerning his travels, miraculous powers, and teachings. According to the best estimates, Pythagoras was born between 580 and 569 B.C. on the Aegean island Samos. He appears to have left Samos permanently as early as his eighteenth year to study in Phoenicia and Egypt, and he may have extended his journeys as far eastward as Babylonia. Some none-too-trustworthy sources say that when Egypt was conquered by the Persian king Cambyses in 525 B.C., Pythagoras was carried back to Babylonia with the other Egyptian captives. Other authorities indicate however, that he followed Cambyses voluntarily. When Pythagoras reappeared after years of wandering (around the age of 50), he sought out a favorable place for a school. Banned from his native Samos by the powerful tyrant Polycrates, he turned westward and finally settled at Croton, a prosperous Dorian colony in southern Italy.

Founding a school was not unusual in the Greek world. The distinctive feature of the school of Pythagoras was that its aims were at once political, philosophical, and religious. Formed of some 300 young aristocrats, the community had the character of a fraternity or a secret society; it was a closely knit order in which all worldly goods were held in common. The school tried rigidly to regulate the diet and way of life of its members, and to invoke a common method of education. Pupils concentrated on four mathemata, or subjects of study: arithmetica (arithmetic, in the sense of number theory as opposed to calculating), harmonia (music), geometria (geometry), and astrologia (astronomy). This fourfold division of knowledge became known in the Middle Ages as the "quadrivium," to which was then added the trivium logic, grammar, and rhetoric—subjects connected with the use of language. These seven liberal arts came to be looked on as the necessary and proper course of study for the educated person.

Pythagoras
(circa 580–500 B.C.)

(The Bettmann Archive.)

Pythagoras divided those who attended his lectures into two grades of disciples: the acoustici (or listeners) and the mathematici. After three years of listening in mute obedience to Pythagoras's voice from behind a curtain, a pupil could be initiated into the inner circle, to whose members were confided the main doctrines of the school. Although women were forbidden by law to attend public meetings, they were admitted to the master's lectures. One source indicates that there were at least 28 women in the select category of mathematici. When Pythagoras was close to 60 years old, he married one of his pupils, Theano. She was a remarkably able mathematician who not only inspired him during the latter years of his life but continued to promulgate his system of thought after his death. (Some contradictory sources say that Theano was Pythagoras's daughter; yet others, that she was only a highly gifted pupil, never his wife.)

Pythagoras followed the custom of Eastern teachers by passing along his views by word of mouth. He seems not to have committed any of his teachings to writing. And furthermore the members of his community were bound not to disclose to outsiders anything taught by the master or discovered by others in the brotherhood as a result of the master's teaching. Legend has it that one talkative disciple was drowned in a shipwreck as the gods' punishment for his public boast that he had added the dodecahedron to the set of regular solids Pythagoras had enumerated. The symbol on which the members of the Pythagorean community swore their oaths was the "tetractys," or holy fourfoldness, which was supposed to stand for the four elements: fire, water, air, earth. The tetractys was represented geometrically by an equilateral triangle made up of ten dots, and arithmetically by the number $1 + 2 + 3 + 4 = 10$.

Strange rituals Pythagoreans?!

According to the Greek writer and satirist Lucian (120–180), Pythagoras asked someone to count; when he had reached 4, Pythagoras interrupted, ''Do you see? What you take to be 4 is 10, a perfect triangle and our oath.''

Like other mystery cults of that time, the Pythagoreans had their strange initiations, rites, and prohibitions. They refused, for example, to eat beans, drink wine, pick up anything that had fallen, or stir a fire with an iron. They insisted, in addition to these curious taboos, on a life of virtue, especially of friendship. From Pythagoreanism comes the story of Damon and Pythias. (Pythias, condemned to death for plotting against the king, was given leave to arrange his affairs after Damon pledged his own life if his friend did not return.) The five-pointed star, or pentagram, was used as a sign whereby members of the brotherhood could recognize one another. It is told that a Pythagorean fell ill while traveling and failed to survive, despite the nursing of a kind-hearted innkeeper. Before dying, he drew the pentagram star on a board and begged his host to hang it outside. Some time later another Pythagorean, passing by, noticed the symbol and after hearing the innkeeper's tale, rewarded him handsomely.

Vegetarians! doctrine of transmigration of souls.

The Pythagoreans fancied that the soul could leave one's body, either temporarily or permanently, and that it could inhabit the body of another person or animal. As a result of this doctrine of transmigration of souls, they would eat no meat or fish lest the animal slaughtered be the abode of a friend. The Pythagoreans would not kill anything except as a gift to the gods, and they would not even wear garments of wool, since wool is an animal product. A story is told in which Pythagoras, coming across a small dog being thrashed, said, ''Stop the beating, for in this dog lives the soul of my friend; I recognize him by his voice.''

Accounts of Pythagoras's death do not agree. What is clear is that political ideas were gradually added to the other doctrines, and for a time, the autocratic Pythagoreans succeeded in dominating the local government in Croton and the other Greek cities in southern Italy. About 500 B.C., there was a violent popular revolt in which the meetinghouse of the Pythagoreans was surrounded and set afire. Only a few of those present survived. In several accounts, Pythagoras himself is said to have perished in the inferno. Those with a sense of drama would have us believe that Pythagoras's disciples made a bridge over the fire with their bodies, so that the master might escape the frenzied mob. It is said in these versions that he fled to nearby Metapontum but in the ensuing flight, having reached a field of sacred beans, chose to die at the hands of his enemies rather than trample down the plants. With the death of Pythagoras, many members of the school emigrated to the Greek mainland; some stayed behind for a time but by the middle of the fourth century B.C. all had left Italy. Although the political influence of

the Pythagoreans was destroyed, they continued to exist for several centuries longer as a philosophical and mathematical order. To the end, the dwindling band of exiles remained a secret society, leaving no written record, and with notable self-denial, ascribing all their discoveries to the master.

What set the Pythagoreans apart from the other sects was the philosophy that "knowledge is the greatest purification," and to them knowledge meant mathematics. Never before or since has mathematics had such an essential part in life and religion as it did with the Pythagoreans. At the heart of their scheme of things was the belief that some sort of an operative reality existed behind the phenomena of nature, and that through the volition of this supreme architect, the universe was created—that beneath the apparent multiplicity and confusion of the world around us there was a fundamental simplicity and stability that reason might discover. They further theorized that everything, physical and spiritual, had been assigned its allotted number and form, the general thesis being "Everything is number." (By "number" was meant a positive integer.) All this culminated in the notion that without the help of mathematics, a rational understanding of the ruling principles at work in the universe would be impossible. Aristotle wrote in the *Metaphysics:*

> The Pythagoreans . . . devoted themselves to mathematics; they were the first to advance this study and having been brought up in it they thought its principles were the principles of all things.

About Pythagoras himself, we are told by another chronicler that "he seems to have attached supreme importance to the study of arithmetic, which he advanced and took out of the domain of commercial utility."

Music provided the Pythagoreans with the best instance of their principle that "number" was the cause of everything in nature. Tradition credits Pythagoras with the discovery that notes sounded by a vibrating string depended on the string's length, and in particular, that a harmonious sound was produced by plucking two equally taut strings, one twice the length of the other. In modern terms, the interval between these two notes is an octave. Similarly, if one string were half again the length of the other, the shorter one would give off a note, called a "fifth," above that emitted by the longer; whereas if one were a third longer than the other a "fourth" would be produced—one note four tones above the other. It was concluded that the most beautiful musical harmonies corresponded to the simplest ratios of whole numbers, namely, the ratios 2:1, 3:2, and 4:3 (the four numbers 1,2,3,4 being enshrined in the famous Pythagorean tetractys, or triangle of dots).

The Pythagorean views on astronomy could be considered an extension of this doctrine of harmonic intervals. Pythagoras held that each of the seven known planets, among which he included the sun and the moon, was carried around the earth on a crystal sphere of its own. Because it was surely impossible for such gigantic spheres to whirl endlessly through space without generating any noise by their motion, each body would have to produce a certain tone according to its distance from the center. The whole system created a celestial harmony, which Pythagoras alone among all mortals could hear. This theory was the basis for the idea of the "music of the spheres," a continually recurring notion in medieval astronomical speculation.

The Pythagorean doctrine was apparently a curious mixture of cosmic philosophy and number mysticism, a sort of supernumerology that assigned to everything material or spiritual a definite integer. Among the writings of the Pythagoreans, we find that 1 represented reason, because reason could produce only one consistent body of truths; 2 stood for man and 3 for woman; 4 was the Pythagorean symbol for justice, because it was the first number to be the product of equals; 5 was identified with marriage, formed as it was by the union of 2 and 3; 6 was the number of creation; and so forth. All the even numbers, after the first even number, were separable into other numbers; hence they were prolific and were considered feminine and earthy—and somewhat less highly regarded in general. And because the Pythagoreans were a predominantly male society, they classified the odd numbers, after the first one, as masculine and divine.

Although these speculations about numbers as models of ''things'' strike us as far-fetched and fanciful today, it must be remembered that the intellectuals of the classical Greek period were largely absorbed in philosophy and that these same men, because they possessed intellectual interest, were the very ones who were laying the foundations for mathematics as a system of thought. To Pythagoras and his followers, mathematics was largely a means to an end, an end in which the human spirit was ennobled through a mystical contemplation of the good and the beautiful. Only with the foundation of the School of Alexandria do we enter a new phase in which mathematics is made into an intellectual exercise pursued for its own sake, independent of its utilitarian applications.

Even though the Pythagoreans first studied numbers less for themselves than for the things they represented, they were nonetheless led to recognize all sorts of new arithmetical properties.

The most complete exposition that has come down to us of the arithmetic of Pythagoras and his immediate successors is contained in the *Introductio Arithmeticae* of Nicomachus of Gerasa (circa 100). Though Nicomachus did not contribute significant new mathematical results, his *Introductio Arithmeticae* is noteworthy as the first systematic work in which arithmetic was treated independent of geometry. The content is much the same as that of the number-theoretic books of Euclid's *Elements* (Books VII, VIII, and IX), but the approach is different. Whereas Euclid represented numbers by straight lines with letters attached—a system that allowed him to work with numbers in general without having to assign them specific values—Nicomachus represented numbers by letters with definite values, thereby having to resort to all sorts of circumlocution to distinguish among undetermined numbers. Euclid always offered proofs of his propositions, a thing wholly lacking in Nicomachus. At times, Nicomachus simply enunciated a general result and gave concrete examples of it; on other occasions he left the general proposition to be inferred from the particular examples presented alone.

Euclid did not share the philosophical proclivities (or more accurately, the Pythagorean tendencies) of Nicomachus but held himself to a more strictly scientific level. Nicomachus's treatise was probably like this because he was not a creative mathematician and because he intended his work to be a popular treatment of arithmetic designed to acquaint the beginner with the important discoveries to that date.

Despite a lack of originality and a mathematical poverty, Nicomachus's *Introductio* became a leading textbook in the Latin West from the time it was written until the 1500s. The Arab world also became acquainted with Greek arithmetic through a translation of the *Introductio* by Thabit-ibn-Korra in the ninth century. Indeed, the influence of Nicomachus's treatise can be judged by the number of versions or commentaries that appeared in ancient times and also by the number of authors who quoted it. An indication of the book's renown is that the second-century Greek writer Lucian, wishing to pay the highest compliment to a calculator, said: ''You reckon like Nicomachus of Gerasa.''

Sometime about 450 B.C., the Greeks adopted an alphabetic notation for representing numbers; the first nine letters of the Greek alphabet were associated with the first nine integers, the next nine letters represented the first nine integral multiples of 10, and the last nine letters were used for the first nine integral multiples of 100. (Three older letters, not found in the present-day Greek alphabet, were introduced to make the required 27.) It is unlikely that the early Pythagoreans had any number symbols, so they must have thought of numbers in a strictly visual way, either as pebbles placed in the sand or as dots in certain geometric patterns. Thus, numbers were classified as triangular, square, pentagonal, and so on, according to the shapes made by the arrangement of the dots. Numbers that can be represented in geometric form are nowadays called figurative, or polygonal, numbers; and such were considered by Nicomachus in the *Introductio*. Pythagoras himself was acquainted at least with the triangular numbers, and very probably with square numbers, and the other polygonal numbers were treated by later members of his school.

The numbers 1, 3, 6, and 10 are examples of triangular numbers, because each of these counts the number of dots that can be arranged evenly in an equilateral triangle.

 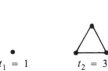 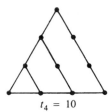

$t_1 = 1$ $\qquad$ $t_2 = 3$ $\qquad$ $t_3 = 6$ $\qquad$ $t_4 = 10$

Similarly, the numbers 1, 4, 9, and 16 are said to be square numbers, because as dots they can be depicted as squares.

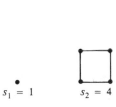 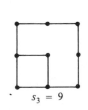 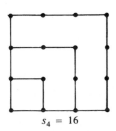

$s_1 = 1$ $\qquad$ $s_2 = 4$ $\qquad$ $s_3 = 9$ $\qquad$ $s_4 = 16$

One can read off some remarkable number-theoretic laws from such configurations. For instance, the sum of two consecutive triangular numbers always equals the square number whose ''side'' is the same as the ''side'' of the larger of the two triangles. This can be confirmed geometrically by separating the dots with a slash and then counting them, as in the accompanying figure. It is just as easy to prove the result by an algebraic argument. However, first notice how the triangular numbers are formed;

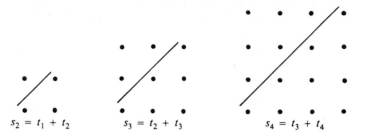

$$s_2 = t_1 + t_2 \qquad s_3 = t_2 + t_3 \qquad s_4 = t_3 + t_4$$

each new one is obtained from the previous triangular number by adding another row containing one more dot than the previous row added. Thus, if t_n designates the nth triangular number, then

$$t_n = t_{n-1} + n$$

$$= t_{n-2} + (n-1) + n$$

$$\vdots$$

$$= t_1 + 2 + 3 + \cdots + (n-1) + n$$

$$= 1 + 2 + 3 + \cdots + (n-1) + n.$$

Our plan is to fit together two triangles, each representing t_n (hence, each consists of n rows of dots), to produce a rectangular array whose sides are n and $n + 1$. In the next figure, for example, $n = 5$.

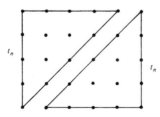

It is clear that such an array contains $n(n + 1)$ dots and so

$$2t_n = n(n + 1),$$

or equivalently,

$$t_n = \frac{n(n + 1)}{2}.$$

With this formula available, one sees easily that the nth square number s_n is the sum of two successive triangular numbers; for

$$s_n = n^2 = \frac{n(n + 1)}{2} + \frac{(n - 1)n}{2} = t_n + t_{n-1}.$$

Gathering up the pieces, we get as a bonus an expression for the sum of the first n numbers:

$$1 + 2 + 3 + \cdots + n = \frac{n(n + 1)}{2}.$$

Likewise, a formula for the sum of the first n odd numbers can be found. The appropriate starting point is the observation that a square made up of n dots on a side can be divided into a smaller square of side $n - 1$ and an L-shaped border (a gnomon).

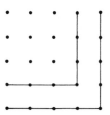

By repeating this subdivision, as in the next diagram, it becomes evident that the differences between successive nested squares produce the sequence of odd numbers; consequently,

$$1 + 3 + 5 + \cdots + (2n - 1) = n^2.$$

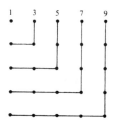

The Pythagoreans may have proved this result by first considering the n equations

$$1^2 \qquad\;\; = 1$$
$$2^2 - 1^2 = 3$$
$$3^2 - 2^2 = 5$$
$$4^2 - 3^2 = 7$$
$$\vdots$$
$$n^2 - (n - 1)^2 = 2n - 1.$$

Adding these equations, we get

$$1^2 + (2^2 - 1^2) + (3^2 - 2^2) + \cdots + (n^2 - (n - 1)^2)$$
$$= 1 + 3 + 5 + \cdots + (2n - 1),$$

which reduces to

$$n^2 = 1 + 3 + 5 + \cdots + (2n - 1).$$

The Pythagoreans could not have expected the theory of figurative numbers to attract the attention of later scholars of the highest rank. In 1665, the mathematician-philosopher Pascal wrote his *Treatise on Figurative Numbers*. In it, he asserted that every positive integer was the sum of three or fewer triangular numbers. For instance,

$$16 = 6 + 10 \qquad\qquad 39 = 3 + 15 + 21$$

$$25 = 1 + 3 + 21 \qquad 150 = 6 + 66 + 78$$

This remarkable result was conjectured by Fermat in a letter to Mersenne dated 1636 and first proved by Gauss in 1801.

Another interesting pattern can be observed from the following equations.

$$1^3 = \quad 1 = t_1{}^2$$

$$1^3 + 2^3 = \quad 9 = t_2{}^2$$

$$1^3 + 2^3 + 3^3 = \quad 36 = t_3{}^2$$

$$1^3 + 2^3 + 3^3 + 4^3 = 100 = t_4{}^2.$$

So far, the right-hand column gives the sequence of squares of the triangular numbers. This pattern leads one to suspect that the sum of the first n cubes equals the square of the nth triangular number. For a formal verification, let us begin by noting that the algebraic identity

$$[k(k-1)+1] + [k(k-1)+3] + [k(k-1)+5]$$
$$+ \cdots + [k(k-1)+(2k-1)] = k^3$$

can be used to produce cubes. Taking successively $k = 1, 2, 3, \ldots, n$ in this formula produces the following set of equations:

$$1 = 1^3$$

$$3 + 5 = 2^3$$

$$7 + 9 + 11 = 3^3$$

$$13 + 15 + 17 + 19 = 4^3$$

$$\vdots$$

$$[n(n-1)+1] + [n(n-1)+3] + \cdots + [n(n-1)+(2n-1)] = n^3.$$

Adding together these last n equations, one finds that

$$1 + 3 + 5 + 7 + 9 + \cdots + [n(n-1)+(2n-1)]$$
$$= 1^3 + 2^3 + 3^3 + \cdots + n^3,$$

where the left-hand side consists of consecutive odd integers. The key to success lies in calculating the number of terms that appear on the left. For this, let us write the last term as

$$n(n-1) + (2n-1) = n^2 + n - 1 = 2\left[\frac{n(n+1)}{2}\right] - 1,$$

so that the expression in question involves the sum of all odd integers from 1 to $2[n(n + 1)/2] - 1$, a total of $n(n + 1)/2$ terms. From what was proved earlier, we know that the sum of the first $n(n + 1)/2$ odd integers equals $[n(n + 1)/2]^2$; it turns out that

$$1^3 + 2^3 + 3^3 + \cdots + n^3 = \left(\frac{n(n + 1)}{2}\right)^2 = t_n^2.$$

This unexpected identity, relating sums of cubes to triangular numbers, goes back to the first century and is usually attributed to Nicomachus himself.

Finding a formula for the sum of the squares of the first n numbers takes a bit more effort. Let us first give a geometric argument for the case $n = 4$, using reasoning that can be generalized for any positive integer n. We begin by placing square arrays containing 1^2, 2^2, 3^2, and 4^2 dots adjacent to each other.

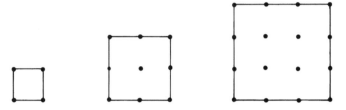

Next let us add horizontal rows consisting of 1, 3, 6, and 10 dots, respectively, to form a rectangle of width $1 + 2 + 3 + 4$ and height $4 + 1$. This is pictured here geometrically.

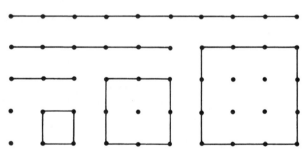

Counting the dots in the squares and rows should make it clear that

$$(1^2 + 2^2 + 3^2 + 4^2) + (1 + 3 + 6 + 10) = (1 + 2 + 3 + 4)(4 + 1)$$

and consequently that

$$(1^2 + 2^2 + 3^2 + 4^2) = 10 \cdot 5 - 20 = 30 = \frac{4 \cdot 5 \cdot 9}{6}.$$

Proceeding along similar lines, one can get a formula for

$$1^2 + 2^2 + 3^2 + \cdots + n^2,$$

where n is arbitrary. Simply place square arrays for 1^2, 2^2, 3^2, . . . , n^2 dots side by side and fit n rows of dots together, beginning with the shortest row on the bottom, to get a rectangle.

The dimensions of the rectangle are $1 + 2 + 3 + \cdots + n$ by $n + 1$, so that it encompasses a total of

$$(1 + 2 + 3 + \cdots + n)(n + 1)$$

dots. This gives one side of the desired identity. For the other side, we add the dots in consecutive squares and rows, to arrive at the sum

$$(1^2 + 2^2 + 3^2 + \cdots + n^2) +$$
$$[1 + (1 + 2) + (1 + 2 + 3) + \cdots + (1 + 2 + 3 + \cdots + n)].$$

In algebraic form, our identity is

$$(1^2 + 2^2 + 3^2 + \cdots + n^2) +$$
$$[1 + (1 + 2) + (1 + 2 + 3) + \cdots + (1 + 2 + 3 + \cdots + n)]$$
$$= (1 + 2 + 3 + \cdots + n)(n + 1).$$

If we let $S = 1^2 + 2^2 + 3^2 + \cdots + n^2$, this becomes

$$S + [1 + (1 + 2) + (1 + 2 + 3) + \cdots + (1 + 2 + 3 + \cdots + n)]$$
$$= (1 + 2 + 3 + \cdots + n)(n + 1).$$

The foregoing expression can be simplified by appealing to the fact that the sum of the first k integers is $k(k + 1)/2$; after making the appropriate substitutions, we get

$$S + \left(\frac{1 \cdot 2}{2} + \frac{2 \cdot 3}{2} + \frac{3 \cdot 4}{2} + \cdots + \frac{n(n + 1)}{2} \right) = \frac{n(n + 1)^2}{2} ,$$

which can be written

$$S + \frac{1}{2}[1(1 + 1) + 2(2 + 1) + 3(3 + 1) + \cdots + n(n + 1)] = \frac{n(n + 1)^2}{2} .$$

This yields

$$S + \frac{1}{2}[(1^2 + 2^2 + 3^2 + \cdots + n^2) + (1 + 2 + 3 + \cdots + n)] = \frac{n(n + 1)^2}{2} ,$$

whence

$$S + \frac{1}{2}\left[S + \frac{n(n + 1)}{2} \right] = \frac{n(n + 1)^2}{2} .$$

It now becomes a matter of solving for S:

$$\frac{3}{2}S = \frac{n(n + 1)^2}{2} - \frac{n(n + 1)}{4}$$

$$= \frac{n(n + 1)(2n + 1)}{4},$$

which leads at once to

$$S = \frac{n(n + 1)(2n + 1)}{6}.$$

All in all, we have shown that the sum of the first n squares has a simple expression in terms of n; namely

$$1^2 + 2^2 + 3^2 + \cdots + n^2 = \frac{n(n + 1)(2n + 1)}{6}.$$

A strikingly original proof of the last result, propounded by the thirteenth-century mathematician Fibonacci, comes from the identity

$$k(k + 1)(2k + 1) = (k - 1)k(2k - 1) + 6k^2.$$

By putting $k = 1, 2, 3, \ldots, n$ in turn into this formula, one gets the set of equations

$$1 \cdot 2 \cdot 3 = 6 \cdot 1^2$$

$$2 \cdot 3 \cdot 5 = 1 \cdot 2 \cdot 3 + 6 \cdot 2^2$$

$$3 \cdot 4 \cdot 7 = 2 \cdot 3 \cdot 5 + 6 \cdot 3^2$$

$$\vdots$$

$$(n - 1)n(2n - 1) = (n - 2)(n - 1)(2n - 3) + 6(n - 1)^2$$
$$n(n + 1)(2n + 1) = (n - 1)n(2n - 1) + 6n^2.$$

What is important is that a common term appears on the left-hand and right-hand sides of successive equations. When these n equations are added and common terms cancelled, it is easily shown that

$$n(n + 1)(2n + 1) = 6(1^2 + 2^2 + 3^2 + \cdots + n^2),$$

leading to the desired conclusion.

Not very far from Croton was the Eleatic school, a philosophical movement challenging the Pythagoreans' doctrine that all natural phenomena can be expressed in some way by whole numbers. This rival school took its name from the Ionian colony at Elea on the western coast of southern Italy and had as its most prominent member Zeno (circa 450 B.C.). We know little of Zeno's life other than Plato's assertion that he went to Athens when nearly 40 years old, where he met with the youthful Socrates. Apparently Zeno was originally a Pythagorean and, like Pythagoras, played an active part in the politics of his native city. There is a widespread legend that he was tortured and killed by a tyrant of Elea whom he had plotted to depose.

Zeno is remembered today for four clever paradoxes—preserved by Aristotle in his *Physics*—about the reality of motion. In these, Zeno pointed out the logical absurdities arising from the concept of ''infinite divisibility'' of time and space. The paradox most often quoted concerns Achilles and a tortoise: Achilles, the swiftest runner in Greece, can never catch a tortoise that has been given a head start. For, by the time Achilles reaches the tortoise's starting point, the animal will have moved to another point; by the time Achilles reaches that point the tortoise will have advanced somewhat further. As the process continues indefinitely, Achilles—though the faster runner—always advances on the slower tortoise yet cannot overtake it.

Although Zeno's argument confounded his contemporaries, a satisfactory explanation incorporates a now-familiar idea, the notion of a ''convergent infinite series.'' The paradox rests partly on the misconception that an infinite number of ever-shorter lengths (and, similarly, time durations) must add up to an infinite total. But an infinite series may have a finite sum. Suppose that Achilles runs ten times as fast as the determined tortoise and gives it an initial start of 100 yards; say, Achilles runs 10 yards per second. Consider the distances he has to cover. They are successively 100 yards, 10 yards, 1 yard, 1/10 yard, and so on. The total number of yards Achilles must travel in order to catch his slower competitor is

$$100 + 10 + 1 + \frac{1}{10} + \frac{1}{100} + \ldots,$$

which forms a convergent geometric series with sum $111\frac{1}{9}$ yards. In the same elapsed time (that is, $11\frac{1}{9}$ seconds) the distance covered by the tortoise will be the sum of the geometric series $10 + 1 + \frac{1}{10} + \frac{1}{100} + \ldots$, which is found to be $11\frac{1}{9}$ yards. Accordingly, when Achilles has traversed $111\frac{1}{9}$ yards he will be dead even with the tortoise, and ahead of him thereafter.

Of course Zeno knew perfectly well that Achilles would win a race with a tortoise, but he was drawing attention to opposing theories on the nature of space and time. (There is a frequently told anecdote that Diogenes the Cynic refuted Zeno's argument, while the latter was lecturing in Athens, simply by getting up and walking; but the story cannot be true because Zeno and Diogenes were not contemporary.) The Eleatic mathematical philosophers held that space and time are undivided wholes, or continua, that cannot be broken down into small indivisible parts. This was at variance with the Pythagorean idea that a line is made up of a series of points—like tiny beads or ''numerical atoms''—and that time is likewise composed of a series of discrete moments. Zeno was partly responsible for the subsequent course of Greek mathematical thought. For at heart, his famous paradoxes were related to the application of infinite processes to geometry. Because of the inability of the Greek geometers to answer them in a clear manner, they banished from mathematics the use of methods that involved the concept of infinity and made a ''horror of the infinite'' part of the Greek mathematical tradition.

3.2 Problems

1. Plutarch (about A.D. 100) stated that if a triangular number is multiplied by 8, and 1 is added, then the result is a square number. Prove that this is fact and illustrate it geometrically in the case of t_2.

2. Prove that the square of any odd multiple of 3 is the difference of two triangular numbers, specifically that

$$[3(2n + 1)]^2 = t_{9n + 4} - t_{3n + 1}.$$

3. Prove that if t_n is a triangular number, then $9t_n + 1$ is also triangular.

4. Write each of the following numbers as the sum of three or fewer triangular numbers:

   (a) 56;   (b) 69;   (c) 185;   (d) 287.

5. For $n \geq 1$, establish the formula

$$(2n + 1)^2 = (4t_n + 1)^2 - (4t_n)^2.$$

6. Verify that 1225 and 41,616 are simultaneously square and triangular numbers. [*Hint:* Finding an integer n such that

$$t_n = \frac{n(n + 1)}{2} = 1225$$

   is equivalent to solving the quadratic equation $n^2 + n - 2450 = 0$.]

7. An oblong number counts the number of dots in a rectangular array having one more row than it has columns; the first few of these numbers are

$$o_1 = 2 \quad o_2 = 6 \quad o_3 = 12 \quad o_4 = 20$$

   and in general, the nth oblong number is given by $o_n = n(n + 1)$. Prove algebraically and geometrically that:

   (a) $o_n = 2 + 4 + 6 + \cdots + 2n.$
   (b) Any oblong number is the sum of two equal triangular numbers.

   (c) $o_n + n^2 = t_{2n}.$
   (d) $o_n - n^2 = n.$
   (e) $n^2 + 2o_n + (n + 1)^2 = (2n + 1)^2.$
   (f) $o_n = 1 + 2 + 3 + \cdots + n + (n + 1) + (n - 1) + (n - 2) + \cdots + 3 + 2.$

8. In 1872, Lebesgue proved that (1) every positive integer is the sum of a square number (possibly 0^2) and two triangular numbers, and (2) every positive integer is the sum of two square numbers and a triangular number. Confirm these results in the cases of the integers 9, 44, 81, and 100.

9. Display the consecutive integers 1 through n in two rows as follows:

1	2	3	$\cdots$	$n - 1$	n
n	$n - 1$	$n - 2$	$\cdots$	2	1

   If the sum obtained by adding the n columns vertically is set equal to the sum obtained by adding the two rows horizontally, what well-known formula results?

10. Derive the identity

$$[n(n - 1) + 1] + [n(n - 1) + 3] + \cdots + [n(n - 1) + (2n - 1)] = n^3,$$

   where n is any positive integer.

11. For any integer $n \geq 1$, prove that:

   (a) $1 + 2 + 3 + \cdots + (n - 1) + n + (n - 1) + \cdots + 3 + 2 + 1 = n^2.$

   (b) $\dfrac{1}{1 \cdot 2} + \dfrac{1}{2 \cdot 3} + \dfrac{1}{3 \cdot 4} + \cdots + \dfrac{1}{n(n + 1)}$
   $= \dfrac{n}{(n + 1)}.$

   [*Hint:* Use the splitting identity

$$1/k - 1/(k + 1) = 1/k(k + 1)$$

   to rewrite the left-hand side.]

   (c) $1 \cdot 2 + 2 \cdot 3 + 3 \cdot 4 + \cdots + n(n + 1)$
   $= \dfrac{n(n + 1)(n + 2)}{3}.$

   [*Hint:* Use the identity $k(k + 1) = k^2 + k$ and collect the squares.]

(d) $\dfrac{1}{1 \cdot 3} + \dfrac{1}{3 \cdot 5} + \dfrac{1}{5 \cdot 7} + \cdots$

$+ \dfrac{1}{(2n - 1)(2n + 1)} = \dfrac{n}{2n + 1}.$

$\left[\text{\textit{Hint:} Use the identity}\right.$

$$\dfrac{1}{(2k - 1)(2k + 1)}$$
$$= \dfrac{1}{2}\left(\dfrac{1}{2k - 1} - \dfrac{1}{2k + 1}\right). \left.\right]$$

(e) $1^3 + 3^3 + 5^3 + \cdots + (2n - 1)^3$
   $= n^2(2n^2 - 1).$

[*Hint:* Separate the left-hand side of the identity

$$1^3 + 2^3 + 3^3 + \cdots + (2n)^3$$
$$= \left[\dfrac{(2n)(2n + 1)}{2}\right]^2$$

into odd and even terms and solve for the sum of odd cubes.]

12. (a) Prove that the sum of a finite arithmetic series equals the product of the number of terms and half the sum of the two extreme terms; in symbols, this reads

$(a + d) + (a + 2d) + (a + 3d) + \cdots$

$+ (a + nd) = n\left[\dfrac{(a + d) + (a + nd)}{2}\right].$

(b) Use the result of part (a) to confirm the identities

$1 + 4 + 7 + \cdots + (3n - 2) = \dfrac{n(3n - 1)}{2}$

and

$1 + 3 + 5 + \cdots + (2n - 1) = n^2.$

13. The identity

$(1 + 2 + 3 + \cdots + n)^2$
$= 1^3 + 2^3 + 3^3 + \cdots + n^3, \qquad n \geq 1$

was known as early as the first century. Provide a derivation of it.

14. Prove the following formula for the sum of triangular numbers, given by the Hindu mathematician Aryabhata (circa 500):

$$t_1 + t_2 + t_3 + \cdots + t_n = \dfrac{n(n + 1)(n + 2)}{6}.$$

[*Hint:* Group the terms on the left-hand side in pairs, replacing $t_{k-1} + t_k$ by k^2; consider the two cases where n is odd and n is even.]

15. Archimedes (287–212 B.C.) also derived the formula

$$1^2 + 2^2 + 3^2 + \cdots + n^2 = \dfrac{n(n + 1)(2n + 1)}{6}$$

for the sum of squares. Fill in any missing details in the following sketch of his proof. In the formula

$n^2 = [k + (n - k)]^2$
$\quad = k^2 + 2k(n - k) + (n - k)^2,$

let k take on the successive values $1, 2, 3, \ldots, n - 1$. Add the resulting $n - 1$ equations, together with the identity $2n^2 = 2n^2$, to arrive at

(*) $(n + 1)n^2 = 2(1^2 + 2^2 + 3^2 + \cdots + n^2)$
$\quad + 2[1(n - 1) + 2(n - 2)$
$\quad + 3(n - 3) + \cdots + (n - 1)1].$

Next, let $k = 1, 2, 3, \ldots, n$ in the formula

$$k^2 = k + 2[1 + 2 + 3 + \cdots + (k - 1)]$$

and add the n equations so obtained to get

(**) $\quad 1^2 + 2^2 + 3^2 + \cdots + n^2$
$\quad = (1 + 2 + 3 + \cdots + n)$
$\quad\quad + 2[1(n - 1) + 2(n - 2)$
$\quad\quad + 3(n - 3) + \cdots + (n - 1)1].$

The desired result follows on combining (*) and (**).

16. The tetrahedral numbers count the number of dots in pyramids built up of triangular numbers. If the base is the triangle of side n, then the pyramid is formed by placing similarly situated triangles upon it, each of which has one less in its sides than that which precedes it.

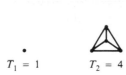

$T_1 = 1$ $T_2 = 4$ $T_3 = 10$

In general, the nth tetrahedral number T_n is given by the formula

$$T_n = t_1 + t_2 + t_3 + \cdots + t_n,$$

where t_n is the kth triangular number. Prove that

$$T_n = \frac{n(n + 1)(n + 2)}{6} = \frac{n + 1}{6}(2t_n + n).$$

[*Hint:* See Problem 14.]

17. Use the following facts to derive the formula for the sum of the squares of the first n integers:

$$1^2 = 1 = t_1,$$
$$1^2 + 2^2 = 5 = 3 + 2 \cdot 1 = t_2 + 2T_1,$$
$$1^2 + 2^2 + 3^2 = 14 = 6 + 2 \cdot 4 = t_3 + 2T_2,$$
$$1^2 + 2^2 + 3^2 + 4^2 = 30 = 10 + 2 \cdot 10$$
$$= t_4 + 2T_3,$$
$$1^2 + 2^2 + 3^2 + 4^2 + 5^2 = 55 = 15 + 2 \cdot 20$$
$$= t_5 + 2T_4,$$

where t_k and T_k are the kth triangular and tetrahedral numbers, respectively.

3.3   The Pythagorean Problem

Although tradition is unanimous in ascribing the so-called Pythagorean theorem to the great teacher himself, we have seen that the Babylonians knew the result for certain specific triangles at least a millennium earlier. We recall the theorem as "the area of the square built upon the hypotenuse of a right triangle is equal to the sum of the areas of the squares upon the remaining sides." Because none of the various Greek writers who attributed the theorem to Pythagoras lived within five centuries of him, there is little convincing evidence to corroborate the general belief that the master, or even one of his immediate disciples, gave the first rigorous proof of this characteristic property of right triangles. Moreover, the persistent legend that when Pythagoras had discovered the theorem, he sacrificed a hundred oxen to the Muses in gratitude for the inspiration appears an unlikely story, because the Pythagorean ritual forbade any sacrifice in which blood was shed. What is certain is that the school Pythagoras founded did much to increase the interest in problems directly connected with the celebrated result that bears his name.

Still more are we in doubt about what line of demonstration the Greeks originally offered for the Pythagorean theorem. If the methods of Book II of Euclid's *Elements* were used, it was probably a dissection type of proof similar to the following. A large square of side $a + b$ is divided into two smaller squares of sides a and b respectively, and two equal rectangles with sides a and b; each of these two rectangles can be split into two equal right triangles by drawing the diagonal c. The four triangles can be arranged within another square of side $a + b$ as shown in the second figure.

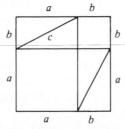
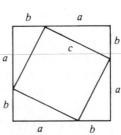

Now the area of the same square can be represented in two ways: as the sum of the areas of two squares and two rectangles,

$$(a + b)^2 = a^2 + b^2 + 2ab;$$

and as the sum of the areas of a square and four triangles,

$$(a + b)^2 = c^2 + 4\left(\frac{ab}{2}\right).$$

When the four triangles are deducted from the larger square in each figure, the resulting areas are equal; or equivalently, $c^2 = a^2 + b^2$. Therefore, the square on c is equal to the sum of the squares on a and b.

   Such proofs by addition of areas are so simple that they may have been made earlier and independently by other cultures (no record of the Pythagorean theorem appears, however, in any of the surviving documents from ancient Egypt). In fact, the contemporary Chinese civilization, which had grown up in effective isolation from both the Greek and Babylonian civilizations, had a neater and possibly much earlier proof than the one just cited. This is found in the oldest extant Chinese text containing formal mathematical theories, the *Arithmetic Classic of the Gnomon and the Circular Paths of Heaven*. Assigning the date of this work is difficult. Astronomical evidence suggests that the oldest parts go back to 600 B.C., but there is reason to believe that it has undergone considerable change since first written. The first firm dates that we can connect with it are over a century later than the dates for *Nine Chapters on the Mathematical Art*. A diagram in the *Arithmetic Classic* represents the oldest known proof of the Pythagorean theorem.

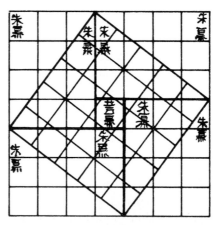

The proof inspired by this figure was much admired for its simple elegance, and it later found its way into the *Vijaganita* (Root Calculations) of the Hindu mathematician Bhaskara, born in 1114. Bhaskara draws the right triangle four times in the square of the hypotenuse, so that in the middle there remains a square whose side equals the difference between the two sides of the right triangle. This last square and the four triangles are then rearranged to make up the areas of two squares, the lengths of whose sides correspond to the legs of the right triangle. "Behold," said Bhaskara, without adding a further word of explanation.

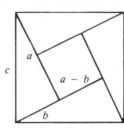 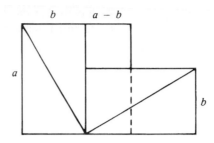

The geometrical discovery that the sides of a right triangle were connected by a law expressible in numbers led naturally to a corresponding arithmetical problem, which we shall call the Pythagorean problem. This problem, one of the earliest problems in the theory of numbers, calls for finding all right triangles whose sides are of integral length, that is, finding all solutions in the positive integers of the Pythagorean equation

$$x^2 + y^2 = z^2.$$

A triple (x, y, z) of positive integers satisfying this equation is said to be a Pythagorean triple.

Ancient tradition attributes to Pythagoras himself a partial solution of the problem, expressed by the numbers

$$x = 2n + 1, \qquad y = 2n^2 + 2n, \qquad z = 2n^2 + 2n + 1,$$

where $n \geq 1$ is an arbitrary integer. As is perhaps more often the rule than the exception in such instances, the attribution of the name may readily be questioned.

Pythagoras presumably arrived at his solution by a relation that produces a square number from the next smaller square number, namely

(1) $$(2k - 1) + (k - 1)^2 = k^2.$$

The strategy was to suppose that $2k - 1$ is a perfect square. (This happens infinitely often; for instance, if $k = 5$, then $2k - 1 = 3^2$.) Letting $2k - 1 = m^2$ and solving for k, we get

$$k = \frac{m^2 + 1}{2} \qquad \text{and} \qquad k - 1 = \frac{m^2 - 1}{2}.$$

When these values are substituted in (1), it follows that

$$m^2 + \left(\frac{m^2 - 1}{2}\right)^2 = \left(\frac{m^2 + 1}{2}\right)^2,$$

whence

(2) $$x = m, \qquad y = \frac{m^2 - 1}{2}, \qquad z = \frac{m^2 + 1}{2},$$

satisfy the Pythagorean equation for any odd integer $m > 1$ (m must be odd, because $m^2 = 2k - 1$ is odd). When $m = 2n + 1$, where $n \geq 1$, the numbers in (2) become

(3) $$x = 2n + 1, \qquad y = 2n^2 + 2n, \qquad z = 2n^2 + 2n + 1,$$

which is Pythagoras's result. Some of the Pythagorean triples that can be obtained from (3) are given in the accompanying table.

n	x	y	z
1	3	4	5
2	5	12	13
3	7	24	25
4	9	40	41
5	11	60	61

As one sees, Pythagoras's solution has the special feature of producing right triangles having the characteristic that the hypotenuse exceeds the larger leg by one.

Another special solution in which the hypotenuse and a leg differ by 2 is ascribed to the Greek philosopher Plato, to wit,

$$(4) \qquad\qquad x = 2n, \qquad y = n^2 - 1, \qquad z = n^2 + 1.$$

This formula can be obtained, like the other, with the help of the relation (1); but now, we apply it twice:

$$(k + 1)^2 = k^2 + (2k + 1)$$

$$= [(k - 1)^2 + (2k - 1)] + 2k + 1 = (k - 1)^2 + 4k.$$

Substituting n^2 for k in order to make $4k$ a square, one arrives at the Platonic formula

$$(2n)^2 + (n^2 - 1)^2 = (n^2 + 1)^2.$$

Observe that from (4) it is possible to produce the Pythagorean triple (8, 15, 17), which cannot be gotten from Pythagoras's formula (3).

Neither of the aforementioned rules accounts for all Pythagorean triples, and it was not until Euclid wrote his *Elements* that a complete solution to the Pythagorean problem appeared. In Book X of the *Elements,* there is geometric wording to the effect that

$$(5) \qquad\qquad x = 2mn, \qquad y = m^2 - n^2, \qquad z = m^2 + n^2,$$

where m and n are positive integers, with $m > n$.

In his *Arithmetica*, Diophantus (third century) also stated that he could get right triangles "with the aid of" two numbers m and n according to the formulas in (5). Diophantus seems to have arrived at these formulas by the following reasoning. Given the equation $x^2 + y^2 = z^2$, put $y = kx - z$, where k is any rational number. Then

$$z^2 - x^2 = y^2 = (kx - z)^2$$

$$= k^2x^2 - 2kxz + z^2,$$

which leads to

$$-x^2 = k^2x^2 - 2kxz$$

or

$$-x = k^2x - 2kz .$$

When this equation is solved for x, we get

$$x = \frac{2k}{k^2 + 1} z .$$

The implication is that

$$y = kx - z = \frac{k^2 - 1}{k^2 + 1} z .$$

But $k = m/n$, with m and n integers (there is no harm in taking $m > n$), so that

$$x = \frac{2mn}{m^2 + n^2} z , \qquad y = \frac{m^2 - n^2}{m^2 + n^2} z .$$

If one sets $z = m^2 + n^2$ in order to obtain a solution in the integers, it is found immediately that

$$x = 2mn, \qquad y = m^2 - n^2, \qquad z = m^2 + n^2.$$

Our argument indicates that x, y, z as defined by the preceding formulas satisfy the Pythagorean equation. The converse problem of showing that any Pythagorean triple is necessarily of this form is much more difficult. The details first appeared in the works of Arab mathematicians around the tenth century.

The most important achievement of the Pythagorean school in its influence on the evolution of the number concept was the discovery of the "irrational." The Pythagoreans felt intuitively that any two line segments had a common measure; that is to say, starting with two line segments, one should be able to find some third segment, perhaps very small, that could be marked off a whole number of times on each of the given segments. From this it would follow that the ratio of the lengths of the original line segments could be expressed as the ratio of integers or as a rational number. (Recall that a rational number is defined as the quotient of two integers a/b, where $b \neq 0$.) One can imagine the shattering effect of the discovery that there exist some ratios that cannot be represented in terms of integers. Who it was that first established this, or whether it was done by arithmetical or geometric methods, will probably remain a mystery forever.

The oldest known proof dealing with incommensurable line segments corresponds in its essentials to the modern proof that $\sqrt{2}$ is irrational. This is the proof of the incommensurability of the diagonal of a square with its side, and it is to be found in the tenth book of Euclid's *Elements*. A reference in one of Aristotle's works, however, makes it clear that the proof was known long before Euclid's time. As in most classical demonstrations, the method of argument was indirect. Thus, the negation of the desired conclusion is assumed, and a contradiction is derived from the assumption.

The reasoning goes as follows. If the diagonal AC and side AB of the square $ABCD$ have a common measure, say δ, then there exist positive integers m and n satisfying

$$AC = m\delta, \qquad AB = n\delta.$$

The ratio of these segments is

$$\frac{AC}{AB} = \frac{m}{n}.$$

To make matters simpler, let us suppose that any common factors of m and n have been cancelled. Now

$$\frac{(AC)^2}{(AB)^2} = \frac{m^2}{n^2}.$$

Applying the Pythagorean theorem to the triangle ABC, one gets $(AC)^2 = 2(AB)^2$, so that the displayed equation becomes

$$2 = \frac{m^2}{n^2},$$

or $2n^2 = m^2$. The task is to show that this cannot happen.

Now $2n^2$, as a multiple of 2, is an even integer; hence m^2 is even. What about m itself? If m were odd, then m^2 would be odd, because the square of any odd integer must be odd. Consequently, m is even; say $m = 2k$. Substituting this value in the equation $m^2 = 2n^2$ and simplifying, we get

$$2k^2 = n^2.$$

By an argument similar to the one above, it can be concluded that n is an even number. The net result is that m and n are both even (that is, each has a factor of 2), which contradicts our initial assumption that they have no common factor whatsoever.

The Pythagoreans were not the first to consider the numerical value of $\sqrt{2}$. An old cuneiform tablet, now in the Yale Babylonian collection, contains the diagram of a square with its diagonals, as shown herewith.

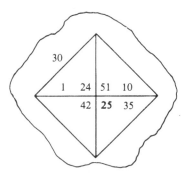

In sexagesimal notation, the number 1;24,51,10 is equal to

$$1 + \frac{24}{60} + \frac{51}{60^2} + \frac{10}{60^3},$$

which gives 1.414213 when translated into the decimal system. You should find this familiar, for it is a very close approximation to $\sqrt{2} = 1.414213562. \ . \ . \ .$ The meaning of the other numbers in the diagram becomes clear when we multiply 1;24,51,10 by

30. The result is 42;25,35, the length of the diagonal of a square of side 30. Thus, the Babylonians not only seemed to know that the diagonal of a square is $\sqrt{2}$ times the length of its side, but also had the arithmetic techniques to accurately approximate $\sqrt{2}$.

Theon of Smyrna (circa 130) devised a procedure for reaching closer and closer approximations of $\sqrt{2}$ by rational numbers. The computations involve two sequences of numbers, the "side numbers" and the "diagonal numbers." We begin with two numbers, one called the first side and denoted by x_1, and the other the first diagonal and indicated by y_1. The second side and diagonal (x_2 and y_2) are formed from the first, the third side and diagonal (x_3 and y_3) from the second, and so on, according to the scheme

$$x_2 = x_1 + y_1, \qquad y_2 = 2x_1 + y_1,$$
$$x_3 = x_2 + y_2, \qquad y_3 = 2x_2 + y_2,$$
$$\vdots \qquad\qquad\qquad \vdots$$

In general, x_n and y_n are obtained from the previous pair of side and diagonal numbers by the formulas

$$x_n = x_{n-1} + y_{n-1}, \qquad y_n = 2x_{n-1} + y_{n-1}.$$

If we take $x_1 = y_1 = 1$ as the initial values, then

$$x_2 = 1 + 1 = 2 \qquad y_2 = 2 \cdot 1 + 1 = 3$$
$$x_3 = 2 + 3 = 5 \qquad y_3 = 2 \cdot 2 + 3 = 7$$
$$x_4 = 5 + 7 = 12 \qquad y_4 = 2 \cdot 5 + 7 = 17$$
$$\vdots \qquad\qquad\qquad \vdots$$

The names *side numbers* and *diagonal numbers* hint that the quotients y_n/x_n of the associated pairs of these numbers come to approximate the ratio of the diagonal of a square to its side:

$$\frac{y_1}{x_1} = 1, \qquad \frac{y_2}{x_2} = \frac{3}{2}, \qquad \frac{y_3}{x_3} = \frac{7}{5}, \qquad \frac{y_4}{x_4} = \frac{17}{12}, \cdots.$$

This follows from the relation

$$(1) \qquad\qquad\qquad y_n^2 = 2x_n^2 \pm 1;$$

for the relation, if true, implies that

$$\left(\frac{y_n}{x_n}\right)^2 = 2 \pm \left(\frac{1}{x_n}\right)^2.$$

Because the value of $(1/x_n)^2$ can be made as small as desired by taking n large enough, it appears that the ratio

$$\frac{y_n}{x_n} = \sqrt{2 \pm \frac{1}{x_n^2}}$$

tends to stay near some fixed number for large n. It can be shown that the fixed "limit" is $\sqrt{2}$. You can see how this works by considering the case $n = 4$. Here,

$$\left(\frac{y_4}{x_4}\right)^2 = \left(\frac{17}{12}\right)^2$$

$$= \frac{289}{144}$$

$$= \frac{288}{144} + \frac{1}{144} = 2 + \left(\frac{1}{12}\right)^2,$$

whence

$$\frac{y_4}{x_4} = \sqrt{2 + \left(\frac{1}{12}\right)^2}.$$

The ratio y_4/x_4 differs from the true value of $\sqrt{2}$ by less than $\frac{1}{7}$ of 1 percent.

Now condition (1), which can be written $y_n^2 - 2x_n^2 = \pm 1$, can be justified by using the algebraic identity

(2) $(2x + y)^2 - 2(x + y)^2 = 2x^2 - y^2.$

If $x = x_0$, $y = y_0$ are any two numbers satisfying the equation $y^2 - 2x^2 = \pm 1$, then we assert that $x = x_0 + y_0$, $y = 2x_0 + y_0$ is also a solution. For by virtue of (2),

$$y^2 - 2x^2 = (2x_0 + y_0)^2 - 2(x_0 + y_0)^2$$

$$= -(y_0^2 - 2x_0^2) = -(\pm 1) = \mp 1.$$

Thus, when one solution of $y^2 - 2x^2 = \pm 1$ is known, it is possible to find infinitely many more solutions by using identity (2).

In the present situation, by the manner in which side and diagonal numbers are formed, this means that if $y_n^2 - 2x_n^2 = \pm 1$ happens to hold for a certain value of n, then it must also hold for $n + 1$, but with opposite sign. Setting $x_1 = y_1 = 1$, we see that $y_n^2 - 2x_n^2 = \pm 1$ holds when $n = 1$, and hence this equation is valid for every value of n thereafter. In consequence, (1) is a correct identity for all $n \geq 1$.

It is natural to raise the question whether the notion of side numbers and diagonal numbers can be used to obtain rational approximations to an arbitrary square root. Theon's original rule of formation was

$$x_n = x_{n-1} + y_{n-1}, \qquad y_n = 2x_{n-1} + y_{n-1}, \qquad n \geq 2.$$

For 2 in the second equation, let us substitute a positive integer a (which is not a perfect square) to develop the following scheme:

$$x_2 = x_1 + y_1, \qquad\qquad y_2 = ax_1 + y_1,$$

$$x_3 = x_2 + y_2, \qquad\qquad y_3 = ax_2 + y_2,$$

$$x_4 = x_3 + y_3, \qquad\qquad y_4 = ax_3 + y_3,$$

$$\vdots \qquad\qquad\qquad\qquad \vdots$$

$$x_n = x_{n-1} + y_{n-1}, \qquad y_n = ax_{n-1} + y_{n-1},$$
$$\vdots \qquad\qquad\qquad \vdots$$

Notice that

$$y_n^2 = (ax_{n-1} + y_{n-1})^2 = a^2x_{n-1}^2 + 2ax_{n-1}y_{n-1} + y_{n-1}^2,$$

$$ax_n^2 = a(x_{n-1} + y_{n-1})^2 = ax_{n-1}^2 + 2ax_{n-1}y_{n-1} + ay_{n-1}^2,$$

and so, on subtraction,

$$y_n^2 - ax_n^2 = (a^2 - a)x_{n-1}^2 + (1 - a)y_{n-1}^2$$
$$= (1 - a)(y_{n-1}^2 - ax_{n-1}^2).$$

The import of this relation is that we have represented $y_n^2 - ax_n^2$ by an expression of the same form, but with n replaced by $n - 1$. Repeating this transformation for the next expression, we evidently arrive at the chain of equalities

$$y_n^2 - ax_n^2 = (1 - a)(y_{n-1}^2 - ax_{n-1}^2)$$
$$= (1 - a)^2(y_{n-2}^2 - ax_{n-2}^2)$$
$$= (1 - a)^3(y_{n-3}^2 - ax_{n-3}^2)$$
$$\vdots$$
$$= (1 - a)^{n-1}(y_1^2 - ax_1^2)$$

and as a result,

$$\left(\frac{y_n}{x_n}\right)^2 = a + \frac{(1 - a)^{n-1}(y_1^2 - ax_1^2)}{x_n^2}, \qquad n \geq 2.$$

From this, it can be concluded that as n increases, the right-hand term tends to zero, whence the values y_n/x_n more and more closely approach the irrational number $\sqrt{a}$.

For an illustration, consider the case of $\sqrt{3}$; that is, $a = 3$. If we take $x_1 = 1$, $y_1 = 2$ as the initial side and diagonal numbers, then the foregoing formula reduces to

$$\left(\frac{y_n}{x_n}\right)^2 = 3 + \frac{(-2)^{n-1}}{x_n^2}, \, n \geq 2.$$

The successive rational approximations of $\sqrt{3}$ are

$$\frac{y_1}{x_1} = \frac{2}{1}, \frac{y_2}{x_2} = \frac{5}{3}, \frac{y_3}{x_3} = \frac{7}{4}, \frac{y_4}{x_4} = \frac{19}{11}, \frac{y_5}{x_5} = \frac{26}{15}, \ldots.$$

A variation of the above theme is afforded by starting with the algebraic identity

(3) $$(y^2 + 3x^2)^2 - 3(2xy)^2 = (y^2 - 3x^2)^2.$$

If one solution, say $x = x_0$, $y = y_0$, of the equation

$$y^2 - 3x^2 = 1$$

is known, then (3) indicates that a second solution can be found simply by letting $x = 2x_0y_0$, $y = y_0{}^2 + 3x_0{}^2$. Indeed, on substitution,

$$y^2 - 3x^2 = (y_0{}^2 + 3x_0{}^2)^2 - 3(2x_0y_0)^2$$

$$= (y_0{}^2 - 3x_0{}^2)^2 = 1^2 = 1.$$

Thus we have a machine for generating solutions of $y^2 - 3x^2 = 1$ from a single solution. By the rule of formation,

$$x_n = 2x_{n-1}y_{n-1}, \qquad y_n = y_{n-1}^2 + 3x_{n-1}^2,$$

a fresh solution x_n, y_n can be derived from a previous one x_{n-1}, y_{n-1}. Because x_n, y_n satisfy

$$y_n{}^2 - 3x_n{}^2 = 1,$$

or what amounts to the same thing,

$$\left(\frac{y_n}{x_n}\right)^2 = 3 + \frac{1}{x_n{}^2},$$

the successive values $(y_n/x_n)^2$ will approach 3 increasingly closely; that is, the sequence y_n/x_n provides a "very good" (in some sense) approximation of $\sqrt{3}$ by rational numbers.

It is clear that the equation $y^2 - 3x^2 = 1$ has at least one solution in the positive integers, namely $x_1 = 1$, $y_1 = 2$. We see then that

$$x_2 = 2x_1y_1 = 2 \cdot 1 \cdot 2 = 4,$$

$$y_2 = y_1{}^2 + 3x_1{}^2 = 2^2 + 3 \cdot 1^2 = 7$$

is also a solution. Thus new solutions are generated out of given ones. The next one is

$$x_3 = 2x_2y_2 = 2 \cdot 4 \cdot 7 = 56,$$

$$y_3 = y_2{}^2 + 3x_2{}^2 = 7^2 + 3 \cdot 4^2 = 97$$

and so on. We have almost finished; for the sequence of rational approximations of the irrational number $\sqrt{3}$ is just

$$\frac{y_1}{x_1} = \frac{2}{1}, \qquad \frac{y_2}{x_2} = \frac{7}{4}, \qquad \frac{y_3}{x_3} = \frac{97}{56}, \qquad \frac{y_4}{x_4} = \frac{18,817}{10,864}, \quad \cdots$$

Let us now view a strictly geometric proof of the incommensurability of the diagonal and side of a square. This argument, apparently older than the first, is in the spirit of the arguments found in Euclid's *Elements*. The basic idea is to show that we can build onto an arbitrary square a sequence of smaller and smaller squares.

In the square $ABCD$, draw the arc BE in order to lay off the side $AB = s_1$ on the diagonal $AC = d_1$. Now draw the line EF perpendicular to d_1, with F the point at which it intersects BC. By one of the congruence theorems, it is easy to prove that the triangles BAF and FAE are congruent; consequently, $FB = FE$, because they are congruent sides. Furthermore CEF is an isosceles right triangle, whence its legs CE and FE are equal.

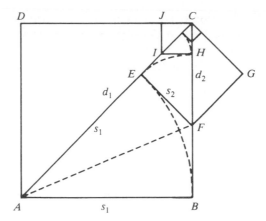

Next, construct a second square $CEFG$ having sides $s_2 = CE = d_1 - s_1$ and diagonal $d_2 = CB - FB = s_1 - s_2$. Laying off the sides $s_2 = FE$ on the diagonal $FC = d_2$, we determine CH, which is then used as s_3, the side of the third square. In this third square, it is seen that $s_3 = d_2 - s_2$ and this diagonal $d_3 = CE - EI = s_2 - s_3$. The process can be repeated over and over, obtaining successively smaller squares whose sides and diagonals satisfy the relations

$$s_n = d_{n-1} - s_{n-1}, \qquad d_n = s_{n-1} - s_n.$$

The geometric preliminaries completed, we assume that the diagonal and side of the original square are commensurable and show that this leads to an impossible situation. If these two lengths are commensurable, then they have a common measure δ, so that there exist integers M_1 and N_1 for which

$$s_1 = M_1\delta, \qquad d_1 = N_1\delta.$$

But then

$$s_2 = d_1 - s_1 = (N_1 - M_1)\delta = M_2\delta,$$

$$d_2 = s_1 - s_2 = (M_1 - M_2)\delta = N_2\delta,$$

where $M_2 < M_1$ and $N_2 < N_1$. Repetition of the argument yields

$$1 \leq \cdots < M_3 < M_2 < M_1, \qquad 1 \leq \cdots < N_3 < N_2 < N_1.$$

We now come to the contradiction. Because there are only finitely many positive integers less than M_1 and N_1, these two sequences must terminate after a finite number of steps. This contradicts the idea that our construction of squares can be carried out indefinitely.

The discovery of irrational numbers caused great consternation among the Pythagoreans, for it challenged the adequacy of their philosophy that number was the essence of all things. This logical scandal encouraged them to maintain the pledge of strict secrecy. Indeed, their resolve is testified to by the very name given to these new entities, "the unutterable." (The Greeks used the term *logos*, meaning "word" or "speech," for the ratio of two integers. Hence, when incommensurable lengths were described as *alogos*, the term carried a double meaning: "not a ratio" and "not to be spoken.")

The knowledge that irrationals existed was a dangerous secret to possess. Popular legend has it that the first Pythagorean to utter the unutterable to an outsider was murdered—thrown off a ship to drown.

It fell on Eudoxus of Cnidos (408–355 B.C.) to resolve the crisis in the foundations of mathematics. His great contribution was a revised theory of proportion applicable to incommensurable as well as commensurable quantities. Everything was based on an elaborate definition of the ratio of magnitudes, but magnitudes themselves were left undefined. Hence, the problem of defining irrational numbers as numbers was avoided entirely. The immediate effect of Eudoxus's approach was to drive mathematics into the hands of the geometers. In the absence of a purely arithmetic theory of irrationals, the primary of the number concept was renounced. Geometry was held to be a more general doctrine than the science of numbers, and for the next two thousand years, it served as the basis of almost all rigorous mathematical reasoning.

The existence of incommensurable geometric quantities necessitated a thorough recasting of the foundations of mathematics, with an increased attention to logical rigor. It was a formidable task and engaged the best efforts of the most notable mathematicians of the fourth century B.C.: Theodorus, Theaetetus, Archytas, and Eudoxus. Theodorus of Cyrene (b. 470 B.C.), the mathematics tutor of the great philosopher Plato, is said to have demonstrated geometrically that the sides of squares represented by $\sqrt{3}$, $\sqrt{5}$, $\sqrt{6}$, $\sqrt{7}$, $\sqrt{8}$, $\sqrt{10}$, $\sqrt{11}$, $\sqrt{12}$, $\sqrt{13}$, $\sqrt{14}$, $\sqrt{15}$, and $\sqrt{17}$ are incommensurable with a unit length. That is, he proved the irrationality of the square roots of nonsquare integers from 3 to 17, "at which point," Plato said, "for some reason he stopped." Theaetetus of Athens (415–369 B.C.), who was a pupil of Theodorus and a member of Plato's school in Athens, extended the result, demonstrating that the square root of any nonsquare integer is irrational. Plato himself added to the theory by showing that a rational number could be the sum of two irrationals. One of the few Pythagoreans to stay behind in southern Italy after the death of Pythagoras, Archytas of Tarentum (428–347 B.C.) is reputed to have been the first to study geometry on a circular cylinder, discovering in the process some of the properties of its oblique section, the ellipse. He also devised an ingenious solution of the problem "to double a cube" by means of cylindrical sections.

Perhaps the most brilliant Greek mathematician before Archimedes was Eudoxus. Born about 408 B.C. in Cnidos on the Black Sea, he set out at the age of 23 to learn geometry from Archytas in Tarentum and for several months, philosophy from Plato in Athens. Eudoxus, too poor to live in Athens, lodged cheaply at the harbor town of Piraeus, where he had first debarked; every day he walked the two miles to Plato's academy. Later he traveled to Egypt, where he remained for sixteen months. Thereafter he earned his living as a teacher, founding a school at Cyzicus in northwestern Asia Minor that attracted many pupils. When he was about 40 years old, Eudoxus made a second visit to Athens accompanied by a considerable following of his own students; there he opened another school, which for a time rivaled Plato's. The reputation of Eudoxus rests on three grounds: his general theory of proportion, the addition of numerous results on the study of the golden section (the division of a line segment in extreme and mean ratio), and the invention of a process known as the method of exhaustion. The procedure Eudoxus proposed was later refined by Archimedes into a powerful tool for determining curvilinear areas, surfaces, and volumes—an important precursor to the integral calculus.

During this period Greek mathematics began to be organized deductively on the basis of explicit axioms. Its final axiomatic form was set forth in the thirteen books of the *Elements* that Euclid wrote about 300 B.C. In compiling the *Elements*, Euclid built on the experience and achievements of his predecessors in the three centuries just past. Theaetetus's elaborate classification of higher types of irrationals is the subject matter of Book X of the *Elements*, although Euclid must be credited with having arranged it into a logical whole. The Eudoxian theory of proportion—which is really a theory of real numbers—is incorporated into Book V; and Book II is mostly a geometric rendition of Pythagorean arithmetic, wherein Euclid represented numbers by line segments instead of the pictorial dot method the early Pythagoreans favored.

3.3 Problems

1. (a) Establish the formula

$$ab + \left(\frac{a - b}{2}\right)^2 = \left(\frac{a + b}{2}\right)^2.$$

   (b) Show that $a = 2n^2$, $b = 2$, gives rise to Plato's formula for Pythagorean triples, whereas $a = (2n + 1)^2$, $b = 1$, yields Pythagoras's own formula.

2. Find all right triangles with sides of integral length whose areas are equal to their perimeters. [*Hint:* The equations $x^2 + y^2 = z^2$ and $x + y + z = \frac{1}{2}xy$ imply that $(x - 4)(y - 4) = 8$.]

3. For $n \geq 3$ a given integer, find a Pythagorean triple having n as one of its members. [*Hint:* For n an odd integer, consider the triple $(n, \frac{1}{2}(n^2 - 1))$; for n even, consider the triple $(n, n^2/4 - 1, n^2/4 + 1)$.]

4. Verify that $(3, 4, 5)$ is the only Pythagorean triple involving consecutive positive integers. [*Hint:* Consider the Pythagorean triple $(x, x + 1, x + 2)$ and show that $x = 3$.]

5. (a) Establish that there are infinitely many Pythagorean triples (x, y, z) in which x and y are consecutive integers. [*Hint:* If $(x, x + 1, z)$ happens to be a Pythagorean triple, so is $(3x + 2z + 1, 3x + 2z + 2, 4x + 3z + 2)$.]

   (b) Find five Pythagorean triples of the form $(x, x + 1, z)$.

6. Consider the sequence of quotients y_n/x_n of Theon's diagonal numbers to side numbers.

   (a) Verify that the first, third, and fifth terms in this sequence are getting successively larger, whereas the second, fourth, and sixth terms are decreasing.

   (b) Compute the difference between 2 and the square of each term, through the first six terms; $(y_n/x_n)^2$ should be getting nearer 2 at each stage, alternating above and below, hence y_n/x_n approximates $\sqrt{2}$.

7. Let two sequences of numbers be formed in accordance with the following rule:

$$x_1 = 2, \qquad y_1 = 3.$$

$$x_n = 3x_{n-1} + 2y_{n-1},$$

$$y_n = 4x_{n-1} + 3y_{n-1} \qquad \text{for } n \geq 2.$$

   (a) Write out the first five numbers in each of the above sequences.

   (b) Show that

$$y_n^2 - 2x_n^2 = y_{n-1}^2 - 2x_{n-1}^2,$$

   whence

$$y_n^2 - 2x_n^2 = y_1^2 - 2x_1^2 = 1.$$

   (c) From part (b), conclude that successive values of y_n/x_n are nearer and nearer approximations of $\sqrt{2}$.

8. Consider the sequence of numbers defined by the following rule:

$$x_1 = 2,$$

$$x_n = \frac{1}{2}\left(x_{n-1} + \frac{2}{x_{n-1}}\right) \text{ for } n > 1.$$

(a) Write out the first four terms of this sequence in decimal form.

(b) Assuming that the terms x_n approach a number L as n increases, show that $L = \sqrt{2}$. [*Hint:* The number L satisfies $L = \frac{1}{2}(L + 2/L)$.]

9. Prove that $\sqrt{3}$ and $\sqrt{2}$ are irrational by assuming that each is rational and arguing until a contradiction is reached.

10. Replace 2 by 3 in Theon's definition of side numbers and diagonal numbers, so that the rule of formation becomes

$$x_n = x_{n-1} + y_{n-1},$$

$$y_n = 3x_{n-1} + y_{n-1} \qquad n \geq 2.$$

(a) Starting with $x_1 = 1$, $y_1 = 2$, write out the first six numbers in each of the resulting sequences.

(b) Confirm for several values of n that when y_n/x_n is in lowest terms,

$$y_n^2 - 3x_n^2 = 1 \text{ or } -2.$$

(c) Assuming that the relation in part (b) holds for all n, show that this implies that the successive ratios y_n/x_n are approaching $\sqrt{3}$.

(d) Write out the first six values of y_n/x_n to get an approximation $\sqrt{3}$.

11. Archimedes (287–212 B.C.) in his book *Measurement of a Circle* presented, without a word of justification, the inequality

$$\frac{1351}{780} > \sqrt{3} > \frac{265}{153}.$$

(a) As an explanation of the probable steps leading to the left-hand bound, show first that

i) $26 - \frac{1}{52} = \sqrt{26^2 - 1 + (\frac{1}{52})^2}$
$$> \sqrt{26^2 - 1},$$

and then

ii) $\frac{1351}{780} = \frac{1}{15}\left(26 - \frac{1}{52}\right)$
$$> \frac{1}{15}\sqrt{26^2 - 1} = \sqrt{3}.$$

(b) Obtain the right-hand bound in a similar manner by replacing $\frac{1}{52}$ with $\frac{1}{51}$.

12. Because $\sqrt{3}$ is approximately $\frac{5}{3}$, one can put $\sqrt{3} = (\frac{5}{3} + 1/x)$, where x is unknown.

(a) Square both sides of this expression, neglect $1/x^2$, and solve the resulting linear equation for x to get a second approximation of $\sqrt{3}$.

(b) Repeat this procedure once more to find a third approximation.

13. (a) Given a positive integer n that is not a perfect square, let a^2 be the nearest square to n (above or below n, as the case may be), so that $n = a^2 \pm b$. Prove that

$$a \pm \frac{b}{2a \pm 1} < \sqrt{n} < a \pm \frac{b}{2a}.$$

(b) Use part (a) to approximate $\sqrt{50}$, $\sqrt{63}$, and $\sqrt{75}$ by rational numbers.

14. Use the inequality of Problem 13 to get Archimedes' bounds on $\sqrt{3}$. [*Hint:* Take $a = 26$ and $b = 1$.]

15. A standard proof of the Pythagorean theorem starts with a right triangle ABC, with its right angle at C, and then draws a perpendicular CD from C to the hypotenuse AB.

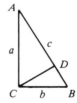

(a) Prove that triangles ACD and CBD are both similar to triangle ABC.

(b) For a triangle ABC with legs of lengths a and b and with hypotenuse of length c, use the proportionality of corresponding sides of similar triangles to establish that $a^2 + b^2 = c^2$.

16. For another proof of the Pythagorean theorem, consider a right triangle ABC (with right angle at C) whose legs have length a and b and whose hypotenuse has length c. On the extension of side BC pick a point D such that BAD is a right angle.

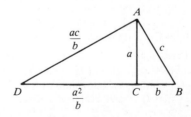

(a) From the similarity of triangles ABC and DBA, show that $AD = ac/b$ and $DC = a^2/b$.

(b) Prove that $a^2 + b^2 = c^2$ by relating the area of triangle ABD to the areas of triangles ABC and ACD.

17. Several years before James Garfield became president of the United States, he devised an original proof of the Pythagorean theorem. It appeared in 1876 in the *New England Journal of Education*. Starting with a right triangle ABC, Garfield placed a congruent triangle EAD as indicated in the figure. He then drew EB so as to form a quadrilateral $EBCD$. Prove that $a^2 + b^2 = c^2$ by relating the area of the quadrilateral to the area of the three triangles ABC, EAD, and EBA.

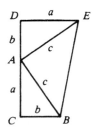

18. The Pythagoreans defined the harmonic mean of a and b, where $a < b$, to be the number h such that

$$\frac{h - a}{b - h} = \frac{a}{b}.$$

For instance, the harmonic mean of 6 and 12 is 8, because $(8 - 6)/(12 - 8) = 6/12$. Prove that h is the harmonic mean of a and b if and only if h satisfies either of the relations:

(a) $\dfrac{1}{a} - \dfrac{1}{h} = \dfrac{1}{h} - \dfrac{1}{b}$.

(b) $h = \dfrac{2ab}{a + b}$.

19. Pappus (circa 320) in his *Mathematical Collection* provided a construction for the harmonic mean of the segments OA and OB as follows. On the perpendicular to OB at B lay off $BC = BD$, and let the perpendicular to OB at A meet OC at the point E. Join ED, and let H be the point at which ED cuts OB. Prove that $h = OH$ is the desired harmonic mean between $a = OA$ and $B = OB$. [*Hint:* From the similarity of triangles OAE and OBC, as well as of triangles HAE and HBD, infer that $a/b = AH/HB$.]

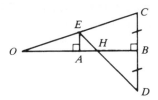

20. Establish the "perfect proportion"

$$\frac{a}{(a + b)/2} = \frac{2ab/(a + b)}{b},$$

between the arithmetic and harmonic means of two numbers a and b.

21. The division of a line segment into two unequal parts so that the whole segment will have the same ratio to its larger part that its larger part has to its smaller part is called the golden section. A classical ruler-and-compass construction for the golden section of a segment AB is as follows. At B erect BC equal and perpendicular to AB. Let M be the midpoint of AB, and with MC as a radius, draw a semicircle cutting AB extended in D and E. Then the segment BE laid off on AB gives P, the golden section.

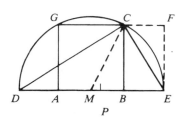

(a) Show that $\triangle DBC$ is similar to $\triangle CBE$, whence $DB/BC = BC/BE$.

(b) Subtract 1 from both sides of the equality in part (a) and substitute equals to conclude that $AB/AP = AP/PB$.

(c) Prove that the value of the common ratio in part (b) is $(\sqrt{5} + 1)/2$, which is the "golden ratio." [*Hint:* Replace PB by $AB - AP$ to see that $AB^2 - AB \cdot AP - AP^2 = 0$. Divide this equation by AP^2 to get a quadratic equation in the ratio AB/AP.]

(d) A golden rectangle is a rectangle whose sides are in the ratio $(\sqrt{5} + 1)/2$. (The golden rectangle has dimensions pleasing to the eye, and was used for the measurements of the facade of the Parthenon and other Greek temples.) Verify that both the rectangles *AEFG* and *BEFC* are golden rectangles.

22. Theodorus of Cyrene (circa 400 B.C.) who was the mathematical teacher of Plato, showed how to construct a line segment of length $\sqrt{n}$ for any positive integer n. Prove the following.

    (a) Given an odd integer n, then $\sqrt{n}$ is represented by the leg of a right triangle whose hypotenuse is $(n + 1)/2$ and whose other leg is $(n - 1)/2$.
    (b) Given an even integer n, then $\sqrt{n}$ is represented by half of the leg of a right triangle whose hypotenuse is $n + 1$ and whose other leg is $n - 1$.

23. It has been suggested that Theodorus also obtained $\sqrt{n}$ ($2 \leq n \leq 17$) by constructing a spiral-like

figure consisting of a sequence of right triangles having a common vertex, so that in each triangle the leg opposite the common vertex had length 1. Show that the hypotenuse of the nth triangle in this sequence has length $\sqrt{n + 1}$. (The reason Theodorus stopped at $\sqrt{17}$ is that at the next step, wherein $\sqrt{18}$ would be constructed, the figure cuts across the initial axis for the first time.)

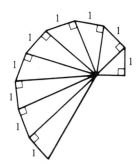

3.4   Three Construction Problems of Antiquity

The mathematician who dominated the second half of the fifth century B.C. was Hippocrates of Chios (460–380 B.C.), who is to be distinguished from his more celebrated contemporary Hippocrates of Cos, the father of Greek medicine. Like Thales, Hippocrates began his life as a merchant and ended as a teacher; but being less shrewd than Thales, Hippocrates was robbed of his money. Accounts differ on whether he was swindled by customhouse collectors at Byzantium or whether his ships were plundered on the high seas by Athenian pirates. At any rate, with his property lost, Hippocrates went to Athens to prosecute the offenders in the law courts. Obliged to stay for many years (perhaps from 450 to 430 B.C.), he attended the lectures of several philosophers. There is good reason to believe that the Pythagoreans were settled in Athens at that time, so he may have come under their influence even though he had no Pythagorean teacher in the formal sense. Ultimately, Hippocrates attained such a proficiency in geometry that he became one of the first to support himself openly by accepting fees for teaching mathematics. If as some say, the Pythagoreans taught him what he knew of arithmetic and geometry, then by the standards of the time he betrayed their trust by selling the secrets of mathematics to anyone who would pay the price. (A more charitable interpretation is that the Pythagoreans, moved by Hippocrates' misfortune, allowed him to earn money by teaching their geometry.) Aristotle spoke unflatteringly of Hippocrates: "It is well known that persons stupid in one respect are by no means so in all others; thus Hippocrates, though a competent geometer, seems in other regards to be stupid and lacking in sense." The Greeks, indeed, were likely to view any man a fool who through his own simplicity was cheated out of his possessions.

By the middle of the fifth century, so many geometric theorems had been established that it became increasingly necessary to tighten the proofs and put all this material in good logical order. Proclus told how Hippocrates composed a work on the elements of geometry, anticipating the better-known *Elements* of Euclid by more than a century. No trace of this first textbook on geometry remains, however (in fact, no mathematical treatise of the fifth century has survived). Although Hippocrates' book may have started a significant tradition, it would have had the shortcomings of a pioneering work, and been rendered obsolete by Euclid's *Elements*. Hippocrates did originate the now familiar pattern of presenting geometry as a chain of propositions, a form in which other propositions can be derived on the basis of earlier ones. Among other innovations, he introduced the use of letters of the alphabet to designate points and lines in geometric figures.

When Hippocrates arrived in Athens, three special problems—the quadrature of the circle, the duplication of the cube, and the trisection of a general angle—were already engaging the attention of geometers. These problems have remained landmarks in the history of mathematics, a source of stimulation and fascination for amateurs and scholars alike through the ages.

The achievement on which Hippocrates' fame chiefly rests has to do with the first of these problems, the quadrature of the circle. This problem, sometimes called the "squaring of the circle," can be stated simply: Is it possible to construct a square whose area shall be equal to the area of a given circle? The problem is much deeper than it first appears, because the important factor is how the square is to be constructed. Tradition has it that Plato (429–348 B.C.) insisted that the task be performed with straightedge and compass only. In this method the assumption is that each instrument will be used for a single, specific operation:

1. With the straightedge, a line can be drawn through two given points.

2. With the compass, a circle with a given center and radius can be drawn.

It is not permissible to use these two instruments in any other way; in particular, neither device is to be used for transferring distances, so that the straightedge cannot be graduated or marked in any way, and the compass must be regarded as collapsing as soon as either point is lifted off the paper. A point or a line is said to be constructible by straightedge and compass if it can be produced from given geometric quantities with these two tools, using them in the prescribed way only a finite number of times.

In the strict Greek sense of construction, the quadrature problem remained unsolved in spite of vigorous efforts by the Greek and other, later geometers. The futility of their attempts was demonstrated in the nineteenth century, when mathematicians were at last able to prove that it is impossible to square the circle by straightedge and compass alone. As it turns out, the test of constructibility under these instrumental limitations uses the ideas of algebra, not geometry, and involves concepts unknown in antiquity or the Middle Ages. Squaring the circle is equivalent to constructing a line segment whose length is $\sqrt{\pi}$ times the radius of the circle. Thus, the impossibility of constructing such a line segment by means laid down by the Greeks would be proved if it could be shown that $\sqrt{\pi}$ is not a constructible length. The argument hinges on the transcendental nature of the number π; that is, π is not the root of any polynomial equation with rational coefficients. (The transcendence of π was established by Lindemann in 1882 in a long and intricate proof.)

Plato
(429–348 B.C.)

(The Bettmann Archive.)

Even early investigators must have suspected that the allowable means were inadequate for solving this quadrature problem; for when they failed to find a construction involving merely circles and straight lines, they introduced special higher curves assumed to be already drawn. Here they were successful. Hippias of Elis (circa 425 B.C.), a near contemporary of Hippocrates, invented a new curve called the quadratrix, for the express purpose of squaring the circle. His solution was perfectly legitimate, but did not satisfy the restriction Plato had laid down. Hearing that Hippias had devised a sliding apparatus by which his curve could be drawn, Plato rejected the solution on the grounds that it was mechanical and not geometrical. Plutarch (*Convivial Questions*) describes Plato as saying: "For in this way the whole good of geometry is set aside and destroyed, since it is reduced to things of the sense and prevented from soaring among eternal images of thought."

Hippocrates' attempts at squaring the circle led him to discover that there are certain plane regions with curved boundaries that are squarable. More specifically, he showed that two lunes (a lune is the moon-shaped figure bounded by two circular arcs of unequal radii) could be drawn, whose areas were together equal to the area of a right triangle. This was accomplished as follows. Starting with an isosceles right triangle *ABC*, he constructed semicircles on the three sides as in the diagram.

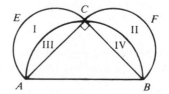

Hippocrates apparently knew that the areas of two circles were proportional to the squares of the lengths of their diameters. Thus,

$$\frac{\text{Area semicircle on } AB}{\text{Area semicircle on } AC} = \frac{AB^2}{AC^2}.$$

This ratio must equal 2; for the Pythagorean theorem, as applied to triangle ABC, allows $AB^2 = AC^2 + CB^2 = 2AC^2$. Hence, the semicircle on AB has twice the area of the semicircle on AC. From this, Hippocrates was led to conclude that the sum of the areas of the two small semicircles equaled the area of the larger one. The next step was to subtract the areas III and IV common to both. The figure shows that the areas remaining—namely, the sum of the areas I and II of the two lunes and the area of triangle ABC—are equal. But triangle ABC has area $\frac{1}{2}(AC \cdot BC) = \frac{1}{2}AC^2$, so that

$$\text{Area lune I} + \text{area lune II} = \frac{1}{2}AC^2.$$

To put it another way, lune I has an area equivalent to half that of triangle ABC,

$$\text{Area lune I} = \frac{1}{2}\left(\frac{1}{2}AC^2\right) = \left(\frac{AC}{2}\right)^2,$$

and the "square of the lune" has been found. Hippocrates thus provided the first example in mathematics of a curvilinear area that admits exact quadrature.

Having shown that the lune could be squared, Hippocrates next tried to square the circle by a similar argument. To this end, he took an isosceles trapezoid $ABCD$ formed by the diameter of a circle and three consecutive sides of half of a regular hexagon inscribed in the circle. Further semicircles were then described, having as diameters the sides AB, BC, and CD of the hexagon, as well as the radius OD of the original circle. Hippocrates proved that the area of the trapezoid $ABCD$ equaled the sum of areas of the three lunes I, II, and III plus the area of the semicircle on OD.

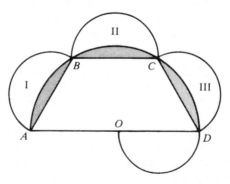

Because the squares of the diameters are to each other as the areas of the respective semicircles,

$$\frac{\text{Area semicircle on } OD}{\text{Area semicircle on } AD} = \frac{OD^2}{AD^2} = \frac{OD^2}{(2OD)^2} = \frac{1}{4}.$$

But each of the sides *AB*, *BC*, and *CD* is equal to the radius *OD*, from which it follows that each of the small semicircles has area a quarter that of the large semicircle. Knowing this, one concludes that the area of the semicircle on *AD* is the same as the total area of four semicircles—the three semicircles on the equal sides of the half-hexagon and the semicircle on the radius *OD*. If the parts common to both of these areas (to wit, the shaded segments lying between the hexagon and the circumference of the semicircle) are removed, the remaining areas will be equal. In other words, the lunes I, II, and III together with the semicircle on *OD* will have an area equivalent to that of the trapezoid *ABCD*:

$$\text{Area trapezoid } ABCD = \text{area lune I} + \text{area lune II} + \text{area lune III}$$
$$+ \text{ area semicircle on } OD.$$

If it were possible to subtract from this sum three squares with areas equal to the areas of the three lunes, then we could construct a rectangle equal in area to the semicircle on *OD*; twice that rectangle would then be equivalent to the circle on *OD*. As any rectangle can be converted to a square having the same area, the circle would have been squared.

Hippocrates' work on lunes has been preserved through the writings of the sixth-century commentator Simplicius and is indeed the only sizable fragment of classical Greek (pre-Alexandrian) mathematics that has been transmitted to us as originally composed. According to Simplicius, Hippocrates believed that he had actually succeeded in obtaining the quadrature of the circle by the argument as we have described it. He did not, needless to say, solve the squaring of the circle.

The mistake lay in assuming that every lune can be squared; whereas this was shown possible only in the special case with which Hippocrates had concerned himself. What he proved for the lune on the side of an inscribed isosceles triangle need not be true for the lune on the side of an inscribed half-hexagon. Actually it is unlikely that Hippocrates, one of the most competent of geometers, would have made such a blunder. He may have hoped that in due course these lune quadratures would lead to the squaring of the circle. But it must have been a mistake on the part of the commentator to think that Hippocrates had claimed to have squared the circle when he had not done so.

Another famous construction problem that concerned geometers of the time was the duplication of the cube; in other words, finding the edge of a cube having a volume twice that of a given cube. Just how the duplication problem originated is a matter of conjecture. Perhaps it dates back to the early Pythagoreans who had succeeded in doubling the square—if upon the diagonal of a given square a new square is constructed, then the new square has exactly twice the area of the original square. After this accomplishment, it would be only natural to extend the problem to three dimensions.

Tradition, however, provides us with a more romantic tale. According to the account that has prevailed most widely, the Athenians appealed to the oracle at Delos in 430 B.C. to learn what they should do to alleviate a devastating plague that had inflicted great suffering on their city and caused the death of their leader, Pericles. The oracle replied that the existing altar of Apollo should be doubled in size. Because the altar was in the form of a cube, the problem was to duplicate the cube. Thoughtless builders merely constructed a cube whose edge was twice as long as the edge of the altar. At this, legend has it, the indignant god made the pestilence even worse than before. When the error was discovered, a deputation of citizens was sent to consult Plato on the matter. Plato told them that "the god has given this oracle, not because he wanted an altar of double the size, but because he wished in setting this task before them to reproach the Greeks for their neglect of mathematics and their contempt of geometry." Whether the plague was actually abated or whether it simply ran its course is not known, but because of the oracle's response, the problem of duplicating the cube is often referred to as the "Delian problem."

History is confused, and there are at least two legends on the subject. We are also told that the poet Euripedes (485–406 B.C.) mentioned the Delian problem in one of his tragedies, now lost. In this version, the origin of the problem is traced to King Minos, who is represented as wishing to erect a tomb to his son Glaucus. Feeling that the dimensions proposed were too undignified for a royal monument, the king exclaimed, "You have enclosed too small a space; quickly double it, without spoiling the beautiful (cubical) form." In each of these accounts, the problem seems to have had its genesis in an architectural difficulty.

Here too, the first real progress in solving the duplication problem was made by Hippocrates. He showed that it can be reduced to finding, between a given line and another line twice as long, two mean proportionals. (That is, two lines are inserted between the given lines so that the four are in geometric proportion.) In our present notation, if a and $2a$ are the two given lines, and x and y are the mean proportionals that could be inserted between them, then the lengths $a, x, y, 2a$ are in geometric progression, which is to say

$$\frac{a}{x} = \frac{x}{y} = \frac{y}{2a}.$$

The first two ratios imply that $x^2 = ay$. From the second pair of ratios, we see that $y^2 = 2ax$. These equations are combined into

$$x^4 = a^2y^2 = 2a^3x,$$

whence it appears that

$$x^3 = 2a^3.$$

In other words, the cube that has edge x will have double the volume of a given cube of edge a.

Hippocrates did not succeed in finding the mean proportionals by constructions using only straightedge and compass, those instruments to which Plato had limited geometry. Nevertheless the reduction of a problem in solid geometry to one in plane geometry was in itself a significant achievement. From this time on, the duplication of the cube was always attacked in the form in which Hippocrates stated it: How may two mean proportionals be found between two given straight lines?

Although Hippocrates advanced two of the three famous construction problems, he made no progress with trisecting an angle. The bisection of an angle with only straightedge and compass is one of the easiest of geometrical constructions, and early investigators had no reason to suspect that dividing an angle into three equal parts under similar restrictions might prove impossible. Some angles can obviously be trisected. In the special case of the right angle POQ, the construction is found as follows.

With O as a center, draw a circle of any radius intersecting the sides of the angle in points A and B. Now draw a circle with center at B and passing through O. The two circles will intersect in two points, one of which will be a point C in the interior of angle POQ.

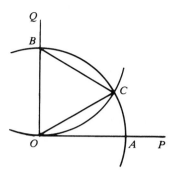

Triangle BOC is equilateral, hence equiangular; therefore $\angle COB = 60°$. But then

$$\angle COA = 90° - 60° = 30° = \tfrac{1}{3}(90°)$$

and line OC is a trisector of the right angle.

For 2000 years mathematicians sought in vain to trisect an arbitrary angle. In 1837, Pierre Wantzel (1814–1848) of the Ecole Polytechnique in Paris supplied the first rigorous proof of the impossibility of trisecting any given angle by straightedge and compass alone. In the same paper, published in Liouville's *Journal de Mathématiques,* Wantzel also demonstrated the futility of duplicating the cube in the manner specified. The key to this conclusion was the conversion of the two geometric problems to questions in the theory of equations. Wantzel obtained simple algebraic criteria that would permit the solution of a polynomial equation with rational coefficients to be geometrically constructed by means of a straightedge and compass. The classical geometric problems of trisection and duplication lead to cubic equations which do not satisfy Wantzel's conditions, and thus the corresponding constructions cannot be carried out.

If the restrictions imposed by the Greeks are relaxed, there are a variety of ways of dividing an angle into three equal parts. The simplest solution of the problem is to allow oneself the liberty of marking the straightedge. The following technique of rotating a marked straightedge until certain conditions are satisfied was devised by Archimedes. Let POQ be the angle to be trisected. With the vertex O as center, draw a circle of any radius r intersecting PO in A and QO in B. Now lay off the distance r on a straightedge. By shifting the straightedge around, you can get a certain position in which it passes through the point B, while the endpoints of the r segment lie on the circle (at C) and the diameter AOA' extended (at D). The line through the points B, C, and D is now drawn with the aid of the straightedge.

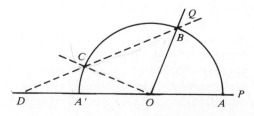

With these preliminaries accomplished, we undertake to show that angle ODC is one-third of angle AOB. First observe that by its construction, $CD = OC = r$, so that triangle ODC is isosceles; hence $\angle COD = \angle ODC$. Because an exterior angle in a triangle is equal to the sum of the nonadjacent interior angles, it follows that in triangle COD,

$$\angle OCB = \angle COD + \angle ODC = 2\angle ODC.$$

Also, in the isosceles triangle OCB, we have $\angle OCB = \angle OBC$. Another appeal to the exterior angle theorem (this time applied to triangle ODB) leads to the equality

$$\angle AOB = \angle ODB + \angle OBC = \angle ODC + \angle OBC.$$

These various observations can be brought together to give

$$\angle AOB = \angle ODC + \angle OBC$$

$$= \angle ODC + \angle OCB$$

$$= \angle ODC + 2\angle ODC = 3\angle ODC,$$

which accomplishes our aim. It is worth emphasizing that the usual rules for straightedge and compass constructions have been violated, because the straightedge was marked. That is, the points C and D were determined by sliding the straightedge to the proper position to make CD equal to r.

3.4 Problems

1. For a variation of Hippocrates' argument that the area of a lune could be reduced to the area of a circle, begin with a square $ABCD$ and construct a semicircle on its diagonal. With the point D as a center and AD as radius, draw a circular arc from A to C, as in the figure. Prove that the area of the lune, shaded in the figure, is equal to the area of triangle ABC. [*Hint:* Similar circular segments (the region between a chord and the arc subtended by the chord) have areas proportional to the squares of the lengths of their chords. Apply this fact to the similar segments I and II.]

2. The following solution to the continued mean proportionals problem is often attributed to Plato, although it could hardly be his in view of his objection to mechanical constructions. Consider two right triangles ABC and BCD, lying on the same side of the common leg BC (see the figure). Suppose that the hypotenuses AC and BD intersect perpendicularly at the point P, and are constructed in such a way that $AP = a$ and $DP = 2a$. Prove that $x = BP$ and $y = CP$ are the required mean proportionals between a and $2a$, that is, that

$$\frac{a}{x} = \frac{x}{y} = \frac{y}{2a}.$$

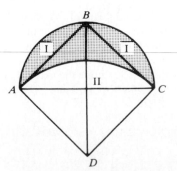

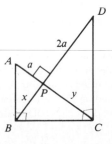

[*Hint:* When parallel lines are cut by a third line, alternate interior angles are equal. Conclude therefore that the triangles *APB*, *CPB*, and *DPC* are similar.]

3. Apollonius (circa 225 B.C.) solved the problem of inserting two mean proportionals between segments of lengths *a* and 2*a*. He first constructed a rectangle *ABCD*, with *AB* = *a* and *AD* = 2*a*, letting *E* be the point at which the diagonals bisected one another. With *E* as a center, he then drew a circle cutting the extensions of *AB* and *AD* at points *P* and *Q*, respectively, so that *P*, *C*, and *Q* all lay on a straight line. (Apollonius is said to have invented a mechanical device by which this last step could be made.) For such a figure, establish that:

  (a) The triangles *PAQ*, *PBC*, and *CDQ* are similar, whence

$$\frac{a}{x} = \frac{y}{2a} = \frac{a+y}{2a+x}.$$

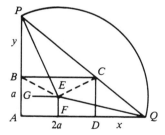

  (b) Triangles *EFQ* and *EGP* are right triangles with equal hypotenuses, whence

$$(a+x)^2 + \left(\frac{a}{2}\right)^2 = a^2 + \left(\frac{a}{2}+y\right)^2$$

or

$$(2a+x)x = (a+y)y.$$

  (c) Segments *DQ* = *x* and *BP* = *y* are the two mean proportionals between *a* and 2*a*;

$$\frac{a}{x} = \frac{x}{y} = \frac{y}{2a}.$$

4. The Greek mathematician Menaechmus (circa 350 B.C.), the tutor of Alexander the Great, obtained a purely theoretical solution to the duplication problem based on finding the point of intersection of certain "conic sections." To duplicate a cube of edge *a*, he constructed two parabolas having a common vertex and perpendicular axes, so that one parabola had a focal chord (latus rectum) of length *a* and the other a chord of length 2*a*. Prove that the abscissa *x* of the point of intersection of the two parabolas satisfies the condition $x^3 = 2a^3$; the sought-for *x*, the cube's edge, is thereby obtained.

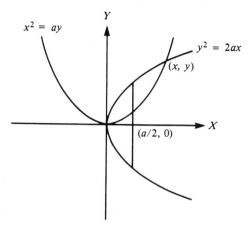

5. The trisection of a given angle can also be accomplished by a construction due to Nicomedes (circa 240 B.C.). Let ∠*AOB* be a given angle. Through the point *B*, draw two lines, one perpendicular to the other side of ∠*AOB* at *C* and one parallel to it. Now mark the length *a* = 2*OB* on a straightedge and slide the straightedge so that it passes through the point *O*, while the endpoints of the *a* segment lie on *BC* and *BD* (at *P* and *Q*, respectively, so that *PQ* = *a*).

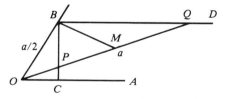

Verify each of the following assertions:

  (a) If *M* is the midpoint of *PQ*, then ∠*MOB* = ∠*BMO*. [*Hint:* The midpoint of the hypotenuse of a right triangle is equidistant from the endpoints of its sides.]

  (b) By the exterior angle theorem, as applied to triangle *BMQ*, we find that ∠*BMP* = ∠*MBQ* + ∠*MQB*.

  (c) ∠*AOQ* = ∠*BQO*.

  (d) ∠*AOB* = ∠*AOQ* + ∠*QOB* = 3∠*BQO*.

6. Nicomedes solved the problem of duplicating the cube by an argument like that of Apollonius. First, construct a rectangle $ABCD$ with $AB = a$ and $AD = 2a$. Let M be the midpoint of AD and N the midpoint of AB, and let the segments CM and BA be extended to meet in G. Take the point F on the perpendicular FN to be such that $FB = a$. Now draw BH parallel to GF and draw FP to cut segment AB produced in P, with P so chosen that $HP = a$. (To accomplish this last step, Nicomedes invented a special plane curve, and even an apparatus that would draw it, called the conchoid.) Prolong the line PC until it meets AD extended in Q.

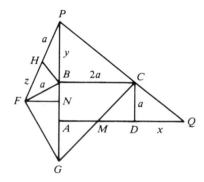

Establish that:

(a) The triangles PAQ, PBC, and CDQ are similar, whence

$$\frac{a}{x} = \frac{y}{2a} = \frac{a + y}{2a + x}.$$

(b) The triangles PBH and PGF are similar, whence

$$\frac{a}{y} = \frac{z + a}{y + 2a},$$

or $a/z = y/2a$, so that $z = x$.

(c) The triangles FNB and FNP have FN as a common side, and so

$$a^2 - \left(\frac{a}{2}\right)^2 = (x + a)^2 - \left(y + \frac{a}{2}\right)^2$$

or

$$\frac{x}{y} = \frac{y + a}{x + 2a}.$$

(d) Segments $DQ = x$ and $BP = y$ are the two mean proportionals between a and $2a$:

$$\frac{a}{x} = \frac{x}{y} = \frac{y}{2a}.$$

7. Isaac Newton (1642–1727) suggested the following construction for duplicating the cube. Given a segment AB, erect a perpendicular BR to AB and draw BT so that angle ABT equals 120°. Let D be the point on BT such that if AD is drawn meeting BR at C, then $CD = AB$.

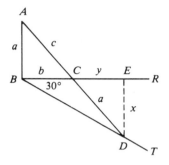

Establish that if DE is drawn perpendicular to BR, each of the following will be true:

(a) Triangles ABC and DEC are similar, whence

$$\frac{a}{x} = \frac{b}{y} = \frac{c}{a}.$$

(b)
$$\frac{1}{\sqrt{3}} = \tan 30° = \frac{DE}{BE}$$
$$= \frac{x}{b + y} = \frac{a^2}{ab + bc}.$$

(c) The result of squaring the last equation and substituting $b^2 = c^2 - a^2$ is

$$c^3(2a + c) = 2a^3(2a + c).$$

(d) Since $c^3 = 2a^3$, the cube of side AC is double the cube of side AB.

8. To find a fourth proportional to given line segments with lengths a, b, and c, first construct two noncollinear rays emanating from a point O. On these rays mark off segments OA and OC of lengths a and c, respectively, and connect the points A and C so as to form a triangle. On the ray on which the length a has been marked, now lay off a segment AB of length b. Finally, construct a line through the point B parallel to the side AC of the triangle constructed earlier, and intersecting the other ray in a point D. If the segment CD has length x, show that x satisfies the proportion

$$\frac{a}{b} = \frac{c}{x}.$$

3.5   The Quadratrix of Hippias

The curve usually called the quadratrix was invented by Hippias of Elis (born about 460 B.C.) in order to trisect an angle. The curve acquired its name from its later use in the quadrature of the circle. Like his contemporary Hippocrates, Hippias was one of the first to teach for money, one of the so-called sophists. The word ''sophist,'' much like the word ''tyrant,'' did not originally have a derogatory meaning although it soon came to receive one. The term first meant ''wise man'' and only later did it take on the connotation of one who reasons adroitly and speciously, rather than soundly.

The sophists were itinerant teachers, usually from Asia Minor or the Aegean Islands, who had acquired learning and experience through wide travel. Whereas the disciples of Pythagoras were forbidden to accept fees for sharing their knowledge, the sophists, less hampered by tradition, had no such qualms. Shortly after the middle of the fifth century B.C. several of these wandering lecturers—some of them reputable scholars, some outright imposters—arrived in Athens to vend their wares. There was a ready market for their talents among the prosperous Athenians, and success there ensured one's reputation throughout Greece, Sicily, and Italy.

The sophists took all knowledge as their province, but their central subject was the art of disputation. They professed to be able to teach their students to speak with clarity and persuasion, with the appearance of logic, on any topic whatever, and to defend either side of a question successfully. This laid them open to the charge of training in cleverness rather than virtue. Their opponents claimed that the sophists taught youth ''to prove that black is white and to make the worse appear the better.'' In spite of the criticisms against them, they were very much in demand. Wealthy people took pride in entrusting the education of their sons to the best and most famous sophists. In the end, their commercialism and the extravagant claims made for their instruction turned Plato and others against them, and gave the term ''sophist'' its present meaning.

Because most of what we know about Hippias's life and character comes from two dialogues of Plato in which sophists are castigated, it is hard to judge him fairly. In the Platonic dialogues named after him, Hippias was pictured as an arrogant, boastful buffoon. He was made to say that he had earned more money than any other two contemporary sophists and had gained, in spite of the competition from the illustrious Protagoras of Abdera (in Thrace), huge sums on his Sicilian lecture tour. His claims were further recounted—that if he had received no lecture fees in Sparta and had not been invited to teach its youth, it was only because Spartan law prohibited foreign teaching. Hippias came from Elis, a small state in the northwest corner of the Peloponnesus, whose inhabitants had charge of the games that took place every fourth year on the plains of Olympia.

In Plato's writing, Hippias boasted that on his previous visit to the Olympic festival everything that he wore was of his own making, not merely his garments, but also his seal ring, oil flask, and sandals. He was said to have brought with him epics, tragedies, and all kinds of prose compositions of his own fashioning, and to have been prepared to lecture on music, letters, and the art of memory. The secret of Hippias's wide knowledge seems to have been his exceptional memory. If he once heard a string of fifty names, for instance, he could repeat them all in correct order. The dialogues *Hippias Major* and *Hippias Minor,* since they were caricatures, are unreliable as portraits—yet they must surely have recorded enough of Hippias's eccentricities that his contemporaries would have recognized him.

Although we know of no other mathematics that we can attribute to him, Hippias's reputation rests securely on his invention of the quadratrix. It is the first example of a curve that could not be drawn by the traditionally required straightedge and compass but had to be plotted point by point. The quadratrix is described by a double motion

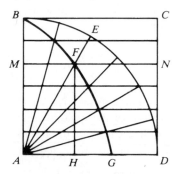

as follows. Let a straight line segment *AE* rotate clockwise about *A* with a constant velocity from the position *AB* to the position *AD,* so that a quadrant *BED* of a circle is described. At the same moment that the radius *AE* leaves its initial position *AB,* a line *MN* leaves *BC* and moves down with a constant velocity towards *AD,* always remaining parallel to *AD.* Both these motions are so timed that *AE* and *MN* will reach their ultimate position *AD* at the same moment. Now, at any given instant in their simultaneous movement, the rotating radius and the moving straight line will intersect at a point (*F* is a typical point). The locus of these points of intersection is the quadratrix. If *FH* is the perpendicular to *AD,* then the property of the quadratrix is that

$$\frac{\angle BAD}{\angle EAD} = \frac{AB}{FH} = \frac{\text{arc } BED}{\text{arc } ED}.$$

It is worth noting that the definition does not actually locate any point of the quadratrix on *AD*. If the rotating radius and moving straight line are made to end their motions together, then they will both coincide with *AD*, hence will not intersect one another at a unique point. The point of the quadratrix on *AD* (namely the point *G*) can be located only as a limit.

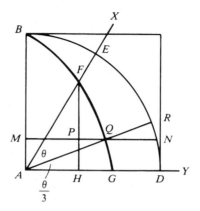

To see the ease with which the quadratrix can be used to trisect an angle, suppose that the given angle is $\angle XAY$. Place this angle at the center of a circle within which the quadratrix is constructed, and let *XA* cut the curve at *F*. Draw *FH* perpendicular to *AD* and trisect *FH*. Through the point *P* of trisection, draw *MN* parallel to *AD*, meeting the quadratrix at *Q*. Now join *AQ* and extend it to meet the quadrant in the point *R*. Then $\angle DAR$ is the required angle. From the definition of the quadratrix, it is easy to prove that

$$\frac{\angle DAR}{\angle DAE} = \frac{PH}{FH} = \frac{(\frac{1}{3}FH)}{FH} = \frac{1}{3},$$

in consequence of which $\angle DAR = \frac{1}{3}\angle DAE = \frac{1}{3}\angle XAY$.

The use of the quadratrix in finding a square equal in area to a given circle is a more sophisticated matter and might not have been obvious to Hippias. Pappus, in his large compendium *Mathematical Collection,* made the statement:

> For the squaring of the circle, there was used by Dinostratus, Nicomedes, and some other more recent geometers a certain curve which took its name from this property; for it is called by them ''square-forming'' [quadratrix].

Hence, any ascription of the curve to Hippias is lacking. And as for Dinostratus (circa 350 B.C.), nothing more is known of his work than is disclosed by this passage, which should remind us of the scantiness of testimony on Greek mathematics and its practitioners. Although there is no universal opinion, Hippias is usually credited with inventing the quadratrix as a device for trisecting angles, and Dinostratus with first applying it to the quadrature of the circle.

Dinostratus's solution of the squaring of the circle, as transmitted to us by Pappus, requires one to know the position of G, the point at which the quadratrix meets the line AD. If it is assumed that G can be found, Pappus's proposition as he established it is

$$\frac{\text{arc } BED}{AB} = \frac{AB}{AG}.$$

This is proved by a double reductio ad absurdum argument, and provides one of the earliest examples in Greek mathematics of the indirect method of reasoning Euclid used so extensively.

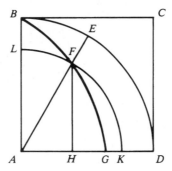

If the ratio (arc BED)/AB is not equal to AB/AG, then it must equal AB/AK, where either $AK > AG$ or $AK < AG$. Let us begin by assuming that $AK > AG$. With A as center and AK as radius, draw a quarter circle KFL, intersecting the quadratrix at F and the side AB at L. Join AF and extend it to meet the circumference BED at E; also, from F draw FH perpendicular to AD. Since corresponding arcs of a circle are proportional to their radii,

$$\frac{\text{arc } BED}{\text{arc } KFL} = \frac{AB}{AK};$$

and if the hypothesis is correct, we must have

$$\frac{\text{arc } BED}{AB} = \frac{AB}{AK},$$

from which it follows that $AB = \text{arc } KFL$. But by the defining property of the quadratrix, it is known that

$$\frac{AB}{FH} = \frac{\text{arc } BED}{\text{arc } ED} = \frac{\text{arc } KFL}{\text{arc } FK}$$

and it was just proved that $AB = \text{arc } KFL$. Therefore, the last relation tells us that $FH = \text{arc } FK$. But this is absurd, for the perpendicular is shorter than any other curve or line from F to AD. Thus the possibility that $AK > AG$ is ruled out. If $AK < AG$, a contradiction is reached in the same manner; hence, we are left with $AK = AG$ and (arc BED)/$AB = AB/AG$.

The quadrature problem just described is the quadrature of a quadrant, and Pappus took for granted that from this, one would be able to arrive at a square equal in area to a circle. For squaring the circle, we shall use Proposition 14 of Book II of Euclid's *Elements:* To construct a square equal to a given rectilinear figure. Let a circle of radius r be given. Using the quadratrix, a line segment of length s can be obtained for which

$$\frac{(C/4)}{r} = \frac{r}{s},$$

where C is the circumference of the circle. Once the length s is available, it is possible to construct a line segment which is the fourth proportional to r, r and s (see Problem 8). The resulting segment will be equal in length to $q = C/4$, the quadrant arc of the circle. Because the area A of the circle is half the product of its radius and its circumference, we have

$$A = \frac{1}{2}rC = 2r\left(\frac{C}{4}\right) = 2rq.$$

A rectangle with $2r$ as one side and q as the other will have area equal to A; a square equal in area to the rectangle is easily constructed by means of a semicircle. This is equivalent to taking the side x of the required square to be the mean proportional between the line segments $2r$ and q,

$$\frac{2r}{x} = \frac{x}{q}.$$

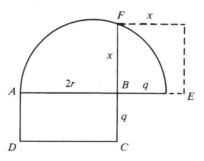

Most sophists had no permanent residence. They engaged their lecture halls, collected their fees for courses of instruction, and then departed. But by the early fourth century B.C., many of them had given up their itinerant practices and established themselves in Athens. The city began to gain a reputation for scholarship that attracted students from near and far. To use Hippias's words—at least those given in Plato's *Dialogues*—Athens had become "the very headquarters of Greek wisdom."

The most celebrated of the new schools to open in Athens was the Academy of Plato, where Aristotle was a student. As a disciple of Socrates, Plato (429–348 B.C.) had found it expedient to leave Athens after his master was sentenced to drink poison. For a dozen years he traveled in the Mediterranean world, stopping in Egypt, Sicily, and southern Italy. In Italy, Plato became familiar with the tenets of the Pythagoreans, which may partly explain his appreciation of the universal value of mathematics. On his way back to Greece he was sold as a slave by the ship's captain but was quickly ransomed by his friends. About 387 B.C., Plato returned to his native city to establish himself as a philosopher. In a grove in the suburbs of Athens, Plato founded a school

A mosaic from Pompeii depicting the Academy of Plato. *(The Bettmann Archive.)*

that became, in a sense, the spiritual ancestor of our Western institutions of higher learning. The land originally belonged to the hero Academos, so that it was called the grove of Academia; and therefore the new school of philosophy was named the Academy. After the fashion of that time, legal recognition was secured by making the Academy a religious brotherhood, dedicated to the worship of the Muses. Accordingly, it had chapels dedicated to these divinities. The Academy was the intellectual center of Greece for 900 years, until permanently closed in 529 A.D. by the Christian Emperor Justinian as a place of pagan and perverse learning.

It is through Plato that mathematics reached the place in higher education that it still holds. He was convinced that the study of mathematics furnished the finest training of the mind and hence was indispensable for philosophers and for those who would govern his ideal state. Because he expected those seeking admission to the Academy to be well grounded in geometry, he caused to be displayed over its portals the warning inscription, "Let no man ignorant of geometry enter here." It is reported that one of Plato's successors as a teacher in the Academy turned away an applicant who knew no geometry, saying, "Depart, for thou hast not the grip of philosophy." Whether or not these stories are true, there is no question that in contrast to the sophists who looked down on the teaching of the abstract concepts of the scientist, Plato gave mathematics a favored place in the curriculum of the Academy. The importance of arithmetical training, in his view, is that "arithmetic has a very great and elevating effect, compelling the mind to reason about abstract number." In speaking of the virtues of mathematics he was, of course, espousing the cause of pure mathematics; by comparison, he thought its practical utility was of no account. Plato carried his dislike of "applied mathematics" to the extreme of protesting the use of mechanical instruments in geometry, restricting the subject to those figures that could be drawn by straightedge and compass.

Plato was primarily a philosopher rather than a mathematician. So far as mathematics is concerned, it is not known that he made any original contribution to the subject matter; but as one who inspired and directed other research workers, he performed as great a service as any of his contemporaries did. According to the Greek commentator Proclus:

> Plato . . . caused mathematics in general, and geometry in particular, to make great advances, by reason of his well-known zeal for the study, for he filled his writings with mathematical discourses, and on every occasion exhibited the remarkable connection between mathematics and philosophy.

Most of the mathematical advances that came during the middle of the fourth century B.C. were made by the friends and pupils of Plato. Proclus, after giving us a list of names of those who contributed to the subject at that time, went on to say, "All these frequented the Academy and conducted their investigations in common." The hand of Plato is also seen in the increased attention given to proof and the methodology of reasoning; accurate definitions were formulated, hypotheses clearly laid down, and logical rigor required. This collective legacy paved the way for the remarkable systemization of mathematics in Euclid's *Elements*.

About 300 B.C., the Platonic Academy found a rival, the Museum, which Ptolemy I set up in Alexandria for teaching and research. The talented mathematicians and scientists for the most part left Athens and adjourned to Alexandria. Although the main center of mathematics had shifted, the direct descendant of Plato's Academy retained its preeminence in philosophy until the Emperor Justinian suppressed the philosophical schools of Athens, decreeing that only those of the orthodox faith should engage in teaching. Edward Gibbon, in *The Decline and Fall of the Roman Empire,* saw Justinian's legislation of 529 as the death knell of classical antiquity, the triumph of Christian ignorance over pagan learning.

> The Gothic arms were less fatal to the schools of Athens than the establishment of a new religion whose minister superseded the exercise of reason, resolved every question by an article of faith, and condemned the infidel or sceptic to eternal flame. . . . The golden chain, as it was fondly styled, continued . . . until the edict of Justinian, which imposed perpetual silence on the schools of Athens.

Beyond 529, the institution of higher learning that Plato had founded ceased to be an instrument of Greek education.

3.5 Problems

1. Complete Dinostratus's proof of the quadrature of the quarter-circle by showing that the assumption $AK < AG$ leads to a contradiction. [*Hint:* Show that in the accompanying figure, arc $PK = FK$, is a contradiction.]

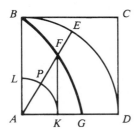 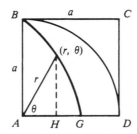

2. (a) Show that in modern polar coordinates, the equation of the quadratrix is

$$r = \frac{2a\theta}{\pi \sin \theta},$$

where θ is the angle made by the radius vector with AD, and r is the length of the radius vector and a the side of the square $ABCD$.

(b) Verify that $AG = \lim_{\theta \to 0} r$ exists, that in fact, $AG = 2a/\pi$.

3. Proposition 14 of Book II of Euclid's *Elements* solves the construction: To describe a square that shall be equal (in area) to a given rectilinear figure. Prove that if $ABCD$ is the given rectangle, AE the diameter of a semicircle, and $BFGH$ a square, then the square is equal in area to the rectangle.

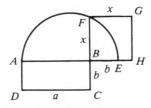

4. Show how Dinostratus, having found a line segment whose length was one-fourth the circumference of a circle, might have used the following theorem— stated by Archimedes in his *Measurement of a Circle*—to help square the circle: The area of any circle is equal to the area of the right triangle that has an altitude equal to radius of the circle and a base equal to the circumference.

5. The tomahawk-shaped instrument shown in the accompanying figure can be used to solve the trisection problem. ($PQ = QR = RS$, with PTR a semicircle on PR as diameter, UR perpendicular to PS.) If $\angle AOB$ is the angle to be trisected, place the tomahawk on the angle so that S lies on OA, the line segment UR passes through O, and the semicircle with diameter PR is tangent to OB at T. Prove that the triangles OTQ, ORQ, and ORS are all congruent, whence $\angle ROA$ is one-third of $\angle AOB$.

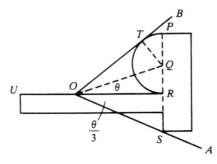

6. The limaçon (from the Latin word for "snail," *limax*) was discovered by Etienne Pascal (1588–1640), father of the better-known Blaise Pascal. The curve is based on the circle C of radius 1 with

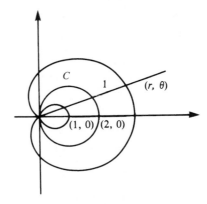

center at (1, 0); for it is defined to be the set of all points whose distance from the circle C measured along a line through the origin is constantly equal to 1, the radius of C. Prove that the equation of the limaçon in polar coordinates is $r = 1 + 2 \cos \theta$, hence in rectangular coordinates is $(x^2 + y^2 - 2x)^2 = x^2 + y^2$. [*Hint:* The polar equation of the circle C is $r = 2 \cos \theta$.]

7. Although the limaçon was invented for other purposes, it was later shown to afford a method for trisecting arbitrary angles. Let ABC be any

central angle in a circle with center $B = (1, 0)$ and radius 1. Draw the limaçon for the circle and let BA extended cut the limaçon in the point D. Let the line from the origin O to D meet the circle at E, as shown in the figure. Prove that angle BDE is one-third as large as angle ABC. [*Hint:* $\angle ABC = \angle BOD + \angle BDO = \angle BEO + \angle BDE = \angle BDE + \angle EBD + \angle BDE = 3\angle BDE$.]

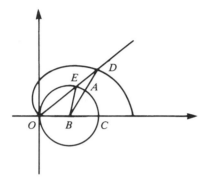

Bibliography

Africa, Thomas W. *Science and the State in Greece and Rome.* New York: Wiley, 1968.

Allman, George. *Greek Geometry from Thales to Euclid.* Dublin: Dublin University Press, 1889.

Anderson, Clifford. *The Fertile Crescent.* Fort Lauderdale, Fla: Sylvester Press, 1968.

Cajori, Florian. "The Purpose of Zeno's Arguments on Motion." *Isis* 3 (1920–21): 7–20.

Clagett, Marshall. *Greek Science in Antiquity.* Freeport, N.Y.: Books for Libraries Press, 1971.

Cohen, Morris, and Drabkin, I. E. *A Source Book in Greek Science.* Cambridge, Mass.: Harvard University Press, 1966.

Dantzig, Tobias. *The Bequest of the Greeks.* New York: Charles Scribner's, 1955.

Dickson, Leonard E. "On the Trisection of an Angle and the Construction of Polygons of 7 and 9 Sides." *American Mathematical Monthly* 21(1914):259–262.

Fowler, D. H. "Ratio in Early Greek Mathematics." *Bulletin (New Series) of the American Mathematical Society* 1(1979):807–847.

———. *The Mathematics of Plato's Academy: A New Reconstruction.* Oxford: Clarendon Press, 1987.

Friedrichs, K. O. *From Pythagoras to Einstein* (New Mathematical Library, No. 16). New York: Random House, 1965.

Gould, S. H. "The Method of Archimedes." *American Mathematical Monthly* 62 (1955):473–476.

Gow, James. *A Short History of Greek Mathematics.* New York: Chelsea, 1968. Reprint.

Heath, Thomas. *A History of Greek Mathematics.* 2 vols. New York: Oxford University Press, 1921.

———. *A Manual of Greek Mathematics.* New York: Oxford University Press, 1931. (Dover reprint, 1963).

Jones, Phillip. "Irrationals and Incommensurables I: Their Discovery and 'Logical Scandal'." *Mathematics Teacher* 49(1956):123–127.

———. "Irrationals and Incommensurables II: The Irrationality of $\sqrt{2}$ and Approximations to It." *Mathematics Teacher* 49 (1956):187–191.

Klein, Jacob. *Greek Mathematical Thought and the Origin of Algebra.* Translated by Eva Brann. Cambridge, Mass.: M.I.T. Press, 1968. (Dover reprint, 1992).

Knorr, Wilbur. *The Ancient Tradition of Geometric Problems.* Boston: Birkhauser, 1986.

Lloyd, G. E. R., *Greek Science After Aristotle.* London: Chatto and Windus, 1973.

Loomis, Elisha Scott. *The Pythagorean Proposition.* 2d ed. Washington: National Council of Teachers of Mathematics, 1968.

Mahoney, Michael. "Another Look at Greek Geometrical Analysis." *Archive for History of Exact Sciences* 5 (1968):318–348.

Maziarz, Edward, and Greenwood, Thomas. *Greek Mathematical Philosophy.* Cambridge, Mass.: M.I.T. Press, 1968.

Salmon, Wesley, ed. *Zeno's Paradoxes.* New York: Bobbs-Merrill, 1970.

Sarton, George. *A History of Science: Ancient Science Through the Golden Age of Greece.* New York: W. W. Norton, 1970.

Sierpinski, Waclaw. *Pythagorean Triangles* (Scripta Mathematica Studies Number Nine). Translated by A. Sharma. New York: Yeshiva University, 1962.

Stapleton, H. E. "Ancient and Modern Aspects of Pythagoreanism." *Osiris* 13 (1958):12–53.

Szabo, Arpad. *The Beginnings of Greek Mathematics.* Dordrecht, Holland: D. Reidel, 1978.

Taton, Rene. *Ancient and Medieval Science: From the Beginning to 1450.* Translated by A. J. Pomeranos. New York: Basic Books, 1963.

Thomas, Ivor, ed. *Selections Illustrating the History of Greek Mathematics.* 2 vols. Cambridge, Mass.: Harvard University Press, 1939–1941.

Vedova, G. C. "Notes on Theon of Smyrna." *American Mathematical Monthly* 58 (1951):675–683.

Weaver, James H. "The Duplication Problem." *American Mathematical Monthly* 23 (1916):106–113.

Zhmud, Leonid. "Pythagoras as a Mathematician." *Historia Mathematica* 16 (1989):249–268.

CHAPTER **4**

The First Alexandrian School: Euclid

It is the glory of geometry that from so few principles, fetched from without, it is able to accomplish so much.

ISAAC NEWTON

4.1 Euclid and the *Elements*

Toward the end of the fourth century B.C., the scene of mathematical activity shifted from Greece to Egypt. The Battle of Chaeronea, won by Philip of Macedon in 338 B.C., saw the extinction of Greek freedom as well as the decay of productive genius on its native soil. Two years later, Philip was murdered by a discontented noble and was succeeded by his 20-year-old son, Alexander the Great. Alexander conquered a great part of the known world within twelve years, from 334 B.C. to his death in 323 B.C., at the age of 33. Because his armies were mainly Greek, he spread Greek culture over wide sections of the Near East. What followed was a new chapter of history, known as the Hellenistic (or Greek-like) Age, which lasted for three centuries, until the Roman Empire was established.

Alexander's great monument in Egypt was the city that still bears his name, Alexandria. Having taken and destroyed the Phoenician seaports in a victorious march down the Eastern Mediterranean, Alexander was quick to see the potential for a new maritime city (a sort of Macedonian Tyre) near the westernmost mouth of the Nile. But he could do little more than lay out the site, because he departed for the conquest of Persia soon afterward. The usual story is that Alexander, with no chalk at hand to mark off the streets, used barley from the commissary instead. This seemed like a good idea until clouds of birds arrived from the delta and ate the grain as fast as it was thrown. Disturbed that this might be a bad omen, Alexander consulted a soothsayer, who concluded that the gods were actually showing that the new city would prosper and give abundant riches.

At Alexander's death, one of his leading generals, Ptolemy, became governor of Egypt and completed the foundation of Alexandria. The city had the advantage of a superb harbor and docking facilities for 1200 ships, so it became with the shortest possible delay the trading center of the world, the commercial junction point of Asia, Africa, and Europe. Alexandria soon outshone and eclipsed Athens, which was reduced

137

to the status of an impoverished provincial town. For nearly a thousand years, it was the center of Hellenistic culture, growing in the later years of the Ptolemaic dynasty to an immense city of a million people. Following its sacking by the Arabs in A.D. 641, the building of Cairo in 969, and the discovery of a shipping route around the Cape of Good Hope, Alexandria withered away, and by the time of the Napoleonic expedition its population had dwindled to a mere 4000.

The early Ptolemies devoted themselves to making Alexandria the center of intellectual life for the whole eastern Mediterranean area. Here they built a great center of learning in the so-called Museum (seat of the Muses), a forerunner of the modern university. The leading scholars of the times—scientists, poets, artists, and writers— came to Alexandria by special invitation of the Ptolemies, who offered them hospitality as long as they wished to stay. At the Museum, they had leisure to pursue their studies, access to the finest libraries, and the opportunity of discussing matters with other resident specialists. Besides free board and exemption from taxes, the members were granted salary stipends, the only demand being that they give regular lectures in return. These fellows of the Museum lived at the king's expense in luxurious conditions, with lecture rooms for their discussions, a colonnaded walkway in which to stroll, and a vast dining hall, where they took their meals together. The poet Theocritus, enjoying the bounty, hailed Ptolemy as ''the best paymaster a free man can have.'' And another sage, Ctesibius of Chalcis, when asked what he gained from philosophy, candidly replied, ''Free dinners.''

Built as a monument to the splendor of the Ptolemies, the Museum was nonetheless a milestone in the history of science, not to mention royal patronage. It was intended as an institution for research and the pursuit of learning, rather than for education; and for two centuries scholars and scientists flocked to Egypt. At its height this center must have had several hundred specialists, whose presence subsequently attracted many pupils eager to develop their own talents. Although one poet of the time contemptuously referred to the Museum as a birdcage in which scholars fattened themselves while engaging in trivial argumentation, science and mathematics flourished with remarkable success. Indeed, it is frequently observed that in the history of mathematics there is only one other span of about 200 years that can be compared for productivity to the period 300–100 B.C., namely the period from Kepler to Gauss (1600–1850).

Scholars could not get along without books, so the first need was to collect manuscripts; when these were sufficiently abundant, a building was required to hold them. Established almost simultaneously with the Museum and adjacent to it was the great Alexandrian library, housing the largest collection of Greek works in existence. There had of course been libraries before it, but not one possessed the resources that belonged to the Ptolemies. Manuscripts were officially sought throughout the world, and their acquisition was vigorously pressed by agents who were commissioned to borrow old works for copying if they could not otherwise be obtained; travelers to Alexandria were required to surrender any books that were not already in the library. Many stories are told of the high-handed methods by which the priceless manuscripts were acquired. One legend has it that Ptolemy III borrowed from Athens the rolls kept by the state containing the authorized texts of the writers Aeschylus, Sophocles, and Euripides. Although he had to make a deposit as a guarantee that the precious volumes would be returned, Ptolemy kept the original rolls and sent back the copies (needless to say, he forfeited the deposit). A staff of trained scribes catalogued the books, edited the texts that were not in good condition, and explained those works of the past that were not easily understood by a new generation of Greeks.

The Alexandrian library was not entirely without rivals in the ancient world. The most prominent rival was in Pergamon, a city in western Asia Minor. To prevent Pergamon from acquiring copies of their literary treasures, the jealous Ptolemies, it is said, prohibited the export of papyrus from Egypt. Early writers were careless with numbers and often exaggerated the size of the library. Some accounts speak of the main collection at the library as having grown to 300,000 or even 500,000 scrolls in Caesar's time (48 B.C.), with an additional 200,000 placed in the annex called the Serapeum. The collection had been built partly by the purchase of private libraries, one of which, according to tradition, was Aristotle's. After the death of Aristotle, his personal papers passed into the hands of a collector who, fearing that they would be confiscated for the library at Pergamon, hid all the manuscripts in a cave. The scrolls were badly damaged by insects and moisture, and the Alexandrian copyists made so many errors when restoring the texts that they no longer agreed with the versions of Aristotle's works already housed in the library.

Before the Museum passed into oblivion in A.D. 641, it produced many distinguished scholars who were to determine the course of mathematics for many centuries: Euclid, Archimedes, Eratosthenes, Apollonius, Pappus, Claudius Ptolemy, and Diophantus. Of these, Euclid (circa 300 B.C.) is in a special class. Posterity has come to know him as the author of the *Elements of Geometry,* the oldest Greek treatise on mathematics to reach us in its entirety. The *Elements* is a compilation of the most important mathematical facts available at that time, organized into thirteen parts, or books, as they were called. (Systematic expositions of geometry had appeared in Greece as far back as the fifth century B.C., but none have been preserved, for the obvious reason that all were supplanted by Euclid's *Elements.*) Although much of the material was drawn from earlier sources, the superbly logical arrangement of the theorems and the development of proofs displays the genius of the author. Euclid unified a collection of isolated discoveries into a single deductive system based on a set of initial postulates, definitions, and axioms.

Few books have been more important to the thought and education of the Western world than Euclid's *Elements.* Scarcely any other book save the Bible has been more widely circulated or studied; for twenty centuries, the first six books were the student's usual introduction to geometry. Over a thousand editions of the *Elements* have appeared since the first printed version in 1482; and before that, manuscript copies dominated much of the teaching of mathematics in Europe. Unfortunately, no copy of the work has been found that actually dates from Euclid's own time. Until the 1800s, most of the Latin and English editions were based ultimately on a Greek revision prepared by Theon of Alexandria (circa 365) some 700 years after the original work had been written. But in 1808, it was discovered that a Vatican manuscript that Napoleon had appropriated for Paris represented a more ancient version than Theon's; from this, scholars were able to reconstruct what appears to be the definitive text.

Although the fame of Euclid, both in antiquity and in modern times, rests almost exclusively on the *Elements,* he was the author of at least ten other works covering a

Euclid
(circa 300 B.C.)

(Smithsonian Institution.)

wide variety of topics. The Greek text of his *Data,* a collection of 95 exercises probably intended for students who had completed the *Elements,* is the only other text by Euclid on pure geometry to have survived. A treatise, *Conic Sections,* which formed the foundation of the first four books of Apollonius's work on the same subject, has been irretrievably lost, and so has a three-volume work called *Porisms* (the term *porism* in Greek mathematics means ''a corollary''). The latter is the most grievous loss, for it apparently was a book on advanced geometry, perhaps an ancient counterpart to analytic geometry.

As with the other great mathematicians of ancient Greece, we know remarkably little about the personal life of Euclid. That Euclid founded a school and taught in Alexandria is certain, but nothing more is known save that, the commentator Proclus has told us, he lived during the reign of Ptolemy I. This would indicate that he was active in the first half of the third century B.C. It is probable that he received his own mathematical training in Athens from the pupils of Plato. Two anecdotes that throw some light on the personality of the man have filtered down to us. Proclus, who wrote a commentary to the *Elements,* related that King Ptolemy once asked him if there was not a shorter way to learning geometry than through the *Elements,* to which he replied that there is ''no royal road to geometry''—implying thereby that mathematics is no respecter of persons. The other story concerns a youth who began to study geometry with Euclid and inquired, after going through the first theorem, ''But what shall I get by learning these things?'' After insisting that knowledge was worth acquiring for its own sake, Euclid called his servant and said, ''Give this man a coin, since he must make a profit from what he learns.'' The rebuke was probably adapted from a maxim of the Pythagorean brotherhood that translates roughly as, ''A diagram and a step (in knowledge), not a diagram and a coin.''

4.2   Euclidean Geometry

For more than two thousand years Euclid has been the honored spokesman of Greek geometry, that most splendid creation of the Greek mind. Since his time, the study of the *Elements,* or parts thereof, has been essential to a liberal education. Generation after generation has regarded this work as the summit and crown of logic, and its study as the best way of developing facility in exact reasoning. Abraham Lincoln at the age of 40, while still a struggling lawyer, mastered the first six books of Euclid, solely as training for his mind. Only within the last hundred years has the *Elements* begun to be supplanted by modern textbooks, which differ from it in logical order, proofs of propositions, and applications, but little in actual content. (The first real pedagogical improvement was by Adrien-Marie Legendre, who in his popular *Eléments de géométrie,* rearranged and simplified the propositions of Euclid. His book ran from an initial edition in 1794 to a twelfth in 1823.) Nevertheless, Euclid's work largely remains the supreme model of a book in pure mathematics.

Anyone familiar with the intellectual process realizes that the content of the *Elements* could not be the effort of a single individual. Unfortunately, Euclid's achievement has so dimmed our view of those who preceded him that it is not possible to say how far he advanced beyond their preparatory work. Few, if any, of the theorems established in the *Elements* are of his own discovery; Euclid's greatness lies not so much in the contribution of original material as in the consummate skill with which he organized a vast body of independent facts into the definitive treatment of Greek geometry and number theory. The particular choice of axioms, the arrangement of the propositions, and the rigor of demonstration are personally his own. One result follows another in strict logical order, with a minimum of assumptions and very little that is superfluous. So vast was the prestige of the *Elements* in the ancient world that its author was seldom referred to by name but rather by the title ''The Writer of the *Elements*'' or sometimes simply ''The Geometer.''

Euclid was aware that to avoid circularity and provide a starting point, certain facts about the nature of the subject had to be assumed without proof. These assumed statements, from which all others are to be deduced as logical consequences, are called the ''axioms'' or ''postulates.'' In the traditional usage, a postulate was viewed as a ''self-evident truth''; the current, more skeptical view is that postulates are arbitrary statements, formulated abstractly with no appeal to their ''truth'' but accepted without further justification as a foundation for reasoning. They are in a sense the ''rules of the game'' from which all deductions may proceed—the foundation on which the whole body of theorems rests.

Euclid tried to build the whole edifice of Greek geometrical knowledge, amassed since the time of Thales, on five postulates of a specifically geometric nature and five axioms that were meant to hold for all mathematics; the latter he called common notions. (The first three postulates are postulates of construction, which assert what we are permitted to draw.) He then deduced from these ten assumptions a logical chain of 465 propositions, using them like stepping-stones in an orderly procession from one proved proposition to another. The marvel is that so much could be obtained from so few sagaciously chosen axioms.

Abruptly and without introductory comment, the first book of the *Elements* opens with a list of 23 definitions. These include, for instance, what a point is (''that which has no parts'') and what a line is (''being without breadth''). The list of definitions concludes: "Parallel lines are straight lines which, being in the same plane and being produced indefinitely in both directions, do not meet one another in either direction." These would not be taken as definitions in a modern sense of the word but rather as naive descriptions of the notions used in the discourse. Although obscure and unhelpful in some respects, they nevertheless suffice to create certain intuitive pictures. Some technical terms that are used, such as *circumference of a circle,* are not defined at all, whereas other terms, like *rhombus,* are included among the definitions but nowhere used in the work. It is curious that Euclid, having defined parallel lines, did not give a formal definition of *parallelogram.*

Euclid then set forth the ten principles of reasoning on which the proofs in the *Elements* were based, introducing them in the following way:

Postulates

Let the following be postulated:

1. A straight line can be drawn from any point to any point.
2. A finite straight line can be produced continuously in a line.
3. A circle may be described with any center and distance.
4. All right angles are equal to one another.
5. If a straight line falling on two straight lines makes the interior angles on the same side less than two right angles, then the two straight lines, if produced indefinitely meet on that side on which are the angles less than two right angles.

Common Notions

1. Things which are equal to the same thing are also equal to one another.
2. If equals are added to equals, the wholes are equal.
3. If equals are subtracted from equals, the remainders are equal.
4. Things which coincide with one another are equal to one another.
5. The whole is greater than the part.

Postulate 5, better known as Euclid's parallel postulate, has become one of the most famous and controversial statements in mathematical history. It asserts that if two lines l and l' are cut by a transversal t so that the angles a and b add up to less than two right angles, then l and l' will meet on that side of t on which these angles lie. The remarkable feature of this postulate is that it makes a positive statement about the whole extent of a straight line, a region for which we have no experience and that is beyond the reach of possible observation.

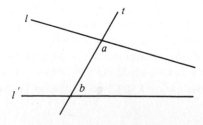

Those geometers who were disturbed by the parallel postulate did not question that its content was a mathematical fact. They questioned only that it was not brief, simple, and self-evident, as postulates were supposed to be; its complexity suggested that it should be a theorem instead of an assumption. The parallel postulate is actually the converse of Euclid's Proposition 27, Book I, the thinking ran, so it should be provable. It was thought impossible for a geometric statement not to be provable if its converse was provable. There is even some suggestion that Euclid was not wholly satisfied with his fifth postulate; he delayed its application until he could advance no further without it, though its earlier use would have simplified some proofs.

Almost from the moment the *Elements* appeared and continuing into the nineteenth century, mathematicians have tried to derive the parallel postulate from the first four postulates, believing that these other axioms were adequate for a complete development of Euclidean geometry. All these attempts to change the status of the famous assertion from ''postulate'' to ''theorem'' ended in failure, for each attempt rested on some hidden assumption that was equivalent to the postulate itself. Futile so far as the main objective was concerned, these efforts led nevertheless to the discovery of non-Euclidean geometries, in which Euclid's axioms except the parallel postulate all hold and in which Euclid's theorems except those based on the parallel postulate all are true. The mark of Euclid's mathematical genius is that he recognized that the fifth postulate demanded explicit statement as an assumption, without a formal proof.

Detailed scrutiny for over 2000 years has revealed numerous flaws in Euclid's treatment of geometry. Most of his definitions are open to criticism on one ground or another. It is curious that while Euclid recognized the necessity for a set of statements to be assumed at the outset of the discourse, he failed to realize the necessity of un-defined terms. A definition, after all, merely gives the meaning of a word in terms of other, simpler words—or words whose meaning is already clear. These words are in their turn defined by even simpler words. Clearly the process of definition in a logical system cannot be continued backward without an end. The only way to avoid the completion of a vicious circle is to allow certain terms to remain undefined.

Euclid mistakenly tried to define the entire technical vocabulary that he used. Inevitably this led him into some curious and unsatisfactory definitions. We are told not what a point and a line are but rather what they are not: ''A point is that which has no parts.'' ''A line is without breadth.'' (What, then, is *part* or *breadth?*) Ideas of ''point'' and ''line'' are the most elementary notions in geometry. They can be de-scribed and explained but cannot satisfactorily be defined by concepts simpler than themselves. There must be a start somewhere in a self-contained system, so they should be accepted without rigorous definition.

Perhaps the greatest objection that has been raised against the author of the *Elements* is the woeful inadequacy of his axioms. He formally postulated some things, yet omitted any mention of others that are equally necessary for his work. Aside from the obvious failure to state that points and lines exist or that the line segment joining two points is unique, Euclid made certain tacit assumptions that were used later in the deductions but not granted by the postulates and not derivable from them. Quite a few of Euclid's proofs were based on reasoning from diagrams, and he was often misled by visual evidence. This is exemplified by the argument used in his very first proposition (more a problem than a theorem). It involved the familiar construction of an equilateral triangle on a given line segment as base.

PROPOSITION   1   *For a line segment AB, there is an equilateral triangle having the segment as one of its sides.*

> **Proof.** Using Postulate 3, describe a circle with center A and radius AB passing through point B. Now, with center B and radius AB, describe a circle passing through A. From the point C, in which the two circles cut one another, draw the segments CA and CB (Postulate 1 allows this), thereby forming a triangle ABC. It is seen that $AC = AB$ and $BC = AB$ because they are radii of the same circle. It then follows from Common Notion 1 that $AB = BC = AC$, and so triangle ABC is equilateral.

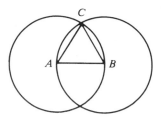

There is only one problem with all this. On the basis of spatial intuition, one feels certain that the two circles will intersect at a point C and will not, somehow or other, slip through each other. Yet the purpose of an axiomatic theory is precisely to provide a system of reasoning free of the dependence on intuition. The whole proposition fails if the circles we are told to construct do not intersect, and there is unhappily nothing in Euclid's postulates that guarantees that they do. To remedy this situation, one must add a postulate that will ensure the ''continuity'' of lines and circles. Later mathematicians satisfactorily filled the gap with the following:

> If a circle or line has one point outside and one point inside another circle, then it has two points in common with the circle.

The mere statement of the postulate involves notions of ''inside'' and ''outside'' that do not explicitly appear in the *Elements*. If geometry is to fulfill its reputation for logical perfection, considerable attention must be paid to the meaning of such terms and to the axioms governing them.

During the last 25 years of the nineteenth century, many mathematicians attempted to give a complete statement of the postulates needed for proving all the long-familiar theorems of Euclidean geometry. They tried, that is, to supply such additional postulates as would give explicitness and form to the ideas that Euclid left intuitive. By far the most influential treatise on geometry of modern times was the work of the renowned German mathematician David Hilbert (1862–1943). Hilbert, who worked in several areas of mathematics during a long career, published in 1899 his main geometrical work, *Grundlagen der Geometrie* (Foundations of Geometry). In it he rested Euclidean geometry on twenty-one postulates involving six undefined terms—with which we should contrast Euclid's five postulates and no undefined terms.

The 48 propositions of the first book of the *Elements* deal mainly with the properties of straight lines, triangles, and parallelograms—what today we should call elementary plane geometry. Much of this material is familiar to any student who has had a traditional high-school course in plane and solid geometry. Although we shall not examine all these results in detail, Proposition 4 is one that deserves a close look. This proposition is called the side-angle-side theorem, for it contains the familiar criterion for congruence of triangles, namely, two triangles are congruent if two sides and the included angle of one are congruent to the corresponding sides and included angle of the other. We have used the word *congruent* where Euclid spoke of *equality*. When he referred to two angles (or for that matter, two line segments) as ''equal,'' he meant that they could be made to coincide. For our purposes, it is safe to think of congruent objects as having the same size and shape.

Euclid tried to give a proof of the side-angle-side theorem by picking up one triangle and superimposing it on the other triangle so that the remaining parts of the two triangles fitted. His argument, which was supposedly valid by Common Notion 4, ran substantially as follows: Given $\triangle ABC$ and $\triangle A'B'C'$, where $AB = A'B'$, $\angle A = \angle A'$, and $AC = A'C'$, move $\triangle ABC$ so as to place point A on point A' and side AB on side $A'B'$. Because $AB = A'B'$, point B must fall on point B'. Because $\angle A = \angle A'$, the side AC has the same direction as side $A'C'$, and because of the equal lengths of AC and $A'C'$, the points C and C' fall on each other. Now, if B and B' coincide and C and C' coincide, so must the connecting line segments BC and $B'C'$. The two triangles coincide in all respects, so it follows that they are congruent.

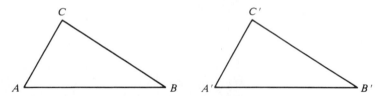

Although this ''principle of superposition'' may seem reasonable enough in dealing with material triangles made of wire or wood, its legitimacy has been questioned for working with conceptual entities whose properties exist only because they have been postulated. Indeed, the prominent British logician Bertrand Russell (1872–1970) spoke of superposition in no uncertain terms as a ''tissue of nonsense.'' The chief criticism is that in assuming that a triangle can be moved about without any alteration in its internal structure, when it is only known that two sides and an included angle remain constant, one is really assuming that these determine the rigidity of the triangle. Thus, in postulating the possibility of movement without change in form or magnitude, congruence itself is actually being postulated. Euclid's proof is therefore a vicious circle of reasoning. It has been conjectured that Euclid felt reluctant to use superposition in proving congruence and did so sparingly in the *Elements* but could not dispense with it entirely, for lack of a better method. Present-day mathematicians avoid the difficulty by taking the side-angle-side theorem as an axiom from which the other congruence theorems are then derived. At any rate, Euclid's approach to the problem of congruence was logically deficient.

A page from the first printed edition of Euclid's Elements. *Published in Latin in 1482. (Courtesy of Burndy Library.)*

Perhaps the most famous of the earlier propositions of Book I is Proposition 5, which states, ''In an isosceles triangle, the angles at the base are congruent to one another.'' (Here, by angles at the base is meant the angles opposite the two congruent sides.) This proposition sometimes marked the limit of the instruction in Euclid in the universities of the Middle Ages. It is historically interesting as having been called ''elefuga,'' a medieval term meaning ''the flight of the fools,'' because at this point the student usually abandoned geometry. Another name commonly used for Proposition 5 is *pons asinorum,* a Latin phrase signifying ''bridge of fools,'' or ''bridge of asses,'' although opinion is not unanimous about the exact implication of the title. The name might have been suggested by the difficulties that poor geometers have with the proposition; anyone unable to proceed beyond it must be a fool. A more generous interpretation is that the diagram that accompanies Euclid's proof resembles a trestle-bridge so steep that a horse could not climb the ramp, though a sure-footed animal such as an ass could. Perhaps only the sure-footed student could proceed beyond this stage in geometry. Here is an abbreviated proof of Euclid's Proposition 5. The contention is that in a triangle ABC, where $AB = AC$, one has $\angle ABC = \angle ACB$. To validate this, select points F and G on the extensions of sides AB and AC such that $AF = AG$.

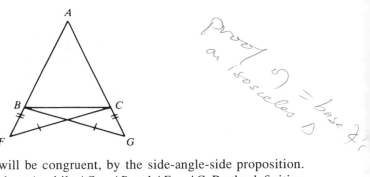

Then triangles AFC and AGB will be congruent, by the side-angle-side proposition. Indeed, they have a common angle at A, while $AC = AB$ and $AF = AG$. By the definition of congruent triangles, all the corresponding parts are equal, so that the bases $FC = GB$, $\angle ACF = \angle ABG$, and $\angle AFC = \angle AGB$. It is worth noticing too that

$$FB = AF - AB = AG - AC = GC.$$

The implication is that triangles BFC and CGB are themselves congruent (also by the side-angle-side proposition), whence as corresponding angles, $\angle BCF = \angle CBG$. This last equality, together with the fact that $\angle ABG = \angle ACF$, tells us that

$$\angle ABG - \angle CBG = \angle ACF - \angle BCF$$

or $\angle ABC = \angle ACB$.

Fortunately, there is a far simpler proof of this proposition (attributed to Pappus of Alexandria, A.D. 300), which requires no auxiliary lines whatever. The pertinent observation is that nowhere in the statement of the side-angle-side proposition is it required that the two triangles be distinct. The details are as follows. Given the isosceles triangle ABC, where $AB = AC$, think of it in two ways, one way as triangle ABC and the other as triangle ACB. Thus, there is a correspondence between $\triangle ABC$ and $\triangle ACB$ with vertices A, B, and C corresponding to vertices A, C, and B, respectively. Under this correspondence, $AB = AC$, $AC = AB$, and $\angle BAC = \angle CAB$. Thus, two sides and an included angle are congruent to the parts that correspond to them, whence the triangles are congruent. This means that all the parts in one triangle are equal to the corresponding parts in the other triangle, and in particular, $\angle ABC = \angle ACB$, which was to be proved.

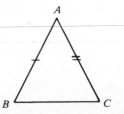
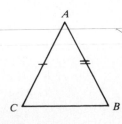

A result about triangles that Euclid found very useful in his development of geometry is the exterior angle theorem. This theorem is the embodiment of practically all the Euclidean axioms, for nearly all are used in its proof.

PROPOSITION 16   *If one of the sides of a triangle is produced, then the exterior angle is greater than either opposite interior angle.*

> **Proof.** Let *ABC* be any triangle and pick *D* to be any point on the extension of side *BC* through *C*. Call *E* the midpoint of *AC;* extend the line segment *BE* to a point *F* so that *BE = EF*. Because *AE = EC, BE = EF*, and ∠*AEB* = ∠*FEC* (vertical angles are equal by Proposition 15), the triangles *AEB* and *FEC* are congruent, from the side-angle-side proposition. The result is that ∠*BAE* = ∠*FCE*. But according to Common Notion 5, the whole is greater than any of its parts, so that ∠*DCA* > ∠*FCE*. Hence the exterior angle ∠*DCA* is greater than ∠*BAE*, which is an opposite interior angle of this triangle.

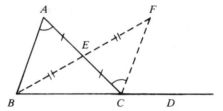

> Likewise, by extending side *AC* to a point *G*, it can be shown that ∠*GCB* > ∠*ABC*. Because ∠*GCB* and ∠*DCA* are vertical angles (hence equal), we immediately have ∠*DCA* greater than ∠*ABC*, the other opposite interior angle.

Aside from the fact that the existence of midpoints must first be established, the main flaw in this argument is Euclid's assumption from his diagram that if the segment *BE* is extended, the point *F* is always "inside" angle *DCA*. On the basis of the postulates he assumes—as distinct from the diagram—there is nothing to justify this conclusion. If the diagram is drawn instead on the curved surface of a sphere, then when *BE* is extended its own length to *F*, the point *F* ends up on the far side of the sphere, and *BF* may be so long that *F* falls "outside" angle *DCA*. Instead of having ∠*DCA* > ∠*FCE*, just the reverse would be true.

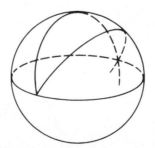

The underlying difficulty is that in making his so-called proof, Euclid took it for granted that a line is infinite. The critical postulate in this regard, Postulate 2, asserts merely that a line can be produced continuously—that it is endless or boundless—but does not necessarily imply that a line is infinite. On a sphere, where the role of a line is played by a great circle (a circle that has the same center as the sphere itself), a line that is produced from a given point will eventually return to that point. Because Euclid was not thinking of such a possibility, he apparently had no misgivings in proceeding on the basis of Postulate 2.

The first 26 propositions of the *Elements* develop theorems on congruent triangles, on isosceles triangles, and on the construction of perpendiculars. One also finds among the results the exterior angle theorem and the fact that the sum of two sides of a triangle is greater than the third side. The subject matter is based mainly on very ancient sources. There is a definite change of character beginning with Proposition 27; here, Euclid introduced the theory of parallels, but still without making use of his parallel postulate.

Euclid defined two lines as parallel if they did not intersect, that is, if no point lay on both of them. Euclid could have used the exterior angle theorem, although he did not do so, to prove the existence of parallel lines. (Or he could have added an extra postulate to the effect that parallel lines actually existed.) To see that this is possible, let l be any line and at each of two distinct points A and B on l erect a perpendicular to l (Proposition 11 allows this). If these perpendiculars were to meet at a point C, then in triangle ABC the exterior angle at B and the opposite interior angle at A, since they are right angles, would be equal. Because Proposition 16 is then violated, the two perpendiculars to l cannot meet; in other words, they are parallel.

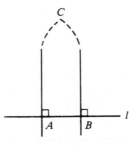

To make the next proposition precise, we require a definition. Suppose that a line t (called a ''transversal'') intersects lines l and l' at two distinct points A and B. In the accompanying figure, angles c, d, e, and f are called interior angles, while a, b, g, h are exterior angles. The usual language is to refer to the pair of angles c and e (d and f) as ''alternate interior angles,'' b and h (a and g) as ''alternate exterior angles.'' The

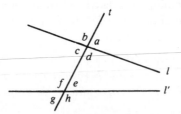

eight angles may also be grouped into four pairs of corresponding angles; angles *a* and *e* form such a pair of corresponding angles, and so do the pairs *b* and *f*, *c*, and *g*, and *d* and *h*.

With this terminology at hand, let us consider another proposition.

PROPOSITION 27   *If two lines are cut by a transversal so as to form a pair of congruent alternate interior angles, then the lines are parallel.*

> **Proof.** Referring to the figure, let the transversal *t* intersect lines *l* and *l'* at points *A* and *B,* so as to form a pair of alternate interior angles, say ∠*b* and ∠*c*, which are equal. To achieve a contradiction, assume that lines *l* and *l'* are not parallel. Then they will meet at a point *C* that lies, let us say, on the right side of *t* so as to form a triangle *ABC*. It can be concluded that an exterior angle (in this case, ∠*b*) is congruent to an opposite interior angle of triangle *ABC* (namely, ∠*c*). But we know that this is impossible, for an exterior angle of a triangle is always greater than either opposite interior angle. In consequence, *l* and *l'* are parallel.

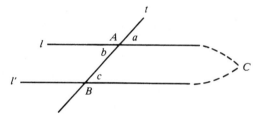

Proposition 27 implies that if two lines are perpendicular to the same line, then the two lines are parallel. From this fact, it is an easy matter to establish that through any point *P* that is not on a given line *l*, there passes a line *l'* that is parallel to *l*. All we need do is drop a perpendicular from *P* to the line *l* with foot at *Q* (Proposition 12 allows this) and at *P* to erect a line *l'* that is perpendicular to *PQ* (the construction is given in Proposition 11). Because *l* and *l'* have a common perpendicular, they must be parallel, with *l'* through *P*.

Let us pass over Proposition 28, which is just a variation of Proposition 27, and next examine Euclid's Proposition 29. It states the converses of the preceding two propositions. To this point, all the results have been obtained without any reference to the parallel postulate. They are, as we say, independent of it and would still be valid if the fifth postulate were deleted, or replaced by another one compatible with the remaining postulates and common notions. To prove Proposition 29, we must use the parallel postulate for the first time.

PROPOSITION 29   *A transversal falling on two parallel lines makes the alternate interior angles congruent to one another, the corresponding angles congruent, and the sum of the interior angles on the same side of the transversal congruent to two right angles.*

Proof. Suppose that the lines and angles are labeled as in the figure. We conclude at once that because $\angle a$ and $\angle b$ are supplementary angles, $\angle a$ and $\angle b$ equals two right angles (this is the content of Proposition 13). If $\angle a > \angle c$, then $\angle a + \angle b > \angle c + \angle b$, and $\angle b + \angle c$ would be less than two right angles. It would follow

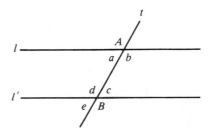

from Postulate 5 that l and l' must meet to the right of t. But this contradicts that l and l' are parallel. Thus, it cannot happen that $\angle a > \angle c$, or to put it in the affirmative, $\angle a \leq \angle c$. A like contradiction arises when we assume that the inequality $\angle a < \angle c$ holds; therefore $\angle a = \angle c$. Because $\angle c$ and $\angle e$ are vertical angles, they are equal, whence $\angle a = \angle e$. Finally, observe that the sum $\angle a + \angle b$ equals two right angles and $\angle a = \angle c$, so that the sum $\angle b + \angle c$ of the interior angles $\angle b$ and $\angle c$ equals two right angles.

■

It is worth noticing that Propositions 27 and 29 both provide proofs by contradiction, sometimes called *reductio ad absurdum* proofs. This is an important form of reasoning that consists in showing that if the conclusion is not accepted, then absurd or impossible results must follow. The element that produces the contradiction is different in each proposition. In Proposition 27, one ends up contradicting the exterior angle theorem, whereas in the case of Proposition 29, it is the parallel postulate that provides the absurdity.

Moving further with these ideas, we look at another important result, namely Proposition 30.

PROPOSITION 30 *Two lines parallel to the same line are parallel to one another.*

Proof. Suppose that each of the lines l and l' is parallel to the line k. We claim that l is also parallel to l'. Let these lines be cut by the transversal t, as indicated in the figure. Because t has fallen on the parallel lines l and k, the angle a equals the angle b by Proposition 29. Likewise, since t has fallen on the parallel lines k and l', the angles b and c are equal. But then $\angle a = \angle c$ (this is Common Notion 1). Because these are alternate interior angles, it is apparent by Proposition 27 that l and l' are parallel.

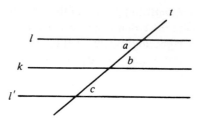

One implication of Proposition 30 is that through a point *P* not on a given line *l*, there cannot be more than one line parallel to *l*. The argument is as follows. Suppose there were two distinct lines through *P*, each parallel to *l*; then from Proposition 30, they would be parallel to each other. This would, by the meaning of *parallel*, contradict that the lines intersect at *P*.

We should stress this last point before temporarily abandoning the subject of parallel lines. Euclid did not require the parallel postulate in order to know that parallel lines exist, or what is more important, that it is possible to construct a parallel to a given line through an external point. The primary effect of Postulate 5 is to ensure that there exists only one line parallel to the given line through a point not on the line.

Throughout Book I, Euclid went forward in a logical chain of propositions until his final goal was reached. The work on parallel lines culminates with the result that the sum of the angles of a triangle is congruent to two right angles. The proof rests on Proposition 29 and hence implicitly involves the parallel postulate. It is surprising how many notable consequences of Euclidean geometry besides the properties of parallel lines stem, directly or indirectly, from this postulate.

PROPOSITION 32   *In any triangle, the sum of the three interior angles is equal to two right angles.*

Proof. Given a triangle *ABC* with angles *a*, *b*, and *c*, extend the side *AB* to a point *D* and through *B* draw a line *l* parallel to side *AC*.

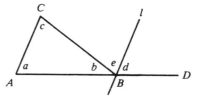

But ∠*c* = ∠*e*, since they are alternate interior angles formed by *l* and *AC* with *BC*. Similarly, Proposition 29 guarantees that ∠*a* = ∠*d*. Now, the sum ∠*b* + ∠*e* + ∠*d* equals two right angles (this is the content of Proposition 13), and so the sum of the interior angles of △*ABC* must equal two right angles.

Book I closes—in Propositions 47 and 48—with a remarkably clever proof of the Pythagorean theorem and its converse. Although few of the propositions and proofs in the *Elements* are Euclid's own discoveries, this proof of the Pythagorean theorem is usually ascribed to Euclid himself. Proclus wrote, ''I admire the writer of the *Elements* not only that he gave a very clear proof of this proposition, but that in the sixth Book, he also explained the more general proposition by means of an irrefutable argument.'' On the surface, this suggests that the proof at the end of Book I was Euclid's own; some authorities contend that it was first advanced by Eudoxus, who antedated Euclid by at least a generation, and the version in which the theory of proportions is applied to the sides of similar triangles bears the mark of Thales.

The proof of the Pythagorean theorem found in Proposition 47 involves the contents of Book I only. The feeling that the reasoning is artificial and unnecessarily intricate led the German philosopher Arthur Schopenhauer (1788–1860) to dismiss the demonstration with the contemptuous remark that it was not an argument but a ''mousetrap.'' Thus, among the many different names applied to Euclid's proof, it is not uncommonly called ''the mousetrap proof.''

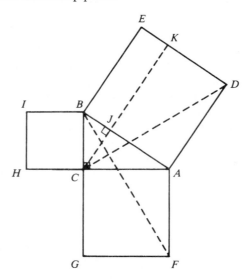

The diagram herewith illustrates Euclid's proof. Given a right triangle *ABC*, with right angle at *C*, erect squares on each of the sides. Next, draw the perpendicular from *C* to *AB* and *DE*, meeting these sides at the points *J* and *K*, respectively. The key observation is that the rectangle *AJKD* has twice the area of the triangle *CAD*:

(1) $AJKD = 2(CAD).$

This is because each figure has the same base *AD* and the same altitude *AJ*. In like manner, since the lower square *AFGC* and the triangle *FAB* have the same base *AF* and the same altitude *AC*, the area of the square is twice the area of the triangle:

(2) $AFGC = 2(FAB).$

Now the two triangles *CAD* and *FAB* are congruent by the side-angle-side theorem ($AC = AF$, $\angle CAD = \angle CAB + \angle DAB = \angle CAB + \angle CAF = \angle FAB$, and $AD = AB$), hence have the same area; that is,

(3) $\triangle CAD = \triangle FAB.$

Putting relations (1) and (2) together, we conclude at once that

(4) $AJKD = AFGC$.

By exactly the same reasoning, it can be demonstrated that the rectangle $BEKJ$ and square $BCHI$ are of equal area:

(5) $BEKJ = BCHI$.

But a glance at the diagram shows that the area of the square on the hypotenuse is the sum of the areas of the two rectangles $AJKD$ and $BEKJ$. Thus,

(6) $ABED = AJKD + BEKJ$

                                        $= AFGC + BCHI$

and, with a change of notation, the theorem obtains:

$$AB^2 = AC^2 + CB^2.$$

The Pythagorean theorem is immediately followed in the *Elements* by a proof of its converse: If in a triangle ABC the square on one of the sides (say BC) is equal to the sum of the squares on the other two sides, the angle contained by these other two sides is a right angle. For the proof, Euclid constructed a right triangle congruent to the given triangle. Specifically, the procedure would be to lay off a line segment AD perpendicular to AC and equal in length to AB.

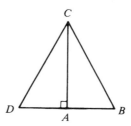

By hypothesis, $AC^2 + AB^2 = BC^2$, and the Pythagorean theorem (as applied to $\triangle CAD$) implies that $AD^2 + AC^2 = CD^2$. Because $AD = AB$, the implication is that $BC^2 = CD^2$, whence $BC = CD$. It follows that triangles CAD and CAB are congruent, for their corresponding sides are congruent. Thus $\angle CAB = \angle CAD$, a right angle.

Euclid's similarity proof of the Pythagorean theorem (Proposition 31 of Book VI) had to be delayed, since the plan of the *Elements* called for the theory of proportion to be expounded in Books V and VI. It depends on a property that is characteristic of right triangles: A perpendicular from the vertex C of the right angle to the hypotenuse divides triangle ABC into two similar right triangles ADC and BDC. Observe that each of the new right triangles so formed and the original triangle are equiangular and hence similar. As regards triangles ABC and ADC, for instance, we have $\angle A = \angle A$, since it is common to both triangles, and $\angle ACB = \angle ADC$, for these are both right angles. The sum of the angles in any triangle equals two right angles, so it is equally clear that $\angle B = \angle ACD$.

Because in Euclidean geometry it is proved that corresponding sides of similar triangles are proportional,

$$\frac{c}{a} = \frac{a}{x} \qquad \text{and} \qquad \frac{c}{b} = \frac{b}{y}.$$

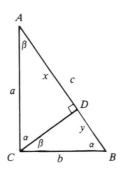

These proportionality relations imply that

$$a^2 = cx \qquad \text{and} \qquad b^2 = cy$$

and, by addition, that

$$a^2 + b^2 = cx + cy = c(x + y) = c^2.$$

You may have gathered that the *Elements* is not a perfect model of mathematical reasoning; critical investigation reveals numerous flaws in its logical structure. The truth is that so far as Euclid's aim was to place geometry on an unimpeachable foundation, he failed in the attempt. This is not to belittle the work; it was a magnificent achievement, a giant step forward marking the real beginning of axiomatic mathematics. Although some of its underpinnings have needed shoring up, Euclid's *Elements* is still a grand work, worthy of study. "This wonderful book," wrote Sir Thomas Heath, "with all its imperfections, which are indeed slight enough when account is taken of the date at which it appeared, is and will remain the greatest mathematical textbook of all time."

The Greek mathematicians of classical times had a unique genius for geometry; no other society of this period developed the subject as an abstract deductive system. Their Chinese counterparts, for instance, concerned themselves with some geometric questions, but always in an empirical, nondemonstrative way. All we find in the early Chinese mathematical handbooks are practical problems connected with everyday life, problems involving the calculation of areas of all kinds of shapes, and volumes of various vessels and dams. Chinese mathematics was profoundly algebraic, so geometric figures served only to transmute numerical information into algebraic form.

The *Nine Chapters on the Mathematical Art* marks the beginning of the mathematical tradition in China. The oldest textbook on arithmetic in existence, it is conspicuously free of the mystic cosmology of the earlier *Arithmetic Classic of the Gnomon and the Circular Paths of Heaven*. The date and origin of the *Nine Chapters* is unknown. We know that it represents the collective effort of many mathematical minds over several centuries. First assembled as a book about the same time that Euclid was drawing up his *Elements*, original copies were destroyed in the famous Burning of the Books of 213 B.C. Fragments of the collection were later recovered, put in order, and augmented by a number of mathematicians before the work received its final form. The text as it survives today is a commentary on the *Nine Chapters*, prepared by Liu Hui in A.D. 263. Liu Hui gave theoretical verifications of each of the problems, at the same time extensively expanding and enriching the material with his own contributions.

In its influence on Chinese mathematical thought, Liu Hui's commentary on the *Nine Chapters* is the most important of all ancient works, studied by generation after generation for more than a thousand years. Indeed, later mathematical writings bear its imprint, both as to ideas and terminology. Chinese mathematics was geared toward proficiency in algebraic manipulation and problem solving, so that there was little incentive to change a procedure that worked; this serves in part to explain the longevity of the *Nine Chapters*.

In the seventh century, the *Nine Chapters* was given wide currency when the government decreed its use throughout the universities as a standard syllabus for students preparing for civil service examinations. It became one of the earliest printed textbooks when a printed version appeared in 1084, the product of a wood-block technique in which each whole page was separately carved from one wooden block. (The oldest known book produced by wood-block printing, the Buddhist *Diamond Sutra*, was cut in 868; the entire Buddhist canon, printed between 972 and 983, required the engraving of 130,000 two-page blocks.)

As its title indicates, the *Nine Chapters* consists of nine distinct sections with a total of 246 problems and their solutions. It may be likened to the *Elements* in being an organization of the mathematical knowledge accumulated by the Chinese up to the middle of the third century. The *Nine Chapters* was not intended as a theoretical work in the Greek style but as a practical handbook with problems that the ruling officials of the state were likely to encounter: measurement of cultivated land, construction of dikes and canals, capacity of granaries, rates of exchange and taxation of foodstuffs. Thus, the chapters bear such titles as "Field Measurement," "Distribution by Proportion," and "Fair Taxes."

The early chapters give computational rules—some correct, others not—for obtaining the areas of rectangles, triangles, trapezoids, and segments of circles; and for the volumes of such familiar solids as spheres, cylinders, pyramids, and circular cones. For instance, the area of a circle is given by $\frac{3}{4}d^2$, where d is the diameter; the result would be correct if the value of π were taken to be 3. The correct formula for the volume of a truncated pyramid, which was also known to the Egyptians, appears here. There are detailed procedures and explanations for the extraction of square and cube roots using counting rods. Special attention is paid to the arithmetic of fractions, with an emphasis on finding common denominators. A chapter is devoted to the solution of linear equations in one unknown by means of the rule of false position.

The *Nine Chapters* provides the first evidence that we have of a systematic method for solving simultaneous linear equations. The method occurs in the 18 problems of the eighth chapter, which is called "The Way of Calculating by Arrays." Some idea of the general procedure may be obtained from the first problem:

There are three grades of corn. After threshing, three bundles of top grade, two bundles of medium grade, and one bundle of low grade make 39 dou [a measure of volume]. Two bundles of top grade, three bundles of medium grade, and one bundle of low grade will produce 34 dou. The yield of one bundle of top grade, two bundles of medium grade, and three bundles of low grade is 26 dou. How many dou are contained in each bundle of each grade?

The relations of the problem are equivalent to a system of three linear equations in three unknowns x, y, z, namely

$$3x + 2y + z = 39$$

$$2x + 3y + z = 34$$

$$x + 2y + 3z = 26$$

The equations were not written in this fashion, but the coefficients of the unknowns and the constants were represented by rods on a counting-board as the array

1	2	3
2	3	2
3	1	1
26	34	39

By performing appropriate multiplications and subtractions, coefficients are eliminated in this array until it is reduced to

0	0	3
0	5	2
36	1	1
99	24	39

We would represent the last array as the system of equations $36z = 99$, $5y + z = 24$, $3x + 2y + z = 39$. From this system we could easily calculate, in turn, the values $z = 2\frac{3}{4}$, $y = 4\frac{1}{4}$, $x = 9\frac{1}{4}$.

During the course of carrying out the array computations, negative numbers might occur within an array. In this earliest accepted use of negative numbers, the Chinese used red rods on the counting-board to represent negative numbers; while black rods stood for positive numbers. A coefficient of zero was indicated by a blank space on the board.

Besides writing a commentary on the *Nine Chapters*, Liu Hui also produced the shorter *Sea Island Mathematical Manual*. A treatise on surveying containing only nine practical problems, it seems to have been intended to supplement the last section of the *Nine Chapters* that dealt with the properties of right triangles. By the seventh century, it was separated from the *Nine Chapters* to become an independent mathematical work. The problems in the *Sea Island Mathematical Manual* involve measuring distances to inaccessible points by using tall poles with sighting-bars fixed at right angles on them. Unlike Thales's technique for finding the distance of a ship at sea from the shore, Liu Hui's problems usually require two observations, and sometimes three or four. The *Manual* most likely takes its name from the first problem of the collection, which begins with the statement, ''There is a sea island that is to be measured.''

There is a sea island that is to be measured. Two poles that are each 30 feet high are erected on the same level, 1000 paces [1 pace = 6 feet] apart, so that the rear pole is in a straight line with the island and the first pole. If a man walks 123 paces back from the first pole, the highest point of the island is just visible through the top of the pole when he views it from ground level. Should he move 127 paces back from the rear pole, the summit of the island is just visible through the top of the pole when seen from a point on ground level. It is required to find the height of the island and its distance from the nearer pole.

The rules given for calculating the required unknowns involve recognizing that corresponding sides of similar triangles are proportional. The problems are geometric, but

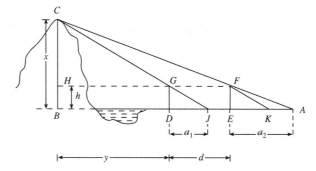

the solution is algebraic. Using modern notation, Hui's explanation of the solution proceeds as follows. Let $EK = DJ$, so that FK is parallel to GJ. Knowing that triangles CHG and FEK are similar, as are triangles CGF and FKA, we have the proportions

$$\frac{CH}{FE} = \frac{HG}{EK} = \frac{CG}{FK} = \frac{GF}{AK}$$

Now, if $CB = x$, $BD = y$, $DE = d$, $GD = FE = h$, $DJ = EK = a_1$, and $EA = a_2$, then these proportions yield

$$\frac{x - h}{h} = \frac{y}{a_1} = \frac{d}{a_2 - a_1}$$

from which the desired distances can be obtained:

$$x = \frac{hd}{a_2 - a_1} + h = 1255 \text{ paces}, \quad y = \frac{a_1 d}{a_2 - a_1} = 30750 \text{ paces}.$$

The other eight problems of the *Manual* also deal with distant measurements, with the solutions always based on properties of similar right triangles.

4.2   Problems

Problems 1–10 contain propositions from Book I of Euclid's *Elements*. In each instance, prove the indicated result.

1. *Proposition 6*. If two angles of a triangle are congruent with one another, then the sides opposite these angles will also be congruent. [*Hint:* Let *ABC*

be a triangle in which $\angle CAB = \angle CBA$. If $AC \neq BC$, say $AC > BC$, then choose a point D on AC such that $AD = BC$.]

2. *Proposition 15*. If two lines cut one another, then they make vertical angles that are equal. [*Hint:* Appeal to Proposition 13, which says that if a ray is drawn from a point on a line, then the sum of the pair of supplementary angles formed is equal to two right angles.]

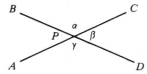

3. *Proposition 17*. In a triangle, the sum of any two angles is less than two right angles. [*Hint:* In △*ABC*, extend segment *BC* to a point *D* and use the exterior angle theorem.]

4. *Proposition 18*. If one side of a triangle is greater than a second side, then the angle opposite the first is greater than the angle opposite the second. [*Hint:* In △*ABC*, for *AC* > *AB*, choose a point *D* on *AC* such that *AD* = *AB*; use the fact that ∠*ADB* is an exterior angle of △*BCD*.]

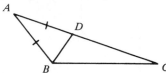

5. *Proposition 26*. Two triangles are congruent if they have one side and two adjacent angles of one congruent with a side and two adjacent angles of the other. [*Hint:* Let △*ABC* and △*DEF* be such that ∠*B* = ∠*E*, ∠*C* = ∠*F*, and *BC* = *EF*. For *AB* ≠ *DE*, say *AB* > *DE*, choose a point *G* on *AB* for which *BG* = *ED*.]

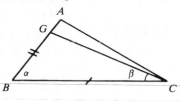

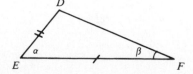

6. *Proposition 28*. Two lines intersected by a third line are parallel if the sum of the two interior angles on the same side of the transversal is equal to two right angles. [*Hint:* In the figure for this problem, α + β = 180°. Use Proposition 13.]

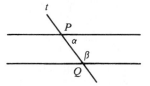

7. *Proposition 33*. If two opposite sides of a quadrilateral are equal and parallel, then the other two sides are also equal and parallel (hence, the quadrilateral is a parallelogram). [*Hint:* In the quadrilateral shown, let *AB* = *DC*, and assume that *AB*, *CD* are parallel. Show that △*ABC* is congruent with △*ADC*.]

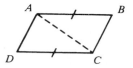

8. *Proposition 35*. Two parallelograms that have the same base and lie between the same parallel lines are equal in area to one another.

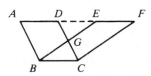

[*Hint:* In the figure, let *ABCD* and *BCFE* be parallelograms, and let *AD*, *EF* lie on a line parallel to *BC*. Show that △*ABE* is congruent with △*DCF*.]

9. *Proposition 37*. Two triangles that have the same base and lie between the same parallel lines are equal in area to one another.

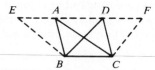

[*Hint:* In the figure, let *ABC* and *DBC* be triangles such that *AD* is parallel to *BC*. Consider the parallelograms *EBCA* and *FCBD*.]

10. *Proposition 41.* If a parallelogram and a triangle have the same base and lie between the same parallel lines, then [the area of] the parallelogram is double [the area of] the triangle.

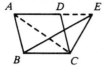

[*Hint:* In the figure, let *ABCD* be a parallelogram and *EBC* be a triangle, with *AD* and *E* on a line parallel to *BC*. Consider the triangles *ABC* and *EBC*.]

11. In *Mathematical Collection,* Pappus (circa 320) gave the following generalization of the Pythagorean theorem, which applies to all

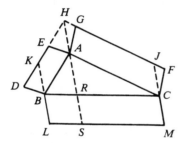

triangles, whether right triangles or not. Let *ABC* be any triangle and *ABDE, ACFG* be arbitrary parallelograms described externally on *AB* and *AC*. Suppose that *DE* and *FG* intersect at the point *H* when extended, and draw *BL* equal and parallel to *HA*. Then (in area)

$$BLMC = ABDE + ACFG.$$

Prove Pappus's theorem. [*Hint:* First extend *HA, BL,* and *MC* until they meet *LM, DE,* and *FG,* respectively. Now, apply Proposition 35 to the parallelograms *ABDE* and *ABKH,* and also to *ACFG* and *ACJH.*]

12. The Greeks constructed a line segment of length $\sqrt{n}$, where *n* is a positive integer, as follows. First write *n* as *n · 1*; then make *AB* = *n* and *BC* = 1.

Draw a semicircle on *AC* as diameter. Erect *BD* perpendicular to *AC* at *B,* meeting the semicircle at the point *D*. By similar triangles, prove that the length of *BD* equals $\sqrt{n}$.

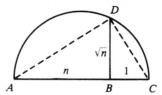

13. Solve the following problems from the last section of the *Nine Chapters:*

(a) A square, walled city measures 200 paces on each side. Gates are located at the centers of each side. If there is a tree 15 paces from the east gate, how far must a man travel out of the south gate to be able to see the tree?

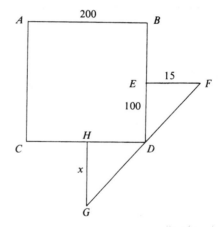

(b) A square, walled city of unknown dimensions has four gates, one at the center of each side. A tree stands 20 paces from the north gate. A man walks 14 paces southward from the south gate and then turns west and walks 1775 paces before he can see the tree. What are the dimensions of the city?

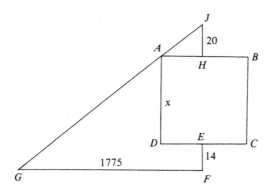

them with a string at eye level. The eastern pole is in a straight line with the northeastern and southeastern corners of the city. By moving northward a_1 feet from the eastern pole, the man's line of observation with the northwestern corner of the city intersects the string at a point b feet from its eastern end. He again goes north a_2 feet from the pole, until the northwestern corner of the city is in line with the western pole. What is the length of a side of the square city?

Show that the dimension of the city wall is

$$x = \frac{(a_2 - a_1)b}{\dfrac{(ba_2)}{d} - a_1}.$$

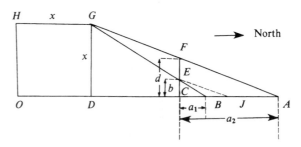

14. The two problems that follow are found in the seventh chapter of the *Nine Chapters;* solve them.

(a) A certain number of people are purchasing some chickens, jointly. If each person contributes 9 wen there is a surplus of 11 wen, and if each person contributes 6 wen there is a deficiency of 16 wen. Find the number of people and the price of the chickens. [*Hint:* If p is the price and n is the number of people, then $9n = p + 11$, and $6n = p - 16$.]

(b) There are 9 equal pieces of gold and 11 equal pieces of silver. The two lots weigh the same. If one piece is removed from each lot and put in the other, the lot containing mainly gold is found to weigh 13 ounces less than the lot containing mainly silver. Find the weight of each piece of gold and silver.

15. Consider the following problem, adapted in modern mathematical language from the *Sea Island Mathematical Manual:*

There is a square, walled city of unknown dimensions. A man erects two poles d feet apart in the east–west direction and joins

[*Hint:* The poles are located at points C and F. Construct EJ parallel to GA. From the similarity of triangles FCA and ECJ it follows that $a_1 + JB = JC = (AC)(EC)/FC$. Now, triangles GBA and EBJ are similar, as are triangles GDB and ECB, whence

$$\frac{AB}{JB} = \frac{BG}{BE} = \frac{GD}{EC}.$$

Thus $x = GD = (AB)(EC)/JB$.]

4.3   The Quadratic Equation

Book II of the *Elements* could be called a treatise on geometric algebra, because it is algebraic in substance but geometric in treatment. Algebraic problems are cast entirely in geometric language and solved by geometric methods. Lacking any adequate algebraic symbolism, Euclid found it necessary to represent numbers by line segments.

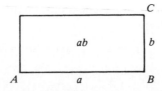

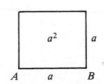

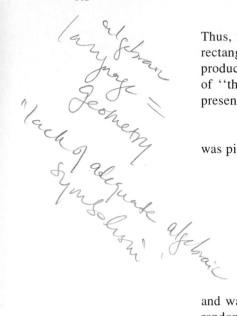

(handwritten margin notes: "Larry Sloman?", "algebraic", "geometry =", "lack of adequate symbolism", "algebraic")

Thus, the product ab (as we write it) of two numbers is thought of as the area of a rectangle with sides whose lengths are the two numbers a and b. Euclid referred to the product as the "rectangle contained by $AB = a$ and $BC = b$"; in place of a^2, he spoke of "the square on AB." Various algebraic identities, even complicated ones, were presented by Euclid in purely geometric form. For instance, the identity

$$(a + b)^2 = a^2 + 2ab + b^2$$

was pictured in terms of the diagram

	a	b
b	ab	b^2
a	a^2	ab

and was quaintly stated in Proposition 4 of Book II as: "If a straight line be cut at random (into two parts a and b), the square on the whole is equal to the square on the two parts and twice the rectangle contained by the parts."

By Euclid's time, Greek geometric algebra had reached a stage of development where it could be used to solve simple equations involving unknown quantities. The equations were given a geometric interpretation and solved by constructive methods; the answers to these constructions were line segments whose lengths corresponded to the unknown values. The linear equation $ax = bc$, for example, was viewed as an equality between areas ax and bc. Consequently, the Greeks would solve this equation by first constructing a rectangle $ABCD$ with sides $AB = b$ and $BC = c$ and then laying off $AE = a$ on the extension of AB. One produces the line segment ED through D to meet the extension of BC in a point F and completes the rectangle $EBFH$. It is clear that $KH = CF$ is the desired quantity x, for the rectangle $KDGH$ (or ax) is equal in area to the rectangle $ABCD$ (or bc); this can be seen by removing equal small triangles from the equal large triangles EHF and EBF.

(handwritten margin: "✳ in class.")

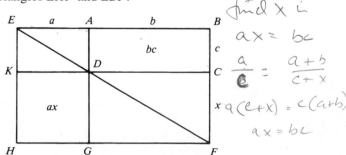

(handwritten notes at right of figure: "find x in", "$ax = bc$", "$\dfrac{a}{c} = \dfrac{a+b}{c+x}$", "$a(c+x) = c(a+b)$", "$ax = bc$")

When it came to quadratic equations, Euclid reduced them to the geometric equivalent of one of the forms

$$x(x + a) = b^2, \qquad x(x - a) = b^2, \qquad x(a - x) = b^2$$

which were then solved by applying theorems on areas. He was not the first to expound on this technique, for according to the *Commentary* of Proclus, "These things are ancient and the discovery of the Muse of the Pythagoreans."

The method of applying areas was fundamental in Euclid's work, and this was, strictly speaking, not so much a case of applying an area as of constructing a figure. In its simplest form, the process consists of constructing a rectangle of unknown height x so that its base lies on a given line segment AB, but in such a way that the area of the rectangle either exceeds a specified value R by the square x^2 or falls short of R by the square x^2.

 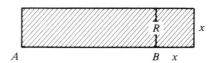

Let us see how Euclid actually used this method. Proposition 5 of Book II of the *Elements* was designed to teach the solution of the quadratic equation

$$x^2 + b^2 = ax.$$

The procedure was disguised by the peculiar geometric garb in which the Greeks were forced to clothe their results. We are told, to a given line segment $AB = a$ apply the rectangle $AQFG$ of known area b^2 in such a way that it shall fall short (from the rectangle of the entire segment AB) by a square figure, say x^2. In brief, this calls for constructing the figure herewith.

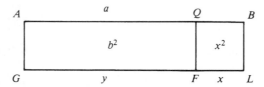

Suppose that the applied rectangle is erected on y as a base and the "deficient" square on x as a base; then the segment AB has length $x + y = a$, while the applied rectangle corresponds to $xy = b^2$. (One should recognize this as an Old Babylonian algebra problem.) Furthermore,

$$x^2 + b^2 = \text{area } ABLG = ax,$$

so that this "application of area" is the geometric equivalent of solving the equation $x^2 + b^2 = ax$.

How does one go about producing the square of area x^2 specified in the quadratic equation? The answer is to be found in Proposition 28 of Euclid's Book VI, a construction proposition, which states: Given a straight line AB, construct along this line a rectangle equal to a given area b^2, assuming that the rectangle falls short of AB by an amount filled out by another rectangle (or square). We are instructed to erect at P, the midpoint of line $AB = a$, a perpendicular PE equal in length to b; then with E as a center and radius $a/2$, we draw an arc cutting AB at the point Q. Then the line segment QB has length equal to the solution of the quadratic equation $x^2 + b^2 = ax$. For it can be proved that $(AQ)(QB) = (PE)^2$, and when QB is set equal to x, this amounts to the statement that $(a - x)x = b^2$.

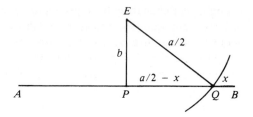

For geometric verification that $(AQ)(QB) = (PE)^2$, construct a rectangle $ABLG$ having width $BL = QB$ and complete the squares $PBDC$ and $QBLF$ on PB and QB as sides. The diagram that Euclid used for this purpose is shown. From various theorems on areas, it can be seen that

$$AQFG + HFKC = (APHG + PQFH) + HFKC$$

$$= PBLH + FLDK + HFKC$$

$$= (PB)^2.$$

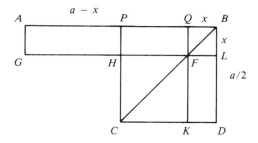

Because the rectangle $AQFG$ has area $(AQ)(QF) = (AQ)(QB)$ and $HFKC = (PQ)^2$, we get

$$(AQ)(QB) + (PQ)^2 = (PB)^2.$$

All of this is, of course, formulated in geometric language. As Euclid expressed it in Proposition 5 of Book II: If a straight line is cut into equal and unequal parts, the rectangle contained by the unequal parts of the whole together with the square on the straight line between the points of section is equal to the square on the half.

All that is needed to complete the argument is an appeal to the Pythagorean theorem. This leads directly to

$$(AQ)(QB) = (PB)^2 - (PQ)^2 = (PE)^2,$$

or with the appropriate substitutions, $(a - x)x = b^2$. The conclusion: $AB = a$ has been divided into two segments AQ and QB, and the length of the segment QB is the number x for which $x^2 + b^2 = ax$.

In the same spirit, Proposition 6 of Book II enables one to solve the quadratic equation

$$x^2 + ax = b^2,$$

or written another way, the equation $(x + a)x = b^2$. The method of solution by application of areas would be to say: To a given line segment $AB = a$, apply the rectangle $AQKF$ of known area b^2 in such a way that it will exceed (the rectangle on the whole segment AB) by a square figure, say x^2. This requires constructing a figure as shown. If the applied rectangle is erected upon the segment $AQ = y$ as a base, then

$$y - x = a, \qquad xy = b^2.$$

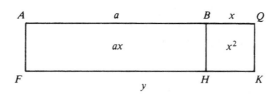

What Euclid wanted to teach is nothing more than the geometric solution of another Babylonian problem.

To get the rectangle $AQKF$, which is equal in area to b^2 and has one side containing the line AB, we use a construction Euclid described in his sixth Book (Proposition 29). At the endpoint B of $AB = a$, erect a perpendicular BE equal in length to b; then with

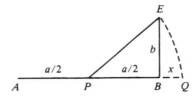

the midpoint P of AB as center and radius PE, draw an arc cutting the extension of AB at the point Q. We maintain that the rectangle with sides AQ and BQ will be equal to the square on BE; that is,

$$(AQ)(BQ) = (BE)^2.$$

The diagram Euclid provided for a demonstration is as shown, where $PQDC$ and $BQKH$ are squares described on PQ and BQ, respectively.

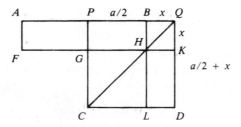

Regarding areas, it is evident that

$$AQKF + GHLC = (APGF + PQKG) + GHLC$$

$$= HKLD + PQKG + GHLC$$

$$= (PQ)^2.$$

Because the rectangle $AQKF$ has an area equal to $(AQ)(QK) = (AQ)(BQ)$ and $GHLC$ is a square of side PB, the foregoing equation can be expressed as

$$(AQ)(BQ) + (PB)^2 = (PQ)^2.$$

Euclid translated all this into ponderous geometric verbiage in Proposition 6: If a straight line is bisected and produced to any point, then the rectangle contained by the whole (with the added straight line) together with the square on half the line bisected is equal to the square on the straight line made up of the half and the part added.

At this point, the Pythagorean theorem comes to the rescue again, for the last-written equation reduces to

$$(AQ)(BQ) = (PQ)^2 - (PB)^2 = (BE)^2 = b^2.$$

We have only to put $AB = a$ and $BQ = x$ to see that the length of the segment BQ is the value required to satisfy the equation

$$(x + a)x = b^2.$$

The special case for which $a = b$ provides us with the opportunity to introduce what the celebrated astronomer Johannes Kepler called "one of the two Jewels of Geometry" (the second is the theorem of Pythagoras). For the construction used in solving the quadratic $(x + a)x = a^2$ amounts to dividing a given line segment AB into what is called the "golden section." Translated into mathematical language, the golden section means that the segment $AB = a$ is cut at a point C so that the whole segment is in the same ratio to the larger part $CB = x$ as CB is to the other part, $AC = a - x$. Stated otherwise, it produces the relation

$$\frac{a}{x} = \frac{x}{a - x}, \qquad x > a - x.$$

This, in turn, leads to the quadratic equation $x(x + a) = a^2$ already mentioned, the positive root of which is

$$x = \tfrac{1}{2}a(\sqrt{5} - 1).$$

When $a = 1$, the value $x = \tfrac{1}{2}(\sqrt{5} - 1)$ is the reciprocal of the "golden ratio"—that is, 0.6180339. . . .

Let us review Euclid's construction for the golden section of a line segment $AB = a$. At the endpoint B of AB, erect a perpendicular BE equal in length to a; with the midpoint P of AB as center and radius PE, draw an arc cutting the extension of AB at the point Q. Take B as center and radius BQ, and draw an arc meeting AB at C. The point C divides the segment AB in the ratio sought.

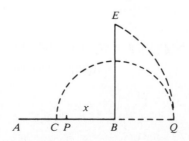

Much of the history of classical mathematics could be written around the idea of the golden section. It appears again in Book IV of the *Elements* with regard to the inscription (with the two traditional instruments, straightedge and compass) of certain regular polygons in a circle. You may recall that a regular polygon is a convex polygon with all its sides equal in length and with equal angles at each vertex. When a regular polygon of n sides is inscribed in a circle, the central angle formed by the radii drawn to two consecutive vertices has measure $360°/n$. The Greeks were able to solve the problem of inscribing in a circle a regular polygon of an assigned number of sides when the number was 3, 4, 5, 6, 15, or twice the number of any inscribable polygon. The first case in which they failed concerned a regular polygon of 7 sides.

The construction of a regular pentagon (polygon of 5 sides), the division of a circle into 5 equal parts and the construction of an angle equal to $360°/5 = 72°$ are equivalent problems. The solution is taught in Propositions 10 and 11 of Book IV; Euclid relied on forming an isosceles triangle having each of the base angles equal to twice the remaining angle. This made the summit angle $36°$ and each of the angles at the base equal to $72°$, thereby permitting the construction of both the regular pentagon and regular decagon (polygon of 10 sides).

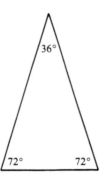

In following the Greek method for constructing regular polygons of 5 and 10 sides, one would proceed as follows. Pick an arbitrary line segment $AB = a$ for the radius of a circle and solve the quadratic equation $x(x + a) = a^2$ to get a line segment whose length is $x = \frac{1}{2}a(\sqrt{5} - 1)$. This is equivalent to cutting AB in golden section by a point C and letting $x = AC$. As we shall presently see, x will be the side of an inscribed decagon, or what amounts to the same thing, x can be stepped off as a chord in the circle of radius $AB = a$ exactly ten times. To confirm this, let us construct the isosceles triangle ABD having as its sides two radii $AB = AD$ of the circle, and as its base BD a segment of length x. Also lay off the segment CD. By virtue of the condition $a/x = x/(a - x)$, we have

$$\frac{AB}{AC} = \frac{AC}{CB}$$

or, since $AB = AD$ and $AC = DB$,

$$\frac{AD}{DB} = \frac{DB}{CB}.$$

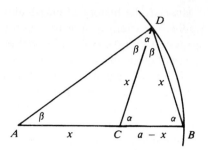

Another point to notice is that $\angle ADB = \angle ABD$, as base angles of an isosceles triangle; these are marked α in the figure. The upshot of all this is that triangles ADB and CBD are similar, for they have two pairs of corresponding sides proportional and the included angles equal. Then $\angle DAB$ equals $\angle CDB$, because these are corresponding angles in similar triangles (they are marked β). A little calculating with angles tells us that $\angle DCB = \angle DBC$, whence $CD = DB = x$. Indeed, more is true: $\angle ADC$ equals $\angle BAD$, since each is a base angle of the isosceles triangle ACD.

With the routine work out of the way, we are now ready to sum up. Because the sum of the angles of triangle DAB must equal two right angles, it can be concluded that

$$180° = \angle DAB + \angle ADB + \angle DBA$$
$$= \beta + \alpha + \alpha$$
$$= \beta + 2\beta + 2\beta$$

and as a result,

$$\beta = \frac{180}{5} = 36°.$$

Segment BD subtends a central angle of 36°, so it is the side of a regular inscribed decagon and will go ten times as a chord within the circle of radius AB. The regular pentagon is drawn by selecting every other point as a vertex.

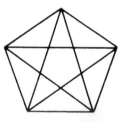

The regular pentagon had a particular appeal to the early Pythagoreans, because its diagonals formed the star pentagram, the sign of recognition of the society. Although it is highly likely that Euclid's method of constructing a pentagon was known to Pythagoras or his immediate disciples, no statement about the extent of their mathematical knowledge can be other than tentative. What is known is that Proclus, whose works inform us concerning the history of Greek geometry, wrote that Eudoxus (circa 370 B.C.) greatly added to the number of theorems that Plato originated concerning the "section," meaning the golden section. This is the first reference we have of the name for such a division of a line segment.

Having taken the opportunity to digress a little, let us now return to the main theme of the section. We saw earlier how the realization that certain geometric magnitudes are not expressible by whole numbers shook the foundations of the Pythagorean doctrine, which maintained that "everything is number." It meant that such a simple equation as $x^2 = 2$ had no solution in their domain of numbers (rational). The dilemma was resolved by putting algebra in a geometric dress; numbers came to be represented by line segments and geometric constructions were substituted for algebraic operations, so that products, for instance, corresponded to rectangular areas. Once the greater applicability of geometry was realized, geometric argument became the basis for all rigorous mathematics.

The geometric algebra, a theory of line segments and areas, of Book II of Euclid's *Elements* was the culmination of the Greek attempt to cope with the irrational through geometry. The book consists of propositions that appear on the surface to belong to geometry, but have content that is entirely algebraic. In particular, the treatment of quadratic problems is reduced to one of the equations

$$x(x + a) = b^2, \qquad x(a - x) = b^2, \qquad x(x - a) = b^2,$$

which are then solved geometrically by means of "application of area," so that the roots appear as line segments. Although the individual solutions by area are awkward, involving as they do intricate constructions of plane figures, they follow exactly the same pattern as the earlier Babylonian algebraic calculations. The geometric algebra of the *Elements* is nothing more than a transposition of an inherited body of Babylonian procedures to geometric form. The chief difference is that, where Babylonian calculations only give a solution to quadratic equations if the square root can be found exactly (otherwise, a convenient approximation is accepted), Greek geometric algebra always gives an answer—a line segment is produced that may very well represent an irrational number.

By embodying all mathematics except the theory of whole numbers in geometry, the Greeks swept the difficulties of the irrational under the rug, so to speak. The cumbersome techniques of geometric algebra allowed the Greeks to solve quadratic equations, but without assuming the existence of irrational numbers. This essentially alien garb, with all its clumsy verbiage and overwhelming diagrams, retarded progress in algebra for many centuries. For although linear and quadratic equations can be expressed clearly in the language of geometric algebra, higher-degree equations are effectively precluded from consideration. It is paradoxical that a religious controversy in the minds of the Pythagoreans, the worshipers of mathematics, should have had such a profoundly deleterious effect on its growth.

Greek geometric algebra had to await a translation into a formal symbolic language before a satisfactory divorce of algebraic calculation from geometry could take place. Historically, the systematic attempt to "symbolize" arithmetic and algebraic operations is a relatively recent phenomenon, the decisive contribution of sixteenth-century mathematics. By the 1500s, negative rational numbers and zero were in regular use in practical calculations, but mathematicians still lacked a clear conception of irrational numbers. The German algebraist Michael Stifel (1486–1567), for instance, in his *Arithmetica Integra* of 1544, argued:

> We are moved and compelled to assert that they truly are numbers, compelled that is, by the results which follow from their use. On the other hand . . . just as an infinite number is not a number, so an irrational number is not a true number, but lies hidden in some sort of cloud of infinity.

Doubt about the soundness of irrational numbers was expressed in the stigma *numerus surdus* (''inaudible number''), the phrase coming from the word *surdus,* ''deaf or mute''—a Latin translation of an Arabic translation of the Greek *alogos* (''irrational number''). Mathematicians such as Stifel pragmatically manipulated irrational numbers uncritically, without seriously questioning their precise meaning or nature, until the late 1800s. Then the question of the logical structure of the real number system was faced squarely. In an epoch-making essay entitled *Continuity and Irrational Numbers* (1872), Richard Dedekind finally established the theory of irrational numbers on a logical foundation, free from the extraneous influence of geometry.

Western Europeans first learned their algebra from the works of Muhammed ibn Mûsâ al-Khowârizmî (circa 820), an astronomer and mathematician of Baghdad who lived in the days portrayed in *The Arabian Nights Tales.* Although we know nothing of al-Khowârizmî's life, he was most probably one of the members of the Baghdad ''House of Wisdom,'' a kind of academy—comparable with the Museum at Alexandria—founded for promoting science. The name ''algebra'' is the European corruption of *al-jabr,* part of the title of al-Khowârizmî's treatise *Hisâb al-jabr w'al muqâbalah.* Apparently the title means ''the science of reunion and reduction.'' The words refer to the two principal operations the Arabs used in solving equations. ''Reunion'' refers to the transference of negative terms from one side of the equation to the other and ''reduction'' to the combination of like terms on the same side into a single term, or the cancellation of like terms on opposite sides of the equation. For example, in the equation (in modern notation) $6x^2 - 4x + 1 = 5x^2 + 3$, ''reunion'' gives

$$6x^2 + 1 = 5x^2 + 4x + 3,$$

and from ''reduction,''

$$x^2 = 4x + 2.$$

In the twelfth century, the book was translated into Latin under the title *Liber Algebrae et Almucabola,* which ultimately gave the name to that part of mathematics dealing with the solution of equations. The influence of al-Khowârizmî is also reflected in the fact that *algorism* (or *algorithm*), a Latin corruption of his name, for a long time meant the art of computing with Hindu-Arabic numerals. Today it is used for any method of calculation according to a set of established rules.

The traditional explanation of the Arabic word *jabr* is that it means ''the setting of a broken bone'' (hence, ''restoring'' or ''reunion''). When the Moors reached Spain in the Middle Ages, they introduced the word *algebra,* and there, in the form *algebrista,* it came to mean ''a bonesetter.'' At one time in Spain, it was not uncommon to see a sign reading *Algebrista y Sangradoe* (''bonesetting and bloodletting'') over the entrance to a barbershop; for until recent times, barbers performed many of the less skilled medical services as a sideline to their regular business.

In speaking of al-Khowârizmî, we do not mean that he personally was the inventor of algebra, for no branch of mathematics sprang up, fully grown, through the work of one person. He was only the representative of an old Persian school who preserved its methods for posterity through his books. This early Arabic algebra was still at the primitive rhetorical stage—a phase characterized by the complete lack of mathematical symbols, in which the calculations were carried out by means of words (even numbers were written out in words rather than presented as symbols). Algebraic rules of procedure were proclaimed as if they were divine revelations, which the reader was to accept and follow as a true believer. Whenever reasons and proofs were given, they were presented as geometric demonstrations; the Arabs, inspired by Euclid's *Elements*, seemed to believe that an argument had to be geometric to be convincing.

In dealing with quadratic equations, al-Khowârizmî divided them into three fundamental types:

(1) $$x^2 + ax = b,$$

(2) $$x^2 + b = ax,$$

(3) $$x^2 = ax + b,$$

with only positive coefficients admitted. (Negative quantities standing alone were still not accepted by these Arabic mathematicians.) All problems were reduced to these standard types and solved according to a few general rules.

Al-Khowârizmî's geometric demonstration of the correctness of his algebraic rules for solving quadratic equations may be illustrated by his discussion of the equation $x^2 + 10x = 39$, a problem he solved by two different methods. This equation reappears frequently in later Arab and Christian texts, running ''like a thread of gold through the algebras of several centuries.'' The first geometrical solution is explained as follows. Given $x^2 + 10x = 39$, construct a square *ABCD* having sides of length x to represent x^2. Now one has to add $10x$ to the x^2. This is accomplished by dividing $10x$ into four parts, each part representing the area $(\frac{10}{4})x$ as a rectangle, and then applying these four rectangles to the four sides of the square.

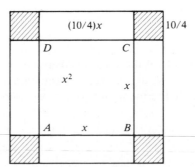

This produces a figure representing $x^2 + 10x = x^2 + 4(\frac{10}{4}x)$. To make the figure into a larger square of sides $x + \frac{10}{2}$, we must add four small squares at the corners, each of which has an area equal to $(\frac{10}{4})^2$. That is, to "complete" the square, we add $4(\frac{10}{4})^2 = (\frac{10}{2})^2$. Then we have

$$\left(x + \frac{10}{2}\right)^2 = (x^2 + 10x) + 4\left(\frac{10}{4}\right)^2 = 39 + \left(\frac{10}{2}\right)^2 = 39 + 25 = 64.$$

Hence, the side of the square must be $x + \frac{10}{2} = 8$, from which it is found that $x = 3$.

In the general setting, the quadratic equation $x^2 + px = q$ is solved by this method of completion of squares by adding four squares, each of area $(p/4)^2$, to the figure representing $x^2 + px$, to get

$$\left(x + \frac{p}{2}\right)^2 = x^2 + px + 4(p/4)^2 = q + \left(\frac{p}{2}\right)^2.$$

This leads to the solution

$$x = \sqrt{\left(\frac{p}{2}\right)^2 + q} - \frac{p}{2}.$$

For al-Khowârizmî's second method of solving $x^2 + 10x = 39$, the starting point is a figure composed of a square of side x (and area x^2) and two rectangles, each having length x and width $\frac{10}{2}$. Because the area of each rectangle is $x(\frac{10}{2})$, the area of the entire figure is $x^2 + 2(\frac{10}{2})x$. To complete this figure so as to form a square, it is necessary to add a new square of area $(\frac{10}{2})^2$. The area of the completed square is $(x + \frac{10}{2})^2$, and consequently

$$\left(x + \frac{10}{2}\right)^2 = x^2 + 2\left(\frac{10}{2}\right)x + \left(\frac{10}{2}\right)^2$$

$$= 64.$$

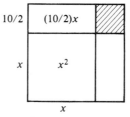

This side of the square is then $x + \frac{10}{2} = 8$, whence the value of the unknown is $x = 3$.

In solving the general equation $x^2 + px = q$ in this manner, a square of side $p/2$ is added to the figure, which represents $x^2 + 2(p/2)x$, thereby making

$$\left(x + \frac{p}{2}\right)^2 = x^2 + 2\left(\frac{p}{2}\right)x + \left(\frac{p}{2}\right)^2$$

$$= q + \left(\frac{p}{2}\right)^2.$$

This yields, as before, the solution

$$x = \sqrt{q + \left(\frac{p}{2}\right)^2} - \frac{p}{2}.$$

In looking at the solution of the quadratic equation, we have seen that geometry was undisputed mistress in Euclid's *Elements;* the algebraic content was clothed in geometric language. With the work of al-Khowârizmî, however, we see the beginning of doing away with this limitation, as geometric explanations come to be auxiliary to a newly predominant algebraic reasoning. The old, ingenious Babylonian tricks and devices for solving individual problems are finally seen as part of al-Khowârizmî's systematic reduction of the quadratics to their standard types, with each type solved according to its own rules. In the work of al-Khowârizmî, we discern a sure progress and an evolution from ancient mathematical practices to improved and more general methods.

The Arabic mathematicians of this period, besides transmitting Hellenistic learning to the West, made lasting contributions of their own. Indeed they revised and reconstructed many fundamental ideas in mathematics. The Arabs recognized, for example, irrational roots of quadratic equations, although these had been disregarded by the Greeks.

Although the Arabs recognized the existence of two solutions of a quadratic equation—something never done by Euclid or the Babylonians—they listed only the positive ones. They did not perceive the reality of negative solutions to an equation. The very idea of a negative root implies the acknowledgment of negative numbers as independent entities having the same mathematical status as positive ones. This understanding is of more recent origins; in the works of al-Khowârizmî and the other Arab algebraists, negative numbers are consistently avoided. The existence and validity of negative as well as positive roots was first affirmed by the Hindu mathematician Bhaskara (born 1114 A.D.). Europeans have admitted them only since the sixteenth or seventeenth century.

Abû Kâmil (circa 850–930), often called "The Reckoner from Egypt," was the second of the great Arabic writers on algebra. Little is known of his life and activities other than that he is apparently of Egyptian descent and wrote in the period following al-Khowârizmî. His *Book of Algebra* (*Kitâb fil-jabr w' al muqâbalah*), a title commonly used by early Muslim algebraists, is essentially a commentary on and elaboration of al-Khowârizmî's work; in part for that reason and in part for its own merit, the book enjoyed widespread popularity in the Muslim world. A much more extensive treatise on algebra than that of al-Khowârizmî, Abû Kâmil's *Algebra* contains a total of 69 problems compared with the 40 of his famous predecessor.

As would be expected with a commentary, Abû Kâmil carried over intact many of the problems that al-Khowârizmî had explained. At the same time, he did not hesitate to add further methods of solution to those presented by the earlier author. This may be seen in Problem 8 of the *Algebra.* As expressed by Abû Kâmil, it reads: "Divide 10 into two parts in such a way that when each of the two parts is divided by the other their sum will be 4½." In modern notation, the problem consists of finding two numbers that satisfy the equations

$$x + y = 10, \qquad \frac{x}{y} + \frac{y}{x} = 4\frac{1}{4}.$$

The algebraic identity

$$\frac{x}{y} + \frac{y}{x} = \frac{x^2 + y^2}{xy}$$

is used to convert the second of these to

$$x^2 + y^2 = 4\frac{1}{4}xy.$$

Abû Kâmil first solves the problem along the lines of al-Khowârizmî. That is, he puts $y = 10 - x$ into the previous equation to obtain a standard type of quadratic equation, namely

$$6\frac{1}{4}x^2 + 100 = 62\frac{1}{2}x,$$

with solution $x = 2$; hence, the corresponding value of y is 8. Abû Kâmil then presents a method of solution of his own, one which involves the old Babylonian procedure of introducing a new unknown quantity z by letting

$$x = 5 - z, \quad y = 5 + z.$$

When these values are substituted, the equation $x^2 + y^2 = 4\frac{1}{4}xy$ becomes

$$50 + 2z^2 = 4\frac{1}{4}(15 - z^2)$$

which yields $z^2 = 9$. This gives $z = 3$ and, in turn, the numbers sought are

$$x = 5 - 3 = 2, \quad y = 5 + 3 = 8.$$

Abû Kâmil developed a calculus of radicals that is quite distinctive. He managed the addition and subtraction of square roots, without using our symbols, by means of the equalities

$$\sqrt{a} \pm \sqrt{b} = \sqrt{a + b \pm 2\sqrt{ab}}.$$

As with al-Khowârizmî's work, the *Algebra* is entirely rhetorical, with all computations (often quite complicated) described in words; the only notation in the text is of integers. For instance, the rule for subtracting the square root of 4 from the square root of 9 is expressed as

> If you wish to subtract the root of 4 from the root of 9 until what remains of the root of 9 is a root of one number, then you add 9 to 4 to give 13. Retain it. Then multiply 9 by 4 to give 36. Take 2 of its roots to give 12. Subtract it from the 13 that was retained. One remains. The root is 1. It is the root of 9 less the root of 4.

This is just a verbal description of what we would write as

$$\sqrt{9} - \sqrt{4} = \sqrt{9 + 4 - 2\sqrt{9 \cdot 4}} = 1.$$

The major advance of Abû Kâmil over earlier writers is in his use of irrational coefficients in indeterminate equations. A case in point is Problem 53 of the *Algebra*. In it he asks for a number such that, if the square root of 3 is added to it and the square root of 2 is added to it, then the product of the two sums will be 20. This, which today would be written as

$$(x + \sqrt{3})(x + \sqrt{2}) = 20,$$

leads to the quadratic equation

$$x^2 + \sqrt{6} + \sqrt{3x^2} + \sqrt{2x^2} = 20.$$

Abû Kâmil gives the correct value

$$x = \sqrt{22\frac{1}{4} - \sqrt{6} + \sqrt{1\frac{1}{2}} - \sqrt{\frac{3}{4}} - \sqrt{\frac{1}{2}}}.$$

The introduction of irrational solutions for some quadratics is another point of departure from the foundation work of al-Khowârizmî.

Abû Kâmil's *Algebra* holds an especially important place in the development of mathematics in the West through its influence on the works of the Italian Leonardo of Pisa, better known as Fibonacci. When Fibonacci wrote his *Liber Abaci* (1202), he drew heavily on the Arabic author, reproducing some 29 problems from the *Algebra* with little or no change. Although Fibonacci was a borrower, he shouldn't be regarded as a plagiarist; Abû Kâmil's methods were so well known at the time that any mathematician felt free to use his results as common property.

4.3   Problems

1. Solve the following quadratic equations by the Arabic method of completing the square.

   (a) $x^2 + 8x = 9$.
   (b) $x^2 + 10x = 144$.
   (c) $x^2 + 12x = 64$.
   (d) $3x^2 + 10x = 32$. [*Hint:* Multiply both sides by 3 and let $y = 3x$.]

2. The French mathematician François Vieta (1540–1603) solved the quadratic equation $x^2 + ax = b$ by substituting $x = y - a/2$. This produces a quadratic in y in which the first-degree term is missing. Use Vieta's method to solve the quadratic equations in Problem 1.

3. To solve the equation $x^2 + ax = b^2$ geometrically, René Descartes (1596–1650) would have used the method as described. Draw a line segment AB of

length b and at A erect a perpendicular AC of length $a/2$. With C as center, construct a circle of radius $a/2$ and draw a line through B and C, intersecting the circle at points D and E. Prove that the length of the segment BE is the value of x that satisfies $x^2 + ax = b^2$.

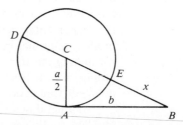

4. Use Abû Kâmil's formula for the difference of two square roots to show that $\sqrt{18} - \sqrt{8} = \sqrt{2}$; then express $\sqrt{18} + \sqrt{8}$ as a single square root.

The next five problems are from the *Algebra* of Abû Kâmil.

5. *Problem 15*. 10 dinar is divided equally among a group of men so that when 6 more men are added to their number and 40 is divided equally among them, then each receives as much as he did previously. Find the original number of men.

6. *Problem 19*. Given that 3 roots of a number plus 4 roots of the difference between the number and the 3 roots equals 20, find the number. [*Hint:* In the equation $3\sqrt{x} + 4\sqrt{x - 3\sqrt{x}} = 20$, let $x = y^2$ to obtain $20 - 3y = 4\sqrt{y^2 - 3y}$; then square both sides.]

7. *Problem 26*. Divide 10 into two parts in such a way that when a certain one of these parts is divided by

the other and the fraction is multiplied by its numerator, the result gives 9.

8. *Problem 54*. Find a number such that if 7 is added to it and the sum multiplied by the root of 3 times the number, then the result is 10 times the number. [*Hint:* To solve the equation $(x + 7)\sqrt{3x} = 10x$, put $x = \frac{1}{3}y^2$; this yields $y^2 + 21 = 10y$.]

9. *Problem 66*. Divide 10 into two parts in such a way that when 50 is divided by one part and 40 by the other, and then the fractions are multiplied, 125 will result. [*Hint:* Algebraically, the equations are $x + y = 10$ and $\frac{50}{x} \cdot \frac{40}{y} = 125$.]

4.4   Euclid's Number Theory

Although Euclid's great work is entitled *Elements of Geometry,* its subject matter extends far beyond what we would now regard as high-school geometry. Three of the books of the *Elements* (namely, VII, VIII, and IX), containing a total of 102 propositions, are devoted to arithmetic in the Greek sense. That is to say, they deal mainly with the nature and properties of what are called the "natural numbers" or the "positive integers." Euclid was building on earlier foundations, because much of the substance of these arithmetical books can be traced to the Pythagoreans. Again he must be accorded the credit of having imposed a logical order on the whole. Many of the results had been long known but not always rigorously proved. Any earlier works on the theory of numbers that may have been written are no longer extant, so that it is impossible to say which proofs were supplied to Euclid and which were his own discoveries.

Euclid was particularly interested in questions pertaining to divisibility, and he properly emphasized the function of the prime numbers. In Book IX, the last of the books on number theory, many significant theorems can be found. Of these the most celebrated is Proposition 20, which reads, "Prime numbers are more than any assigned multitude of prime numbers." What we have here is the famous assertion that there are infinitely many primes. Proposition 14 contains the essence of what today is called the fundamental theorem of arithmetic—any integer greater than 1 can be written as a product of primes in exactly one way. Proposition 35 gives a derivation of the formula for finding the sum of numbers in geometric progression; and the following, and last, proposition in Book IX establishes a criterion for forming "perfect numbers" (the nomenclature is no doubt Pythagorean).

As Euclid possessed no algebraic symbolism, he was forced to represent arbitrary numbers by line segments marked by one letter, or by two letters placed at the ends of the segment. His proofs, which were given in a verbal form, as opposed to the modern symbolic form, did not make use of geometry. In Books VII, VIII, and IX, no geometrical figures were used for indeed none were necessary. Although Euclid may have

Euclid's proof that the base angles of an isosceles triangle are equal. From Isaac Barrow's edition of Euclid's Elements *(1665). (From* An Introduction to the History of Mathematics, 6/E, *copyright © 1990 by Saunders College Publishing, a division of Holt, Rinehart and Winston, Inc., reprinted by permission of the publisher.)*

adopted the language "plane numbers" and "solid numbers" to refer to products of two and three numbers, these were represented throughout the text not by rectangles or volumes but by segments.

Book VII begins with a variety of definitions that serve all three arithmetical books, including those of prime and composite numbers. Where Euclid phrased these in terms of line segments, we shall use modern notation and wording.

Definition

An integer b is said to be divisible by an integer a ≠ 0, in symbols a|b, if there exists some integer c such that b = ac. One writes a∤b to indicate that b is not divisible by a.

Thus, 39 is divisible by 13, since 39 = 13 · 3. However, 10 is not divisible by 3; for there is no integer c that makes the statement 10 = 3c true.

There is other language for expressing the divisibility relation $a|b$. We might say that *a divides b, a* is a *divisor* of *b*, that *a* is a *factor* of *b*, or that *b* is a *multiple* of *a*. Notice too that in the definition given there is a restriction on the divisor *a;* whenever the notation $a|b$ is used, the understanding is that *a* is different from zero. Because Euclid always represented numbers by line segments, he did not use the phrases "is a divisor of" or "is a multiple of." Instead, he replaced these by "measures" and "is measured by," respectively. For Euclid, a number *b* was measured by another number *a* if *b = ac* for some third number *c*.

Euclid, in representing numbers by line segments, would never have considered a negative number. And in the modern view, the divisors of an integer always occur in pairs. If *a* is a divisor of *b*, then so is *−a;* indeed, *b = ac* implies that *b = (−a)(−c)*. To find all the divisors of a given integer, it suffices to obtain the positive divisors and then adjoin to them their negatives. For this reason we shall usually limit ourselves—as Euclid did for his own reasons—to positive divisors.

It will be helpful to list a number of simple facts involving the concept of divisor. For integers *a, b, c,* the following hold:

1. $a|0$, $1|a$, $a|a$.
2. $a|1$ if and only if $a = ±1$.
3. If $a|b$ and $c|d$, then $ac|bd$.
4. If $a|b$ and $b|c$, then $a|c$.
5. $a|b$ and $b|a$ if and only if $a = ±b$.
6. If $a|b$ and $a|c$, then $a|(bx + cy)$ for any integers x and y.

We shall establish assertion (6), leaving the verification of the other parts as an exercise. Now the relations $a|b$ and $a|c$ ensure that there exist integers r and s satisfying $b = ar$ and $c = as$. But then

$$bx + cy = arx + asy = a(rx + sy)$$

whatever the choice of x and y. Because $rx + sy$ is itself an integer, the last-written equation says simply that $a|(bx + cy)$, as desired.

It is convenient to call an expression of the form $bx + cy$, where x and y are integers, a linear combination of b and c. Note that $b + c$ and $b − c$ are both linear combinations of b and c (in the first instance take $x = y = 1$; in the second let $x = 1$, $y = −1$). Hence, as a special case of (6), we see that if $a|b$ and $a|c$, then $a|(b + c)$ and $a|(b − c)$.

Classifying positive integers greater than 1 as either prime or composite is very important in number theory; because of the fundamental theorem of arithmetic, many properties of integers can be deduced from properties of primes. Fact (1) tells us that any integer $a > 1$ is divisible by ±1 and by ±a, divisors that are frequently named improper divisors. If they exhaust the divisors of a, then a is said to be a prime number. Put somewhat differently is this definition.

Definition

An integer p >1 is called a prime number, *or simply a* prime, *if its only positive divisors are* 1 *and p. An integer that is greater than* 1 *and not a prime is termed* composite.

Among the first ten positive integers, 2, 3, 5, and 7 are all primes, whereas 4, 6, 8, 9, and 10 are composite numbers. Note that the integer 2 is the only even prime, and according to our definition, the number 1 is distinguished in the sense of being neither prime nor composite. To illustrate Euclid's language, let us record his way of defining a prime: ''A prime number is that which is measured by a unit (that is, by 1) alone.''

It is often of interest to find out whether two given numbers have any factors in common, and if so which ones.

Definition

If a and b are arbitrary integers, then an integer d is said to be a common divisor *of a and b if we have both d|a and d|b.*

Because 1 divides each integer, 1 is a common divisor of a and b. Hence, any pair of integers possesses at least one positive common divisor. In fact, if either a and b is nonzero, then a finite number of positive common divisors exist. Among these, there is one that is the largest, called the *greatest common divisor of a and b,* and denoted by the symbol gcd (a, b).

> **Example.** The positive divisors of 12 are 1, 2, 3, 4, 6, and 12, and the positive divisors of 30 are 1, 2, 3, 5, 6, 10, 15, and 30; hence, the positive common divisors of 12 and 30 are 1, 2, 3, 6. Because 6 is the largest of these integers, it follows that gcd $(12, 30) = 6$.

To obtain the greatest common divisor of two integers, we could always proceed as in the last example by listing all their positive divisors and picking out the largest one common to each; but this is cumbersome for large numbers. A more efficient process is given early in the seventh book of the *Elements*. Although there is historical evidence that this method predates Euclid by at least a century, it today goes under the name ''Euclidean algorithm.''

Euclid's procedure relies on a result so basic that it is often taken for granted: the division theorem. Roughly, the theorem asserts that an integer a can be divided by a positive integer b in such a way that the remainder is smaller than b. An exact statement of this fact follows.

DIVISION THEOREM

For integers a and b, with b > 0, there exist unique integers q and r satisfying

$$a = qb + r, \qquad 0 \le r < b.$$

The integers q and r are called the *quotient* and the *remainder* in the division of a by b. We accept the division theorem without proof, noting that b is a divisor of a if and only if the remainder r in the division of a by b is zero.

In examining the division theorem, let us take $b = 7$. Then, for the choices $a = 1, -2, 28$, and -59, one gets the representations

$$1 = 0 \cdot 7 + 1$$

$$-2 = (-1) \cdot 7 + 5$$

$$28 = 4 \cdot 7 + 0$$

$$-59 = (-9) \cdot 7 + 4.$$

The aim is to focus attention not so much on the division theorem as on its use in finding greatest common divisors. To this end, let a and b be two integers whose greatest common divisor is desired; there is no harm in assuming that $a \geq b > 0$. The first step is to apply the division theorem to a and b, to get

$$a = q_1 b + r_1, \qquad 0 \leq r_1 < b.$$

If it happens that $r_1 = 0$, then $b \mid a$, and since $b \mid b$ also, gcd $(a, b) = b$. If $r_1 \neq 0$, divide b by r_1 to produce integers q_2 and r_2 satisfying

$$b = q_2 r_1 + r_2, \qquad 0 \leq r_2 < r_1.$$

If $r_2 = 0$, then we stop; otherwise, we go on as before, dividing r_1 by r_2, to obtain

$$r_1 = q_3 r_2 + r_3, \qquad 0 \leq r_3 < r_2.$$

This division continues until some zero remainder appears, say at the $(n + 1)$st stage, at which r_{n-1} is divided by r_n. A zero remainder must occur sooner or later, since the decreasing sequence $b > r_1 > r_2 > \cdots \geq 0$ cannot contain more than b integers.

The result is the following system of equations:

$$a = q_1 b + r_1, \qquad\qquad 0 < r_1 < b,$$

$$b = q_2 r_1 + r_2, \qquad\qquad 0 < r_2 < r_1,$$

$$r_1 = q_3 r_2 + r_3, \qquad\qquad 0 < r_3 < r_2,$$

$$\vdots$$

$$r_{n-3} = q_{n-1} r_{n-2} + r_{n-1}, \quad 0 < r_{n-1} < r_{n-2},$$

$$r_{n-2} = q_n r_{n-1} + r_n, \qquad 0 < r_n < r_{n-1},$$

$$r_{n-1} = q_{n+1} r_n + 0.$$

We argue that r_n, the last nonzero remainder that appears in this algorithm, is equal to gcd (a, b). Now $r_n \mid r_{n-1}$ by the last equation of the above system. From the equation immediately preceding, it follows that $r_n \mid r_{n-2}$; for r_{n-2} is a linear combination of r_n and r_{n-1}, both of which are divisible by r_n. Working backward through these equations, we find that r_n divides each of the preceding remainders r_k. Finally $r_n \mid b$, and from the first equation $a = q_1 b + r_1$, we get $r_n \mid a$. Therefore, r_n is a positive common divisor of a and b.

Next, suppose that d is an arbitrary positive common divisor of a and b. The first of the equations tells us that $d|r_1$. It is clear, in going down the list of the equations, that d divides $r_2, r_3, \ldots$ and ultimately r_n also. But $d|r_n$, with d and r_n both positive integers, implies that $d \leq r_n$. In consequence, r_n is the largest of the positive common divisors of a and b; that is, gcd $(a, b) = r_n$.

There is another important point that deserves mention. Namely, gcd (a, b) can always be expressed as a linear combination of the integers a and b. To verify this, we fall back on the Euclidean algorithm. Starting with the next-to-last equation arising from the algorithm, we write r_n as

$$r_n = r_{n-2} - q_n r_{n-1},$$

a linear combination of r_{n-1} and r_{n-2}. Now solve the preceding equation in the algorithm for r_{n-1} and substitute to

$$r_n = r_{n-2} - q_n(r_{n-3} - q_{n-1}r_{n-2})$$

$$= (1 + q_n q_{n-1})r_{n-2} + (-q_n)r_{n-3}.$$

This eliminates r_{n-1} and represents r_n as a linear combination of r_{n-2} and r_{n-3}. Continuing backward through the system of equations, we successively eliminate the remainders $r_{n-1}, r_{n-2}, \ldots, r_2, r_1$ until a stage is reached at which $r_n = $ gcd (a, b) is expressed as a linear combination of a and b.

To summarize, what we have obtained is the following:

THEOREM

For integers a and b, of which both are not zero, there exist integers x and y such that gcd $(a, b) = ax + by$.

Example. Let us see how the Euclidean algorithm works in a concrete case by calculating, say, gcd (12,378, 3054). The appropriate applications of the division algorithm produce the equations

$$12{,}378 = 4 \cdot 3054 + 162$$

$$3054 = 18 \cdot 162 + 138$$

$$162 = 1 \cdot 138 + 24$$

$$138 = 5 \cdot 24 + 18$$

$$24 = 1 \cdot 18 + 6$$

$$18 = 3 \cdot 6 + 0.$$

Our previous discussion tells us that the last nonzero remainder appearing above, namely the integer 6, is the greatest common divisor of 12,378 and 3054:

$$6 = \text{gcd } (12{,}378, 3054).$$

To represent 6 as a linear combination of the integers 12,378 and 3054, we start with the next-to-last of the displayed equations and successively eliminate the remainders 18, 24, 138, and 162:

$$6 = 24 - 18$$
$$= 24 - (138 - 5 \cdot 24)$$
$$= 6 \cdot 24 - 138$$
$$= 6(162 - 138) - 138$$
$$= 6 \cdot 162 - 7 \cdot 138$$
$$= 6 \cdot 162 - 7(3054 - 18 \cdot 162)$$
$$= 132 \cdot 162 - 7 \cdot 3054$$
$$= 132(12{,}378 - 4 \cdot 3054) - 7 \cdot 3054$$
$$= 132 \cdot 12{,}378 + (-535)3054.$$

Thus, we have

$$6 = \gcd(12{,}378, 3054) = 12{,}378x + 3054y,$$

where $x = 132$ and $y = -535$. It might be well to record that this is not the only way to express the integer 6 as a linear combination of 12,378 and 3054. Among other possibilities, one could add and subtract $3054 \cdot 12{,}378$ to get

$$6 = (132 + 3054)12{,}378 + (-535 - 12{,}378)3054$$
$$= 3186 \cdot 12{,}378 + (-12{,}913)3054.$$

It may happen that 1 and -1 are the only common divisors of a given pair of integers, whence $\gcd(a, b) = 1$. For example,

$$\gcd(2, 5) = \gcd(9, 16) = \gcd(27, 35) = 1.$$

This situation occurs often enough to prompt a definition:

Definition

Two integers a and b are said to be relatively prime, *or* prime to each other, *whenever* $\gcd(a, b) = 1$.

We should emphasize that it is possible for a pair of integers to be relatively prime without either integer being a prime. On the other hand, if p is a prime number, then $\gcd(a, p) = 1$ if and only if $p{\nmid}a$. This is true because the only positive divisors of p are 1 and p itself, so that either $\gcd(a, p) = 1$ or $\gcd(a, p) = p$. The latter case holds provided that $p\,|\,a$.

The next theorem characterizes relatively prime integers in terms of linear combinations.

THEOREM

Let a and b be integers, of which both are not zero. Then a and b are relatively prime if and only if there exist integers x and y such that $1 = ax + by$.

Proof. If a and b are relatively prime, so that gcd $(a, b) = 1$, then our last theorem guarantees the existence of integers x and y satisfying $1 = ax + by$. As for the other direction, suppose that $1 = ax + by$ for some choices of x and y, and that $d =$ gcd (a, b). Because $d|a$ and $d|b$, we must have $d|(ax +by)$ or $d|1$. Because d is a positive integer, this last divisibility condition forces $d = 1$, and the desired conclusion follows. ∎

This result leads to an observation that is useful in certain situations.

COROLLARY 1

If gcd (a, b) = d, then gcd (a/d, b/d) = 1.

Proof. Before starting with the proof proper, we should observe that although a/d and b/d have the appearance of fractions, they are in fact integers, since d is a divisor of both a and b. Because gcd $(a, b) = d$, it is possible to find integers x and y such that $d = ax + by$. On dividing both sides of this equation by d, one obtains the expression

$$1 = (a/d)x + (b/d)y.$$

Because a/d and b/d are integers, an appeal to the theorem is legitimate. The conclusion is that a/d and b/d are relatively prime. ∎

In illustration of the corollary, we observe that gcd $(12, 30) = 6$ and

$$\gcd (12/6, 30/6) = \gcd (2, 5) = 1,$$

as expected.

It is not true, without imposing an extra condition, that $a|c$ and $b|c$ together yield $ab|c$. For instance, $6|24$ and $8|24$, but clearly $6 \cdot 8 \nmid 24$. Were 6 and 8 relatively prime, of course, the situation would be altered. This brings us to another corollary.

COROLLARY 2

If $a|c$ and $b|c$, with gcd $(a, b) = 1$, then $ab|c$.

Proof. Because $a|c$ and $b|c$, there exist integers r and s for which $c = ar = bs$. Also, the condition gcd $(a, b) = 1$ allows us to write $1 = ax + by$ for suitable choices of integers x and y. If this last equation is multiplied by c, it appears that

$$c = c \cdot 1 = c(ax + by) = acx + bcy.$$

If the appropriate substitutions are now made on the right-hand side, then

$$c = a(bs)x + b(ar)y = ab(sx + ry)$$

or as a divisibility statement, $ab \mid c$. ∎

Proposition 24 of Book VII of Euclid's *Elements* seems mild enough, but it is fundamentally important in number theory. In modern notation, it may be stated as follows.

EUCLID'S
LEMMA

If $a \mid bc$, with gcd $(a, b) = 1$, then $a \mid c$.

Proof. We start again by writing $1 = ax + by$, where x and y are integers. Multiplication of this equation by c produces

$$c = 1 \cdot c = (ax + by)c = acx + bcy.$$

Because $a \mid ac$ and $a \mid bc$, it follows that $a \mid (acx + bcy)$, which may be restated as $a \mid c$. ∎

If a and b are not relatively prime, then the conclusion of Euclid's lemma may fail to hold. A specific example: $12 \mid 9 \cdot 8$, but $12 \nmid 9$ and $12 \nmid 8$.

The fundamental theorem of arithmetic, otherwise known as the "unique factorization theorem," asserts that any integer greater than 1 can be represented as a product of primes, and that the product is unique apart from the order in which the factors appear. Although this theorem is sometimes attributed to Euclid, it apparently was not expressly stated before 1801, when Gauss featured it in his *Disquisitiones Arithmeticae*. The nearest that Euclid himself came to this result was Proposition 14 of Book IX: "If a number be the least that is measured by prime numbers, it will not be measured by any other prime number except those originally measuring it." Some authorities argue that Euclid's failure to "discover" the fundamental theorem stems from his inability to form products wherein the number of factors is unspecified. Others argue that the theorem asserts the existence of a certain representation, and that the Greeks could not conceive of the existence of anything that was not constructible by elementary geometry.

Because every number either is a prime or, by the fundamental theorem, can be broken down into unique prime factors and no further, the primes serve as the "building blocks" from which all other integers can be made. Accordingly, the prime numbers have intrigued mathematicians through the ages, and although many remarkable theorems relating to their distribution in the sequence of positive integers have been proved, even more remarkable is what remains unproved. The open questions can be counted among the outstanding unsolved problems of all mathematics.

To begin on a simple note, we observe that the prime 3 divides the integer 36. We may write 36 as the product

$$6 \cdot 6, \text{ or } 9 \cdot 4, \text{ or } 12 \cdot 3, \text{ or } 18 \cdot 2;$$

and in each instance, 3 divides at least one of the factors involved in the product. This is typical of the general situation, and the precise result can be stated.

THEOREM

If p is a prime and $p \mid ab$, then $p \mid a$ or $p \mid b$.

> **Proof.** If $p \mid a$, then we need go no further, so let us assume that $p \nmid a$. Since the only positive divisors of p (hence, the only candidates for the value of gcd (a, p)) are 1 and p itself, this implies that gcd $(a, p) = 1$. Citing Euclid's lemma, it follows immediately that $p \mid b$.
>
> ■

This theorem extends to products with more than two factors. We state the result without proof.

COROLLARY

If p is a prime and $p \mid a_1 a_2 \cdots a_n$, then $p \mid a_k$ for some k, where $1 \leq k \leq n$.

Let us next show that any composite number is divisible by a prime (Proposition 31, Book VII). For a composite number n, there exists an integer d satisfying the conditions $d \mid n$ and $1 < d < n$. Among all such integers d, choose p to be the smallest. Then p must be a prime number. Otherwise, it too would possess a divisor q with $1 < q < p$; but $q \mid p$ and $p \mid n$ imply that $q \mid n$, which contradicts our choice of p as the smallest divisor, not equal to 1, of n. Thus, there exists a prime p with $p \mid n$.

With this preparation we arrive at the fundamental theorem of arithmetic. As indicated earlier, the theorem asserts that every integer larger than 1 can be factored into primes in essentially one way; the linguistic ambiguity "essentially" means that the representation $2 \cdot 3 \cdot 2$ is not considered different from $2 \cdot 2 \cdot 3$ as a factorization of 12. The precise formulation is given as follows.

FUNDAMENTAL THEOREM OF ARITHMETIC

Every positive integer $n > 1$ can be expressed as a product of primes; this representation is unique, apart from the order in which the factors occur.

> **Proof.** Either n is a prime or it is composite. In the first case there is nothing to prove. If n is composite, then there exists a prime divisor of n, as we have shown. Thus, n may be written as $n = p_1 n_1$, where p_1 is prime and $1 < n_1 < n$. If n_1 is prime, then we have our representation. In the contrary case, the argument is repeated to produce a second prime number p_2 such that $n_1 = p_2 n_2$; that is,
>
> $$n = p_1 p_2 n_2, \qquad 1 < n_2 < n_1.$$
>
> If n_2 is a prime, then it is not necessary to go further. Otherwise, write $n_2 = p_3 n_3$, with p_3 a prime; hence,
>
> $$n = p_1 p_2 p_3 n_3, \qquad 1 < n_3 < n_2.$$

The decreasing sequence

$$n > n_1 > n_2 > \cdots > 1$$

cannot continue indefinitely, so that after a finite number of steps n_k is a prime, say p_k. This leads to the prime factorization

$$n = p_1 p_2 \cdots p_k.$$

The second part of the proof—the uniqueness of the prime factorization—is more difficult. To this purpose let us suppose that the integer n can be represented as a product of primes in two ways; say,

$$n = p_1 p_2 \cdots p_r = q_1 q_2 \cdots q_s, \qquad r \le s,$$

where the p_i and q_j are all primes, written in increasing order, so that

$$p_1 \le p_2 \le \cdots \le p_r \qquad \text{and} \qquad q_1 \le q_2 \le \cdots \le q_s.$$

Because $p_1 | q_1 q_2 \cdots q_s$, we know that $p_1 | q_k$ for some value of k. Being a prime, q_k has only two divisors, 1 and itself. Because p_1 is greater than 1, we must conclude that $p_1 = q_k$, but then it must be that $p_1 \ge q_1$. An entirely similar argument (starting with q_1 rather than p_1) yields $q_1 \ge p_1$, so that in fact $p_1 = q_1$. We can cancel this common factor and obtain

$$p_2 p_3 \cdots p_r = q_2 q_3 \cdots q_s.$$

Now repeat the process to get $p_2 = q_2$; cancel again, to see that

$$p_3 p_4 \cdots p_r = q_3 q_4 \cdots q_s.$$

Continue in this fashion. If the inequality $r < s$ held, we should eventually arrive at the equation

$$1 = q_{r+1} q_{r+2} \cdots q_s$$

which is absurd, since each $q_i > 1$. It follows that $r = s$ and that

$$p_1 = q_1, p_2 = q_2, \quad \ldots, p_r = q_r$$

making the two factorizations of n identical. The proof is now complete. ∎

Of course, several of the primes that appear in the factorization of a given integer n may be repeated (as is the case with $360 = 2 \cdot 2 \cdot 2 \cdot 3 \cdot 3 \cdot 5$). By collecting the equal primes and replacing them by a single factor, we could write n in the so-called *standard form*

$$n = p_1^{k_1} p_2^{k_2} \cdots p_r^{k_r},$$

where each k_i is a positive integer and each p_i is a prime with $p_1 < p_2 < \cdots < p_r$.

To illustrate: The standard form of the integer 360 is $360 = 2^2 \cdot 3^2 \cdot 5$. Further examples are

$$4725 = 3^3 \cdot 5^2 \cdot 7 \qquad \text{and} \qquad 17,640 = 2^3 \cdot 3^2 \cdot 5 \cdot 7^2.$$

We cannot resist giving another proof of the irrationality of $\sqrt{2}$, this time using the fundamental theorem of arithmetic.

THEOREM

The number $\sqrt{2}$ is irrational.

Proof. Suppose to the contrary that $\sqrt{2}$ is a rational number, say, $\sqrt{2} = a/b$, where a and b are both integers with gcd $(a, b) = 1$. Squaring, we get $a^2 = 2b^2$, so that $b \mid a^2$. If $b > 1$, then the fundamental theorem guarantees the existence of a prime p such that $p \mid b$. From $p \mid b$ and $b \mid a^2$, it follows that $p \mid a^2$; but then $p \mid a$, hence gcd $(a, b) \geq p$. We therefore arrive at a contradiction, unless $b = 1$. If this happens, then $a^2 = 2$, which is impossible (we assume you are willing to grant that no integer can be multiplied by itself to give 2). Our original supposition that $\sqrt{2}$ is a rational number is untenable; so it must be an irrational number. ∎

By this time, you are probably asking, Is there a prime number that is the largest, or do the primes go on forever? The answer is to be found in a very ingenious (yet quite simple) proof given by Euclid in Book IX of his *Elements*. In general terms what he showed is that beyond each prime another and larger prime can be found. The actual details follow; the argument is Euclid's, although the words and modern notation are not.

THEOREM

There are an infinite number of primes.

Proof. Write the primes 2, 3, 5, 7, 11, . . . in ascending order. For any particular prime p, consider the number

$$N = (2 \cdot 3 \cdot 5 \cdot 7 \cdot 11 \cdots p) + 1.$$

That is, form the product of all the primes from 2 to p, and increase this product by one. Because $N > 1$, we can use the fundamental theorem to conclude that N is divisible by some prime q. But none of the primes 2, 3, 5, . . . , p divides N. For if q were one of these primes, then on combining the relation $q \mid 2 \cdot 3 \cdot 5 \cdots p$ with $q \mid N$, we would get $q \mid (N - 2 \cdot 3 \cdot 5 \cdots p)$, or what is the same thing, $q \mid 1$. The only positive divisor of the integer 1 is 1 itself, and since $q > 1$, the contradiction is obvious. Consequently, there exists a new prime q larger than p. ∎

Euclid's proof demonstrates the existence of some prime larger than p; but we do not necessarily arrive at the very next prime after p when we use the method indicated by his proof. For example, this process yields 59 as a prime beyond 13:

$$N = (2 \cdot 3 \cdot 5 \cdot 7 \cdot 11 \cdot 13) + 1 = 30031 = 59 \cdot 509$$

Frequently, there are a great many primes between the prime p considered and the one obtained in the manner the proof suggests.

How can we determine, given a particular integer, whether it is prime or composite, and if it is composite, how can we actually find a nontrivial divisor? The most obvious approach is successive division of the integer in question by each of the numbers preceding it; if none of them (except 1) serves as a divisor, then the integer must be a prime. Although this method is very simple, it cannot be regarded as useful in practice. For even if one is undaunted by large calculations, the amount of work involved may be prohibitive.

Composite numbers have a property that enables us to reduce materially the necessary computations. If an integer $a > 1$ is composite, it can be written as $a = bc$, where $1 < b < a$ and $1 < c < a$. Assuming that $b \leq c$, we get $b^2 \leq bc = a$, and so $b \leq \sqrt{a}$. Because $b > 1$, there is for b at least one prime factor p. Then $p \leq b \leq \sqrt{a}$; furthermore, because $p \mid b$ and $b \mid a$, it follows that $p \mid a$. The point is simply this: A composite number a will always possess a prime divisor p satisfying $p \leq \sqrt{a}$.

In testing the primality of a specific integer $a > 1$, it therefore suffices to divide a by those primes not exceeding $\sqrt{a}$ (presuming, of course, the availability of a list of primes up to $\sqrt{a}$). This can be clarified by considering the integer $a = 509$. Because $22 < \sqrt{509} < 23$, we need only try out the primes that are not larger than 22 as possible divisors, namely, the primes 2, 3, 5, 7, 11, 13, 17, and 19. Dividing 509 by each of these in turn, we find that none serves as a divisor of 509. The conclusion is that 509 is a prime number.

Example. The foregoing technique provides a practical means for determining the canonical form of an integer, say $a = 2093$. Because $45 < \sqrt{2093} < 46$, it is enough to examine the primes 2, 3, 5, 7, 11, 13, 17, 19, 23, 29, 31, 37, 41, and 43. By trial, the first of these to divide 2093 is 7, with $2093 = 7 \cdot 299$. As regards the integer 299, the seven primes less than 18 (note that $17 < \sqrt{299} < 18$) are 2, 3, 5, 7, 11, 13, and 17. The first prime divisor of 299 is 13, and carrying out the required division, we obtain $299 = 13 \cdot 23$. But 23 is itself a prime, whence 2093 has exactly three prime factors, namely 7, 13, and 23:

$$2093 = 7 \cdot 13 \cdot 23.$$

4.4   Problems

1. Given integers a, b, c, verify that:

   (a) If $a \mid b$, then $a \mid bc$.
   (b) If $a \mid b$ and $a \mid c$, then $a^2 \mid bc$.
   (c) $a \mid b$ if and only if $ac \mid bc$, provided $c \neq 0$.
   (d) If $a \mid (a + b)$, then $a \mid b$.

2. Show that if $a \mid b$, then $(-a) \mid b$, $a \mid (-b)$, and $(-a) \mid (-b)$.

3. For any positive number n, it can be shown that there exists an even integer a that is representable as the sum of two odd primes in n different ways. Confirm that the integers 66, 96, and 108 can be written as the sum of two primes in six, seven, and eight ways, respectively.

4. A conjecture of Lagrange (1775) asserts that every odd integer greater than 5 can be written as a sum $p + 2q$, where p and q are both primes. Verify that this holds for all odd integers through 75.

5. Find an example to show that the following conjecture is not true: Every positive integer can be written in the form $p + a^2$, where p is a prime (or else equal to 1) and $a \geq 0$.

6. Prove that the only prime of the form $n^3 - 1$ is 7. [*Hint:* Factor $n^3 - 1$ as $(n - 1)(n^2 + n + 1)$.]

7. Find a set of four consecutive odd integers of which three are primes, and a set of five consecutive odd integers of which four are primes.

8. Although the answer is not known, it appears that each positive multiple of 6 can be written as the difference of two primes. Confirm this as far as 90.

9. Consider the primes arranged in their natural order 2, 3, 5, 7, ... It is conjectured that beginning with 3, every other prime can be composed of the addition and subtraction of all smaller primes (and 1), each taken once. For example:

$$3 = 1 + 2, \qquad 7 = 1 - 2 + 3 + 5,$$
$$13 = 1 + 2 - 3 - 5 + 7 + 11$$
$$= -1 + 2 + 3 + 5 - 7 + 11.$$

Show that this also holds for 19, 29, 37, and 43.

10. Establish each of these statements.

   (a) The square of any integer is of the form either $4n$ or $4n + 1$.
   (b) The square of any odd integer is of the form $8n + 1$. [*Hint:* Any odd integer is of the form $4k + 1$ or $4k + 3$.]
   (c) The square of any integer not divisible by 2 or 3 is of the form $12n + 1$. [*Hint:* By the division theorem, an integer can be represented in one of the forms $6k$, $6k + 1$, $6k + 2$, $6k + 3$, $6k + 4$, or $6k + 5$.]

11. For any arbitrary integer a, show that $2 | a(a + 1)$ and $3 | a(a + 1)(a + 2)$.

12. Prove that if a is an integer not divisible by 3, then $3 | (a^2 - 1)$.

13. Verify that the difference of two consecutive squares is never divisible by 2; that is, 2 does not divide $(a + 1)^2 - a^2$ for any choice of a.

14. For a positive integer a, show that gcd $(a, 0) = a$, gcd $(a, 1) = 1$, and gcd $(a, a) = a$.

15. Find gcd $(143, 277)$, gcd $(136, 232)$, and gcd $(272, 1479)$.

16. Use the Euclidean algorithm to obtain integers x and y satisfying:

   (a) gcd $(56, 72) = 56x + 72y$.
   (b) gcd $(24, 138) = 24x + 138y$.
   (c) gcd $(119, 272) = 119x + 272y$.
   (d) gcd $(1769, 2378) = 1769x + 2378y$.

17. Prove that any two consecutive integers are relatively prime, that is, gcd $(a, a + 1) = 1$ for any integer a.

18. Establish that the product of any three consecutive integers is divisible by 6, and the product of any four consecutive integers is divisible by 24.

19. Given that p is a prime and $p | a^n$, show that $p^n | a^n$.

20. (a) Find all prime numbers that divide 40! (recall that $40! = 1 \cdot 2 \cdot 3 \cdot 4 \cdots 40$).
   (b) Find the prime factorization of the integers 1234; 10,140; and 36,000.

21. (a) An unanswered question is whether there are infinitely many primes that are 1 more than a power of 2, such as $5 = 2^2 + 1$. Find two more of these primes.
   (b) It is equally uncertain whether there are infinitely many primes that are 1 less than a power of 2, such as $3 = 2^2 - 1$. Find four more of these primes.

22. Prove that the only prime p for which $3p + 1$ is a perfect square is $p = 5$. [*Hint:* If $3p + 1 = a^2$, then $3p = a^2 - 1 = (a + 1)(a - 1)$.]

23. It has been conjectured that every even integer can be written as the difference of two consecutive primes in an infinite number of ways. For example,

$$4 = 11 - 7 = 17 - 13 = 23 - 19$$
$$= 47 - 43 = 131 - 127 = \cdots.$$

Express the integer 6 as the difference of two consecutive primes in ten ways.

24. Determine whether the integer 701 is prime by testing all primes $p \le \sqrt{701}$ as possible divisors. Do the same for the integer 1009.

25. Prove that $\sqrt{p}$ is irrational for any prime p.

26. Use the division theorem to show that every prime except 2 and 3 is of the form $6n + 1$ or $6n + 5$.

4.5   Eratosthenes, the Wise Man of Alexandria

Another Alexandrian mathematician whose work in number theory remains significant is Eratosthenes (276–194 B.C.). Eratosthenes was born in Cyrene, a Greek colony just west of Egypt and under Ptolemaic domination, but spent most of his working days in Alexandria. At some time during his early life he studied at Plato's school in Athens. When about 30 years of age Eratosthenes was invited to Alexandria by King Ptolemy III to serve as tutor for his son and heir. Later, Eratosthenes assumed the most prestigious position in the Hellenistic world, chief librarian at the Museum, a post he was to hold for the last 40 years of his life. It is reported that in old age he lost his sight, and unwilling to live when he was no longer able to read, he committed suicide by refusing to eat.

Eratosthenes was acknowledged to be the foremost scholar of his day and was undoubtedly one of the most learned men of antiquity. An author of extraordinary versatility, he wrote works (of which only some fragments and summaries remain) on geography, philosophy, history, astronomy, mathematics, and literary criticism; and he also composed poetry. Eratosthenes was given two nicknames that are significant in light of the prodigious range of his interests. In honor of his varied accomplishments, his friends called him Pentathis, a name applied to the champion in five athletic events—hence, to men who tried their hands at everything. His detractors felt that in attempting too many specialities, Eratosthenes failed to surpass his contemporaries in any one of them. They dubbed him Beta (the second letter of the Greek alphabet), insinuating that while Eratosthenes stood at least second in all fields, he was first in none. Perhaps a kinder explanation of this second nickname is that certain lecture halls in the Museum were marked with letters, and Eratosthenes was given the name of the room in which he taught.

Although Eratosthenes could be regarded as among a second echelon in many endeavors, he was certainly not *beta* in the fields of geography and mathematics. His three-volume *Geographica,* now lost except for fragments, was the first scientific attempt to put geographical studies on a sound mathematical basis. In this work, he discussed the arguments for a spherical earth and described the position of various land masses in the known world. Eratosthenes' actual mapping of the populated quarters of the earth was based on hearsay and speculation, but it was the most accurate map of the world that had yet appeared and the first to use a grid of meridians of longitude and parallels of latitude. He regarded the inhabited lands as placed wholly in the northern hemisphere, surrounded by a continuous body of ocean. Eratosthenes made the first suggestion for the circumnavigation of the globe when he observed: "If it were not for the vast extent of the Atlantic Sea one might sail from Iberia (Spain) to India along one and the same parallel." The vast amount of quantitative data accumulated by Eratosthenes as head of the largest library of antiquity made his *Geographica* the prime authority for centuries; the longitude and latitude of 8000 places on earth were given, as well as numerous estimates of distances between locations.

As a mathematician, Eratosthenes produced as his chief work a solution of the famous Delian problem of doubling the cube and the invention of a method for finding prime numbers. His mechanical contrivance for effecting duplication, called a

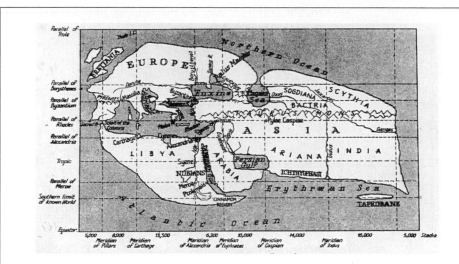

The habitable world according to Eratosthenes. (*From* A Short History of Scientific Ideas *by Charles Singer. Reproduced by permission of Oxford University Press.*)

''mesolabium,'' or mean-finder, consisted of a rectangular framework along which three rectangular plates (marked with their diagonals) of height equal to the width of the frame slide in three grooves, moving independently of one another and able to overlap. Suppose that the original positions of the rectangular plates are shown as in the figure, where *AP, FQ* are the sides of the frame and *ARGF, RSHG, STIH* are the

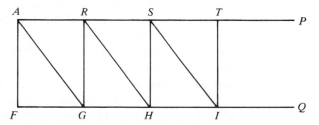

plates that slide. If the first plate remains stationary while the second slides under the first, and the third under the second, to a position in which the points *A, B, C, D*

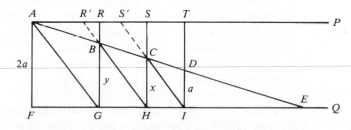

are brought into line, then the result looks like the preceding figure. Draw a straight line through the collinear points A, B, C, and D, meeting the side FQ at E. From the theory of similar triangles, we then obtain

$$\frac{HE}{GE} = \frac{BE}{AE} = \frac{GE}{FE},$$

while

$$\frac{BG}{AF} = \frac{GE}{FE} \quad \text{and} \quad \frac{CH}{BG} = \frac{HE}{GE}.$$

Tying the various relations together, we see that

$$\frac{CH}{BG} = \frac{BG}{AF}.$$

By similar reasoning,

$$\frac{DI}{CH} = \frac{CH}{BG},$$

and so DI, CH, BG, AF are in continued proportion. On setting $DI = a$, $AF = 2a$, $CH = x$, and $BG = y$, we get

$$\frac{a}{x} = \frac{x}{y} = \frac{y}{2a},$$

which makes apparent the conclusion that x and y are the required mean proportionals between the lengths a and $2a$. Put another way: If a is the length of the edge of a given cube, the cube that has edge x will have volume double the original one's.

Eratosthenes was so pleased with his contrivance for solving the Delian problem that he had a monument erected to Ptolemy III on which the proof was inscribed, and he also caused the mean-finder to be cast in bronze. What could be more curious behavior—the best way Eratosthenes could think of to thank and flatter the king was to dedicate the solution of an esoteric mathematical problem to him! Of course, any mechanical solution was not as "pure" as straightedge and compass constructions would be, and as such would be abhorrent to the principles of Plato.

We have seen that if an integer $a > 1$ is not divisible by a prime $p \leq \sqrt{a}$, then a itself is necessarily a prime. Eratosthenes used this fact as the basis of a clever technique, called the sieve of Eratosthenes, for finding all primes less than a given integer n. The scheme calls for writing down the integers from 2 to n in their natural order and then systematically eliminating all the composite numbers by striking out all multiples $2p$, $3p$, $4p$, . . . of the primes $p \leq \sqrt{n}$. The integers that are left on the list—that fall through the "sieve"— are primes.

To see by example how this works, suppose that we want to find all primes not exceeding 100. Recognizing that 2 is a prime, we begin by crossing out all even integers from our listing, except 2 itself. The first of the remaining integers is 3, which must be a prime. We keep 3, but strike out all higher multiples of 3, so that 6, 9, 12, . . . are now removed. The smallest integer after 3 not yet deleted is 5. It is not divisible by either 2 or 3 (otherwise it would have been cancelled), hence is also a prime. Because all proper multiples of 5 are composite numbers, we next remove 10, 15, 20, . . . , retaining 5 itself. The first surviving integer 7 is a prime, for it is not divisible by 2, 3, or 5, the only primes that precede it. After the proper multiples of 7, the largest prime less than $\sqrt{100} = 10$, have been eliminated, all composite integers in the sequence 2, 3, 4, . . . , 100 have fallen through the sieve. The positive integers that remain, to wit 2, 3, 5, 7, 11, 13, 17, 19, 23, 29, 31, 37, 41, 43, 47, 53, 59, 61, 67, 71, 73, 79, 83, 89, 97 are all the primes less than 100.

The accompanying table represents the result of the completed sieve. The multiples of 2 are crossed out by \; the multiples of 3 are crossed out by /; the multiples of 5 are crossed out by —; the multiples of 7 are crossed out by ~.

	2	3	4	5	6	7	8	9	10
11	12	13	14	15	16	17	18	19	20
21	22	23	24	25	26	27	28	29	30
31	32	33	34	35	36	37	38	39	40
41	42	43	44	45	46	47	48	49	50
51	52	53	54	55	56	57	58	59	60
61	62	63	64	65	66	67	68	69	70
71	72	73	74	75	76	77	78	79	80
81	82	83	84	85	86	87	88	89	90
91	92	93	94	95	96	97	98	99	100

Today Eratosthenes is best remembered for having devised a practical method for calculating the earth's circumference. Although his was not the first or last such estimate made in antiquity, it was far more accurate than all previous estimates. The extraordinary thing about Eratosthenes' achievement is its simplicity. His procedure was based on estimates of the arc of the great circle through Alexandria and Syene, the city that today is called Aswan. The two cities had certain advantages. They were thought to be on the same meridian; the distance between them had been measured by a bematistes, or surveyor, trained to walk with equal steps and count them, and had been found to be 5000 stadia; and travelers had commented on the curious fact that in Syene, at the time of the summer solstice, the sun at noon cast no shadow from an upright stick. This meant that Syene was directly under the Tropic of Cancer, or at least, nearly so. Story has it that Eratosthenes confirmed the position of the tropic by observing the water in a deep well. At noontime of the summer solstice, the bottom was completely illuminated by the sun's rays, the edge of the well casting no shadow at all on the water below.

Because the sun is so vastly distant from the earth, its rays may be regarded as striking the earth in parallel lines. Eratosthenes argued that at noon on the day of the summer solstice, the continuation of a line through the well at Syene would pass through the center of the earth, the sun being directly overhead. At the same time at Alexandria, the sun was found to cast a shadow indicating that the sun's angular position from zenith was $\alpha = 7° 12' = \frac{360°}{50}$, or $\frac{1}{50}$ of a complete circle. In making this determination, Eratosthenes apparently used a sundial consisting of a hemispherical bowl with a vertical pointer at its center to cast a shadow; the direction and height of the sun could be read off by observing the sun's shadow with lines drawn on the concave interior. Now an imaginary line drawn through the vertical pointer of the sundial would pass through the center of the earth and there form an angle with the line through the well at Syene. This central angle would have to equal α, according to the theorem that asserts that the alternate interior angles formed by a transversal cutting a pair of parallel lines are equal. In brief, the angle the sun's rays would make with the pointer of the sundial would equal the angle subtended at the earth's center by the arc connecting Alexandria and Syene.

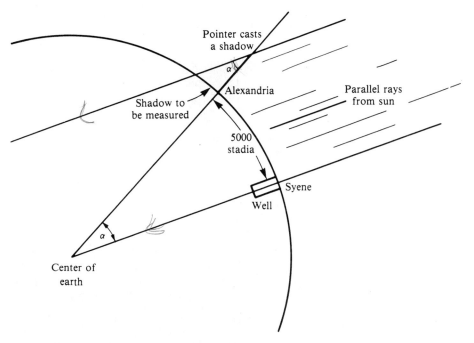

Assuming that the sundial at Alexandria, the well at Syene, the center of the earth, and the center of the sun when directly over Syene all lay in the same plane, Eratosthenes inferred that

$$\frac{\alpha}{360°} = \frac{5000}{\text{circumference}}, \qquad \alpha = \frac{360°}{50},$$

and there was but one unknown (the earth's circumference) in the equation. This gave him 50 times the 5000 stadia, or 250,000 stadia, for the entire circumference of the earth. For some reason not known to us (perhaps to account for any error that existed in measuring the distance between Alexandria and Syene), he added an extra 2000 to this figure to conclude that the desired circumference was 252,000 stadia. Unfortunately, there was more than one kind of stadium used for measuring distance. If it is assumed that Eratosthenes used Egyptian stadia of 516.73 feet each, then his 252,000 stadia work out to the incredibly excellent value of 24,662 miles, just 245 miles less than the true value. The ancient world certainly accepted Eratosthenes' measurement as the best possible. Pliny (A.D. 23–79), the Roman naturalist, said it was so bold and subtle a feat that it would be a shame not to accept the figure, and he even recorded divine sanction of it.

Such a close estimate must, however, be regarded as somewhat accidental. Although the method was sound in theory, the accuracy of the answer would have to depend on the precision with which the basic data could be determined. Eratosthenes made several compensating errors. The figure of $\frac{1}{50}$ of the circle for the difference in latitude is near the truth, but Syene is not directly on the tropic, Alexandria is not on the same meridian (it lies about 3° to the west of Syene), and the direct distance between the two places is 4530 stadia, not 5000. This does not matter very much, however. Eratosthenes' achievement lies in his method; for a man who was regarded as a ''second-stringer'' in the Alexandrian era of Greek mathematics, it showed the touch of genius.

Any discussion of Alexandria must take into account the advances made in astronomy, a branch of science completely dependent on mathematics. For fourteen centuries, the accepted blueprint of the solar system was that of the Alexandrian Claudius Ptolemy (A.D. 100–170). Ptolemy did for astronomy what Euclid did for geometry; by incorporating a brilliant power of synthesis and exposition with original genius, he reduced the works of his predecessors to a matter of ''historical interest'' with little chance of survival. His great treatise *Syntaxis Mathematica* (The Mathematical System), or the *Almagest,* as it became known to the Arabs and medieval Europeans, was destined to remain the supreme authority on astronomy until the publication of Copernicus's *De Revolutionibus* (1543). We are ignorant of most of the events in Ptolemy's life, except for the knowledge that he was a native of Egypt and that his numerous astronomical observations were made in the period between A.D. 127 and 151, probably at the Museum.

The very name of Ptolemy's masterpiece has its own curious history. The Greeks called it *Megale Syntaxis* (the Great Collection). Later translators from Greek into Arabic, either through admiration or carelessness, combined the Arabic article *al* with the superlative *megiste* to form the hybrid word *almagisti,* ''The greatest,'' whence the Latin *Almagestum* and colloquial *Almagest,* by which name it has been known ever since.

Ptolemy came at the end of a long line of Greek thinkers who viewed the earth as the fixed and immovable center of the universe, around which the planets swung in concentric circles. To assert that the earth was at any place other than the center of the heavens was to deny humans their position of supremacy in the universe, to believe that human affairs were no more significant to the gods than those of other planets. Some astronomers, notably Aristarchus of Samos, proposed the heliocentric hypothesis—that the earth and the planets all revolved in circles about a fixed sun—but it was rejected for various reasons. One did not have to be trained in astronomy to observe that the earth seemed stable under the feet, that lighter bodies did not fly into the air, or that projectiles shot straight upward did not fall farther to the west. Archimedes advanced the more scientific argument that if the earth were in motion, its distance from the stars would vary, and this apparently was not so.

Claudius Ptolemy
(circa 145)

(The Bettmann Archive.)

According to the Pythagorean prejudice for the beauty and perfection of the circle, the motion of the sun and planets had to be circular. However, their deviation from circular orbits was great enough to have been observed and to require explanation. To reduce celestial motion to combinations of circular movements, the Greek astronomer Apollonius had worked out an ingenious scheme of epicycles, or small circles having their centers on the circumferences of other circles. In the epicycle system, each planet travels around the earth in a large circle, called a ''deferent''; this circle does not represent the true path of the planet, but rather the path of the center of a small circle, the epicycle, around which the planet revolves. Claudius Ptolemy, to rationalize these ideas with his accumulated observations, proposed the notion of eccentric solar motion.

Planet

Epicycle
(orbit of planet)

Moving center

Deferent
(orbit of moving center)

Earth
(fixed in space)

Equant

His system as described by the *Almagest* was perhaps as complicated, relative to his own time, as Einstein's relativity theory is to our time. It will be enough for our purposes to say that Ptolemy set the earth eccentrically within the main circle representing the deferent of the planet and made the center of the epicycle move with uniform velocity, not about the center of the deferent, but about an offset point. This latter point, called the "equant," or equalizing point, lay at an equal distance from the earth on the opposite side of the circle. The equant was a remarkable invention that not only allowed Ptolemy to describe important features of planetary motion in terms of circles but also fitted the observational data available in the second century. It obviously had the motion appear the fastest when the deferent was near the terrestrial observer and slowest at the opposite point; and that was the explanation of why the sun appeared sometimes near the earth and sometimes farther away.

The chief flaw in Ptolemy's system lay in its mistaken premise of an earth-centered universe. Yet the heliocentric theory was not ignored. Ptolemy devoted a column or two to the refutation of this theory, thereby preserving it for the ages to ponder on and for Copernicus to develop. Copernicus was still plagued by epicycles and the matter was not resolved until Kepler (1609) observed that the planets moved, not in Pythagoras's ideal circle, but in elliptical orbits. As soon as Kepler made this radical break with tradition, everything fell into place.

A work that exerted almost as much influence on succeeding centuries as the *Almagest* did was Ptolemy's *Geographike Syntaxis* (Geographical Directory). Written in eight books, it is an attempt to summarize the geographical knowledge of the habitable world as known at that time, that is, the continents of Europe, Asia, and Africa. The *Geography* was accompanied by a collection of maps, a general map of the world and 26 others showing regional details. Ptolemy developed his own manner of representing the curved surface of the earth on a plane surface. He divided the circumference of the globe into 360 parts, or degrees, as they came to be called, and covered the surface with a network of meridians and parallels. In choosing an arbitrary prime meridian, Ptolemy drew a line passing through the westernmost of the Fortunate Islands (the Canaries), but was mistaken by about 7° in his idea of the distance of these islands from the mainland. On his world map, he sought to reproduce on a flat surface the contour of the globe by representing the parallels and meridians as curved lines, with the meridians converging to the poles; for the smaller regional maps, a simple rectangular grid was considered sufficient.

A glance at Ptolemy's map will reveal a somewhat misleading picture of the known world. Its length from his own zero meridian in the Fortunate Islands to the city of Sera in China covers 180° (as against 126° in reality), with the result that the westward distance from western Europe to eastern Asia is much less than it should be. He was ignorant of the peninsular shape of India, so Ptolemy completely distorted the southern coastline of Asia; and the island of Ceylon is exaggerated to 14 times its actual size. He somehow assumed that the land mass of China ran far to the south and then to the west until it joined the east coast of Africa, thereby making the Indian Ocean a landlocked sea. The distortion of Ptolemy's world map is partly due to his rejection of Eratosthenes' estimate of the earth's circumference, and his adoption of the less appropriate estimate of 180,000 stadia. This figure is too small by nearly 5000 miles, or about one-quarter of the correct distance.

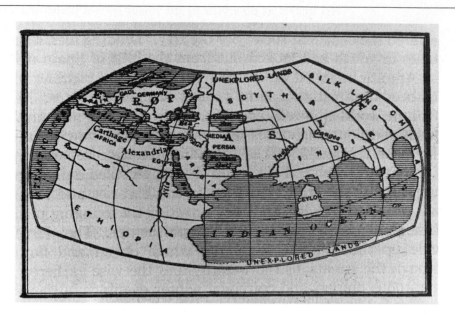

The habitable world according to Ptolemy. (From Ancient Times *by James Henry Breasted, © copyright 1916 by James Henry Breasted. Reproduced by permission of Ginn and Company.)*

The main part of the *Geography* is an exhaustive gazetteer of some 8000 places, arranged by regions, with their supposed latitudes and longitudes. Although Ptolemy gave the impression that his coordinates were based on astronomical observation, he relied largely on Roman road-itineraries (official lists of stopping-places on the roads of the empire, with distances between them) and on reports accumulated from traders and travelers who came to Alexandria. Because he worked from this sketchy data, it is not surprising that the positions he gave for many localities outside the well-known Mediterranean area were grossly inaccurate. Paris, for instance, was put opposite the mouth of the Loire River. But Ptolemy came remarkably close to the truth when he described the Nile as formed by two rivers flowing from two lakes a little south of the equator (these are Victoria and Albert Nyanza), a fact of geography that was not confirmed until the nineteenth century.

Ptolemy's geographical treatise had its effect on western Europe much later than his *Almagest* did. It was translated into Latin in 1409, not from an Arabic manuscript but from a Greek one brought from Constantinople. Although initially printed in 1475, the first printed edition to be accompanied by maps, drawn by medieval cartographers from coordinates contained in the text, was published in Rome in 1478. Columbus possessed a copy of this latter edition. The Latin *Geography* was received with great deference, partly because the author represented the world approximately as it had been known for many centuries and partly because of the mistaken conception that he had used rigorous mathematical methods for determining places. Besides, the scholars of the early fifteenth century had no reliable criteria for criticizing Ptolemy. The maps based on his information, despite their many errors, were vastly superior to those previously available and covered many areas not usually touched by marine charts of the day.

Ptolemy's diminution of the distance between Europe and Asia by some 50° latitude fortified Columbus's belief that he could easily reach the Orient by sailing westward across the Atlantic—perhaps even induced him to undertake his great voyage of discovery. Indeed, Columbus died in the conviction that the land he had first sighted was an outlying island of southeastern India; and the error is perpetuated in the application of the name "Indian" to the natives of the American continents.

4.5 Problems

1. In the *Almagest*, Ptolemy proved a geometrical result known today as "Ptolemy's theorem." If *ABCD* is a (convex) quadrilateral inscribed in a circle, then the product of the diagonals is equal to the sum of the products of the two pairs of opposite sides. In symbols;

$$AC \cdot BD = AB \cdot CD + BC \cdot AD.$$

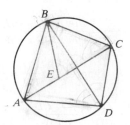

If *BE* is drawn so that $\angle ABE = \angle DBC$, complete the details of the following proof of Ptolemy's theorem:

(a) The triangles *ABE* and *DBC* are similar, whence

$$\frac{AB}{BD} = \frac{AE}{CD}.$$

(b) $\angle ABD = \angle ABE + \angle EBD$
$$= \angle DBC + EBD = \angle EBC.$$

(c) The triangles *ABD* and *EBC* are similar, whence

$$\frac{AD}{EC} = \frac{BD}{BC}.$$

(d) The result of adding $AB \cdot CD = AE \cdot BD$ and $BC \cdot AD = EC \cdot BD$ is

$$AC \cdot BD = AB \cdot CD + BC \cdot AD.$$

2. Let *AB* and *AC*, where $AB < AC$, be two chords of a circle terminating at an endpoint *A* of the diameter *AD*.

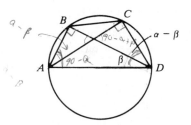

If $\angle CDA = \alpha$ and $\angle BDA = \beta$, show that Ptolemy's theorem leads to

$$BC = \sin \alpha \cos \beta - \cos \alpha \sin \beta,$$

a result which is reminiscent of the trigonometric formula for $\sin(\alpha - \beta)$.

3. Use Ptolemy's theorem to prove that if *P* lies on the arc *AB* of the circumcircle of the equilateral triangle *ABC*, then $PC = PA + PB$.

4. Like other Greek geometers, Ptolemy used chords of angles rather than sines. Sines were invented much later, around the fifth century, by the Hindu astronomers. Book I of the *Almagest* contains a table giving the lengths of the chords of central angles in a circle of radius 60, increasing by half a degree at a time from 1/2° to 180°.

(a) Derive the relation

$$\text{chord } 2\alpha = 120 \sin \alpha$$

between Ptolemy's value for the length of a chord corresponding to angle 2α and the sine of α.

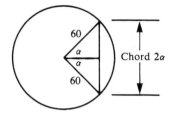

(b) From Ptolemy's value chord 1° = 1;2,50 and using an inscribed 360-gon to approximate the circumference of a circle, obtain his approximation to π. [*Hint:* π = circumference/diameter ≈ (360 chord 1°)/diameter.]

(c) From Ptolemy's value chord 120° = 103;55,23 and using the fact that $\sqrt{3} = 2 \sin 60°$, obtain his approximation to $\sqrt{3}$.

5. Supply the missing details in the following proof of the formula for the area K of a triangle in terms of its sides a, b, c, namely

$$K = \sqrt{s(s - a)(s - b)(s - c)},$$
$$s = \tfrac{1}{2}(a + b + c).$$

(This formula appears in Heron's *Metrica*, and a proof is worked out in his *Dioptra*. According to Arabic tradition the result was known earlier to Archimedes, who undoubtedly had a proof of it.)

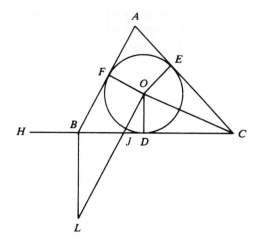

In triangle ABC, inscribe a circle with center O, touching the sides BC, AC, AB at points D, E, F, respectively. Extend segment CB to H so that HB = AF; also draw OL perpendicular to OC to cut BC at J and meet the perpendicular to BC at B in the point L. Then:

(a) $K = \tfrac{1}{2}(BC)\ ((OD) + \tfrac{1}{2}\ AC)(OE) + \tfrac{1}{2}(AB)(OF)$
$= (HC)(OD);$

(b) $\angle CLB + \angle BOC = 180°$ and $\angle BOC + \angle AOF$ $= 180°$, so that $\angle CLB = \angle AOF;$

(c) triangles AOF and CLB are similar, hence

$$\frac{BC}{BH} = \frac{BC}{AF} = \frac{BL}{OF} = \frac{BL}{OD} = \frac{BJ}{JD};$$

(d) $\dfrac{BC}{BH} + 1 = \dfrac{BJ}{JD} + 1$ implies that $\dfrac{CH}{BH} = \dfrac{BJ}{JD};$

(e) $\dfrac{(CH)^2}{CH \cdot HB} = \dfrac{BC \cdot CD}{JD \cdot CD} = \dfrac{BD \cdot CD}{(OD)^2};$

(f) $K^2 = (CH)^2(OD)^2$
$= CH \cdot HB \cdot BD \cdot DC$
$= s(s - a)(s - b)(s - c),$

where $a = BC$, $b = AC$, $c = AB$.

6. The Hindu mathematician Brahmagupta (c. 600) discovered a formula for the area K of a quadrilateral inscribed in a circle:

$$K = \sqrt{(s - a)(s - b)(s - c)(s - d)},$$

where a, b, c, d are the sides of the quadrilateral and $s = \frac{1}{2}(a + b + c + d)$ is its semiperimeter. Prove that Heron's formula is a special case of Brahmagupta's formula.

7. If a quadrilateral with sides a, b, c, d is inscribed in one circle and circumscribed about another, show that its area K is given by

$$K = \sqrt{abcd}.$$

[*Hint:* Use the fact that the tangents to a circle from an external point are equal in length to conclude that $s = a + c = b + d$. Now apply Brahmagupta's formula.]

8. Establish the following result due to Brahmagupta: If a quadrilateral inscribed in a circle has perpendicular diagonals meeting at a point P, then

any line through P which is perpendicular to a side of the quadrilateral will bisect the opposite side.

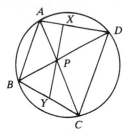

[*Hint:* If XY is perpendicular to BC, then

$$\angle DPX = \angle BPY = \angle PCY = \angle ACB$$
$$= \angle ADB = \angle XDP,$$

so that triangle XPD is isosceles. Similarly, triangle XPA is isosceles.]

4.6 Archimedes

The work of Archimedes (about 287–212 B.C.) epitomizes Alexandrian mathematics. Considered the greatest creative genius of the ancient world, Archimedes lived a generation or two after Euclid and was a contemporary of Eratosthenes. We know few details of his life, though several fanciful stories have clustered around his name. Archimedes was the son of the astronomer Phidias and was born in Syracuse, a Greek settlement on the southeastern coast of Sicily. At the time, it was the largest city in the Hellenistic world. According to Plutarch, Archimedes came from the same royal family as the city's ruler, King Hieron II. This enlightened dictator reigned, according to the historian Polybius, for 54 years "without killing, exiling, or injuring a single citizen, which is indeed the most remarkable of all things." Archimedes almost certainly visited Egypt, and because he corresponded regularly with several scholars at the Museum in Alexandria, it is likely that he studied at that center of Greek science. He spent most of his productive years in Syracuse, however, where under Hieron's protection and patronage, he devoted himself wholeheartedly to study and experiment. Archimedes earned great renown in antiquity for his mathematical writings, his mechanical inventions, and the brilliant way in which he conducted the defense of his native city during the Second Punic War (218–201 B.C.). It is well attested that he perished in the indiscriminate slaughter that followed the sacking of Syracuse by Roman troops.

Archimedes' mechanical skill together with his theoretical knowledge enabled him to devise a series of ingenious contrivances. Of these the most famous is the Archimedean screw, a pump still used in parts of the world. Archimedes apparently invented it during his visit to Egypt for the purpose of raising canal water over levees into irrigated fields. It was later used for pumping water out of mines and from ships' holds. The simple and useful device consists of a long tube, open at both ends and containing a continuous screw or spiral piece of metal of the same length as the cylinder. When the lower end of the tube is tilted into the standing water and the spiral insert is rotated, water is carried to the top and flows out of the cylinder's upper opening.

Several of the stories about Archimedes that have come down to us relate to his skill as an engineer, for it is natural that his mechanical inventions would have a broader appeal than his more specialized mathematical achievements. One familiar legend concerns his exploit in launching a large ship. When King Hieron was amazed at the great weights that Archimedes could move by means of levers, cogwheels, and pulleys, Archimedes is reported to have boasted that if he had a fixed fulcrum to work with he could move anything: ''Give me a place to stand and I will move the earth.'' Hieron asked Archimedes to reduce the problem to practice, and pointed out the difficulty that his men were experiencing with a ship so heavy that it could not be launched from the slips in the usual way. Archimedes designed a combination of levers and pulleys that (in the words of that man of letters, Plutarch) he alone ''while sitting far off, with no great effort, but only holding the end of a compound pulley quietly in his hand and pulling at it, drew the ship along smoothly and safely as if she were moving through the water.'' The same story was told by Proclus, who represented Hieron as operating the pulley himself and crying out in amazement, ''From this day forth Archimedes is to be believed in everything that he may say.''

Despite his mechanical talents, Archimedes was far more concerned with theoretical studies than with discoveries connected with practical needs, regarding these as the ''diversions of geometry at play.'' In *The Life of Marcellus,* Plutarch went on to say:

> Though these inventions had obtained for him the reputation of more than human sagacity, he yet would not deign to leave behind him any written work on these subjects, but, regarding as ignoble and vulgar the business of mechanics and every sort of art which is directed towards use and profit, he placed his whole ambition in those speculations whose beauty and subtlety are untainted by any admixture of the common needs of life.

Although Archimedes was not greatly interested in the practical applications of his knowledge, he was usually willing to help his admiring friend and patron, King Hieron, with a problem. One of the best-known stories tells of his success in determining the purity of a golden crown. It appears that Hieron, on gaining power in Syracuse, had a crown of pure gold made as an offering to the gods. The weight of the completed crown matched the weight of the gold that had been assigned to the goldsmith; yet Hieron suspected that the maker had appropriated some of the gold, replacing it with an equal weight of silver. Being unable to verify his suspicion, Hieron consulted Archimedes. The story has it that the great scientist suddenly realized how to settle the question while he was at the public baths of the city. Getting into the tub, he observed that the lower his body submerged into the water the more water overflowed the top of the tub. This gave him the idea that if the goldsmith had actually debased the crown by alloying it with silver, the crown would displace a greater volume when immersed in water than would a weight of gold equal to the weight of the crown; for pure gold would be more dense than an alloy of gold and the lighter metal silver. The Roman architect Vitruvius related that Archimedes, recognizing the value of this method of solution,

> without a moment's delay and transported with joy . . . jumped out of the tub and rushed home naked, crying out in a loud voice that he had found what he was seeking; for as he ran, he shouted repeatedly in Greek, ''Eureka, eureka!'' [''I have found it, I have found it!'']

Whether Archimedes actually dashed naked through the streets of Syracuse, as alleged, is a matter of speculation; but the common people cheerfully believed such a story, because it made a great man look ridiculous.

The widest fame Archimedes enjoyed in the classical world came from the active part he took in defending his city against the Romans. During the third century B.C., Rome and the African city-state Carthage were locked in the bitter Punic wars. It was clear to the Romans that their mastery of southern Italy would be threatened if ever a hostile power controlled Sicily. While King Hieron was still alive, Syracuse remained Rome's loyal ally; but Hieron died in 215 B.C. and was succeeded by his 15-year-old grandson, who fell under the influence of courtiers in the pay of Carthage. Roman forces under a tough and businesslike general named Marcellus, seizing the opportunity to annex the whole of Sicily, attacked Syracuse by land and sea. Geographically the site was a natural fortress, and Archimedes, then an old man of 75, personally directed the defense.

A vivid account of this famous siege was given by Plutarch in his writing on the life of Marcellus. He told how Archimedes used his engineering skill to construct ingenious war machines, by which he inflicted great losses on the Romans. The city walls were fortified with a series of powerful catapults and crossbows set to throw a hail of missiles at specified ranges, so that however close the attackers came, they were always under fire. The assault by sea was repulsed by devices that could be run out from the walls to drop huge stones or masses of lead through the planking of the galleys beneath. Cranes caught the bows of the vessels with grapnels, lifted them out of the water, and dropped them stern-first from a height. Plutarch wrote that the Roman soldiers were in abject terror and refused to advance:

> If they only saw a rope or piece of wood extending beyond the walls, they took flight exclaiming that Archimedes had once again invented a new machine for their destruction.

But the tale that Archimedes set the enemy ships on fire by concentrating the sun's rays on them through the use of great concave mirrors, though repeated by many later writers, is probably not true. (Such a device was, however, used in defending Constantinople in 514.) After a two-year-siege, the Romans temporarily withdrew their forces and the overconfident Syracusans relaxed their vigilance. When the defenders had feasted and drunk their fill at a religious festival, pro-Roman sympathizers inside the city directed the enemy to a weak point in the walls. Marcellus gave explicit orders to his officers that the life and house of Archimedes should be spared; but before they could locate the great scientist, he had been slain by a common soldier.

The account of how Archimedes met his death has been told in various forms. According to the traditional story, he was absorbed in a geometrical problem whose diagram was drawn in the sand. As the shadow of the approaching Roman soldier fell over his diagrams, the agitated mathematician called out, "Don't spoil my circles!" The soldier, insulted at having orders thus given to him, retaliated by drawing his sword. Another legend has it that Archimedes was slain by looters who supposed that his astronomical instruments, constructed of polished brass, were actually made of gold.

The death of Archimedes during the siege of Syracuse. (The Bettmann Archive.)

Marcellus deeply regretted the death of Archimedes and erected an elaborate monument in his honor. Archimedes had expressed the wish to friends that his tomb should bear the figure of a sphere inscribed in a right cylinder, in memory of his discovery of the relation between the two bodies (the volume of the sphere is equal to two-thirds that of the circumscribing cylinder). In building his tomb, the Romans complied with his wish. Many centuries later, the Roman orator Cicero identified the monument by means of this inscription. His account in *Tuscalan Disputations* of how he found it in a ruined state, neglected by the people of Syracuse, is worth repeating:

> When I was questor [B.C. 75] I hunted out his grave, which was unknown to the people of Syracuse, since they entirely denied its existence, and I found it completely covered and surrounded by brambles and thorn-bushes. . . . Slaves sent in with sickles cleared and uncovered the place. When a passage had been made to it, we approached the pedestal facing us: the epigram was apparent with about half of the little verse worn away. And thus one of the noblest cities of Greece, once indeed a very great seat of learning, would have been ignorant of the monument of its most brilliant citizen, except that it was revealed by a man of Arpinum [Cicero].

The tomb has since disappeared and its exact location is unknown.

A survey of the contents of a few of Archimedes' principal works is enough to reveal the wide range of subjects he studied and the surprising ingenuity with which he treated them. The dozen items that have come down to us were preserved by a school of Byzantine mathematicians in Constantinople; between the sixth and tenth centuries, they made it their objective to collect and copy the dispersed treatises of Archimedes. These have greatly lost their original form, having suffered the linguistic transformation from the Sicilian-Doric dialect into Attic Greek. Unlike the *Elements* of Euclid, the works that have immortalized Archimedes were never popular in antiquity; where Euclid worked up existing material into systematic treatises that any educated student would understand, Archimedes aimed at producing small tracts of limited scope addressed to the most eminent mathematicians of the day. "It is not possible," wrote Plutarch several centuries later, "to find in all geometry more difficult and more intricate questions, or more simple and lucid explanations."

It was Archimedes' practice first to send statements of his results, with the request that the other mathematicians discover the proofs for themselves; the complete treatise, with its supporting evidence, would follow thereafter. He was not above enunciating theorems he knew to be false so that "those vain mathematicians who claim to discover everything, without ever giving their proofs, may be deceived into saying that they have discovered the impossible."

Of all his mathematical achievements, Archimedes seems to have taken chief pride in those contained in *On the Sphere and Cylinder*. Written in two books, some 53 propositions in all, it begins with a prefatory letter announcing the main results obtained. Archimedes indicated that he was publishing them for the first time so that expert mathematicians could examine the proofs and judge their value. Those propositions selected for mention included:

1. The surface of a sphere is four times the area of a great circle of the sphere [or as we would say, $S = 4\pi r^2$].

2. If about a sphere there is circumscribed a cylinder whose height is equal to the diameter of the sphere, then the volume of the cylinder is three halves of the volume of the sphere; and the surface of the circumscribing cylinder, including its bases, is three halves of the surface of the sphere.

Then follow some definitions and assumptions. Of the five assumptions, there is a famous one, a property that Archimedes himself attributed to Eudoxus. This is usually known today as the postulate of Archimedes: Of two unequal line segments, some finite multiple of the shorter one will exceed the longer. Using this, Archimedes derived the above results, plus numerous others relative to the area or volume of figures bounded by curved lines or surfaces.

Book II of *On the Sphere and Cylinder* treats some problems and theorems suggested by the first book. In his work on segments of a sphere, Archimedes was confronted with the solution of a cubic equation. This occurs in Proposition 4 of Book II, which poses one of the great problems of Greek geometry—to pass a plane through a sphere in such a way that the volumes of the segments cut off are in a given ratio.

The problem can be analyzed as follows. Suppose $2r$ is the diameter of the given sphere. It is required to find a plane cutting this diameter at right angles so that the segments into which the sphere is divided have their volumes in a given ratio, say m/n. Because the volume of a spherical segment of height h, cut from a sphere of radius r, is given by the formula $V = \pi h^2(r - h/3)$, we must have

$$\frac{h^2(3r - h)}{k^2(3r - k)} = \frac{m}{n} \, .$$

If k is eliminated by the relation $h + k = 2r$, this becomes

$$nh^2(3r - h) = m(2r - h)^2(r + h)$$
$$= m(h^3 - 3h^2r + 4r^3),$$

or what amounts to the same thing,

$$(m + n)h^3 - 3r(m + n)h^2 + 4mr^3 = 0,$$

a cubic equation in which the term containing h is missing. This can be written

$$\frac{3r - h}{mr/(m + n)} = \frac{4r^2}{h^2} \, ,$$

and Archimedes treated it as a particular instance of the more general equation

$$\frac{a - x}{b} = \frac{c^2}{x^2} \, .$$

Archimedes promised to provide a complete solution to the equation and then to apply it to the particular case at hand; but either the explanation was omitted or else this part of the text has been lost. The details were found centuries later in a fragment of a manuscript, which is usually attributed to Archimedes because it was written in the Sicilian-Doric dialect he used. The reconstructed solution proceeds in much the same way that the geometer Menaechmus attacked the Delian problem—by finding the intersection of conics. That is, both members of $(a - x)/b = c^2/x^2$ are equated to a/y. This leads to two equations,

$$x^2 = \left(\frac{c^2}{a}\right)y, \qquad (a - x)y = ab,$$

which represent, respectively, a parabola and a hyperbola. The points of intersection of these two conics will furnish the solutions of $x^2(a - x) = bc^2$. The fragment also proves that if $bc^2 = 4a^3/27$, then the curves touch at the point for which $x = 2a/3$, while if $bc^2 < 4a^3/27$, there are two solutions. Except for a simple cubic encountered by Diophantus of Alexandria in the first half of the fourth century, interest in cubic equations disappeared after Archimedes, not to reappear in the history of European mathematics for more than a thousand years.

Of the works of Archimedes known in the Middle Ages, the most popular, and the first to be translated into Latin, was *The Measurement of a Circle*. It is a short treatise, perhaps a part of a longer work, comprising only three propositions. The object of the first is to show that the area of a circle can be calculated as soon as its circumference is known.

PROPOSITION 1   *The area of any circle is equal to the area of a right triangle in which one of the sides about the right angle is equal to the radius, and the other to the circumference of the circle.*

The next proposition (whose proof we include) establishes that if the circumference of a circle is $3\frac{1}{7}$ of the diameter, then the area of the circle is to the square of its diameter as 11 is to 14. Archimedes could not have originally placed it before Proposition 3, because the approximation depends on the result of that proposition.

PROPOSITION 2   *The area of a circle is to the square on its diameter as* 11 *to* 14, *very nearly.*

Proof. Take a circle with diameter *AB* and let a square *CDEF* be circumscribed about it. Produce the side *CD* so that *DG* is twice *CD* and *GH* is one-seventh *CD*. Because the areas of triangle *ACG* and *ACD* are in the ratio 21:7 and *ACD* and *AGH* are in the ratio 7:1, triangle *ACH* and triangle *ACD* are in the ratio 22:7. But the square *CDEF* is four times the triangle *ACD*, and therefore the triangle *ACH* is to the square *CDEF* as 22:28, or 11:14. The triangle *ACH* equals the circle, since *AC* equals the radius and *CH* equals the circumference (which will be shown in Proposition 3 to be very nearly $3\frac{1}{7}$ of the diameter). Thus the circle and the square *CDEF* are in the ratio 11:14, very nearly.

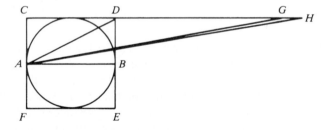

The most important proposition in *The Measurement of a Circle* contains Archimedes' estimate of the numerical value of π. He did not call it π. The symbol π for the ratio of the circumference of a circle to its diameter was not used by Archimedes or any other Greek mathematician. It was introduced in 1706 by an obscure English writer, William Jones, in his *Synopsis Palmariorum Matheseos, or a New Introduction to the Mathematics*. In this book for beginners, Jones published the circumference-to-diameter ratio to 100 decimal places, all correct. It was not until the usage given it by Leonhard Euler in the famous *Introductio in Analysin Infinitorum* (1748) that the letter π was definitely adopted for this ratio, no doubt because it is the first letter of the Greek word *perimetros* (perimeter).

    The approach Archimedes took in obtaining a value for π was based on the following fact: the circumference of a circle lies between the perimeters of the inscribed and circumscribed regular polygons of *n* sides, and as *n* increases, the deviation of the circumference from the two perimeters becomes smaller. This type of demonstration has since become known as the "method of exhaustion"—not for what it does to the user, but because the difference in area between the polygons and the circle is gradually exhausted. Although it amounts to considering the circle as the limit of the inscribed (or circumscribed) polygons as the number of sides increases indefinitely, there is no direct passage to the limit. For the Greek mathematician never thought of the process as continued for an infinite number of steps; he considered it only carried out in finite stages to a desired degree of accuracy.

In calculating a suitable approximation for π, Archimedes successively inscribed and circumscribed regular polygons of 6, 12, 24, 48, and 96 sides within and without the circle. The choice for the number of sides was natural. Of all the regular polygons, the hexagon is most easily inscribed. Simply mark off from any point on the circumference chords of a length equal to the radius of the circle until all six vertices, say A, B, C, D, E, and F, are obtained. When tangents are drawn to the circle A, B, C, D, E, and F, another regular hexagon is produced, one that circumscribes the circle.

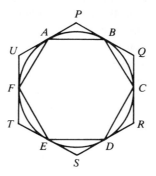

From the regular hexagon, the regular inscribed 12-sided polygon is constructed by bisecting the arc subtended on the circumscribed circle by each side of the hexagon, using the additional points thus found and the original vertices to form the required dodecagon. Continuing in this way, by repeated bisection of arcs, Archimedes obtained the regular polygons of 12, 24, 48, and 96 sides from the hexagon.

If p_n and P_n represent the perimeters of the inscribed and circumscribed regular polygons of n sides, and C the circumference of the circle, it follows that

$$p_6 < p_{12} < p_{24} < p_{48} < p_{96} < \cdots < p_n < C$$
$$< P_n < \cdots < P_{96} < P_{48} < P_{24} < P_{12} < P_6.$$

Both of these sequences are bounded monotonic sequences, and hence each has a limit; and it can be proved that the limits are the same, with C their common value. Moreover, P_{2n} is the harmonic mean of p_n and P_n, and p_{2n} is the geometric mean of p_n and P_{2n}:

$$P_{2n} = \frac{2p_n P_n}{p_n + P_n}, \qquad p_{2n} = \sqrt{p_n P_{2n}}.$$

Starting from the perimeters $p_6 = 3d$ and $P_6 = 2\sqrt{3}d$, where d is the diameter of the circle, one can use these recursion relations to compute P_{2n} and p_{2n} successively until the values P_{96} and p_{96} required by Archimedes are reached. Assuming the inequality

$$\frac{265}{153} < \sqrt{3} < \frac{1351}{780}$$

as known without further explanation, Archimedes found that

$$\left(3 + \frac{10}{71}\right)d < \frac{96 \cdot 66}{2077\frac{1}{4}}d < p_{96} \quad \text{and} \quad P_{96} < \frac{96 \cdot 153}{4673\frac{1}{2}} < \left(3 + \frac{10}{70}\right)d,$$

whence the final result

$$3\frac{10}{71} < \pi < 3\frac{1}{7}.$$

The result of Archimedes' computation was expressed as this proposition.

PROPOSITION 3 *The circumference of any circle exceeds three times its diameter by a part that is less than $\frac{1}{7}$ but more than $\frac{10}{71}$ of the diameter.*

The approximation of $\frac{22}{7}$ is often called the Archimedean value of π. Because $\frac{22}{7} \approx 3.1429$ is less than 0.2 percent larger than the actual value of π and is such a simple number for ordinary calculation, it was good enough for most purposes in antiquity. Archimedes could theoretically have provided a better estimate of π using polygons of 192 or 384 sides, but the arithmetic—made difficult in any case by the clumsy Greek alphabetic number symbols—would have been prohibitive.

Historians of science have focused considerable attention on the attempts of early societies to arrive at an approximate value for the ratio of a circle's circumference to its diameter (that is, the number π), perhaps because the increasing accuracy of the results seems to offer a measure of the mathematical skill of the culture at that time. The ancient Chinese were considerably more advanced in arithmetic calculation than their Western contemporaries, so it is not surprising that they obtained remarkably accurate values for π. Texts from the pre-Christian era generally used 3 as an approximation for π, but from the first century mathematicians in China were searching for better estimates. Liu Hsin (circa 23) employed 3.1547, and Chang Heng (78–139) used the value $\sqrt{10}$, whose decimal approximation is 3.1622; or the fraction 92/29, whose decimal approximation is 3.1724.

By taking the ratio of the perimeter of a regular inscribed polygon to the diameter of a circle enclosing the polygon, third century mathematicians obtained more accurate approximations. Liu Hui, in his commentary on the *Nine Chapters of the Mathematical Art,* used a polygon of 384 sides to derive for π the bounds

$$3.141024 < \pi < 3.142904,$$

and with a 3072-sided polygon found his best value for π, namely 3.14159. In the fifth century, the brilliant mathematician and astronomer Tsu Chung-Chi (430–501) refined the method to obtain

$$3.1415926 < \pi < 3.1415927;$$

and, from these, gave the fraction 22/7 as an "inaccurate" value for π and 355/113 as the "accurate" value. This latter value yields π correct to six decimal places. Comparable rational approximations were not attained in the Western world until the sixteenth century when the Dutch fortress engineer Adriaan Anthonizoon (1527–1607) derived anew the ratio 355/113. No fraction with denominator less than 113 gives a closer approximation to π; in fact, 355/113 is such a good rational estimate that no better one is reached until 52163/16604. By using the Archimedean method on a polygon of 2^{62} sides, the indefatigable Ludolph van Ceulen (1540–1610) carried the value of π correctly to 35 decimal places. (This computational feat was considered so extraordinary that his widow had all 35 digits of the ''Ludolphine number'' carved upon his tombstone.) His was one of the last major attempts to evaluate π by the method of perimeters; thereafter, the techniques of calculus prevailed.

The Sand-Reckoner of Archimedes was a computational accomplishment of another kind. It contained a new system of notation for expressing numbers in excess of one hundred million, for which Greek mathematics had not yet developed any characters. Archimedes contrived a procedure for counting in units of ten thousand myriads, 10^8 in our notation, and used exponents for ordering his classes of magnitudes. To demonstrate that his system would adequately describe enormously large numbers, he undertook to enumerate the grains of sand that the finite universe, bounded by the sphere of the fixed stars, could hold. (Like other astronomers of the time, Archimedes believed the universe to be a sphere whose center was the immobile earth and whose radius equaled the distance from the earth to the sun.)

To give a reasonable maximum bound on the dimension of the universe, Archimedes quoted certain earlier views on the size of the celestial bodies. Like most earlier astronomers, he assumed that the earth had a diameter greater than that of the moon but less than that of the sun, and that the diameter of the sun was 30 times the diameter of the moon. (The factor 30 was a convenient exaggeration of the traditional estimate of 20.) If the diameters of the sun, moon, earth, and universe are represented by D with suitable subscripts, this means postulating that

$$D_{sun} = 30D_{moon} < 30D_{earth}.$$

By a clever geometric argument, Archimedes proved that the perimeter of a regular polygon of 1000 sides inscribed in a circle of diameter D_{univ} was greater than $3D_{univ}$ and at the same time less than $1000D_{sun}$; hence,

$$3D_{univ} < 1000D_{sun} < 30,000D_{earth}.$$

For the circumference of the earth, he took a then accepted value of 300,000 stadia, but in order to be on the safe side multiplied by a factor of 10, thereby assuming that

$$D_{earth} < 1,000,000 \text{ stadia}.$$

Archimedes concluded from these assumptions that for the diameter of the universe as far as the sun,

$$D_{univ} < 10^{10} \text{ stadia}.$$

To make good his boast, Archimedes next supposed that a grain of sand had minute but definite size. Underestimating the size of a grain of sand, he proposed that 10,000 grains of sand would be needed to fill the space of a poppy seed and that 40 poppy seeds lined up in a row would exceed one finger-breadth. Therefore (using $V = \frac{1}{6}\pi D^3 < D^3$) a sphere of diameter one finger-breadth would contain at most 64,000 poppy seeds, consequently at most 640 million grains of sand—in any event, no more than 1 billion $= 10^9$ grains. Taking one stadium to be less than $10,000 = 10^4$ finger-breadths, Archimedes then found the number of grains of sand in a sphere of diameter 1 stadium to be fewer than $10^9(10^4)^3 = 10^{21}$. A secure upper bound for the grains in a sphere with diameter 10^{10} stadia was $10^{21}(10^{10})^3 = 10^{51}$, or as Archimedes put it, "one thousand units of the seventh order of numbers."

The figure just mentioned gives the number of grains of sand needed to fill up the "conventional universe." To demonstrate the practicality of his method beyond any doubt, Archimedes also referred to the view of Aristarchus of Samos (sometimes called the Copernicus of antiquity) that the universe was heliocentric, with the earth revolving around the sun. He showed that a universe of the dimensions Aristarchus proposed in *On the Size and Distance of the Sun and Moon* had room for only fewer than 10^{63} grains of sand. Archimedes concluded the discussion with the following words:

> These things will appear incredible to the numerous persons who have not studied mathematics; but to those who are conversant therewith and have given thought to the distances and the sizes of the earth, the sun, and the moon, and of the whole universe, the proof will carry conviction.

The treatise *On Spirals* contains twenty-eight propositions dealing with the properties of the curve now known appropriately as the spiral of Archimedes. It is described in the words of the inventor himself:

> If a straight line [half-ray] one extremity of which remains fixed be made to revolve at a uniform rate in the plane until it returns to the position from which it started, and if, at the same time as the straight line is revolving, a point moves at a uniform rate along the straight line, starting from the fixed extremity, the point will describe a spiral in the plane.

In modern polar coordinates, the equation connecting the length r of the radius vector with the angle θ through which the line has revolved from its initial position is $r = a\theta$, where $a > 0$ is some constant. For let OA be the revolving half-line, O the fixed extremity, and P the point that moves away from O along OA. If $OP = r$ and $AOP = \theta$; then the characteristic property of the Archimedean spiral requires r/θ to be constant.

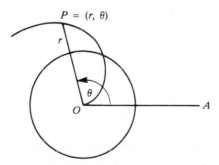

In the view of the modern mathematician, perhaps the greatest mathematical achievement of Archimedes, and certainly one of the most fascinating results, was his calculation of the area enclosed by the first loop of the spiral (corresponding to $0 \leq \theta \leq 2\pi$) and the fixed line. As he put it: "The space bounded by the spiral and the initial line after one complete revolution is equal to one-third of the circle described from the fixed extremity as center, with radius that part of the initial line over which the moving point advances in one revolution.'' This is equivalent to the modern formulation $A = \frac{1}{3}\pi(2\pi a)^2$. Nowadays, a problem of this kind is made easy by the use of integral calculus. Archimedes, in its stead, used the method of exhaustion; he divided the spiral curve into numerous equal parts and circumscribed and inscribed circular sectors, adding up their areas.

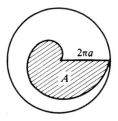

The method of exhaustion is traditionally attributed to Eudoxus of Cnidos (390–337 B.C.), although Euclid and Archimedes used it most frequently and to greatest advantage. The method plays a leading part in Book XII of the *Elements,* where it was used to prove that the areas of circles are to one another as the squares of their diameters, and also that the volumes of pyramids that are of the same height and have triangular bases are proportional to the areas of their bases. Archimedes subsequently exploited exhaustive techniques in finding the areas of curvilinear plane figures and volumes bounded by curved surfaces. The method is encountered in Archimedes' work in two main forms. One version consists in enclosing the geometric figure whose area or volume is sought between two others, which can be calculated and can be shown to approach each other indefinitely. The essence of the other approach is to inscribe suitably chosen figures within the figure for which the area or volume is required; then in some fashion the area or volumes of the inscribed figures are increased until the difference between them and the quantity to be calculated becomes arbitrarily small. The phrase ''method of exhaustion'' was not used by the ancient Greeks to describe this procedure but introduced by the Jesuit mathematician Gregory St. Vincent in his *Opus Geometricum* (1647).

Archimedes used the method in the *Quadrature of a Parabola* to find the area of the segment formed by drawing any chord of the parabola. Archimedes begins to ''exhaust'' the area of the parabolic segment by inscribing in it a triangle of the same base as the segment and of a height equal to the height of the segment. (By ''height of a parabolic segment'' we mean the distance from the chord to the point on the parabola at which the tangent is parallel to the chord.) The other two sides of the inscribed triangle provide two new parabolic segments; in each of these another triangle is inscribed in the same way, with the process continued as far as desired to build up an inscribed polygon as the sum of a sequence of triangles. In this way Archimedes found that the segment cut off by the chord had an area equal to $\frac{4}{3}$ the area of the first triangle constructed.

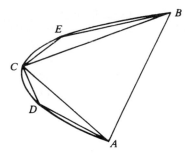

Archimedes' argument is typical of his general approach in determining areas or volumes by exhaustion, so it is worth looking at more closely. In the parabolic segment bounded by the chord AB, Archimedes constructed a triangle ABC having AB for its base and the point C for the third vertex. At C, the tangent to the parabola was parallel to the chord. (It is proved that C is the point on the curve that has the greatest perpendicular distance from the base AB.) Let the area of triangle ABC be denoted by $\triangle$. In each of the two smaller segments cut off by the chords AC and CB, Archimedes similarly inscribed triangles ADC and CEB. From the properties of the parabola, he demonstrated that each of the two new triangles had an area equal to $\frac{1}{8}\triangle$; hence, the area of ADC and CEB together equaled $\frac{1}{4}\triangle$. Next, more triangles were constructed with vertices on the parabola and bases on the new chords AD, DC, CE, and EB. Each of these four triangles had an area equal to $\frac{1}{8}$ that of triangle ADC, or equal to $(1/8^2)\triangle$, so that this set of triangles added $(1/4^2)\triangle$ to the area of the inscribed figure. Continuing, Archimedes obtained a sequence of polygonal figures by adding an ever-increasing number of triangles to the original triangle ABC. The area of the nth such polygon is given by

$$\triangle\left(1 + \frac{1}{4} + \frac{1}{4^2} + \frac{1}{4^3} + \cdots + \frac{1}{4^n}\right).$$

This is a finite geometric progression of ratio $\frac{1}{4}$ whose sum,

$$\triangle\left(\frac{4}{3} - \frac{1}{3}\left(\frac{1}{4}\right)^n\right),$$

measures areas closer and closer to the area required. At this point, the modern mathematician would use the limit concept to conclude that the parabolic segment has an area of $\frac{4}{3}\triangle$. Archimedes, who did not have command of this notion, instead proved by a double *reductio ad absurdum* argument that if the polygons exhausted the parabolic segment, then its area could be neither greater nor less than $\frac{4}{3}\triangle$.

In 1906, the Greek text of yet another work by Archimedes was discovered almost by accident in the library of a monastery in Constantinople. A Danish philologist, Johan Ludvig Heiberg, was drawn there by the report of a tenth-century parchment manuscript that seemed originally to have had mathematical content (a so-called palimpsest). Sometime between the twelfth and fourteenth centuries, monks had washed off the earlier text to provide space for a collection of prayers and liturgies, a not uncommon practice caused by the high cost of parchment. Fortunately, most of the expunged contents could be deciphered with a magnifying glass. The manuscript contained

The Method

method of exhaustion

(integral calculus)

fragments of many treatises of Archimedes that had had sufficiently wide circulation to be preserved elsewhere; it also contained the only surviving copy of a largely unknown work entitled *The Method*. Historians had been aware of the existence of *The Method* through allusions by ancient writers, such as Heron, but it had been believed irretrievably lost. Sent as a letter to Eratosthenes, it recalled certain mathematical results that Archimedes had propounded without proof on a former occasion; and it went on to acquaint Eratosthenes with the method that had been used in reaching these and many other conclusions. Anticipating the view of modern integral calculus, Archimedes asserted that surfaces were to be considered ''made up'' of an infinity of parallel lines and that solids of revolution were ''filled up'' by circles. But Archimedes did not regard such intuitive reasoning as a proof, only as an investigation preliminary to a rigorous demonstration by the method of exhaustion. By this ingenious method, he found the surface areas, volumes, and centers of gravity of numerous solids of revolution. Although these achievements are remarkable anticipations of results found later in the integral calculus, we must be careful not to impute to Archimedes the idea expressed in the calculus; for the concept of limit, which lies at the very heart of the subject, was entirely alien to his arguments.

In the preface to *The Method,* Archimedes said, ''I presume there will be some among the present as well as future generations who by means of the method here explained will be enabled to find other theorems which have not yet fallen to our share.'' Unfortunately, his hope of finding successors to continue his work remained unfulfilled. After Archimedes' time, the trend of Greek mathematics was in other directions; and more than eighteen centuries were to pass before Newton and Leibniz took up the task of developing the classical method of exhaustion into the principles constituting the calculus.

4.6 Problems

1. Verify the following results from Book I of Archimedes' *On the Sphere and Cylinder:*

    (a) *Proposition 13.* The surface area of any right circular cylinder, excluding its bases, is equal to the area of a circle whose radius is the mean proportional between the side of the cylinder and the diameter of the base of the cylinder.

    (b) *Proposition 14.* The lateral area of any isosceles cone, excluding the base, is equal to the area of the circle whose radius is the mean proportional between the side of the cone and the radius of the circle that is the base of the cone.

    (c) *Proposition 15.* The lateral area of any isosceles cone has the same ratio to the area of its base as the side of the cone has to the radius of the circle that is the base of the cone.

    (d) *Proposition 23.* The surface area of any sphere is equal to four times the area of a great circle of the sphere.

    (e) *Proposition 24.* The volume of any sphere is equal to four times the volume of the cone whose base equals a great circle of the sphere, and whose height equals the radius of the sphere.

2. Prove that if a sphere is inscribed in a right circular cylinder whose height is equal to the diameter of the sphere, then:

    (a) The volume of the cylinder is $\frac{3}{2}$ the volume of the sphere.

    (b) The surface area of the cylinder, including its bases, is $\frac{3}{2}$ the surface area of the sphere.

3. Prove Archimedes' ''theorem of the broken chord'': If *AB* and *BC* make up any broken chord in a circle (where $BC > AB$), and *M* is the midpoint of the arc

ABC and MF the perpendicular to the longer chord, then F is the midpoint of the broken chord. That is, $AB + BF = FC$. [*Hint:* Extend chord BC to D, so that $FD = FC$; then $\triangle MBA$ is similar to $\triangle MBD$.]

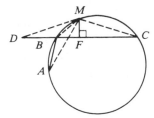

4. To find a formula for the length of the side of a regular inscribed polygon of $2n$ sides in terms of the length of the side of the regular polygon of n sides, proceed as follows. Let $PR = S_n$ be the side of a regular n-gon inscribed in a circle of radius 1. Through the center O of the circle, draw a perpendicular to PR, bisecting PR at T and meeting the circle at Q; then $PQ = QR = S_{2n}$ are sides of the inscribed regular $2n$-gon. Prove that;

(a) $OT^2 = OR^2 - TR^2 = 1 - \dfrac{S_n^2}{4}$,

(b) $QT^2 = (1 - OT)^2 = \left(1 - \dfrac{\sqrt{4 - S_n^2}}{2}\right)^2$,

(c) $S_{2n}^2 = QT^2 + TR^2 = 2 - \sqrt{4 - S_n^2}$.

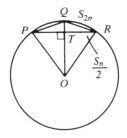

5. For regular polygons inscribed in a circle of radius 1, use $S_6 = 1$ to conclude that

$$S_{12} = \sqrt{2 - \sqrt{3}},$$

$$S_{24} = \sqrt{2 - \sqrt{2 + \sqrt{3}}},$$

$$S_{48} = \sqrt{2 - \sqrt{2 + \sqrt{2 + \sqrt{3}}}},$$

$$S_{96} = \sqrt{2 - \sqrt{2 + \sqrt{2 + \sqrt{2 + \sqrt{3}}}}},$$

and hence that $\pi \approx 48S_{96} \approx 3.14103 \approx \frac{22}{7}$, which was the value Archimedes found.

6. In the *Book of Lemmas* (a collection of 13 geometrical propositions that has come down to us only in an Arabic translation), Archimedes introduced a figure that, owing to its shape, is known as the arbelos, or "shoemaker's knife." If

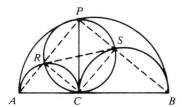

a straight line AB is divided into two parts at C and if on one side of AB are described semicircles with AB, AC, and CB as diameters, then the region included between the circumferences of the three semicircles is the shoemaker's knife. Prove that if PC is the straight line perpendicular to AB at C, then the area of the shoemaker's knife equals the area of the circle whose diameter is PC. [*Hint:* $AB^2 = AC^2 + BC^2 + 2AC \cdot BC = AC^2 + BC^2 + 2PC^2$.]

7. Prove that if the common external tangent to the two smaller semicircles in the shoemaker's knife touches these curves at R and S, then RS and PC bisect each other, and R, S, P, C lie on the circle whose diameter is PC.

8. The *Book of Lemmas* also contains a geometrical figure called the "salinon," or "salt cellar." Take $AC = DB$ on the diameter AB of a semicircle. Then describe semicircles, with AC and DB as the diameters, on the same side of AB as the given semicircle; also describe a semicircle, with CD as the diameter, on the other side of the given semicircle. The region bounded by the circumference of the semicircles is the salt cellar. Prove that if PQ is the line of symmetry of the figure, then the area of the salt cellar equals the area of the circle whose diameter is PQ.

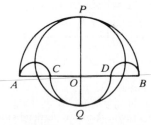

9. Use the techniques of calculus to show that the area bounded by the first complete turn of the spiral $r = a\theta$ and the initial line is equal to one-third of the "first circle" (that is, the circle with radius $2\pi a$).

10. Like Hippias's quadratrix, the spiral of Archimedes can be used to trisect an angle and square the circle. Given a spiral, place the angle to be trisected so that the vertex and the initial side of the angle coincide with the initial point of the spiral and the initial position OA of the rotating ray. Let the terminal side of the angle intersect the spiral of P. Trisect the segment OP at the points Q and R, and draw circles with center at O and with OQ and OR as radii. Prove that if these circles meet the spiral in points U and V, then the lines OU and OV will trisect $\angle AOP$.

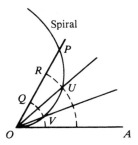

11. A clever solution to the problem of the quadrature of the circle is achieved by means of the spiral of Archimedes. Given a circle with center at O and radius a, draw the spiral whose equation in polar coordinates is $r = a\theta$ and whose initial point is O. Prove that when the rotating ray is revolved perpendicular to its initial position OA, the segment OP will have a length equal to one-fourth the circumference of the circle. Show how this resolves the quadrature problem.

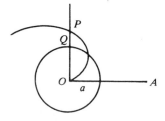

12. If OA is the initial line and A the end of the first revolution of the spiral, and if the tangent to the spiral at A is drawn, then the perpendicular to OA at O will meet the tangent at some point B. Establish that the length of the segment OB is equal to the circumference of the circle with radius OA; hence, the area of $\triangle AOB$ is equal to the area of this circle. [*Hint:* The slope of the tangent at A is 2π.]

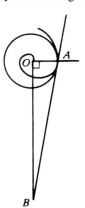

Bibliography

Ang, Tian Se, and Swetz, Frank. "A Chinese Mathematical Classic of the Third Century: The Sea Island Mathematical Manual of Liu Hui." *Historia Mathematica* 13(1986): 99–117.

Archibald, Raymond C. "The First Translation of Euclid's *Elements* into English and its Source." *American Mathematical Monthly* 57(1950): 443–452.

Archimedes. "The Sand Reckoner." In *The World of Mathematics,* vol. 1, J. W. Newman, ed. New York: Simon & Schuster, 1956.

Berggren, J. L. *Episodes in the Mathematics of Medieval Islam.* New York: Springer-Verlag, 1986.

Bose, D. M., ed. *A Concise History of Science in India.* New Delhi: Indian National Science Academy, 1971.

Clagett, Marshall. *Archimedes in the Middle Ages.* Madison: University of Wisconsin Press, 1964.

Collison, Mary. "The Unique Factorization Theorem: From Euclid to Gauss." *Mathematics Magazine* 53(1980): 96–100.

Daus, Paul. "Why and How We Should Correct the Mistakes in Euclid." *Mathematics Teacher* 53(1960): 576–581.

Davis, Harold T. *Alexandria, the Golden City.* 2 vols. Evanston: Principia Press of Illinois, 1957.

Dijksterhuis, E. J. *Archimedes.* New York: Humanities Press, 1957.

Diller, Aubrey. "The Ancient Measurement of the Earth." *Isis,* 40(1939): 6–9.

Durant, Will. *The Story of Civilization.* Vol. 2, *The Life of Greece.* New York: Simon & Schuster, 1939.

Eells, Walter C. "Greek Methods of Solving Quadratic Equations." *American Mathematical Monthly* 18(1911): 3–14.

Erhardt, Erika von, and Erhardt, Rudolf von. "Archimedes' Sand-Reckoner." *Isis* 33(1942): 578–602.

Fisher, Irene. "How Far Is It from Here to There?" *Mathematics Teacher* 58(1965): 123–130.

———. "The Shape and Size of the Earth." *Mathematics Teacher* 60(1967): 508–516.

Ganz, Solomon. "The Sources of Al-Khwārizmī's Algebra." *Osiris* 1(1936): 264–277.

———. "The Origin and Development of the Quadratic Equations in Babylonian, Greek and Early Arabic Algebra." *Osiris* 3(1938): 405–557.

Gould, S. H. "The Method of Archimedes." *American Mathematical Monthly* 62(1955): 264–277.

Heath, Thomas. *The Works of Archimedes.* Cambridge: Cambridge University Press, 1897. (Dover reprint, 1953).

———. *The Thirteen Books of Euclid's Elements.* 3 vols. Cambridge: Cambridge University Press, 1908. (Dover reprint, 1956).

Huxley, G. L. *Anthemius of Tralles: A Study in Later Greek Geometry.* Watertown, Mass.: Eaton Press, 1959.

Ibn Lablān, Kūshyār. *Principles of Hindu Reckoning.* Madison: University of Wisconsin Press, 1966.

Joseph, George. *The Crest of the Peacock: Non-European Roots of Mathematics.* London: Tauris, 1991.

Karpinsky, Louis. "The Algebra of Abu Kamil." *American Mathematical Monthly* 21(1914): 37–47.

Knorr, Wilbur. *The Evolution of the Euclidean Elements.* Dordrecht, Holland: D. Reidel, 1975.

———. "Problems in the Interpretation of Greek Number Theory: Euclid and the Fundamental Theorem of Arithmetic." *Studies in the History and Philosophy of Science* 7(1976): 353–368.

———. "Archimedes and the Measurement of the Circle: A New Interpretation." *Archive for History of Exact Sciences* 15(1976): 115–140.

———. "Archimedes' Lost Treatise of the Centers of Gravity of Solids." *Mathematics Intelligencer* 1(No. 2, 1978): 102–109.

Langer, R. E. "Alexandria—Shrine of Mathematics." *American Mathematical Monthly* 48(1941): 109–125.

Levey, Martin. *The Algebra of Abū Kāmil.* Madison: University of Wisconsin Press, 1966.

Li, Yan, and Shiran, Du. *Chinese Mathematics: A Concise History.* Translated by John Crossley and Anthony Lun. Oxford: Clarendon Press, 1987.

Lynch, John. *Aristotle's School: A Study of a Greek Educational Institution.* Berkeley: University of California Press, 1972.

Meder, A. E., Jr. "What Is Wrong with Euclid?" *Mathematics Teacher* 51(1958): 578–584.

Neugebauer, Otto. "Archimedes and Aristarchus." *Isis* 34(1942): 251–263.

Pixley, Loren. "Archimedes." *Mathematics Teacher* 58(1965): 634–636.

Proclus. *A Commentary on the First Book of Euclid's Elements.* Translated by Glenn Morrow. Princeton, N.J.: Princeton University Press, 1970.

Sarton, George. *A History of Science: Hellenistic Science and Culture in the Last Three Centuries* B.C. New York: W. W. Norton, 1970.

Swetz, Frank. *The Sea Island Mathematical Manual: Surveying and Mathematics in Ancient China.* University Park, Pa.: Pennsylvania State University Press, 1992.

Swetz, Frank, and Kao, T. I. *Was Pythagoras Chinese? An Examination of Right Triangle Theory in Ancient China.* Reston, Va.: National Council of Teachers of Mathematics, 1977.

Yong, Lam Lay, and Se, Ang Tian. *Fleeting Footsteps: Tracing the Conception of Arithmetic and Algebra in Ancient China.* Singapore: World Scientific, 1992.

CHAPTER 5

The Second Alexandrian School: Diophantus

When we cannot use the compass of mathematics or the touch of experience . . . . it is certain that we cannot take a single step forward.

VOLTAIRE

5.1   The Decline of Alexandrian Mathematics

The end of the third century B.C. saw the close of the Golden Age of Greek mathematics. As the next century wore on, political strife and anarchic conditions in Egypt proved more and more stifling to original scientific work and scholarship at the Alexandrian Museum. Ptolemy VII, the victor in a power struggle in 146 B.C.—unheedful of his predecessors' enlightened policies toward the arts and sciences—banished from Egypt all those scientists and scholars who had not demonstrated their loyalty to him. Alexandria's loss enriched the rest of the Mediterranean world, for learning was noticeably stimulated in those places to which the exiled Alexandrian scholars fled. According to Athenaeus of Naucratis:

> The King sent many Alexandrians into exile, filling the islands and towns with men who had been close to his brother—philologists, philosophers, mathematicians, musicians, painters, physicians and other professional men. The refugees, reduced by poverty to teaching what they knew, instructed many other men.

Until Diophantus once more brought fame to the Museum, Alexandria no longer enjoyed the primacy that it had once held over leading Eastern centers of learning.

The last two centuries of the pre-Christian era saw the steady and relentless growth of Roman power. When Rome began to expand outside of peninsular Italy, it first gained mastery over the western half of the Mediterranean basin. Syracuse, though protected by ingenious military machines that the mathematician Archimedes had devised, yielded to siege in 212 B.C., as Carthage did in 202 B.C. Then, after 200 B.C., the Roman armies turned eastward into Greece and Asia Minor. Greece proper was conquered in 146 B.C., and by 64 B.C. Mesopotamia had fallen before the Roman legions. On the Ides of March in 44 B.C., the daggers of Brutus, Cassius, and their fellow conspirators brought an abrupt end to the reign of Julius Caesar. After Caesar's death the Roman world was ruled by Caesar's grandnephew Octavian (who later received the honorific title Augustus) in the West; and in the East by Mark Antony in association with Cleopatra, Queen of Egypt. In the inevitable clash with Antony, Octavian's general Agrippa won a decisive naval battle at Actium off the west coast of Greece in 31 B.C. The suicides of Antony and Cleopatra in the following year ended the Ptolemaic dynasty. Nothing remained for Octavian but to incorporate Egypt into the dominions of the Roman people.

On August 1, 30 B.C., Octavian entered Alexandria in triumph. He visited the tomb of Alexander the Great, laying a crown of gold upon the glass coffin and scattering flowers to pay his respects. The Macedonian king whose body lay before him had lived only to the age of 32. Octavian at 32 was now the sole ruler of a world-state stretching from the Euphrates to Scotland and from the Danube to the Sahara.

With the passing of Cleopatra, Egypt was reduced to the status of a province in the Roman Empire. During Octavian's reign the empire consisted of Italy and more than thirty provinces of varying size and importance. Egypt was a Roman province of a peculiar kind; it was like a vast private estate of the emperor. With the annual sailing of the grain fleet from Alexandria, the country could send enough grain to satisfy Italy's needs for four months of every year. Because an ambitious Egyptian governor might try to starve out Rome itself, Octavian decided that it would be unsafe to put such manifest temptation in the hands of a senator. He determined instead, against all tradition, to rule the land through a military commander, whom he titled the Prefect of Alexandria and Egypt. Further, he ordained that no senator should set foot in the new province without the emperor's express permission.

The beginning of Roman rule brought a period of tranquility to Alexandria, in which the city enjoyed reasonable prosperity. It was the second city of the empire and still the greatest port on the Mediterranean Sea, with an active trade reaching westward and northward to Italy, Greece, and Asia Minor, and eastward as far as India. With some justification, Edward Gibbon, in his six-volume *Decline and Fall of the Roman Empire* (1776) could say, "If a man were called to fix the period in the history of the world during which the condition of the human race was most happy and prosperous, he would, without hesitation, name that which elapsed from the death of Domitian to the accession of Commodus (96–180 A.D.)." For Rome at its height brought to the Mediterranean peoples the blessings of Pax Romana, a durable peace the like of which had not previously been seen over so large an area and has never been seen again. With the passage of time, unfortunately, this sense of security was rarely to exist in Alexandria. The story of Roman Egypt is a sad record of short-sighted exploitation by an absentee landlord, leading inevitably to economic distress, mismanagement, and constant civil unrest. The population of Alexandria was a mixture of different cultures and ethnic groups—Greeks, Christians, Jews, and native Egyptians—who, it became increasingly clear, were unable to live together in one society without the subjugation of one group by another. By A.D. 200, the city was plagued by large, unruly mobs who at the slightest provocation sought to vent their frustrations in brawls and bloodshed. The relative stability of the 300-year reign of the Ptolemies had given way to an era of street riots and political confusion, during which the commercial and intellectual glories of Alexandria slowly but surely deteriorated.

The question of when and why Greek mathematics began to wane is both controversial and complex. Although it is always perilous to fix dividing lines in the study of history, one may safely say that under Roman rule the overall picture was one of

Roman v. Greek.

Romans—
more practical,
utilitarian—
not Θ'l

— Engineering

Agrippa's survey
team — had specialists
from Alexandria

declining mathematical activity and originality. The new masters of the Mediterranean were a practical and utilitarian people, who never showed any inclination or aptitude for extensive theoretical studies. It is remarkable that although the Roman and Greek civilizations existed over roughly the same centuries—750 B.C. to A.D. 450—in all that time there appeared no Roman mathematician of note. The chief Roman concern was the application of arithmetic and geometry to impressive engineering projects: viaducts, bridges, roads that survive even today, public buildings, and land surveys. Even among the Roman engineers, the small amount of mathematics they required could be applied in practice without any grasp of the theory behind it. Agrippa for instance, in carrying out Julius Caesar's plan of surveying the empire, was obliged to call in specialists from Alexandria to carry out the measurements. Cicero's attitude illustrated the Roman intellectuals' contempt for theoretical knowledge. In *Tuscalan Disputations* he recorded:

> The Greeks held the geometer in the highest honor; accordingly nothing made more brilliant progress among them than mathematics. But we have established as the limits of this art its usefulness in measuring and counting.

Small glimmers
Diophantus/Pappus
"silver age"

It would be wrong to conclude that Alexandrian mathematics immediately deteriorated with Roman neglect, or that the intellectual stagnation could not be temporarily arrested by exceptional individuals working in particular fields. There were occasional rallies, as in the period 250–350, when the extraordinary talents of Diophantus and Pappus succeeded in making their age a "silver age" of Greek mathematics. But cultural interests in the Roman world were by this time so completely alienated from mathematics that their brilliant work aroused but slight and passing attention.

Christianity;
it's
beginnings.

Soon after the foundation of the Roman Empire a new movement developed in Alexandria, and also in many other parts of the empire, which was to accelerate the demise of Greek learning. This was the development of Christianity. The new religion began as a sect within Palestinian Judaism, spread throughout the Roman world in spite of sporadic but repeated imperial repression, and finally won official recognition as the religion of the empire. This reversal in condition, from enemy of the government to subsidized state religion subordinate to the emperor, was to transform the future of Europe and the Mediterranean world.

It seems that initially the Christians were merely an annoyance to the Roman state in their stiff-necked refusal to acknowledge the divinity of the emperor, and the movement was allowed to develop with little interference. In the second and third centuries, as the Roman Empire was racked with internal crises and frequent invasions from without, the Church became a scapegoat on which to blame these catastrophes. As one Church father of the time, Tertulian, observed:

> If the Tiber reaches the walls, if the Nile fails to reach the fields, if the heaven withholds its rain, if the earth quakes, if there is famine, if there is pestilence, at once the cry is raised, "The Christians to the lions!"

When in 249–250 the Germanic tribes momentarily broke through the frontier defenses (in 268 even taking Athens for a short time), the emperor issued an order that all citizens should worship the traditional gods of the Roman state to gain divine support in this time of trouble. The Christians could not make the necessary sacrifices; the result was a series of savage outbursts of violence against them. The Church was still relatively small and uninfluential, comprising not more than one-third of the population in the Greek-speaking eastern part of the empire and less than 10 percent of the Latin-speaking inhabitants in the west. Had the repressions continued for a longer time the growth of the Christian movement might well have been slowed or even stopped. As it was, most emperors felt that in desperate times it was better to conciliate factions than to identify scapegoats. Even the most extended and sweeping persecution, the Great Persecution (303) initiated by Diocletian, was almost entirely restricted to the eastern empire, lasting for eight years in its European provinces and ten years in North Africa and Asia.

The fourth century saw the conversion to Christianity of a Roman emperor and the subsequent imposition of Christianity as the single official religion in the entire empire. One of the principal instigators of the Great Persecution, the Emperor Galerius, who died in 311, repented while mortally ill. Apparently thinking that the god of the Christians was punishing him, he issued an edict of universal toleration, which not only ended active persecution but also made Christianity a legal religion for the first time. Constantine the Great, who came to the throne in 312, went further; he became the first emperor to adhere personally to the Christian faith. Later in life Constantine recounted that while crossing the Alps, some time before his conquest of Italy, he had seen a flaming cross in the sky with the words, "By this sign you shall conquer." It was also reported that the day before his victorious battle of the Milvian Bridge outside Rome he was bidden, in a dream, to mark the shields of his troops with some symbol of Christianity. Although Constantine made Christianity a favored religion, he realized that the vast majority of his subjects were pagan, and he did not try to make his religion the only recognized one. Like many Christians of the time, Constantine himself put off his baptism until he lay on his deathbed—when presumably he could sin no more. In 392, Emperor Theodosius, a devout Christian, promulgated laws closing all the pagan temples in the empire and forbidding the exercise of pagan ceremonies of any kind, even those conducted in the privacy of the home. By the time Theodosius died, in 395, the empire was officially Christian.

By the fourth century, the great days of Greek mathematical thought were past. Scholars were beginning to turn their intellectual interests and energies to the debates on theological questions. The spirit of the early Church was not a spirit of scientific inquiry, for doctrines of faith were not demonstrable in terms of logic. Christianity looked inward to the mysteries of the soul, not outward to the mysteries of the natural world. Most of the significant Christian thinkers of the fourth century ridiculed physical science and mathematics, promoting the Bible as the source of all knowledge. The position taken by Saint Augustine was emblematic of an age that preferred revelation to reason: "The words of the Scripture have more authority than the whole human intellect." Certainly the idea that truth depends on divine revelation is not uniquely Christian; nevertheless, the recent success of the new religion created a climate of opinion increasingly hostile to pagan scientists and scholars. Whereas Christians were formerly persecuted, they now took steps to apply against paganism the proscriptions once enforced against them. Unfortunately, all Greek learning was identified with paganism, and Alexandrian mobs could rely on the encouragement of the Roman emperors as they looted libraries as well as pagan temples. In a period of growing antirationalism, the destruction of ancient learning was of little consequence to the majority of the people. The days of the Museum as an island of reason in a sea of ignorance were finally at an end.

The next few centuries were unhappy times for the empire as a whole. There were constant civil wars as one usurper after another rose to claim the title of emperor. Few successful claimants maintained themselves on the throne for as long as ten years. No single emperor was strong enough to deal with external threats and internal usurpations in every part of the empire at the same time, so Constantine was forced in 330 to found a new Christian capital on the old site of Byzantium. He renamed it Constantinople, which remained its name until 1930; the city then became known as Istanbul. After 330, the empire was more or less permanently divided into an eastern and a western half. In the fifth century, the Roman state in the west disintegrated before the onslaught of the invading Germanic peoples, the so-called barbarian invasions. First, Britain was overrun by the invading Saxons. Then the Vandals and kindred tribes ravaged Gaul and moved into Spain. Finally the Visigoths, followed by the Huns under Attila, sacked Italy. By this time the Church, in the person of the bishop of Rome, had taken the place of the emperor as the defender of the eternal city; twice, in 452 and again in 455, the pope went out from Rome to negotiate with the barbarian chiefs and implore them to spare the capital. The year 476 is taken by most historians as the symbolic end of the western empire. For then, the imperial forces (by now entirely German) elected one of their own generals to replace the reigning emperor and to rule under the title King of the Germans in Italy. In truth the death knell of the empire had sounded years before; there was a visible lack of loyalty to empire and emperor, and by the fifth century, few cared to save the Roman state in the West.

The eastern territories around Constantinople, which had been largely spared these invasions, remained independent and isolated for nearly a thousand years after the empire in the West had slipped into the hands of the Germans. While Europe was blanketed with barbarism and general illiteracy, the spark of Greek learning was kept alive in the Eastern Roman, or Byzantine, Empire. Science and mathematics, to be sure, were as dormant in one half of the empire as the other. But a knowledge of the Alexandrian tradition never completely died out in the East; although Byzantine scholars did not attempt original research on their own account, they were actively engaged in preserving and multiplying copies of the works of antiquity. Eight centuries would elapse before Western Europe had a second opportunity to acquaint itself with the treasures of Greek civilization. Without the efforts of the Byzantine copyists, most of the ancient scientific and literary texts would have been lost forever. There might never have been a Renaissance.

5.2   The *Arithmetica*

From the time of the discovery of irrational numbers, Greek mathematics had veered away from the purely arithmetical approach. One result was that all algebraic problems, even to the solution of simple equations, were cast in a clumsy and inflexible geometric mold. With Diophantus, next to Pappus the last great mathematician of classical antiquity, came an emancipation of algebra.

Practically nothing is known of Diophantus as an individual, save that he lived in Alexandria about the year 250. Although his works were written in Greek and he

Woodcut of the temple of knowledge, showing the gradations from the Seven Liberal Arts to the theology of Peter Lombard. (From Margarita Philosophica *(1508) of Gregor Reisch.)*

displayed the Greek genius for theoretical abstraction, Diophantus was most likely a Hellenized Babylonian. What personal particulars we have of his career come from the wording of an epigram problem (apparently dating from the fourth century) to the effect: His boyhood lasted for $\frac{1}{6}$ of his life; his beard grew after $\frac{1}{12}$ more; after $\frac{1}{7}$ more he married, and his son was born five years later; the son lived to half his father's age and the father died four years after his son. If x was the age at which Diophantus died,

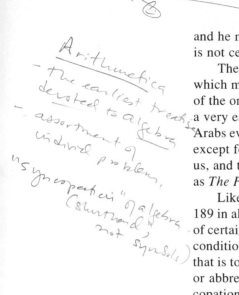

the equation becomes

$$\tfrac{1}{6}x + \tfrac{1}{12}x + \tfrac{1}{7}x + 5 + \tfrac{1}{2}x + 4 = x$$

and he must have reached an age of $x = 84$, but in what year or even in what century is not certain.

The great work on which the reputation of Diophantus rests in his *Arithmetica,* which may be described as the earliest treatise devoted to algebra. Only six books out of the original thirteen have been preserved; the missing books were apparently lost at a very early date, probably before the tenth century, for there is no indication that the Arabs ever possessed them. Of the other works attributed to Diophantus, we know little except for their titles. Fragments of a tract on polygonal numbers have come down to us, and the *Arithmetica* alludes to the existence of a collection of theorems referred to as *The Porisms,* but this is lost in its entirety.

Like the Rhind Papyrus, the *Arithmetica* is an assortment of individual problems, 189 in all, with their solutions. The apparent object was to teach the method of solution of certain problems in which it is required to find rational numbers satisfying prescribed conditions. First, a word about the notation. Before Diophantus, algebra was rhetorical, that is to say, the results were reached by verbal argument without recourse to symbols or abbreviations of any kind. One of Diophantus's main contributions was the "syncopation" of algebra. "Syncopated algebra," as it is called, is more a case of shorthand for expressing much used quantities and operations than of abstract symbolism in our sense. Diophantus had stenographic abbreviations for the unknown, successive powers of the unknown up through the sixth, equality, subtraction, and reciprocals.

Instead of our customary $x,$ he used the symbol ς for unknown quantities; this is perhaps a fusion of $\alpha\rho,$ the first two letters of *arithmos,* the Greek word for "number." The square of the unknown was denoted by $\Delta^{\Upsilon},$ the first two letters of the word *dunamis,* meaning "power." Similarly, K^{Υ} represented the cube of the unknown quantity, coming from the Greek word *kubos,* for "cube." For higher powers, he used the following abbreviating symbols:

$$\Delta^{\Upsilon}\Delta \text{ (for square-square) indicates } x^4,$$

$$\Delta K^{\Upsilon} \text{ (for square-cube) indicates } x^5,$$

$$KK^{\Upsilon} \text{ (for cube-cube) indicates } x^6.$$

Diophantus did not go beyond the sixth power, since he had no occasion to use a higher power in solving any of his problems.

The sign for subtraction was $\pitchfork$, something like an inverted ψ; and ι acted as an equals sign, connecting two sides of an equation. He had no symbol for addition but relied on juxtaposition, that is, putting terms alongside one another. In Diophantus's system of notation, the coefficients of the different powers of the unknown were represented by ordinary numerals following the power symbol:

$$K^{\Upsilon}35 \qquad \text{means} \qquad 35x^3.$$

(To avoid confusion, we have retained Arabic numerals; Diophantus would have written $K^{\Upsilon}\lambda\varepsilon$ for $35x^3$, where $\lambda\varepsilon$ stands for 35.) When there were units in the addition, they were expressed by M—an abbreviation for the Greek work *monades*, meaning "units"—with the appropriate numeral:

$$K^T35M12 \quad \text{means} \quad 35x^3 + 12.$$

Because Diophantus had no addition symbol, in an expression containing several terms with different signs, he had to place all negative terms together after the sign for subtraction. Thus, the expression $x^3 - 5x^2 + 8x - 2$ would appear as

$$K^T1 \;\varsigma8 \;\psi\; \Delta^T5M2.$$

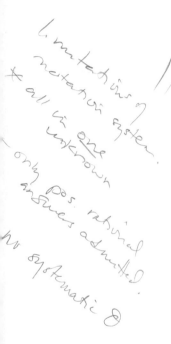

Since most of the problems in the *Arithmetica* require the determination of several quantities, Diophantus worked under a serious notational handicap. For want of other symbols besides ς to represent variables, he was compelled to reduce all his problems, no matter how complicated, to equations in one unknown. Either he expressed the other unknown quantities in terms of the one symbol, or he assigned them arbitrary values consistent with the conditions of the problem. All these eliminations were done beforehand, as a preliminary to the actual work.

Only positive rational answers were admitted, and Diophantus felt satisfied when he had found a single solution. (It made no difference to him whether the solution was integral or rational.) Diophantus had no concept of negative quantities, although he allowed for subtraction as an operation. Thus, in Problem 2 of Book V, we find his description of the equation $4x + 20 = 4$ as "absurd," because it would lead to the "impossible" solution $x = -4$. As he said, "The 4 ought to be some number greater than 20." It will be seen that his methods varied from case to case, and there was not a trace in his work of a systematic theory. Each question required its own special technique, which would often not serve for the most closely related problems.

We shall now describe several typical problems from the *Arithmetica*, though in modern notation. These will tell you more about the ingenuity of Diophantus's methods than any summary of this work could hope to do.

1. *Book I, Problem 17.* Find four numbers such that when any three of them are added together, their sum is one of four given numbers. Say the given sums are 20, 22, 24, and 27.

   Let x be the sum of all four numbers. Then the numbers are just $x - 20$, $x - 22$, $x - 24$, and $x - 27$. (For instance, if (1) + (2) + (3) = 20, then when (4) is added to both sides of this equation, $x = (1) + (2) + (3) + (4) = 20 + (4)$ or (4) $= x - 20$.) It follows that

   $$x = (x - 20) + (x - 22) + (x - 24) + (x - 27)$$

   and so $3x = 93$, or $x = 31$. The required numbers are therefore 11, 9, 7, and 4.

2. *Book II, Problem 8.* Divide a given square number, say 16, into the sum of two squares.

   Let one of the required squares be x^2. Then $16 - x^2$ must be equal to a square. Here Diophantus was satisfied to choose a particular instance of a perfect square, in this case the number $(2x - 4)^2$, so that

   $$16 - x^2 = (2x - 4)^2.$$

Diophantus's choice of $(2x - 4)^2$ was designed to eliminate the constant terms from the foregoing equation; he could just as well have picked $(3x - 4)^2$. The result was the equation

$$5x^2 = 16x,$$

with (positive) solution $x = \frac{16}{5}$. Therefore one square would be $\frac{256}{25}$, and the other, $16 - \frac{256}{25} = \frac{144}{25}$.

3. *Book II, Problem 20*. Find two numbers such that the square of either added to the other gives a square.

   Diophantus chose the numbers to be x and $2x + 1$. If these are used, the square of the first plus the second automatically becomes a square, no matter what the value of x, thereby satisfying one condition:

$$x^2 + (2x + 1) = (x + 1)^2.$$

   The square of the second number plus the first is

$$(2x + 1)^2 + x = 4x^2 + 5x + 1.$$

   To make this expression into a square, Diophantus assumed that it would equal $(2x - 2)^2$. The effect would be to produce a linear equation in x, which would also happen if one used $(2x - 3)^2$ or $(2x - 4)^2$ instead of $(2x - 2)^2$. Then

$$4x^2 + 5x + 1 = (2x - 2)^2 = 4x^2 - 8x + 4,$$

   leading to the equation $13x = 3$, or $x = \frac{3}{13}$. The desired numbers are $\frac{3}{13}$ and $\frac{19}{13}$.

4. *Book II, Problem 13*. Find a number such that if two given numbers, say 6 and 7, are subtracted from it, both remainders are squares.

   Call the number x, so that the problem is one of making $x - 6$ and $x - 7$ into perfect squares. Let

$$x - 6 = a^2 \qquad \text{and} \qquad x - 7 = b^2.$$

   Here, we see an approach that comes close to a ''method'' in Diophantus's work: the use of the algebraic identity

$$a^2 - b^2 = (a + b)(a - b).$$

   The difference $a^2 - b^2 = (x - 6) - (x - 7) = 1$ is resolved into two suitably chosen factors, from which a and b can be obtained. If one takes 2 and $\frac{1}{2}$ as the factors, setting

$$a + b = 2 \qquad \text{and} \qquad a - b = \frac{1}{2},$$

   then $a = \frac{5}{4}$, $b = \frac{3}{4}$. It follows that

$$x - 6 = \frac{25}{16}, \, x - 7 = \frac{9}{16},$$

   whence $x = \frac{121}{16}$ is the number sought.

5. *Book III, Problem 17*. Find two numbers such that their product added to either one or to their sum gives a square.

Call the numbers in question x and $4x - 1$. Then

$$x(4x - 1) + x = 4x^2 = (2x)^2,$$

so that one condition is satisfied immediately. Now it is also required that each of the expressions

$$x(4x - 1) + (4x - 1) + x = 4x^2 + 4x - 1$$

and

$$x(4x - 1) + (4x - 1) = 4x^2 + 3x - 1$$

has to be a square. Diophantus's method of solution again depends on using the identity

$$a^2 - b^2 = (a + b)(a - b).$$

It involves taking the difference between $4x^2 + 4x - 1$ and $4x^2 + 3x - 1$, namely x, separating this into the two factors $4x$ and $\frac{1}{4}$, and equating one factor with $a + b$ and the other with $a - b$. But

$$a + b = 4x, \qquad a - b = \tfrac{1}{4}$$

implies that $a = \frac{1}{2}(4x + \frac{1}{4})$ and $b = \frac{1}{2}(4x - \frac{1}{4})$. Thus

$$4x^2 + 4x - 1 = [\tfrac{1}{2}(4x + \tfrac{1}{4})]^2$$

$$4x^2 + 3x - 1 = [\tfrac{1}{2}(4x - \tfrac{1}{4})]^2$$

from either of which equations we arrive at the value $x = \frac{65}{224}$. The two numbers are therefore $\frac{65}{224}$ and $\frac{36}{224}$.

6. *Book III, Problem 21*. Divide a given number, for instance 20, into two parts and find a square whose addition to either of the parts produces a square.

Let (1) and (2) be the two parts of 20, and take

$$x^2 + 2x + 1 = (x + 1)^2$$

to be the added square. The conditions require that each of the expressions

$$(1) + (x^2 + 2x + 1)$$

and

$$(2) + (x^2 + 2x + 1)$$

should be squares. Diophantus observed that when $x^2 + 2x + 1$ was added to either $2x + 3$ or $4x + 8$, a perfect square resulted:

$$(2x + 3) + (x^2 + 2x + 1) = (x + 2)^2,$$

and

$$(4x + 8) + (x^2 + 2x + 1) = (x + 3)^2.$$

Taking $2x + 3$ and $4x + 8$ as the two parts of 20 gives $6x + 11 = 20$, whence $x = \frac{3}{2}$. The two parts of 20 are therefore 6 and 14, while the added square is $\frac{25}{4}$.

There are other possibilities. If Diophantus had called the square to be added x^2 and used the relations

$$(4x + 2) + x^2 = (x + 2)^2$$
$$(6x + 9) + x^2 = (x + 3)^2$$

then the two required parts of 20 would be $\frac{68}{20}$ and $\frac{132}{10}$.

7. *Book VI, Problem 19*. Find a right triangle such that its area added to one of its legs gives a square and its perimeter is a cube.

Using the formula for right triangles attributed to Pythagoras, Diophantus called the sides

$$2x + 1, \qquad 2x^2 + 2x, \qquad 2x^2 + 2x + 1.$$

The perimeter of the triangle would then be

$$4x^2 + 6x + 2 = 2(2x + 1)(x + 1).$$

It is difficult to make a quadratic a cube, and Diophantus, noticing the factor $x + 1$ in the expression for the perimeter, considered in turn the triangle

$$\frac{2x + 1}{x + 1}, \qquad 2x, \qquad \frac{2x^2 + 2x + 1}{x + 1}$$

obtained by dividing each of the sides by $x + 1$. This new triangle would have perimeter $2(2x + 1)$ and area $(2x^2 + x)/(x + 1)$. Adding the latter value to $(2x + 1)/(x + 1)$, one finds that

$$\frac{2x^2 + x}{x + 1} + \frac{2x + 1}{x + 1} = \frac{(2x + 1)(x + 1)}{x + 1} = 2x + 1.$$

The problem requires x to be chosen so that $2x + 1$ is a square and $2(2x + 1)$ is a cube, that is, finding a cube that is twice a square. The obvious choice is $2(2x + 1) = 8$, or $x = \frac{3}{2}$, which leads to the triangle with sides $\frac{8}{5}$, 3, and $\frac{17}{5}$.

5.3   Diophantine Equations

Diophantus was not the first to propose or solve indeterminate problems of second degree. Arithmetical problems clothed in poetic garb were a common type of mathematical recreation long before his time. Perhaps the most difficult of these—in the sense that it leads to excessively large numbers—is the famous "cattle problem." This appears in a memorandum that, according to its heading, Archimedes sent to Eratosthenes with instructions that it "be solved by those in Alexandria who occupy themselves with such matters." In essence, the problem is to calculate "the number of oxen of the Sun, which once grazed upon the isle Thrinacia [Sicily]." The wording appears to hark back to the twelfth book of Homer's *Odyssey*, in which the following line occurs: "Next you will reach the island of Thrinacia, where in great numbers feed many oxen and fat sheep of the Sun."

The cattle problem requires that one find the number of bulls and cows of each of four colors—eight unknown quantities. The first part of the problem connects the unknowns by seven simple linear equations. To add to the problem's complexity, the second part subjects the unknowns to the additional conditions that the sum of a certain pair must be a perfect square while the sum of another certain pair must be a triangular number. To be specific, if W, X, Y, Z denote the numbers of white, black, spotted, and brown bulls, and if w, x, y, z are the numbers of cows of the corresponding colors, then the relations among the numbers of bulls are

$$W = \tfrac{5}{6}X + Z, \qquad X = \tfrac{9}{20}Y + Z, \qquad Y = \tfrac{13}{42}W + Z,$$

and among the numbers of cows,

$$w = \tfrac{7}{12}(X + x), \qquad x = \tfrac{9}{20}(Y + y), \qquad y = \tfrac{11}{30}(Z + z), \qquad z = \tfrac{13}{42}(W + w);$$

and also $W + X$ is a square number, $Y + Z$ is a triangular number. When reduced to a single equation, the problem involves solving the equation

$$x^2 - 4{,}729{,}494y^2 = 1,$$

y a multiple of 9314. The problem led to what would later be known as the Pell equation. The name originated in the mistaken notion of Leonhard Euler that the English mathematician John Pell (1611–1685) was the author of the method of solution that was really the work of his countryman Lord Brouncker. Although the historical error has long been recognized, Pell's name is the one that is indelibly attached to the equation.

Many tried to solve the cattle problem, but the large numbers required to satisfy the nine conditions discouraged investigators. It was not until an article published by A. Amthor in 1880 that there was serious progress. By expanding $\sqrt{4{,}729{,}494}$ as a continued fraction, Amthor concluded that the number of cattle must be 776 . . . , where the dots represent 206,542 unknown decimal digits. In 1889 a surveyor and civil engineer, A. H. Bell, undertook to determine the exact figures needed to express Amthor's result. After nearly four years of computation by himself and two others who constituted the Hillsboro Mathematics Club of Hillsboro, Illinois, Bell specified what he believed to be 32 of the leftmost digits and 12 of the rightmost digits. The first complete solution of Archimedes' problem was given by H. C. Williams, R. A. German, and C. R. Zarnke in 1965, using a computer. They confirmed that the total number of the "cattle of the sun" is an enormous integer written in 206,545 digits, the first 30 and last 12 of which Bell correctly calculated.

A clearer idea of the magnitude of the answer can be obtained by considering the space it would take to print it. If we assume that 15 printed digits take up 1 inch of space, the number would be over $\tfrac{1}{5}$ of a mile long. The resulting value is so large that the island of Sicily, whose area is about 7 million acres, could not contain all the cattle. Moreover, there are 1397 bulls for each cow, a ratio that could lead to serious difficulties in herd management.

We have seen that Archimedes speculated about very large numbers, for the *Sand-Reckoner* was an attempt to prove that his system could be used to express the number of grains of sand in a sphere the size of the earth. But given the magnitude of the values required to fulfill the conditions of the cattle problem and the great difficulty inherent in the work, it is hardly likely that the famous geometer of Syracuse or the Alexandrian mathematicians came anywhere near its solution. They probably displayed the equations involved and left the matter at that.

Owing to its geometric significance, it is not surprising that the Pythagorean equation $x^2 + y^2 = z^2$ received attention earlier than the conceptually simpler first-degree equation $ax + by = c$, where a, b, c are known integers. Although the theory required for solving the latter equation is found in Euclid's *Elements,* it does not appear in the extant works of subsequent Greek writers. Possibly Diophantus considered the equation too trivial to be included in the *Arithmetica.* Most of his problems involved making expressions of first- or second-degree terms into squares or cubes. The earliest attempts to solve the indeterminate equation $ax + by = c$ by a general method were made in India, beginning about the fifth century, in the work of the Hindu mathematicians Aryabhata (born 476), Brahmagupta (circa 600), Mahavira (circa 850), and Bhaskara (1114–1185).

Alexander's invasion of India, and the founding of Greek kingdoms within India and on its borders, immensely stimulated the communication of ideas between Asia and the Mediterranean world. It seems likely that Indian mathematics was directly influenced and inspired by the Greeks at an early stage and affected by Chinese traditions at a later time. The whole question of which methods were evolved by the Indians themselves is the subject of much conjecture. Initially, their mathematics developed as an outgrowth of astronomy, and it is no accident that a substantial part of what has come down to us appeared as chapters in works on astronomy. Indeed there seem to have been no separate mathematical texts. Because the writers lacked algebraic symbolism, they expressed problems in verse and with a flowery style. This both pleased and attracted readers and aided the memory. Little emphasis was placed on demonstrations, so that sometimes there would be only an illustrating figure and the author's comment, ''Behold.''

In the period from 400 to 1200, the Indians developed a system of mathematics superior, in everything except geometry, to that of the Greeks. Among those who contributed to the subject, the noted astronomer Aryabhata investigated the summation of arithmetic and geometric series, drew up a table of sines of angles in the first quadrant, and tried to solve quadratic and linear indeterminate equations. In the *Aryabhatiya,* he calculated the value of π as follows:

> Add four to one hundred, multiply by eight and then add sixty-two thousand; the result is approximately the circumference of a circle of diameter twenty thousand. By this rule the relation of the circumference to diameter is given.

In other words,

$$\pi = \frac{\text{circumference}}{\text{diameter}} \approx \frac{8(100 + 4) + 62000}{20,000} = \frac{62,832}{20,000} = 3.1416,$$

a remarkably close approximation. Brahmagupta, who lived more than a century after Aryabhata, based his work largely on what his illustrious predecessor had done. His practice, however, of taking $\sqrt{10}$ as the ''neat value'' of π was somewhat of a step backward. He introduced negative numbers (the term mentioned was equivalent to our word *negative*) and developed a satisfactory rule for obtaining two roots of a quadratic equation, even in cases in which one of them was negative. Brahmagupta also gave the formula $A = \sqrt{(s - a)(s - b)(s - c)(s - d)}$ for the area of a cyclic quadrilateral whose sides are a, b, c, d and whose semiperimeter is s.

The most enduring contribution of Aryabhata and Brahmagupta was to the study of indeterminate equations, the favorite subject of Diophantus. Although they repeated many of Diophantus's problems, the approach was different. Where Diophantus sought to solve equations in the rational numbers, the Indian mathematicians admitted only positive integers as solutions. Nowadays, in honor of Diophantus, any equation in one or more unknowns that is to be solved for integral values of the unknowns is called a diophantine equation. The term is somewhat misleading, for its seems to imply that a particular equation is under consideration, whereas what is important is the nature of the required solutions.

Although Aryabhata apparently knew of a method for finding a solution of the linear diophantine equation $ax + by = c$, Brahmagupta was the first to get all possible integral solutions. In this he advanced beyond Diophantus, who had been content to give one particular solution of an indeterminate equation.

The condition for solvability of this equation is easy to state; the diophantine equation $ax + by = c$ admits a solution if and only if $d \mid c$, where $d = \gcd(a, b)$. We know that there are integers r and s for which $a = dr$ and $b = ds$. If a solution of $ax + by = c$ exists, so that $ax_0 + by_0 = c$ for suitable x_0 and y_0, then

$$c = ax_0 + by_0 = drx_0 + dsy_0 = d(rx_0 + sy_0),$$

which simply says that $d \mid c$. Conversely, assume that $d \mid c$, say $c = dt$. Now, integers x_0 and y_0 can be found satisfying $d = ax_0 + by_0$. When this relation is multiplied by t, we get

$$c = dt = (ax_0 + by_0)t = a(tx_0) + b(ty_0).$$

Hence, the diophantine equation $ax + by = c$ has $x = tx_0$ and $y = ty_0$ as a particular solution. This proves part of the following theorem.

THEOREM   *The linear diophantine equation $ax + by = c$ has a solution if and only if $d \mid c$, where $d = \gcd(a, b)$. If x_0, y_0 is any particular solution of this equation, then all other solutions are given by*

$$x = x_0 + \frac{b}{d}t, \qquad y = y_0 - \frac{a}{d}t$$

for some integer t.

Proof. To establish the second assertion of the theorem, let us suppose that a solution x_0, y_0 of the given equation is known. If x', y' is any other solution, then

$$ax_0 + by_0 = c = ax' + by',$$

which is equivalent to

$$a(x' - x_0) = b(y_0 - y').$$

There exist integers r and s such that $a = dr$, $b = ds$, and

$$\gcd(r, s) = \gcd\left(\frac{a}{d}, \frac{b}{d}\right) = 1.$$

Substituting these values into the last-written equation and cancelling the common factor d, we find that

$$r(x' - x_0) = s(y_0 - y').$$

The situation is now this: $r \mid s(y_0 - y')$, with r and s relatively prime. By Euclid's lemma, it must be the case that $r \mid (y_0 - y')$; that is, $y_0 - y' = rt$ for some integer t. Substituting, we obtain

$$x' - x_0 = st.$$

This leads us to the formulas

$$x' = x_0 + st = x_0 + \frac{b}{d}t,$$

$$y' = y_0 - rt = y_0 - \frac{a}{d}t.$$

It is easy to see that these values satisfy the diophantine equation regardless of the choice of the integer t; for,

$$ax' + by' = a\left(x_0 + \frac{b}{d}t\right) + b\left(y_0 - \frac{a}{d}t\right)$$

$$= (ax_0 + by_0) + \left(\frac{ab}{d} - \frac{ab}{d}\right)t$$

$$= c + 0 \cdot t = c.$$

Thus, there are infinitely numerous solutions of the given equation, one for each integral value of t.

Bhaskara (1114–1185) was the leading Indian mathematician of the twelfth century. His most celebrated work is the *Siddhanta Siromani* (Head Jewel of an Astronomical System), written in 1150. The contents became known to western Europe through its Arabic translation in 1587. The *Siddhanta Siromani* is arranged in four parts, of which the first two, the *Lilavati* (The Beautiful) and the *Vijaganita* (Root Extractions), deal with arithmetic and algebra, respectively. The first part is named after Bhaskara's daughter, and many of his fanciful problems are propounded in the form of questions addressed to her. For instance:

> One-fifth of a swarm of bees is resting on a kadaba bush and a third on a silindha bush; one-third of the difference between these two numbers is on a kutaja, and a single bee has flown off in the breeze, drawn by the odor of a jasmine and a pandam. Tell me, beautiful maiden, how many bees are there?

Bhaskara was celebrated as an astrologer no less than as a mathematician. There is a legend that astrologers predicted that his daughter Lilavati would never wed, but he calculated a lucky day and hour for her marriage. As the hour for this event approached, the young girl was bending over a water clock, when a pearl dropped unnoticed from her wedding headdress and chanced to stop the outflow of water. So the propitious moment passed, and since any other time was prophesied as being sure to bring misfortune, Lilavati never married. To console the unhappy girl, Bhaskara promised to give her name to a book which, "will last to the latest times."

By Bhaskara's time, mathematics in India had long since evolved from its purely utilitarian function, and problems were often posed simply for pleasure. The following is a typical problem taken from the *Lilavati:*

> Say quickly, mathematician, what is the multiplier by which 221 being multiplied, and 65 added to the product, the sum divided by 195 becomes exhausted [leaves no remainder]?

In present-day notation, the problem is equivalent to finding integers x and y that will satisfy the linear diophantine equation $221y + 65 = 195x$, or

$$195x - 221y = 65.$$

Applying Euclid's algorithm to the evaluation of gcd (195, 221), we find that

$$221 = 1 \cdot 195 + 26,$$

$$195 = 7 \cdot 26 + 13,$$

$$26 = 2 \cdot 13,$$

whence gcd (195, 221) = 13. Because $13 \mid 65$, a solution of this equation exists. To obtain 13 as a linear combination of 195 and 221, we work backward through the preceding calculations:

$$13 = 195 - 7 \cdot 26$$

$$= 195 - 7(221 - 195)$$

$$= 8 \cdot 195 + (-7)221.$$

On multiplying this relation by 5, we obtain

$$65 = 40 \cdot 195 + (-35)221,$$

so that $x = 40$ and $y = 35$ provides one solution to the diophantine equation in question. All other solutions are expressed by

$$x = 40 + (\tfrac{-221}{13})t = 40 - 17t,$$

$$y = 35 - (\tfrac{195}{13})t = 35 - 15t,$$

for any integer t. In the *Lilavati,* Bhaskara arrived at the values 6 and 5 for x and y, respectively, and noted that there were many solutions; the equation, he added, was also satisfied by $x = 57$, $y = 50$.

There is evidence of a close acquaintance between Hindu and Chinese mathematics at this time, although the question of priority has been disputed. A testimony to the algebraic abilities of Chinese scholars is provided by the contents of the *Mathematical Classic* of Chang Ch'iu-chien (sixth century), a contemporary of Aryabhata. This elaborate treatise contains one of the most famous problems in indeterminate equations, in the sense of its transmission to other societies—the problem of the "hundred fowls." This problem, which occurs in the works of Mahavira and Bhaskara, states:

> If a cock is worth 5 coins, a hen 3 coins, and three chickens together 1 coin, how many cocks, hens and chickens, totaling 100, can be bought for 100 coins?

In terms of equations, the problem would be written (if x equals the number of cocks, y the number of hens, z the number of chickens):

$$5x + 3y + \tfrac{1}{3}z = 100, \qquad x + y + z = 100.$$

Eliminating one of the unknowns, we shall wind up with a linear diophantine equation in the other two unknowns, which is the case just discussed. Specifically, since $z = 100 - x - y$, we have $5x + 3y + \tfrac{1}{3}(100 - x - y) = 100$, or

$$7x + 4y = 100.$$

This equation has the general solution $x = 4t$, $y = 25 - 7t$, which then makes $z = 75 + 3t$, where t is an arbitrary integer. Chang himself gave several answers:

$$x = 4, \qquad y = 18, \qquad z = 78;$$
$$x = 8, \qquad y = 11, \qquad z = 81;$$
$$x = 12, \qquad y = 4, \qquad z = 84.$$

A little further effort produces all solutions in the positive integers. For this, t must be chosen to satisfy simultaneously the inequalities

$$4t > 0, \qquad 25 - 7t > 0, \qquad 75 + 3t > 0.$$

The last two of these are equivalent to the requirement $-25 < t < 3\tfrac{4}{7}$. Because t must have a positive value, we conclude that $t = 1, 2, 3$, leading to precisely the values Chang obtained.

The type of word puzzle that involves simultaneous linear diophantine equations has a long history, appearing in the Chinese literature as early as the first century. In what is the oldest known instance of the "remainder theorem," Sun-Tsu asked:

> There are certain things whose number is unknown. When divided by 3, the remainder is 2; when divided by 5, the remainder is 3; when divided by 7, the remainder is 2. What will be the number of things?

Thus, we are to find an integer N that simultaneously satisfies the three equations:

$$N = 3x + 2,$$
$$N = 5y + 3,$$
$$N = 7z + 2,$$

where x, y, and z are integers. With regard to the first equation, $N - 3x = 2$, our theorem tells us that

$$N = 8 - 3t, \qquad x = 2 - t,$$

for any integer t. If this value of N is substituted in the second equation of the system, we obtain the relation

$$5y + 3t = 5.$$

Here, the general solution is given by

$$y = -5 + 3s, \qquad t = 10 - 5s,$$

where s is arbitrary. The implication is that N will be of the form

$$N = 8 - 3t = 8 - 3(10 - 5s) = -22 + 15s.$$

For N to satisfy the last equation of the system, we must have

$$7z - 15s = -24,$$

which leads to

$$z = 48 - 15r, \qquad s = 24 - 7r.$$

This yields in turn

$$N = -22 + 15s = -22 + 15(24 - 7r) = 338 - 105r.$$

All in all, $N = 338 - 105r$ provides a solution to the system of diophantine equations for any integer r. Sun-Tsu seems not to have been aware that there were an infinite number of solutions to this indeterminate problem, for $N = 23$ is the only value he gave. In fact, it is not at all certain that he had a general method of solution in mind.

To conclude, let us mention that the Indians devoted considerable effort to solving indeterminate quadratic equations, particularly the misnamed Pell equation $x^2 = 1 + ay^2$, and more generally $x^2 = c + ay^2$, where a is a nonsquare integer. Diophantus was frequently led to special cases of this equation in solving problems in his *Arithmetica;* in Problem 28 of Book II, for instance, he made $9 + 9y^2$ equal to a square x^2 by taking $x = 3y - 4$. Brahmagupta discussed the equation $x^2 = 1 + ay^2$, but its solution was first effected, as far as we know, by Bhaskara. Brahmagupta said that a person who could, within a year, solve the equation $x^2 = 1 + 92y^2$ would be a good mathematician; for those times he must at least have been an efficient arithmetician, because $x = 1151$, $y = 120$ is the smallest solution in the positive integers. In his *Lilavati,* Bhaskara found particular solutions of $x^2 = 1 + ay^2$ for the five cases $a = 8, 11, 32, 61$, and 67. In the case of $x^2 = 1 + 61y^2$, for example, the answers that were given,

$$x = 1,776,319,049, \qquad y = 22,615,390,$$

were the least positive solution. From one solution, an infinite number of integral solutions can readily be obtained using a rule Brahmagupta discovered. It amounts to the following. If p and q are one set of values of x and y satisfying $x^2 = 1 + ay^2$ and p' and q' are the same or another set, then $x = pp' + aqq'$ and $y = pq' + p'q$ give another solution. Thus, the solution $x = 17, y = 6$ of $x^2 = 1 + 8y^2$ leads to a second pair of values $x = 577, y = 204$, satisfying the equation.

5.3   Problems

Solve problems 1–12, which are from the *Arithmetica* of Diophantus.

1. *Book I, Problem 16.* Find three numbers such that when any two of them are added, the sum is one of three given numbers. Say the given sums are 20, 30, and 40.

2. *Book I, Problem 18.* Find three numbers such that the sum of any pair exceeds the third by a given amount; say the given excesses are 20, 30, and 40. [*Hint:* Let the sum of all three numbers be $2x$. Add number (3) to both sides of the equation (1) + (2) = (3) + 20 to get (3) = $x - 10$. Obtain expressions for (1) and (2) similarly.]

3. *Book I, Problem 27.* Find two numbers such that their sum and product are given numbers; say their sum is 20 and their product is 96. [*Hint:* Call the numbers $10 + x$ and $10 - x$. Then one condition is already satisfied.]

4. *Book I, Problem 28.* Find two numbers such that their sum and the sum of their squares are given numbers; say their sum is 20 and the sum of their squares is 208.

5. *Book II, Problem 10.* Find two square numbers having a given difference; say their difference is 60. [*Hint:* Take x^2 for one of the squares and $(x + a)^2$ for the other, where a is an integer chosen so that a^2 is not greater than 60.]

6. *Book II, Problem 12.* Find a number whose subtraction from two given numbers (say 9 and 21) allows both remainders to be squares. [*Hint:* Call the required number $9 - x^2$, so that one condition holds automatically.]

7. *Book II, Problem 22.* Find two numbers such that the square of either added to the sum of both gives a square. [*Hint:* If the numbers are taken to be x and $x + 1$, then one condition is satisfied.]

8. *Book III, Problem 12.* Find three numbers such that the product of any two added to the third gives a square. [*Hint:* Let the numbers be x, $x + 6$, and 9, so that one condition is satisfied.]

9. *Book III, Problem 14.* Find three numbers such that the product of any two added to the square of the third gives a square. [*Hint:* Let the numbers be x, $4x + 4$, and 1, so that two of the conditions are satisfied.]

10. *Book IV, Problem 2.* Find two numbers such that their difference and also the difference of their cubes are given numbers; say their difference is 6 and the difference of their cubes is 504. [*Hint:* Call the numbers $x + 3$ and $x - 3$.]

11. *Book IV, Problem 26.* Find two numbers such that their product added to either one gives a cube. [*Hint:* If the numbers are called $8x$ and $x^2 - 1$, then one condition holds, for $8x(x^2 - 1) + 8x = (2x)^3$.]

12. *Book VI, Problem 1.* Find a right triangle such that the hypotenuse minus either side gives a cube. [*Hint:* Consider the triangle with sides $x^2 - 4$, $4x$, and $x^2 + 4$.]

13. Which of the following diophantine equations cannot be solved?

   (a) $6x + 51y = 22$.
   (b) $33x + 14y = 115$.
   (c) $14x + 35y = 93$.

14. Determine all solutions in the integers of the following diophantine equations.

   (a) $56x + 72y = 40$.
   (b) $24x + 138y = 18$.
   (c) $221x + 35y = 11$.

15. Determine all solutions in the positive integers of the following diophantine equations.

   (a) $18x + 5y = 48$.
   (b) $123x + 57y = 30$.
   (c) $123x + 360y = 99$.

16. *Alcuin of York, 775.* A hundred bushels of grain are distributed among 100 persons in such a way that each man receives 3 bushels, each woman 2 bushels, and each child half a bushel. How many men, women, and children are there?

17. *Mahavira, 850.* There were 63 equal piles of plantain fruit put together and 7 single fruits. They were divided evenly among 23 travelers. What is the number of fruits in each pile? [*Hint:* Consider the diophantine equation $63x + 7 = 23y$.]

18. *Yen Kung, 1372.* We have an unknown number of coins. If you make 77 strings of them, you are 50 coins short; but if you make 78 strings, it is exact.

How many coins are there? [*Hint:* If N is the number of coins, then $N = 77x + 27 = 78y$ for integers x and y.]

19. *Christoff Rudolff, 1526.* Find the number of men, women, and children in a company of 20 persons, if together they pay 20 coins, each man paying 3, each woman 2, and each child $\frac{1}{2}$.

20. *Euler, 1770.* Divide 100 into two summands such that one is divisible by 7 and the other by 11.

21. *Bhaskara, 1150.* What number divided by 6 leaves a remainder of 5, divided by 5 leaves a remainder of 4, divided by 4 leaves a remainder of 3, and divided by 3 leaves a remainder of 2?

22. *Fibonacci, 1202.* Find a multiple of 7 having the remainders of 1, 2, 3, 4, 5 when divided by 2, 3, 4, 5, 6.

23. *Regiomontanus, 1436–1476.* Find a number having remainders 3, 11, 15 when divided by 10, 13, 17.

5.4   The Later Commentators

The time of Diophantus brings us to the final stages of Hellenistic mathematics. Although nominally the School of Alexandria continued to exist for several hundred years more, the days of creative scholarship were over. Those who follow Diophantus are known mainly for their commentaries on earlier treatises. There are, however, a few mathematicians of this period whose names deserve particular mention. The most notable, the last in a long line of accomplished geometers, is Pappus of Alexandria. Although nowhere near the equal of Archimedes, Apollonius, or Euclid, who flourished five centuries earlier, Pappus towered above his contemporaries.

The great work on which his reputation rests is the *Mathematical Collection,* originally written in eight books, of which the first and part of the second are missing. Only one of Pappus's other writings has survived, and that in fragmentary form, namely his commentary on Ptolemy's *Almagest.* We can fairly well date when Pappus lived, for in his commentary on the *Almagest,* Pappus referred to an eclipse of the sun that took place in the year 320, and he spoke as though it were an eclipse he had recently seen. Proclus (who died in Athens in 484) quoted Pappus several times in his *Commentary on the First Book of Euclid's Elements,* so that it is reasonable to infer that Pappus thrived in the first half of the fourth century A.D.

The *Mathematical Collection* of Pappus was intended to be a consolidation of the geometric knowledge of its time. The books contain theorems of all kinds about proportion, solid geometry, and higher plane curves, and also contributions to mechanics—which was at that period regarded as part of mathematics. The design of the *Collection* was to give a synopsis of the contents on the great mathematical works of the past and then to clarify any obscure passages through various alternative proofs and supplementary lemmas. Not content with mere description of an earlier treatise, Pappus went on at several points to extend and generalize the results of his predecessor. The first proposition of Book IV, for instance, contains Pappus's generalization of the Pythagorean theorem on the square of the hypotenuse. Pappus himself made many notable contributions. One of these was the discovery that the quadratrix of Hippias could be obtained as the intersection of a cone of revolution with a right cylinder whose base was the spiral of Archimedes. Most striking is his theorem on the generation of a solid by the revolution of a plane area about an axis: The volume of the solid of revolution is equal to the product of the rotating plane area and the distance traversed by its center of gravity.

Part of Book III of the *Collection* is devoted to the three classical problems of antiquity: the quadrature of the circle, the duplication of the cube, and the trisection of an angle. After reproducing the "solutions" of the various ancient geometers, Pappus virtually stated that the problems were impossible of solution under the terms in which they had been formulated by the Greeks; for they did not belong among the "plane problems," nor among problems solvable by straightedge and compass:

> The earlier geometers were not able to solve the aforementioned problem about the angle, when they sought to do so by means of planes [plane methods], because by nature it is solid; for they were not familiar with the sections of a cone, and for this reason were at a loss.

The nineteenth century saw the curtain fall on the three famous problems of Greek geometry with formal proof of their insolubility under Platonic conditions.

Though the *Collection* of Pappus is not of the same order as the earlier classics of the Alexandrian School, it is an invaluable record of parts of mathematics that would otherwise be unknown. Of all the extant Greek works, the *Collection* is richest in information on the lost treatises of the ancient geometers and particularly on the missing books of Euclid and Apollonius. Many results of ancient authors are available to us only in the form in which Pappus preserved them. Despite Pappus's attempt to arouse interest in the traditional geometry of the Greeks, this study was never effectively revived and it practically ceased to be a living interest.

An unhappy consequence of the conversion of the fourth-century Roman emperors to Christianity was that the role of the persecuted was now shifted to the pagans. When the Greek temples were ordered razed, the immense library housed in the Temple of Serapis, consisting of over 300,000 rolls of manuscripts, fell prey to the vandalism of fanatics. With it, a painfully accumulated record of centuries of genius was heedlessly wiped away. Not content with eradicating "pagan science" by the torch, Christian mobs murdered many of the Museum's scholars in the streets of Alexandria. Such was the fate of the first prominent woman mathematician, Hypatia, daughter and pupil of Theon of Alexandria.

Hypatia (370–415) was distinguished in mathematics, medicine, and philosophy and is reported to have written a commentary on the first six books of Diophantus's *Arithmetica,* as well as a treatise on Apollonius's *Conic Sections.* From her father she had obtained a knowledge of the astronomical discoveries of Claudius Ptolemy, and she edited the *Almagest* of this great astronomer. She was also the unquestioned leader of the neo-Platonic school of philosophy and took part in the last attempt to oppose the Christian religion. As a living symbol of the old culture, she was destined to be a pawn in a struggle for political mastery of Alexandria.

Hypatia lectured at the Museum on mathematics and philosophy, and her classes attracted many distinguished listeners. Among these was the philosopher Synesius of Cyrene, who was later to become bishop of Ptolemais. In the letters of Synesius that have come down to us, he always spoke of Hypatia in the highest terms, calling her "mother, sister, reverend teacher," and praising both her learning and her virtue. In spite of support from Synesius and other Christians, the Christian leaders regarded Hypatia's neo-Platonic philosophy as heretical. Her position was further threatened by her friendship with Orestes, Roman governor of the city and the only countervailing force to Cyril, Bishop of Alexandria. The followers of Cyril spread rumors that Hypatia was drawing large crowds to her lecture hall, where under the guise of scholarship, she was expounding paganism—and that moreover, her influence over Orestes was the only obstacle to a reconciliation between the governor and himself. As she returned one day from her classes, Hypatia was waylaid by a mob of religious zealots, slashed by sharp oyster shells, and finally torn limb from limb, her remains delivered to the flames. With the death of Hypatia, the long and glorious history of Greek mathematics was at an end.

Although it seems likely that the greater part of the Museum and its library was plundered by the Christians well before Alexandria was taken by the Moslems in 641, the Moslems burned what books were still left. Accounts say that the Arab military governor, perplexed by the hoard of writings gathering dust in the library, referred the matter to Mecca for advice. According to the Christian writer Bar-Hebiaeus, the reply came: "Either the manuscripts contain what is in the Koran, in which case we do not have to read them, or they contain what is contrary to the Koran, in which case we must not read them." In either event, their destruction was decreed; the contents of the library were distributed among the public baths, of which there were some 4000, where they served to supply the fires for the next six months. The story is probably based on truth, but the library was then only a ghost of the past. A great stock of writings appears to have been destroyed at the time of Julius Caesar's siege of Alexandria in 48 B.C. Caesar, fearing that he would be cut off by sea, sent an incendiary crew to set fire to the Egyptian fleet, which had been left undefended in the harbor. The conflagration spread to the wharves and warehouses, and before it could be brought under control, consumed the original library building. Part of the loss was recouped when Marc Antony presented 200,000 volumes from the library of Pergamon to Cleopatra.

The successors of Diophantus are noted mainly for translating and commenting on the writings of earlier scholars. The traditions of Alexandrian mathematics became more remote as one commentator after another skimmed the surface of his predecessors' work to produce a volume that would gain a wider audience. At a time of general decay of learning, originality often gave way to fraudulent scholarship—the indiscriminate appropriation of materials without acknowledgment or the falsification of sources to gain the appearance of quoting Greek treatises. There were, to be sure, reputable commentators who showed some knowledge of the mathematics of the Golden Age, most notably Proclus (410–485) and Boethius (475–524); their own work, even though a pale reflection of what Greek mathematics was at its highest moment, furnishes a link between classical and medieval learning.

Proclus received his early training in Alexandria but spent most of his life at Athens, where he was head of the Academy of Plato. The range and volume of his production was enormous, extending from philosophy and theology through mathematics, physics, and astronomy, to literary criticism and poetry. Although Proclus lived a good thousand years after the inception of Greek mathematics, he had access to numerous historical and critical works that have since vanished completely. The historical work whose loss is most deeply to be deplored is the great *History of Geometry* by Eudemus, the pupil of Aristotle. Luckily a brief outline (called the Eudemian Summary) has been preserved by Proclus in his *Commentary on the First Book of Euclid's Elements*. Even in its fragmentary form, the history is of incomparable value as our main source of information on Greek geometry from Thales to Euclid.

Woodcut showing a contest between the old and new arithmetic, symbolized by Boethius (left) using the Hindu-Arabic numerals and Pythagoras (right) still reckoning with a counting board. (From Margarita Philosophica *(1508) of Gregor Reisch.)*

For those who are interested in tracing developments in theoretical mathematics, the Roman period is singularly barren of interest. However excellent the Romans may have been in the arts, literature, and law, they showed no disposition to master the Greek sciences, let alone to add to them. As the early Church emerged from the catacombs, people argued less about mathematics and more about salvation. Learning of any kind was deemed useful as it was necessary for the proper understanding of the

Scriptures and the writings of the Church fathers. The trivium of liberal arts became the accepted format of Christian education; and within the trivium, the study of grammar and rhetoric received far more attention than what was devoted to logic. Contributing to this lack of interest in theoretical studies was the fact that a knowledge of the Greek language, in which much of the scientific learning of antiquity remained, gradually faded in the Latin-speaking West.

Perhaps the best known Roman commentator to interest himself in the Greek works then available was Anicius Boethius (circa 475–524). Memorably characterized as "the last of the Romans and the first of the Scholastics," Boethius provided a bridge between Antiquity and the Middle Ages. Born into one of the wealthy and illustrious families of senatorial rank, he received the best education to be had in those troubled times. Scholars disagree over where Boethius was educated, some favoring Athens and others Alexandria. As a young man Boethius entered the Roman administrative system: The Ostrogothic king Theodoric, who had ruled Italy since 493, needed the experience of the old Roman aristocracy in his task of governing. Holding a number of trusted positions, Boethius reached the height of his political power in 522 when he became Master of the Offices, a post in which he functioned virtually as the king's prime minister. Soon afterwards, Boethius fell out of Theodoric's favor and was accused of treasonable conduct. The official charge—widely accepted now as unjust—was that he corresponded with the Byzantine Emperor Justinian in a conspiracy to overthrow Theodoric. In prison awaiting execution, Boethius wrote *The Consolation of Philosophy,* one of the classics of Western thought.

Boethius, realizing the sad state of the sciences and aware of his own command of Greek, had previously embarked on the ambitious program of providing the scholars of his day with textbooks on all four subjects of the quadrivium (arithmetic, geometry, music, and astronomy). His geometry consisted of nothing more than definitions and statements of theorems—with no proofs—from Books I, III, and IV of the *Elements,* along with various practical applications. Boethius's popular work, *De Institutione Arithmetica,* is actually a paraphrase, bordering on a translation, of the *Introductio Arithmeticae* of Nicomachus. Although occasionally adding material and condensing portions of the original, Boethius contributed nothing really new. He was not an expert mathematician, and in his departures from Nicomachus he was trying to exhibit his own talents, after the fashion of Latin commentators. Yet such was the poverty of mathematical learning of the time that it is mainly through Boethius that the Middle Ages came to know the principles of formal arithmetic. His *Arithmetica* remained for over a thousand years the authoritative text on the subject in monastic schools (that the Church proclaimed him a martyr no doubt helped too). Indeed, the last known edition of Boethius's *Arithmetica* was published in Paris in 1521. In the East, meanwhile, the Greek masterpieces were being zealously preserved, studied, and recopied by each generation. After the original texts were rediscovered by the Latin West in the fifteenth century, Boethius sank into an obscurity that became as great as his reputation once was.

Magnus Aurelius Cassiodorus (circa 480–575), a younger friend of Boethius and not so great a scholar, yet made a more substantial contribution to the preservation of the classical heritage. Like Boethius, he was a Roman aristocrat who rose to high position in the government of Theodoric. Upon retiring from public life to his estate

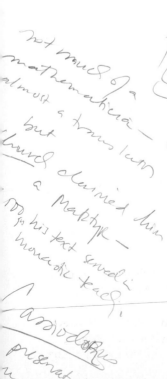

at Vivarium, Cassiodorus founded a large monastery with the conscious aim of making it a center of Christian learning and scholarship—the first education-oriented monastic house. This involved the creation of a scriptorium for the translation into Latin of the classical texts to be studied; copies were made for their own library and to be sent to monasteries that were less well equipped. For the education of his monks and to facilitate their teaching of others, Cassiodorus composed the *Introduction to Divine and Human Writings*. Most of it was devoted to holy scripture and the works of the church fathers, but he did offer a brief discussion of each of the seven liberal arts. This primitive textbook served as the basis of the curriculum of the church schools in the early Middle Ages.

5.4 Problems

1. Let the cubic equation $ax^3 + bx^2 + cx + d = 0$ have integral coefficients a, b, c, d.

    (a) Prove that if this equation has a rational root r/s, where r and s are relatively prime, then r divides d, and s divides a. [*Hint:* Substitute $x = r/s$ in the equation, clear of fractions, and use Euclid's lemma.]
    (b) Show that if $a = 1$, every rational root of the cubic must be an integer that divides the constant term d.

2. Find the rational roots of the following cubic equations.

    (a) $2x^3 - 5x^2 - 2x + 15 = 0$.
    (b) $32x^3 - 6x - 1 = 0$.
    (c) $6x^3 - x^2 - 4x - 1 = 0$.
    (d) $x^3 - 7x^2 + 20x - 24 = 0$.
    (e) $x^3 - 2x^2 + 7x + 2 = 0$.

3. A real number r is said to be constructible if there exists a line segment of length $|r|$ that can be constructed by straightedge and compass from a given line segment of unit length. (Because there is a one-to-one correspondence between constructions by straightedge and compass and algebraic operations that are purely rational or involve real square roots, this translates into: A real number r is constructible if it can be calculated from 0 and 1 by a finite number of additions, subtractions,

multiplications, and divisions, and extractions of square roots.) The following theorem is well known: If the cubic equation

$$ax^3 + bx^2 + cx + d = 0$$
$$a, b, c, d \text{ integers,}$$

has a constructible real number as a root, then it has a rational root. Show the impossibility of constructing, with straightedge and compass, the side of a cube of a volume twice that of a given cube. [*Hint:* If the original cube has side of length 1, the edge of the desired cube must satisfy $x^3 - 2 = 0$. Show that this equation has no constructible real number as its root.]

4. The relation between the angle α (that is, the angle whose measure is α) and the numbers

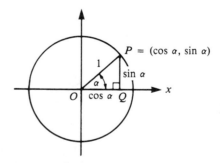

sin α and cos α is exhibited in the accompanying figure. Prove that it is possible to construct an angle α with the aid of a straightedge and compass if and only if either sin α or cos α is a constructible real number.

5. Establish that it is impossible by straightedge and compass alone to trisect the angle 60°. [*Hint:* A 60° angle can be trisected if and only if it is possible to construct a 20° angle. From the trigonometric identity

$$\cos 3\alpha = 4 \cos^3 \alpha - 3 \cos \alpha$$

applied to $\alpha = 20°$, it is found that

$$\tfrac{1}{2} = 4 \cos^3 20° - 3 \cos 20°.$$

If we let $x = \cos 20°$, this can be rewritten as a cubic equation $8x^3 - 6x - 1 = 0$.]

6. (a) Show by means of the trigonometric identity $\cos 3\alpha = 4 \cos^3 \alpha - 3 \cos \alpha$ that the angle 90° can be trisected using only a straightedge and compass.

    (b) Show by means of the trigonometric identity $\sin 3\alpha = 3 \sin \alpha - 4 \sin^3 \alpha$ that the angle 30° cannot be trisected.

Bibliography

Cantor, Norman. *Medieval History: The Life and Death of a Civilization.* 2d ed. London: Macmillan, 1970.

Davis, Harold T. *Alexandria, the Golden City.* 2 vols. Evanston: Principia Press of Illinois, 1957.

Deakin, Michael. ''Hypatia and Her Mathematics.'' *American Mathematical Monthly* 101(1994): 234–243.

Gibbon, Edward. *The Decline and Fall of the Roman Empire.* 3 vols. New York: Modern Library (Random House), 1977.

Heath, Thomas. *Diophantus of Alexandria: A Study in the History of Greek Algebra.* 2d ed. New York: Cambridge University Press, 1910. (Dover reprint, 1964).

Hughes, Barnabas. ''Rhetoric, Anyone?'' *Mathematics Teacher* 63(1970): 267–270.

Kingsley, Charles. *Hypatia, or New Foes with Old Faces.* Chicago: W. B. Conkley, 1853.

Schrader, Dorothy. ''*De Arithmetica, Book I,* of Boethius.'' *Mathematics Teacher* 61(1968): 615–628.

Stahl, W. H. *Roman Science: Origins, Development and Influence to the Late Middle Ages.* Madison: University of Wisconsin Press, 1962.

Swift, J. D. ''Diophantus of Alexandria.'' *American Mathematical Monthly* 63(1956): 163–170.

CHAPTER **6**

The First Awakening: Fibonacci

Algebra is generous, she often gives more than is asked of her.

D'ALEMBERT

6.1 The Decline and Revival of Learning

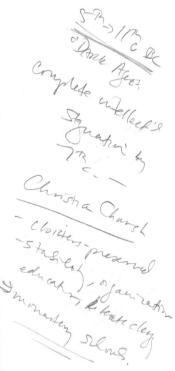

In western Europe the period from the barbarian invasions of the fifth century until the eleventh century, often called the Dark Ages, marked the low ebb of mathematics. Roman culture and thought persisted in the early part of the sixth century—at least in certain places, such as Italy and Southern Gaul, where the Latin language was still in use. But by the beginning of the seventh century almost nothing remained of the civilization that had flourished for a millenium in the Mediterranean lands. This was the darkest part of Europe's Dark Ages. The time was one of complete intellectual stagnation, no less so in mathematics than in science and philosophy. As a cloud of ignorance began to settle over the greater part of Europe, the Christian Church became the sole custodian of intellectual life, preserving in the cloisters of a few monasteries what feeble spark of learning remained.

The part played by the Church at this time cannot be too greatly stressed; for as the Roman Empire in the West collapsed, the Church emerged as the one stable institution among the ruins. It alone had the organization, the dedication, and the educated men to provide the leadership that was so badly needed by the new society that was coming into being. Although the Church did not immediately step into the place of the state as the provider and director of learning, it was not long before the Church was compelled, if only to provide a literate clergy, to concern itself with education. There ultimately grew up a system of monastery schools that by the end of the Dark Ages was almost as complete and comprehensive as the municipal system that had passed away with the Roman Empire.

It cannot be said that the early monastery schools were ideal centers of learning. The essential aim of the schools was to produce ecclesiastical leaders; and intellectual life there was nourished less by the great writers of antiquity than by the works of the fathers of the Church. Thus they taught, within the boundaries fixed by the Church's interests and doctrines, the bare elements of reading and writing rudimentary Latin, and summary explanations of Biblical texts. Suspicion of the ancient pagan authors, which had contributed to the decline of learning in the early Christian era, lingered. Secular books were no longer read, except possibly in the form of extracts illustrating some moral or grammatical point. Even Latin literature was studied more as a means of educating the student in the writing of serviceable Latin than for the ideas involved.

Yet the intellectual heritage of the Roman world did survive all the negligence of the darkest ages. The contents of old manuscripts were preserved for later times only because they were copied during this period, and many others became better known because they were reproduced in different regions by monastic scribes. Although the scribe may have lacked a deep appreciation of the classical texts he was copying, he nonetheless preserved nearly all that was valuable in the Latin writings of the ancient world. By rescuing the remains of classical literature from destruction, he became the conduit that preserved the ancient culture from extinction and enabled its emergence into a European civilization.

Certain kings gave to educational work in the Dark Ages a support second in importance only to that given by the Church. For a brief spell at the close of the eighth century, and during the whole of the ninth, Europe witnessed a remarkable resurgence of its intellectual strength brought on by a combination of favorable circumstances and extraordinary individuals. This revival of learning, usually described as the Carolingian Renaissance, had its origin and focus at the court of the Frankish king Charlemagne (742–814). On Christmas Day of 800, while Charlemagne was kneeling in prayer at St. Peter's, the pope suddenly placed on his head a golden crown and hailed him as Holy Roman Emperor. The acclamation was well deserved, for Charlemagne was not only the most powerful ruler in Europe, for all practical purposes he was the only ruler. His dominion encompassed what is now France, western Germany, parts of Austria, and Italy as far south as Rome.

Early in his reign Charlemagne realized that drastic reforms would be necessary to alleviate the pitiful conditions of ignorance among the clergy and the civil servants in his government. An important necessity for this reconstruction was a new curriculum and a new educational system. Charlemagne therefore invited the most renowned scholar of the day, the Englishman Alcuin of York, to become his educational advisor. Alcuin was eminently successful in accomplishing the tasks Charlemagne set before him. About 789, he ordained that every abbey and monastery throughout the realm should have its own school, with the seven liberal arts as divided into the quadrivium (arithmetic, geometry, astronomy, and music) and the trivium (grammar, rhetoric, and logic) a firm part of the curriculum. Alcuin even dictated how these subjects should be taught, by writing elementary textbooks for each of them.

Because a necessary condition for a revival of learning was the wide distribution of manuscripts, he became an energetic searcher after books. Emissaries were sent to Ireland, Spain, and Italy for texts that could be copied for the students to use. He also encouraged a form of writing by the introduction of a rounded, well-proportioned, and perfectly legible script that could be easily written and read. This was known as "Carolingian minuscule," and its advantages were so great that it was adopted by virtually all the Italian printers of the fifteenth century—and is the source of our printed alphabet today. Under Alcuin's tutelage, the palace school at Aachen was transformed from a prominent center of court etiquette into a genuine place of learning. According to tradition, the king himself attended classes there, along with all the members of his family and the young nobles that he had marked for high position in church and state. In 796 Alcuin retired from the court and became abbot of the preeminent monastery of St. Martin at Tours, where he continued to teach and collect manuscripts.

Mathematics was relatively nonessential to the needs of a society that was still struggling with basic literacy, so that learning imparted under the heading of the quadrivium, with its strong mathematical base, was vulnerable to neglect in Charlemagne's schools. Indeed, within the trivium the study of grammar and rhetoric received far more attention than logic did. Thus it is not surprising that while the ninth century's "little Renaissance" produced marked educational advances, there was no notable change in the mathematical climate of western Europe. Even Euclid's *Elements* was lacking in Western libraries, so that the standard authority of the day on geometry was Boethius, whose shortcomings have already been noted. On the practical side, little was taught beyond the arithmetical operations needed for calculating the ecclesiastical calendar, the most pressing problem of which was establishing the exact date of Easter and other movable feast days. The school curriculum also included lessons on logical and mathematical thinking. A work ascribed to Alcuin, *Propositions for Sharpening Youthful Minds,* presented 53 puzzles in arithmetic. Although some could be solved through elaborate calculations, many required mathematical ingenuity. The best-known puzzle is the problem of three men and their three sisters having to cross a river in a boat holding only two people, where it is assumed that to be safe each girl must have no other companion than her brother.

The Carolingian revival was short-lived, for just when it seemed that Charlemagne had solved the problem of European political disunity, the final wave of barbarian invasion broke over the West. In the ninth and tenth centuries, Vikings from the North, Magyars from the East, and Saracens from the South simultaneously plundered the coasts, plains, and river valleys of the Frankish kingdom. Weakened internally by the question of royal succession and assaulted from without by these new invaders, the Carolingian empire fell apart into pieces that would one day become the separate nations of France, Germany, and Italy. Of the three incursions, the one by the Vikings was the most persistent and the most serious; for 200 years the Northmen kept the whole of the West in a state of turmoil, laying waste the lands on the seaboard. A strong government might have repelled or lessened this evil, but the rising nations of western Europe were everywhere still weak and completely incapable of dealing with the marauders.

The failure of Christian Europe to realize political unity under the Carolingian empire did not mean that just when it had started to develop as a cultural entity, it sank back into complete barbarism. Generally the educational establishment Charlemagne created continued to function during this period of torment, so that there always remained centers where learning was cherished. The losses that took place through the failure of some monastery schools were constantly made good by the fresh efforts of others. Never again would Europe face the possible extinction of literacy that had been the danger in the seventh century.

The slow rise of science and mathematics to renewed prominence during the eleventh century corresponded to another transformation, that of the schools themselves. The monastic and palace schools of Carolingian Europe were the intellectual arenas for the revival of learning in the ninth century; here, all education aimed at a better understanding of Scripture and other sacred texts. Anyway, the monastery schools, irregularly and unpredictably staffed, were never intended for educating a large segment of society. The orientation of education changed as the foremost teachers and students of the time were attracted instead to the famous cathedral schools—among them Cologne, Tours, Liege, Chartres, Reims, and Paris. It was inevitable that in time the cathedral schools would themselves prove inadequate for the numbers who wished to attend them. Teachers who were not members of the school settled in its vicinity and, with the sanction of the authorities, gave lectures on subjects that had no place in the circumscribed intellectual world of the Church. At first a system of private initiative prevailed; teachers dispensed instruction in return for fees. Students transferred from one master to another at will, and bitter competition for students often took place. These early associations of students grouped around individual teachers paved the way for the universities of the thirteenth and fourteenth centuries. The acquisition of increased numbers of Latin versions of Arabic and Greek scientific texts transformed the curriculum, so that the trivium steadily declined in importance as logic and mathematics came to occupy the most prominent place in the scheme of study. But these radical changes in both the substance and social character of learning carry us somewhat ahead of our story. For in the eleventh century the immediate future of learning lay not in the West but in the East, where the increasing splendor of Arabic civilization was set off against the continued intellectual darkness of Europe.

Two far-reaching movements of peoples had destroyed the last remnants of Mediterranean unity. The first was the continuing influx, across the Rhine and the Danube, of the Germanic tribes; the second, the rise of a new religious grouping of the Arab world, the religion of Islam. The expansion of Islam occurred over exactly one hundred years—from the death of Mohammed in 632 until the battle of Tours in 732, when the Arab armies, having penetrated the very heartland of France, were checked by Charles Martel (the grandfather of Charlemagne). The defeat at Tours put a stop to further Arab advances to the north, and the Arabs remained satisfied with bringing all Spain under their rule. In the year (711) in which they landed in Spain, the Arabs were battering, less successfully, at the gates of Constantinople. The capital of the Byzantine Empire managed to survive Arab assaults until the fifteenth century and thereby saved western Europe from Moslem conquest via the Balkan peninsula. As a result of the Arabic conquests, three sharply contrasting civilizations arose within the Mediterranean basin: the Byzantine, the Latin-European, and the Islamic. In varying degrees, each of these civilizations was heir to the late Roman Empire.

The Arabs who overran the southern and eastern shores of the Mediterranean brought with them nothing that could be called scholarship; their science and philosophy, like their arts, came from the lands they had conquered. The Arabs, eager to absorb new ideas, began to collect these old manuscripts that had been reproduced in sufficient numbers to survive the wars attendant on the breakup of the Roman Empire and the lack of interest of the early Christians in antique learning. Thus the Arabs met the ideas of Aristotle, Euclid, Archimedes, and Ptolemy. They rendered a lasting service to Europe by industriously translating into their own tongue what one Arabic scribe after another would devoutly call the science of the Greeks. By the tenth century, nearly all the texts of Greek science and mathematics that were to become known to Western Christendom were available in Arabic copies. A complete version of Euclid's

Elements was obtained and translated about the year 800; and Ptolemy's *Megale Syntaxis*—which became a preeminent, almost divine book—appeared in Arabic in 827 under the generally accepted name of the *Almagest*. The Arabs, by hastening to acquire the accumulated heritage of late antiquity, preserved many a classic Greek work that would otherwise have been irretrievably lost to the Latin-speaking West. This, more than anything else, was Islam's great and enduring contribution to the advancement of knowledge.

Adding significant material from Persia and India to the extensive foundation of Greek learning, the Arabs were able to build a structure of scientific and philosophical thought that was to make them the great scholars of the time. Baghdad, a new capital city established by the eastern Mohammedans, became for centuries one of the greatest centers of learning, quite surpassing any city in Western Christendom. At Baghdad were a library of immense proportions, an academy (known as the House of Wisdom) for teaching and study, and a host of translators able to take the writings of the classical past and turn them into accurate Arabic. By the tenth and eleventh centuries, mathematics was almost exclusively regarded as an Arabic science, as the perpetuation of the terms *algebra* and *Arabic numerals* indicates.

The scholars of Islam were not so much making an original contribution, however, as they were more widely disseminating the developments in mathematics that had taken place among the Persians and Hindus. Hindu mathematics had evolved independently of the influence of Greek mathematics; and unlike the Greeks who favored geometry, the Hindus had a lively interest in arithmetic and algebra. The so-called Arabic numerals, with the introduction of the all-important zero, constituted the most significant mathematical idea the Arabs borrowed from the East. The vast improvement their "new arithmetic" was over the arithmetic of the Latin world will be realized by anyone who tries to add, subtract, multiply, or divide using only Roman numerals. Arabic mathematicians also developed trigonometry for astronomical purposes, using the ratios we now call the trigonometric functions instead of the "chords of an angle" Ptolemy and the Alexandrians used. (The chord of an angle is the length of the chord standing on the arc of a circle whose radius is 60 and subtending a given angle at the center.)

The roads by which Arabic learning came to the West ran not through the eastern Mediterranean (where the Christian Crusaders captured Jerusalem in 1099), but rather through Spain and Sicily. The scientific tradition was established later in Western Islam than in the East. The Moors (western Mohammedans from that part of North Africa once known as Mauritania) crossed over into Spain early in the seventh century, bringing with them the cultural resources of the Arab world. Cordoba, with its 600 mosques and its library of 600,000 volumes, by the middle of the century had risen to be the intellectual center of the western part of the Mohammedan empire, a counterpart of Baghdad in the East. The Greek scientific writings moved westward through the Islamic world and reached Spain by the ninth century. Thus, at a time when most learned men in Christian Europe were painfully studying secondhand abstracts— sometimes clear, more often confused accounts—of Greek works, students at the great Moorish schools of Cordoba, Toledo, Seville, and Granada were jealously guarding the originals.

The impulse that Charlemagne gave to education, though losing force as time went on, sufficed to maintain a continuity of learning in Europe until the greater revival of the eleventh and twelfth centuries. The two centuries from 1050 until 1250 were ones of great intellectual excitement and social dynamism. Christendom, swollen with an increased population, armed with new feudal institutions, and inspired by the ideal of the Crusades, was everywhere pressing forward—over the Pyrenees into Moorish Spain, and into the Byzantine Empire and Palestine. Unlike the Carolingian renaissance, which was imposed artificially from above, the renaissance of the twelfth century grew spontaneously along with greatly changed material conditions. A passion for learning superseded the previous intellectual stagnation, as Europeans began to add to their inherited knowledge. The immediate problem for a Western scholar was to find out where learning was to be had and to make the effort to go and get it. This was often a difficult adventure, involving hardship, travel to remote and dangerous places, and perhaps an abjuration of faith. The discovery by Crusaders that the Moslems possessed a great store of knowledge set Europe buzzing, and to tap this new source of information scholars set out for those places at which contact between the Christian and Islamic civilizations was most intimate.

The most obvious point of contact from which the Arabic materials were passed to the Latin West was the Spanish peninsula. Spain's doors were opened by the Christian recovery of Toledo in 1085. Western students flocked to its centers of learning, eager to learn science as it was transmitted by the Arabs. As soon as it became known that the masterpieces of antiquity were locked up in the Arabic, many zealous scholars undertook to get access to them and render them into Latin. (It is useful to remember that Latin had become the exclusive vehicle for technical and intellectual subjects in the West, and remained the academic language until the eighteenth century.) The recovery of ancient science in the eleventh and twelfth centuries, augmented by what the Arabs themselves had contributed, marked a turning point in European intellectual history.

At Toledo there arose a regular school of translation of Arabic books of science, drawing from many lands those who thirsted for this knowledge. Toledo was not the only intellectual clearinghouse. There were whole regions, such as the Norman kingdom of Southern Italy and Sicily, that became, because of their open character, forums for exchanging ideas and texts. The work of translation was extremely awkward. First the Arabic text had to be read aloud, then rendered into Hebrew or current Spanish idiom; and finally a Christian translator turned it into Latin. The process was neither rapid nor free of error or misunderstanding, especially considering the intricacy of the scientific treatises. Moreover, medieval Latin was not yet equipped with an adequate supply of technical terms, so that the meaning of some of these in Arabic was imperfectly known to the translators themselves. At best, the Latin translations, having passed through the medium of two wholly different languages, were slavishly literal and reasonably accurate. At worst, the versions that finally reached the medieval student, with accumulated errors, bore but slight resemblance to the Greek originals.

The second half of the twelfth century saw the work of the most industrious and prolific of these pioneer translators from the Arabic, Gerard of Cremona (1114–1187). Although he had studied all the arts in Italy, he was especially interested in astronomy. Ptolemy's works were not available to him, so Gerard was drawn to Toledo, where he learned Arabic from a native Christian teacher. There he produced a Latin version of

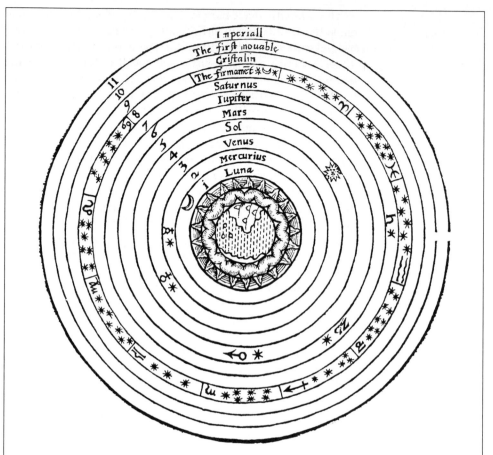

The pre-Copernican universe showing the earth as the center. From John Blagrave's The Mathematical Jewell *(1585). (Courtesy of Theatrum Orbis Terrarum Ltd.)*

Ptolemy's great work on astronomy, the *Almagest*—probably with the new Arabic numerals. Gerard devoted his life to translating scientific works from the Arabic, and it is said that more of Arabic science passed into western Europe at his hands than in any other way. He is credited with having produced Latin versions of no fewer than 90 complete Arabic texts, among them Archimedes' *Measurement of a Circle,* Apollonius's *Conic Sections,* and al-Khowârizmî's works on algebra. (It was known for some time that Gerard had translated Euclid's *Elements,* but not until 1901 was the first extant trace, Books X–XIII, found in the library of the Vatican.) The direction and scope of mathematical activity in the Middle Ages was very largely based on these translations.

Another pioneer translator was the English monk Adelard of Bath (1090–1150), who traveled far and wide—to Spain, Southern Italy, Sicily, Greece, Syria, and Palestine—seeking out the knowledge he had heard of. In the disguise of a Mohammedan student, Adelard attended lectures at Cordoba (about 1120) and succeeded in gaining an Arabic copy of Euclid's *Elements,* which he subsequently translated. In this way he made the geometry of the great Alexandrian known for the first time in the Latin West. Some 150 years after Adelard, Johannes Campanus brought out a new translation, which because of its clarity and completeness, drove the earlier Latinized versions from the field; it followed the original Greek text more closely than its predecessors, but still at some distance. This became the basis for the first printed edition of Euclid's *Elements,* which coming out in 1482, was the first mathematical book of any importance to appear in print. (By then, books by ancient and modern authors were being printed daily, yet because of the difficulty of typesetting the figures, little or nothing mathematical had appeared.)

By the late twelfth century, what amounted to a torrent of translations from Arabic works had reached Europe; and by the thirteenth century, many Greek works also had been translated. Of all that was obtained from Arabic sources, the philosophy of "the new Aristotle," that is, the scientific works of Aristotle not previously available in Latin, was prized most. Aristotle's *Physics, Metaphysics,* and *New Logic* (four advanced works on logic) had been translated and were beginning to be circulated. These writings were chiefly responsible for a shift in educational interest toward speculative philosophy and science, which the churchman and scholar John of Salisbury (1115–1180) complained were becoming preferred to the history and poetry of his youth. In 1210, the teaching of Aristotle was forbidden at the University of Paris, under pain of excommunication of the offending master.

Arabic being the new language of science, it enjoyed a greater prestige during this period than Greek. Moreover, spoken Arabic was more accessible than spoken Greek, a knowledge of which had gradually faded in the Latin-speaking West. Consequently, the practice was to make translations from the Arabic versions of Greek works and not from the original Greek. Once Sicily had fallen into Norman hands (after Arab rule from 902 until 1091), it provided a point of contact by which the original Greek classics could find their way into Europe. The region still retained a considerable Arabic-speaking population and had never broken off commercial relations with Constantinople, so conditions especially favored an exchange of ideas among Arabic, Greek, and Latin scholars. Thus there appeared in Sicily, besides translations from the Arabic, some of the earliest retranslations to be made directly from the Greek. Ptolemy's *Almagest* was first translated into Latin from Greek in Sicily in 1163, some twelve years before it was rendered from Arabic by Gerard of Cremona at Toledo. Unfortunately, this version from the Greek gained no currency, and only the version from the Arabic was available in Europe until the fifteenth century.

These struggling translators received little or no remuneration and with few exceptions enjoyed little or no fame. The only motive for their work was a devotion to truth and knowledge. Yet they accomplished a great feat; they renewed Greek science and philosophy in the West, adding to it the treasures of Arabic mathematics and medicine. Europe had never had this material before; the vast scientific and mathematical body of antique thought, from the Ionian philosophers and Aristotle to the Alexandrian mathematicians and Ptolemy, had never been translated into Latin at all. The late Roman Empire had almost abandoned the study of the Greek language, in which many of the masterpieces of antique learning remained, and Charlemagne's

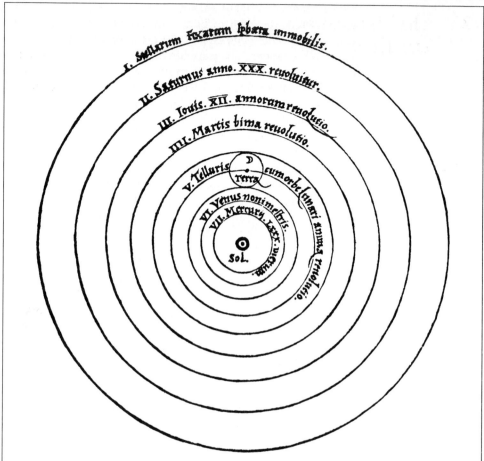

The Copernican universe with the sun as the center. From Copernicus's De Revolutionibus Orbium Coelestium *(1543). (By courtesy of Editions Culture et Civilisation.)*

scholars had been fully occupied in saving the Latin and Christian classics. The impossibility of drawing on the wellsprings of Greek culture had led to an impoverishment of knowledge and thought. But by the middle 1200s, when all that was worthwhile in Arabic learning had been transmitted to Europe through Latin translations, Western scholars stood once again on the solid foundation of Hellenistic thought. It is wonderfully fortunate that the decline of Arabic scholarship and creativity did not occur before Europe's own intellectual reawakening. When the real revival of learning came, and a genuine Renaissance took place in the 1400s, Islam had spent itself as a great force. But by then western Europe was prepared to accept the intellectual legacy bequeathed to it by earlier ages.

[handwritten margin note: Greek language had been all but lost, only access was Arab to Latin translation]

6.2 The *Liber Abaci* and *Liber Quadratorum*

The greatest mathematician of the Middle Ages was Leonardo of Pisa, better known by his other name, Fibonacci (a contraction of *filius Bonaccio,* "son of Bonaccio"). It is safe to say that the mathematical renaissance of the West dates from him. Fibonacci was born in Pisa about 1175 and educated in North Africa, where his father was in charge of a customshouse. As a young man, he traveled widely in the countries of the Mediterranean, observing and analyzing the arithmetical systems used in the commerce of the different countries. He quickly recognized the enormous advantages of the Hindu-Arabic decimal system, with its positional notation and zero symbol, over the clumsy Roman system still used in his own country. Returning to Pisa in 1202, Fibonacci wrote his famous *Liber Abaci* (Book of Counting), in which he explained the virtues of this number system "in order that the Latin race might no longer be deficient in that knowledge." The first chapter opens with the sentence:

These are the nine figures of the Indians:

9   8   7   6   5   4   3   2   1.

With these nine figures, and with this sign 0 . . . any number may be written, as will be demonstrated below.

It was chiefly by means of the second edition of this work, which appeared in 1228, that Christian Europe became acquainted with the Arabic numerals.

Arabic numerals were not entirely new to Europe; Gerard had brought the system from Spain a half-century earlier. However, no book previously produced had shown by such a wealth of examples from every field their superiority to the traditional Roman numeration. The *Liber Abaci* embodied virtually all the arithmetical knowledge of Fibonacci's time, including much Arabic science, and gave original interpretations of all this material. As the mathematical masterwork of the Middle Ages, it remained a model and source for the next several hundred years. (Curiously, although the *Liber Abaci* circulated widely in manuscript, it was not printed in Italy until 1857, nor was it translated into English.)

Although we have referred to our present number system as Hindu-Arabic, its origins are obscure and much disputed. The most widely accepted theory is that it originated in India about the third century, was carried to Baghdad in the eighth century, and finally was transmitted to Western Europe by way of Moorish Spain.

Written number symbols appeared in India before the dawn of the Christian era. One of the earliest preserved examples is found in records cut on the walls of a cave in a hill called Nana Ghat, near Bombay. If correctly interpreted, these include

—   =   ∓   ∔   Ϙ   7   ?

1   2   4?   6   7   9   (Third century B.C.)

The next important trace of numeral appears in carved inscriptions at Nasik, India. These Brahmi numerals of the second century A.D. form a ciphered system with the following first nine symbols:

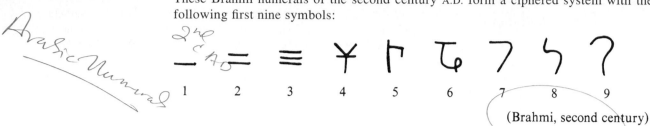

(Brahmi, second century)

Historical evidence indicates that the idea of positional notation with a zero was known in India by the fifth century, if not a century earlier. (It is clear that the form for a symbol for zero underwent changes from a mere dot to a small circle.) The numerals used in the eighth century are termed ''Devanagari,'' or ''sacred,'' numerals and the characters are essentially as shown here:

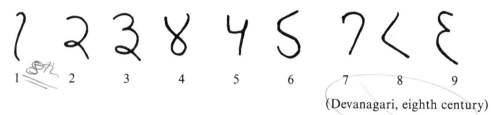

(Devanagari, eighth century)

How and when these numerals first reached the Arabs is a question that has never been satisfactorily settled. During the early Arabic expansion, public decrees were written in Greek as well as Arabic, because Greek was widely understood in the Near East. The ruling Caliph, to promote his own language, passed a law in 706 that forbade the use of Greek in favor of Arabic, but nonetheless decreed that the Greek alphabetic system could be used in writing out numbers. This indicates that the Hindu symbols had not yet penetrated as far as Damascus, the seat of the caliphs. Around 800, the system was definitely known to the Arabs. The mathematician al-Khowârizmî prepared a small book explaining the use of the Hindu numbers, including the use of zero as a place holder. When this was translated into Latin by Adelard of Bath in the 1100s, the numerals were incorrectly assumed to be of Arabic origin.

The outward appearance of the Hindu numerals went through a series of changes in transit from India, and the Arabs selected from the various shapes those most suitable for handwriting. The symbols ultimately adopted by the Western Arabs, or Moors, are the so-called Gobar numerals, from the Arab word for ''dust.'' They acquired their peculiar name from the custom of the Arab pupils who, lacking other writing materials, sprinkled white dust on a black tablet and made their computations with a stylus. It will be seen that the Gobar numerals resemble our modern numerals much more closely than the Hindu ones do:

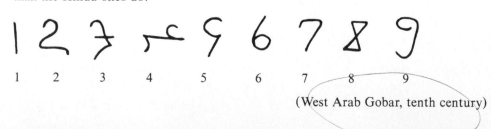

(West Arab Gobar, tenth century)

These primitive western forms appear in a tenth century edition of Boethius's *Geometry*. Because their introduction breaks the continuity of the text, probably they were not part of the original work but inserted by a copyist at a later date.

Coming closer to our present-day notation, the oldest definitely dated European manuscript known to contain the Hindu-Arabic numerals is the *Codex Vigilanus,* written in Spain in 976. The nine symbols used are

1	2	3	4	5	6	7	8	9

(Spain, 976)

What is interesting is that during their long migration from culture to culture, the Indian number signs remained astonishingly constant in form (look at the shapes for 6, 7, and 9).

At first there was stubborn resistance to the spread of the new numerals. In 1299, the city of Florence issued an ordinance forbidding merchants from using Arabic numerals in bookkeeping, ordering them either to use Roman numerals or to write out the numerical adjectives in full. This decree was probably due to the great variety of shapes of certain digits, some quite different from those now in use, and the consequent opportunity for ambiguity, misunderstanding, and outright fraud. A 0 can be changed to a 6 or 9 without difficulty, but it is not so easy to falsify Roman numerals. If we add to this the confusion and insecurity that the zero produced in the minds of ordinary people (who could understand a symbol that meant nothing at all?), and the scarcity of scrap paper cheap enough to be thrown away after the computation was finished, it is easy to see why it took so long for Arabic numerals to come into general use.

It did take a few more centuries, but the Arabic symbols were bound to win out in the end. Calculating with an abacus or a counting board and registering the results in Roman numerals was simply too slow a procedure. For the final victory no certain date can be set. Outside of Italy accounts were kept in Roman numerals until about 1550 and, in the more conservative monasteries and universities, for a hundred years longer. After printed books were introduced in 1450, the form of the Arabic numerals became standardized. Indeed, so great was the stabilizing influence of printing that the digits of today have essentially the same appearance as the digits of the fifteenth century.

Fibonacci compiled another work of note, the *Liber Quadratorum* (Book of Squares). Although the *Liber Abaci* contains a few diophantine problems, the *Liber Quadratorum* is devoted entirely to diophantine equations of second degree. In the dedication, Fibonacci related that he had been presented to the Emperor Frederick II at court and that one of Frederick's retinue, a certain John of Palermo, on that occasion propounded several problems as a test of Fibonacci's mathematical skill. One problem required that he find a number for which increasing or decreasing its square by 5 would give also a square as the result. It should be said that the problem was not original with John of Palermo, having been investigated by Arab writers with whom Fibonacci was unquestionably familiar. Fibonacci gave a correct answer, namely $\frac{41}{12}$:

$$\left(\tfrac{41}{12}\right)^2 + 5 = \left(\tfrac{49}{12}\right)^2, \qquad \left(\tfrac{41}{12}\right)^2 - 5 = \left(\tfrac{31}{12}\right)^2.$$

Leonardo of Pisa (Fibonacci)
(circa 1175–1250)

(*By courtesy of Columbia University, David Eugene Smith Collection.*)

Through considering this problem and others allied to it, Fibonacci was led to write the *Liber Quadratorum* (1225).

For some idea of the contents of this remarkable work, let us consider a typical problem from it. Solve, in the rational numbers, the pair of equations

$$x^2 + x = u^2,$$

$$x^2 - x = v^2,$$

where x, u, v are unknowns. A solution is obtained by taking any three squares that are in arithmetic progression, say the squares a^2, b^2, and c^2, and letting the common difference be d. Then

$$a^2 = b^2 - d, \qquad c^2 = b^2 + d.$$

Fibonacci proposed a solution to the problem by giving x the value b^2/d. For

$$x^2 + x = \frac{b^4}{d^2} + \frac{b^2}{d} = \frac{b^2(b^2 + d)}{d^2} = \frac{b^2 c^2}{d^2} = \left(\frac{bc}{d}\right)^2,$$

$$x^2 - x = \frac{b^4}{d^2} - \frac{b^2}{d} = \frac{b^2(b^2 - d)}{d^2} = \frac{b^2 a^2}{d^2} = \left(\frac{ba}{d}\right)^2.$$

The simplest numerical example would be $a^2 = 1$, $b^2 = 25$, $c^2 = 49$ (here, the common difference is 24), and this illustration was furnished by Fibonacci. It leads to the solution $x = 25/24$:

$$x^2 + x = (\tfrac{35}{24})^2, \qquad x^2 - x = (\tfrac{5}{24})^2.$$

At no time did it seem to occur to Fibonacci that the real question in diophantine analysis was to find all solutions, not just one.

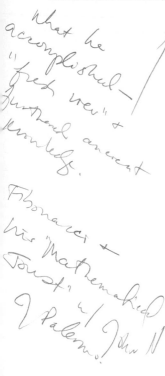

In surveying Fibonacci's activity, one must view him as a pioneer in the revival of mathematics in the Christian West. In mathematical content, his work does not surpass the work of his Arab predecessors. Fibonacci, far from being a slavish imitator of others, gave fresh consideration to the ancient knowledge and independently furthered it. Many of his proofs were original and, in some cases, his results were original also.

Fibonacci's work indicates a combination of inventive genius and a profound knowledge of earlier writers on mathematics. A striking illustration of Fibonacci's ability was his observation that the classification of irrationals given by Euclid in Book X of the *Elements* did not include all irrationals. This exists in a small treatise entitled *Flos* (meaning "blossom," or "flower"). Sent to Emperor Frederick II, the *Flos* was stimulated by the mathematical disputation held in the emperor's presence at Pisa about 1224. It is an analysis of fifteen indeterminate problems, including two of the three questions posed by John of Palermo. Fibonacci stated that the second challenge put to him was finding a cube that with two squares and ten roots should be equal to 20; in other words, the problem is to solve the equation

$$x^3 + 2x^2 + 10x = 20.$$

It is especially interesting that the first mention of a cubic in Europe after the time of the Greeks should be the result of a mathematical joust, for (as we shall see) the solution of the general cubic equation came about in connection with another problem-solving contest.

The specific cubic mentioned above can be written in the form

$$10\left(x + \frac{x^2}{5} + \frac{x^3}{10}\right) = 20,$$

so that any root x of it must satisfy

$$x + \frac{x^2}{5} + \frac{x^3}{10} = 2.$$

To see that x cannot be a rational number, we use Fibonacci's argument, which was substantially as follows. Suppose to the contrary that x were rational, say $x = a/b$, where gcd $(a, b) = 1$. The expression

$$\frac{a}{b} + \frac{a^2}{5b^2} + \frac{a^3}{10b^3} = \frac{a(10b^2 + 2ab + a^2)}{10b^3}$$

will not be an integer unless b^3 (and in turn b itself) divides $10b^2 + 2ab + a^2$. But this means that b must divide the difference

$$(10b^2 + 2ab + a^2) - (10b^2 + 2ab) = a^2,$$

which leads to the conclusion that b divides a. This contradicts the condition that gcd $(a, b) = 1$, so no rational root of the cubic equation exists.

By checking each of the cases, Fibonacci next demonstrated that a root of the equation could not be represented by any of the Euclidean irrational magnitudes

$$a \pm \sqrt{b}, \ \sqrt{a} \pm \sqrt{b}, \ \sqrt{a \pm \sqrt{b}}, \ \text{or} \ \sqrt{\sqrt{a} \pm \sqrt{b}},$$

where a and b denoted rational numbers. Hence, its construction could not be carried out with straightedge and compass only. This was the first indication that there was more to the number system than what could be constructed using the geometric algebra of the Greeks. Fibonacci contented himself with finding an accurate approximation to the required root. He gave it in sexagesimal notation, simply making the statement

$$x = 1;22,7,42,33,4,40,$$

whose value, in decimal form, is 1.3688081075. . . . This was a remarkable estimate of the only real root of the cubic equation, correct to nine decimal places; and it was the most accurate European approximation to an irrational root of an algebraic equation that would exist for the next 300 years. But we are not told how the result was found. Although Fibonacci never revealed his sources, the possibility cannot be excluded that he had learned the solution in his travels. The same problem appears in the algebra of the great Persian poet and mathematician Omar Khayyam (circa 1050–1130), where it was solved geometrically by intersecting a circle and a hyperbola.

Fibonacci, like the Arabic mathematicians before him, recognized that a quadratic equation can be satisfied by two values; yet he habitually rejected negative numbers as solutions. He took a step forward, however, in his *Flos,* when he interpreted a negative number in a financial problem to mean a loss instead of a gain.

Brahmagupta (circa 600) and Bhaskara (circa 1150), in writing common fractions, had used the scheme of placing the numerator above the denominator, without any line of separation. The Arabs at first copied the Hindu notation, but later improved on it by inserting a horizontal bar between the two numbers. Fibonacci followed the Arab practice in the *Liber Abaci.* He habitually put the fractional part of a mixed number before the integral part, with juxtaposition used to imply their addition. A kind of ascending continued fraction, which he called *fractiones in gradibus* ("step fractions") was introduced by Fibonacci. His notation $\dfrac{7\ \ 1}{10\ 10}\,8$, for example was meant to be read

$$8 + \frac{1}{10} + \frac{7}{10 \cdot 10}.$$

In the same way, the expression $\dfrac{1\ 5\ 7}{2\ 6\ 10}$ signifies

$$\frac{7}{10} + \frac{5}{10 \cdot 6} + \frac{1}{10 \cdot 6 \cdot 2}.$$

His habit of indicating numbers from right to left was influenced by the Arabs.

A significant, if less gifted, contemporary of Fibonacci was Jordanus Nemorarius, or Jordanus de Nemore (circa 1225). Virtually nothing is known with any certainty of his life or even his identity. His name appears four times in the *Biblionomia,* a library catalog compiled around 1250, so that it is reasonable to assume that he wrote during the first part of the thirteenth century. A manuscript sometimes attributed to Jordanus contains the marginal note, "This is enough to say for the instruction of the students at Toulouse"; hence, he may have lectured at the University of Toulouse, which was founded in 1229.

Such speculations aside, Jordanus is known to us only through his written works. Six of these are strictly mathematical treatises, dealing with arithmetic (number theory), algebra, and astronomy. His *De Triangulus* in particular represents medieval geometry at its highest level by giving rigorous—and frequently new—proofs of Euclidean theorems. The proofs are derived largely from Arabic sources, which were themselves based on Greek mathematical texts. Worth noting are the three proofs Jordanus gives for the classical problem of trisecting an angle, two constructions for finding continued mean proportionals between two given lines, and a proof of Heron's formula for the area of a triangle in terms of its sides (that is, $A = \sqrt{s(s - a)(s - b)(s - c)}$, where s is the semiperimeter).

The largest and most original of Jordanus's works is the *De Numeris Datis*. It is a text on advanced algebra that complements the Arabic treatises of al-Khowarîzmî and Abû Kâmil. There are 115 problems, divided into four books, offering a development of quadratic, simultaneous, and proportional equations; for the most part the material had not appeared elsewhere. The *De Datis* is wholly rhetorical, with letters of the alphabet used to represent general numbers. The format usually consists of a formal statement of the problem, a proof that more often than not appears as a series of instructions (tantamount to constructing equations), and then a specific numerical illustration. The numbers occurring in the example are written in cumbersome Roman numerals. Proposition 6 of Book IV illustrates Jordanus's approach:

> If the ratio of two numbers together with the sum of their squares is given, then each of them is known. [Proof] Let the ratio of x and y be given. Let d be the square of x and c be the square of $y;$ and let $d + c$ be known. Now the ratio of d to c is the square of the ratio of x and y. Hence the former is known. Consequently d and c are known.

This can be expressed in modern algebraic notation as follows: If $x/y = a$, $x^2 + y^2 = b$ are given and $x^2 = d$, $y^2 = c$, then $d/c = x^2/y^2 = a^2$. But $x^2 + y^2 = b$ implies that $(d/c + 1)y^2 = b$, which leads to $y = \sqrt{b/(a^2 + 1)}$. After giving his proof, Jordanus offers the example $x/y = 2$, $x^2 + y^2 = 500$. His rules provide the solution $y = \sqrt{500/(2^2 + 1)} = 10$, and so $x = 20$.

Another problem is this: If the sum of the squares of the two parts of a given number added to their difference is known, then the two parts can be found. In modern notation, the two equations are $x + y = a$, $x^2 + y^2 + x - y = b$. Here, Jordanus's example is $x + y = 10$, $x^2 + y^2 + x - y = 62$, with solution $x = 7$, $y = 3$. The text's single cubic equation occurs in the concluding proposition: $a/x^2 = b$ and $a^2/x = c$ produces the cubic $x^3 = c/b^2$.

As the first Western mathematician consistently to employ letters of the alphabet to designate quantities, known as well as unknown, Jordanus advanced the evolution of algebraic symbolism. Yet this practice was overlooked by subsequent writers in algebra for some 350 years before Francois Vièta realized the facility to be gained through Jordanus's lettering scheme.

The two central mathematical figures of the European Middle Ages, Fibonacci and Jordanus, had a notable lack of successors during the next two centuries. Although the study of mathematics was not entirely abandoned in this so-called barren period, the subject was in the hands of lesser talents who did not contribute work of lasting importance. For many of these, mathematics was a mere sideline to activities concerning the Church: in England, there was Thomas Bradwardine (1290–1349), who became Archbishop of Canterbury only a month before falling victim to the plague; in France, Nicole Oresme (1323–1349), whose career carried him from a professorship in Paris to a bishopric in Brittany; and in Germany, Nicholas Cusa (1401–1465), appointed a cardinal by Pope Nicholas V. Mathematics was concerned with practical applications during this period, so the powerful mercantile cities fostered a growing use of the Hindu-Arabic numerals and the new arithmetic that went with them, whereas interest in the more advanced algebra promoted by Fibonacci and Jordanus languished. Contemporary Western scholars, more inclined to theology and metaphysics, did not care to invest in the labor required to learn mathematics. We shall shortly see that the ideas of Fibonacci and Jordanus were to enjoy a second life when revived by the Italian algebraists during the time that has come to be called the Renaissance.

6.2   Problems

The first three problems appear in Fibonacci's *Liber Abaci*.

1. Two birds start flying from the tops of two towers 50 feet apart; one tower is 30 feet high and the other 40 feet high. Starting at the same time and flying at the same rate, the birds reach a fountain between the bases of the towers at the same moment. How far is the fountain from each tower?

2. A merchant doing business in Lucca doubled his money there and then spent 12 denarii. On leaving, he went to Florence, where he also doubled his money and spent 12 denarii. Returning home to Pisa, he there doubled his money and again spent 12 denarii, nothing remaining. How much did he have in the beginning?

3. Three men, each having denarii, found a purse containing 23 denarii. The first man said to the second, "If I take this purse, I will have twice as much as you." The second said to the third, "If I take this purse, I will have three times as much as you." The third man said to the first, "If I take this purse, I will have four times as much as you." How many denarii did each man have?

The next three problems are taken from Fibonacci's *Liber Quadratorum*.

4. Given the squares of three successive odd numbers, show that the largest square exceeds the middle square by eight more than the middle square exceeds the smallest.

5. Assuming that x and y are integers:

    (a) Find a number of the form

    $$4xy(x + y)(x - y)$$

    that is divisible by 5, the quotient being a square.

    (b) Prove that if $x + y$ is even, then the product $xy(x + y)(x - y)$ is divisible by 24, and that without this restriction, $4xy(x - y)(x + y)$ is divisible by 24. [*Hint:* Consider that any integer is of the form $3k$, $3k + 1$, or $3k + 2$ in showing that $3 \mid xy(x + y)(x - y)$. Similarly, because any integer is of the form $8k$, $8k + 1$, . . . , or $8k + 7$, then $8 \mid xy(x - y)(x + y)$.]

6. (a) Find a square number such that when twice its root is added to it or subtracted from it, one obtains other square numbers. In other words, solve a problem of the type

    $$x^2 + 2x = u^2, \qquad x^2 - 2x = v^2$$

    in the rational numbers.

(b) Find three square numbers such that the addition of the first and second, and also the addition of all three squares, produces square numbers. In other words, solve a problem of the type

$$x^2 + y^2 = u^2, \qquad x^2 + y^2 + z^2 = v^2$$

in the rational numbers. [*Hint:* Let x and y be two relatively prime integers such that $x^2 + y^2$ equals a square, say $x^2 + y^2 = u^2$. Now note the identity

$$u^2 + \left(\frac{u^2 - 1}{2}\right)^2 = \left(\frac{u^2 + 1}{2}\right)^2 .]$$

7. Fibonacci proved that if the sum of two consecutive integers is a square (that is, if $n + (n - 1) = u^2$ for some u), then the square of the larger integer will equal the sum of two nonzero squares. Verify this result and furnish several numerical examples.

8. The algebraic identity

$$(a^2 + b^2)(c^2 + d^2) = (ac + bd)^2 + (ad - bc)^2$$
$$= (ad + bc)^2 + (ac - bd)^2$$

appears in the *Liber Quadratorum*. Establish this identity and use it to express the integer $481 = 13 \cdot 37$ as the sum of two squares in two different ways.

9. (a) Given rational numbers a and b, find two other rational numbers x and y such that $a^2 + b^2 = x^2 + y^2$. [*Hint:* Choose any two integers c, d for which $c^2 + d^2$ is a square; now write $(a^2 + b^2)(c^2 + d^2)$ as a sum of two squares.]

(b) Illustrate part (a) by expressing $61 = 5^2 + 6^2$ as the sum of squares of two rational numbers.

10. Solve the following problem, which is one of the tournament problems that John of Palermo posed to Fibonacci. Each of three men owned a share in a pile of money, their shares being $\frac{1}{2}$, $\frac{1}{3}$, and $\frac{1}{6}$ of the total. Each man took some money at random until nothing was left. The first man afterward returned $\frac{1}{2}$ of what he had taken, the second $\frac{1}{3}$, and the third $\frac{1}{6}$. When the amount thus returned was divided into three equal parts and given to each man, each one had what he was originally entitled to. How much money was there in the pile at the start, and how much did each man take? [*Hint:* Let t denote the original sum, u the amount each man received when the money left in the pile was divided equally, and x, y, z the amounts the men took. Then

$$u = \frac{1}{3}\left(\frac{x}{2} + \frac{y}{3} + \frac{z}{6}\right)$$

and

$$\frac{x}{2} + u = \frac{t}{2},$$

$$\frac{2y}{3} + u = \frac{t}{3},$$

$$\frac{5z}{6} + u = \frac{t}{6},$$

which implies that $47u = 7t$.]

11. The famous French scholar Gerbert d'Aurillac (940–1003), who was later elected to the papal throne as Sylvester II, solved a problem that was considered remarkably difficult for the time, namely, to determine the sides of a right triangle whose hypotenuse a and area b^2 were given numbers. Use the technique employed in the Cairo papyrus (see page 76) to get Gerbert's answer of $\frac{1}{2}(\sqrt{a^2 + 4b^2} \pm \sqrt{a^2 - 4b^2})$.

12. Gerbert, in his *Geometry*, determined the area of an equilateral triangle of side s to be $3s^2/7$. Show that he arrived at this conclusion by taking the value of $\sqrt{3}$ to be equal to $\frac{12}{7}$.

13. The trisection of an angle can be accomplished by a construction described by Jordanus in his *De Triangulis*. Let $\angle POQ$ be a given angle. With its vertex as center, draw a circle with any radius r intersecting PO in A and QO in B. From O, draw a radius OC perpendicular to OB; now construct a chord AD cutting OC in a point E in such a way that $DE = r$. Finally, through O draw a line OF parallel to DA.

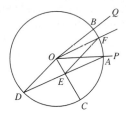

To see that $\angle FOB$ is one-third $\angle POQ$, establish the following assertions:
  (a) $OFED$ is a parallelogram, whence triangle OFE is isosceles;
  (b) $\angle OAD = \angle ODA = \angle OFE = \angle FOA = \alpha$;
  (c) The sum of the angles of triangle OFE equals

$$2(90° - \angle AOB + \alpha) + \alpha = 180°,$$

or $\alpha = (2/3)\angle AOB$.

Problems 14–17 are found in Jordanus's *De Numeris Datis*.

14. Book I, Problem 28. Solve the system of equations

$$x + y = a, \quad c/x + c/y = b.$$

For example, if each of the two parts of 10 divide 40 so that the sum is 25, what are the parts? [Hint: Notice that $xy = (ca)/b$.]

15. Book I, Problem 15. Obtain x and y if

$$x + y = a, \quad x^2 + y^2 + x - y = b.$$

For example, if the sum of the squares of the parts of 10 when increased by the difference of the parts equals 62, what are the parts? [Hint: First show that $(x - y)^2 + 2(x - y) = 2b - a^2$.]

16. Book II, Problem 22. Find the solution of the system of equations

$$x + y = a, \quad (x + c)/y = b.$$

For example, if 12 is separated into two parts so that the first increased by 2 is $\frac{3}{4}$ of the other, what are the parts?

17. Book II, Problem 16. Solve the following equations for x and y:

$$a/x = y/b, \quad x/y = c.$$

For example, if the first and fourth term of a given ratio are 18 and 2, and the ratio of the second and third equals 4, what are the second and third terms?

6.3   The Fibonacci Sequence

It is ironic that Fibonacci is remembered mainly today because a nineteenth century French number theorist, Edouard Lucas, attached his name to a sequence that appeared in a trivial problem in the *Liber Abaci*. Fibonacci posed the following problem dealing with the number of offspring of a pair of rabbits.

> A man put one pair of rabbits in a certain place entirely surrounded by a wall. How many pairs of rabbits can be produced from that pair in a year, if the nature of these rabbits is such that every month each pair bears a new pair which from the second month on becomes productive?

On the basis that none of the rabbits die, a pair is born during the first month, so that there are two pairs present. During the second month, the original pair has produced another pair. One month later, both the original pair and the firstborn pair have produced new pairs, so that two adult and three young pairs are present, and so on. The figures are tabulated in the chart.

Growth of Rabbit Colony			
Months	Adult Pairs	Young Pairs	Total
1	1	1	2
2	2	1	3
3	3	2	5
4	5	3	8
5	8	5	13
6	13	8	21
7	21	13	34
8	34	21	55
9	55	34	89
10	89	55	144
11	144	89	233
12	233	144	377

The point to remember is that each month the young pairs grow up and become adult pairs, making the new "adult" entry the previous one plus the previous "young" entry. Each of the pairs that was adult last month produces one young pair, so that the new young entry is equal to the previous adult entry.

When continued indefinitely, the sequence

$$1, 1, 2, 3, 5, 8, 13, 21, 34, 55, 89, 144, 233, 377, \ldots$$

is called the Fibonacci sequence and its terms the Fibonacci numbers. If we let F_n denote the nth Fibonacci number, then we can write this remarkable sequence as follows:

$$2 = 1 + 1 \quad \text{or} \quad F_3 = F_1 + F_2,$$
$$3 = 1 + 2 \quad \text{or} \quad F_4 = F_2 + F_3,$$
$$5 = 2 + 3 \quad \text{or} \quad F_5 = F_3 + F_4,$$
$$8 = 3 + 5 \quad \text{or} \quad F_6 = F_4 + F_5,$$
$$\vdots \qquad\qquad \vdots$$

In general, the rule for formation is easily discernible:

$$F_1 = F_2 = 1, \qquad F_n = F_{n-2} + F_{n-1} \qquad \text{for } n \geq 3.$$

That is, each term in the sequence (after the second) is the sum of the two that immediately precede it. Such sequences, in which from a certain point forward every term can be represented as a linear combination of preceding terms, are "recursive sequences.'' The Fibonacci sequence is one of the earliest recursive sequences in mathematical work. Fibonacci himself was probably aware of the recursive nature of his sequence, but not until 1634—by which time mathematical notation had made sufficient progress—did Albert Girard write the formula in his posthumously published work *L'Arithmetique de Simon Stevin de Bruges.*

It may not have escaped your attention that successive terms of the Fibonacci sequence are relatively prime. We will establish this fact next.

THEOREM *No two consecutive Fibonacci numbers F_n and F_{n+1} have a factor $d > 1$ in common.*

> ***Proof.*** Suppose that $d > 1$ divides F_n and F_{n+1}. Then their difference $F_{n+1} - F_n = F_{n-1}$ will also be divisible by d. From this and the formula $F_n - F_{n-1} = F_{n-2}$, it can be concluded that $d \mid F_{n-2}$. Working backward, we can show that $F_{n-3}, F_{n-4}, \ldots$, and finally F_1 are all divisible by d. But $F_1 = 1$, which is certainly not divisible by any $d > 1$. This contradiction invalidates our supposition and therefore proves the theorem. ∎

Because $F_3 = 2$, $F_5 = 5$, $F_7 = 13$, and $F_{11} = 89$ are all prime numbers, one might be tempted to guess that F_n is prime whatever $n > 2$ is a prime. The conjecture fails at an early stage, for a little figuring shows that

$$F_{19} = 4181 = 37 \cdot 113.$$

Not only is there no known device for predicting which F_n are prime but it is not even known whether the number of prime Fibonacci numbers is infinite.

On the positive side, one can prove that for any prime p, there are infinitely many Fibonacci numbers that are divisible by p and that lie at equal distances from one another in the Fibonacci sequence. For instance, 3 divides every fourth term in the Fibonacci sequence, 5 divides every fifth term, and 7 divides every eighth term.

We saw earlier that by the Euclidean algorithm, the greatest common divisor of two positive integers can be found after finitely many divisions. When the integers are suitably chosen, the number of divisions required can be made arbitrarily large. The precise result is this. For $n > 0$, there exist integers a and b for which exactly n divisions are needed in calculating gcd (a, b) by the Euclidean algorithm.

To verify our contention, let us take $a = F_{n+2}$ and $b = F_{n+1}$. The Euclidean algorithm for obtaining gcd (F_{n+2}, F_{n+1}) leads to the following system of equations:

$$F_{n+2} = 1 \cdot F_{n+1} + F_n$$

$$F_{n+1} = 1 \cdot F_n + F_{n-1}$$

$$\vdots$$

$$F_4 = 1 \cdot F_3 + F_2$$

$$F_3 = 2 \cdot F_2 + 0.$$

Evidently the number of necessary divisions is n. For example, to find the greatest common divisor of the numbers $F_8 = 21$ and $F_7 = 13$ by the Euclidean algorithm, one needs 6 divisions:

$$21 = 1 \cdot 13 + 8$$
$$13 = 1 \cdot 8 + 5$$
$$8 = 1 \cdot 5 + 3$$
$$5 = 1 \cdot 3 + 2$$
$$3 = 1 \cdot 2 + 1$$
$$2 = 2 \cdot 1 + 0.$$

You no doubt recall that the last nonzero remainder appearing in the Euclidean algorithm for F_{n+2} and F_{n+1} furnishes the value of gcd (F_{n+2}, F_{n+1}). Hence,

$$\gcd (F_{n+2}, F_{n+1}) = F_2 = 1,$$

which shows anew that consecutive Fibonacci numbers are relatively prime.

The Fibonacci numbers have numerous easily derivable properties. One of the simplest is that the sum of the first n Fibonacci numbers equals $F_{n+2} - 1$. For instance, when we add the first eight Fibonacci numbers together, we get

$$1 + 1 + 2 + 3 + 5 + 8 + 13 + 21 = 54 = 55 - 1,$$

where $55 = F_{10}$.

That this is typical of the general situation follows from the relations

$$F_1 = F_3 - F_2$$
$$F_2 = F_4 - F_3$$
$$F_3 = F_5 - F_4$$
$$\vdots$$
$$F_{n-1} = F_{n+1} - F_n$$
$$F_n = F_{n+2} - F_{n+1}.$$

When these equations are added, the left-hand side gives the sum of the first n Fibonacci numbers, and on the right-hand side the terms cancel in pairs, leaving us only with $F_{n+2} - F_2$. The conclusion:

$$F_1 + F_2 + F_3 + \cdots + F_n = F_{n+2} - F_2 = F_{n+2} - 1.$$

Another Fibonacci property of interest is the identity

(1) $$F_n^2 = F_{n-1}F_{n+1} + (-1)^{n-1}, \qquad n \geq 2.$$

The last term means that the sign in front of the final 1 alternates. This can be illustrated by taking, say, $n = 6$ and $n = 7$:

$$F_6^2 = 8^2 = 5 \cdot 13 - 1 = F_5 F_7 - 1,$$
$$F_7^2 = 13^2 = 8 \cdot 21 + 1 = F_6 F_8 + 1.$$

To establish identity (1), let us start with the equation

$$F_n{}^2 - F_{n-1}F_{n+1} = F_n(F_{n-1} + F_{n-2}) - F_{n-1}F_{n+1}$$

$$= (F_n - F_{n+1})F_{n-1} + F_nF_{n-2}.$$

Recalling the rule of formation of the Fibonacci sequence gives $F_{n+1} = F_n + F_{n-1}$, so the expression in parentheses can be replaced by $-F_{n-1}$ to give

$$F_n{}^2 - F_{n-1}F_{n+1} = (-1)(F_{n-1}^2 - F_nF_{n-2}).$$

Except for the initial sign, the right-hand side of this equation is the same as the left-hand side but with all the subscripts decreased by 1. By an entirely similar argument, $F_{n-1}^2 - F_nF_{n-2}$ can be shown to equal $(-1)(F_{n-2}^2 - F_{n-1}F_{n-3})$, whence

$$F_n{}^2 - F_{n-1}F_{n+1} = (-1)^2(F_{n-2}^2 - F_{n-1}F_{n-3}).$$

After $n - 2$ such steps, we eventually arrive at

$$F_n{}^2 - F_{n-1}F_{n+1} = (-1)^{n-2}(F_2{}^2 - F_3F_1)$$

$$= (-1)^{n-2}(1^2 - 2 \cdot 1)$$

$$= (-1)^{n-2}(-1)$$

$$= (-1)^{n-1},$$

which is what we sought to prove.

For $n = 2k$, where k is an integer, relation (1) becomes

(2) $$F_{2k}^2 = F_{2k-1}F_{2k+1} - 1.$$

This identity is the basis of a well-known geometric deception whereby a square 8 units by 8 can be broken into pieces that seemingly fit together to form a rectangle 5 by 13. To accomplish this, divide the square into four parts as shown in the left-hand diagram and rearrange them as indicated on the right.

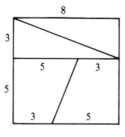
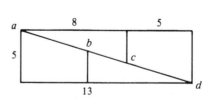

The area of the square is $8^2 = 64$, whereas the rectangle, which seems to have the same constituent parts, has an area $5 \cdot 13 = 65$, and so the area has apparently been increased by one square unit. The puzzle is easy to explain. The points a, b, c, d do not all lie on the diagonal of the rectangle, but instead are the vertices of a parallelogram whose area is exactly equal to the extra unit of area.

The construction can be carried out with any square whose sides are equal to the Fibonacci number F_{2k}. When the square is partitioned as in the diagram, the pieces

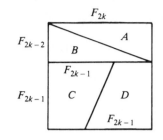

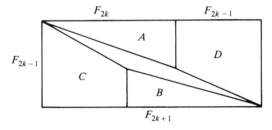

can be re-formed to produce a rectangle having a slot in the shape of a slim parallelogram (our figure is exaggerated). The identity $F_{2k-1}F_{2k+1} - 1 = F_{2k}^2$ can be interpreted as asserting that the area of the rectangle minus the area of the parallelogram is precisely equal to the area of the original square. It can be shown that the height of the parallelogram—that is, the width of the slot at its widest point—is

$$\frac{1}{\sqrt{F_{2k}^2 + F_{2k-1}^2}} .$$

When F_{2k} is reasonably large (say, $F_{2k} = 144$, so that $F_{2k-2} = 55$), the slot is so narrow as to be almost imperceptible to the eye.

This is a convenient place to examine a remarkable connection between the Fibonacci numbers and what the Greeks called the golden ratio. We start by forming the sequence

$$u_n = \frac{F_{n+1}}{F_n} , \qquad (n \geq 1)$$

of the ratios of consecutive Fibonacci numbers. The first few terms are

$$u_1 = \tfrac{1}{1} = 1 \qquad\qquad u_5 = \tfrac{8}{5} = 1.60$$

$$u_2 = \tfrac{2}{1} = 2 \qquad\qquad u_6 = \tfrac{13}{8} = 1.625$$

$$u_3 = \tfrac{3}{2} = 1.5 \qquad\qquad u_7 = \tfrac{21}{13} = 1.615\ldots$$

$$u_4 = \tfrac{5}{3} = 1.66\ldots \qquad\qquad u_8 = \tfrac{34}{21} = 1.619.\ldots$$

As the index increases, the sequence seems to tend to a number that falls between 1.61 and 1.62. Let us assume that the limiting value actually exists; call it α. For any $n \geq 1$, we have

$$\frac{F_{n+1}}{F_n} = \frac{F_n + F_{n-1}}{F_n} = 1 + \frac{F_{n-1}}{F_n},$$

which by virtue of our definition of the u_n's, can be replaced by

$$u_n = 1 + \frac{1}{u_{n-1}}.$$

As n increases, the left- and right-hand sides of the foregoing equation are getting closer and closer to α and $1 + 1/\alpha$, respectively, so that the equation as a whole is approaching

$$\alpha = 1 + \frac{1}{\alpha} \quad \text{or} \quad \alpha^2 - \alpha - 1 = 0.$$

But the only positive root of this quadratic equation is

$$\alpha = \tfrac{1}{2}(1 + \sqrt{5}) = 1.618033989 \ldots,$$

the so-called golden ratio. Thus, the sequence of the ratios of consecutive Fibonacci numbers gives an approximation of the golden ratio, and the further out we go, the better the approximation becomes.

6.3 Problems

1. It can be established that each positive integer is representable as a sum of Fibonacci numbers, none taken more than once; for example,

$$5 = F_3 + F_4,$$

$$6 = F_1 + F_3 + F_4,$$

$$7 = F_1 + F_2 + F_3 + F_4.$$

Write the integers 50, 75, 100, and 125 in this manner.

2. (a) Show that the sum of the first n Fibonacci numbers with odd indices is given by the formula

$$F_1 + F_3 + F_5 + \cdots + F_{2n-1} = F_{2n}.$$

[Hint: Add the equalities $F_1 = F_2$, $F_3 = F_4 - F_2$, $F_5 = F_6 - F_4$, ...]

(b) Show that the sum of the first n Fibonacci numbers with even indices is given by the formula

$$F_2 + F_4 + F_6 + \cdots + F_{2n}$$
$$= F_{2n+1} - 1.$$

[Hint: Use part (a) and the identity $F_1 + F_2 + F_3 + \cdots + F_{2n} = F_{2n+2} - 1$.]

(c) Obtain the following formula for the alternating sum of Fibonacci numbers:

$$F_1 - F_2 + F_3 - F_4 + \cdots + (-1)^{n+1} F_n$$
$$= (-1)^{n+1} F_{n-1} + 1.$$

3. For any prime $p \neq 2$ or 5, it is known that either F_{p-1} or F_{p+1} is divisible by p. Confirm this in the case of the primes 7, 11, 13, and 17.

4. From the formula $F_{n+1} F_{n-1} - F_n^2 = (-1)^n$, conclude that consecutive Fibonacci numbers are relatively prime.

5. One can prove that the greatest common divisor of two Fibonacci numbers is also a Fibonacci number; specifically,

$$\gcd(F_n, F_m) = F_d, \quad \text{where } d = \gcd(n, m).$$

Verify this identity in the case of $\gcd(F_9, F_{12})$ and $\gcd(F_{15}, F_{20})$.

6. Use Problem 5 to prove that for $n > 2$, $F_n | F_m$ if and only if $n | m$.

7. Establish each of the following assertions.

    (a) $2 | F_n$ (that is, F_n is even) if and only if $3 | n$.

    (b) $3 | F_n$ if and only if $4 | n$.

    (c) $4 | F_n$ if and only if $6 | n$.

    (d) $5 | F_n$ if and only if $5 | n$.

[*Hint:* All these require the aid of the previous problem.]

8. Show that the sum of the squares of the first n Fibonacci numbers is given by the formula

$$F_1^2 + F_2^2 + F_3^2 + \cdots + F_n^2 = F_n F_{n+1}.$$

[*Hint:* Note that $F_n^2 = F_n(F_{n+1} - F_{n-1})$
$$= F_n F_{n+1} - F_n F_{n-1}.]$$

6.4   Fibonacci and the Pythagorean Problem

In Chapter 3 we mentioned the ancient problem of finding all right triangles whose sides are of integral length, the so-called Pythagorean problem. From a number-theoretic point of view, solving this problem amounts to determining formulas giving all triples (x, y, z) of positive integers that satisfy the equation $x^2 + y^2 = z^2$. Such a triple of integers x, y, and z is referred to as a Pythagorean triple. Both Euclid's *Elements* (clothed in its geometric language) and the *Arithmetica* of Diophantus indicate a rule for making as many Pythagorean triples as you like. Choose any pair of integers, call them s and t, and let

$$x = 2st, \qquad y = s^2 - t^2, \qquad z = s^2 + t^2.$$

It was left to later Arab mathematicians to show that all Pythagorean triples can be produced from these formulas, a result available to the well-traveled Fibonacci.

Before giving Fibonacci's argument, let us make several observations. First, notice that if (x, y, z) is a Pythagorean triple and k is any positive integer, then the triple (kx, ky, kz) arrived at by multiplying each of the entries by k is also a Pythagorean triple; for the relation $x^2 + y^2 = z^2$ implies that $k^2x^2 + k^2y^2 = k^2z^2$, or what is the same thing, $(kx)^2 + (ky)^2 = (kz)^2$. Thus, from the triple $(3, 4, 5)$, we could get the triples $(6, 8, 10)$, $(9, 12, 15)$, $(12, 16, 20)$, and infinitely many more. But none of these is essentially different from the triple $(3, 4, 5)$. It is more interesting to find the basic Pythagorean triples—those that cannot be gotten by multiplying some other one by a suitable positive integer. These are termed primitive Pythagorean triples.

Definition

A Pythagorean triple (x, y, z) is said to be primitive if the three numbers x, y, and z have no common divisor $d > 1$.

Some examples of primitive Pythagorean triples are $(3, 4, 5)$, $(5, 12, 13)$, and $(8, 15, 17)$, whereas $(10, 24, 26)$ is not primitive.

According to our definition of a primitive Pythagorean triple (x, y, z) there is no divisor common to all three numbers. Actually much more is true, no two of the numbers x, y, and z can have a common divisor $d > 1$. Phrased somewhat differently: the integers x, y, z are relatively prime in pairs. To see this, let us suppose that gcd $(x, y) = d > 1$. By the fundamental theorem of arithmetic, there must exist some prime p

with $p|d$. Since $d|x$ and $d|y$, we should then have $p|x$ and $p|y$, which in turn imply $p|x^2$ and $p|y^2$. But then $p|(x^2 + y^2)$ or $p|z^2$. An appeal to Euclid's lemma now gives $p|z$. The implication of all this is that p is a common divisor of the three integers x, y, and z, a contradiction that (x, y, z) is a primitive triple. Because this contradiction arose out of the assumption that $d > 1$, we must conclude that $d = 1$. In the same way, one can verify that gcd $(y, z) = 1$ and gcd $(x, z) = 1$.

Here is another property of primitive Pythagorean triples.

LEMMA

If (x, y, z) is a primitive Pythagorean triple, then one of the integers x and y is even, and the other is odd.

Proof. By the result of the last paragraph, x and y cannot both be even, so that all we need to show here is that they cannot both be odd. As is well known, any odd number can be put in the form $2n + 1$, where n is an integer. Thus, if x and y are odd, there exist appropriate choices of h and k for which

$$x = 2h + 1 \qquad \text{and} \qquad y = 2k + 1.$$

Then

$$z^2 = x^2 + y^2 = (2h + 1)^2 + (2k + 1)^2$$

$$= 4h^2 + 4h + 1 + 4k^2 + 4k + 1$$

$$= 4(h^2 + h + k^2 + k) + 2 = 4m + 2.$$

Because z^2 is of the form $4m + 2 = 2(2m + 1)$, it is an even number. This, in its turn, forces z to be even (z cannot be odd, since the square of an odd number is also odd). But the square of any even number is divisible by 4, and $4m + 2$ is clearly not divisible by 4. This situation being impossible, we see that x and y cannot both be odd. ∎

By virtue of this lemma, there exist no primitive Pythagorean triples (x, y, z) all of whose values are prime numbers (you may supply your own argument). There are primitive Pythagorean triples in which z and one of x or y is prime, for instance, the triples $(3, 4, 5)$, $(11, 60, 61)$, and $(19, 180, 181)$. It is not known whether infinitely many such triples exist.

In a primitive Pythagorean triple (x, y, z), exactly one of x and y is an even integer. We shall hereafter write our triples so that x is even and y odd; then z must be odd (otherwise, gcd $(x, z) \geq 2$).

With the routine work out of the way, all primitive Pythagorean triples can be described in a straightforward manner.

THEOREM

The triple (x,y,z) is a primitive Pythagorean triple if and only if there exist relatively prime integers $s > t > 0$ such that

$$x = 2st, \qquad y = s^2 - t^2, \qquad z = s^2 + t^2$$

where one of s and t is even and the other is odd.

Proof. To start, let (x, y, z) be any primitive Pythagorean triple. We have agreed to take x even, and y and z both odd, so it follows that $z + y$ and $z - y$ are even integers; say, $z + y = 2u$ and $z - y = 2v$. Now the equation $x^2 + y^2 = z^2$ can be rewritten

$$x^2 = z^2 - y^2 = (z + y)(z - y),$$

whence on division by 4,

$$\left(\frac{x}{2}\right)^2 = \left(\frac{z + y}{2}\right)\left(\frac{z - y}{2}\right) = uv.$$

Notice that u and v are relatively prime integers; for if gcd $(u, v) = d > 1$, then $d \mid (u - v)$ and $d \mid (u + v)$, or equivalently, $d \mid y$ and $d \mid z$, which violates the condition that gcd $(y, z) = 1$.

It can be proved that if the product of two relatively prime integers equals the square of an integer, then each of them is itself a perfect square. Granting this fact, we conclude that u and v are each perfect squares. To be definite, let us write

$$u = s^2, \qquad v = t^2$$

where s and t are positive integers. The result of substituting these values of u and v is

$$z = u + v = s^2 + t^2,$$
$$y = u - v = s^2 - t^2,$$
$$x^2 = 4uv = 4s^2t^2,$$

or in the last case, $x = 2st$. Because any common divisor of s and t divides both y and z, the relation gcd $(y, z) = 1$ forces gcd $(s, t) = 1$. It remains for us to observe that if s and t were both even, or both odd, then this would make each of y and z even, an impossibility. Hence, exactly one of the pair s, t is even, and the other is odd.

It is easy to see that any triple (x, y, z) satisfying the conditions of this theorem is a Pythagorean triple. For if $x = 2st$, $y = s^2 - t^2$, $z = s^2 + t^2$, then the following identity holds:

$$x^2 + y^2 = (2st)^2 + (s^2 - t^2)^2$$
$$= 4s^2t^2 + s^4 - 2s^2t^2 + t^4$$
$$= s^4 + 2s^2t^2 + t^4 = (s^2 + t^2)^2 = z^2.$$

Let us next show that the triple (x, y, z) is primitive. We assume to the contrary that x, y, z have a common divisor $d > 1$ and argue until a contradiction is reached. Consider any prime divisor p of d. Observe first that $p \neq 2$, since it divides the odd integer z (one of s or t is odd, and the other is even; hence, $s^2 + t^2 = z$ must be odd). From $p \mid y$ and $p \mid z$, we obtain $p \mid (z + y)$ and $p \mid (z - y)$, or put otherwise, $p \mid 2s^2$ and $p \mid 2t^2$. But then $p \mid s$ and $p \mid t$, which is incompatible with gcd $(s, t) = 1$. In consequence, $d = 1$ and (x, y, z) is a primitive Pythagorean triple. ∎

We have a method for producing primitive Pythagorean triples, namely by means of the formulas

$$x = 2st, \qquad y = s^2 - t^2, \qquad z = s^2 + t^2,$$

and the theorem indicates that all primitive Pythagorean triples can be so obtained. The accompanying table lists some primitive Pythagorean triples arising from small values of s and t. For each value of $s = 2, 3, 4, \ldots, 7$, we have taken those values of t that are relatively prime to s, less than s, and even whenever s is odd.

s	t	x	y	z
2	1	4	3	5
3	2	12	5	13
4	1	8	15	17
4	3	24	7	25
5	2	20	21	29
5	4	40	9	41
6	1	12	35	37
6	5	60	11	61
7	2	28	45	53
7	4	56	33	65
7	6	84	13	85

From this or a more extensive table, you might be led to suspect that if (x, y, z) is a primitive Pythagorean triple, then exactly one of the integers x or y is divisible by 3. Let us show that this is indeed the case. As members of a primitive Pythagorean triple, $x, y,$ and z can be written

$$x = 2st, \qquad y = s^2 - t^2, \qquad z = s^2 + t^2$$

for suitable integers s and t. Recall that the square of any integer has either the form $3k$ or the form $3k + 1$. If either s^2 or t^2 happens to be of the form $3k$ (this is to say, if either $3 \mid s^2$ or $3 \mid t^2$), then $3 \mid s$ or $3 \mid t$, in which case $3 \mid x$, and there is nothing more to prove. Thus, it suffices to assume that both s^2 and t^2 take the form $3k + 1$; to be specific, let $s^2 = 3k + 1$ and $t^2 = 3h + 1$. Substituting, we get

$$y = s^2 - t^2 = (3k + 1) - (3h + 1) = 3(k - h).$$

This is simply the statement that $3 \mid y$. All in all, what we have proved is as follows.

THEOREM

In a primitive Pythagorean triple (x, y, z), either x or y is divisible by 3.

Let us turn to another of the famous tournament problems Fibonacci solved, namely, one equivalent to finding a number x such that both $x^2 + 5$ and $x^2 - 5$ are squares of rational numbers; say,

(1) $$x^2 + 5 = a^2 \quad \text{and} \quad x^2 - 5 = b^2.$$

We shall see that the solution depends ultimately on knowing the general form of primitive Pythagorean triples.

A solution is sought in the rational numbers, so let us express x, a, and b as fractions with a common denominator:

$$x = \frac{x_1}{d}, \qquad a = \frac{a_1}{d}, \qquad b = \frac{b_1}{d}.$$

Substituting these values in equation (1) and clearing fractions gives

(2) $$x_1^2 + 5d^2 = a_1^2, \qquad x_1^2 - 5d^2 = b_1^2.$$

When the second equation is subtracted from the first, we get

$$10d^2 = a_1^2 - b_1^2 = (a_1 + b_1)(a_1 - b_1).$$

The left-hand side is even, so that a_1 and b_1 must both be even or both odd. In either event, $a_1 - b_1$ is an even integer, say $a_1 - b_1 = 2k$, from which it can be inferred that $a_1 + b_1 = 5d^2/k$. Now solve the last two equations simultaneously for a_1 and b_1 to obtain

$$a_1 = \frac{5d^2}{2k} + k, \qquad b_1 = \frac{5d^2}{2k} - k.$$

If these two expressions are now substituted in equations (2), then one arrives at

$$x_1^2 + 5d^2 = \left(\frac{5d^2}{2k} + k\right)^2 = \left(\frac{5d^2}{2k}\right)^2 + 5d^2 + k^2,$$

$$x_1^2 - 5d^2 = \left(\frac{5d^2}{2k} - k\right)^2 = \left(\frac{5d^2}{2k}\right)^2 - 5d^2 + k^2,$$

which on addition yield the single condition

$$k^2 + \left(\frac{5d^2}{2k}\right)^2 = x_1^2.$$

The point is precisely this: The three numbers k, $5d^2/2k$, x_1 form a Pythagorean triple. As such, they must arise from a primitive Pythagorean triple and so can be written as

$$k = (2mn)t, \qquad \frac{5d^2}{2k} = (m^2 - n^2)t, \qquad x_1 = (m^2 + n^2)t$$

for some choice of t.

To eliminate k, let us take the product of the first two of these equations. The result is

$$5d^2 = 4mn(m^2 - n^2)t^2.$$

We are seeking values for the integers m and n that will make the right-hand side of this equation 5 times a perfect square. As a first attempt, it is reasonable to set $m = 5$, so that the condition reduces to

$$d^2 = 4n(5^2 - n^2)t^2.$$

Evidently the right-hand side becomes a square when $n = 4$:

$$d^2 = 4 \cdot 4(5^2 - 4^2)t^2 = 16 \cdot 9t^2 = (12t)^2.$$

These values for m and n lead to

$$x_1 = (m^2 + n^2)t = (5^2 + 4^2)t = 41t.$$

Putting the pieces together, we get

$$x = \frac{x_1}{d} = \frac{41t}{12t} = \frac{41}{12}$$

as a solution to Fibonacci's tournament problem.

Bibliography

Boussard, Jacques. *The Civilization of Charlemagne.* Translated by Francis Partridge. New York: World University Library (McGraw-Hill), 1968.

Duckett, E. S. *Alcuin, Friend of Charlemagne.* New York: Macmillan, 1951.

Gandz, Solomon. "The Origin of Ghubār Numerals." *Isis* 16(1931): 393–424.

Gies, Joseph, and Gies, Frances. *Leonard of Pisa and the New Mathematics of the Middle Ages.* New York: Thomas Y. Crowell, 1969.

Grimm, Richard. "The Autobiography of Leonardo Pisano." *Fibonacci Quarterly* 11(1973): 99–104.

Hill, G. F. *The Development of Arabic Numerals in Europe.* Oxford: Oxford University Press, 1915.

Hoggatt, V. E., Jr. *Fibonacci and Lucas Numbers.* Boston: Houghton Mifflin, 1969.

Hoyrup, Jens. "Jordanus de Nemore, 13th Century Mathematical Innovator." *Archive for History of Exact Sciences* 37 (1988): 307–363.

Hughes, Bernard. *Jordanus de Nemore: De Numeris Datis.* Berkeley: University of California Press, 1981.

McClenon, R. B. "Leonardo of Pisa and His *Liber Quadratorum.*" *American Mathematical Monthly* 26(1919): 1–8.

Mahoney, Michael. "Mathematics." In *Science in the Middle Ages.* Edited by David Lindberg. Chicago: University of Chicago Press, 1978.

Pisano, Leonardo (Fibonacci). *The Book of Squares.* Translated by L. E. Signer. Orlando, Fla.: Academic Press, 1987.

Smith, D. E., and Karpinski, L. C. *The Hindu-Arabic Numerals.* Boston: Ginn, 1911.

Sullivan, J. W. N. *The History of Mathematics in Europe from the Fall of Greek Science to the Rise of the Conception of Mathematical Rigor.* New York: Oxford University Press, 1925.

Vorobyov, N. *The Fibonacci Numbers.* Boston: D.C. Heath, 1963.

Weinberg, Joseph. "The Disputation Between Leonardo of Pisa and John of Palermo." *Scripta Mathematica* 3(1935): 279–281.

West, Andrew. *Alcuin and the Rise of the Christian Schools.* New York: Charles Scribner's Sons, 1901.

CHAPTER 7

The Cubic Controversy: Cardan and Tartaglia

A mathematical problem should be difficult in order to entice us, yet not completely inaccessible, lest it mock at our efforts.

DAVID HILBERT

7.1 Europe in the Fourteenth and Fifteenth Centuries

If the thirteenth century can be seen as the highest point of medieval Europe, then perhaps the fourteenth century was the lowest. Although the thirteenth century had given abundant promise for the future, many events conspired to make the following century a period almost as dark as what followed the collapse of Rome. The afflictions were those classic riders of the Apocalypse: famine, plague, war, and death. The fourteenth century opened with a series of heavy rainfalls so constant and so widespread that chroniclers of the time compared it with the great flood of Genesis. Not only did the climate become wetter, but it turned significantly colder also, in what has been called the Little Ice Age. The cumulative effect was a disastrous crop failure and an attendant famine in which mortality increased alarmingly in the towns, some losing 10 percent of their inhabitants in six months. Those who suffered malnutrition lacked resistance to disease. Upon a people weakened by hunger fell a worse calamity, the Black Death. The Black Death was bubonic plague, carried by brown rats—specifically by a flea parasitic on brown rats—and easily spread in the crowded, dirty conditions of the medieval towns. The outbreak of the plague reached the Mediterranean in 1347, via Italian ships from the Crimea, the port center in the Black Sea. (Because the Crimea was the terminus of the greatest of the caravan routes, it is probable that the seeds of the epidemic were brought from China.) The disease then swept in a great arc through western Europe, striking France in 1348 and afflicting England a year later. Medical knowledge was hopelessly inadequate; nothing could be done to resist the attack. The Black Death raged at its worst for three years, and even when the worst was over it returned with lesser virulence at intervals of twelve to fifteen years until the late seventeenth century. The Great Plague of London in 1665 was the last English eruption. In the absence of trustworthy vital statistics, it is impossible to make firm estimates of the terrible mortality. At Paris, it is said, over 800 people died of it each day, and at Avignon 10,000 people were buried in a single mass grave in the first six weeks. The few figures that wc have indicate that in some towns half, in general perhaps a third, of the population were carried away, whereas other regions were completely depopulated. Food shortages were aggravated by sickness in the agricultural districts. At Montpellier in France, so many inhabitants died that the town fathers invited repopulation from as far away as Italy. Peculiarly at peril were those whose occupations called for them to remain in the stricken towns: officials who tried to preserve order, doctors and priests who stayed to aid and console the dying, scholars who continued their studies. These also perished in great numbers; and society, deprived of its natural leaders, was shaken and unstable for decades following.

The smoke of war hung over the whole sad century. The most famous of these wars was that series of English invasions of France extending from 1338 until 1452 and known to us as the Hundred Years' War. It dragged on for generations before either side won a permanent victory. Even the brief interludes of peace were far from tranquil. Thousands of soldiers refused to lay down their arms and instead formed wandering bands of brigands, the Free Companies of mercenaries, who pillaged the countryside and held for ransom those whom they captured. To this litany of afflictions one must add the first social revolts by the rural peasantry and the urban poor. Savage rebellions occurred in Flanders in 1323–1328, in northern France in 1358 (the famous Jacquerie, which gave its name to all other purely peasant risings), and in England, with the Peasants' Revolt of 1381.

People of the fourteenth century saw the future as an endless succession of evils; despair and defeat everywhere overwhelmed confidence and hope. The depressed mood of the time is preserved for us in the Danse Macabre, or Dance of Death, an actual dance in pantomime performed with public sermons, in which a figure from every walk of life confronts the corpse he must become.

Yet the ultimate ruin by which Western civilization was threatened never materialized. By approximately 1450 the calamities of war, plague, and famine had tapered off, with the result that population increased, compensating for the losses from 1300 on, and the towns began growing rapidly. Prosperity was once again possible, provided that public order could be restored. The great majority of the people of western Europe had become convinced that the ills of a strong monarchy were less to be feared than weakness of government, that rebellion was more dangerous to society than was royal tyranny. Thus, after two centuries of chaos, political security returned with the advent of the "new monarchies" of Louis XI in France (1461), Ferdinand and Isabella in Spain (1477), and Henry VII in England (1485). The rise of these strong national states marked the demise of feudalism, and provided the solid foundation upon which a new European civilization could be built.

As the long-stagnant economy responded to the stimulus of the dramatic growth in population, western Europe experienced a recovery that seemed to many a remarkable rebirth. Not only did Europeans succeed in restoring order, stability, and prosperity but also embarked on a series of undertakings that vastly expanded their literary and artistic horizons. To later generations this reawakening of the human intellect is known as the Renaissance. The word is the legacy of the great nineteenth-century historian Jacob Burkhardt, who in *The Civilization of the Renaissance in Italy* (1860)

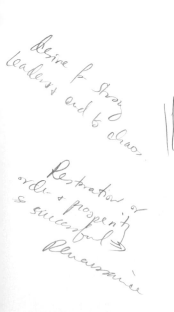

popularized the idea of the Italian Renaissance as a distinct epoch in cultural history, differentiated clearly from the preceding period and from the contemporary culture north of the Alps. In recent years, the whole concept of a "renaissance" has come under suspicion by those who claim that the greater period of cultural achievement came in the twelfth century. There is no longer any general agreement about the character of the Renaissance, its causes, or even its geographical or chronological limits. Ultimately, the Renaissance cannot be disregarded; for medieval civilization—founded as it was upon a basis of land tenure and an almost purely agricultural economy— could not continue indefinitely to absorb an expanding urban population and accommodate a money economy founded on trade without changing into something recognizably different. Thus, depending on context, we shall use the term *Renaissance* in either of its current senses: as a great revival of literature and the arts, with its reverence for classical culture, or as that period of transition (roughly, 1350–1550) in which the decisive change from a largely feudal and ecclesiastical culture to a predominantly secular, lay, urban, and national culture took place.

The reason that a cultural rebirth was experienced and nurtured first in Italy was doubtless that Italy had not been as seriously affected by war and economic dislocation as the northern countries. (It had experienced many small wars but no great conflict.) At the beginning of the fifteenth century, feudalism had disappeared in central and northern Italy, giving place to a vigorous urban society of politically independent city-states. The intellectuals and artists of these prosperous territorial states, hoping to bolster or replace the tottering traditions of medieval culture, thought they had found a model for their secular, individualistic society in the classical past. A cultivation of the Latin and Greek classics flourished with an intensity unknown since the decline of Rome. This "revival of classical culture" was one of the distinguishing characteristics of the Renaissance and one of the chief forces in its changing civilization.

Two events helped to hasten this upsurge of interest in the literary remains of antiquity: the fall of Constantinople to the Turks (1453) and Johann Gutenberg's invention of printing with movable, metallic type (about 1450). Long before the Arabs had subjugated Egypt, fugitive scholars from Alexandria had reached Constantinople with their books, making the fortress city the chief resting place of what was left of classical literature in the original Greek. On May 29, 1453, the Ottoman Turks seized the great city; even though Constantinople had long been a mere enclave in Turkish territory, its fall stunned Christendom. This final collapse of the Byzantine empire drove a host of Greek scholars to seek refuge on Italian soil, bringing with them a precious store of classical manuscripts. Many of the treasures of Greek learning, hitherto known indirectly through Arabic translations, could now be studied from the original sources.

The invention of printing revolutionized the transmission and dissemination of ideas, thereby making the newly acquired knowledge accessible to a larger audience. Handwritten books were scarce and dear, and they had necessarily been the monopoly of the wealthy and scholars under their patronage. Those few books that were available to the public had to be chained down, and as further insurance against their loss, many bore maledictions damning anyone who stole, mutilated, or even approached them without washing his hands. When it became possible to issue books not in single copies, but in the hundreds or even thousands, the world of letters and learning was opened up to the moderately well-to-do classes everywhere.

There is no need to labor the importance of printing with movable type. Still, it should be stressed that the first printing presses were made in the early fifteenth century in medieval Germany, not in Renaissance Italy, and that Italian scholars for a long time scorned the new process. Moreover, the stimulus that had led to the invention of printing was the typically medieval desire for quicker and cheaper ways of producing religious texts. There had been printing before Gutenberg, but Gutenberg's Bible certainly heralded a new day.

The first form of printing in Europe, perhaps the first form of printing on paper, was the block printing (the transference of ink from carved wooden blocks) of playing cards in the latter decades of the fourteenth century. Among the many block books produced, some attained considerable popularity. The best-known was the *Poor Man's Bible,* a 40-page book of religious pictures with a minimum of inscriptions, intended for the instruction of the uneducated in the principal lessons of the Bible. What came in the fifteenth century therefore was not the invention of printing but the notion of separate metal type for each letter. Of course, the production of paper made from linen helped popularize the new discovery; there would have been little use in a cheap method of duplication if the only material available had been expensive parchment.

On this last point, a digression may be permitted. By the eighth century, when the advance of Islam produced the final separation of East and West, Egyptian papyrus was no longer available. The monastic scholars therefore wrote on parchment made from the skins of animals, usually sheepskin or goatskin. Parchment was prepared for scribal use by a slow process that involved soaking the skin in a lye solution to dissolve organic materials, stretching it on a frame and rubbing it with a pumice stone for smoothness, and finally pressing the skin and cutting it to size. Parchment had many advantages over papyrus; it resisted dampness, and if a text were no longer required, it could be scraped off and the same writing surface could be used again. Even so, parchment was expensive, and without a cheaper material to print on, the invention of printing would not have been so useful and significant.

The first use of paper from hemp, tree bark, fish nets, and rags is carefully dated in Chinese dynastic records as belonging to the year 105, but this discovery like most was probably a gradual process. The secret of its manufacture was taught by Chinese prisoners to their Arab captors at Samarkand in the eighth century. For the next 500 years, papermaking was an Arab monopoly until passed on by the Moors in Spain to the Christian conquerors. At the opening of the fourteenth century, paper was still a fairly rare material in Europe, imported from Damascus and turned out in small quantities from several newly established mills in Italy. By the end of the century, it was manufactured in Italy, Spain, France, and southern Germany and had largely displaced parchment as the standard writing material of all but the wealthy. Gutenberg's famous Bible was one of the few early books printed on parchment, and each of his Bibles is said to have required the skins of 300 sheep.

Once invented, the "divine art" of printing from cast movable type spread like wildfire through central and western Europe, so that by the end of the century the names of 1500 printers were known. To ascertain accurately the number of books that all these presses produced before the year 1500 is impossible. According to the titles collected in various catalogs of incunabula, about 30,000 printed works appeared. Assuming that the editions were small, averaging about 300 copies, there would have been nearly 9 million books (including pamphlets) in Europe by 1500, as against the few score thousand manuscripts that lately had held all the irrecoverable lore of the past.

The first printed books were little concerned with mathematics. Many mathematical works written in the mid-1400s, such as Regiomontanus's treatise *De Triangulis,* did not appear in print until very much later. The principal standbys of the earlier printers were the Bible (which appears in many editions, both in Latin and in the popular languages), books of meditation, and religious tracts of various sorts. Those mathematical works that did come off the presses were unoriginal, falling far below the level of the great thirteenth- and fourteenth-century mathematicians. The first popular textbook, the *Treviso Arithmetic,* was published in 1478 at Treviso, an important mercantile town not far to the north of Venice. Essentially a list of rules for performing common calculations, it was written, claims the anonymous author, at the request of young people preparing to enter commercial careers. The *Treviso Arithmetic* was significant not so much for its content as for initiating a remarkable movement. Before the close of the fifteenth century, over 200 mathematical books had been printed in Italy alone. Euclid's *Elements,* with the Latin commentary by Campanus of Novara, was published in 1482 at Venice and again in 1491 at Vicenza. Campanus lacked linguistic competence in Arabic, so this version contained numerous errors and barbarous terminology. In 1505, Zamberti brought out a new translation, working from a recovered Greek manuscript.

One of the earliest European scholars to take advantage of the recovery of the original Greek texts was the mathematician-astronomer Johannes Müller (1436–1476), better known as Regiomontanus, from the Latin name of his native town of Königsberg. The most distinguished scientific man of his time, Regiomontanus was active in translating and publishing the classical manuscripts available, including Ptolemy's treatise on astronomy, the *Almagest.* The fruits of this study were shown in his greatest publication, *De Triangulis Omnimodis* (On Triangles of All Kinds). The work was finished about 1464 but remained unprinted until 1533. Trigonometry was one of the few branches of mathematics to receive substantial development at the hands of the Greeks and the Arabs. In the *De Triangulis,* Regiomontanus systematically summed up the work of these pioneers and went on to solve all sorts of problems relating to plane and spherical triangles. The only trigonometric functions introduced were the sine and the cosine, but at a later date Regiomontanus computed a table of tangents. For all practical purposes, *De Triangulis* established trigonometry as a separate branch of mathematics, independent of astronomy.

It is difficult if not impossible to assess the influence of this new trigonometric learning on the great voyages of discovery in the late 1400s. At one time, historians thought that the Portuguese navigators in venturing south of the equator along the coast of Africa had used the tables of solar declination in Regiomontanus's almanac, the *Ephemerides Astronomicae;* but it appears that the first editions (1474) of this work contain no such tables. What is known is that Columbus carried a copy of the *Ephemerides* with him on his four trips to the New World. On one occasion, having read that Regiomontanus predicted a total eclipse of the moon for February 29, 1504, Columbus took advantage of this knowledge to frighten the natives into reprovisioning his ships.

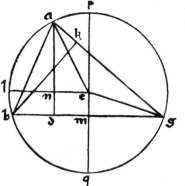

XXVI.

Data area trianguli cum eo, quod fub duobus lateribus continetur rectangulo, angulus quem bafis refpicit, aut cognitus emerget, aut cū angulo cognito duobus rectis æquipollebit.

Refumptis figurationibus præcedētis, fi perpendicularis b k uerfus lineā a g ,p cedens, extra triangulum ceciderit, erit per ea, quæ in præcedenti commemorauimus, proportio b k ad b a nota, & ideo per primi huius angulum b a k notum accipiemus, fic angulus b a g cum angulo b a k noto duobus rectis æquiualebūt. Si ue ro perpendicularis b k intra triangulum ceciderit, quemadmodū in tertia figuratione præcedentis cernitur, erit ut prius a b ad b k notā habens proportionē, & ideo angulus b a k fiue b a g notus conclude tur. At fi perpendicularis b k coinciderit lateri a b, necefse eft angulum b a g fuiffe rectum, & ideo cognitum, quod quidem accidit, quando area trianguli propofiti æquatur ei, quod fub duobus lateribus eius continetur rectangulo.

The area of a triangle: from Regiomontanus's De Triangulis Omnimodis *(1533 edition). (From* A Short History of Mathematics *by Vera Sanford. Reproduced by permission of the publisher, Houghton Mifflin Company.)*

The period of Regiomontanus was also the time of Luca Pacioli (1445–1514), a Franciscan friar who was commonly called Fra Luca di Borga. Many scholars of this time felt the compulsive urge to bring together, within the pages of a large book, all known information in some given field. There was a systematic compendium, or ''summa,'' for every interest and taste. Pacioli's *Summa de Arithmetica Geometria Proportioni et Proportionalita,* published in Venice in 1494, was the most influential mathematical book of that time. The first comprehensive work to appear after the *Liber Abaci* of Fibonacci, it contained almost nothing that could not be found in Fibonacci's treatise, which indicates how little European mathematics had progressed in nearly 300 years. But as an encyclopedic account of the main mathematical facts inherited from the Middle Ages, the *Summa* goes far beyond what was taught in the universities. Written carelessly in Italian, it is notable historically for its wide circulation (perhaps due to the author's explanation of the mechanics of double-entry bookkeeping).

The universities that were being established were to become prominent in the cultivation and spread of learning. The Latin *universitas* was originally a mere synonym for *communitas,* a general work indicating a collection of individuals loosely associated for communicating ideas. Initially, the only educational centers were monasteries. Their primary function was religious service, not intellectual, and they were disinclined to teach outsiders. They preserved, rather than added to, literature. As the number of laymen seeking education grew, schools attached to the churches of bishops became prominent as centers of learning. Cathedral schools were provided mainly for those who would enter the ranks of the ''secular clergy'' and carry on the work of the Church in the world, not apart from it. Such schools flourished as a sideline to the work of the bishop and were prone to be affected by the reputation of the local teachers, waxing and waning with the comings and goings of particular personalities.

Cathedral schools were of course hardly conducive to the free flow of ideas. Thus, long before the formal beginnings of universities, assemblies of students gathered around an individual master or two who had no connection with the Church—who, however, still needed the permission of the bishop to teach. An excellent teacher became a celebrated figure, and students traveled from town to town in pursuit of some famous scholar whose reputation had reached their homelands. The force of personality of Peter Abelard (1079–1142) is said to have attracted students from every corner of Europe to his crowded lecture hall in Paris. Twenty of his pupils subsequently became cardinals, and more than fifty became bishops. There is an oft-told tale that when the theological writings of Abelard were condemned by the Church, the king of France suddenly forbade Abelard to teach in his lands. On hearing the news, Abelard climbed a tree and his students flocked to hear him from below. When the king then prohibited him from teaching in the air, Abelard began lecturing in a boat; at this point the king relented. Abelard is especially important because his brilliance as a teacher popularized the cathedral school at Notre Dame as a center of higher learning, thus opening the way for the foundation of the University of Paris.

The growth of the universities in the twelfth and thirteenth centuries was a natural consequence of a demand that the older cathedral and monastery schools were unable to satisfy. A developing body of secular knowledge that had a marked professional value (medicine, and especially law) and that required for its mastery protracted study under an eminent specialist began to make the university an indispensable institution. Students living in the centers made famous by the cathedral schools began to find it necessary to organize in order to regulate their own conduct, to protect themselves from extortion by local citizens, and—because many were not native to the area—to secure legal rights. Thus the students in voluntary association tended, like the merchants and craftsmen of those days, to form self-governing guilds and eventually to gain legal recognition through the charter of a king or a pope. The universities at Bologna (1158), Paris (1200), Padua (1222), Oxford (1214), and Cambridge (1231) can all trace their inception to this period. These embryonic universities bore little physical resemblance to what they later became, and there were great variations among the institutions of different towns. Not until the fifteenth century did the universities acquire permanent buildings. Before this, teachers lectured in their own quarters or in rented halls, and general meetings took place in churches or monastic halls. Competition for eminent scholars gradually led to contracted salaries, so that as early as 1180 Bologna paid several professors from municipal funds; the selection of professors, however, remained a student prerogative. Students individually paid the master who taught liberal arts, because teaching skill was equivalent to the skill of any other tradesman. Teachers of theology, on the other hand, were forbidden to stipulate charges in advance—theology being a ''spiritual gift''—but were allowed to accept donations after a lecture was concluded.

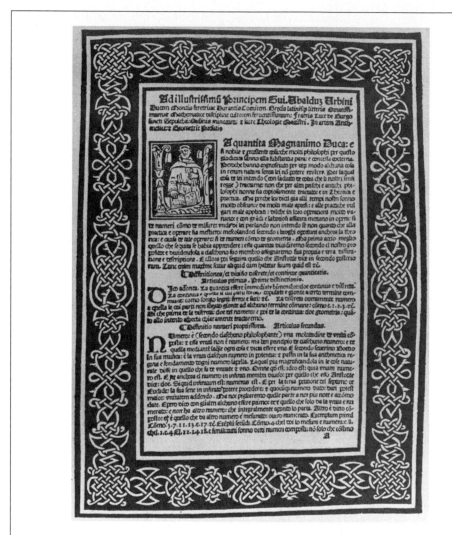

First page of Pacioli's Summa *(1523 edition). (From* Rara Arithmetica, *by David Eugene Smith, published by Chelsea Publishing Co., 1970.)*

Paris and Bologna were the great "mother" universities, serving as models for the later universities that sprang up in every part of Europe during the next two centuries. The universities of Italy and southern France followed the academic pattern of Bologna, whereas those in northern Europe looked to Paris as the standard. Both schools developed much the same methods of teaching and came to grant the same degrees, but they emphasized different studies and were organized differently. The rise of secular administrative governments in Italy made legal studies the door to high civil office and profitable employment. Thus, at Bologna jurisprudence always dominated

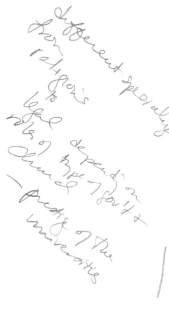

and little attention was given to theology and philosophy. For these subjects, the student went by preference to Paris, where canon law was secondary and civil law was not taught at all. Education in the North was everywhere still in the hands of the Church, so it was a matter of course that ecclesiastical studies should predominate at Paris, and that the church authorities should claim a large share in university government. In Bologna, the university was a union of student guilds, which gained control over all academic affairs, save only the bestowing of degrees—which were licenses to teach. In Paris the system of organization was the reverse, with the governance of the university in the hands of the masters. One reason for this difference will be found in the differing ages of the students. At Bologna, with its interest in the "lucrative science" of law, many students were mature men who had already attained high civil position. The students at the faculty of arts at Paris, much the largest faculty there, were too young (possibly 12 or 14 years old) and too poor to assert themselves in any similar way.

Universities enjoyed an enormous prestige as custodians of learning in an age in which education was esteemed as at almost no other period in history. A city's trade, population, and notability depended on the presence of a university, so that cities without schools were willing to underwrite universities that might secede from their established seats. The medieval universities had no permanent buildings and little corporate property, so it was simple for students to migrate to another city when for any reason they were dissatisfied. The masters, because they were entirely dependent for their livelihood on meager tuition fees, had no choice but to follow. (Cambridge, for example, was raised to university status by a migration from Oxford in 1209.)

Students obtained numerous privileges—including the right of trying their own members in practically all civil and criminal cases—through the potent threat of withdrawing to rival cities. A threatened secession from Bologna in 1321 was withdrawn on the terms that an offending magistrate should be publicly flogged and that the city should erect a chapel for the university. The students, having humbled the municipal authorities by a ruthless boycotting of recalcitrant teachers, went on to enforce a series of statutes governing all phases of instruction. Each master had to give a certain number of lectures covering a prescribed minimum of work, might be fined for tardiness or for evading difficult material on which he was supposed to expound, could not leave town without the permission of the student rector, and even on the occasion of his wedding was only allowed one day off. In the long run, the strength of the student body proved its undoing. As the local authorities began to pay teachers' salaries in order to propitiate the students and keep up the reputation of the local university, the state gradually gained the responsibility for appointments and supervision of faculty.

Because the academic base consisted of the seven arts of the traditional trivium (grammar, logic, and rhetoric) and quadrivium (arithmetic, music, geometry, and astronomy), superficially mathematics seemed to be important. Little attention was paid to the quadrivium, however, ostensibly because these studies had practical applications. Paris, Oxford, and Cambridge systematically discouraged all technical instruction, holding that a university education should be general and not technical. The real reason seems to have been that distinction could be more easily attained in theology and philosophy than in the sciences. By 1336, in an effort to stimulate interest in mathematics, a statute was passed at the University of Paris that no student could graduate without attending lectures on "some mathematical books." It also appears that after 1452, candidates for the degree master of arts at Paris had to take an oath that they had read the first six books of Euclid. Although the Renaissance was to prove to have been as much a landmark in mathematics as in other branches of learning, the university curriculum continued to provide for a literary rather than a scientific education.

As the revival of commerce and the growth of town life in the fourteenth century gradually altered medieval culture, many efforts were made to shore it up or to replace it with something new. When neither the feudal nor the ecclesiastical tradition of the earlier period proved adequate, intellectuals of the Italian city-states looked to a more remote past to find a congenial civilization. Most of the Latin authors—and as they later discovered, Greek authors—had written for an urban, secular, and individualistic society not unlike their own. Italian scholars devoted themselves with a passionate zeal to the study of classical writings, interpreting them in the light of the present age. Behind this "cult of the classics" lay the belief that antiquity, both Latin and Greek, offered a model of perfection by which to judge all civilizations; and in its literature could be found new solutions to all political, social, and ethical problems. There began a systematic and astonishingly successful search for mislaid or forgotten manuscripts, many of which still existed in only a few scattered copies. From one end of Europe to another, scholars rummaged through old libraries in towns and monasteries. The collecting, copying (at first by hand and later, when the printing trade had developed, by press), and diffusion of the treasures they had unearthed was just the beginning.

Manuscripts had to be edited to purify them from the many errors medieval copyists had made and to secure the correct form for each passage. Bibliophiles compiled grammars and lexicons and composed guides to ancient works, and commentaries on them. A tradition of critical judgment in dealing with authoritative texts emerged—a development quite impossible when the Church had had a monopoly on learning; this would be of great value once the interests of the educated turned toward scientific research.

The Renaissance thirst for antique culture inspired a growing fashion of collecting libraries. This enthusiasm pervaded all branches of society, as the princes of the church, state, and commerce vied with one another in assembling books. Renaissance men venerated manuscripts just as their grandfathers had adored the relics of the Holy Land. (A sort of snobbery existed among some wealthy owners, who boasted that their collections contained no printed works.) The development of the Vatican Library in Rome during this period was largely the work of Pope Nicholas V (1395–1455), who had been, before his elevation to the papal throne, librarian to Cosimo de Medici. It cannot be said that the Vatican Library had substantial reality at this time, containing as it did a mere 350 volumes in various states of repair. Nicholas dispatched agents all over Europe to collect manuscripts, with the authority to excommunicate those who refused to give them up. At the same time, some of the most distinguished scholars in Rome were set to making translations of the Greek works into Latin. By the time of his death, Nicholas had built the library to over 5000 volumes and made it one of the finest in Italy. While the primary purpose was to collect and preserve works on the history and doctrines of the Church, an increasing number of secular works found their way into the collection. Vespasiano da Bisticci, a writer of the time, said with some exaggeration, "Never since the time of Ptolemy had half so large a number of books of every kind been brought together."

For a time it seemed that the people of the Renaissance had far less intention of creating something new than of reviving something old, less an idea of moving forward to the future than of returning to the past. Like every fad, this exaltation of ancient life was carried to absurd extremes by some of its devotees. Literary clubs, called ''academies'' in the ancient Greek fashion, were formed, at which discourses on classical subjects were read and followed by discussion and debate. Greek was the language of the meetings and Greek names were adopted by the members. In imitation of the ancient custom, successful poets were crowned with wreaths of laurel. Classical ways of feeling, thinking, and writing cast such a deep spell over some scholars that they slipped into the habit of pretending to be Greeks or Romans, even going through the motions of reviving pagan religious rituals. Despite these excesses, the Renaissance Italians of the fifteenth century performed an invaluable service to future generations by restoring the whole surviving heritage of Greek literature, editing all of it, and finally bringing out printed editions of the entirety. The accomplishment becomes even more impressive when we recall that the knowledge of ancient Greek script had almost disappeared in the West during the Middle Ages. As the range of Hellenistic prose and verse was brought back into the mainstream of Western scholarship, there developed an ideal of education for general human cultivation. This new attitude toward learning differed so markedly from the strictly utilitarian or professional objective of study that had dominated the centuries just preceding that it engendered an entirely new experience.

All this activity on behalf of the classics directly influenced the universities, gradually transforming the prevailing curriculum to the humanities. The term ''humanities'' is simply a translation of the ancient Latin phrase *studia humanitatis* and was used in the Renaissance to mean a clearly defined set of scholarly disciplines (grammar, rhetoric, poetry, history, and moral philosophy) based on the study of the classics of Greece and Rome. The humanities of the Renaissance were not the seven liberal arts of the Middle Ages under another name; for the humanities omitted not only the mathematical disciplines of the quadrivium, but also logic, adding three subjects that are best implied in the trivium, namely poetry, history, and moral philosophy. Thus, the Renaissance thinkers created a version of the classical curriculum that in all its variations was to become one of the great staples of the university, until pushed aside in the nineteenth and twentieth centuries by science, modern languages, and social science.

Another sign of the breakup of the traditional disciplines was the conscious and deliberate creation of a new educational model, the gentleman. Gentlemanly training demanded that one be schooled in the classical writings, graceful in deportment, proper in style of dress, and of discriminating taste in music, painting, and the literary arts. In the universities an atmosphere of largely verbal scholarship arose, resting primarily on grammar—which meant reading, writing, and rigorous analysis of language and style of literary works—and on rhetoric, the art of persuasion and eloquence in speaking. Elegant Latin was regarded as essential for public documents, and Ciceronian phrases were henceforth reckoned among the tools of diplomacy. Close study and imitation of the ancients were held necessary to achieve this style.

What distinguished the Greek revival of the Renaissance from its medieval forerunners was not simply that Greek became part of the general curriculum of studies, but that the whole focus of interest was on the literary and historical masterpieces of Greek literature. By emphasizing the scholarly worth of the humanities as a molder of the gentlemanly character, the Renaissance educators subordinated, and sometimes even impugned, learning from experience and direct observation. The effect was to impede the study of the physical sciences and mathematics, which were beyond the scope of literary treatment, and if anything, offended the aesthetic senses of these men of letters. Although the Renaissance movement as a whole made relatively little progress in science, it nevertheless indirectly opened the way to the Scientific Revolution of the 1600s by recovering more of the ancient learning than the medieval scholars had possessed. Although Euclid and Ptolemy, and even much of Archimedes, were known in the Middle Ages, such advanced authors as Diophantus and Pappus were first translated during the Renaissance. By the 1600s, almost all the extant corpus of Greek mathematics was easily available to those interested in the subject. The result, apparent from the middle 1500s on, was a rapid and noticeable rise in the level of sophistication of European mathematics.

7.2   The Battle of the Scholars

The Renaissance produced little brilliant mathematics commensurate with the achievements in literature, painting, and architecture. The generally low level of prevailing mathematical knowledge stood in the way of any intellectual breakthrough. Although mathematics was included in the curriculum of most universities, it was maintained only in a half-hearted manner. Indeed, during the late 1400s, Bologna was practically the only place where the teaching of the subject was properly organized, and even there it appeared chiefly as a sideline to astronomy. There were few university chairs in mathematics, and no mathematician could command respect from the learned world without also being a teacher, scholar, or patron of the Renaissance humanities.

Regiomontanus set the pattern for combining mathematics with humanistic learning. At the University of Vienna, he lectured enthusiastically on the classical Latin poets Virgil, Juvenal, and Horace, drawing a larger audience than if his subject had been astronomy or mathematics. On a visit to Rome he copied the tragedies of Seneca while learning Greek in order to undertake a more comprehensive translation of the *Almagest* and subsequently the *Conic Sections* of Apollonius. To make the Greek tradition generally available, Regiomontanus became an ardent advocate of the new craft of printing, even installing a printing press in his house for publishing his and other people's manuscripts. Starting with Regiomontanus, mathematicians displayed an astute appreciation of the power of printing. A recurrent feature of the mathematical revival was an ambitious printing program designed to achieve a rapid dissemination of texts and translations.

Mathematics benefited immensely from the humanist passion—almost missionary zeal—for the discovery, translation, and circulation of ancient Greek texts. Though their main interest was in the literary classics, the humanists took all classical learning as their province, and mathematical works were cherished equally with literary ones in their retrieval. These manuscript collectors were responsible for assembling in Italy an almost complete corpus of Greek mathematical writings. The medieval scholar had generally been limited to Euclid, Ptolemy, and sometimes Archimedes, all in translation from the Arabic. By the fifteenth century, typical holdings encompassed not only the works of the aforementioned authors in both Latin and Greek, but also Diophantus,

Apollonius, Pappus, and Proclus. The mathematician, like many of his Renaissance contemporaries, often tried little more than to comprehend what the ancients had done, certain that this was the most that could be known. Although much of this effort was wasted, the return to original sources made a first step toward an intellectual advance. It would then be only a matter of time before mathematicians were stimulated to go beyond the strict letter of the texts to develop new concepts and results strictly unforeseen by the Greeks. Besides, it was far better to read an author, say Euclid, directly than to read what some commentator thought an Arabic paraphrase of the author meant.

By 1500 the situation had changed radically. The newly translated works had been absorbed, and scholars, discontented with looking backward to antiquity, were prepared to go beyond the mathematical knowledge possessed by the Greeks. It came as an enormous and exhilarating surprise when the Italian algebraists of the early 1500s showed how to solve the cubic equation, something the ancient Greeks and the Arabs had missed. (The advance in algebra, however, that proved to be the most significant was the introduction of better symbolism.) In arithmetic, developing commercial and banking interests stimulated improved methods of computation, such as the use of decimal fractions and logarithms. Trigonometry, in connection with its increasing use in navigation, surveying, and military engineering, began to break away from astronomy and acquire a status as a separate branch of mathematics. Refined astronomical instruments necessitated the computation of more extended tables of trigonometric functions. In a monument to German diligence and perseverance, Georg Joachim (generally called Rhaeticus, 1514–1576) worked out a table of sines for every ten seconds to fifteen decimal places. Only in geometry was the progress less pronounced. Renaissance geometers tended to accept the elementary properties found in Euclid's *Elements* as an exclusive model for their conduct and to ignore developments that could not claim Greek paternity.

Mathematicians were eager to make known the newly discovered ways in which they could aid the ordinary person, from teaching the merchant how to reckon profits to showing the mapmaker the principles underlying the projection of a spherical surface onto a plane. Even the sixteenth century found the rules of simple arithmetic and geometry difficult to comprehend, and long division was truly long in the time required to accomplish it. Thus, the middle 1500s saw an increasing number of books of elementary instruction, written in plain and simple language. These practical textbooks, although producing nothing new, were important in diffusing mathematics to an ever-increasing public. The great majority were in Latin, but a good many appeared in the vernacular. Fair examples are the works of the English mathematician Robert Recorde (1510–1558): *The Grounde of Artes* (1542, a popular arithmetic text that ran through 29 editions), *The Pathewaie of Knowledge* (1551, a geometry containing an abridgment of the *Elements*), and *The Whetstone of Witte* (1557, on algebra). Books on algebra became so numerous in Germany that the subject was long known in Europe as the "cossic art," after the German word *coss* for "unknown" (literally, "thing"). Through the trend of producing textbooks in the popular languages, mathematics assumed increasing importance in the education of all cultured people and not just of specialists training for an occupation.

Algebraists, who had floundered under the weight of a cumbersome syncopated notation, began to introduce a symbolism that would make algebraic writing more

The Arte

as their woꝛkes doe extende) to diſtincte it onely into twoo partes. Whereof the firſte is, *when one number is equalle vnto one other.* And the seconde is, *when one number is compared as equalle vnto.2.other nombers.*

Alwaies willyng you to remēber, that you reduce your nombers, to their leaſte denominations, and ſmalleſte foꝛmes, befoꝛe you pꝛocede any farther.

And again, if your *equation* be ſoche, that the greateſte denomination *Coſſike,* be ioined to any parte of a compounde nomber, you ſhall tourne it ſo, that the nomber of the greateſte ſigne alone, maie ſtande as equalle to the reſte.

And this is all that neadeth to be taughte, concernyng this wooꝛke.

Howbeit, foꝛ eaſie alteratiõ of *equations.* I will pꝛopounde a fewe exāples, bicauſe the extraction of their rootes, maie the moꝛe aptly bee wꝛoughte. And to auoide the tediouſe repetition of theſe wooꝛdes: is equalle to: I will ſette as I doe often in wooꝛke vſe, a paire of paralleles, oꝛ Gemowe lines of one lengthe, thus:======, bicauſe noe.2. thynges, can be moare equalle. And now marke theſe nombers.

1.   14.℥.——.15.q.======71.q.

2.   20.℥.———.18.q.=====.102.q.

3.   26.ʒ.——10℥======9.ʒ.——10℥——213.q.

4.   19.℥.——192.q.===10ʒ.——108q——19℥

5.   18.℥.——24.q.=====8.ʒ.——2.℥.

6.   34ʒ.———12℥.====40℥——480q——9.ʒ.

1.   In the firſte there appeareth.2. nombers, that is 14.℥.

Extract from Recorde's The Whetstone of Witte *(1557). (Courtesy of Theatrum Orbis Terrarum Ltd.)*

efficient and compact, one that was also better suited to the needs of typography. These improvements came intermittently, and there was a lack of uniformity in symbols, even for common arithmetic operations (the present division sign ÷ was often used to indicate subtraction). Also, different symbols were proposed in different countries, tried, and often discarded. The Italian algebraists were slow in taking up new notation, preferring the initial letters *p* and *m* for ''plus'' and ''minus'' at a time when the Germans, less fettered by tradition, were adopting the familiar mathematical signs + and −. Although the development of symbols for operations in algebra was proceeding rapidly, the quantities described in equations were still represented by actual instead of general numbers. As a result, there could not be a complete treatment of, say, the quadratic equation. Instead, methods of solution were described and direct solutions were offered for many special cases, each illustrated by equations having appropriately chosen particular numerical coefficients.

The liberation of algebra from the necessity of dealing only with concrete examples was largely the work of the great French mathematician Francois Vieta (1540–1603), who initiated using consonants to represent known quantities and vowels for the unknowns. This one step marked a decisive change, not only in convenience of notation but also in the abstraction of mathematical thought. In moving from varied but specific examples such as $3x^2 + 5x + 10 = 0$ to the general $ax^2 + bx + c = 0$, an entire class of equations could be considered at once, so that a solution to the abstract equation would solve all the specific equations at one fell swoop.

Italian mathematics of the 1500s can be summarized in the names of del Ferro, Tartaglia, Cardan, Ferrari, and Bombelli. The collective achievement of the first four was the solution of the cubic and biquadratic equations and implicitly a deeper understanding of equations in general. This feat was perhaps the greatest contribution of algebra since the work of the Babylonians some 3000 years earlier. Third-degree, or cubic, equations were in no sense peculiar to the Renaissance, attempts at their solution going back to classical antiquity. We have seen that the problem of duplicating the cube, the so-called Delian problem, attained special celebrity among the Greeks. This problem is nothing more than the attempt to find two mean proportionals between a (the length of the edge of the given cube) and $2a$; that is, to solve

$$\frac{a}{x} = \frac{x}{y} = \frac{y}{2a},$$

which requires substantially the solution to the cubic equation $x^3 = 2a^3$. Another noteworthy cubic equation is encountered in Diophantus's *Arithmetica* in connection with Problem 17 of Book VI:

> Find a right triangle such that the area added to the hypotenuse gives a square, while the perimeter is a cube.

The manner in which Diophantus set up the problem leads to the cubic $x^3 + x = 4x^2 + 4$. We do not know how the solution was obtained, for he said simply that x was found to be 4. Perhaps he reduced the equation to the form $x(x^2 + 1) = 4(x^2 + 1)$ and saw that it was satisfied by $x = 4$. Arab writers contributed solutions to special cubics but seem to have believed that many cases could not be solved. Part of the poet Omar Khayyam's (circa 1100) fame as a mathematician rests on his claim of being the first to handle any type of cubic having a positive root. In the thirteenth century, John of Palermo proposed solving the equation $x^3 + 2x^2 + 10x = 20$ as one of his challenge problems to Fibonacci in their contests. Fibonacci showed by geometry that no rational solution was possible, but he gave an approximate value for a root. Over the next several hundred years, mathematicians searched for a "cubic formula" that could be used to solve cubic equations in much the same way the quadratic formula was used for quadratic equations. The credit for finally discovering such a formula belongs to the Italian mathematical school at Bologna during the 1500s.

Fra Luca Pacioli
(circa 1445–1514)

(The Bettmann Archive.)

The most complete and detailed fifteenth century mathematical treatise was the *Summa de Arithmetica, Geometria, Proportioni, et Proportionalita* (1494) of Fra Luca Pacioli, a work in which the author borrowed shamelessly from earlier writers. The main contribution of the *Summa* (which was, after all, a summary) was to lay out the boundaries of contemporary mathematical knowledge and so to supply a program of sorts for the renaissance of mathematics. Pacioli ended his *Summa* by asserting that the solution of the cubic equation was as impossible as the quadrature of the circle. This put off some mathematicians from the attempt but only induced others to try. In the first or second decade of the sixteenth century, Scipione del Ferro (1465–1526) of the University of Bologna shattered Pacioli's prediction by solving the cubic equation for the special case $x^3 + px = q$, where p and q are positive. Pacioli may have personally stimulated this first great achievement of Renaissance algebra, for in 1501–2 he lectured at the University of Bologna, where one of his colleagues was del Ferro. (Pope Nicholas V had, in 1450, proclaimed a general reorganization of the university and allocated four chairs to the mathematical sciences. By 1500 there were as many as eight professors at a time teaching mathematics there.)

It was the practice in those days to treat mathematical discoveries as personal properties, disclosing neither method nor proof, to prevent their application by others to similar problems. This was because scholarly reputation was largely based on public contests. Not only could an immediate monetary prize be gained by proposing problems beyond the reach of one's rival, but the outcomes of these challenges strongly influenced academic appointments; at that time, university positions were temporary and subject to renewal based on demonstrated achievement. (As the printing of scientific periodicals became commonplace, this attitude of secrecy gradually shifted to the view that publication of results was the scholar's best path to recognition.) At any rate, loath to surrender an advantage over other competitors, del Ferro never published his solution and divulged the secret only to a few close friends, among them his pupil and successor Antonio Maria Fiore. This exchange was to lead to one of the most famous of mathematical disputes, its origin being a problem-solving contest at Venice in 1535 in which Fiore challenged Nicolo Tartaglia to solve various kinds of cubics.

Nicolo Tartaglia
(circa 1500–1557)

(Source: Princeton University Press.)

One of the most important restorers of the algebraic tradition, Nicolo Tartaglia (1500–1557), was also one of the least influential. Tartaglia (whose actual family name was Fontana) was born in Brescia, in northern Italy. When the French sacked Brescia in 1512, many of the inhabitants sought refuge in the local cathedral. The soldiers however violated the cathedral's sanctuary and massacred the townspeople. The boy Nicolo's father was among those killed in the butchery, and he himself was left for dead after receiving a severe sabre cut that cleft his jaw and palate. Although his mother found the lad and treated the wounds as best she knew, he was left with an impediment in his speech that earned him the cruel nickname Tartaglia, "the stammerer." Later in life he used the nickname formally in his published works; he wore a long beard to cover the monstrous scars, but he could never overcome the stuttering.

Although his early years were spent in direst poverty, Tartaglia was determined to educate himself. His widowed mother had accumulated a small sum of money so that he might be tutored by a writing-master. The funds ran out after fifteen days, but the boy stole a copybook from which he subsequently learned to read and write. It is said that lacking the means to buy paper, Tartaglia made use of the tombstones in the cemetery as slates on which to work out his exercises. Possessing a mind of extraordinary power, he eventually acquired such proficiency in mathematics that he earned his livelihood by teaching the subject in Verona and Venice. It is ironic that Tartaglia, a man disfigured by a sabre, contributed to the ultimate obsolescence of the sabre by his pioneering work *Nova Scientia* (1537), on the application of mathematics to artillery fire. Tartaglia's "new science" was, of course, ballistics. Even though the theories he developed were often completely wrong, he was the first to offer a theoretical discussion as against the so-called experience of gunners. Anticipating Galileo, Tartaglia taught that falling bodies of different weights traverse equal distances in equal times.

Tartaglia's unfortunate early experiences may have encouraged a suspicious character. Self-taught, he was jealous of his prerogatives and constantly impelled to try to establish his intellectual credentials. Either through intent or simple ignorance of the literature, he had a habit of claiming other people's discoveries as his own. An instance of this is the ''arithmetic triangle'' commonly attributed to Pascal, which Tartaglia asserted was his invention although it had previously appeared in print. Tartaglia seems to have felt that his lack of a classical education placed him at a disadvantage as a humanist; and in his *General Trattato di Numeri et Misure* (1556–60), intended to replace Pacioli's *Summa,* he adorned the preface with quotations from both Cicero and Ptolemy.

In 1530, Tartaglia was sent two problems by a friend, namely:

1. Find a number whose cube added to three times its square makes 5; that is, find a value of x satisfying the equation $x^3 + 3x^2 = 5$.

2. Find three numbers, the second of which exceeds the first by 2, and the third of which exceeds the second by 2 also, and whose product is 1000; that is, solve the equation $x(x + 2)(x + 4) = 1000$, or equivalently, $x^3 + 6x^2 + 8x = 1000$.

For some time Tartaglia was unable to solve these problems, but in 1535 he finally managed to do so, and he also announced that he could effect the solution of any equation of the type $x^3 + px^2 = q$. Fiore, believing Tartaglia's claim to be a bluff, challenged him to a public problem-solving contest. Each contestant was to propose 30 problems, the victor being the one who could solve the greatest number within 50 days. Tartaglia was aware that his rival had inherited the solution of some form of cubic equation from a deceased master, and he worked frantically to find the general procedure. Shortly before the appointed date, he devised a scheme for solving cubics that lacked the second-degree term. Thus, Tartaglia entered the competition prepared to handle two types of cubics, whereas his opponent was equipped for but one. Within two hours, Tartaglia had reduced all 30 problems posed to him to particular cases of the equation $x^3 + px = q$, for which he knew the answer. Of the problems he himself put to Fiore, the latter failed to master a single one (most of which led to equations of the form $x^3 + px^2 = q$).

At this point, Girolamo Cardano (1501–1576), better known as Cardan, appeared on the scene. Cardan's life was deplorable even by the standards of the times. He saw one son executed for wife-poisoning; he personally cropped the ears of a second son who attempted the same offense; he was imprisoned for heresy after having published the horoscope of Christ; and in general he divided his time between intensive study and extensive debauchery. Yet in his range of interests as well as vices, Cardan was a true Renaissance man: physician, philosopher, mathematician, astrologer, dabbler in the occult, and prolific writer.

After a frivolous youth devoted mainly to gambling, Cardan began his university studies at Pavia and completed them at Padua in 1525 with a doctorate in medicine. Ostensibly on the grounds of his illegitimate birth but more likely owing to his reputation as a gambler, Cardan's repeated applications to the College of Physicians in Milan were all turned down. It is not surprising that his first published work, *De Malo Recentiorum Medicorum Medendi Usu Libellus* (On the Bad Practices of Medicine in Common Use), ridiculed the practitioners in Milan. By the time he was 50 years old, Cardan stood second only to Vesalius among European physicians and traveled widely to treat the well-known. So great was his fame that the archbishop of Scotland was among his patients. The archbishop was believed to be suffering from consumption; and Cardan, on the strength of a statement—later admitted to be false—that he could cure this complaint, journeyed to Edinburgh to treat the archbishop. Fortunately for the patient, and also for Cardan's reputation, it turned out that he was suffering from attacks of asthma. When Cardan passed through London on the return trip, he was received by the young King Edward VI, whose horoscope he obligingly cast. The comfortable predictions of a long life and prosperous future proved to be a great embarrassment when the boy died shortly thereafter. At various times, Cardan was professor of mathematics at the universities of Milan, Pavia, and Bologna, resigning each position as a result of some new scandal connected with his name. Forbidden to lecture publicly or to write or publish books, he finally settled in Rome, where for some strange reason, he obtained a handsome pension as astrologer to the papal court. According to various accounts, having predicted that he would die on a certain day, Cardan felt obliged to commit suicide to authenticate the prediction.

When the news of the mathematical joust between Tartaglia and Fiore eventually reached Cardan in Milan, Cardan begged Tartaglia for the cubic solution, offering to include the result in his forthcoming book *Practica Arithmeticae* (1539) under Tartaglia's name. Tartaglia refused on the grounds that in due time he intended to publish his own discourse on algebra. Being credited for a formula is not the same thing as having a treatise, an original work, under your own name; it is the book, not the footnote reference, that history will cite. Cardan, in the hope of learning the secret, invited Tartaglia to visit him. After many entreaties and much flattery, Tartaglia revealed his method of solution on the promise, probably given under oath, that Cardan would keep it confidential. Rumors began to circulate, however, that Tartaglia was not the first discoverer of the cubic formula, and in 1543 Cardan journeyed to Bologna to try to verify these reports. After examining the posthumous papers of del Ferro, he concluded that del Ferro was the one who had made the breakthrough. Cardan no longer felt bound by his promise to Tartaglia, and when Cardan's work *Ars Magna* appeared in 1545, the formula and method of proof were fully disclosed. Cardan candidly admitted (at three places in the text) that he had gotten the solution to the special cubic equation $x^3 + px = q$ from his "friend" Tartaglia, but claimed to have carried out for himself the proof that the formula he had received was correct. Angered at this apparent breach of a solemn oath and feeling cheated out of the rewards of his monumental work, Tartaglia accused Cardan of lying. Thus began one of the bitterest feuds in the history of science, carried on with name-calling and mudslinging of the lowest order.

7.3   Cardan's *Ars Magna*

Cardan wrote on a wide variety of subjects, including mathematics, astrology, music, philosophy, and medicine. When he died, 131 of his works had been published and 111 existed in manuscript form, and he had claimed to have burned 170 others that were unsatisfactory. These ran the gamut from *Practica Arithmeticae* (1539), a book on numerical calculation based largely on Pacioli's work of 1494, to *Liber de Vita Propria* (1575), an autobiography in which he did not spare the most shameful revelations. His passion for the games of chess, dice, and cards inspired Cardan to write *Liber de Ludo Aleae* (Book on Games of Chance). Found among his papers after his death and published in 1663, this work broke the ground for a theory of probability more than 50 years before

Girolamo Cardano
(1501–1576)

(Source: Princeton University Press.)

Fermat and Pascal, to whom the first steps are usually attributed. In it he even gives advice on how to cheat, no doubt gained from personal experience. One of the ironic twists of fate in Cardan's life is that his excessive gambling, which had cost him time, money, and reputation, should have helped him earn a place in the history of mathematics.

In permanent significance, the *Ars Magna* (The Great Art) undoubtedly stands at the head of the entire body of Cardan's writings, mathematical or otherwise. This work, which was first printed in 1545, today would be classified as a text on algebraic equations. It makes very clear that Cardan was no mere plagiarist but one who combined a measure of honest toil with his piracy. Although negative numbers had become known in Europe through Arabic texts, most Western algebraists did not accept them as bona fide numbers and preferred to write their equations so that only positive terms appeared. Thus, there was no one cubic equation at the time, but rather thirteen of them, according to whether the terms of the various degrees appeared on the same side of the equality sign or on opposite sides. In giving Cardan the formula for $x^3 + px = q$, Tartaglia did not automatically provide solutions for all the other forms that the cubic might take. Cardan was forced to expand Tartaglia's discovery to cover these other cases, devising and proving the rule separately in each instance.

Hitherto, mathematicians had confined their attention to those roots of equations that were positive numbers. Cardan was the first to take notice of negative roots, although he called them fictitious, and the first to recognize that a cubic might have three roots. Another notable aspect of Cardan's discussion was the clear realization of the existence of what we now call complex or imaginary numbers (the ghosts of real numbers, as Napier was later to call them). Cardan kept these numbers out of the *Ars Magna* except in one case, when he considered the problem of dividing 10 into two parts whose product was 40. He obtained the roots $5 + \sqrt{-15}$ and $5 - \sqrt{-15}$ as solutions of the quadratic equation $x(10 - x) = 40$, and then stated, "Putting aside the mental tortures involved, multiply $5 + \sqrt{-15}$ by $5 - \sqrt{-15}$, making $25 - (-15)$, whence the product is 40." Cardan somehow felt obliged to accept these solutions yet hastened to add that there was no interpretation for them, remarking, "So progresses arithmetic subtlety the end of which, as is said, is as refined as it is useless." But merely writing down the meaningless gave it a symbolic meaning, and Cardan deserves credit for having paid attention to the situation.

Among the innovations that Cardan introduced in the *Ars Magna* was the trick of changing a cubic equation to one in which the second-degree term was absent. If one starts with the equation

$$x^3 + ax^2 + bx + c = 0,$$

all that is needed is to make the substitution $x = y - a/3$. With this new variable, the given equation becomes

$$0 = \left(y - \frac{a}{3}\right)^3 + a\left(y - \frac{a}{3}\right)^2 + b\left(y - \frac{a}{3}\right) - c$$

$$= \left[y^3 - 3y^2\left(\frac{a}{3}\right) + 3y\left(\frac{a}{3}\right)^2 - \left(\frac{a}{3}\right)^3\right]$$

$$+ a\left[y^2 - 2y\left(\frac{a}{3}\right) + \left(\frac{a}{3}\right)^2\right] + b\left(y - \frac{a}{3}\right) + c$$

$$= y^3 + \left(b - \frac{a^2}{3}\right)y + \left(\frac{2a^3}{27} - \frac{ab}{3} + c\right).$$

If one sets

$$p = b - \frac{a^2}{3} \quad \text{and} \quad q = -\left(\frac{2a^3}{27} - \frac{ab}{3} + c\right),$$

then the last equation can be written

$$y^3 + py = q,$$

which is the so-called reduced form of the cubic. It lacks a term in y^2, but otherwise the coefficients are arbitrary.

Cardan solved the cubic equation $x^3 + 20x = 6x^2 + 33$ by this reduction technique. Through the substitution $x = y - (-6)/3 = y + 2$, it is transformed to the equation

$$(y^3 + 6y^2 + 12y + 8) + 20(y + 2) = 6(y^2 + 4y + 4) + 33,$$

or simplified,

$$y^3 + 8y = 9.$$

This last equation has one obvious solution, namely $y = 1$; hence, $x = y + 2 = 3$ will satisfy the original cubic.

Let us examine how Cardan managed to arrive at the general solution of the reduced cubic. Because the Renaissance was a period of the highest veneration of Greek

mathematics, it is not unexpected that his proofs should be based on geometric arguments, emulating (as Cardan himself emphasized) the reasoning of Euclid. The technique for dealing with the cubic

(1) $$x^3 + px = q, \qquad p > 0, q > 0,$$

although geometric, is equivalent to using the algebraic identity

(2) $$(a - b)^3 + 3ab(a - b) = a^3 - b^3.$$

If a and b are chosen so that $3ab = p$ and $a^3 - b^3 = q$, then identity (2) becomes

$$(a - b)^3 + p(a - b) = q,$$

which shows that $x = a - b$ will furnish a solution to the cubic (1). The problem
therefore involves solving the pair of simultaneous equations

$$a^3 - b^3 = q,$$

$$ab = \frac{p}{3},$$

for a and b. To do so, one squares the first equation and cubes the second, to get

$$a^6 - 2a^3 b^3 + b^6 = q^2,$$

$$4a^3 b^3 = \frac{4p^3}{27}.$$

When the equations are added, it follows that

$$(a^3 + b^3)^2 = a^6 + 2a^3 b^3 + b^6 = q^2 + \frac{4p^3}{27}$$

and so

$$a^3 + b^3 = \sqrt{q^2 + \frac{4p^3}{27}}.$$

If the equations

$$a^3 - b^3 = q \qquad \text{and} \qquad a^3 + b^3 = \sqrt{q^2 + \frac{4p^3}{27}}$$

are now solved simultaneously, then a^3 and b^3 can be determined; the result is

$$a^3 = \frac{1}{2}\left(q + \sqrt{q^2 + \frac{4p^3}{27}}\right) = \frac{q}{2} + \sqrt{\frac{q^2}{4} + \frac{p^3}{27}},$$

$$b^3 = \frac{1}{2}\left(-q + \sqrt{q^2 + \frac{4p^3}{27}}\right) = -\frac{q}{2} + \sqrt{\frac{q^2}{4} + \frac{p^3}{27}}.$$

But then

$$a = \sqrt[3]{\frac{q}{2} + \sqrt{\frac{q^2}{4} + \frac{p^3}{27}}},$$

$$b = \sqrt[3]{-\frac{q}{2} + \sqrt{\frac{q^2}{4} + \frac{p^3}{27}}},$$

and consequently,

$$x = a - b = \sqrt[3]{\frac{q}{2} + \sqrt{\frac{q^2}{4} + \frac{p^3}{27}}} - \sqrt[3]{-\frac{q}{2} + \sqrt{\frac{q^2}{4} + \frac{p^3}{27}}}.$$

As Tartaglia feared, this last formula has forever since been known as Cardan's formula for the solution of the cubic equation. The mathematician to whom we owe the chief contribution made to algebra in the sixteenth century is largely forgotten, and the discovery goes by the name of a scoundrel.

Cardan illustrated his method by solving the equation

$$x^3 + 6x = 20.$$

In this case, $p = 6$ and $q = 20$, so that $p^3/27 = 8$ and $q^2/4 = 100$; whence the formula yields

$$x = \sqrt[3]{\sqrt{108} + 10} - \sqrt[3]{\sqrt{108} - 10}.$$

As remarked earlier, Cardan was forced to treat an elaborate list of equation types, produced largely by his failure to allow negative coefficients. In solving the equation

$$x^3 = px + q, \qquad p > 0, q > 0,$$

he used a geometric argument corresponding to the identity

$$(a + b)^3 = a^3 + b^3 + 3ab(a + b),$$

to arrive at the solution

$$x = \sqrt[3]{\frac{q}{2} + \sqrt{\frac{q^2}{4} - \frac{p^3}{27}}} + \sqrt[3]{\frac{q}{2} - \sqrt{\frac{q^2}{4} - \frac{p^3}{27}}}.$$

There is one difficulty connected with this last formula, which Cardan observed but could not resolve. When $(q/2)^2 < (p/3)^3$, the formula leads inevitably to square roots of negative numbers. That is, $\sqrt{q^2/4 - p^3/27}$ involves "imaginary numbers."

Consider, for example, the historic equation

$$x^3 = 15x + 4,$$

treated by Rafael Bombelli, the last great sixteenth-century Bolognese mathematician, in his *Algebra* (1572). A direct application of the Cardan-Tartaglia formula would lead to

$$x = \sqrt[3]{2 + \sqrt{-121}} + \sqrt[3]{2 - \sqrt{-121}}.$$

Bombelli knew, nevertheless, that the equation had three real solutions, namely 4, $-2 + \sqrt{3}$, and $-2 - \sqrt{3}$. One is left in the paradoxical situation in which the formula produces a result useless for most purposes, yet in other ways three perfectly good solutions can be found. This impasse, which arises when all three roots are real and different from zero, is known as the ''irreducible case'' of the cubic equation.

Bombelli was the first mathematician bold enough to accept the existence of imaginary numbers, and hence to throw some light on the puzzle of irreducible cubic equations. A native of Bologna, he himself had not received any formal instruction in mathematics and did not teach at the university. He was the son of a wool merchant and by profession an engineer-architect. Bombelli felt that only Cardan among his predecessors had explored algebra in depth, and that Cardan had not been clear in his exposition. He therefore decided to write a systematic treatment of algebra to be a successor to Cardan's *Ars Magna*. Bombelli composed the first draft of his treatise about 1560, but it remained in manuscript form until 1572, shortly before his death. The preparation of his *Algebra* took considerably longer than Bombelli had foreseen, for as he wrote in the work:

> A Greek manuscript in this science was found in the Vatican Library, composed by Diophantus. . . . We set to translating it and have already done five of the seven (sic) extant books. The rest we have not been able to finish because of other commitments.

Tremendously enthusiastic over the rediscovery of the *Arithmetica*, Bombelli took 143 problems and their solutions from its first four books and embodied them in his *Algebra,* interspersing them with his own contributions. Although Bombelli did not distinguish among the problems, he nonetheless acknowledged that he had borrowed freely from Diophantus. (A manuscript of the *Algebra* was found in 1923; the absence of the 143 problems borrowed from the *Arithmetica* suggests that Bombelli had not seen the Vatican copy when he first wrote the work.) Whereas the works of Pacioli and Cardan contained many problems of applied arithmetic, Bombelli's problems were all abstract. He claimed that while others wrote for a practical, rather than a scientific purpose, he had ''restored the effectiveness of arithmetic, imitating the ancient writers.'' The publication of Bombelli's *Algebra* completed a movement that began in Italy about 1200, when Fibonacci introduced the rules of algebra in the *Liber Abaci.*

Bombelli's skill in operating with imaginary numbers enabled him to demonstrate the applicability of Cardan's formula, even in the irreducible case (all roots real) of the cubic equation. Assuming that the complex numbers behaved like other numbers in calculations, he made a circuitous passage into, and out of, the complex domain and ended by showing that the apparently imaginary expression for the root of the equation $x^3 = 15x + 4$ gave a real value. Bombelli had the ingenious idea that the complex values of the radicals

$$\sqrt[3]{2 + \sqrt{-121}} \quad \text{and} \quad \sqrt[3]{2 - \sqrt{-121}}$$

might be related much as the radicals themselves; that is, they might differ only in a sign. This prompted him to set

$$\sqrt[3]{2 + \sqrt{-121}} = a + b\sqrt{-1} \quad \text{and} \quad \sqrt[3]{2 - \sqrt{-121}} = a - b\sqrt{-1},$$

where $a > 0$ and $b > 0$ are to be determined. As Bombelli said:

> It was a wild thought in the judgment of many; and I too for a long time was of the same opinion. The whole matter seemed to rest on sophistry rather than on truth. Yet I sought so long, until I actually proved this to be the case.

Now the relation $\sqrt[3]{2 + \sqrt{-121}} = a + b\sqrt{-1}$ implies that

$$2 + \sqrt{-121} = (a + b\sqrt{-1})^3$$
$$= a^3 + 3a^2b\sqrt{-1} + 3ab^2(\sqrt{-1})^2 + b^3(\sqrt{-1})^3$$
$$= a(a^2 - 3b^2) + b(3a^2 - b^2)\sqrt{-1}.$$

This equality would hold provided that

$$a(a^2 - 3b^2) = 2 \quad \text{and} \quad b(3a^2 - b^2) = 11.$$

If solutions are sought in the integers, then the first of these conditions tells us that a must be equal to 1 or 2, and the second condition asserts that b has the value 1 or 11; only the choices $a = 2$ and $b = 1$ satisfy both conditions. Therefore,

$$2 + \sqrt{-121} = (2 + \sqrt{-1})^3 \quad \text{and} \quad 2 - \sqrt{-121} = (2 - \sqrt{-1})^3.$$

Bombelli concluded that one solution to the cubic equation $x^3 = 15x + 4$ was

$$x = \sqrt[3]{2 + \sqrt{-121}} + \sqrt[3]{2 - \sqrt{-121}}$$
$$= \sqrt[3]{(2 + \sqrt{-1})^3} + \sqrt[3]{(2 - \sqrt{-1})^3}$$
$$= (2 + \sqrt{-1}) + (2 - \sqrt{-1}) = 4.$$

In proving the reality of the roots of the cubic $x^3 = 15x + 4$, he demonstrated the extraordinary fact that real numbers could be engendered by imaginary numbers. From this time on, imaginary numbers lost some of their mystical character, although their full acceptance as bona fide numbers came only in the 1800s.

7.3 Problems

1. Find all three roots of each of the following cubic equations by first reducing them to cubics that lack a term in x^2.

(a) $x^3 + 11x = 6x^2 + 6$.

(b) $x^3 + 6x^2 + x = 14$.

(c) $x^3 + 6x^2 = 20x + 56$.

(d) $x^3 + 64 = 6x^2 + 24x$.

2. Derive Cardan's formula

$$x = \sqrt[3]{\frac{q}{2} + \sqrt{\frac{q^2}{4} - \frac{p^3}{27}}} + \sqrt[3]{\frac{q}{2} - \sqrt{\frac{q^2}{4} - \frac{p^3}{27}}}$$

for solving the cubic equation $x^3 = px + q$, where $p > 0$, $q > 0$.

3. Using Cardan's formula, obtain one root of each of the following cubic equations.

  (a) $x^3 + 24x = 16$.
  (b) $x^3 + 15x = 6x^2 + 18$.
  (c) $x^3 + 27x = 6x^2 + 58$.
  (d) $x^3 = 9x + 12$.
  (e) $x^3 = 6x^2 + 15x + 8$.
  (f) $x^3 = 3x^2 + 27x + 41$.

4. Solve the cubic equation $x^3 + 6x^2 + x = 14$.

Problems 5–11 appear in Cardan's *Ars Magna*.

5. *Chapter 5, Problem 2.* There were two leaders, each of whom divided 48 aurei among his soldiers. One of these had two more soldiers than the other. The one who had two soldiers fewer had four aurei more for each soldier. Find how many soldiers each had.

6. *Chapter 37, Problem 1.* The dowry of Francis's wife is 100 aurei more than Francis's own property is worth, and the square of the dowry is 400 more than the square of his property's value. Find the dowry and the property value.

7. *Chapter 31, Problem 1.* Divide 8 into two parts, the product of the cubes of which is 16. [*Hint:* If the two parts are $4 + x$ and $4 - x$, it follows that $(4 + x)(4 - x) = \sqrt[3]{16}$.]

8. *Chapter 5, Problem 4.* There is a number for which adding twice its square root to it and twice its square root to this sum gives 10. What is the number? [*Hint:* Call the number x^2; if $y^2 = x^2 + 2x$, then $y^2 + 2y = 10$.]

9. *Chapter 17, Problem 3.* An oracle ordered a prince to build a sacred building whose space would be 400 cubits, the length being 6 cubits more than the width and the width 3 cubits more than the height. Find these quantities. [*Hint:* If x is the height, then $x(x + 3)(x + 9) = 400$.]

10. *Chapter 32, Problem 3.* Divide 6 into two parts, the sum of the squares of which is equal to the difference between their cubes. [*Hint:* Calling the two parts $3 + x$ and $3 - x$ leads to the equation $x^3 + 27x = x^2 + 9$.]

11. *Chapter 38, Problem 1.* Find two numbers whose difference is 8 and for which the sum of the cube of one and the square of the other is 100. [*Hint:* Let the numbers be called $x + 2$ and $x - 6$, so that $(x + 2)^3 + (x - 6)^2 = 100$.]

12. The following method of Vieta (1540–1603) is useful in solving the reduced cubic $x^3 + ax = b$. By substitution of $x = a/3y - y$, the given equation becomes $y^6 + by^3 - a^3/27 = 0$, a quadratic in y^3. By the quadratic formula,

$$y^3 = \frac{1}{2}\left(-b \pm \sqrt{b^2 + \frac{4a^3}{27}}\right),$$

from which y and then x can be determined. Use this method to find a root of the cubics $x^3 + 81x = 702$ and $x^3 + 6x^2 + 18x + 13 = 0$. [*Hint:* $\sqrt{142{,}884} = 378$.]

13. By making the substitution $x = y + 5/y$, find a root of the cubic equation $x^3 = 15x + 126$.

14. Use Cardan's formula to find, in these examples of the irreducible case in cubics, a root of the given equations.

  (a) $x^3 = 63x + 162$.
    [*Hint:* $81 \pm 30\sqrt{-3} = (-3 \pm 2\sqrt{-3})^3$.]

  (b) $x^3 = 7x + 6$.
    $$\left[\text{Hint: } 3 \pm \frac{10}{9}\sqrt{-3} = \left(\frac{3}{2} \pm \frac{1}{6}\sqrt{-3}\right)^3.\right]$$

  (c) $x^3 + 6 = 2x^2 + 5x$.
    $$\left[\text{Hint: } -\frac{28}{27} - \frac{5}{3}\sqrt{-3} = \left(\frac{1}{6} + \frac{5}{6}\sqrt{-3}\right)^3.\right]$$

15. The great Persian poet, Omar Khayyam (circa 1050–1130), found a geometric solution of the cubic equation $x^3 + a^2x = b$ by using a pair of intersecting conic sections. In modern notation, he first constructed the parabola $x^2 = ay$. Then he drew a semicircle with diameter $AC = b/a^2$ on the x-axis, and let P be the point of intersection of the semicircle with the parabola (see the figure). A perpendicular is dropped from P to the x-axis to produce a point Q.

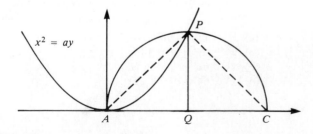

Complete the details in the following proof that the x-coordinate of P, that is, the length of segment AQ, is the root of the given cubic.

(a) $AQ^2 = a(PQ)$.

(b) Triangles AQP and PQC are similar, so that

$$\frac{AQ}{PQ} = \frac{PQ}{QC}, \quad \text{or} \quad PQ^2 = AQ\left(\frac{b}{a^2} - AQ\right).$$

(c) Substitution gives $AQ^3 + a^2AQ = b$.

16. It is also possible to use the parabola $y = x^2$ for duplicating a cube of edge a. Draw a circle with center $(a/2, 1/2)$ that passes through the origin $(0,0)$. Then the x-coordinate of the point of intersection of the circle and the parabola $y = x^2$ will serve as the edge of a cube double in volume to the given cube. Prove this conclusion.

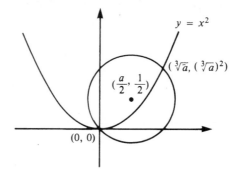

7.4   Ferrari's Solution of the Quartic Equation

After the cubic had been solved, it was only natural that mathematicians should attack the quartic (fourth-degree) equation. The solution was discovered during work on a problem proposed to Cardan in 1540. Divide the number 10 into three proportional parts so that the product of the first and second parts is 6. If the numbers are called $6/x, x, x^3/6$, the conditions laid down are clearly fulfilled. In particular, the requirement that

$$\frac{6}{x} + x + \frac{x^3}{6} = 10$$

is equivalent to the quartic

$$x^4 + 6x^2 + 36 = 60x.$$

After an unsuccessful attempt at solving this equation, Cardan turned it over to his disciple Ludovico Ferrari (1522–1565). Ferrari, using the rules for solving the cubic, eventually succeeded where his master had failed. At least, Cardan had the pleasure of incorporating the result in the *Ars Magna,* with due credit given Ferrari.

Ferrari, the son of poor parents, was taken into Cardan's household as a servant boy at the age of 14. Although he had not received any formal education, Ferrari was exceptionally gifted, and Cardan undertook to instruct him in Latin, Greek, and mathematics. Cardan soon made him his personal secretary, and after four years of service, Ferrari left to become public lecturer in mathematics at the University of Milan. He became professor of mathematics at Bologna in 1565 and died in the same year, having been poisoned with white arsenic—by his own sister, as rumor had it.

Ferrari joined the fray surrounding the solution of the cubic by swearing that he had been present at the fateful meeting between Cardan and Tartaglia and that there had been no oath of secrecy involved. Always eager to defend his old master, Ferrari then challenged Tartaglia to a public disputation on mathematics and related disciplines, writing in a widely distributed manifesto: "You have written things that falsely and unworthily slander Signor Cardan, compared with whom you are hardly worth mentioning." Tartaglia's counterstatement asked Ferrari either to let Cardan fight his

HIERONYMI CAR

DANI, PRÆSTANTISSIMI MATHE
MATICI, PHILOSOPHI, AC MEDICI,

ARTIS MAGNÆ,

SIVE DE REGVLIS ALGEBRAICIS,
Lib.unus. Qui & totius operis de Arithmetica, quod
OPVS PERFECTVM
inscripsit, est in ordine Decimus.

HAbes in hoc libro, studiose Lector, Regulas Algebraicas (Itali, de la Cof
fa uocant) nouis adinuentionibus, ac demonstrationibus ab Authore ita
locupletatas, ut pro pauculis antea uulgo tritis, iam septuaginta euaserint. Ne
q; solum, ubi unus numerus alteri, aut duo uni, uerum etiam, ubi duo duobus,
aut tres uni æquales fuerint, nodum explicant.   Hunc aut librum ideo seor-
sim edere placuit, ut hoc abstrusissimo, & plane inexhausto totius Arithmeti-
cæ thesauro in lucem eruto, & quasi in theatro quodam omnibus ad spectan
dum exposito, Lectores incitarentur, ut reliquos Operis Perfecti libros, qui per
Tomos edentur, tanto auidius amplectantur, ac minore fastidio perdiscant.

Title page of Cardan's Ars Magna *(1545). (Source: M.I.T. Press.)*

own battles or to admit that he was acting on Cardan's behalf; the challenge would be accepted if Cardan were willing to countersign Ferrari's letter and if (because Tartaglia feared some sort of trickery) topics from the *Ars Magna* were excluded. Another acrimonious dispute ensued in which twelve letters were exchanged, full of charges and insults, with each party trying to justify his own position. In one of these retorts, Ferrari made the mistake of calling himself Cardan's creation, allowing Tartaglia the satisfaction of thereafter referring to him as ''Cardan's creature.''

The contest finally took place in Ferrari's hometown of Milan in 1548 before a large and distinguished gathering. Perhaps aware of his own limitations, Cardan had the foresight to leave Milan for several days. There is no record of the proceedings except for a few statements to the effect that the meeting soon deteriorated into a shouting match over a problem of Ferrari's that Tartaglia had been unable to resolve. The altercation ran into the dinner hour, at which time everyone felt compelled to leave. Tartaglia departed the next morning claiming to have come off the better in the dispute, but it seems more likely that Ferrari was declared the winner. The best evidence of this is that Tartaglia lost his teaching post in Brescia, and Ferrari received a host of flattering offers, among them an invitation to lecture in Venice, Tartaglia's stronghold.

Ferrari's method for solving the general quartic could, in modern notation, be summarized as follows. First, reduce the equation

$$x^4 + ax^3 + bx^2 + cx + d = 0$$

to the special form

$$y^4 + py^2 + qy + r = 0,$$

in which the term in y^3 is missing, by substituting $x = y - a/4$. Now the left-hand side of

$$y^4 + py^2 = -qy - r$$

contains two of the terms of the square of $y^2 + p$. Let us complete the square by adding $py^2 + p^2$ to each side to get

$$(y^2 + p)^2 = y^4 + 2py^2 + p^2 = py^2 + p^2 - qy - r.$$

We now introduce another unknown for the purpose of converting the left member of this equation into $(y^2 + p + z)^2$. This is done by adding $2(y^2 + p)z + z^2$ to each side, and leads to

$$(y^2 + p + z)^2 = py^2 + p^2 - qy - r + 2(y^2 + p)z + z^2$$
$$= (p + 2z)y^2 - qy + (p^2 - r + 2pz + z^2).$$

The problem now reduces to finding a value of z that makes the right-hand side, a quadratic in y, a perfect square. This will be the case when the discriminant of the quadratic is zero; that is, when

$$4(p + 2z)(p^2 - r + 2pz + z^2) = q^2,$$

which requires solving a cubic in z; namely,

$$8z^3 + 20pz^2 + (16p^2 - 8r)z + (4p^3 - 4pr - q^2) = 0.$$

The last equation is known as the *resolvent cubic* of the given quartic equation, and it can be solved in the usual way. There are in general three solutions of the resolvent cubic, and y can be determined from any one of them by extracting square roots. Once a value of y is known, the solution of the original quartic is readily reached.

If the procedure sounds complicated, an example from the *Ars Magna* might help to clarify the sequence of steps. Cardan considered (Chapter 39, Problem 9) the quartic equation $x^4 + 4x + 8 = 10x^2$, or equivalently,

$$x^4 - 10x^2 = -4x - 8.$$

Completing the square on the left-hand side, one gets

$$(x^2 - 10)^2 = -10x^2 - 4x + 92.$$

By adding the quantity $2z(x^2 - 10) + z^2$ to each side, this equation is changed to

(1) $$(x^2 - 10 + z)^2 = (2z - 10)x^2 - 4x + (92 - 20z + z^2),$$

where z is a new unknown. Now the right-hand expression is a perfect square if z is chosen to satisfy the condition

$$4(2z - 10)(92 - 20z + z^2) = 16,$$

or after simplification,

$$z^3 - 25z^2 + 192z = 462.$$

This is a cubic equation from which z can be found. We start by letting $z = u + \frac{25}{3}$; this substitution reduces the equation to the form

$$u^3 = \frac{49}{3}u + \frac{524}{27}.$$

A solution is $u = -\frac{4}{3}$, so that $z = 7$. It is this value of z that should give squares on both sides of equation (1). The result of substituting $z = 7$ is

$$(x^2 - 3)^2 = 4x^2 - 4x + 1 = (2x - 1)^2,$$

whence $x^2 - 3 = \pm(2x - 1)$. The positive sign gives

$$x^2 - 2x - 2 = 0,$$

and the negative sign yields

$$x^2 + 2x - 4 = 0.$$

In solving these equations by the quadratic formula, it is found that the four solutions of the original quartic equation are

$$1 + \sqrt{3}, \ 1 - \sqrt{3}, \ -1 + \sqrt{5}, \ -1 - \sqrt{5}.$$

Our story has a postscript. We have seen that in the case of quadratic, cubic, and quartic equations, explicit formulas for the roots were found that were formed from the coefficients of the equation by using the four operations of arithmetic (addition, multiplication, subtraction, and division) and by taking radicals of various sorts. The next natural step was to seek similar solutions of equations of higher degrees, the presumption being that an equation of degree n should be capable of formal solution by means of radicals and probably by radicals of an exponent not larger than n. For close to 300 years, algebraists wrestled with the general equation of fifth degree (the quintic equation) and made almost no progress. But these repeated failures at least had the effect of suggesting the possibility, startling at the time, that the quintic equation might not be solvable in this way.

Paolo Ruffini (1765–1822), an Italian physician who taught mathematics as well as medicine at the University of Modena, confirmed the suspicion of the impossibility of finding an algebraic solution for the general fifth-degree equation. Ruffini's proof, which appeared in his two-volume *Teorie generale delle equazioni* of 1799, was sound in general outline although faulty in some details. The Norwegian genius Niels Henrik Abel (1802–1829), when he was about 19 years old, made a study of the same problem. At first he thought he had found a solution of the general quintic by radicals, but later he established the unsolvability of the equation, using a more rigorous argument than Ruffini's. Abel fully realized the importance of his discovery and had it published in 1824, at his own expense, in a pamphlet that bore the title *Memoire sur les equations algébriques où on démontre l'impossibilité de la résolution de l'equation generale du cinquième degré.* So that expenses could be kept down, the whole pamphlet had to be condensed to six pages of actual print, making it difficult to follow the reasoning. Thus, the significance of Abel's masterpiece went unnoticed by contemporary scholars. When Europe's leading mathematician, Carl Friedrich Gauss, duly received his copy, he tossed it aside unread with the disgusted exclamation, "Here is another of those monstrosities!"

Abel's opportunity came when he had the great good fortune to make the acquaintance of August Leopold Crelle, a German civil engineer and enthusiastic mathematical amateur. At this time, Crelle was making plans to launch a new journal, which would be the first periodical devoted exclusively to mathematical research. Abel eagerly accepted the invitation to submit articles, and the first three volumes of the *Journal für die reine und angewandte mathematik* (Journal for Pure and Applied Mathematics), or *Crelle's Journal* as it is commonly called, contained 22 papers by Abel. In the founding volume (1826), he expanded his earlier research into what now is known as the Abel-Ruffini theorem: It is impossible to find a general formula for the roots of a polynomial equation of degree five or higher if the formula for the solution is allowed to use only arithmetic operations and extraction of roots. When Abel composed his paper, he was not aware that he had a precursor. He was later to write, however, in a manuscript *Sur la résolution algébraique des equations* (dated 1828 but only published after his death): "The only one before me, if I am not mistaken, who has tried to prove the impossibility of the algebraic solution of the general equation is the mathematician Ruffini, but his paper is so complicated that it is difficult to judge the correctness of his arguments."

Abel's theorem on the unsolvability of higher equations applied to general equations only. Many special equations existed that were solvable by radicals, and the characterization of these remained an open question. It was reserved for another young mathematician, Evariste Galois (1811–1832) to definitively answer what specific equations of a given degree admit an algebraic solution. The posthumous publication of Galois's manuscripts in Liouville's *Journal de Mathématiques* in 1846 represented both the completion of Abel's research and the foundation of group theory, one of the most important branches of modern mathematics. Considering the significance of his discovery, one naturally asks why it required 14 years after Galois's death for the essential elements of his work to become available in print. The reason is a combination of sheer bad luck and negligence. The original memoir was mislaid by the editor appointed to examine it, and after resubmission, it was returned by a second editor, who judged the contents incomprehensible.

The sequence of events seems to be this. Galois first submitted his results on the algebraic solution of equations to the Academy of Sciences in May 1829, while he was still only 17 years old. Augustin-Louis Cauchy (1789–1857), a member of the Academy and a professor at the Ecole Polytechnique, was appointed referee. Cauchy either forgot or lost the communication, as well as another presented a week later. Galois then (February 1830) submitted a new version of his investigations to the Academy, hoping to enter it in the competition for the Grand Prize in Mathematics, the pinnacle of mathematical honor. This time it was entrusted to the permanent secretary, Joseph Fourier (1768–1830), who died shortly thereafter, before examining the manuscript. It was never retrieved from among his papers. A further disappointment awaited for Galois. In January 1831, he submitted his paper for the third time under the title *"Une mémoire sur les conditions de résolubilité des equations par radicaux."* After a delay of some six months, during which Galois wrote to the president of the Academy asking what had happened, it was rejected by the referee Simeon-Denis Poisson (1781–1840). At the conclusion of his report, Poisson remarked:

> His arguments are not sufficiently clear, nor developed enough for us to judge their correctness. . . . It is hoped that the author would publish his work in its entirety so that we can form a definite opinion.

In May 1832, Galois was provoked into a duel in unclear circumstances. (The theory has been advanced that the challenger was hired by the police, who arranged the confrontation in order to eliminate what they considered to be a dangerous radical.) On the eve of the duel, apparently certain of death, Galois wrote a letter to a friend describing the contents of the memoir Poisson had rejected. Its seven pages, hastily written, contain a summary of the discoveries he had been unable to develop. The letter ends with the plea;

> Eventually there will be, I hope, some people who will find it profitable to decipher this mess.

Galois spent the rest of the night annotating and making corrections to some of his papers; next to a theorem, he scrawled:

> There are a few things left to be completed in this proof. I do not have time.

The duel took place on May 30, 1832, early in the morning. Galois was grievously wounded by a shot in the abdomen, and lay where he had fallen until found by a passing peasant, who took him to the hospital. He died the next morning of peritonitis, attended by his younger brother. Galois tried to console him, saying, "Do not cry. I need all my courage to die at twenty." He was buried in a common ditch at the cemetery of Montparnasse; the exact location is unknown.

By 1843, Galois's manuscripts had found their way to Joseph Liouville (1809–1882), who after spending several months in the attempt to understand them, became convinced of their importance. He addressed the Academy of Sciences on July 4, 1843, opening with the words:

> I hope to interest the Academy in announcing that among the papers of Evariste Galois I have found a solution, as precise as it is profound, of this beautiful problem: whether or not it [the general equation of fifth degree] is solvable by radicals.

Liouville announced that he would publish Galois's papers in the December 1843 issue in his recently founded periodical *Journal de Mathématiques Pures et Appliquées*. But for some reason, publication of the heavily edited version of the celebrated 1831 memoir did not occur until the October–November 1846 issue.

Although no trace of Galois's grave remains, his enduring monument lies in his ideas. During the late 1800s, Galois's theory—as well as a new topic it brought to life, group theory—became an integral and accepted part of mathematics. Galois's theory appears to have been taught in the German universities for the first time by Richard Dedekind, who lectured on the topic at Göttingen in the winter of 1856–1857; it is said that only two students came to hear him. The first full and clear presentation of the Galois theory was given by Camille Jordan in his book *Traité des substitutions et des equations algébraiques* (1870).

7.4 Problems

1. Solve the following quartic equations by Ferrari's method.

   (a) $x^4 + 3 = 12x$.
   (b) $x^4 + 6x^2 + 8x + 21 = 0$.
      [*Hint:* The reduced form of the resolvent cubic is $u^3 - 24u + 32 = 0$, with $u = 4$ as a solution.]
   (c) $x^4 + 9x + 4 = 4x^2$. [*Hint:* The resolvent cubic $z^3 - 10z^2 + 28z - \frac{273}{8} = 0$ has $z = \frac{13}{2}$ as a solution.]

2. Solve the quartic $x^4 + 4x^3 + 8x^2 + 7x + 4 = 0$.
   [*Hint:* First replace the given quartic by $y^4 + 2y^2 - y + 2 = 0$. The resolvent cubic of this last equation is $z^3 + 5z^2 + 6z + \frac{15}{8} = 0$, with $z = -\frac{1}{2}$ as a solution.]

3. Solve the quartic $x^4 + 8x^3 + 15x^2 = 8x + 16$.
   [*Hint:* First replace the given quartic by $y^4 - 9y^2 - 4y + 12 = 0$. The reduced form of the resolvent cubic of this last equation is $u^3 = \frac{75}{4}u + \frac{125}{4}$, with $u = \frac{25}{2}$ as a solution.]

4. Use Ferrari's method to show that the quartic equation

   $$x^4 + 6x^3 = 6x^2 + 30x + 11$$

   has the four real roots $1 + \sqrt{2}$, $1 - \sqrt{2}$, $-4 + \sqrt{5}$, and $-4 - \sqrt{5}$.

5. Find a solution to the following problem from the *Ars Magna*.

   Chapter 26, Problem 1. Four men form an organization. The first deposits a given quantity of aurei; the second deposits the fourth power of one-tenth of the first; the third, five times the square of

one-tenth the first; and the fourth, 5. Let the sum of the first and second equal the sum of the third and fourth. How much did each deposit? [*Hint:* If it is assumed that the first deposited $10x$, then the conditions imply that $x^4 + 10x = 5x^2 + 5$.]

6. The following method of Vieta was a notable improvement in Ferrari's technique for solving the quartic $y^4 + py^2 + qy + r = 0$. To both sides of the equation, add $y^2z^2 + \frac{1}{4}z^4$, where z is a new unknown, so that

   $$(y^2 + \tfrac{1}{2}z^2)^2 = y^2z^2 + \tfrac{1}{4}z^4 - r - qy - py^2$$

   $$= y^2(z^2 - p) - qy + (\tfrac{1}{4}z^4 - r).$$

   The right-hand side is a perfect square if z is chosen to satisfy

   $$q^2 = 4(z^2 - p)(\tfrac{1}{4}z^4 - r)$$

   $$= z^6 - pz^4 - 4rz^2 + 4rp,$$

   which is a cubic in z^2 and therefore solvable. Use Vieta's procedure to find one root of the quartic equation

   $$y^4 - y^3 + y^2 - y = 10.$$

Bibliography

Bidwell, James, and Lange, Bernard. "Girolamo Cardano: A Defense of His Character." *Mathematics Teacher* 64(1971): 25–31.

Boas, Marie. *The Scientific Renaissance.* New York: Harper, 1962.

Cardano, Girolamo. *The Great Art, or the Rules of Algebra.* Translated by Richard Witmer. Cambridge, Mass.: M.I.T. Press, 1968. (Dover reprint, 1993).

Crombie, A. C. *Medieval and Early Modern Science.* 2 vols. Cambridge, Mass.: Harvard University Press, 1963.

Dales, Richard. *The Scientific Achievement of the Middle Ages.* Philadelphia: University of Pennsylvania Press, 1973.

Feldman, Richard, Jr. ''The Cardano-Tartaglia Dispute.'' *Mathematics Teacher* 54(1961): 160–163.

Grant, Edward, ed. *A Source Book in Medieval Science.* Cambridge, Mass.: Harvard University Press, 1974.

Haskins, Charles. *Studies in the History of Medieval Science.* Cambridge, Mass.: Harvard University Press, 1924.

Hay, Cynthia, ed. *Mathematics from Manuscript to Print: 1300–1600.* Oxford: Clarendon Press, 1988.

Kearney, Hugh. *Science and Change 1500–1700.* New York: World University Library, 1971.

Lindberg, David. *The Beginnings of Western Science: 600 B.C. to 1450 A.D.* Chicago: University of Chicago Press, 1992.

Ore, Oystein. *Cardano, the Gambling Scholar.* Princeton, N.J.: Princeton University Press, 1953. (Dover reprint, 1965).

Randall, John. *The School of Padua, and the Emergence of Modern Science.* Padua: Editrice Antenore, 1961.

Rose, Paul. *The Italian Renaissance of Mathematics.* Geneva: Librairie Droz, 1975.

Sarton, George. *Six Wings: Men of Science in the Renaissance.* Bloomington: Indiana University Press, 1957.

Schrader, Dorothy. ''The Arithmetic of Medieval Universities.'' *Mathematics Teacher* 60(1967): 264–275.

Smith, David E. ''The First Printed Arithmetic (Treviso, 1478).'' *Isis* 6(1924): 311–313.

Swetz, Frank. *Capitalism and Arithmetic: The New Math of the 15th Century.* La Salle, Ill.: Open Court, 1987.

Taton, René. ''Evariste Galois and his Contemporaries.'' *Bulletin of the London Mathematical Society* 15(1983): 107–118.

Thorndike, Lynn. *Scientific Thought in the Fifteenth Century.* New York: Columbia University Press, 1929.

Wightman, W. P. D. *Science and the Renaissance.* Vol. 1. Edinburgh: Oliver and Boyd, 1962.

Zimmer, E. *Regiomontanus: His Life and Work.* Translated by E. Brown. New York: North-Holland/Elsevier, 1990.

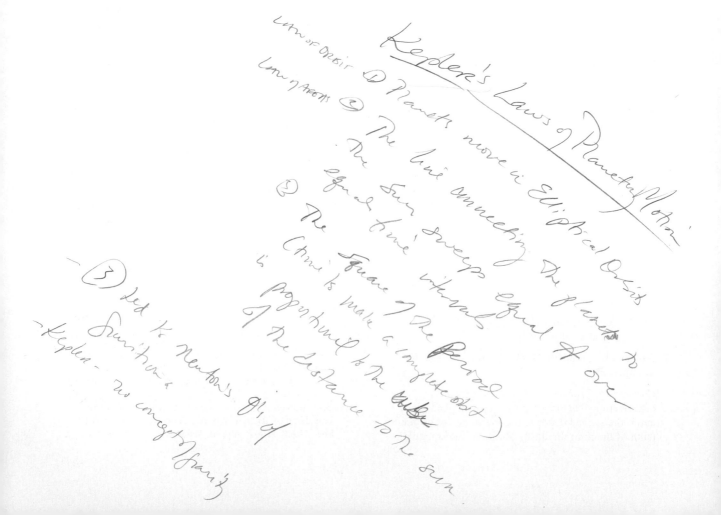

CHAPTER 8

The Mechanical World: Descartes and Newton

The discoveries of Newton have done more for England and for the race, than has been done by whole dynasties of British monarchs.

THOMAS HILL

8.1 The Dawn of Modern Mathematics

The Renaissance, which by the sixteenth century was well under way in Italy, soon spread north and west, first to Germany, then to France and the Low Countries, and finally to England. By the late 1600s, scientific, technological, and economic leadership centered on the English Channel—in those countries that had been galvanized by the commerce arising from the great voyages of discovery. At the start, the revival was mainly literary, but gradually scholars began to pay less attention to what was written in ancient books and to place more reliance on their own observations. The age was characterized by an eagerness to experiment, and above all to determine how things happened. Seventeenth-century science may be said to have begun with the appearance of William Gilbert's *De Magnete* in 1600, the first treatise on physical science whose content was based entirely on experimentation; and the culmination would have been Isaac Newton's *Opticks* in 1704.

In between the *De Magnete* and the *Opticks* came the contributions of Johannes Kepler, who was convinced that planetary bodies moved not in Aristotle's "ideal circles" but in elliptical orbits, and he thereby formulated the laws of terrestrial motion (1619). Also there were the demonstration by William Harvey (1628) of the circulatory route of the blood from the heart through arteries and veins by way of the lungs; the laying down of the principles of modern chemistry by Robert Boyle in his *Sceptical Chymist* (1661); and the publication of Robert Hooke's *Micrographia* (1665), the earliest large-scale work on the microscopic observation of cellular structure. However, no brief summing can do justice to the achievements of a period that saw so many new discoveries and so many advances in scientific methods.

309

Whereas the Renaissance marked a return to classical concepts, the seventeenth century set mathematics on entirely new foundations. So extensive and radical were the changes that historians have come to regard the half-century from 1637 to 1687 as the fountainhead of modern mathematics—the first date alluding to the publication of Descartes's *La Géométrie* and the second to the date of publication of Newton's *Principia Mathematica*.

Renaissance mathematics had added little to the geometry of the ancient Greeks, but 1600 ushered in an unexpected revival in the subject. In 1637 the French mathematical community witnessed one of those strange coincidences, once thought rare but which the history of science has shown to be frequent. Two men, Pierre de Fermat and René Descartes, simultaneously wedded algebra to geometry, to produce a remarkable innovation, analytic geometry. About the time when Fermat and Descartes were laying the foundations of a coordinate geometry, two other equally original mathematicians, Pascal and Desargues, were rendering a similar service in the area of synthetic projective geometry. But it was not only on account of the far-reaching developments in geometry that the seventeenth century has become illustrious in the history of mathematics, for the activities of the mathematicians of the period stretched into many fields, new and old. Number mysticism gave way to number theory, in Fermat's reflections on diophantine analysis. The mathematical theory of probability, a subject to which Cardan contributed in his book *Liber de Ludo Aleae,* took its first full steps in an exchange of letters between Pascal and Fermat concerning the calculation of probabilities. Leibniz's attempt to reduce logical discussion to systematic form was the forerunner of modern symbolic logic; but it was so far in advance of its time that not until 200 years later was the idea realized through the work of the English mathematician George Boole. Hardly less important were the studies of Galileo, Descartes, Torricelli, and Newton, which were to turn mechanics into an exact science during the next two centuries.

During the middle years of the Renaissance, trigonometry had become a systematic branch of mathematics in its own right in place of serving as handmaiden to astronomy. The aim of facilitating work with complicated trigonometric tables was responsible for one of the greatest computational improvements in arithmetic, the invention of logarithms, by John Napier (1550–1617). Napier worked at least twenty years on the theory, which he explained in his book *Mirifici Logarithmorum Canonis Descriptio* (A Description of an Admirable Table of Logarithms, 1614). Seldom has a new discovery won such universal acclaim and acceptance. With logarithms, the operations of multiplication and division can be reduced to addition and subtraction, thereby saving an immense amount of calculation, especially when large numbers are involved. Astronomy was notorious for the time-consuming computations it imposed; the French mathematician Pierre de Laplace was later to assert that the invention of logarithms "by shortening the labors, doubled the life of the astronomer."

Above all, for mathematics the seventeenth century was the century of the rise of calculus. Although we normally ascribe the invention of calculus to two brilliant contemporaries, Isaac Newton (1642–1727) and Gottfried Leibniz (1646–1716), great advances in mathematics are seldom the work of single individuals. Cavalieri, Torricelli, Barrow, Descartes, Fermat, and Wallis had all paved the way to the threshold but had hesitated when it came to crossing it. By the second half of the seventeenth century, the raw materials lay at hand out of which the calculus would emerge. All that remained was for a Leibniz or a Newton to fuse these ideas in a tremendous synthesis. Newton's well-known statement to Hooke, "If I have seen <u>farther than others, it is because I have stood on the shoulders of giants</u>," shows his appreciation of this cumulative and progressive growth of mathematics.

Probably no single figure from the 1600s is as well known as the mathematician-physicist-astronomer Galileo Galilei (1564–1642). His name is associated with events of profound significance: with the birth of modern science, with the Copernican revolution, with the dethronement of Aristotle as the supreme authority in the schools, and with the struggle against external restrictions on scientific inquiry. Galileo's original intention was to enter the lucrative profession of medicine, and in 1581, he enrolled at the University of Pisa as a medical student. While a student at Pisa, Galileo is supposed to have made his first independent discovery, the isochronism (equality of time) of the pendulum. Tradition has it that this came about through his observation that a chandelier in the cathedral, set in motion while being lit, performed all its swings at equal intervals of time although its successive swings gradually grew narrower in amplitude.

Galileo's formal introduction to mathematics came late. There is a story that Galileo, after listening at the door of a classroom in which Ostilio Ricci (a pupil of the famous Italian mathematician Tartaglia) was lecturing on the geometry of Euclid, became so fascinated with mathematics that he abandoned his medical plans. Indeed, he appears to have had little fondness for medicine and left the university in 1585 without a degree. Under the tutelage of Ricci, Galileo spent the next year pursuing the study of Euclid; he then went on to the other Greek geometers, winding up with the mechanical works of Archimedes. From 1585 to 1589, he earned money by giving private lessons in mathematics. Then, at the age of 25, Galileo succeeded in obtaining a lectureship at the University of Pisa. The appointment was only for three years, and the salary a mere pittance, but he gained academic standing.

Galileo is alleged to have performed, during his stay at Pisa, a public demonstration at the Leaning Tower to show that bodies of the same material but different weights fall with equal speed. This was an open challenge to the prevailing Aristotelian physics, according to which "the downward movement of a mass of gold or lead, or any body endowed with weight, is quicker in proportion to its size;" that is, the heavier the body, the faster the fall. Aristotelians claimed that the simultaneous arrival of the two weights—if the demonstration was actually carried out—was the effect of sorcery and not a refutation of Aristotle. They managed to pack the young professor's lectures and hiss at his every word. Because he had aroused antagonism in the faculty, Galileo had little hope of reappointment at Pisa at the end of his three-year contract. He left, in 1592, to become professor of mathematics at the famed University of Padua, a post he held for the next 18 years.

In 1609, Galileo heard rumors that Dutch spectacle-makers had invented a remarkable contrivance for making distant objects appear quite close. Surmising how such a device, called a "telescope," might be constructed, he set to work to fashion one for himself. (It is noteworthy that the Dutch instrument was of a totally different type from what Galileo designed.) Although Galileo by no means invented the telescope, he seems to have been the first to look at the sky systematically with one and to publish findings.

Galileo Galilei
(1564–1642)

(From A Short History of Astronomy *by Arthur Berry, 1961, Dover Publications, Inc., N.Y.)*

In a series of observations made in 1610, Galileo was able to distinguish the four satellites revolving about Jupiter—perhaps the most dramatic disproof of the Aristotelian view that the earth is at the center of all astronomical motions. Within a month he published this truly earth-shattering news in a 29-page booklet entitled the *Sidereus Nuncius* (The Starry Messenger), ''unfolding great and marvelous sights'' such as the existence of unknown stars, the nature of the Milky Way, and the rugged surface of the moon. Such ideas were so disturbing that there were professors at Padua who refused to credit Galileo's discoveries, refused even to look into his telescope for fear of seeing in it things that would discredit the infallibility of Aristotle and Ptolemy, and even the Church. His open publication of Copernican views made Galileo's position as a teacher at Padua, a stronghold of Aristotelianism, untenable. Later in the year, he accepted an appointment as ''First Mathematician'' of the University of Pisa, and also the post of court mathematician to the Grand Duke of Tuscany.

At the beginning of the sixteenth century, most people still believed in the ancient description of the universe. As conceived by Aristotle and elaborated by Ptolemy, this system placed the earth at the center; and then at increasing distances from it came nine crystalline and concentric spheres. The first seven spheres carried the sun, the moon, and the five known planets, and the fixed stars were attached to the eighth one, often called the ''firmament.'' An elaborate theory of epicycles, deferents, equants, and eccentrics accounted for each planet's motion within its own sphere. On the outside lay the ninth sphere, known as the ''primum mobile'' and representing the Prime Mover, or God; this was held to provide in some inexplicable fashion the motive power for all the others. Beyond this last sphere, there was nothing, no matter, no space, nothing at all. The Aristotelian universe was a finite one contained within the primum mobile.

From the standpoint of Aristotle, the earth was the main body in the universe, and everything else existed for its sake and the sake of its inhabitants. In the new cosmology produced by Nicolaus Copernicus (1473–1543), the sun changed places with the earth; the sun became the unique central body and the earth merely one of several planets revolving about the stationary sun. The ancient theory, because it made the earth the center, is known as the "geocentric theory," and the Copernican, because it treated the sun as central, is called "heliocentric." (In Greek, the words for "earth" and "sun" are *ge* and *helios*.) From the theory's inception, theologians—both Protestant and Catholic—viewed with extreme dislike a theory in which the earth became a comparatively insignificant part of Creation. Had not God created the universe for man's enjoyment and put the earth at the center to prove this? Indeed the psalmist declared, in the 93rd Psalm, "He hath made the round world so sure, that it cannot be moved." Moving the earth was like displacing God's throne.

Underlying the issue whether the Copernican pattern of celestial motion was physically correct was the matter of authority. Copernicanism was so incompatible with the traditional interpretation of various passages in the Bible that if it should prevail, it seemed that the Bible would lose authority, and Christianity would suffer. Besides, if freedom of judgment could be exercised to the extent of deciding between rival astronomical theories, it was but a short step to questioning authority itself.

Over the next several years, Galileo found himself involved in disputes about the relation of his astronomical views to the Bible. A question frequently raised to confound the adherents of the heliocentric theory was how to explain the "Miracle of Joshua." The tenth chapter of the Book of Joshua relates that God, at Joshua's prayer, made the sun stand still and lengthened the day so that the Israelites could pursue their enemies; had the sun gone down, the victory would not have been total. In several widely circulated letters (1613), Galileo maintained that "this passage shows manifestly the impossibility of the Aristotelian and Ptolemaic world systems and on the other hand accords very well with the Copernican." With a brilliant dialectic turn, Galileo pointed out that if one accepted the traditional cosmology of Aristotle, the account of Joshua stopping the sun could not be understood literally. For it was admitted that the originative source of the sun's motion, as well as that of the planets and stars, was the primum mobile. Therefore, if the whole of the heavenly movement was not to be disarranged, Joshua must have stopped not the sun but the outermost celestial sphere. On the other hand, by accepting Copernican theory, one could take the story literally. For if it was assumed that the revolution of the planets was impressed on them by the sun, which is in the center of the universe, then by stopping the sun Joshua was able to stop the whole solar system without disordering the other parts. Quite simply, if one were going to interpret scriptural language in its strict meaning, it would be better in this case to be a Copernican.

Galileo went on to state that the Holy Scriptures did not have as their aim the teaching of science, and that the words of the Bible were not to be taken literally. Where the sun was described as moving around the earth, this was a reflection of the incomplete knowledge of those times; certainly it was not meant as an endorsement of a given astronomical theory. He quoted "an ecclesiastic of the most eminent degree" who once said, "The intention of the Holy Ghost is to teach us not how the heavens go, but how to go to Heaven." Further on, he added that "before a physical proposition is condemned, it must be shown to be not rigorously demonstrated." In implying that it was the Church that ought to give scientific proof if Galileo were to be faulted, he provided exactly the opportunity his enemies wanted. They proclaimed everywhere

that Galileo had assailed the authority of the Scriptures as a privileged source of knowledge and had tried, as an outsider, to meddle in religious matters. Mathematics was
denounced from the pulpit as a devilish art and all mathematicians as enemies of the
true religion.

Toward the beginning of 1616, the pope submitted the following two propositions
to the Holy Office for examination: (1) "The sun is the center of the world and entirely
motionless as regards spatial motion," and (2) "The earth is not the center of the world
and is not motionless, but moves with regard to itself and in daily motion." After a
day's deliberation, a special commission of theologians ruled that the first of these was
"foolish and absurd, philosophically and formally heretical, inasmuch as it expressly
contradicts the doctrine of the Holy Scripture in many passages." As for the second
proposition, it could be "equally censured philosophically and was at least erroneous
in faith."

As an avowed protagonist of a moving earth, Galileo also had to be disciplined.
He was summoned to the palace of Cardinal Bellarmine and in the presence of witnesses, admonished to "abstain altogether from teaching or defending this opinion and
doctrine, and even from discussing it." If he did not acquiesce, he was to be imprisoned. Galileo declared that he submitted. Immediately afterwards, the Holy Office
proceeded against the writings of Copernicus. The work that had caused the upheaval
in astronomical thought, the *De Revolutionibus Orbium Coelestium* (On the Revolution
of the Heavenly Spheres, 1543), and all other texts that affirmed the earth's motion
were put on the Index of Prohibited Books "pending correction" of various passages.

Galileo was more or less silenced until 1623, when a new pope, a longtime admirer
of his, was elected and took the name Urban VIII. After several friendly audiences,
Urban granted Galileo permission to write about Copernicus's theory provided that he
represent it not as reality but as a convenient scientific hypothesis, and provided that
arguments for the Ptolemaic view were given equal and impartial discussion. Galileo
began work on a book that he believed would comply with these instructions. The
writing went slowly, as Galileo was in ill health, but by 1630 he had finally completed
the manuscript of the astronomical treatise that was to lead to his celebrated trial:
Dialogo Sopra Due Massimi Sistemi del Mondo (Dialogue Concerning the Two Chief
World Systems). After many arguments and delays in getting the necessary license for
printing it, Galileo had the book approved by the Church authorities and it was published in Florence in 1632.

To reach the widest possible audience, Galileo wrote the *Dialogue* not in Latin,
the academic language of the universities, but in Italian. It is addressed "To the Dis-
cerning Reader," whom he wished to win over to his cause. The immediate response
from the public was enormous, with the work sold out as it came off the presses. One
reason is that the *Dialogue* was the most readable of the three great masterpieces of
contemporary astronomical literature, of which the other two were the *De Revolutionibus* of Copernicus and the *Principia* of Newton. It was not a severely technical treatise
like the others—mathematics would have been out of place in it—but a piece of brilliant
polemic, directed at the clerical establishment. As far as the text itself was concerned,
the *Dialogue* consisted of a lively conversation extending over four successive days in
which three people discussed the arguments for and against Copernicanism, though
coming to no definite conclusion. Of the three speakers, Salviati, the Copernican

Nicolaus Copernicus
(1473–1543)

(From A Short History of Astronomy *by Arthur Berry, 1961, Dover Publications, Inc., N.Y.)*

scholar, represented Galileo; Simplicio, the archetype of the bumbling Aristotelian philosopher, stood for authority; and Sagredo, the intelligent and cultivated layman, acted as moderator. The form of a dialogue was chosen partly for literary reasons, but still more because it would enable Galileo to claim that certain views expressed were not really his own but those of an imaginary character. Needless to say, the Aristotelian cause came out a miserable second best, as Simplicio was made to look foolish and forced to withdraw from the conversation.

With the publication of the *Dialogue,* Galileo's enemies in science as well as in the Church redoubled their denunciation of him. An ecclesiastical commission that examined the work reported that he had transgressed orders by treating Copernicanism not as hypothesis but as fact. Moveover, their search of the Inquisition's records for 1616 disclosed a notary's unsigned statement to the effect that Galileo had been personally ordered not to teach Copernicanism in any manner, orally or in writing. The pope, who had known nothing of this injunction, felt that he had been tricked into the granting of permission for Galileo to write about the topic. Worst of all, Urban became convinced that Galileo had held him up to ridicule in the character of Simplicio, the clumsy defender of the geocentric theory. Urban could not have failed to recognize his own arguments put into Simplicio's mouth almost verbatim. (After hearing one of Simplicio's objections to the new theory, Salviati commented sarcastically, ''An admirable and truly angelic argument.'') His vanity bitterly wounded, the pope came to support the Jesuit position that the teachings expounded in the *Dialogue* were potentially more dangerous to Christianity than all the heresies of Luther and Calvin.

Although 70 years old and seriously ill, the author of the *Dialogue* was summoned to Rome to stand trial before a tribunal of the dreaded Inquisition. The line of questioning was designed to elicit from Galileo an admission that he had broken his promise

Frontpiece of Galileo's Dialogo *(1632): Aristotle, Ptolemy, and Copernicus (from left to right) debate the structure of the universe. (By courtesy Editions Culture et Civilisation.)*

to obey the injunction issued in 1616. His defense was that far from violating the decree by advocating Copernicanism, he had written the *Dialogue* to show that the arguments for it were "invalid and inconclusive." At a second examination, Galileo declared that he had just reread his book for the first time in three years and did admit that many sections presented the case for Copernicanism in an extremely favorable light. He was never actually put to torture during his stay in prison, but the ever-present threat of such questioning induced Galileo to confess his "errors" and to sign a recognition of

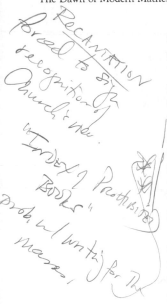

the Church's authority in astronomical matters. (There is a persistent legend to the effect that Galileo, on making his public abjuration, muttered to himself the words, "Nevertheless it does move.") The full text of the recantation was read from every pulpit, and to all university students by their professors. As an additional safeguard against Copernican ideas, the *Dialogue* was placed on the Index of Prohibited Books, where it remained until 1822. In 1744, for the first time, the Church permitted it to be printed, and then only with the sentence on Galileo and his recantation included in the same volume. Galileo's trial illustrates one of the grave problems that the early seventeenth-century scientists had to confront. They were allowed to pursue their experiments provided that they kept the results to themselves. Because the *Dialogue* was written in a popular style accessible to a wide public, Galileo's prosecutors charged him not so much for holding his own views as for openly proclaiming them.

Apparently broken in spirit, Galileo was sentenced to permanent house arrest, under surveillance of officers of the Inquisition. His devotion to scientific study remained unabated, however. Before his astronomical discoveries with the newly devised telescope had made him world-famous, he had been about to publish a great work on mechanics. Galileo turned again to this project of a quarter-century earlier, and though nearly blind, went on to write his most important work, *Discorsi e Dimostrzioni Matematiche Intorno a Due Nuove Scienze* (Discourses and Mathematical Demonstrations Concerning Two New Sciences), a treatise analyzing projectile motion and gravitational acceleration. From this book the science of dynamics can be said to have taken its origin. Because the Inquisition would allow no work of Galileo's to be published in Italy, the manuscript had to be smuggled out of the country, and it was printed in Holland in 1638.

One is tempted to say of Galileo what had once been said of Plato: He was the maker of mathematicians rather than the author of mathematical treatises. Although Galileo had exceptional mathematical ability, it showed itself episodically and not in any magnum opus on some branch of the subject. (His pupils Cavalieri and Torricelli went on to achieve a degree of fame.) His great contribution was the revolutionary idea that mathematics was the vehicle of scientific explanation, or as he himself put it, "Nature is written in mathematical language." Thus, the emphasis was changed from the Pythagorean view that Nature was mathematical to the view that Nature lent itself to mathematization. The underlying principle of this "new Pythagoreanism" was that the universe behaves in a logical way that could be understood through a combination of direct observation and mathematical reasoning.

It is difficult to say precisely why there should have been such spectacular mathematical achievements at this particular time, but the improvement in the means of mathematical expression was a necessity for the transition from ancient to modern conceptions. New results have often become possible only because of a different mode of writing. Certainly, by 1600, new mathematical notations were blossoming in Europe like flowers in the springtime. The signs + and − first appeared in print in Johann Widmann's *Mercantile Arithmetic* (1489), where they refer, not to addition or subtraction or to positive or negative numbers, but to surpluses and deficits in business problems. The first person to use them in writing an algebraic expression was the Dutch mathematician Vander Hoecke (1514). In the earliest English treatise on algebra, published under the alluring title *The Whetstone of Witte* (1557), Robert Recorde

DISCORSI
E
DIMOSTRAZIONI
MATEMATICHE,
intorno à due nuoue ſcienze
Attenenti alla
MECANICA & I MOVIMENTI LOCALI,
del Signor
GALILEO GALILEI LINCEO,
Filoſofo e Matematico primario del Sereniſſimo
Grand Duca di Toſcana.
Con vna Appendice del centro di grauità d'alcuni Solidi.

IN LEIDA,
Appreſſo gli Elſevirii. M. D. C. XXXVIII.

Title page of Galileo's Discorsi *(1638). (By courtesy Editions Culture et Civilisation.)*

introduced the symbol =, but with longer lines, to denote equality. Recorde selected the sign because, so he says, ''Noe 2 thynges can be moare equalle,'' than two parallel straight lines. However, this notation was not immediately popular. Xylander (1575) preferred two upright parallel lines; and the use of the sign $\propto$ for equality (possibly a corruption of the first two letters of the word *aequalis*) continued well into the 1700s. Symbols for multiplication developed much more slowly than the symbols for addition and subtraction. Thomas Harriot, in his *Artis Analyticae Praxis* (Practice of the Analytic Art, 1631), denoted multiplication by a dot, but the *Clavis Mathematicae* (Key to Mathematics, 1631) of another English algebraist, William Oughtred, used the cross sign ×. Harriot was also responsible for the symbols > and < for ''greater'' and ''less,'' which were an improvement over the symbols $\sqsubset$ and $\sqsupset$, invented simultaneously by Oughtred for the same purpose. (The lack of symmetry in Oughtred's notation made it difficult to remember.) In Rahn's *Teutsche Algebra,* published in 1659, the symbol ÷ for division was encountered in print for the first time. The square root sign $\sqrt{}$ is traceable to Christoff Rudolff's *Die Coss* (1525), where it had only two strokes. It is frequently said that Rudolff chose $\vee$ because it resembled a small *r,* the initial letter of the word *radix.* Typesetters were inclined to improvise on these notations, turning the root $\sqrt{}$ into the position $>$ for ''greater than'' and into the position $<$ for ''less than.''

In the 1500s, dominance in algebra passed from Italy to other countries, notably France. It was there that François Vièta (1540–1603), councillor to the court of Henry IV and the leading mathematician of his day, took a decisive step in perfecting algebraic symbolism. Since the time of Euclid, letters had been used to represent the quantities that entered into an equation, but there had been no way of distinguishing quantities assumed to be known from those unknown quantities that were to be found. Vièta suggested using letters of the alphabet (capitals) as symbols for quantities, both known and unknown—vowels were to designate unknown quantities, what we now call variables, and consonants to represent numbers assumed to be given. Simple though it was, this convention had enormous consequences in liberating algebra from having to deal with particular examples involving specific numerical coefficients. Before the introduction of Vièta's literal notation (that is, a notation in which letters stand in place of numbers), attention had focused on specific equations only. Each equation, such as $3x + 2 = 0$ or $6x^2 + 5x + 1 = 0$, had an individuality all its own and had to be handled on its own merits. Literal notation made it possible to build up a general theory of equations—to study not an equation like $6x^2 + 5x + 1 = 0$ but the quadratic equation $ax^2 + bx + c = 0$. Previously the idea of an equation in which the coefficients might be of either sign, each independent of the sign preceding it, was wholly lacking.

The vowel-consonant notation of Vièta had a short existence; for within a half-century of Vièta's death, Descartes's *Géométrie* appeared. In this work, letters at the beginning of the alphabet were used for given quantities, and those near the end (especially *x*) for the unknown. This rule was rapidly assimilated into seventeenth-century practice and has survived to modern times.

Vièta retained the last vestige of verbal algebra by writing *A quadratus, A cubus,* and so on, for the different powers of a quantity *A.* The immensely convenient idea of using exponents to indicate the powers to which a quantity is raised was another contribution of Descartes and occurred for the first time in his *Discours de la Méthode,* published in 1637. The successive powers of *x* were denoted, much as is still done, by $x, xx, x^3, x^4, \ldots$. For some strange reason, however, Descartes invariably wrote the expression *xx* in place of x^2. This use of a repeated letter for the second power continued for many years, certain writers preferring *xx* on the unmathematical grounds that it occupied no more space than x^2 did.

The chief difference between Diophantus's abbreviated symbolism and Descartes's notation is that Diophantus used no sign for addition, denoting the operation instead by simple juxtaposition. Furthermore, there was no clearly displayed relationship between Δ^{Υ} and the symbol ς of which it is the square, as there is with x^2 and x. Whereas Diophantus would have written

$$K^{\Upsilon}4\ \Delta^{\Upsilon}6\varsigma 2M3,$$

this became, according to Descartes's symbolism,

$$4x^3 - 6xx \propto 2x + 3$$

(both correspond to our $4x^3 - 6x^2 = 2x + 3$). A century earlier, the Italian algebraist Bombelli might have indicated the same equation by

4Cm. 6Q aeqtur 2R p. 3.

Here, a syncopated algebra is met in which the letters p and m are used as abbreviations for addition (*piu*) and subtraction (*meno*), while each power is represented by its own symbol, the unknown denoted by R, its square by Q, and its cube by C. Bombelli had another form of expression, namely

4⏞m. 6⏝   aeqtur 2⏝   p. 3

where the power of the unknown quantity is represented as a numeral above a short circular arc. On the other hand, Cardan, who did not accept negative coefficients, would probably have written the equation in question as

4 cubus aequantur 6 quadratus & 2 res & 3

(modern form: $4x^3 = 6x^2 + 2x + 3$). For "unknown," Cardan had the Latin term *res*, literally "thing;" the ligature &, or the word *et*, stands for addition.

Thus, we have seen algebra pass through three stages: the rhetorical, in which all statements and equations were written out in ordinary language; the syncopated, in which familiar terms were abbreviated; and the symbolic, in which every part of an expression was characterized by an *ad hoc* symbol.

Decimal fractions were the most important innovation in arithmetic since the general introduction of the Hindu-Arabic numeration system. Such fractions, along with a more or less convenient notation for them, were inevitable; it is only surprising that this "admirable invention" took so long to appear. Although the new numerals had gained ascendency in Western Europe by 1500, sexagesimal notation continued to be employed for representing fractions. Decimal and sexagesimal bases were actually used in combination in individual numbers, with decimal notation being used for a number's integral part and sexagesimal notation for its fractional part. This dual usage is illustrated in the calculation of tables of square, cube, and higher roots. John of Meurs (circa 1343), for example, extracted $\sqrt{2}$ by the method

$$\sqrt{2} = \frac{1}{1000} \sqrt{2000000} \approx \frac{1}{1000} \, 1414 = 1 + \frac{414}{1000},$$

which he has clearly done using a ten-based computation. He expressed the result, however, in sexagesimal terms as $1°24'50''24'''$. If it now seems odd to use mixed notations in this way, recall that trigonometric tables still appear in this format.

Although the idea was maturing in the work of various mathematicians—particularly the Frenchman Francois Vièta—the first person to give a systematic exposition of the rules of operation of decimal fractions was Simon Stevin (1548–1620), a native of Bruges. In his younger days Stevin worked as a bookkeeper in Antwerp, leaving the Low Countries during the period of unrest preceding the Dutch revolt against Spanish rule. After traveling in Prussia, Poland, and Norway for ten years, he finally established

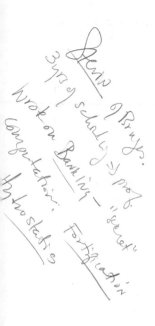

himself in the northern part of the Netherlands, an area which by then had shaken off the domination of Spain. In 1583 Stevin entered the newly founded University of Leiden. He was 35 years of age by then—at that time, rather late in life to become a student—but he did so well that within a few years he was teaching mathematics at the university. On the recommendation of a former pupil, Prince Maurice of Nassau, Stevin was appointed quartermaster general for the Dutch armies and inspector of dikes and canals; he held these powerful positions until the time of his death.

Eager to explain his views, Stevin wrote on a wide variety of topics. His earliest publication, *Table of Interest Rates* (1582), gave the common reader rules for computing simple and compound interest. It also offered tables, previously kept secret by bankers, for speedily figuring discounts and annuities. *The Art of Fortification* (1586) was put to practical use by military engineers during the wars of the next two centuries. In mathematics, Stevin brought out his *Arithmetic* (1585), elaborating on the best from older arithmetic and algebraic writings; he also prepared a French version of the first four books of Diophantus's *Arithmetica;* the latter work, the first translation of Diophantus into any European vernacular, was probably based on the Latin text of Xylander. Stevin's fame in the history of science rests mainly on the *Principles of Hydrostatics* of 1586, the first substantial advance in the subject beyond the work of Archimedes.

Calculation with decimal fractions was the theme of Stevin's popular pamphlet *De Thiende* (*The Tenth*), published in Flemish in 1585; a French translation under the title *La Disme* was printed in the same year. The English rendition in 1608 introduced the word ''decimal'' into our language. The 29-page booklet carried the subtitle ''Teaching how all computations that are met in business may be performed by integers alone without the aid of fractions.'' That is, as soon as fractions are put into decimal form they may be treated as integers; the operations of ordinary arithmetic are done in the same manner as for integers, with the decimal point merely having to be placed correctly in the final result. Stevin did not think of using a single sign to separate the two parts of a number, as we would use a decimal point. Instead, he suggested a cumbersome notation using encircled numbers to indicate powers of ten. Each succeeding digit in the fractional part of a number was to be followed by an encircled number noting the corresponding power of ten in the denominator; the integral part of the number would be followed by an encircled zero. For instance, the number that we would represent by 34.567 was written by Stevin as either

$$34 \; ⓪ \; 5 \; ① \; 6 \; ② \; 7 \; ③ \quad \text{or} \quad \begin{array}{cccc} ⓪ & ① & ② & ③ \\ 34 & 5 & 6 & 7, \end{array}$$

as the circumstances required. In multiplying, say, 0.000378 by 0.54 he arranged the computation as

$$\begin{array}{ccccc} & ④ & ⑤ & ⑥ & \\ & 3 & 7 & 8 & \\ & & 5 & 4 & ② \\ \hline 1 & 5 & 1 & 2 & \\ 1 & 8 & 9 & 0 & \\ \hline 2 & 0 & 4 & 1 & 2 \\ ④ & ⑤ & ⑥ & ⑦ & ⑧ \end{array}$$

The *La Disme* gained rapid and wide circulation, but Stevin's clumsy circle notation—taken from Bombelli's *Algebra*—was short-lived. The first writer to use a period as a separator between the integral and fractional parts of a number seems to have been the Vatican astronomer Christoph Clavius; this device appears in his *Astrolabium* of 1593. The main introducer of the decimal point into computational practice was the Scottish nobleman John Napier. Napier's new mathematical instrument, common logarithms, became the natural vehicle for the decimal idea. In the preface to his influential treatise on logarithms Napier stated, "In numbers distinguished by a period in their midst, whatever is written after the period is a fraction." Within 50 years of Stevin's little tract, decimal numeration was widely adopted throughout Europe.

During the half-century after Copernicus, few eminent scientists were bold enough to champion a theory that pushed the sun out of its honored place at the center of the universe. Then in the early 1600s, a youthful Johannes Kepler, with his intense faith in the mathematical simplicity of Nature, set forth a chain of daring conclusions that would strengthen and purify Copernicus's scheme. Although Kepler is chiefly remembered for his discovery of three great laws concerning the motion of the planets, his achievements in pure mathematics would have been enough to win him enduring recognition. He was the first to enunciate clearly the principle of continuity—the continuous change of a mathematical entity from one state to another—treating the parabola as the limiting case of either an ellipse or a hyperbola in which one of the two foci moves off to infinity. (He introduced the word "focus.") Kepler found the area of a circle by imagining the circle to be made up of an infinite number of triangles having their common vertex at the center and their infinitely small bases lying along the circumference. Likewise, he showed the volume of a sphere to be one-third the product of the radius and the surface area by regarding the sphere as consisting of infinitely many small cones, each with its vertex at the center of the sphere and its base on the surface. This identification of curvilinear areas and volumes with the sum of an infinite number of infinitesimal elements of the same dimension is a remarkable anticipation of results found later in integral calculus. Kepler applied these conceptions to the volume of some 92 solids obtained by revolving conic sections about their diameters, chords, or tangents. His interest in volume problems apparently came after noting the inaccuracy of the methods then in use in estimating, for purposes of taxation, the capacity of wine casks.

Johannes Kepler (1571–1630) was born in Weil, a city near Stuttgart in southern Germany, to impoverished parents. His father was a drunkard who repeatedly went off to fight as a mercenary soldier in the Low Countries, before settling down as a tavern-keeper. His mother was equally erratic; an ignorant woman of violent temper, her taste for sorcery eventually led to imprisonment on the charge of practicing witchcraft. Kepler himself was a sickly child unable to attend school regularly. An attack of smallpox, which nearly killed him at the age of 4, left him with crippled hands and impaired vision. Owing to the kindly patronage of the duke of Württemberg, the young Kepler was able to enroll at the University of Tübingen, one of the centers of Protestant theology, receiving the degree master of arts in 1591. He was fortunate enough to study with Michael Mastlin, the professor of mathematics, who taught him privately the work of Copernicus while hardly daring to recognize it openly in his professorial lectures.

Johannes Kepler
(1571–1630)

(From A Short History of Astronomy *by Arthur Berry, 1961, Dover Publications, Inc., N.Y.)*

In 1593, at age 22, Kepler prepared a disputation enthusiastically supporting the Copernican theory of a sun-centered universe; but it was never presented, because the professor in charge of those academic exercises at Tübingen was so unalterably opposed to Copernicanism that he refused to permit the dissertation to be heard. Kepler, feeling that his views were incompatible with dogma that then prevailed, abandoned his intention of entering the Lutheran ministry and instead accepted the position of ''provincial mathematician'' in the Protestant seminary at Graz (the capital of the Austrian province of Styria).

It could hardly be said that Kepler was a success in this humble teaching post. Lecturing beyond the capacity of his pupils, he managed to drive most of them away. In the first year, his mathematics classes were attended by only twelve students and in the second year by none. To keep his salary from being wasted, Kepler was assigned the duty of preparing a yearly almanac, which was expected to contain not merely the usual meteorological predictions but prophecies on the remarkable events of the coming year. When his predictions for the year 1595—a serious cold spell, peasant uprisings, and invasion by the Turks—were all fulfilled, Kepler gained a considerable reputation as a prophet and astrologer.

From the time of his earliest research, Kepler had an almost Pythagorean belief in the simplicity and harmonious unity of the universe. He was convinced that the heavenly bodies were arranged according to some simple geometric law. (''Geometry provided God with a model for the Creation and was implanted into man, and not merely conveyed to his mind through his eyes,'' he declared in the *Harmonices Mundi.*) While at Graz, he made his first attempt to discover hidden mathematical regularities embedded in the solar system. There were at that time six known planets—Mercury, Venus, Earth, Mars, Jupiter, and Saturn—and these were known to be at successively greater distances from the sun. Casting around for a pattern from which the Creator might have worked, Kepler arrived at the idea of relating the distances between the planetary orbits to the five geometric bodies known as the regular solids: cube,

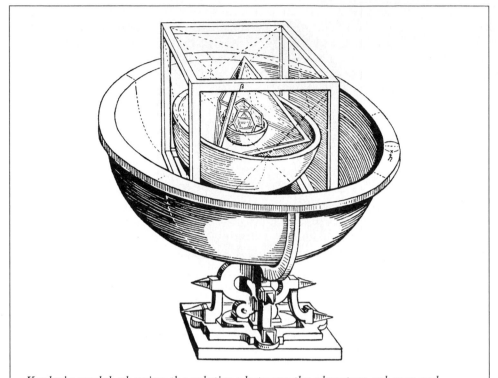

Kepler's model, showing the relations between the planetary spheres and regular geometric solids. From his Mysterium Cosmographicum *(1596).*
(Extract taken from A History of Science, Technology and Philosophy in the 16th and 17th Centuries, *by A. Wolf. Reproduced by kind permission of Unwin Hyman Ltd.)*

tetrahedron, dodecahedron, icosahedron, and octohedron. He theorized that the hypothetical spheres of the planets could be inscribed in and circumscribed about the five regular solids properly distributed in succession between them. The result satisfied Kepler that he had uncovered one of the fundamental secrets of the universe—the cosmos was constructed so that there were six and only six planets. In a letter penned shortly thereafter he recorded his inordinate delight, writing ''The joy which this discovery gave me can never be described; I regret no more the time wasted.'' The fruits of these intellectual speculations were set forth in 1596 in a book whose title may be abridged to *Mysterium Cosmographicum* (The Mystery of the Universe). The work attracted considerable attention, not only because of the novelty of the ideas, but because it ardently advocated the Copernican planetary system as opposed to Ptolemy's. Of course, the discovery of new planets quite upset the underlying assumptions of the *Mysterium Cosmographicum.*

As a Lutheran, Kepler was subject to all kinds of disciplinary measures in the Catholic city of Graz; but his reputation mostly kept him free from the persecution that Galileo suffered as a result of his knowledge. Then, one day in 1598, all Protestant

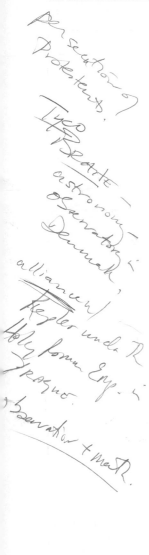

clergy and teachers were abruptly ordered to leave town by sunset. Although Kepler was allowed to return within a month, he saw that he could not hope to be left in peace much longer. By a happy chance for science, the *Mysterium Cosmographicum* had caught the eye of the Danish astronomer Tycho Brahe, who needed a junior partner in his astronomical research and offered Kepler the position as his chief assistant.

Tycho Brahe (1546–1601) was the pioneer of accurate astronomical observation. With the help of the king of Denmark, he had constructed in 1576 a splendid observatory on a 2000-acre island near Copenhagen. He equipped this with the most accurate instruments possible, including a 37-foot quadrant for measuring altitudes. None of the instruments had lenses, for the telescope was not invented until around 1600. With immense patience and skill, he labored for some 20 years compiling a vastly more extensive and incomparably more precise set of records than any of his predecessors had possessed. When the king died, the patronage was not extended by his successor, so that Tycho moved to Prague in 1599, taking his most portable instruments along. There he entered the service of the eccentric Rudolf II, Holy Roman Emperor and the greatest patron of astrologers and alchemists in Europe. Kepler accepted Tycho's invitation to join him and arrived in Prague early in the following year. It was a fortunate alliance. Tycho was a splendid observer but a poor mathematician, whereas Kepler was a splendid mathematician but a poor observer.

Toward the end of 1601, after drinking copiously at a dinner party, Tycho was suddenly felled by a burst bladder. Adhering to the strict etiquette of the day, he remained at table through the rest of the meal, not wishing to leave before the other guests did. He died 11 days later, after intense suffering. On his deathbed, he turned to Kepler in particular and begged him to complete some of his tables on planetary motion as quickly as possible. It is also said that, in the delirium that preceded his death, Tycho repeated several times, "I hope that I will not appear to have died in vain." Kepler did not gain possession of Tycho's instruments, and they were inadvertently burned. Because Kepler's poor eyesight made him an indifferent observer, the loss was of little practical consequence. The most important scientific inheritance Tycho left him was a vast wealth of astronomical observations of unparalleled accuracy. In the hands of Kepler, this store of information was to produce the next great advance in mathematical astronomy.

Shortly before Tycho's death, Kepler received the title of Imperial Mathematician from the Emperor Rudolf; and now he succeeded to Tycho's position as official astronomer to the emperor. But his royal master had the habit of paying his salary only rarely or in part, so that Kepler was forced to earn additional income by casting the horoscopes of eminent men. "Mother Astronomy would certainly starve if the daughter Astrology did not earn their bread," he is reported to have said.

The riddle of the orbit of Mars engrossed Kepler's attention for the next eight years. Tycho's very accurate measures of the position of Mars relative to the sun enabled Kepler to test various hypotheses and cast them aside when they proved incompatible with the observed movement. With a Pythagorean craving for simplicity, he felt sure that the orbit was a circle. It was only after many failures to fit the data to a circular orbit that he began to suspect that it must be some other closed path. For a long time, Kepler was inclined to believe that it was an oval, shaped like an egg. He tried various sorts of ovals, but none eliminated the discrepancies between his tentative theories and Tycho's observations. Years of work and disappointment finally forced

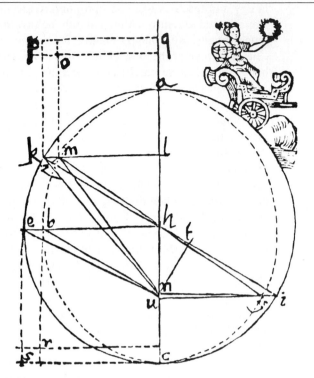

Diagram used by Kepler to demonstrate the elliptical orbit of Mars. From his
Astronomia Nova *(1609). (From* A Short History of Astronomy *by Arthur Berry, 1961,*
Dover Publications, Inc., N.Y.)

him to the conclusion that only an elliptical orbit, with the sun occupying one of the
two foci, satisfied Tycho's data. The same was presumably true for all other planets,
because the harmony of nature demanded that all "have similar habits." This was
Kepler's celebrated first law. Another conclusion he extracted from the astronomical
data was that the speed with which a planet traversed its elliptical orbit varied in a
regular pattern, accelerating with approach to the sun and decelerating with departure
from the sun. From this he was led to another pillar of celestial mechanics, Kepler's
second law. The line drawn from the sun to a planet sweeps over equal areas in equal
times.

The full history of his investigation of Mars, together with the laws just stated,
was published in 1609 in a long book called *Astronomia Nova*. After ten years' further
effort, Kepler arrived at a relation, his third and last great law of planetary motion,
connecting the times of revolution of any two planets with their respective distances
from the sun. The ground was thus prepared for the later achievements of Isaac Newton,
who was able to prove mathematically not only that the behavior implicitly extended
by Kepler to all the planets agreed with observation but also that no other behavior
was possible.

Kepler's celebrated results may be described thus:

1. The planets move in elliptical orbits with the sun at one focus.

2. Each planet moves around its orbit, not uniformly, but in such a way that a straight line drawn from the sun to the planet sweeps out equal areas in equal time intervals.

3. The squares of the times required for any two planets to make complete orbits about the sun is proportional to the cubes of their mean distances from the sun.

Archimedes is reported to have boasted, "Give me a place to stand and I will move the world." The observations Tycho Brahe had gathered became for Kepler a place to stand, and he did move the world. His three laws, which established the first correct principles of planetary mechanics, overturned medieval cosmology and much of Aristotelian physics. From ancient times through the days of Copernicus and Tycho the idea that a planet's orbit must be a circle had been unchallenged. Indeed, a circle was accepted as the perfect geometrical form, and perfection was accepted as the normal state of heavenly affairs. That the real orbits of planets were ellipses was a triumph for the new astronomy and the right of Western scientists to pursue their investigations independent of theological doctrines.

One fundamental question could not yet be answered: What held the planets in their courses? As we shall see, the vortex theory of Descartes and Newton's theory of universal gravitation were responses to this challenge.

8.2   Descartes: The *Discours de la Méthode*

Among the principal movers in the seventeenth-century scientific revolution, René Descartes must certainly be included. Through the publication of *La Géométrie,* which made analytic geometry known to his contemporaries, Descartes is generally acknowledged to have laid the foundations for the growth of mathematics in modern times. This first really great advance beyond the techniques known to the ancients changed the face of mathematics and led, within a generation, to the development of the calculus by Newton and Leibniz. It is not too much to say that Descartes's career marks the turning point between medieval and modern mathematics.

René Descartes (1596–1650) was born at La Haye, a small town about 200 miles southwest of Paris, in the province of Touraine. His father belonged to the lesser nobility. He was a councilor at the Parlement of Brittany—in effect, a provincial judge. Descartes went through the normal upbringing of a gentleman of that time. At age 8 he was placed in the lately founded Jesuit College of La Flèche, perhaps the most illustrious school in which a student could enroll. There he came to know Marin Mersenne, who was 7 or 8 years older. The first five years of the curriculum at La Flèche were devoted to the traditional course in languages and the humanities. The final three years embraced logic, philosophy, physics, and mathematics. Mathematics, because of the certainty of its demonstrations, was the only subject that really satisfied Descartes, even at this early age.

Descartes's health was delicate during infancy and childhood, and he was not expected to live long. His teachers at La Flèche, recognizing this physical weakness, treated him with exceptional consideration; regular attendance at lectures was not

René Descartes
(1596–1650)

(Smithsonian Institution.)

required of him and he was allowed to lie in bed each morning as late as he pleased. He never lost this habit; throughout the rest of his life (except for one unfortunate incident that may well have hastened his death), Descartes preferred to rise late, spending the early hours in bed meditating and writing. Indeed, when he visited Pascal in 1647, Descartes stated that the only way he could do good work in mathematics and preserve his health was never to allow anyone to get him up in the morning before he felt so inclined.

On leaving school in 1612, Descartes followed the usual path of a young man of wealth living in France by going to Paris to taste the pleasures of its social life. This phase did not last long, for in Paris, he renewed his schoolboy friendship with that most indefatigable of learned gossips, the good Father Mersenne. Mersenne soon rekindled Descartes's interest in serious study, and in almost cloistral retirement they devoted two years to mathematical investigation. Although the younger Descartes had no deep inclination to follow his father's profession, he then entered the University of Poitiers, where he earned a degree in law in 1616.

In 1617 Descartes, then 21 years old and tired of textbooks, decided to learn more about the world at first hand. He enlisted in the army as a gentleman volunteer, first joining the troops of Prince Maurice of Nassau in Holland and afterwards taking service under the duke of Bavaria. There is no evidence of any real soldiering on Descartes's part, only years of leisure, in which he had time to pursue his favorite studies. The night of November 10, 1619, while in winter quarters with the Bavarian army along the Danube, was critical in Descartes's life. He escaped the cold by shutting himself up alone all day in a "*poêle*"—literally a stove, actually an overheated room. Tired from the heat, he dreamed three feverish dreams, in which he discovered "the foundations of a marvelous science." At the same time his future career as a mathematician and philosopher was revealed to him. (Near the close of the final dream, as Descartes tells us, he saw a book opened at a passage of the Latin poet Ausonius, containing the words ''Which way shall I follow?'' As the dream continued, an unknown man handed him a bit of verse beginning, ''Is and is not,'' which he understood as representing truth and falsehood in human knowledge.) Descartes neglected to specify the exact nature of the marvelous science whose foundations he found on this memorable day. Some authorities are inclined to believe that he formulated the principles of analytic geometry; others feel that Descartes conceived a complete reform of philosophy based on the methods of mathematics. As Bertrand Russell observed, ''Socrates used to meditate all day in the snow, but Descartes's mind only worked when he was warm.''

By 1628, having grown weary from years of aimless wandering through Holland, Germany, Hungary, and Italy, Descartes settled down to what might be called the productive period of his life. Holland, which had recently won independence after a protracted struggle with Spain, seemed the country best fitted to offer the tolerance and tranquility Descartes needed to pursue his researches. There, in great seclusion (barring three brief visits to France to look after family affairs), he meditated and wrote for twenty years. Until then he had published nothing. Descartes conceived therefore of writing an almost encyclopedic treatise on physics, which he chose to call *Le Monde* (The World). The time from 1629 to 1633 was occupied with building up a cosmological theory of vortices to explain all natural phenomena. On the eve of the completion of *Le Monde,* he learned that Galileo's *Dialogue on the Two Chief Systems of the World,* published the previous year, had earned the censure of the Church. It was clear that the earth was not to be summarily dismissed from its position as the immovable center of the solar system. His own work, affirming as it did the heliocentric hypothesis, would have made him equally guilty with Galileo, so Descartes prudently abandoned the project. The publication of *Le Monde* had to wait until 1664, well after his death.

It was not moral weakness that forced Descartes to suspend publication of *Le Monde,* but rather that he never ceased to regard himself as a sincere and devoted Roman Catholic. He wrote sadly to Mersenne, ''This has so strongly affected me that I have almost resolved to burn all my manuscript, or at least show it to no one. But on no account will I publish anything that contains a word that might displease the Church.'' Not that the fruits of his labor were withheld from the world, for Descartes did not destroy his papers as he first threatened to do. The ideas contained therein, modified but not abandoned, had their presentation to the public in his first principal published work, the *Discours de la Méthode* (1637). Although the *Discours* included a summary of *Le Monde,* Descartes so sidestepped the controversy over Copernicanism that one could glean little from it concerning his cosmology; in particular, any mention of vortices was studiously avoided. Finally, in 1644, the *Principia Philosophiae* was issued, in which he explained at some length the formation of the physical world, by ''gradual and natural means'' out of matter and motion. Descartes's new ''mechanical philosophy'' quickly became the rage, a dominant feature of discussion in intellectual circles.

By 1649, Descartes's reputation had been established throughout Europe, and he was invited by Queen Christina of Sweden, the daughter of Gustavus Adolphus, to visit her court to tutor her in philosophy. She also suggested that he might help her in planning an academy of sciences that would rival the best in Europe. When Descartes had misgivings about living in ''the land of bears amongst rocks and ice,'' the young queen (she was then but 22 years old) dispatched an admiral to coax him and then a Swedish warship to fetch him. Accepting the invitation was a fatal decision on Descartes's part. It is even said that a presentiment of death came over him as he prepared for the journey.

Descartes was received with every honor and had no cause for complaint until the time drew near for his personal instruction of the queen to begin. From childhood on, Christina had slept no more than five hours a night, and she was indifferent to heat or cold. She proposed to Descartes that they meet three times a week, always at five o'clock in the morning, when her mind was unfatigued and she felt the most energetic. For two months Descartes conformed to his royal pupil's schedule, walking in the winter dawn from his rooms to the ice-cold library. His own lifelong routine was radically changed; as a foreign Catholic at a Lutheran court he felt isolated and homesick. "It seems to me," he wrote to his friend the Comte de Brégy, "that men's thoughts freeze here in winter just like water." The rigors of one of the bitterest winters in memory proved too much for Descartes's constitution, which had never been robust. On February 1, 1650, he caught a cold that rapidly developed into pneumonia, and he died after ten days of suffering and delirium. He was buried where Catholics were usually interred, in a cemetery set aside for infants who died before baptism. Fifteen years later his remains (except for the right hand, which was kept as a memento by the official who arranged the transaction) were conveyed back to France, where a magnificent monument was erected to his memory in the Church of Saint Genevieve. As Descartes's doctrines were by then under the ban of both the Church and the universities, the funeral oration was prohibited by a court order, which arrived during the funeral service.

The year 1637 saw the publication of the work that is considered the most significant of Descartes's writings: *Discours de la Méthode pour bien conduire sa Raison et chercher la Vérité dans les Sciences* (Discourse on the Method of Rightly Conducting the Reason in the Search for Truth in the Sciences), with its scientific appendages *La Dioptrique, Les Météores,* and *La Géométrie.* The *Discours* is not, as commonly supposed, a formal philosophical treatise but a short autobiographical résumé of Descartes's progress in arriving at his method. Its first edition had 78 pages, roughly a sixth of the entire work. It was written in his native tongue, though traditionally Latin was used for learned subjects, and it showed at once the power and precision of the vernacular as a vehicle for expressing highly complicated philosophical and scientific thoughts. (In *Principia Philosophiae,* Descartes reverted to Latin to make the work more acceptable to the Church and the universities.) Descartes's use of the French language speeded the diffusion of his ideas. The work was widely read; but though the *Discours* brought fame to its author, the fortune went to the book's printer. The printer paid a small price indeed for one of the landmarks in Western thought. Descartes had asked only to be given, instead of royalties, 200 free copies of the new book for distribution to his friends.

The whole of Descartes's philosophy of "systematic doubt" as expounded in the *Discours* is dominated by his pursuit of certainty. The certainty of mathematics, he delighted to repeat, consists of this—it starts with the simplest elements whose truth is recognized, and then proceeds by the process of deduction from one evident proposition to another. Mathematics should therefore be a model for other branches of study. To let Descartes speak for himself:

> The long chains of simple and easy reasonings by means of which geometers are accustomed to reach the conclusions of their most difficult demonstrations led me to imagine that all things, to the knowledge of which man is competent, are mutually connected in the same way, and that there is nothing so far removed from us as to be beyond our reach, or so hidden that we cannot discover it, provided only we abstain from accepting the false for the true, and always preserve in our thoughts the order necessary for the deduction of one truth from another.

The character of the reasoning of mathematics rather than the results was what so impressed Descartes. And he was anxious to see whether, by arguing in a mathematical fashion from the most universal principles, it would be possible to deduce everything rationally knowable.

The starting point for Descartes's thought was to discover the simplest ideas or principles, those of which there could be no doubt. Because he had lost all confidence in traditional teachings, Descartes began by breaking away from authority altogether in matters of science and philosophy, deliberately rejecting all entrenched dogmas and doctrines. In his own words from the *Discours:*

> I thought that I ought to reject as absolutely false all opinions in regard to which I could suppose the least ground of doubt, in order to ascertain whether after that there remained anything in my belief that was wholly indubitable.

Descartes was thus led to one proposition so sound that it could not be doubted, the certainty of his own existence; for doubt itself is an act of thought and thought does not take place without a thinker. He enunciated this in the most famous sentence in philosophy, one that has been the subject of numerous commentaries: "Je pense, donc je suis." [I think, therefore I am.] Having satisfied himself of the existence of a thinking being, Descartes passed on to a search for other propositions that appeared equally self-evident and irrefutable. For him, there was no greater guarantee of the truth of a proposition than that it should survive the most careful scrutiny of his own independent criticism. "We ought never to allow ourselves to be persuaded of the truth of anything unless on the evidence of our reason," Descartes wrote. This unbounded confidence in the capacity of human reason helped launch the Great Debate between faith and reason that was to preoccupy most western Europeans in the century to come.

Three appendixes to the *Discours* were actual illustrations of Descartes's new method of discovering scientific truths. Although the *Discours* was intended to be a preface to *La Dioptrique, Les Météores,* and *La Géométrie,* history has completely reversed the sequence; and today the *Discours* is studied by students of modern philosophy, while these works on science are virtually ignored. *La Dioptrique* (Dioptrics) deals with the nature and properties of light, including an account of the law of refraction, the anatomy of the human eye, and the shape of lenses best adapted for the newly invented telescope. *Les Météores* (Meteorology) aims at a scientific explanation of atmospheric phenomena; it is concerned with such topics as how snow crystals are formed, the size of raindrops, the cause of thunder and lightning, and the formation of the rainbow. Of the three essays accompanying the *Discours,* the third, *La Géométrie* (Geometry), is the one in which Descartes made his great and lasting contribution to

LA
GEOMETRIE.
LIVRE PREMIER.

Des problesmes qu'on peut construire sans y employer que des cercles & des lignes droites.

TOus les Problesmes de Geometrie se peuuent facilement reduire a tels termes, qu'il n'est besoin par aprés que de connoistre la longeur de quelques lignes droites, pour les construire.

Et comme toute l'Arithmetique n'est composée, que de quatre ou cinq operations, qui sont l'Addition, la Soustraction, la Multiplication, la Diuision, & l'Extraction des racines, qu'on peut prendre pour vne espece de Diuision : Ainsi n'at'on autre chose a faire en Geometrie touchant les lignes qu'on cherche, pour les preparer a estre connuës, que leur en adiouster d'autres, ou en oster, Oubien en ayant vne, que ie nommeray l'vnité pour la rapporter d'autant mieux aux nombres, & qui peut ordinairement estre prise a discretion, puis en ayant encore deux autres, en trouuer vne quatriesme, qui soit à l'vne de ces deux, comme l'autre est a l'vnité, ce qui est le mesme que la Multiplication ; oubien en trouuer vne quatriesme, qui soit a l'vne de ces deux, comme l'vnité

Commēt le calcul d'Arithmetique se rapporte aux operations de Geometrie.

First page of Descartes's La Géométrie (1637). *(Reprinted by permission of Open Court Publishing Company, La Salle, Illinois, from* The Geometry of René Descartes, *translated by David Eugene Smith and Marcia L. Latham.)*

pure mathematics. In the *Géométrie*, he combined the methods of algebra and geometry to produce the new field of analytic geometry. The English philosopher John Stuart Mill called this "the greatest single step ever made in the progress of the exact sciences." Tradition holds that the idea of analytic geometry came to Descartes while he watched a fly crawl along the ceiling of his room near a corner; his immediate problem became expressing the path of the fly in terms of its distance from the adjacent walls. The story is more agreeable fable than fact.

Of the three parts of *La Géométrie*, the first two are devoted mainly to applying algebra to geometry, while the third treats the theory of equations. Book I bears the

title *Problems Which Can Be Constructed by Means of Circles and Straight Lines Only.* At the threshold of the work, Descartes introduced the algebraic notation still in use today. The last letters of the alphabet, x, y, and z, designate unknown quantities, and the first letters of the alphabet stand for constants. Descartes was perhaps the first to use the same letter for both positive and negative quantities. Our modern exponential notation for powers is also found here. In a more radical step, Descartes broke with Greek tradition by divorcing numbers from reference to physical quantity. Instead of interpreting a^2 (or aa as he wrote it) and a^3, for example, as an area and a volume, he considered them nothing more than lines. For Descartes, a^2 was simply the fourth term in the proportion $1:a = a:a^2$ and as such could be represented by a line once a was given. To devise a construction that corresponded to the proportion $1:a = a:a^2$, he arbitrarily chose a unit length 1, to which all other lengths were referred. Then a triangle with sides 1 and a were drawn; in a similar triangle, in which the side corresponding to 1 was a, the other side would be a^2.

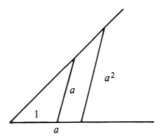

The problems that formed the central theme in *La Géométrie* were generalizations of the three- and four-line locus problems Pappus had propounded in his commentary on the *Conics* of Apollonius. In their original form the Pappus problems run: Given four lines in a plane, find the locus of a point that moves so that the product of the distances from two fixed lines (along specified directions) is proportional to the square of the distance from the third line [three-line locus problem], or proportional to the product of the distances from the other two lines [four-line locus problem]. Whereas Pappus had stated without proof that the required locus was one of the conic sections, Descartes showed this algebraically. Subsequently, Newton solved the problem geometrically in his *Principia* (1687).

Descartes began his attack on the problem by choosing one of the given lines, say *AB*, and a fixed point on it, say *A* (he selected what would later be called an axis of coordinates and an origin). From an arbitrary point *C* of the locus sought, a straight line *CB* was drawn to *AB*, meeting it at a given angle. The lines *AB* and *BC* were then the quantities that would determine the position of *C*, and he called them x and y:

> I would simplify matters by considering one of the given lines and one of those to be drawn (for example, *AB* and *BC*) as the principal lines to which I shall try to refer all others. Call the segment of the line *AB* between *A* and *B*, x and call *BC*, y.

The lengths of the other lines were expressed in terms of x and y; and by the conditions of the problem, Descartes combined these, to arrive at an equation of the curve upon which *C* would have to lie.

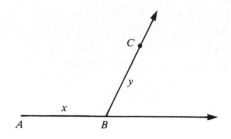

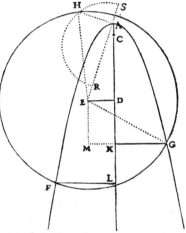

394   LA GEOMETRIE.

pour le quarré de la ligne G E, a caufe qu'elle eft la baze
du triangle rectangle E M G.

Mais a caufe que cete mefme ligne G E eft le demi-
diametre du cercle F G, elle fe peut encore expliquer en
d'autres termes, a fçauoir E D eftant $\frac{1}{2}q$, & A D eftant
$\frac{1}{2}p + \frac{1}{2}$, E A eft $\sqrt{\frac{1}{4}qq + \frac{1}{4}pp + \frac{1}{2}p + \frac{1}{4}}$ a caufe de l'angle droit A D E, puis H A eftant moyene proportionelle
entre A S qui eft 1 & A R qui eft r, elle eft $\sqrt{r}$. & à caufe de l'angle droit E A H, le quarré de H E, on E G eft
$\frac{1}{4}qq + \frac{1}{4}pp + \frac{1}{2}p \div \frac{1}{4} + r$: fibienque il y a Equation
                                                    entre

Extract from Descartes's La Géométrie *(1637). (Reprinted by permission of Open Court Publishing Company, La Salle, Illinois, from* The Geometry of René Descartes, *translated by David Eugene Smith and Marcia L. Latham.)*

What we have here is the germinal idea of a coordinate system in which the position of a point in the plane is defined by its distances, x and y, from two fixed axes. Descartes was choosing what in current language is an oblique coordinate system, although he did not formally introduce a second axis, the y-axis. Nowhere in *La Géométrie* does the modern rectangular coordinate system appear. Descartes's presentation differed from that now current also in his use of only positive values of x and y, that is, by his restriction of curves to the first quadrant. The name "coordinate" does not appear in the work of Descartes. This term is due to Leibniz, and so are "abscissa" and "ordinate" (1692).

In the second book of *La Géométrie,* called *On the Nature of Curved Lines*, Descartes distinguished between two kinds of curves, geometrical and mechanical (or as Leibniz later preferred to call them, algebraic and transcendental). He insisted that curves more complex than lines, circles, and conic sections were proper objects of geometric investigation provided that they were securely defined. Descartes's criterion for the acceptability of curves was that their points should be determined by the intersection of two lines, each moving parallel to one coordinate axis with "commensurable" velocities. These are the geometric curves for which there is an algebraic equation in two variables defining all their points. As Descartes said,

> All points of those curves which we may call "geometric," that is, those which admit of precise and exact measurement, must bear a definite relation to all points of a straight line, and this relation must be expressed by means of a single equation.

On the other hand, curves like the quadratrix and spiral, which arise from "two simultaneous motions whose relation does not admit of precise determination" were rejected from geometry and stigmatized as mechanical curves. Such curves by their nature allow direct construction of only certain of their points. In this way Descartes widened the scope of elementary geometry, which until then had been restricted to constructions by straightedge and compass, by giving full geometric status to many curves formerly excluded.

Descartes, in Book II, devised a purely algebraic method for finding the normal at any given point on a curve whose equation was known; the tangent line could be taken as the perpendicular through the point to the normal. Apart from the isolated attempt of Archimedes to draw the tangent to his spiral, constructing tangents to curves had not been seriously studied until the early 1600s, when it exercised the minds of some of the ablest mathematicians. Descartes, who was led to the "tangent problem" by his optical studies, called attention to it in Book II with the statement, "I dare say that this is not only the most useful and the most general problem in geometry that I know, but even that I have ever desired to know."

The approach Descartes used in constructing the normal—and so indirectly, the tangent—at any specified point P on a curve took as unknown the point of intersection of the normal and the x-axis. In other words, according to Descartes, let $f(x, y) = 0$ be the equation of the curve and (x_0, y_0) the coordinates of the point P. Suppose further that the normal to P is already drawn and that Q, with coordinates $(x_1, 0)$, is the point at which the normal meets the x-axis. Then the equation of the circle with center Q passing through P is

$$(x - x_1)^2 + y^2 = (x_0 - x_1)^2 + y_0^2.$$

The result of eliminating one of the variables, say y, between this equation and the equation $f(x, y) = 0$ of the curve is a new relation of the form $g(x, x_1) = 0$. In general, a circle described with center Q and radius PQ will cut the curve in two points; but if PQ is the desired normal, the two points will coincide and the circle will be tangent to the curve at P. Therefore, in the equation $g(x, x_1) = 0$ leading to the x-coordinates of the points at which the circle intersects the curve, we impose the condition that will render a pair of equal roots. This condition will give the correct value of x_1 for Q to be the point at which the normal cuts the x-axis. Thus, the problem is reduced to determining a double root of an algebraic equation.

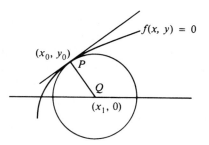

Descartes used his method of tangents on many different curves in *La Géométrie*, each selected to show the process to best advantage. For an illustration, consider the construction of the tangent to the parabola $y^2 = ax$ at the point (a, a). The equation of the circle having its center on the x-axis and going through (a, a) is

$$(x - x_1)^2 + y^2 = (a - x_1)^2 + a^2.$$

The elimination of y by substituting ax for y^2 gives

$$(x - x_1)^2 + ax = (a - x_1)^2 + a^2$$

or

$$x^2 + (a - 2x_1)x + 2a(x_1 - a) = 0.$$

Because this is to be an equation with equal roots, Descartes would have compared it with $(x - r)^2 = 0$. Equating corresponding coefficients in

$$x^2 + (a - 2x_1)x + 2a(x_1 - a) = x^2 - 2rx + r^2$$

gives $a - 2x_1 = -2r$ and $2a(x_1 - a) = r^2$; whence

$$(a - 2x_1)^2 = 4r^2 = 8a(x_1 - a),$$

or $x_1 = 3a/2$. Thus, the point on the x-axis through which the normal to the parabola at (a, a) passes is $(3a/2, 0)$. The required tangent is then the line through (a, a) perpendicular to this normal.

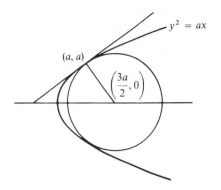

Missing in Descartes's exposition was a general scheme for determining conditions under which a polynomial equation

$$a_0x^n + a_1x^{n-1} + \cdots + a_{n-1}x + a_n = 0$$

would have a double root r. In the examples worked out in *La Géométrie*, he usually found the condition by equating the polynomial as shown term by term with the polynomial

$$(x - r)^2(b_2x^{n-2} + b_3x^{n-3} + \cdots + b_{n-1}x + b_n).$$

This was frequently tedious, because a lot of time was spent calculating the coefficients b_k in the equated polynomial, which were later themselves to be eliminated from the calculation. Thus, Descartes's tangent method was soon superseded by Fermat's.

The third and last book of the *Géométrie* belongs more properly to algebra than to geometry, concerning as it does the nature of equations and the principles underlying their solution. In the summary of the main properties of equations with which the book opens, Descartes recommended that all terms should be taken to one side and equated with zero. Though he was not the first to suggest this, he was the earliest writer to realize the advantage to be gained. He pointed out that a polynomial $f(x)$ was divisible by $(x - a)$ if and only if a was a root of $f(x)$. We find too an intuitive proof of the theorem that any equation of degree n has n roots. Descartes's words are: ''Every equation can have as many distinct roots (values of the unknown quantity) as the number of dimensions of the unknown quantity in the equation.'' His proof relied on the principle that every root a must appear in the corresponding linear factor $(x - a)$ of $f(x)$, and that it requires n such factors to achieve x^n as the highest power of x in $f(x)$.

Apart from a reform of notation, which gives this part of the work a modern look, the lasting contributions are a systematic use of negative and imaginary roots (Descartes himself was among the first to use this latter designation) and a result still known as Descartes's rule of signs. This remarkable rule—implicit in the work of several earlier writers, including Cardan and Harriot, but first used explicitly by Descartes—enables one merely by looking at a given equation to assign an upper bound to the number of its positive roots.

Two consecutive terms of an equation are said to present a variation in sign if their coefficients have opposite signs; thus, in the equation $x^3 + x^2 - x + 2 = 0$, there are two variations. With this idea, Descartes's rule of signs may be stated as follows:

The number of positive roots (each root counted as often as its multiplicity) of an equation

$$f(x) = a_0 x^n + a_1 x^{n-1} + \cdots + a_{n-1}x + a_n = 0, \qquad a_0 > 0$$

with real coefficients is equal either to the number of variations in the signs of its coefficients or to this number decreased by a positive even integer.

For example, the equation $x^3 + x^2 - x + 2 = 0$ has either two positive roots or none, the exact number not being found by Descartes's rule. The two roots may coincide to give a repeated root, but if they are distinct then neither is repeated.

As originally formulated by Descartes, the rule stated that an equation could have as many "true" (real, positive) roots as its successive terms showed changes in sign. John Wallis claimed in his *Algebra* (1685) that Descartes failed to notice that the rule breaks down in the case of imaginary roots, observing, "This rule is either a mistake or an inadvertence, for it must be taken with this caution, that is, that the roots are real, not imaginary." But Descartes said not that the equation always had to have so many roots, only that it might have. There is ample evidence that he was aware of the failure of his rule if the equation contained imaginary roots, and it is unfortunate that he did not express the fact more clearly. Newton formulated it more precisely in his *Arithmetica Universalis* (1707).

Because the negative roots of $f(x) = 0$ are simply the positive roots of $f(-x) = 0$, Descartes also gave a law of signs for the number of negative roots: No equation can have more negative roots than there are variations of sign in the coefficients of the polynomial $f(-x)$. Thus, $x^6 - 10x^2 + x + 1 = 0$ has either two negative roots or none, since $x^6 - 10x^2 - x + 1 = 0$ has two sign changes.

If it should happen that the largest possible number of positive roots added to the largest possible number of negative roots gives a sum less than the degree of the equation, then Descartes was prepared to recognize the existence of imaginary roots to fill out the number. Take, for instance, the equation $x^6 + 3x^3 + x - 1 = 0$. This equation, having only one sign variation, cannot have more than one positive root. Changing x to $-x$, we get $x^6 - 3x^3 - x - 1 = 0$, and because there is one variation, the original equation cannot have more than one negative root; hence, it cannot have more than two real roots. The proposed equation is of degree six, so it has at least four imaginary roots.

In this book, Descartes carefully considered the problem of solving equations of the third degree and higher, and he succeeded in effecting a simple solution of the quartic. Starting with a general quartic $x^4 + ax^3 + bx^2 + cx + d = 0$, he replaced x by $z - a/4$ to obtain a reduced quartic equation

$$z^4 + pz^2 + qz + r = 0$$

lacking a term in z^3. The left-hand member is expressed as the product of two quadratic factors

$$(z^2 + kz + m)(z^2 - kz + n) = z^4 + (m + n - k^2)z^2 + k(n - m)z + mn.$$

Comparing the coefficients in the two forms of the equation leads to the relations

$$p = m + n - k^2, \qquad q = k(n - m), \qquad r = mn.$$

If $k \neq 0$, the first two give

$$2n = p + k^2 + q/k, \qquad 2m = p + k^2 - q/k.$$

Substituting these values in $mn = r$, one gets the equation

$$k^6 + 2pk^4 + (p^2 - 4r)k^2 - q^2 = 0.$$

The latter, a cubic in k^2, can be solved by methods already obtained. Any root $k^2 \neq 0$ produces a pair of quadratic equations

$$z^2 + kz + \frac{1}{2}\left(p + k^2 - \frac{q}{k}\right) = 0$$

and

$$z^2 - kz + \frac{1}{2}\left(p + k^2 + \frac{q}{k}\right) = 0,$$

whose four roots are the roots of the reduced quartic.

Descartes considered as an example of the technique the equation

$$x^4 - 17x^2 - 20x - 6 = 0.$$

This led him to the cubic

$$k^6 - 34k^4 + 313k^2 - 400 = 0.$$

Because $k^2 = 16$ is a root, the solution of the given quartic is reduced to solving two quadratics, namely

$$z^2 + 4z + \tfrac{1}{2}(16 - 17 + \tfrac{20}{4}) = 0$$

and

$$z^2 - 4z + \tfrac{1}{2}(16 - 17 - \tfrac{20}{4}) = 0,$$

which is to say,

$$z^2 + 4z + 2 = 0 \qquad \text{and} \qquad z^2 - 4z - 3 = 0.$$

Solving these equations, Descartes got the roots

$$-2 + \sqrt{2}, \qquad -2 - \sqrt{2}, \qquad 2 + \sqrt{7}, \qquad 2 - \sqrt{7};$$

and these are the roots of the original quartic.

Descartes's exposition in *La Géométrie* is so far from clear that the work has the appearance of an early draft of itself. He did not arrange the material in an orderly and systematic manner, and as a rule, he gave only indications of proofs, gladly leaving their detailed execution to the reader. Such statements as ''I did not undertake to say everything,'' or ''It already wearies me to write so much about it,'' occur frequently in the text. In concluding the work, Descartes justified the omissions and obscurities he affected with the remark that much was deliberately omitted ''in order to give others the pleasure of discovering [it] for themselves.''

This mathematical sloth was remedied when *La Géométrie* was translated into Latin and published with an explanatory commentary by Frans van Schooten, a professor of mathematics at Leyden. The Latin version underwent a total of four editions, appearing first in 1649, then in 1659–61, 1683, and 1693. It is safe to say that these editions established the place of coordinate geometry in the university mathematical courses in western Europe. For they not only made Descartes's work available in the scholarly language of the time but included a large amount of supplementary material, clarifying the original account with which readers were having difficulty. In the greatly expanded two-volume version of 1659–61, the commentary of van Schooten was more than twice the length of *La Géométrie* itself. The next generation of mathematicians all studied *La Géométrie* in one version or another, Wallis in the 1649 edition, and Newton in both the 1649 and 1659–61 editions, and Huygens used Descartes in the original French.

The 1659–61 edition of *La Géométrie* contained a convenient means of determining the double roots that Descartes's tangent method called for. This was supplied by the Dutch mathematician Johann Hudde, one of van Schooten's most capable students and the burgomaster of Amsterdam for some 30 years. Hudde's rule, as it came to be known, asserted that if r was a double root of the polynomial equation

$$a_0x^n + a_1x^{n-1} + \cdots + a_{n-1}x + a_n = 0,$$

then r was also a root of the equation

$$a_0b_0x^n + a_1b_1x^{n-1} + \cdots + a_{n-1}b_{n-1}x + a_nb_n = 0,$$

where $b_0, b_1, \ldots, b_n$ were any n numbers in arithmetic progression. The rule can be illustrated by the equation $x^3 - 5x^2 + 8x - 4 = 0$. Taking the arithmetic progression 3, 2, 1, 0 (where the largest term is equal to the degree of the equation) and multiplying each coefficient of the equation by the corresponding term in the progression, one gets $3x^3 - 10x^2 + 8x = 0$, or $3x^2 - 10x + 8 = 0$. This equation has 2 as a root, and since 2 is also a root of the original equation, it is a double root. Had there been no common roots, the original equation would not have possessed a double root.

In addition to the attempt to transfer the methods of mathematical thought to the physical sciences, Descartes's great endeavor was to replace the medieval world-picture—with its blend of Aristotelian physics and Ptolemaic astronomy—by a scientific system in which all physical explanations could be sought. Descartes believed that the principal phenomena of nature could be framed in terms of mathematical and mechanical laws alone, laws of which one could be absolutely certain. He told us that ''the rules of mechanics . . . are the same as those of nature.'' His conception of a mechanical universe, in which physical objects or events in one part exerted an influence on all others, was to attain such popularity that references to the solar system as a gigantic piece of clockwork machinery became commonplace.

In the *Principia Philosophiae,* which appeared in Amsterdam in 1644, Descartes used the notion of vortices as a creative mechanism to account for the origin and the current state of the universe. Rejecting the notion of a void or vacuum in nature, he postulated that the space occupied by the solar system was filled with vaguely defined primordial matter, the so-called plenum or ether. This all-pervading material, having been endowed by God's hand at the beginning with a fixed and finite quantity of motion, evolved thereafter by the laws of mechanics, without interference. The motion of the stellar mass set up an immense whirlpool or vortex, in the middle of which was the sun. Within this vast ethereal vortex there were lesser vortices fitting together like a mass of soap bubbles; these exerted local influences much as eddies arose in streams and were carried along by the current. To escape censure by the Church on the question of the earth's motion, Descartes asserted that the earth was at rest within its vortex and therefore stationary relative to it. At the same time, it was borne by its surrounding celestial matter around the sun, just as a boat at repose in the middle of a sea could be carried imperceptibly by the tides. Thus, by a stretch of the imagination, Descartes was able formally to declare that the earth could be considered not to move. One critic has claimed that this theory of a motionless earth carried along by an enveloping ether was like a worm in a Dutch cheese sent from Amsterdam to Batavia; the worm has traveled 6000 miles without changing its location.

By comparing heavenly bodies in motion with terrestrial objects caught up in whirlpools of water, Descartes was able very plausibly to explain the workings of the solar system in readily understandable terms. Everyone had seen eddies form in the course of a river, trapping and carrying along a passing leaf, and the mental picture carried conviction. It could not be said of Descartes's ideas on the creation, as was written of the cosmology of Robert Fludd (1574–1637), ''The obscurity of the style is only equalled by the absurdity of the matter.''

Though the theory of vortices was an ingenious scheme, it was unsupported by any experimental evidence and was irreconcilable with a multitude of known celestial phenomena—including Kepler's law that the motions of the planets were not circular but elliptical. Despite the inherent defects of Descartes's new doctrine, there was something in it that satisfied the philosophical hunger of the times. It attracted a host of enthusiastic adherents, who found it wholly intelligible, and it withstood for half a century all efforts to dislodge it (probably because it had no real competition). But then came Isaac Newton, with his mathematical proofs involving gravitation, which could not be explained by vortical motion. By curious coincidence, the publication of Descartes's *Principia Philosophiae* coincided almost exactly with the birth of Newton, who before many years was to demolish the Cartesian concept of ethereal vortices to make way for his greater theory of universal gravitation. Not that Descartes's views were easily displaced by the celebrated alternative, particularly in France where national pride led to an unwillingness to accept Newtonian physics. For a generation or so after the publication of Newton's *Principia Mathematica,* the spiritual heirs of Descartes tried to remove some of the glaring blemishes in the original theory. Not before the second half of the eighteenth century did the Cartesian scientific ideas altogether lose their command. Descartes, for all his fantasies, made the first serious effort to emancipate natural knowledge from theology, that is, to explain the architecture of the universe by mechanical principles without recourse to divine intervention or occult forces.

It was the general policy of the Church to allow new scientific positions to be stated hypothetically, provided that they were formally declared to be conditional on the Church's interpretation of divine revealed truth. (Galileo was condemned mainly because the acid mockery of his dialogue too obviously belied his formal declaration.) To avoid ecclesiastical censure, Descartes explicitly put forward the vortex theory as

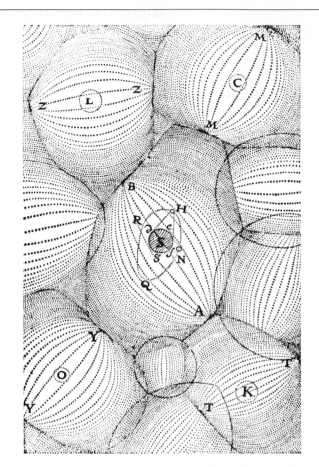

Diagram of Descartes's system of celestial vortices. From his Principia
Philosophiae *(1644).* (Extract taken from A History of Science, Technology and Philosophy
in the 16th and 17th Centuries, by A. Wolf. Reproduced by kind permission of Unwin Hyman Ltd.)

speculative. He stated in the *Principia Philosophiae* itself, ''I want what I have written
to be taken simply as an hypothesis which is perhaps far removed from the truth.''
Descartes did not hesitate to say that though his theory might be wrong in detail, it
had, nevertheless, the great merit of providing a plausible explanation of gravity, light,
magnetism, and other physical properties hitherto regarded as ''innate'' to the world.
This wily maneuver could hardly have deceived the trained censors of the Holy Office,
and Descartes's works were placed on the Index of Prohibited Books in 1663. This,
however, did not detract from their growing popularity—although it was enough to
prevent a funeral oration for him. By 1740, his works were removed from the Paris
Index in order to provide an alternative to the Newtonian synthesis, which was then
gaining gradual acceptance in France.

Descartes's achievements in mathematics, philosophy, optics, meteorology, and science leave no doubt that he was the dominant thinker of the 1600s. Descartes was only incidentally a mathematician, and his analytic geometry—or as Voltaire described it, "the method of giving algebraic equations to curves"—was just an episode in a career devoted to many innovations. It is curious that this man who brought such luster to France lived nearly all his productive years beyond her borders, taught in none of her schools, and even as a soldier fought in none of her foreign wars.

8.2 Problems

1. To multiply two numbers geometrically, Descartes said:

Let *AB* be taken as unity, and let it be required to multiply *BD* by *BC*. I have only to join the points *A* and *C*, and draw *DE* parallel to *CA*; then *BE* is the product of *BD* and *BC*.

Show that the length of *BE* is the product of the lengths of *BD* and *BC*.

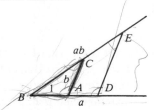

2. In *La Géométrie*, Descartes constructed the positive solutions to the quadratic equation $x^2 = ax - b^2$, where $b < a/2$. Given a circle of radius $NL = a/2$, draw a tangent to *L* and lay off from the point of contact a length $LM = b$. Then, through *M*, draw a line parallel to *NL*,

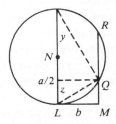

cutting the circle in the points *Q* and *R*. Prove that the lengths *MQ* and *MR* represent the two positive solutions to $x^2 = ax - b^2$. [*Hint:* If the parallel to

LM through *Q* cuts the diameter in segments of length *y* and *z*, then $y + z = a$ and $yz = b^2$.]

3. Assume that in the five-line Pappus problem, four of the lines l_1, l_2, l_3, l_4 are parallel and an equal distance apart, and that the fifth line l_5 is perpendicular to the others. Prove that if l_5 and l_2 are taken as the *x*-axis and *y*-axis,

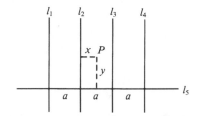

respectively, and if p_k denotes the distance of a point $P = (x, y)$ from the line l_k, then the locus of all points satisfying $p_1 p_3 p_4 = a p_2 p_5$ is given by

$$(a + x)(a - x)(2a - x) = axy.$$

This locus, which occurs in *La Géométrie*, was later called the Cartesian parabola, or trident, by Newton.

4. Show that the equation $x^3 - x^2 + 2x + 1 = 0$ has no positive roots. [*Hint:* Multiply by $x + 1$, which does not change the number of positive roots.]

5. Find the number of positive roots of the equation $x^5 + 2x^3 - x^2 + x - 1 = 0$.

6. From Descartes's rule of signs, conclude that the equation $x^{2n} - 1 = 0$ has $2n - 2$ imaginary roots.

7. Without actually obtaining these roots, show that

(a) $x^3 + 3x + 7 = 0$ and
(b) $x^6 - 5x^5 - 7x^2 + 8x + 20 = 0$

both possess imaginary roots.

8. Verify the following assertions.

(a) If all the coefficients of an equation are positive and the equation involves no odd powers of x, then all its roots are imaginary.

(b) If all the coefficients of an equation are positive and all terms involve odd powers of x, then zero is the only real root of the equation.

(c) An equation with only positive coefficients cannot have a positive root.

9. Prove that:

(a) The equation $x^3 + a^2x + b^2 = 0$ has one negative and two imaginary roots if $b \neq 0$.

(b) The equation $x^3 - a^2x + b^2 = 0$ has one negative root, while the other two roots are either imaginary or both positive.

(c) The equation $x^4 + a^2x^2 + b^2x - c^2 = 0$ has just two imaginary roots if $c \neq 0$.

10. Use Descartes's method to find the solutions of the following quartic equations:

(a) $x^4 - 3x^2 + 6x - 2 = 0$. [*Hint:* The sextic $k^6 - 6k^4 + 17k^2 - 36 = 0$ has $k^2 = 4$ as a solution.]

(b) $x^4 - 2x^2 - 8x - 3 = 0$.

11. The logarithmic spiral, whose equation in polar coordinates is $r = ae^{c\theta}$, where $-\infty < \theta < \infty$ ($a > 0$, $c > 0$), was invented by Descartes in 1638. Prove that if $P = (r_1, \theta_1)$ is any point on

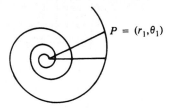

the spiral other than the origin, then the part of the spiral from the origin to P is an unending curve of finite length. [*Hint:* The length of the arc in question is given by

$$\int_{-\infty}^{\theta_1} \sqrt{r^2 + \left(\frac{dr}{d\theta}\right)^2}\, d\theta.]$$

8.3 Newton: The *Principia Mathematica*

Descartes's vortex theory, which offered a general explanation of the motions of the planets in mechanical terms, envisioned physical nature as a machine and man as the reasoning mind capable of comprehending it. Descartes began with a program of scientific rationalism—"There is nothing in my physics that is not in geometry," he wrote to Mersenne—but ended by building what Huygens called a philosophical romance. The inconsistency of vortex motion with Kepler's laws eventually led to a search for a different kind of scientific explanation for the fabric of the heavens. It remained for a still greater mind, Isaac Newton, to give the scholarly world the synthesis for which it yearned. Newton's *Philosophiae Naturalis Principia Mathematica* (1687) was the climax of the soaring intellectual thought that marked the seventeenth century, the Century of Genius. Probably the most momentous scientific treatise ever printed, it aimed, in Newton's words, "to subject the phenomena of Nature to the laws of mathematics." In realizing this ideal, the *Principia* laid the foundations of modern celestial mechanics, the principles of which were to dominate the mathematical physics of the eighteenth and nineteenth centuries.

In the year that Galileo died, on Christmas Day of 1642, Isaac Newton was born in the village of Woolsthorpe, in Lincolnshire near Cambridge. His father, a farmer of moderate means, had died a few months before Newton was born, so that the responsibility for the boy's upbringing devolved on his mother. Newton was born prematurely and was so small and frail at birth that no one thought he would survive beyond a few

Isaac Newton
(1642–1727)

(From A Short History of Astronomy *by Arthur Berry, 1961, Dover Publications, Inc. N.Y.)*

days. But the infant, so undersized that his mother said he could have been fitted into a quart pot, must soon have thrown off these "symptoms of early death," for he lived to be 85 years old.

Newton's life may be conveniently divided into three parts, each of which was largely confined to a particular locality. The first (1643–1669) covers his boyhood in Lincolnshire and his undergraduate days; the second (1669–1687), his life as Lucasian professor at Cambridge, during which he produced most of his work in mathematics; and the third (1687–1727), lasting nearly as long as the other two periods combined, his career in London as a highly paid government official.

The first eighteen years of Newton's life were times of turmoil and torment for his native land. In the year of his birth, the Great Civil War (often called the Puritan Revolution) broke out between Charles I and Parliament. By 1650 England had beheaded its monarch—"tyrant, traitor, and enemy of the Commonwealth"—and was experimenting with a republican form of government, under the leadership of Oliver Cromwell. Cromwell's death in 1658 weakened the Commonwealth sufficiently to allow the restoration of Charles II to the English throne in 1660. Newton used to tell of the great storm that swept over England on the day Cromwell passed away. The force of the gale was so strong that he tried to measure it by first jumping with the wind, then against it, and comparing the difference in the measurements of the longest jumps. In later years, he liked to call this the first of his experiments.

For one who became so great, Newton showed little sign of the massive talent that was to change the current of scientific thought. At age 12, he was sent to a public school in the neighboring town of Grantham. He seems to have been relatively inconspicuous in his academic work, being reported as "idle" and "inattentive." The young

Newton was apt to neglect the prescribed studies to spend most of his time creating mechanical contrivances such as sundials, waterwheels, and homemade clocks. After two years at Grantham, he was called home at the death of his stepfather to help with the farm, but he showed little aptitude or interest in the agricultural pursuits of his forefathers. Fortunately an uncle recognized Newton's quickness, sent him back to school at Grantham, and arranged for him to go to the college the uncle had attended, which happened to be Trinity College, Cambridge. Newton arrived at Cambridge in 1661, as a subsizar, a student who had some of his fees remitted in return for performing various menial duties.

Cambridge was an excellent place for someone wanting to pursue a law degree, and it appears that Newton enrolled with that intention. The seventeenth century was close enough to the Middle Ages that medieval modes of thought persisted in education. The curriculum was still cast in the old humanistic tradition, with its exclusive diet of reading the ancients and their commentators. The main courses for undergraduates were logic, rhetoric (the art of literary composition), and moral philosophy. The Latin and Greek languages were studied less for their own sake than as ancillary to other subjects. Although English had become the prevailing language for teaching and examining, the educated person nonetheless required Latin for writing and formal oration; academic occasions were adorned with ornate Latin speeches. A knowledge of Greek was essential among scholars, not only because of its use in the New Testament but because it was the language of the ancient authorities in philosophy. As for science and mathematics, the most significant advances took place outside of the universities and barely affected their curriculum. Not until the establishment (1663) of the Lucasian professorship, endowed by a Mr. Henry Lucas in his will, were undergraduates at Cambridge provided any formal instruction in mathematics by the university. King Charles II granted a dispensation that would allow the chair to be held by someone not in Holy Orders, and he furthermore ordered all undergraduates past the second year to attend its holder's lectures.

Newton's mathematical genius blossomed suddenly and unexpectedly. He probably entered Cambridge more backward than most undergraduates, knowing little more than the bare rudiments of computation, which he picked up from the elementary arithmetics of his day. It is not clear when or how he was first introduced to advanced mathematics. There is a dubious story that Newton's mathematical awakening came in 1663 when he purchased a book on astronomy at a country fair and found that he could not understand the diagrams in it without knowing trigonometry. With a view to throwing some light on the trigonometry, he then secured an English edition of Euclid's *Elements,* but reportedly abandoned it as "a trifling book" and turned to more advanced learning. He thereupon managed to master by himself van Schooten's richly annotated Latin edition of Descartes's *Géométrie*—which was about eight times the size of the original—and passed on to a careful study of John Wallis's *Arithmetica Infinitorum* (published in 1656). Whatever Newton's mathematical taste as a young man, there is ample documentation that he pored over the texts of Euclid and Descartes, deriving continuous inspiration from the latter's *Géométrie*. We are told that he always regretted that "he had applied himself to the works of Descartes and other algebraic writers before he had considered the *Elements* of Euclid with that attention which so excellent a writer deserves." Although Newton was a "late bloomer" in mathematics, by 1664 his reading had taken him to the frontiers of contemporary mathematical knowledge. The absence of precocity was more than counterbalanced by the rapidity with which his formidable talent matured, once stimulated.

The arrival in 1663 of Isaac Barrow (1630–1677) as the first occupant of the Lucasian chair seems both to have inspired Newton's developing mathematical power and led him to adopt an academic career. Barrow, then only 33 years old and an early member of the Royal Society, was already regarded as one of the foremost mathematicians of the period. His researches on drawing tangents to curves and on determining areas bounded by curves had very nearly led him to the invention of the calculus. Barrow had also come to mathematics by a circuitous route. As a boy he so plagued his teachers, and was so rebellious at home, that his father was heard to pray that if it pleased God to take any of his children, he could best spare Isaac. He entered Trinity College in the year 1644, received his bachelor's degree in 1648, and then stayed on as a fellow of the college. An excellent scholar in Greek, theology, physics, and astronomy, Barrow edited various works of Euclid, Archimedes, and Apollonius. He was nominated by a former teacher for the Greek professorship at Cambridge, but because of his staunch loyalist sentiments was denied the position. Driven out of Cambridge in 1655 in the Puritan purge, he sold his personal library and set out on an adventurous four-year tour of eastern Europe, which included fighting off a pirate attack during a sea voyage from Italy to Turkey.

While living in Constantinople, Barrow developed an interest in the writings of the early church fathers. On his return to England in 1660 (which happened to coincide with the restoration of Charles II to the throne), he took Holy Orders and was rewarded with the Greek professorship previously denied him. The stipend for this post was small, and two years later he augmented it by accepting the appointment as professor of geometry at Gresham College, London, a position he held only a short time before transferring to the Lucasian chair. As he explained, there were many scholars who could worthily undertake the duties of the Greek professorship, so he saw no reason why he should not retire from the "grammatical mill" and follow his own inclination.

The new chair in mathematics had few obligations connected with it; the occupant need merely work during one term of the academic year, lecturing once a week and conferring with students twice a week. Barrow, however, instituted a regular series of general introductory lectures during his tenure as Lucasian professor. One of those attending was Isaac Newton, a student destined to become the greatest mathematician England has ever produced. Barrow's lectures for the years 1664–1666 were published posthumously in 1683 as *Lectiones Mathematicae*. However, his *Lectiones Opticae* (1669) and *Lectiones Geometricae* (1670) were printed almost immediately, with Newton assisting in the preparation.

Barrow's *Lectiones Geometricae* presented, in thirteen lectures, a collection of theorems concerned with drawing tangents to curves and finding lengths of curves and the areas bounded by them. His method for determining the tangent to a point P on a curve given by the polynomial equation $f(x, y) = 0$ closely resembled that used in our modern calculus textbooks. He observed that the tangent could be obtained if some other point on it were known, for instance the point T at which the desired tangent should meet the x-axis. For this, Barrow took a point Q on the curve, close to the first point P, and by drawing parallels to the coordinate axes constructed a small right triangle PQR (which he called the differential triangle). The closer the point Q to the

Isaac Barrow
(1630–1677)

(Extract taken from A History of Science,
Technology and Philosophy in the 16th and 17th
Centuries, *by A. Wolf. Reproduced by kind
permission of Unwin Hyman Ltd.)*

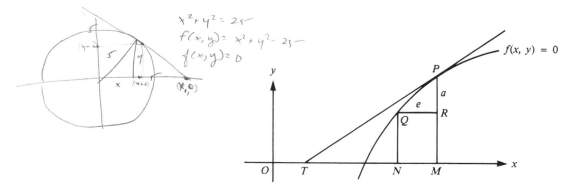

point P, the more nearly similar the triangles PQR and PTM. Barrow thought it reasonable to take them as similar on the grounds of the "indefinite smallness" of that part of the curve, whence

$$\frac{TM}{MP} = \frac{QR}{RP}.$$

On setting $QR = e$ and $RP = a$, it follows that if (x, y) are the coordinates of the point P, then Q has coordinates $(x - e, y - a)$. Now substitute these coordinates into the equation $f(x, y) = 0$, and in Barrow's own words,

> reject all terms in which there is no a or e (for they destroy each other by the nature of the curve); reject all terms in which a and e are above the first power, or are multiplied together (for they are no value with the rest, as being infinitely small).

The ratio a/e can then be found. Finally, T can be determined by using the length of the line segment TM:

$$OT = OM - TM = OM - MP\left(\frac{QR}{RP}\right) = x - y\left(\frac{e}{a}\right).$$

Barrow gave five examples of this method of the differential triangle, including

i) $x^3 + y^3 = r^3$, ii) $x^3 + y^3 = rxy$, iii) $y = (r - x)\tan\dfrac{\pi x}{2r}$.

To see the process clearly, it will suffice to consider the case $x^3 + y^3 = r^3$. Since Q is taken to be on the curve, its coordinates satisfy the equation, so that

$$(x - e)^3 + (y - a)^3 = r^3$$

or

$$x^3 - 3x^2e + 3xe^2 - e^3 + y^3 - 3y^2a + 3ya^2 - a^3 = r^3.$$

Because $x^3 + y^3 = r^3$, the last-written equation becomes

$$-3x^2e + 3xe^2 - e^3 - 3y^2a + 3ya^2 - a^3 = 0.$$

Discarding the terms that contain powers of a and e beyond the first, we get

$$3x^2e + 3y^2a = 0$$

and from this,

$$\frac{a}{e} = -\frac{x^2}{y^2}.$$

The *Lectiones Geometricae* was the culmination of the seventeenth-century investigations leading toward the calculus. Although the tangent method of Barrow resembled the process of differentiation, he apparently did not perceive the deeper significance. Nor was he able to justify why the higher powers of a and e should be neglected in his calculations, which for a rigorous foundation can be explained only in terms of limits. Thus, we can better describe Barrow as a signal precursor of the differential calculus, and Newton as its first inventor.

All through the Middle Ages, and right on until the second half of the seventeenth century, outbreaks of bubonic plague were a recurrent feature of English life. The years 1665–1666 are best remembered for the Great Plague, which swept through closely packed London, leaving no fewer than 68,500 dead—one in four of the estimated population. In June, 1665, the diarist Samuel Pepys (1633–1703) wrote:

> This day, much against my will, I did in Drury Lane see two or three houses marked with a red cross upon the doors, and "Lord have mercy upon us" writ there; which was a sad sight to me, being the first of the kind that, to my remembrance, I ever saw.

Then, in September, Pepys recorded in his *Diary,* "I have stayed in the city till about 7400 died in one week and of them about 6000 of the plague, and little noise heard day or night but tolling of bells." As a precaution, Cambridge University was closed for the better part of two years and Newton took refuge at the isolated family farm in Woolsthorpe. In 1666 plague-ridden England was shaken by another disaster, the Great Fire of London; the smoldering ruins covered 436 acres or approximately half of the city. The fire, although it destroyed some 13,200 homes and 90 parish churches, did at least mitigate the plague.

While Newton was forced to live in seclusion at home, he began to lay the foundations for his future accomplishments in those fields with which his name is associated—pure mathematics, optics, and astronomy. During these two ''golden years'' at Woolsthorpe, Newton made three discoveries, each of which by itself would have made him an outstanding figure in the history of modern science. The first was the invention of the mathematical method he called fluxions, but which today is known as the differential calculus; the second was the analysis of white light (sunlight) into lights of different colors, separated in the visible spectrum according to their different refrangibilities; the third discovery was the conception of the law of universal gravitation. These three discoveries were made before he was 25 years old. Referring to this period of leisure and quiet, Newton later wrote, ''All this was in the two plague years of 1665 and 1666, for in those days I was in the prime of my age for invention and minded Mathematics and Philosophy [physics] more than at any time since.''

Newton returned to Cambridge early in 1667, the virulence of the plague having subsided. In 1668, he was elected a fellow of Trinity College, and a few months thereafter, was created a master of arts. One of the highest honors in the scientific world, the Lucasian professorship, became his a year later, when Barrow was called to London as chaplain to Charles II and resigned the chair of mathematics. Barrow proposed his incomparable pupil, Isaac Newton, as his successor. Whether Barrow made way for Newton in recognition of the latter's superior powers or simply out of the desire to make theology his full-time study remains a question, but the effect was the same. Newton accepted the appointment and held the chair actively until 1696. (Newton was denied further promotions at the university because he was a Unitarian and did not accept the doctrine of the Trinity.) As for Barrow, he did no further mathematical work. After Barrow became a doctor of divinity, the king appointed him (1672) master, then vice-chancellor, of Trinity College, saying that he was giving the position to the best scholar in England.

On assuming the Lucasian chair, Newton chose optics for his inaugural course of lectures, conscious that he would eclipse Barrow's *Lectiones Opticae* both in subject matter and methodology. Owing to the novelty of the subject and his rigorous treatment of it, Newton's lectures did not attract a large audience. Often finding no one present—despite an official letter from Charles II confirming the statutory attendance requirement—he would leave the lecture hall and return to his private research. By the end of 1669, he had worked out the details of his discovery of the heterogeneous character of white light, though these were not published until many years later under the title *Opticks* (1704).

Newton was led to the study of optics by the imperfections of the contemporary lens telescopes, in which chromatic aberration produced colored edges around the images seen. Accordingly, in 1668, he designed and constructed the first reflecting telescope, that is, a telescope in which the rays of light from the object viewed are concentrated by means of a concave mirror instead of the convex lens of Galileo's refracting telescope. Knowledge of Newton's telescope was confined to a limited circle of Cambridge friends. Eventually word reached the Royal Society whose fellows, their curiosity aroused, asked Newton to send the instrument for inspection. Its appearance at the meeting of January 1672 caused a sensation, and Newton suddenly became a celebrated figure. Among those who examined the telescope were King Charles II, Robert Hooke, and Christopher Wren (the architect of St. Paul's in London, and at this time Savilian professor of astronomy at Oxford). At the same meeting, Newton was elected a fellow of the society. He immediately responded by sending the society an article embodying the results of his lectures on optics. Printed in the *Philosophical Transactions* of the Royal Society under the title *New Theory about Light and Color,* this was his first scientific paper to be published. It contained the first formal announcement, beyond the bounds of his Cambridge lecture hall, of the ingenious experiment of 1666 in which he demonstrated that white light is composed of the various colors of the spectrum. Light from a tiny hole in a shutter had been passed through a triangular glass prism. Newton wrote that as he observed the refracted colors on the wall, ''I was surprised to see them in oblong form; which according to the received laws of refraction, I expected would have been circular.'' The band of light varied in color from red at the bottom, through orange, yellow, green, and blue, to violet at the top—the order of colors in the rainbow. Newton concluded that sunlight, and white light generally, is composed of rays of every color and that some colors are more sharply bent, or refracted, than others. This was in direct contrast to the current theory that white light was basic, whereas colors were due to various mixtures of light and ''darkness.''

Newton's paper on the nature of light provoked much criticism from those with established reputations, particularly from Robert Hooke, a genius second to Newton but to few others. Senior to Newton in age by seven years, and one of the original fellows of the Royal Society, Hooke viewed himself as the society's authoritative spokesman on experimental physics. Hooke had recorded in his great work *Micrographia* (1665) an experiment similar to Newton's with the prism. He felt slighted by Newton's ignoring his priority and joined other members of the society in condemning Newton for asserting conclusions that they felt did not necessarily follow from the experiments described. Hooke maintained that Newton did not prove that all colors were actually in every ray of light before it suffered a refraction.

Another point of contention concerned the medium through which light is transmitted. Hooke, recognizing the analogy between light and sound, favored the hypothesis that light traveled on the waves of a weightless invisible agent called ''lumeniferous ether'' and argued that this doctrine explained most of the optical phenomena then known. Newton, for his part, claimed that such a theory could not be reconciled with the fact that light traveled in a straight line. He proposed that light was composed of a stream of tiny particles, or corpuscles, of different sizes (the size corresponding with the color) and moving with different velocities. The increasingly acrimonious dispute was carried on for four years. The friction with Hooke only confirmed Newton's tendency to secrecy and isolation and dimmed his early enthusiasm for the Royal Society. ''I am so persecuted,'' he wrote, ''with discussions arising out of my theory of light that I blame my own imprudence for parting with so substantial a blessing to run after a shadow.'' Having neither the time nor the desire to engage in controversy, Newton retired from the public world of science. The decision led him to refrain from publishing a general account of his optics until after the death of Hooke in 1703. There was even a greater delay in the appearance of Newton's mathematical writings, few of which were voluntarily given to the world by himself.

During the next few years Newton spent most of his time at optics and mathematics. As a consequence of his study of Wallis's *Arithmetica Infinitorum* in the winter of 1664–1665, he had discovered the general binomial theorem, or expansion of

$(a + b)^n$, where n may be a fractional or a negative exponent. (Except when n is a positive integer, the resulting expansion is an infinite series.) He first enunciated the formula, and tried to recapture his original train of thought leading to it, twelve years later (1676) in two letters written to Henry Oldenburg, the multilingual secretary of the Royal Society. These letters were to be translated into Latin and forwarded to Leibniz, who in his early struggles with his version of the calculus had asked for information about Newton's work on infinite series. As given in the first letter to Oldenburg (the *Epistola Prior* of June 1676), the formula, or rule, as Newton called it, was written in the form

$$(P + PQ)^{m/n} = P^{m/n} + \frac{m}{n}AQ + \frac{m - n}{2n}BQ + \frac{m - 2n}{3n}CQ + \frac{m - 3n}{4n}DQ + \cdots,$$

where each of A, B, C, D denotes the term immediately preceding it; that is, A represents $P^{m/n}$, B represents $(m/n)AQ$, and so on. Newton's letter, and his calculations in general, employed negative and fractional exponents, which after this time became a universally recognized practice. He wrote, "For as analysts, instead of aa, aaa, etc., are accustomed to write a^2, a^3, etc., so instead of $\sqrt{a}$, $\sqrt{a^3}$, $\sqrt{c:a^5}$, etc., I write $a^{1/2}$, $a^{3/2}$, $a^{5/3}$, and instead of $1/a$, $1/aa$, $1/a^3$, I write a^{-1}, a^{-2}, a^{-3}."

Wallis had earlier constructed a table of values of what today would be written $\int_0^1 (1 - x^2)^n dx$ for certain positive integers n, but the evaluation of $\int_0^1 (1 - x^2)^{1/2} dx$ had eluded him. Nonetheless, by a highly elaborate and difficult analysis, he arrived at a remarkable expression for $4/\pi$ in the form of an infinite product:

$$\frac{4}{\pi} = \frac{1}{\int_0^1 (1 - x^2)^{1/2} dx} = \frac{3 \cdot 3 \cdot 5 \cdot 5 \cdot 7 \cdot 7 \cdots}{2 \cdot 4 \cdot 4 \cdot 6 \cdot 6 \cdot 8 \cdots}.$$

Newton had the insight to change Wallis's fixed upper bound in the integral to a free variable x (he had no symbol for the integral, but defined the integral as a limit of a sequence of sums), and then to look for a general pattern that seemed to run through a set of particular instances. Considering the expansions whose modern equivalents are

$$\int_0^x (1 - t^2) dt = x - \frac{1}{3}x^3,$$

$$\int_0^x (1 - t^2)^2 dt = x - \frac{2}{3}x^3 + \frac{1}{5}x^5,$$

$$\int_0^x (1 - t^2)^3 dt = x - \frac{3}{3}x^3 + \frac{3}{5}x^5 - \frac{1}{7}x^7,$$

$$\int_0^x (1 - t^2)^4 dt = x - \frac{4}{3}x^3 + \frac{6}{5}x^5 - \frac{4}{7}x^7 + \frac{1}{9}x^9,$$

Newton observed that the first term of each expression is x, that x increases in odd powers, that the algebraic signs of the terms alternate, and that the second terms $\frac{1}{3}x^3$, $\frac{2}{3}x^3$, $\frac{3}{3}x^3$, $\frac{4}{3}x^3$ are in arithmetical progression. Reasoning by analogy, he assumed that the first two terms of $\int_0^x (1 - t^2)^{1/2} dt$ should be

$$x - \frac{\frac{1}{2}}{3}x^3.$$

Further attempts to recognize a pattern by interpolating from specific cases led him to

$$\int_0^x (1 - t^2)^{1/2} dt = x - \frac{\frac{1}{2}}{3}x^3 - \frac{\frac{1}{8}}{5}x^5 - \frac{\frac{1}{16}}{7}x^7 - \frac{\frac{5}{128}}{9}x^9 - \cdots.$$

The successive powers of x revealed for the first time the binomial character of the sequence of coefficients. The numerators $\frac{1}{2}$, $-\frac{1}{8}$, $\frac{1}{16}$, $-\frac{5}{128}$ are just $\binom{n}{1}$, $\binom{n}{2}$, $\binom{n}{3}$, $\binom{n}{4}$ for the particular case $n = \frac{1}{2}$ in the general formula

$$\int_0^x (1 - t^2)^n dt = x - \binom{n}{1}\frac{1}{3}x^3 + \binom{n}{2}\frac{1}{5}x^5 - \binom{n}{3}\frac{1}{7}x^7 + \binom{n}{5}\frac{1}{9}x^9 - \cdots,$$

where

$$\binom{n}{k} = \frac{n(n - 1)(n - 2) \cdots (n - k + 1)}{1 \cdot 2 \cdot 3 \cdots k}.$$

In other words, Newton found that the binomial coefficient form held for nonintegral values, and in particular, he used the coefficient

$$\binom{\frac{1}{2}}{k} = \frac{\frac{1}{2}(-\frac{1}{2})(-\frac{3}{2}) \cdots - (k - \frac{3}{2})}{1 \cdot 2 \cdot 3 \cdots k}.$$

From this, Newton went on to deduce by differentiation the expansion

$$(1 - x^2)^{1/2} = 1 - \frac{1}{2}x^2 - \frac{1}{8}x^4 - \frac{1}{16}x^6 - \frac{5}{128}x^8 - \cdots,$$

and checked its correctness by multiplying the preceding series by itself, term by term, to get $1 - x^2$.

It is odd that Newton derived the binomial expansion in an integral form, namely $\int_0^x (1 - t^2)^{1/2} dt$, before he realized that the same form is preserved in $(1 - x^2)^{1/2}$ if from the expression for $\int_0^x (1 - t^2)^{1/2} dt$ the denominators 1, 3, 5, 7, 9, . . . are omitted and each exponent lowered by 1.

Newton was aware that deriving the binomial theorem by looking for structural patterns in tabulated cases was not a rigorous method of reasoning. He therefore checked that the particular expansion of $(1 - x^2)^{-1}$ arising from his formula was equivalent term by term with the expansion obtained by ''brute-force'' long division. Newton never did publish his binomial theorem, nor did he prove it in generality. It became widely known through private circulation of his tract *De Analysi* (1669), but no account appeared in a printed text until 1685, when Wallis's *De Algebra Tractatus* (Treatise of Algebra) quoted extracts of Newton's letters to Oldenburg.

The well-known anecdote that the problems of gravitation were brought home to Newton by the fall of a ripened apple in the orchard at Woolsthorpe seems to rest on good authority. A friend of Newton, William Stukeley, a fellow of the Royal Society, related:

> Amidst other discourse, he told me, he was just in the same situation [under the shade of some apple trees], as when formerly, the notion of gravitation came into his mind. It was

occasioned by the fall of an apple, as he sat in a contemplative mood. Why should the apple always descend perpendicularly to the ground, thought he to himself. Why should it not go sideways or upwards, but constantly to the earth's center? Assuredly, the reason is that the earth draws it.

Newton began to wonder how high gravity extended. Skyward to the moon perhaps? If so, could the same force that pulled the apple to the ground be the force that retained the moon in its curvilinear orbit? To settle the question, he needed to know the ratio by which the gravitational pull decreased with distance from earth. And for this, Kepler's celebrated third law of planetary motion—which stated that the squares of the times of revolution of any two planets (including the earth) about the sun are proportional to the cubes of their mean distances from the sun—provided valuable information.

Newton and several other physicists of the day (notably Huygens, Halley, and Wren) had all shown independently that if Kepler's third law were true—and they were not quite certain that it was—then the attractive force of gravity would diminish with the square of the distance. The argument probably went as follows. For a planet of mass m moving with velocity v in a circular orbit of radius r, the centrifugal force is

$$F = \frac{mv^2}{r}.$$

But if T is the time of one revolution, then

$$v = \frac{2\pi r}{T}.$$

On substituting this value of v, one gets

$$F = \frac{4\pi^2 mr}{T^2}$$

as an expression for the constant force required to hold the planet in its circular orbit. By Kepler's third law, $T^2/r^3 = c$, where c is a constant; whence

$$F = \left(\frac{4\pi^2 m}{c}\right)\frac{1}{r^2}.$$

Thus, if the earth's gravity provided the force maintaining the moon in its orbit, this force would be inversely proportional to the square of the separating distance.

During the plague years, Newton carried out a calculation to see whether a force of attraction that varied with the inverse square of the distance between two bodies would account for the motion of the moon around the earth. Unfortunately, Fate played a trick on Newton in this enterprise, and his test was at first a disappointment. Through Galileo's experiments with falling bodies, afterwards repeated more accurately by others, it was generally known that the rate of fall at the surface of the earth was $16 \cdot 60^2$ feet in one minute. The accepted value for the distance of the moon from the earth was $60r$, where r was the earth's radius. Hence, if the inverse-square law held, the gravitational attraction the earth exerted on the moon would be $1/60^2$ of the attraction the earth exerted on an object on its own surface. The moon would therefore descend a distance of 16 feet in one minute toward the center of the earth.

The next stage in Newton's calculation, determining the distance over which the moon actually fell in one minute toward the earth's center, required an accurate value of the earth's radius and the mean time of the moon's revolution around the earth. The value of the latter was very nearly 27 days, 7 hours, and 43 minutes, or 39,343 minutes. "Being away from books," Newton took for his calculations the standard local estimate, used by seamen and old geographers, that there were 60 miles to a degree of latitude along the earth's equator. This led him to infer that the circumference of the earth was $60 \cdot 360$ miles, so that its radius would be $60 \cdot 360/2\pi$, or 3438 miles. The moon's distance from the center of the earth was known to be 60 times the earth's radius, so the moon's orbit, taken to be circular, would be $60^2 \cdot 360$ miles long. If this were assumed to be the usual statute mile, whose length had been defined in 1593 to be 5280 feet, then the orbit would be $60^2 \cdot 360 \cdot 5280$ feet long. (Some historians argue that Newton was more likely to have set a mile equal to 5000 feet.) Hence the moon's velocity in feet per minute at any point such as P would be

$$\frac{60^2 \cdot 360 \cdot 5280 \text{ feet}}{39,343 \text{ minutes}} = 173,930 \text{ feet/minute}.$$

Constrained to follow its curved path, the moon would have traveled from P to Q, an arc of length 173,930 feet, in one minute. Moreover, in that time it had "fallen" the distance $PS = RQ$ toward the earth. Because triangles PSQ and QSP' are similar, it follows that

$$PS \cdot SP' = (SQ)^2.$$

$$\frac{60^2 \cdot 360 \cdot 5280 \text{ feet}}{39,343 \text{ minutes}} = 173,930 \text{ feet/minute}$$

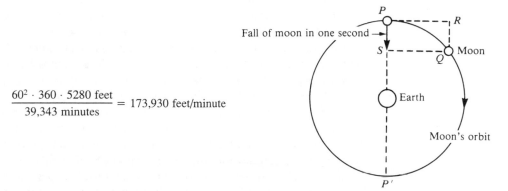

If PP' is used as an approximation to SP', and arc PQ as an approximation to SQ, this last relation becomes

$$PS = \frac{(\text{arc } PQ)^2}{PP'} = \frac{(173,930)^2\pi}{60 \cdot 60 \cdot 360 \cdot 5280} \text{ feet,}$$

or $PS = RQ = 13.89$ feet (or 14.67 feet if one takes a mile to be 5000 feet). Thus there was a serious discrepancy between the two values for the fall of the moon— the 16 feet per minute as determined from the inverse-square law of gravity, and on the other hand, the 13.89 feet per minute as deduced from the moon's mean period and the size of the orbit. Although Newton said that he found his calculations to "answer pretty nearly,'' they did not match well enough to be convincing. Somewhat discouraged that the results did not answer expectation, he abandoned all work on the gravitational problem, without bothering to publish any account of it.

During the dozen years from 1667 to 1679, when Newton pushed the idea of gravitation to the back of his mind, others began to duplicate much of his first work. In 1673 the great Dutch scientist Christiaan Huygens published his mathematical analysis of the motion of the pendulum, *Horologium Oscillatorium sive de Motu Pendulorum,* a work in which he derived the law of centrifugal force for uniform circular motions. As a result, the inverse-square law for gravitational attraction was formulated independently by that versatile but jealous physicist Robert Hooke, by the astronomer Edmund Halley, by the architect and astronomer Christopher Wren, and by Huygens himself. Hooke, Halley and Wren—all brilliant young members of the newly founded Royal Society—were greatly interested in the problem of gravitation for noncircular orbits but were unable to handle the mathematics involved. Hooke, acting formally in his new capacity as secretary of the society, wrote (1679) a conciliatory letter to Newton, begging him to renew his correspondence with its members on scientific matters. Stimulated by Hooke's opinions on the dynamics of planetary motions, Newton's attention was drawn to his calculations on the moon that had lain neglected for twelve years.

The error in Newton's original computation arose from taking the wrong value for the length of a degree of terrestrial latitude, which then yielded an incorrect value for the radius of the earth. The reason commonly given for Newton's careless adoption of an inaccurate value is that because he was in the country at the time, he used the figure given in the only book at hand, a sailor's manual. Several years later, Jean Picard obtained a very accurate value for the length of a degree on the earth's meridian; his figure of 69.1 miles varied greatly from the 60 miles Newton had used. Picard's determination of the size of the earth was carried out in 1669 and published in his *Measure de la Terre* in 1671. The figures were communicated to the Royal Society in 1672, but Newton apparently paid no attention to them.

When Newton finally became acquainted with Picard's results (it is possible, even probable, that this took place in 1684), he began again to calculate how far the moon would fall toward the earth in one minute according to the new data. His biographers tell us that as he made the final calculation his emotions overcame him, and his hand froze when he foresaw the result. He cried out to his assistant, ''Work it out for me, I cannot complete it; what do you get?'' ''Sixteen feet per minute,'' was the answer. The eagerly expected agreement in the figures confirmed Newton's conjecture that the earth's gravitation attraction does indeed provide the force maintaining the moon in its orbit.

At this point, according to his own account, Newton was led to the discovery that if a planet were acted on by an attractive force varying according to the inverse-square law, then it would have to describe an elliptical orbit, with the attractive force residing in one of the ellipse's foci. Despite the magnificence of this accomplishment, he was silent about his discovery, ''being upon other matters.'' No doubt Newton, with his morbid sensitivity to any kind of criticism or questioning, however small, was leery of publicity. But the real sticking point seems to have been his trouble in proving that the gravitational pull of a spherical body is the same as if the sphere's whole mass were located at its center. (This is true for precisely those systems showing an inverse-square law of attraction.) An essential link in his line of reasoning was missing, and it apparently was not produced by Newton until 1685.

The last act of the drama began several years later when Christopher Wren, in a sporting humor, made a gentlemanly wager with Hooke and Halley. In January 1684, Wren offered the prize of a book, up to the cost of 40 shillings, if either of his two friends could deduce within two months' time the orbit of a planet acting under an inverse-square law of force. Halley failed to do so. Hooke claimed ''that upon that principle [the inverse-square law] all the laws of the celestial motions were to be demonstrated,'' and that he himself had done so; but he made an excuse for not putting forward his alleged explanation just then—and it was never forthcoming. In August of that year, the young and energetic Halley set out for Cambridge (or as he says, did himself ''the honour to visit Newton'') to see what suggestions Newton had to offer. Halley put the question to him: What path would the planets describe about the sun if they are attracted by a force that varies inversely as the square of the distance? Newton answered immediately, ''An ellipse.'' The amazed and overjoyed Halley asked him how he knew. ''Why,'' replied Newton, ''I have calculated it.'' Somewhat characteristically, Newton was unable to find the notes of the demonstration that he had effected some years earlier but promised to reconstruct the solution and send it to Halley. This was accordingly communicated to Halley in November 1684.

The reworking of the old problem seems to have aroused Newton's interest in the whole question of planetary motion. He realized that his calculation could hardly stand by itself and therefore worked out enough new material to serve as a course of nine lectures on celestial mechanics. These were delivered, under the title *De Motu Corporum* (On the Motion of Bodies), when Cambridge opened in the autumn. Halley, on receiving Newton's promised demonstration, was excited enough to pay a second visit to Cambridge to persuade him to make his work public. On this trip, he studied Newton's manuscript lecture notes and realized their immense importance. Halley begged that the results might be published, but had to be content with a pledge that they would be forwarded to the Royal Society to secure priority. Halley then reported to the body that he had lately seen Newton at Cambridge, and that there Newton had shown him a curious treatise, *De Motu,* promising to send it to the society to be entered on their register. The Royal Society not only published its *Philosophical Transactions* but from time to time undertook to bring out meritorious scientific works. Thus the society authorized the printing of *De Motu Corporum* and appointed Halley ''to put Newton in mind of his promise.''

Newton threw himself into the task of devising a rigorous mathematical formulation of the whole system of planetary motion, a formulation dependent only on his law of universal gravitation. Early in 1685, he was finally able to prove that the gravitational attraction exerted by a solid sphere, whose density varies only with distance to the center, acts as though the sphere's mass were concentrated at its center. With the essential link in the argument then complete, it became possible for Newton to elucidate a whole series of cosmic movements that had heretofore been enigmatic. Under the personal encouragement of Halley, who kept in constant touch with the progress of the work, *De Motu Corporum* broadened in scope to the bulky *Philosophiae Naturalis Principia Mathematica* (The Mathematical Principles of

Edmund Halley
(1656–1742)

(Extract taken from A History of Science,
Technology and Philosophy in the 16th and 17th
Centuries, *by A. Wolf. Reproduced by kind
permission of Unwin Hyman Ltd.)*

Natural Philosophy). At the time the *Principia* was published, *natural philosophy* meant science in general and physics and astronomy in particular.

Newton worked at its composition with a speed that was little short of fanatical; by his own testimony, "*The Book of Principles* was writ in about 17 or 18 months." In March 1686, he duly presented the manuscript of Book I, *De Motu Corporum* (The Motion of Bodies) of the *Principia* to the Royal Society. Hooke at once came forth with the claim that he had been first in discovering the inverse-square law of gravitation and had been the prime mover in the whole series of results in the *Principia*. Hooke had no doubt surmised that the motions of the planets implied an inverse-square law; but because he had been unable to verify that such a law accounted for elliptical orbits, he had little claim to a scientific discovery. Halley acted as peacemaker and convinced Newton that he should soothe Hooke by inserting in a scholium a suitable acknowledgment, in which he should state that the inverse-square law "obtains in the celestial bodies, as Sir Christopher Wren, Dr. Hooke, and Dr. Halley have severally observed."

Newton was so annoyed by Hooke's galling priority claims that when he announced to Halley (June 1687) that Book II was completed, he added that he was ready to eliminate the proposed third book:

> The third I now design to suppress. Philosophy is such an impertinently litigious lady that a man had as good be engaged in lawsuits as have to do with her. I found her so formerly, and now I can no sooner come near her again, but she gives me warning.

Fortunately for the scientific world, Halley's persuasive powers convinced the irritated author that he should continue.

The Royal Society went virtually bankrupt when one of its books, Willoughby's *History of Fishes,* failed to sell as expected. And although the society wanted to pay for the publication of the *Principia,* it could not enter on any fresh printing expenses. Rather than let the work remain unprinted, Halley agreed to defray the costs entirely out of his own pocket, no slight burden for a man of meager means. (In return, the society voted to give Halley fifty copies of the unsalable *History of Fishes*.) In September 1687, the complete *Principia* was published, in Latin, under the imprint of the Royal Society and its then president, Samuel Pepys, although it is to be doubted that the pompous and prosy diarist could have understood as much as a single sentence of it.

If it were not for Halley, there might never have been a *Principia*. Not only did he furnish the funds for its prompt issue and act as a conciliator between Newton and the Royal Society, but he set aside his own research to keep an editorial eye on what is regarded as the greatest scientific work of all time. Halley gathered the necessary astronomical data, corrected obscurities in the text, and superintended the illustration and printing. The mathematician Augustus De Morgan (1806–1871) wrote in his *Gallery of British Worthies* of Halley's efforts, ''But for him, in all human probability, the work would never have been thought of, nor when thought of written, nor when written printed.'' Newton generously acknowledges in the preface of the *Principia* that it was Halley who had persuaded him to deal with the subject and communicate his results to the Royal Society. Halley on his part prefixed a 48-line ode to the *Principia* eulogizing the discoveries therein. Certain verses were constantly quoted for their appropriate praise of Newton:

> Then ye who now on heavenly nectar fare,
> Come celebrate with me in song the name
> Of Newton, to the Muses dear; for he
> Unlocked the hidden treasuries of the Truth:
> So richly through his mind had Phoebus cast
> The radiance of his own divinity.
> Nearer the gods no mortal may approach.

Newton's great work shows by its very title that it was intended to be a rebuttal to Descartes's theory of vortices. It is the *Mathematical Principles of Natural Philosophy* as opposed to Descartes's *Principles of Philosophy*—principles that Newton held to be thoroughly unreliable. Legend has it that Newton repeatedly wrote the pencil note ''error'' in the margin of his copy of Descartes's book; and tired of having to write the same note again and again, he threw the book away never to reread it.

Newton's *Principia* tried to explain all the motions of the heavens according to the law of universal gravitation. The concept of gravitational attraction was introduced long before Newton's time. For those who took the geocentric view, wherein the earth was thought to occupy a special position at the center of the universe, it was natural to believe that gravity was associated with the earth alone. When this ceased to make sense in a Copernican universe, gravity came to be regarded as an attribute of any large material body, and was spoken of as the tendency by which various celestial bodies tend to unite and come together. Newton's master stroke lay in generalizing this to a law of universal gravitation, wherein every particle of matter, however small, attracts every other particle. In its modern form, the principle states:

> Any two material particles attract each other with a force varying directly with the product of their masses and inversely with the square of the distance between them.

Nowhere in the *Principia* is the law found in these words, but different parts of it exist in different passages. In one place, Newton wrote ''There is a power of gravity tending to all bodies, proportional to the several quantities of matter which they contain. . . .''

And in another, he stated that "gravity . . . operates . . . according to the quantity of solid matter which they contain . . . decreasing always with the inverse square of the distances." All this is perhaps more understandable in terms of symbols. What Newton had conjectured is that two particles attract each other with a force given by the formula

$$F = G \frac{Mm}{d^2},$$

where M and m are the masses of the particles, d is the distance between them, and G is a constant called the constant of gravitation.

Although the results were first found by means of the new fluxional calculus, Newton was careful in the *Principia* to recast all his demonstrations in the form of classical Greek geometry with an almost complete lack of analytical calculations. He probably felt that the time-honored approach provided convincing arguments in a language other mathematicians and astronomers would understand. The fluxional calculus was then unpublished, and had Newton used it to arrive at results that were opposed to many of the theories prevalent at the time (such as Descartes's vortex theory of the universe), the controversy about the truth of his findings would have been complicated by disputes over the validity of the methods. Because the printed pages of the *Principia* have the appearance of Greek geometry, it is to be expected that there are few references to the "new analysis" through which Newton claimed to have discovered the propositions; in fact, the word *fluxion* does not appear at any place in the work except in one lemma in which Newton appears to have forgotten to change over from the initial analysis to the geometric style of presentation.

The *Principia* consists of three books (containing 53, 42, and 48 propositions, respectively) as well as 25 pages of introductory matter. In the prefatory section, Newton defined such concepts as mass, inertia, momentum, and centripetal force; and he laid down the three famous "Axioms or Laws of Motion," which had to precede the mathematical propositions. These laws are, in his own words:

1. Every body continues in its state of rest, or of uniform motion in a straight line, unless it is compelled to change that state by impressed forces.
2. The change of motion [rate of change of momentum] is proportional to the impressed force and takes place in the direction of the straight line in which the force is impressed.
3. To every action there is always an opposed and equal reaction.

The first two laws are deductions from the historical experiments of Galileo, but the principle expressed in the third law is one Newton himself introduced.

Book I, containing the first of two parts of *Du Motu Corporum,* is a mathematical treatment of the laws of motion under the influence of impressed forces, there being no resisting medium; it consists of what we should today call theoretical mechanics. The book is written in generality, and in the earlier propositions a geometrical point-mass plays the role of a physical body. Newton's apparent aim was to develop the subject in such a way that he could apply it to the motion of comets, or planets and their satellites, in Book III. The outstanding result of the book was the proof that if a body describes an elliptical path (or for that matter, any path that is a conic section) under the influence of an attractive force at one focus, then the force must vary inversely with the square of the body's distance from the focus; conversely, if the attractive force varies inversely with the square of distance, then the orbit describes a conic section, with the center of attraction at a focus of the conic. Turning to results on attraction of spheres, Newton showed that if every particle attracts every other particle according to an inverse-square law, then the whole mass of a uniformly dense spherical body can be considered concentrated at its center.

The second book of the *Principia, De Motu Corporum Liber Secundus,* deals with the motion of bodies—particularly of pendulums—in resisting media such as air and water. The effect of the resistance was regarded as proportional to the body's velocity, to the square of its velocity, or to both combined. The book contains the mathematics of fluid dynamics, in a wide sense of the term: calculation of the density and compression of gases and liquids, the first printed analysis of wave motion in a fluid, and an examination of the flow of liquids through orifices. One of Newton's motives in the book was to show mathematically that Descartes's theory of vortices, with its set of whirlpools in which each planet was supposedly held, was dynamically unsound and hence not a possible explanation of celestial motion. He demonstrated that a vortex would impart to a planet a motion that could not be reconciled with Kepler's laws ("it is manifest that the planets are not carried around in corporeal vortices").

Book III, which bears the special title *De Systemate Mundi,* was the crowning achievement of Newton's work. It contains the application of the general theory developed in Book I to the solar system; many important phenomena of motion, terrestrial as well as celestial, are explained by the law of universal gravitation. By reference to gravitational principles, Newton could compare the ratios of the masses and mean densities of the sun, the earth, and any planet that had a satellite. He then made the remarkable surmise that the earth had a mean density between 5 and 6 times that of water (current calculations put the figure at 5.517). He showed that the earth is not exactly spherical as had been supposed since ancient times, and calculated the oblateness, or "flattening," at the poles; his numerical estimate of the ellipticity of the earth as an oblate spheroid was $\frac{1}{230}$ (today's figure is about $\frac{1}{297}$). Newton ascribed the tides to the solar and lunar attraction for the seas, demonstrating that very high tides would occur at new and full moon when the gravitational pull of the sun and moon act together, and low tides at quarters when the pulls tend to neutralize each other. Finally, he investigated at some length the behavior of comets. Whereas Kepler had supposed that comets run through space along straight lines, Newton, by treating them as planets with highly eccentric orbits, deduced that comets must describe conic sections with the sun in one focus. This suggests that some comets, instead of being seen only once, travel on long elliptical paths, a suggestion Newton's friend Halley seized on.

The unpredictability of comets had always made them objects of wonder. Ancient chroniclers saw the fiery rush of comets as harbingers of evil—war, plague, earthquake—a natural view, because some calamitous event was bound to be taking place in some country where the comet was visible. A final blow to the superstitious fear of comets was given by Halley, who showed that their sudden appearances and disappearances were in accord with gravitational law. Using Newton's methods, Halley calculated the orbit of the Great Comet of 1682 and found it nearly identical with the orbits of a comet that had attracted Kepler's attention in 1607 and of a bright comet observed by Peter Apian in 1531. He rightly concluded that they were all appearances of a single comet describing a highly flattened elliptical orbit about the sun, within a period of some 75.5 years. This led Halley to make the daring prediction, which could not be fulfilled until after his death, that the comet would again be seen in 1759. Its arrival, exactly as forecast, was an independent confirmation of Newton's mathematical astronomy. ''Halley's Comet'' has made subsequent returns to our sky, at the same intervals, in 1835, 1910 and 1986, and its previous visits have been traced back to dates before the Christian era. One famous appearance was before the battle of Hastings in 1066, and it was interpreted by King Harold as indicating his defeat by William the Conqueror. In the Bayeux Tapestry, the comet appears on one of the corners with the inscription: ''Here they marvel at a star.''

The first edition of the *Principia* was very small, probably numbering no more than 300 copies, which were fairly soon dispersed by sale or gift. By 1691, it was impossible to get a copy of the work, and there was already talk of a new edition, possibly to be supervised by some younger mathematician. But not until 1709 was Newton able to enlist the services of someone. This was Roger Cotes (1682–1716), a brilliant Cambridge mathematician who was able to criticize and correct the *Principia* with confidence. Of this second edition, 750 copies appeared from the press in 1713. No fundamental work in the physical sciences met with such a demand as the *Principia* did. It is reported that one reader, unable to secure a hardbound copy for himself, transcribed the entire work longhand. This should not be interpreted as an indication of general enthusiasm, or even that there were many people in England capable of understanding the contents. Newton told a friend that ''to avoid being baited by little smatterers in mathematics,'' he purposely made the *Principia* abstruse. He was confident that competent mathematicians would comprehend his concise reasoning and accept his conclusions. Still, there was a wide public interest in the *Principia* that led to a flood of popularizations of it in the vernacular. Within the next century, there had appeared 40 of these in English, 17 in French, 11 in Latin, 3 in German, and 1 each in Italian and Portuguese. Eighteen editions of the *Principia* itself were published by 1789.

The immediate continental reception to Newton's principles of natural philosophy did not measure up to the admiration of the English. The scarcity of first-edition copies of the *Principia* naturally did not favor wide dissemination of Newton's ideas. Moreover, in France, where the Cartesian idea of vortices for explaining planetary motion was uncritically accepted, Newton's constitution of the universe was dismissed on the ground that it rested on assuming an ''action at a distance'' without intervention of material substance; that is, on some unintelligible, even occult, quantity of the sort natural science had so recently rejected. Even so able a man as Huygens regarded the idea of gravity as ''absurd,'' and Leibniz wrote, ''I do not see how he [Newton] conceived gravity or attraction.'' Newton would have dearly liked to give a mechanical explanation, but the remote force of gravity was a mystery to him also. In the famous *General Scholium* appended to the second edition of the *Principia* to refute criticism that he had introduced occult quantities, Newton was still reluctant to commit himself: ''I have not yet been able to deduce the reason of these properties of gravity from phenomena, and I do not invent hypotheses.'' (This was not meant as a condemnation of all hypotheses in science, but only of those of speculative character that could not be proved or disproved by mathematically treated experiment.) ''It is enough,'' he added with a certain arrogance, ''to say that gravity really does exist and act according to the laws which we have explained.''

PHILOSOPHIÆ

NATURALIS

PRINCIPIA

MATHEMATICA·

Autore *JS. NEWTON*, *Trin. Coll. Cantab. Soc.* Matheseos
Professore *Lucasiano*, & Societatis Regalis Sodali.

IMPRIMATUR·

S. PEPYS, *Reg. Soc.* PRÆSES.

Julii 5. 1686.

LONDINI,

Jussu *Societatis Regiæ* ac Typis *Josephi Streater.* Prostat apud
plures Bibliopolas. *Anno* MDCLXXXVII.

Title page of Newton's Principia Mathematica *(1687). (By courtesy Editions Culture et Civilisation.)*

Newton had shown that the motions of the planetary bodies were not capricious but subject to precise calculation. The phenomena of the universe were under "natural law," to be interpreted not by the Church, but by reason. These feelings were summed up in Alexander Pope's lines:

Nature and Nature's law lay hid at night,
God said, 'Let Newton be,' and all was light.

God had let others be also, but Newton reaped the fame, for the coming age of science was called the Age of Newton.

Whereas Newton's physical laws led some to view the universe as a self-sufficient mechanism, a gigantic piece of clockwork machinery, Newton himself insisted that the solar system was not a godless creation. This admirable ordering of the universe was precisely what confirmed in Newton his belief in a divine controller. The feeling that he was the man destined to unveil the ultimate truth about God's creation led him to try his hand at theology and biblical studies. In his later years, he wrote at length on prophecies and predictions of future events. Newton's main manuscript on religion, *Observations Upon the Prophecies of Daniel, and the Apocalypse of John,* was published (1733) after his death. What he called his mystic reveries probably consumed as much time and effort as the *Principia* did.

Although by Newton's death in 1727 Newtonian attraction was still regarded as obscure and inadmissible, there was a growing challenge to Cartesian attempts to explain the details of planetary motion. Not until the late 1730s, when Voltaire took up his worshipful defense of Newton in his *Letters Concerning the English Nation,* did Newton's gravitational view of the cosmic scheme begin to have advocates in France. Voltaire convinced his mistress Emilie Breteuil, Marquise du Chatelet, a competent scientist and mathematician, that she should prepare a French translation of the *Principia,* enriched by a commentary on the work. This account, which did not appear in print until 1759, remains the sole French translation of Newton's masterpiece. Half a century elapsed before opposition to Newton's views began to crumble and the leading mathematicians of Europe took up the task of perfecting the structure of the Newtonian synthesis.

When Newton published the *Principia,* he reached the zenith of his genius. Whatever else he might do would pale in comparison with what had been achieved in that period of unparalleled concentration in the middle of the 1680s. As it was, Newton lived another 40 years, but a continually increasing portion of his time was devoted to public business of one sort or another and almost none to scientific work. The severe mental exertion of composing the *Principia* took its toll. Newton began to suffer from insomnia and lack of appetite, and by 1692 his mental health had deteriorated to the point where he was afflicted with some sort of nervous illness.

There have been few cases of such a dramatic rechanneling of the energies of a scientist as befell Newton on his recovery. A royal appointment (1696) as warden, and subsequently master, of the British mint made it possible for Newton effectively to sever his academic ties with Cambridge and pass the remaining years of his life in London. The post at the mint was not a sinecure, although it had been the king's intention to provide one. At the time Newton became warden, the general debasement of the currency had become a national calamity. Counterfeiting and adulteration of coins were so prevalent that pieces of full value were rare. Newton supervised the minting of new coinage, using the knowledge obtained through his numerous experiments with alloys to establish new standards of purity and weight. An important part of his duties involved organizing a campaign against counterfeiters and clippers, setting up a network of informers to track them down, and interrogating the chief offenders. Counterfeiting was stopped by raising the relief designs so high that coins could be struck by only the mightiest of presses, whereas grooving the rims of coins prevented undetected clipping.

The news of Newton's mental crisis led some to suppose that his intellectual powers had vanished forever. The effort spent on the *Principia* may have overstrained

S E C T. III.

De motu Corporum in Conicis Sectionibus excentricis.

Prop. XI. Prob. VI.

Revolvatur corpus in Ellipsi: Requiritur lex vis centripetæ tendentis ad umbilicum Ellipseos.

Esto Ellipseos superioris umbilicus S. Agatur *S P* secans Ellipseos tum diametrum *D K* in *E*, tum ordinatim applicatam *Q v* in *x*, & compleatur parallelogrammum *Q x P R*. Patet *E P* æqualem esse semiaxi majori *A C*, co quod acta ab altero Ellipseos umbilico *H* linea *H I* ipsi *E C* parallela, (ob æquales *C S, C H*) æquentur *E S, E I*, adeo ut *E P* semisumma sit ipsarum *P S, P I*, id est (ob parallelas *H I, P R* & angulos æquales *I P R, H P Z*) ipsorum *P S, P H*, quæ

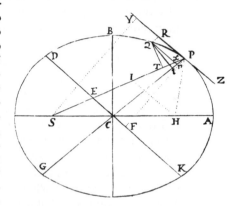

conjunctim axem totum 2 *A C* adæquant. Ad *S P* demittatur perpendicularis *Q T*, & Ellipseos latere recto principali (seu $\frac{2\,BC\ quad.}{AC}$) dicto *L*, erit *L x Q R* ad *L x P v* ut *Q R* ad *P v*; id est ut *P E* (seu *A C*) ad *P C*: & *L x P v* ad *G v P* ut *L* ad *G v*;

Extract from Newton's Principia Mathematica *(1687). (By courtesy Editions Culture et Civilisation.)*

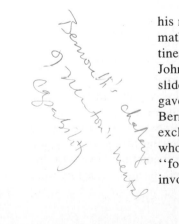

his mind and diminished his passion for scientific research, but his genius for solving mathematical problems with incredible speed was by no means impaired, as the Continental mathematicians learned when they dared to test him. In 1697, for example, John Bernoulli proposed the problem of finding the curve along which an object would slide from one point to another not directly beneath it in the shortest time. Newton gave the solution within a day, and when he permitted it to be published anonymously, Bernoulli recognized the author from the sheer power and originality of the work, exclaiming, ''The lion is known by its paw.'' On another occasion (1716) Leibniz, who was stung by the accusation of plagiarism from friends of Newton, sent a problem ''for the purpose of feeling the pulse of the English analysts.'' The challenge, which involved finding the orthogonal trajectories of the family of hyperbolas having the same vertices, was presumably meant to be a severe test for the most experienced mathematicians. It is reported that Newton, aged 64, received the problem around five o'clock, after returning home weary with the day's business at the mint, and solved it before going to bed.

During the latter years of his life, Newton began to take more interest in the Royal Society. Without opposition, he was elected its president in 1703, an office to which he was reelected annually until the close of his life more than 24 years later. Newton became something of a national figure, one of the principal sights of London for all visiting foreign intellectuals. Queen Anne knighted him, a farmer's son—the first scientist so honored—in a ceremony at Cambridge in 1705.

Newton was well past 80 before he began to suffer seriously from the complaints of old age. Taken ill while presiding at a meeting of the Royal Society, he died less than three weeks later, on March 20, 1727, in his eighty-fifth year. Newton's passing seemed to arouse in the British people a national consciousness that so great a scientist had been one of their countrymen. His body was interred in Westminster Abbey, where the greatest of England's departed lay. Dukes and earls, all fellows of the Royal Society, carried his coffin. It was a great occasion. Voltaire, who attended the funeral, was moved to tell, ''I have seen a professor of mathematics, only because he was great in his vocation, buried like a king who had done good to his subjects.'' Inscribed in Latin on Newton's magnificent monument are the words: ''Let mortals rejoice that there has existed such and so great an ornament of the human race.''

The exalted esteem in which Newton's genius was held reached almost absurd heights in later times. When the queen of Prussia asked Leibniz what he thought of Newton, he replied: ''Taking mathematicians from the beginning of the world to the time when Newton lived, what he had done was much the better half.'' Competent critics agreed that the *Principia,* on which the fame of Newton chiefly rests, surpassed all other scientific works in power and originality. A century later, Lagrange called the *Principia* the greatest production of the human mind, and his contemporary Laplace felt that it was assured a preeminence above all other productions of human genius. With more than a touch of envy, Lagrange remarked that because there was only one solar system it could be granted to only one man to discover its fundamental laws: ''Newton was the greatest genius that ever existed, and the most fortunate, for we cannot find more than once a system of the world to establish.''

Beside these magnificent eulogies, it is pleasant to consider Newton's evaluation of his own work. Shortly before his death, he told some friends:

> I do not know what I may appear to the world; but to myself I seem to have been only like a boy playing on the seashore, and diverting myself in now and then finding a smoother pebble or prettier shell than ordinary, while the great ocean of truth lay all undiscovered before me.

Newton was aware that he had traveled along a broad highway prepared for him by others. His well-known statement, ''If I have seen farther than others, it is because I have stood on the shoulders of giants,'' shows his appreciation of the cumulative and progressive growth of science. Yet it took the genius of Newton to find the key and show, once and for all, just how Nature is regulated by mathematical law. With the *Principia,* all the pieces of the puzzle suddenly fell into place, all obscurities were made plain.

8.4   Gottfried Leibniz: The Calculus Controversy

The invention of the calculus was one of the great intellectual achievements of the 1600s. By one of those curious coincidences of mathematical history not one, but two men devised the idea—and almost simultaneously. The methods of the calculus of Newton in England and Leibniz on the Continent were so similar that the question whether Leibniz borrowed the crucial concepts from Newton or discovered them independently gave rise to a long and bitter controversy. The tactics of the principal protagonists were so unworthy of these two titans, and the violence of the accusations and counteraccusations so injurious, that neither escaped with his reputation untarnished. When inferences of plagiarism became public charges, a committee of the Royal Society, called to adjudicate this most notorious of scientific disputes, found—not surprisingly—in favor of the society's own president against one of its oldest foreign members.

Gottfried Wilhelm Leibniz (1646–1716) was born in the university town of Leipzig some two years before the Peace of Westphalia put an end to the Thirty Years' War. His father, a jurist and professor of moral philosophy at the university, died when the boy was 6 years old. As a result, the young Leibniz was left almost without direction in his studies. The boy's world was the world of books. A precocious child, he taught himself Latin from an illustrated copy of Livy's history of Rome when he was about eight, and had begun the study of Greek by the time he was twelve. This led to his being given unhampered access to his father's library, which had previously been kept under lock and key. Here, according to his own testimony, he became acquainted with a wide range of classical writers. Leibniz wrote in later life: ''I began to think when I was very young; and before I was fifteen I used to go for long walks by myself in the woods, comparing and contrasting the principles of Aristotle with those of Democritus.''

In the fall of 1661, the same date that Newton entered Cambridge, Leibniz became a student at the university of his native city, Leipzig. Only 15 at the time, he was regarded as something of a prodigy and soon outstripped all his contemporaries. The education Leibniz received at Leipzig followed traditional, conservative lines, with its emphasis on religion (orthodox Lutheran doctrine) and philosophy. Arithmetic was taught at an elementary level, prescribed by the textbooks of the German Jesuit Christoph Clavius (1537–1612). And although lectures were given on Euclid's *Elements,* Leibniz did not pay them sufficient attention. Van Schooten's edition of Descartes's *Géométrie,* which he tried to read on his own, seemed much too complicated for him. Hardly 17 years old, Leibniz graduated from Leipzig in 1663 after defending a thesis on a point of philosophy. He passed the summer term at the University of Jena, where he attended mathematics lectures, then returned to Leipzig to concentrate on legal studies; he earned his master's degree the following year. Leibniz was given a teaching position in the philosophical faculty at Leipzig, for which he qualified by writing *Disputatio Arithmetica de Complexionibus;* this work, which was expanded into *Ars Combinatoria* (1666), extensively develops the theory of permutations and combinations for the purpose of making logical deductions. Leibniz was ignorant of the mathematical literature, so the *Ars Combinatoria* contained little that was new. Leibniz later called it the work of a young man just out of school. Nonetheless, it is a work of great

Gottfried Wilhelm
Leibniz
(1646–1716)

(Smithsonian Institution.)

interest in that it discusses the establishment of a new mathematics-like language of reasoning (*characteristica universalis*) in which all scientific concepts could be formed by combinations from a ''basic alphabet of human thoughts.'' Moreover, Leibniz suggested that a calculus of reasoning could be devised that would provide an automatic method of solution for all problems that could be expressed in his scientific language.

In 1666, Leibniz applied for the degree of doctor of law but was refused by the Leipzig faculty on the threadbare grounds that he was too young; a more likely explanation is that they were jealous of his ability. Disgusted, Leibniz quit his native city, never to return except in passing, in order to enroll at the University of Altdorf (Nuremberg). The following year, he received a doctorate from Altdorf, probably using a thesis he had already completed in Leipzig. His dissertation made such a favorable impression that he was offered a professorship on the strength of it; but Leibniz declined, having, as he said, ''very different things in view.'' He chose instead to be near the center of political power. An essay that he wrote on the study of law gained him a post (1667) in the service of the archbishop-elector of Mainz, where he was charged with reforming the current statutes. Except for a four-year sojourn in Paris, the rest of Leibniz's life was spent in residence at the courts of Mainz and Hanover—at Mainz until 1672 and at Hanover from 1676 until his death in 1716.

Leibniz's first problem was to acquaint himself with the entangled legal and political position of the German states. Throughout the seventeenth century, Europe had been involved in a political revolution no less significant than the scientific one. While England was undergoing the Puritan Revolution and Civil War, the Continent was shattered by the terrible Thirty Years' War. What began in 1618 as a rebellion by a group of Bohemian leaders against the House of Hapsburg was followed by a bitter war of reprisal that spread rapidly to involve all of Germany, then at various times Scandinavia, the Netherlands, France, Spain, Hungary, Italy, Poland, Russia, and

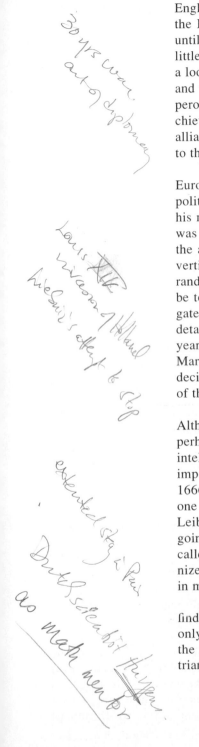

England. For 30 consecutive years Germany became the battlefield of Europe, until the Peace of Westphalia (1648) finally ended the Continent's most destructive war until the twentieth century. From 1648 onward, the medieval Holy Roman Empire was little more than a polite anachronism. There remained only the titles and trappings and a loosely knit union of 360 states (or nearly 2000 enclaves if all the imperial knights and their estates were counted) nominally held together by a mere shadow of an emperor in Vienna. The Peace of Westphalia granted the various German states the mischievous privilege of maintaining their own armies and entering into independent alliances for their own welfare and protection. It was this bargaining power that gave to the art of diplomacy its current importance.

With the disintegration of the empire, France, now unified, became the center of European power and the symbol of its culture. French now became the language of polite society and of diplomacy everywhere. Louis XIV, the Sun King, who came into his royal inheritance in 1661, was the true ''arbiter of Europe.'' At the time Leibniz was appointed legal counsel at Mainz, both Germany and Holland were threatened by the aggressive policy of France. The young lawyer conceived (1670) the idea of diverting Louis's attention from Germany by proposing to him—in a famous memorandum, *Consilium Aegyptiacum*—that a worthier objective for a Christian king would be to seize Egypt from the Turks, who were then massing their forces at the eastern gate of Europe. The elector agreed to send the author himself to Paris to explain the details of the plan. (This bears a curious resemblance to a plan Napoleon devised 128 years later, and it is sometimes supposed that he read Leibniz's memorandum.) In March 1672, Leibniz left Mainz for Paris but arrived there too late. Louis had already decided to invade the ''nation of fishwives and tradesmen,'' as he called Holland. One of the king's ministers observed that Crusades had gone out of style.

Leibniz's journey to Paris turned into a long stay, lasting from 1672 to 1676. Although the political ends for which it was undertaken were not realized, it was perhaps the most important event in the future mathematician's life. Paris was the intellectual capital of Europe, and Leibniz came into contact with many scholars. Most important was the great Dutch scientist Christiaan Huygens, who lived in Paris from 1666 until 1681. When Newton went to Cambridge, he studied under Isaac Barrow, one of the foremost mathematicians of the period; but the mathematical instruction Leibniz had received at Leipzig was far from adequate. Leibniz himself said that before going to Paris he was ''not even a novice in mathematics.'' So far as Leibniz can be called the mathematics pupil of anyone, he was a pupil of Huygens. Huygens recognized the brilliance of the studious young German, and realizing that he was lacking in mathematical training, undertook to guide his first investigations.

During one of his early meetings with Huygens, Leibniz asserted that he could find the sum of any infinite series whose terms were formed by some rule (provided only that the series should converge). Huygens, wishing to make an immediate test of the young man, suggested that he try to determine the sum of the reciprocals of the triangular numbers:

$$\frac{1}{1} + \frac{1}{3} + \frac{1}{6} + \frac{1}{10} + \frac{1}{15} + \cdots + \frac{2}{n(n+1)} + \cdots.$$

By the clever device of writing each term as the sum of two others, using

$$\frac{2}{n(n+1)} = 2\left(\frac{1}{n} - \frac{1}{n+1}\right),$$

Leibniz was able to obtain the sum demanded by Huygens. For

$$\frac{1}{1} + \frac{1}{3} + \frac{1}{6} + \frac{1}{10} + \frac{1}{15} + \cdots$$

$$= \frac{2}{1 \cdot 2} + \frac{2}{2 \cdot 3} + \frac{2}{3 \cdot 4} + \frac{2}{4 \cdot 5} + \cdots$$

$$= 2\left(1 - \frac{1}{2}\right) + 2\left(\frac{1}{2} - \frac{1}{3}\right) + 2\left(\frac{1}{3} - \frac{1}{4}\right) + \cdots$$

$$= 2.$$

Also at this time, Leibniz constructed a working model of a new calculating machine that was an improvement on the machine Pascal had already invented. Whereas Pascal's device performed only addition and subtraction, Leibniz's machine accomplished multiplication and division by repeated addition and subtraction. "It is unworthy of excellent men," wrote Leibniz, "to lose hours like slaves in the labor of calculation. . . ."

It was impossible for Leibniz to find much time for mathematical studies just then. In 1673, he crossed the channel to England on a diplomatic mission for the elector of Mainz to encourage peace negotiations between France and Holland. While in London, he made the acquaintance of his fellow countryman Henry Oldenburg, the permanent secretary of the Royal Society, who introduced him to various mathematicians and scientists connected with the society—Pell, Collins, Boyle, and Hooke. Leibniz was unanimously elected a member of that illustrious body, owing in part to an exhibition of his calculating machine at a meeting of the society and in part to the friendly offices of Oldenburg.

The diplomatic mission from Mainz met little success; and after two months' stay in London, Leibniz returned to Paris, where he was able to find time to pursue his studies without hindrance. Huygens was in the midst of publishing his great work, *Horologium Oscillatorium,* which dealt with numerous mechanical problems arising out of pendular motion. He gave his young friend a copy as a present, but Leibniz's mathematical knowledge was insufficient to enable him to understand the contents of the book. Looking on it as something of a disgrace to be ignorant of such matters, Leibniz proceeded with the greatest fervor to work his way through the standard mathematical works of the period. He devoted himself to van Schooten's two-volume edition of Descartes's *Géométrie* (which heretofore had been beyond him), thus, as he said, "entering the house of geometry truly as it were by the back door." Guided by the friendly advice of Huygens, he also made himself acquainted with the manuscripts of Pascal. On several occasions Leibniz was to declare that he was led to the invention of calculus more by studying Pascal's writings than anything else. These few years in Paris were the most intensely creative period of Leibniz's life, during which time he grew from a beginner to a mature mathematician.

As the first fruits of these studies, Leibniz was able to obtain the celebrated alternating series that bears his name,

$$\frac{\Rightarrow}{4} = 1 - \frac{1}{3} + \frac{1}{5} - \frac{1}{7} + \frac{1}{9} - \frac{1}{11} + \neq\neq\neq.$$

Leibniz seems to have found this series in 1673, but the formula was known to the young Scottish mathematician James Gregory (1638–1675) in 1671. It is an elegant formula for $\Rightarrow$ but the series converges too slowly for computational purposes. Newton pointed this out when in a neat counterblast, he sent Leibniz (through Oldenburg) the variant expression

$$\frac{\Rightarrow}{2\sqrt{2}} = 1 + \frac{1}{3} - \frac{1}{5} - \frac{1}{7} + \frac{1}{9} + \frac{1}{11} - \frac{1}{13} - \frac{1}{15} + \neq\neq\neq.$$

In the midst of the London negotiations, Leibniz's patron, the elector of Mainz, died unexpectedly. Leibniz, hoping to gain a permanent foothold in Paris, sought a seat in the Académie des Sciences with a royal pension, but all his efforts failed (not until 1700 was Leibniz admitted to the Académie, in the same year as Newton). With reluctance, Leibniz looked around for a new political office. He finally settled on the position of counsel and librarian to the Duke of Hanover, a post he was to fill for the remainder of his life.

In the years from 1672 to 1676, spent in Paris, Leibniz's slowly flowering mathematical genius matured. (This concentrated period of creativity is reminiscent of Newton's ''golden years'' 1664–1666 at Woolsthorpe.) During this time, he developed the principal features and notation of his version of the calculus. Various methods had been invented for determining the tangent lines to certain classes of curves, but as yet nobody had made known similar procedures for solving the inverse problem, that is, deriving the equation of the curve itself from the properties of its tangents. Leibniz stated the inverse tangent problem thus: ''To find the locus of the function, provided the locus which determines the subtangent is known.'' By the middle of 1673, he had settled down to an exploration of this problem, fully recognizing that ''almost the whole of the theory of the inverse method of tangents is reducible to quadratures [integrations].''

Because Leibniz was still struggling with the notation for his calculus, it is not surprising that these early calculations were clumsy. Either he expressed his results in rhetorical form or else used abbreviations, such as ''omn.'' for the Latin *omnia* (''all'') to mean ''sum.'' The letter *l* was used to symbolize what we should write as *dy,* the ''difference'' of two neighboring ordinates. In a notable manuscript dated October 29, 1675, but never published, Leibniz made his symbolic connection of the direct and inverse tangent problems. In this essay Leibniz wrote of a theorem he had obtained by a geometrical argument:

We have a theorem that to me seems admirable, and one that will be of great service to this new calculus, namely,

$$\frac{\overline{\text{omn. } l^2}}{2} = \text{omn. } \overline{\text{omn.} l\frac{l}{a}}, \text{ whatever } l \text{ may be.}$$

The horizontal overbars were used in place of our parentheses, and the constant a (which would be dx) was taken equal to 1. Leibniz remarked, "This is a very fine theorem, and one that is not at all obvious." Immediately afterward, he stated another theorem of the same kind,

$$\text{omn. } xl = x \text{ omn. } l - \text{omn. omn. } l.$$

These equations are historically important, because it was here that Leibniz first introduced the symbol $\int$, an elongated form of the letter S for "sum." In the middle of the paper, he said:

It will be useful to write $\int$ for omn., so that $\int l = $ omn. l, or the sum of the l's. Thus,

$$\frac{\int \overline{l^2}}{2} = \int \int \overline{l \frac{l}{a}} \quad \text{and} \quad \int \overline{xl} = x \int \overline{l} - \int \int l.$$

Thus, in our notation of the calculus, he has shown that

$$\frac{1}{2}\left(\int dy\right)^2 = \int\left(\int dy\right)dy \quad \text{and} \quad \int x \, dy = xy - \int y \, dx.$$

Leibniz was not yet using the differential under the integral sign; there exists another manuscript, written several weeks thereafter (November 21), in which he improved his notation by writing $\int f(x)dx$, the direct ancestor of the modern form.

Later on, in the same manuscript of October 29, Leibniz explored the contrary calculus, perceiving the dual nature of the integration and differentiation processes:

Given l, and its relation to x, to find $\int l$. This is to be obtained from the contrary calculus, that is to say, suppose that $\int l = ya$. Let $l = ya/d$; then just as $\int$ will increase, so d will diminish dimensions. But $\int$ means a sum, and d a difference. From the given y, we can always find y/d or l, that is, the difference of the y's. Hence one equation can be transformed into another.

The symbol d was at first placed by Leibniz in the denominator, probably by analogy with the division process; in a paper written on November 1, three days later, he replaced y/d by the familiar dy, which then seemed more appropriate to him and which he kept in all his future work.

Leibniz next investigated the basic algorithms of calculus, especially the question whether $d(xy)$ was equal to $dx \, dy$ and whether $d(x/y)$ was equal to dx/dy. In the manuscript of November 11, he concluded that the expressions were not the same, though he could not give the true value of each. Ten days later, he correctly determined the product rule, and in July 1677 gave a statement of the quotient rule. In the case of the product, Leibniz subtracted xy from $(x + dx)(y + dy)$ and discarded the term $dx \, dy$ with the remark that "the omission of the quantity $dx \, dy$, which is infinitely small in comparison with the rest, for it is supposed that dx and dy are infinitely small, will leave $x \, dy + y \, dx$." All this is, of course, without sound justification. To find the differential of the quotient $z = x/y$, Leibniz set $x = zy$ and used the product rule: $dx = z \, dy + y \, dz$ leads to

$$dz = \frac{dx - z \, dy}{y} = \frac{dx - (x/y)dy}{y} = \frac{y \, dx - x \, dy}{y^2}.$$

By November 1676, he was also able to state the rule $d(x^n) = nx^{n-1} dx$ for integral and fractional values of n.

Leibniz's investigation into the calculus was based on what he called the "characteristic triangle." Isaac Barrow had used the characteristic triangle in England, but according to Leibniz's own account, the inspiration for its use came from reading the work of Pascal. Referring to himself in the third person, Leibniz wrote that on running across a figure in Pascal "a light suddenly burst upon him." For a curve $y = f(x)$, the characteristic triangle is the right triangle whose sides consist of $PQ(= dx)$, $QR(= dy)$, and PR, part of the tangent to the curve at a typical point P. Leibniz noted

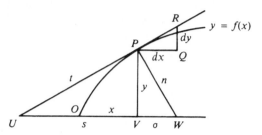

that the characteristic triangle was similar to the triangle PVW formed by the normal n, the subnormal σ, and the ordinate y at the point of contact, and also similar to the triangle PVU formed by the tangent t, the subtangent s, and the ordinate y. From the similarity of the characteristic triangle and the triangle PVW, he got

$$\frac{dy}{\sigma} = \frac{dx}{y} \qquad \text{or} \qquad \sigma\, dx = y\, dy.$$

Regarding dx and dy as infinitely small and summing up, Leibniz came to the result

$$\int \sigma\, dx = \int y\, dy.$$

To solve a definite problem (November 11), he supposed the subnormal to be inversely proportional to the ordinate—that is, $\sigma = a^2/y$—and found that $y^3/3 = a^2x$; and hence the curve with the given property was the cubic parabola.

For another application, Leibniz used the fact that when the characteristic triangle was very small the chord PR could be considered of the same length as the length ds of the curve. Because the characteristic triangle was similar to triangle PVW,

$$\frac{n}{ds} = \frac{y}{dx} \qquad \text{or} \qquad y\, ds = n\, dx.$$

Summation then gave Leibniz

$$\int y\, ds = \int n\, dx,$$

a formula for the surface of revolution obtained by rotating the original curve about the x-axis.

In this memorable series of manuscripts, the symbols dx (first as x/d) and $\int f(x)dx$ of Leibniz's calculus came into being. There were few if any new discoveries here—Newton's disparaging judgment was that "not a single previously unsolved problem was solved''—but a formalism was developed that helped systematize and generalize the diverse geometric results of old. His newly contrived symbolism freed calculus of its bondage to geometry and allowed Leibniz to achieve many results virtually without effort.

Leibniz gradually elaborated his differential-integral calculus but never actually founded it on the limit concept; his differential ratio dy/dx was always thought of as a quotient of ''differences'' and his integral simply as a sum. The first prominent mathematician to suggest that the theory of limits was fundamental in calculus was Jean d'Alembert (1717–1783). D'Alembert wrote most of the mathematical articles in that cardinal document of the Enlightenment, the *Encyclopédie* (28 volumes, 1751–1772), and in an article entitled ''Différentiel'' (volume 4, 1754) said ''the differentiation of equations consists simply in finding the limits of the ratio of finite differences of two variables in the equation.'' In other words, he came to the expression of the derivative as the limit of a quotient of increments, or as we should write it,

$$\frac{dy}{dx} = \lim_{\Delta x \to 0} \frac{\Delta y}{\Delta x} .$$

Unfortunately, d'Alembert's elaboration of the limit concept itself lacked precision. Therefore, a conscientious mathematician of the 1700s would have been no more satisfied with this definition than with currently available interpretations of the derivative.

Before Leibniz left Paris in the autumn of 1676, he found himself in possession of the rules and notation of his calculus. He suspected, but could not be sure, that Newton had developed an equivalent approach, one far more geometrically slanted. This was precisely the case. That development had taken place as early as 1665–1666, when Newton was in his twenties, during the same period in which he had discovered the binomial theorem. The bulk of Newton's early work on calculus was condensed into a small treatise of some 30 crowded pages, covering such things as tangency, curvature, centers of gravity, and area. The work, which Newton seems never to have given a definite title, is known in the learned literature as the October 1666 Tract. By seeking a pattern from tabulated values of $\int_0^x (1 + t)^{-1} \, dt$, he was able to show at that time that the area of the rectangular hyperbola $(x + 1)y = 1$ was

$$z = x - \frac{x^2}{2} + \frac{x^3}{3} - \frac{x^4}{4} + \frac{x^5}{5} - \cdots ,$$

which is the series expansion for the natural logarithm of $1 + x$. With boyish enthusiasm, Newton demonstrated his newly found numerical facility by calculating this expression, for particular values of x, to impractically large numbers of decimal places (as many as 68 decimal places, using the series through terms involving x^{25}).

In mid-1669, Newton came across Nicholas Mercator's *Logarithmotechnia* (1668). The first two parts of that book were devoted to a table of common logarithms; the third part contained various approximation formulas for logarithms, one of which was Newton's own reduction of log $(1 + x)$ to an infinite series,

$$\log(1 + x) = x - \frac{x^2}{2} + \frac{x^3}{3} - \frac{x^4}{4} + \frac{x^5}{5} - \cdots .$$

Feeling crestfallen that Mercator had anticipated him in publication, Newton, spurred by a desire to protect his priority in individual topics, hurriedly set to work to write up his earlier research in series expansions. The resulting compendium was submitted to Isaac Barrow, then Lucasian professor of mathematics, for his approval. By the summer of that year, Barrow confided enthusiastically to the mathematician John Collins, "A friend of mine here that hath a very excellent genius to those things, brought me the other day some papers, wherein he set down methods of calculating the dimension of magnitudes like that of Mr. Mercator concerning the hyperbola, but very general. . . ." These "papers" turned out to be the short tract *De Analysi per Aequationes Numero Terminorum Infinitas* (On the Analysis by Equations Unlimited in the Number of Their Terms). The receipt of this work is supposed to have persuaded Barrow to recommend Newton, "an unparalleled genius," as his successor in the Lucasian professorship.

The *De Analysi* opens with a rule, stated without proof, for computing the area under the curve $y = ax^{m/n}$:

> To the base *AB* of some curve *AD*, let the ordinate *BD* be perpendicular and let *AB* be called *x* and *BD*, *y*. Let again *a, b, c, . . .* be given quantities and *m, n* integers. Then

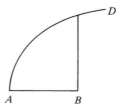

Rule 1. If $ax^{m/n} = y$, then will $\dfrac{na}{m + n} x^{(m + n)/n} = $ area *ABD*

Because the *x*-coordinate of *A* is zero, Newton correctly evaluated the integral $\int_0^x at^{m/n}dt$. Later in the *De Analysi,* he elaborated the demonstration of Rule 1. Newton assumed that he had a curve and that the area under the curve was given by

$$(1) \qquad z = \left(\frac{n}{m + n}\right)ax^{(m + n)/n}.$$

If *o* is an infinitesimal increase in *x*, then the new abscissa would be $x + o$, and the increase in area (the area bounded by the curve, the *x*-axis, and the ordinate at $x + o$) would be

$$(2) \qquad z + oy = \left(\frac{n}{m + n}\right)a(x + o)^{(m + n)/n},$$

where *oy* is the increment by which the area increases. Newton then applied the binomial expansion to the right-hand side of this equation, subtracted (1) from (2), divided through by *o*, and discarded terms still containing *o*, to arrive at the result

$$y = ax^{m/n}.$$

He said that conversely, if the curve were $y = ax^{m/n}$, the area under it would be

$$z = \left(\frac{n}{m + n}\right) ax^{(m + n)/n}.$$

Although Newton seemed to grasp the relation between differentiation and integration, he argued obscurely. There was no attempt to explain the logic by which terms involving powers of o could be neglected in the calculation.

After much persuasion, Barrow convinced his ''pupil'' that he should let Collins see the *De Analysi*. Collins quickly circulated copies among eminent mathematicians of his acquaintance, retaining a transcript of it for himself, before returning the original to Newton. It was Collins's desire to offer the tract to the presses as soon as possible, but Newton had already conceived the bolder scheme of expanding *De Analysi* into a comprehensive account of his fluxional—that is, calculus—methods. By 1671, the project had grown into the *De Methodis Serierum et Fluxionum* (On the Methods of Series and Fluxions). Between 1671 and 1676, Newton tried to arrange for the printing of this important work, either as a treatise complete in itself or as an appendix to one of his Lucasian lectures; but all the attempts failed, because booksellers found little market for severely mathematical publications. It first appeared 65 years later in 1736 in John Colson's English translation, *The Method of Fluxions and Infinite Series;* it was retranslated (1744) into Latin under the title *Methodus Fluxionum et Serierum Infinitarum*. Meanwhile, the substance of the manuscript was revealed privately to interested parties, and updated into the *De Quadratura Curvarum* (prepared in the period 1691–1693), which formed an appendix to the 1704 edition of Newton's *Opticks*.

In the *De Methodis Fluxionum*, Newton abandoned his use of infinitesimals in favor of fluxions. Newton conceived of mathematical quantities as generated by a continuous motion analogous to that of a point tracing out a curve. Each of these flowing quantities (variables) was called a ''fluent,'' and its rate of generation was known as the ''fluxion of the fluent'' and designated by a letter with a dot over it. Thus, if the fluent was represented by x, Newton denoted its fluxion by $\dot{x}$; and denoted the fluxion of $\dot{x}$, the second fluxion, by $\ddot{x}$, and so on. (In modern language, the fluxion of the variable x relative to an independent time-variable t would be its velocity dx/dt.) The infinitely small part by which a fluent was increased in a small interval of time, designated o, was called the moment of the fluent, and was denoted in fluxional notation by $\dot{x}o$. Newton did not actually introduce the familiar dot for fluxions until 1691. Earlier he used the literal symbols p, q, r, for the fluxions of x, y, z.

In the *De Methodis Fluxionum*, Newton stated the fundamental task of the calculus: ''The relation of the fluents being given, to find the relation of their fluxions [and conversely].'' He then illustrated this by several examples. The following extract shows the similarity between Newton's approach and the modern method of differentiating a function:

Thus let any equation $x^3 - ax^2 + axy - y^3 = 0$ be given and substitute $x + \dot{x}o$ for x, $y + \dot{y}o$ for y, and there will arise

$$x^3 + 3x^2\dot{x}o + 3x\dot{x}o\dot{x}o + \dot{x}^3o^3 - ax^2 - 2ax\dot{x}o - a\dot{x}o\dot{x}o + axy + ax\dot{y}o$$
$$+ ay\dot{x}o + a\dot{x}o\dot{y}o - y^3 - 3y^2\dot{y}o - 3y\dot{y}o\dot{y}o - \dot{y}^3 = 0.$$

OF

ANALYSIS

BY

Equations of an infinite Number of Terms.

1. *THE General Method, which I had devised some considerable Time ago, for measuring the Quantity of Curves, by Means of Series, infinite in the Number of Terms, is rather shortly explained, than accurately demonstrated in what follows.*

2. Let the Base AB of any Curve AD have BD for it's perpendicular Ordinate; and call AB=*x*, BD=*y*, and let *a*, *b*, *c*, &c. be given Quantities, and *m* and *n* whole Numbers. Then

The Quadrature of Simple Curves,

R U L E I.

3. If $ax^{\frac{m}{n}} = y$; it shall be $\dfrac{an}{m+n}x^{\frac{m+n}{n}} =$ Area ABD.

The thing will be evident by an Example.

1. If $x^2 (= 1x^2) = y$, that is $a = 1 = n$, and $m = 2$; it shall be $\frac{1}{3}x^3 =$ ABD.

T t   2. Suppose

First page of the English translation (1745) of Newton's De Analysi. *(Printed by permission of the Johnson Reprint Corporation.)*

Now, by supposition, $x^3 - ax^2 + axy - y^3 = 0$, which therefore, being expunged and the remaining terms divided by o, there will remain

$$3x^2\dot{x} - 2ax\dot{x} + ay\dot{x} + ax\dot{y} - 3y^2\dot{y} + 3x\dot{x}\dot{x}o - a\dot{x}\dot{x}o + a\dot{x}\dot{y}o - 3y\dot{y}\dot{y}o + \dot{x}^3oo - \dot{y}^3oo = 0.$$

But whereas o is supposed to be infinitely little, that it may represent the moments of quantities, the terms that are multiplied by it will be nothing in respect to the rest; I therefore reject them, and there remains

$$3x^2\dot{x} - 2ax\dot{x} + ay\dot{x} + ax\dot{y} - 3y^2\dot{y} = 0.$$

From this, it is clear that the method of fluxions was not essentially different from that used in the *De Analysi,* nor were the obscurities of the earlier work removed. Newton still dropped all terms containing o once he allowed himself to divide through by o—yet o could not be treated as zero, because that would render the division illegitimate. The real advance was in treating moments $\dot{x}o$ and $\dot{y}o$ as varying with time, where earlier the moments o were fixed bits of x and y.

The difficulties in understanding Newton's creation were due in part to the change of approach in each of his three works on the calculus. Infinitesimals were emphasized in the *De Analysi* but abandoned in favor of a theory of fluxions in *De Methodis Fluxionum;* and still later, that theory came to rest on prime and ultimate ratios in the *De Quadratura Curvarum.* The *De Quadratura Curvarum,* the last-written of Newton's trio but the first published, was the climax of his effort to establish the calculus on sound foundations. In earlier works, Newton neglected terms involving the quantity *o* with the doubtful justification that they were infinitely small compared with the other terms. Yet as long as *o* was a quantity, however small, it could not be rejected without affecting the result. Newton seems to have been aware of this, for in *De Quadratura Curvarum,* he declared that ''in mathematics the minutest errors are not to be neglected.'' Thus, he removed all traces of infinitely small quantities (''in the method of fluxions there is no necessity of introducing figures infinitely small''), replacing them with a new doctrine of ultimate ratios. According to Newton: ''By the ultimate ratio of evanescent quantities is to be understood the ratio of quantities, not before they vanish, nor afterwards, but with which they vanish.'' Although he may have been flirting with the limit concept, Newton's definition was far from clear; even among his ablest admirers the word *vanish* caused much confusion and severe criticism.

After Leibniz had returned to Paris in 1673, he maintained an active correspondence with Oldenburg, through which he was kept informed of the latest work of the English mathematicians. Oldenburg himself had never done any serious mathematics; thus, he was advised on all mathematical questions by John Collins (1625–1683), a self-educated man who developed a wide correspondence with the leading scholars of his time. In a letter (July 1676) from Oldenburg to Leibniz, based on a draft by Collins, mention was made that Newton had already developed a rule for finding tangents to algebraic curves. Probably as a result of this communication, Leibniz was moved, in the middle of October 1676, to make his return trip to Hanover by way of London and Amsterdam. During his week's stay in London, he was permitted by Collins to examine a copy of Newton's tract *De Analysi,* from which he made copious extracts, and also various letters of mathematicians that Collins had collected. There is much difference of opinion among the partisans of Newton and Leibniz about whether the fluxional method, briefly exposed in the *De Analysi,* caught Leibniz's eye or whether his interest was limited to the solution of problems by infinite series. But in transcribing portions of manuscripts confided to Collins, he opened himself to the suspicion that he had illicitly appropriated Newton's findings.

Throughout the year, Leibniz had become keenly interested in the mathematical writings of Newton and had asked Oldenburg for further information about them. Under the united persuasions of Oldenburg and Collins, Newton wrote Oldenburg two letters giving the status of his current research, with the request that Latin versions be transmitted to Leibniz. These celebrated letters, known now as the *Epistola Prior* of June 13, 1676, and the *Epistola Posterior* of October 24, 1676, established Newton's priority for numerous results in his mathematical hoard. The first letter contained the binomial theorem, as well as a summary of Newton's more important work on series. Although nothing was said of his secret discovery of fluxions, there was a vague hint that Newton had an important method in mind. Leibniz must have guessed that Newton's methods were essentially similar to his own, for he wrote back to Oldenburg asking that Newton amplify the more decisive points in the letter.

Newton could not refuse Oldenburg's request to give Leibniz the information he wanted but was still disinclined to divulge his secret. The response to Leibniz's pleas was a letter of 15 closely written pages sent through Oldenburg, a veritable treatise on the construction and application of infinite series. Having reached a point in the letter where an explanation of the method of fluxions would logically follow, he carefully concealed the basic principle in a Latin anagram:

$$6accd\mathit{æ}13\mathit{eff}7i3l9n4o4qrr4s8t12vx.$$

Anyone who could unscramble this jumble of six a's, two c's, one d, and so on, into a sentence might arrive at "Data æquatione quotcunque fluentes quantitates involvent fluxiones invenire et vice versa." Freely translated, this means "Given an equation involving any number of fluent quantities, to find the fluxions, and conversely." By this anagram Newton, without disclosing his method, could later establish a claim to priority in the invention of the calculus.

Anagrams were frequently used by scientists in Newton's time, and indeed long before that, as a means of establishing priority of discovery without revealing what had been found. The writer thereby had time to publish at his leisure. A notable instance occurred in 1610, when Galileo adopted the device to communicate to Kepler the news that the planet Saturn appeared to have close satellites, one on each side. (Actually, his telescope was too weak, and a better one would have revealed that the "handles," as Galileo called them, were Saturn's rings.) The triple nature of Saturn was announced by a jumble of letters that read, when properly disentangled, "I have observed the most distant planet is triple." The mystery of Saturn was solved, nearly a half-century after Galileo's first observation, by Christiaan Huygens. Recognizing that the planet's peculiarity was due to its encirclement by a thin plane ring, Huygens published (1656) his discovery in the cryptic form $a^7c^5d^1e^5g^1h^1i^7l^4m^2n^9o^4p^2q^1r^2s^1t^5u^5$, meaning in English, "It is surrounded by a thin flat ring, nowhere touching, and inclined to the elliptic."

The *Epistola Posterior* later acquired great importance when it became part of the documentary foundation for the Royal Society's report on the calculus priority dispute. It is now generally agreed that Newton's anagram of transposed letters could not have given Leibniz much of a clue to the invention of the calculus; even the enunciation contained therein is so brief and obscure as to be practically useless. "If Leibniz," wrote the mathematician Augustus De Morgan, "could have taken a hint, either from the preceding letters in alphabetical order [in the anagram], or, had he known it, in their significant arrangement, he would have derived as much credit as if he had made the invention independently." In any event, the forwarding of the *Epistola Posterior* was delayed for months, not to be conveyed to Leibniz before the end of June 1677. Leibniz replied immediately to the letter-packet, hoping to keep the correspondence alive. Addressed to Oldenburg, the response imparted a full and complete statement of the principles of his form of the differential calculus—notation apart, all but identical with Newton's. Although Leibniz's "noble frankness" appeared in marked contrast to Newton's basic reluctance to discuss his thoughts, it did not extend to informing Newton of the week only lately spent examining Collins's copy of the *De Analysi*. For Newton, however, the matter was closed; he had no further interest in continuing a correspondence that might steal his time and draw him into another controversy. He had already written to Oldenburg:

I.

NOVA METHODUS PRO MAXIMIS ET MINIMIS, ITEMQUE TAN-GENTIBUS, QUAE NEC FRACTAS NEC IRRATIONALES QUANTITATES MORATUR, ET SINGULARE PRO ILLIS CALCULI GENUS *).

Sit (fig. 111) axis AX, et curvae plures, ut VV, WW, YY, ZZ, quarum ordinatae ad axem normales, VX, WX, YX, ZX, quae vocentur respective v, w, y, x, et ipsa AX, abscissa ab axe, vocetur x. Tangentes sint VB, WC, YD, ZE, axi occurrentes respective in punctis B, C, D, E. Jam recta aliqua pro arbitrio assumta vocetur dx, et recta, quae sit ad dx, ut v (vel w, vel y, vel z) est ad XB (vel XC, vel XD, vel XE) vocetur dv (vel dw, vel dy, vel dz) sive differentia ipsarum v (vel ipsarum w, vel y, vel z). His positis, calculi regulae erunt tales.

Sit a quantitas data constans, erit da aequalis 0, et $\overline{dax}$ erit aequalis adx. Si sit y aequ. v (seu ordinata quaevis curvae YY aequalis cuivis ordinatae respondenti curvae VV) erit dy aequ. dv. Jam *Additio et Subtractio*: si sit z — y + w + x aequ. v, erit dz — y + w + x seu dv aequ. dz — dy + dw + dx. *Multiplicatio*: $d\overline{xv}$ aequ. xdv + vdx, seu posito y aequ. xv, fiet dy aequ. xdv + vdx. In arbitrio enim est vel formulam, ut xv, vel compendio pro ea literam, ut y, adhibere. Notandum, et x et dx eodem modo in hoc calculo tractari, ut y et dy, vel aliam literam indeterminatam cum sua differentiali. Notandum etiam, non dari semper regressum a differentiali Aequatione, nisi cum quadam cautione, de quo alibi.

Porro *Divisio*: $d\dfrac{v}{y}$ vel (posito z aequ. $\dfrac{v}{y}$) dz aequ. $\dfrac{\pm vdy \mp ydv}{yy}$.

Quoad *Signa* hoc probe notandum, cum in calculo pro litera substituitur simpliciter ejus differentialis, servari quidem eadem signa, et pro + z scribi + dz, pro — z scribi — dz, ut ex addi-

---

*) Act. Erud. Lips. an. 1684.

A page from Leibniz's first page on the differential calculus. Published in the Acta Eruditorum *(1694). (From* A Concise History of Mathematics *by Dirk Struik, 1967, Dover Publications, Inc., N.Y.)*

I hope that this will so far satisfy Mr. Leibniz that it will not be necessary for me to write any more about the subject. For having other things in my head, it proves an unwelcome interruption to me to be at this point of time put upon considering these things.

During the next several years matters stood thus. In 1684, Leibniz published the particulars of his differential calculus in a newly established scientific monthly, the *Acta Eruditorum*. (The *Acta Eruditorum* was established in Leipzig in 1682 and did in Germany what the *Journal des Savants* had been doing so well in France since 1665. Because it was published in Latin, the journal soon gained an international audience.)

The paper, entitled *Nova Methodus pro Maximis et Minimis, itemque Tangentibus . . .* (A New Method for Maxima and Minima, as well as Tangents . . .), contained mechanical rules, without proof, for computing the differentials—which Leibniz always called differences—of powers, products, and quotients. It also had the familiar *d* notation for differentials. The first published account of the calculus was too barren and obscure to make the subject generally understood. It was so unenlightening that the Bernoulli brothers called it an enigma rather than an explanation. Two years later, in another paper in the *Acta Eruditorum,* Leibniz dealt briefly with the integral calculus; this marked the first appearance in print of the integral notation $\int x$ (later $\int x\ dx$) for summation.

Thus it came about that although Newton had invented his method of fluxions many years before Leibniz had invented his rival method of "differences," Leibniz was the first of the two to publish and to make known to the learned world his results. Had Newton secured publication for his *De Methodis Fluxionum* when it was originally written in 1671, he would have had no competitor, and mathematical history might well have taken a different course.

At this moment, relations between the two great contemporaries were harmonious, with no apparent trace of enmity. Even the first edition (1687) of the *Principia* contained no accusation on Newton's part, nothing but an ambiguous acknowledgment of Leibniz's achievement in the field. A scholium to Book II reads:

> In letters which passed between me and that most excellent geometer, G. W. Leibniz, ten years ago, when I signified that I knew a method of determining maxima and minima, of drawing tangents and the like, and when I concealed it in transposed letters . . . the most distinguished man wrote back that he had also fallen upon a method of the same kind, and communicated his method, which hardly differed from mine, except in his forms of words and symbols.

In 1713, when the second edition of the *Principia* appeared, the priority dispute over the invention of the calculus was at its height. The scholium of the first edition was allowed to stand, but contained the added phrase "and the concept of the generation of quantities," so that the last clause read "which hardly differed from mine, except in his forms of words and symbols, and the concept of the generation of quantities." Soon afterward, Newton denied that the scholium "allowed him [Leibniz] the invention of the calculus differentialis independently of my own." When in 1726 the third edition of the *Principia* was issued, a new scholium appeared in the place of the one just quoted; neither Leibniz's name nor his work was mentioned. With regard to the admission in the scholium of Leibniz's rights as a second or simultaneous inventor, De Morgan remarks that Newton was weak enough "first to deny the plain and obvious meaning, and secondly, to omit it entirely from the third edition of the *Principia.*"

The death of Oldenburg in 1677 ended Leibniz's tenuous link with the English mathematicians, because there was no other intermediary sufficiently interested in carrying on the correspondence. By this time, Leibniz had arrived in Hanover, where he was to serve under three successive dukes—John Friedrich, Ernst August, and Georg Ludwig—in various official capacities. Residence at the Court of Hanover (a Versailles in miniature, with Handel conducting the chamber concerts) meant intellectual isolation for Leibniz and the end of his mathematical aspirations. The first of his masters, Duke John Friedrich, used him as a kind of all-purpose civil servant: diplomat, counselor, and librarian. With his flair for engineering, Leibniz also acted as a consultant on many technical projects, one of which was a scheme to increase the production of the Harz Mountain silver mines by using wind-powered pumps to clear the mine passages of seeping groundwater. But Leibniz's functions and duties became increasingly trivial, with much time devoted to business of no permanent value. When the duke was succeeded by his brother Ernst August in 1679, Leibniz was appointed court historian and given the task of writing a history of the House of Brunswick. There was nothing that did not arouse a fruitful interest in him. In pursuit of genealogical lines, he made a series of travels (1687–1690) throughout the German states and Italy, ransacking archives to find original documents. Leibniz declined, during his stay in Italy, an offer to be Vatican librarian, for its acceptance would have required that he change his religion.

With the passing of Ernst August in 1698, the most brilliant period in Leibniz's life came to a close. The old duke's son, Georg Ludwig (afterwards George I of England), made little use of Leibniz's services except to urge him to complete the history of the Hanoverian royal house. In 1700, Leibniz succeeded in persuading the elector of Brandenburg to found the Berlin Academy of Sciences, and was named its first president; however, ten years' further effort was required to get the learned body properly running, with a meeting place of its own (Frederick the Great later remarked of Leibniz's many-sided activities, ''Leibniz himself comprises a whole Academy.'') Aside from occasional visits to Berlin—and to Vienna where he attempted to establish a similar academy—Leibniz grew into a recluse, not so much from personal inclination as from a lack of interest on the part of his masters. In his last years, Hanover became his prison, and his life there a misery.

We have seen that by 1693, Newton had written out three accounts of his approach to calculus. Although he had circulated manuscripts on the subject among friends in England, never once after the early 1670s did he evince any desire to make public his invention. This excessive reluctance to publish and apparent indifference to popular approval was not an indication of modesty, but a strong dislike for anything bordering on controversy. It was an ironic misfortune that Newton's hesitation to secure priority of invention was punished by raising up a formidable rival, one who was to give him more trouble than what he had sought to avoid by withholding publication.

The first printed account of Newton's long-hidden treasure appeared in the writings of John Wallis (1616–1703), the venerable Savilian professor of geometry at Oxford. Volume II of Wallis's *Opera Mathematica* (1693), an augmented Latin translation of his *Algebra,* contained a direct statement of Newton's method of fluxions with its use of dots, ''pricked letters,'' $\dot{x}, \ddot{x}, \ldots$ for velocities, or fluxions. In 1695, Wallis warned his young colleague at Cambridge not to delay further in putting forward his ideas:

> Your notions of fluxions pass there [on the Continent] with great applause, by the name of Leibniz's calculus differentialis. . . . You are not so kind to your reputation, and that of the Nation, as you might be, when you let things of worth lie by you so long, till others carry away the reputation that is due you. I have endeavoured to do you justice in that point.

The ''justice'' that Wallis had done was to state in the preface of Volume I of his collected *Opera Mathematica* (which did not appear until 1695, two years after the second volume) that Newton's method of fluxions was similar to Leibniz's differential

John Wallis
(1616–1703)

(Extract taken from A History of Science, Technology and Philosophy in the 16th and 17th Centuries, *by A. Wolf. Reproduced by kind permission of Unwin Hyman Ltd.)*

calculus, and that Newton's method had been sent to Oldenburg in 1676 to be communicated to Leibniz. With that, the seeds of the great controversy on the invention of the calculus were already being sown.

The first serious volley in the battle between Leibniz and Newton and their respective supporters was fired by Nicolas Fatio de Duiller, an obscure Swiss mystic and mathematician living in London. Fatio felt slighted at having been omitted from a list, compiled by Leibniz, of mathematicians capable of solving John Bernoulli's problem of the path of quickest descent. He was determined to have his revenge. When he emigrated to London in 1687 he soon ingratiated himself with Newton and his inner circle of followers to such an extent that he was considered a possible editor for a new edition of the *Principia*. (Fatio is the author of the epigram on Newton's tomb: ''Let mortals rejoice that there has existed such and so great an ornament of the human race.'') In 1699, when Fatio prepared a tract on a *Geometrical Investigation of the Solid of Least Resistance,* he made a thinly veiled charge that Leibniz's ideas were obtained from Fatio's idol, Newton:

> I am now fully convinced by the evidence itself on the subject that Newton is the first inventor of this calculus, and the earliest by many years; whether Leibniz, its second inventor, may have borrowed anything from him, I should rather leave to the judgement of those who had seen the letters of Newton, and his original manuscripts.

Leibniz was indignant, but pointed out to friends that because he had been given credit in the first edition of the *Principia* as an independent inventor of the calculus, Newton would doubtless vindicate him. When Newton maintained a haughty silence, Leibniz defended himself in the *Acta Eruditorum,* whose editors then refused to publish a reply written by Fatio.

For almost five years nothing more happened, until Newton's *Opticks* came out in 1704. Published with the main body of the work were two mathematical tracts; one was *Tractatus de Quadratura Curvarum,* dealing with his method of fluxions. In the preface, Newton gave the following reason for including these tracts:

> In a letter written to Mr. Leibniz in the year 1676, and published by Dr. Wallis [in volume III (1699) of the *Opera*], I mentioned a method by which I found some general theorems about squaring curvilinear figures. . . . And some years ago I lent out a manuscript containing such theorems; and having since met with some things copied out of it, I have on this occasion made it public.

The reference to material purloined from a manuscript concerns Leibniz's second visit to London. Newton became convinced that Leibniz had in some way gotten the key to the calculus from perusing *De Analysi* and other papers entrusted to Collins.

The next January, Leibniz countered with an anonymous review in the *Acta Eruditorum* of Newton's two mathematical tracts. After criticizing the mathematical contents, Leibniz went on to claim that Newton's method of fluxions was his own calculus, merely renamed, in slightly different notation:

> The elements of this calculus have been given to the public by its inventor, Dr. Wilhelm Leibniz, in these *Acts*. . . . Instead of the Leibnizian differences, then, Dr. Newton employs, and has always employed, fluxions, which are very much the same as the arguments of fluents produced in the least intervals of time; and these fluxions he has used elegantly in his *Mathematical Principles of Nature* and in other later publications, just as Honoratus Fabri, in his *Synopsis of Geometry* substituted progressive motions for the methods of Cavalieri.

This innocent-sounding statement provoked a storm of wrath in England. For Fabri was a notorious plagiarist, who had greatly damaged his reputation by his *Synopsis of Geometry,* in which he had published Cavalieri's work as his own. The likening of Newton to Fabri was an outrageous insult, and together with the rest of the passage, it was taken as a direct accusation of plagiarism.

This charge of plagiarism was hurled back at Leibniz in 1708 by John Keill, a Scottish mathematician soon to be professor of astronomy at Oxford. In a paper on *The Laws of Centripetal Force,* addressed as a letter to Halley, Newton's ''avowed champion'' inserted the following:

> All these laws follow from that very celebrated arithmetic of fluxions which, without any doubt, Dr. Newton invented first, as can readily be proved by anyone who reads the letters about it published by Wallis; yet the same arithmetic afterwards, under a changed name and method of notation, was published by Dr. Leibniz in *Acta Eruditorum*.

The paper was published in the *Philosophical Transactions* for 1710, and the copy sent to Leibniz was delayed until 1711. Leibniz demanded that the Royal Society intervene, asking that ''such vain and vociferous clamours'' be suppressed and that a public apology be extracted from Keill.

In the early spring of 1712, the Royal Society responded to Leibniz's appeal by appointing a committee of eleven to examine all documents in the society's possession that bore on the matter. Grandiloquently called ''a large committee of distinguished persons of several nations,'' the group comprised only two foreigners, and only one of those two, de Moivre, was a mathematician. Six members of the committee were mathematicians, and at least seven were intimate personal friends of Newton, who had been president of the society since 1703. Halley was one of the committee, and as senior member, probably acted as its secretary. No one was the official representative of Leibniz, although the Prussian minister to England served on this ostensibly independent body.

After a month's research of the question, the committee delivered its report to the Royal Society. The report is a statement of the case for Newton, presented as if it were a set of impartial findings. It left the question whether Leibniz was guilty of conscious plagiarism completely unresolved, and asserted gratuitously that Newton was the first inventor of the infinitesimal calculus. The committee's conclusion runs:

> That the differential method is one and the same with the method of fluxions, excepting the name and the mode of notation; Mr. Leibniz calling those quantities differences, which Mr. Newton calls moments or fluxions; and marking them with the letter *d,* a mark not used by Mr. Newton. And therefore we take the proper question to be, not who invented this or that method, but who was the first inventor of the method. And we believe that those who reputed Mr. Leibniz the first inventor, knew little or nothing of his correspondence with Mr. Collins or Mr. Oldenburg long before; nor of Mr. Newton's having the method above fifteen years before Mr. Leibniz began to publish it in the *Acta Eruditorum* of Leipzig.
>
> For which reasons, we reckon Mr. Newton the first inventor; and are of the opinion, that Mr. Keill, in asserting the same, has been no ways injurious to Mr. Leibniz.

Leibniz had never seriously questioned that Newton invented the method of fluxions, but had claimed an independent discovery of the calculus. He had asked the society to vindicate him of the charge of having stolen the idea from Newton, merely fabricating another notation for it. Although the report affirmed that Newton was not a plagiarist (because he had been adjudged the first inventor of the calculus), it failed to pronounce on Leibniz's originality, and therefore on the truth of Keill's accusation. The dispute could not, nor did it, terminate with the utterly illogical conclusion that Keill had not injured Leibniz by his abusive attack.

The biased investigators based their judgment on a superficial examination of the available evidence. They assumed that Leibniz had seen a letter of December 1672 from Newton to that scientific intermediary, John Collins, in which "the method of fluxions was sufficiently described to any intelligent person." Newton would always maintain that by reading this letter Leibniz hit on the invention of his differential calculus. It has been established that the Latin version that Oldenburg passed on to Leibniz was not an accurate rendering of the essential content of the letter, but only a résumé. The whole portion in which Newton explained his tangent method by an illustrative example was missing in the copy dispatched to Leibniz. Even allowing the possibility that Leibniz had inspected the original during his short visit to London in the autumn of 1676, this would have taken place six months after he had devised his own version of the calculus independent of Newton's previous investigations.

The finished report was approved unanimously at a meeting of the society in April 1712 and ordered printed. No one bothered to inform Leibniz of the findings. A subcommittee of three, including Halley as a member, was named to see it through the presses. The report was published the following January and appeared in Latin for the benefit of continental readers. It is known for short as the *Commercium Epistolicum,* the first two words of its title *Commercium Epistolicum D. Johanns Collinsii et Aliorum de Analysi Promota.* Few copies were printed and these were not for sale but given as gifts to important individuals and institutions.

Although Leibniz had his adherents among continental mathematicians, few were willing to risk their reputations for him. A notable exception, the bellicose John Bernoulli, dared to intervene on Leibniz's behalf. On receiving a copy of the *Commercium Epistolicum,* Bernoulli wrote to Leibniz, "You are, at the start, accused before a tribunal which, it appears, is composed of the plaintiffs themselves and their witnesses. . . . You fall under the law, you are condemned." He attempted indirectly to weaken the evidence by charging that "Newton was ignorant of the correct way of taking second differentials a long time after it was known to us." Leibniz had little choice but to present his case in public, for that supposedly impartial body, the Royal Society, had heard no testimony from him. Without waiting to see the *Commercium Espitolicum,* he responded by publishing, under the cover of anonymity, a broadsheet known as the *Charta Volans.* Bernoulli's letter was printed in it, without naming names, as the judgment of a "very eminent mathematician." The *Charta Volans* raised the wrath of the English; Keill, Newton's gladiator-in-chief, exclaimed, "They have thrown all the dirt and scandal they could without proving anything," and referred to Leibniz and his colleagues as those Leipzig rogues.

Because Newton implacably resented any reflection on his honor, the dispute became yet more heated. Following the printing of the *Commercium Espitolicum,* there appeared an anonymous review of it (which has passed into history under the name of the *Recensio*) in the *Philosophical Transactions* of 1715. This too was the handiwork of Newton, though it was deceptively palmed off as Keill's. The review, *An Account of the Book Entitled Commercium Epistolicum,* purported to be a rectification of the imperfect summaries published abroad. Full of unsupported innuendoes, it was a polemic against Leibniz and revealed the lengths to which Newton would go to destroy an enemy. It is reported that Newton once "had pleasantly" said to a friend that "he had broken Leibniz's heart with his reply to him."

By waiting nearly 30 years before taking up his case, Newton had placed his rival at a disadvantage for which Leibniz was not prepared. Leibniz, when the battle was joined, lacked the textual evidence that would prove that he was not a latecomer to calculus (the Royal Society refused him access to the documents in its possession). Newton had more numerous allies; he had become something of a national hero in his old age and was fully supported by the members of the Royal Society, with their official organ the *Philosophical Transactions.* Leibniz commanded no such backing. As time wore on, his influence at court had decreased more and more, so that toward the end of his life he found little consideration and at last only neglect from the duke of Hanover. When his employer left Hanover (1714) to assume the crown of England as George I, Leibniz hoped to be given the post of royal historian; instead he was ordered to stay behind like a lazy schoolboy until he had completed the history of the House of Brunswick.

Leibniz spent the last decades of his life in philosophical speculation, with his thoughts set out for the most part in short occasional papers. The crown of his philosophical thinking, the *Theodicy* (1710), was the only lengthy work published in his lifetime. Dealing with the question of whether the claims of rationalism were at odds with Christian doctrine, it became one of the most popular books of the eighteenth-century Enlightenment. Leibniz's contention was that although any created world is bound to be imperfect, the evil that we see about us is the least possible amount of evil. This "best of all possible worlds" optimism was ridiculed later in the figure of Dr. Pangloss in Voltaire's novel *Candide* (1759).

Leibniz died in 1716 from a noxious potion he took during an attack of the gout. Unlike Newton, who was buried in Westminster Abbey amid great pomp and surrounded by eminent persons, Leibniz was buried in near solitude. There was not a single representative of the Court of Hanover at his funeral; the only mourner at the graveside was his faithful secretary. A friend wrote in his memoirs that Leibniz "was buried more like a robber than what he really was, the ornament of his country." Little notice was taken either in Berlin or in London of the passing of the greatest mathematician Germany had produced in the seventeenth century, and only in Paris was an adequate eulogy to his memory pronounced before the Académie des Sciences. Diderot, the editor of the great *Encyclopédie,* said of Leibniz that he brought as much fame to Germany as Plato, Aristotle, and Archimedes together had brought to Greece. Yet not until 50 years later was his grave at Court Church in Hanover made recognizable by a suitable inscription.

Newton, in his vendetta, continued to pursue Leibniz even beyond the pale. He superintended a second edition of the *Commercium Epistolicum* in 1722, with the *Recensio* prefixed as a kind of introduction and an anonymous new preface "To the Reader" (written by Newton himself). The title page appeared under the old date of 1712, which was tantamount to a declaration that this edition was an exact reprinting of the old; but numerous insertions and omissions were apparent when the edition was compared with the original text. Among the most notable variations was the addition of the sentence, "This *Collectio* was sent to Dr. Leibniz on June 26, 1676." What gave significance to this statement was that the *Collectio* of John Collins—a 50-page set of memoranda and extracts of letters presenting the history of the development of series expansions—contained the full text of Newton's celebrated letter of December, 1672, the letter the *Commercium Epistolicum* states "sufficiently described the method of fluxions to any intelligent person." We know now that because the *Collectio* was too extensive a document for communication, Oldenburg sent Leibniz only a drastically curtailed survey of it. This is just a further instance of the carelessness on the part of those stooping to frame an accusation of plagiarism against Leibniz.

The bitterness of the calculus priority dispute materially affected the history of mathematics in Western Europe. In England, it was regarded as an attempt by impudent foreigners to defraud Newton, her most eminent son, of the fruits of his genius. The natural reaction was that the purely geometrical methods preferred by Newton over the analytic ones were alone studied and used. English mathematicians were also blinded to the obvious advantages of Leibniz's *d* notation over Newton's "dottage" (the use of dotted or pricked letters to denote differential operations). For more than a century, excessive loyalty to the great man's reputation kept them out of touch with developments on the Continent. The loss of cooperation was detrimental to both English and continental mathematics, but most especially to the English school. England produced few creative mathematicians during the eighteenth-century "Age of Newton," and none who could rightly be called great.

Few readers understood the elements of the new calculus contained in Newton's *Principia* or the *Acta Eruditorum* articles written by Leibniz: Neither of these was a proper textbook on the subject. Such a work, confined simply to the differential

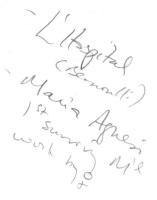

calculus, did appear in 1696 when the Marquis de L'Hospital published the lectures of his private teacher John Bernoulli, in a volume entitled *Analyse des Infiniment Petits.* Enjoying a wide circulation through its numerous editions, L'Hospital's *Analyse* dominated the field for more than fifty years. In the middle of the eighteenth century, it found a worthy rival in the *Instituzioni Analitiche* of Maria Agnesi.

Maria Gaetana Agnesi (1718–1799), a brilliant linguist, philosopher, and mathematician, was the first of twenty-one children of a professor of mathematics at the University of Bologna. Educated by private tutors, she early developed an exceptional gift for languages; by her eleventh year she was fluent in Latin, Greek, Hebrew, French, Spanish, and German, as well as her native Italian. Her father gathered in his home the most distinguished intellectuals from Italy and abroad, with his young daughter acting as hostess. She would engage these scholars in academic disputations on a variety of scientific and philosophical topics, usually speaking in Latin; and in ordinary conversation with a foreign visitor would respond in his own language. Her *Propositions Philosophicae,* based on these domestic discussions, appeared when Agnesi was 20 years old.

Thereafter followed a decade of concentrated work—undertaken at first for her own amusement and subsequently for the instruction of her younger brothers—which resulted in the publication of the *Instituzioni Analitiche* (1748). It is the first surviving mathematical work written by a woman. In two huge volumes containing more than 1000 pages, the *Instituzioni Analitiche* attempted a complete and unified treatment of algebra and analysis, one that would keep the reader abreast of the most recent advances in the differential and integral calculus. Widely acclaimed as a model of clarity and exposition, and translated from Italian into French (1775) and English (1801), it was soon superseded by the classic textbooks on the calculus published by Euler in 1775 and 1768–1770. The English translation of Agnesi's book was done by John Colson, the Lucasian Professor of Mathematics at Cambridge who had also previously (1736) rendered Newton's *Principia* from Latin into English. Colson went to the trouble of learning the Italian language late in life specifically to translate the *Instituzioni Analitiche,* ''so that British Youth might have the benefit of it as well as the Youth of Italy.''

Of the many curves investigated in the *Instituzioni Analitiche,* Agnesi's name is most frequently associated with the versed sine curve or versiera (from the Latin *vertere,* ''to turn''), whose Cartesian equation is $yx^2 = a^2(a - y)$. Through confusion with the Italian word *avversiera,* which means ''wife of the devil'' or ''witch,'' this cubic curve has subsequently come to be known as the ''witch of Agnesi.''

Maria Agnesi is often said to have been the first woman professor of mathematics on a university faculty. In recognition of her exceptional accomplishments, Pope Benedict XIV appointed her (1750) to the chair of mathematics and natural philosophy at the University of Bologna, but there is a difference of opinion as to whether she actually held that position. Some writers believe that she taught at Bologna from 1750 until 1752 while her father was seriously ill; other sources insist that Agnesi declined the Pontiff's offer, having no ambition for public life and finding such a display of her talent distasteful. After her father's death in 1752, she withdrew from all mathematical activity, devoting the rest of her long life to charitable projects and religious meditation. In 1771, Agnesi became directress of a home for the poor and infirm, a post she held until her death at 81 years of age.

8.4 Problems

1. Assuming Leibniz's series

$$\frac{\pi}{4} = 1 - \frac{1}{3} + \frac{1}{5} - \frac{1}{7} + \cdots,$$

prove that

$$\frac{\pi}{8} = \frac{1}{1 \cdot 3} + \frac{1}{5 \cdot 7} + \frac{1}{9 \cdot 11} + \cdots$$

$$= \frac{1}{2^2 - 1} + \frac{1}{6^2 - 1} + \frac{1}{10^2 - 1} + \cdots.$$

2. Given that $P_n = n!$ denotes the number of permutations of n objects, establish the following identities from Leibniz's *Ars Combinatoria*:

$$2P_n - (n - 1)P_{n-1} = P_n + P_{n-1}$$

and

$$P_n^2 = P_{n-1}(P_{n+1} - P_n).$$

3. Verify Leibniz's famous identity,

$$\sqrt{6} = \sqrt{1 + \sqrt{-3}} + \sqrt{1 - \sqrt{-3}},$$

which gives an imaginary decomposition of the real number $\sqrt{6}$.

4. Obtain Mercator's logarithmic series

$$\log (1 + x) = x - \frac{x^2}{2} + \frac{x^3}{3} - \frac{x^4}{4} + \cdots,$$

for $-1 < x \leq 1$, by first calculating by long division the series

$$\frac{1}{1 + x} = 1 - x + x^2 - x^3 + \cdots$$

and then integrating termwise between 0 and x.

5. Prove that

$$\log\left(\frac{1 + x}{1 - x}\right) = 2\left(x + \frac{x^3}{3} + \frac{x^5}{5} + \frac{x^7}{7} + \cdots\right),$$

for $-1 < x < 1$, and hence

$$\log 2 = 2\left(\frac{1}{3} + \frac{1}{3} \cdot \frac{1}{3^3} + \frac{1}{5} \cdot \frac{1}{3^5} + \cdots\right).$$

6. Supply the details of the following derivation, due to Euler, of the infinite series expansion for $\log (1 + x)$:

(a) Show that $\log (1 + x)$ can be given by the limit

$$\log (1 + x) = \lim_{n \to \infty} n[(1 + x)^{1/n} - 1].$$

[*Hint:* Since

$$\frac{d}{du}(1 + x)^u = \frac{d}{du}[e^{u \log (1 + x)}]$$
$$= (1 + x)^u \log (1 + x),$$

the above limit is the value of the derivative of $(1 + x)^u$ at $u = 0$.]

(b) Next use the binomial series expansion of $(1 + x)^{1/n}$ to obtain

$$\log (1 + x) = \lim_{n \to \infty}\left[x + \frac{(1/n - 1)}{2!}x^2\right.$$
$$+ \frac{(1/n - 1)(1/n - 2)}{3!}x^3$$
$$\left. + \frac{(1/n - 1)(1/n - 2)(1/n - 3)}{4!}x^4 + \cdots\right].$$

(c) From the fact $\lim_{n \to \infty}\left(\frac{1}{n} - k\right) = -k,$

conclude that

$$\log (1 + x) = x - \frac{x^2}{2} + \frac{x^3}{3} - \frac{x^4}{4} + \cdots.$$

7. Show that the binomial theorem, as stated by Newton in his letter to Oldenburg, is equivalent to the more familiar form

$$(1 + x)^r = 1 + rx + \frac{r(r - 1)}{2!}x^2$$
$$+ \frac{r(r - 1)(r - 2)}{3!}x^3 + \cdots,$$

where r is an arbitrary integral or fractional exponent. The necessary condition $|x| < 1$ for convergence was not stated by Newton.

8. Use the binomial theorem to obtain the following series expansions.

(a) $(1 + x)^{-1} = 1 - x + x^2 - x^3 + \cdots$
$+ (-1)^n x^n + \cdots$.

(b) $(1 - x)^{-2} = 1 + 2x + 3x^2 + \cdots$
$+ (n + 1)x^n + \cdots$.

(c) $(1 + x)^{1/2} = 1 \dfrac{1}{2}x - \dfrac{1}{2 \cdot 4}x^2$

$+ \dfrac{1 \cdot 3}{2 \cdot 4 \cdot 6}x^3 + \cdots$

$+ (-1)^n \dfrac{1 \cdot 3 \cdot 5 \cdots (2n - 3)}{2 \cdot 4 \cdot 6 \cdots (2n)}x^n$

$+ \cdots$.

(d) $(1 + x)^{-1/2} = 1 - \dfrac{1}{2}x + \dfrac{1 \cdot 3}{2 \cdot 4}x^2 + \cdots$

$+ (-1)^n \dfrac{1 \cdot 3 \cdot 5 \cdots (2n - 1)}{2 \cdot 4 \cdot 6 \cdots (2n)}x^n + \cdots$.

9. Prove that the inequality $2 < (1 + 1/n)^n < 3$ holds for all integers $n \geq 1$. [*Hint:* By the binomial theorem,

$$\left(1 + \frac{1}{n}\right)^n < 1 + 1 + \frac{1}{2} + \frac{1}{2 \cdot 3}$$

$$+ \cdots + \frac{1}{2 \cdot 3 \cdot 4 \cdots n}$$

$$< 1 + 1 + \frac{1}{2} + \frac{1}{2^2} + \cdots$$

$$+ \frac{1}{2^{n-1}}.]$$

10. Supply the missing details in the following derivation of Newton's series

$$\frac{\pi}{2\sqrt{2}} = 1 + \frac{1}{3} - \frac{1}{5} - \frac{1}{7} + \frac{1}{9} + \frac{1}{11} + \cdots.$$

(a) From the expansion

$$\frac{1}{1 + x^4} = 1 - x^4 + x^8$$
$$- x^{12} + x^{16} + \cdots$$

and termwise integration, show that

$$\int_0^1 \left(\frac{1 + x^2}{1 + x^4}\right)dx = 1 + \frac{1}{3} - \frac{1}{5}$$

$$- \frac{1}{7} + \frac{1}{9} + \frac{1}{11} - \cdots.$$

(b) Verify that

$$\int_0^1 \left(\frac{1 + x^2}{1 + x^4}\right)dx = \frac{1}{2}\int_0^1 \left(\frac{1}{1 + \sqrt{2}x + x^2}\right.$$

$$\left. + \frac{1}{1 - \sqrt{2}x + x^2}\right)dx$$

$$= \frac{1}{2}\int_{-1}^1 \frac{dx}{1 + \sqrt{2}x + x^2}.$$

(c) Use the integration formula

$$\int \frac{dx}{a^2 + x^2} = \frac{1}{a}\arctan\frac{x}{a}$$

to get

$$\frac{1}{2}\int_{-1}^1 \frac{dx}{1 + \sqrt{2}x + x^2} = \frac{1}{\sqrt{2}}[\arctan(1 + \sqrt{2})$$

$$- \arctan(1 - \sqrt{2})]$$

$$= \frac{1}{\sqrt{2}}\left(\frac{3\pi}{8} - \left(-\frac{\pi}{8}\right)\right)$$

$$= \frac{\pi}{2\sqrt{2}}.$$

11. Newton's method for approximating a real root of the equation $y = f(x)$ is to start with an estimated value x_1 and in place of the curve use the line $y - f(x_1) = f'(x_1)(x - x_1)$, which is tangent to the curve at the point $(x_1, f(x_1))$. The intersection of this tangent with the x-axis,

$$x_2 = x_1 - \frac{f(x_1)}{f'(x_1)},$$

gives the next approximation of the desired root. Newton applied (1669) his method of tangents to $x^3 - 2x - 5 = 0$. Find to five decimal places the root of this cubic lying between 2 and 3.

12. (a) Apply Newton's iterative method to the polynomial $f(x) = x^n - a$, where n is a positive integer, to show that if x_1 is an approximation to $\sqrt[n]{a}$, then

$$x_2 = \frac{(n - 1)x_1 + a/x_1^{n-1}}{n}$$

is a better approximation, and so on.

(b) Find $\sqrt[3]{2}$ correct to five decimal places. [*Hint:* Take $x_1 = 1$ as a first approximation and use part (a).]

13. The construction of the witch of Agnesi curve can be accomplished as follows: Consider a circle, with diameter $OA = a$, lying between two parallel tangents. Draw a secant through the lower point O of tangency and meeting the upper tangent at B and the circle at C. Let the point P be the intersection of lines CP and BP parallel and perpendicular, respectively, to the lower tangent.

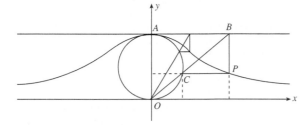

To see that the locus of the point P as the secant varies is the desired curve, show that:

(a) Triangles BPC and OAB are similar, whence $(BP)(AB) = (OA)(PC)$.

(b) If O is the origin of a rectangular coordinate system and P and C have coordinates (x,y) and (u,v), respectively, then $(a - y)x = a(x - u)$ and so $u = (xy)/a$, $v = y$.

(c) Substituting the values u and v into the equation

$$u^2 + \left(v - \frac{a}{2}\right)^2 = \left(\frac{a}{2}\right)^2$$

yields $y(x^2 + a^2) = a^3$.

Bibliography

Adamczewski, Jan. *Nicolaus Copernicus and His Epoch.* Philadelphia: Copernicus Society of America, 1973.

Aiton, E. J. *The Vortex Theory of Planetary Motions.* New York: American Elsevier, 1972.

Baron, Margaret. *The Origins of the Infinitesimal Calculus.* London: Pergamon, 1969. (Dover reprint, 1993).

Basalla, George, ed. *The Rise of Modern Science.* Lexington, Mass.: D. C. Heath, 1968.

Bell, E. T. "Newton After Three Centuries." *American Mathematical Monthly* 49(1942): 553–575.

Bissell, Christopher. "Cartesian Geometry: The Dutch Contribution." *Mathematical Intelligencer* 9(No. 4, 1987): 39–44.

Boyer, Carl. "Analytic Geometry: The Discovery of Fermat and Descartes." *Mathematics Teacher* 37(1944): 99–105.

———. "The Invention of Analytic Geometry." *Scientific American* 180(Jan. 1949): 40–45.

———. *The History of Calculus and Its Conceptual Development.* New York: Dover, 1959.

———. "Galileo's Place in the History of Mathematics." In *Galileo: Man of Science,* ed. Ernan McMullin. New York: Basic Books, 1967.

———. "The Making of a Mathematician. Essay Review of *The Mathematical Papers of Isaac Newton.*" *History of Science* 6(1967): 97–106.

Broad, William. "Priority War: Discord in the Pursuit of Glory." *Science* 211(1981): 465–467.

Bullough, Vern, ed. *The Scientific Revolution.* New York: Holt, Rinehart & Winston, 1970.

Burt, Edwin. *The Metaphysical Foundations of Modern Physical Science.* New York: Harcourt Brace, 1925.

Cajori, Florian. "Discussion of Fluxions: From Berkeley to Woodhouse." *American Mathematical Monthly* 24(1917): 145–154.

———. "Grafting of the Theory of Limits on the Calculus of Leibniz." *American Mathematical Monthly* 30(1923): 223–234.

———. "The History of the Notations of Calculus." *Annals of Mathematics* (2) 25(1924): 1–46.

———. "Leibniz, the Master Builder of Notations." *Isis* 7 (1925): 412–429.

Charon, Jean. *Cosmology.* Translated by Patrick Moore. New York: World University Library, 1970.

Child, J. M. *The Early Mathematical Manuscripts of Leibniz.* Chicago: Open Court, 1920.

Christianson, Gale. *In the Presence of the Creator: Isaac Newton and His Times.* New York: The Free Press, 1984.

Clarke, M. L. *Classical Education in Britain 1500–1900.* Cambridge: Cambridge University Press, 1959.

Cohen, I. Bernard. "Newton and the Modern World." *American Scholar* 11(1941–42): 328–338.

———. "Newton in the Light of Recent Scholarship." *Isis* 51(1960): 489–514.

———. *Introduction to Newton's "Principia."* Cambridge, Mass.: Harvard University Press, 1971.

Crombie, A. C. "Descartes." *Scientific American* 201(Oct. 1959): 160–166.

Crowther, J. G. *Founders of British Science.* London: Cresset Press, 1960.

De Santillana, Giorgio. *The Crime of Galileo.* Chicago: University of Chicago Press, 1955.

Descartes, René. *The Geometry of René Descartes.* Translated by D. E. Smith and Marcia Latham. New York: Dover, 1954.

Dijksterhuis, E. J. *The Mechanization of the World Picture.* Oxford: Oxford University Press, 1961.

Drake, Stillman. *Galileo at Work, His Scientific Biography.* Chicago: University of Chicago Press, 1978.

Dreyer, J. L. E. *Tycho Brahe: A Picture of the Scientific Life and Work in the Sixteenth Century.* New York: Dover, 1963.

Edwards, C. H., Jr. *The Historical Development of the Calculus*. New York: Springer-Verlag, 1979.

Fauvel, John, ed. *Let Newton Be!* Oxford: Oxford University Press, 1988.

Fermi, Laura, and Fermi, Bernardini. *Galileo and the Scientific Revolution*. New York: Basic Books, 1961.

Field, J. V. "A Lutheran Astronomer: Johannes Kepler." *Archive for the History of Exact Sciences* 31(1985): 189–272.

Frank, Robert. "Science, Medicine, and the Universities of Early Modern England." *History of Science* 11(1973): 194–216, 239–269.

Gade, John Allyne. *The Life and Times of Tycho Brahe*. Princeton, N. J.: Princeton University Press, 1970.

Galilei, Galileo. *Dialogue Concerning Two New Sciences*. Translated by Henry Crew and Alfonso de Salvio. New York: Macmillan, 1914.

———. *Discourses on the Two Chief World Systems*. Translated by Stillman Drake. Berkeley: University of California Press, 1967.

Gunther, R. T. *Early Science in Cambridge*. London: Dawsons of Pall Mall, 1969.

Haldane, Elizabeth. *Descartes: His Life and Times*. New York: American Scholar Publications, 1966.

Hall, A. Rupert, *The Scientific Revolution 1500–1800*. Boston: Beacon Press, 1956.

———. *From Galileo to Newton 1630–1720*. New York: Harper & Row, 1963.

———. *Philosophers at War: The Quarrel Between Newton and Leibniz*. Cambridge: Cambridge University Press, 1980.

Hall, A. Rupert, and Hall, Marie Boas. *Unpublished Scientific Papers of Newton*. Cambridge: Cambridge University Press, 1962.

Harre, R., ed. *Early Seventeenth Century Scientists*. Oxford: Pergamon, 1965.

Herivel, John. *The Background to Newton's Principia*. Oxford: Oxford University Press, 1965.

Hoffman, Joseph E. *Leibniz in Paris 1672–1676*. Cambridge: Cambridge University Press, 1964.

Hutchings, Donald, ed. *Late Seventeenth Century Scientists*. Oxford: Pergamon, 1969.

Jones, Phillip. "Sir Isaac Newton: 1642–1727." *Mathematics Teacher* 51(1958): 124–127.

Kitcher, Philip. "Fluxions, Limits and Infinite Littleness—A Study of Newton's Presentation of the Calculus." *Isis* 64(1973): 33–49.

Kreiling, Frederick. "Leibniz." *Scientific American* 218(May 1968): 94–100.

Langer, R. E. "René Descartes." *American Mathematical Monthly* 44(1937): 495–512.

MacPike, Eugene. *Hevellius, Flamsteed and Halley: Three Contemporary Astronomers and Their Mutual Relations*. London: Taylor and Francis, 1937.

Manuel, Frank E. *A Portrait of Isaac Newton*. Cambridge, Mass.: Harvard University Press, 1968.

———. "Newton as Autocrat of Science." *Daedalus* 97(1968): 969–1001.

Mellone, S. H. *The Dawn of Modern Thought—Descartes, Spinoza, and Newton*. London: Oxford University Press, 1930.

Merz, John. *Leibniz*. New York: Hacker Press, 1948.

Meyer, R. W. *Leibniz and the Seventeenth Century Revolution*. Translated by J. P. Stern. Cambridge: Bowes and Bowes, 1952.

Molland, A. G. "Shifting the Foundations: Descartes' Transformation of Ancient Geometry." *Historia Mathematica* 3(1976): 21–49.

Moore, Louis Trenchard. *Isaac Newton: A Biography*. New York: Charles Scribner's Sons, 1934.

Newton, Isaac. *Mathematical Principles of Natural Philosophy*. Translated by Andrew Motte. Edited by Florian Cajori. 2 vols. Berkeley: University of California Press, 1934.

———. *Mathematical Works*. Edited by D. T. Whiteside. 2 vols. New York: Johnson Reprint, 1964–67.

———. *Mathematical Papers*. Edited by D. T. Whiteside. 7 vols. Cambridge: Cambridge University Press, 1967–76.

Rickey, Frederick. "Isaac Newton: Man, Myth and Mathematics." *College Mathematics Journal* 18 (1987): 362–389.

Ronan, Colin A. *Edmund Halley, Genius in Eclipse*. Garden City, N.Y.: Doubleday, 1969.

———. *Galileo*. New York: C. P. Putnam's Sons. 1974.

Rosenfeld, L. "Newton and the Law of Gravitation." *Archive for History of Exact Sciences* 2(1965): 365–386.

———. "Newton's Views on Aether and Gravitation." *Archive for History of Exact Sciences* 6(1969): 29–37.

Rosenthal, Arthur. "The History of Calculus." *American Mathematical Monthly* 58(1951): 75–86.

Sabra, A. I. *Theories of Light, from Descartes to Newton*. London: Oldburne Book Company, 1967.

Sanford, Vera. "The Decimal Point." *Mathematics Teacher* 45(1952): 71, 73.

Sarton, George. "Simon Stevin of Bruges (1548–1620)." *Isis* 21(1934): 241–290.

———. "The First Explanation of Decimal Fractions and Measures." *Isis* 23(1935): 153–218.

Schrader, Dorothy. "The Newton-Leibniz Controversy Concerning the Discovery of the Calculus." *Mathematics Teacher* 55(1962): 385–396.

Scott, J. F. *The Mathematical Works of John Wallis, D. D., F. R. S. (1616–1703)*. London: Taylor and Francis, 1938.

———. *The Scientific Work of René Descartes*. London: Taylor and Francis, 1952.

Scriba, C. J. "The Inverse Method of Tangents: A Dialogue Between Leibniz and Newton (1675–1677)." *Archive for History of Exact Sciences* 2(1965): 113–137.

Shea, William. *The Magic of Numbers and Motion: The Scientific Career of René Descartes*. Canton, Mass.: Science History Publications, 1991.

Slichter, C. S. "The Principia and the Modern Age." *American Mathematical Monthly* 44(1937): 433–444.

Stimson, Dorothy. *Scientists and Amateurs: A History of the Royal Society*. New York: Henry Schuman, 1948.

———. *The Gradual Acceptance of the Copernican Theory of the Universe*. Gloucester, Mass.: Peter Smith, 1972.

Struik, Dirk. "Simon Stevin and Decimal Fractions." *Mathematics Teacher* 52(1959): 474–478.

Sullivan, J. W. N. *Isaac Newton 1642–1727*. New York: Macmillan, 1938.

Thomson, Thomas. *History of the Royal Society from Its Institution to the End of the 18th Century*. Ann Arbor, Mich.: University Microfilms, 1967.

Toulmin, Stephen, and Goodfield, June. *The Fabric of the Heavens*. New York: Harper & Row, 1961.

Truesdell, C. "Maria Gaetana Agnesi." *Archive for History of Exact Sciences* 40(1989): 113–142.

Tuck, Bernard. "Life and Times of Johann Kepler." *Mathematics Teacher* 60(1967): 58–65.

Turnbull, H. W., ed. *The Correspondence of Isaac Newton*. 7 vols. Cambridge: Cambridge University Press, 1959–77.

Veitch, John. *The Method, Meditation and Philosophy of René Descartes*. New York: Tudor, 1901.

Vièta, François. *The Analytic Art*. Translated by T. Richard Witmer. Kent, Ohio: Kent State University Press, 1983.

Vrooman, Jack. *René Descartes, A Biography*. New York: C. P. Putnam's Sons, 1970.

Weld, Charles. *A History of the Royal Society*. Reprint edition. New York: Arno Press, 1975.

Westfall, Richard. "The Achievement of Isaac Newton." *Mathematical Intelligencer* 9(No. 4, 1987): 45–49.

Whiteside, D. T. "Patterns of Mathematical Thought in the Late Seventeenth Century." *Archive for History of Exact Sciences* 1(1961): 179–388.

Wilson, C. A. "From Kepler's Laws, So-Called, to Universal Gravitation: Empirical Factors." *Archive for History of Exact Sciences* 6(1973): 33–49.

Wolf, A. *A History of Science, Technology and Philosophy in the 16th and 17th Centuries*. 2d ed. 2 vols. New York: Harper, 1959.

Yoder, Joella. *Unrolling Time: Christiaan Huygens and the Mathematization of Nature*. Cambridge: Cambridge University Press, 1988.

CHAPTER 9

The Development of Probability Theory: Pascal, Bernoulli, and Laplace

The theory of probabilities is at bottom nothing but common sense reduced to calculus; it enables us to appreciate with exactness that which accurate minds feel with a sort of instinct for which ofttimes they are unable to account.

LAPLACE

9.1 The Origins of Probability Theory

According to legend, probability theory began as a branch of mathematics with the correspondence between Blaise Pascal and Pierre de Fermat in 1654. In fine detail this is wrong. Years before Pascal and Fermat ever thought of defining the "true worth of a chance," isolated problems of a probablistic nature had been tackled by some mathematicians. It would be more appropriate to say that Pascal and Fermat supplied vital links in a chain of reasoning that gave us probability theory as we now know it. The difficulty in trying to trace this chain to its origin is that probability theory started essentially as an empirical science and developed only later on the mathematical side. The subject had its twin roots in two fairly distinct lines of investigation: the solution of wagering problems connected with games of chance, and the processing of statistical data for such matters as insurance rates and mortality tables.

Straightforward counting of population and economic wealth goes back to antiquity. The numbering of the people of Israel, Emperor Augustus's balance sheets on the Roman Empire, and William the Conqueror's inventory of his possessions, the *Doomsday Book* (so named by the English because there was no appeal from it) are a few of the most outstanding historical examples of gathering data. Although quantitative facts were collected at various times, such information was of less value to a feudal society than to a society that based its economy on trade and manufacture. Systematic and extensive statistical investigations began only when various institutions of the new merchant class, particularly insurance companies, called for the probabilistic estimation of certain events.

The oldest type of insurance seems to have been designed for protecting merchant vessels and was in vogue even in Roman times. In the fourteenth century, the first marine insurance companies were established in Italy and Holland; and the idea spread to other countries by the sixteenth century. (The famous Lloyd's of London was founded sometime before 1688.) Although a life-insurance policy was known to have been underwritten by a small group of men in London in 1583, a well-organized company for this purpose was not established until 1699. Certainly some sort of scientific determination of probabilities was required to put these operations on a sound actuarial footing. Yet for many years, the Dutch financed public business through annuities—loans paying an annual interest for life to the lender—which were so badly calculated that the towns were regularly losing money; despite a brisk commerce in these contracts, no government appeared to have made the cost of an annuity a function of the age of the purchaser.

The London merchant John Graunt (1620–1674) was the first to draw an extensive set of statistical inferences from mass data. In 1662, Graunt produced a tract entitled *Natural and Political Observations Made upon the Bills of Mortality,* a work that may be said to have launched the discipline we now call mathematical statistics. The Bills of Mortality, on which Graunt based his deductions, were originally weekly and yearly returns of the number of burials in several London parishes. The practice of keeping such tallies seems to have arisen as early as 1532 out of a desire to know about the current state of the Plague. The bills were regularly made out for all of London starting in 1563, although they were not actually published until 1625; they gradually became more explicit in the information they imparted. An attempt was made to classify all deaths in London with respect to several causes (only two, disease and accident, were listed at first). The sex of the victim was distinguished, but the bills did not specify an age at the time of death. The data was obtained by searchers, usually "ancient matrons," sworn to their office, and their work included viewing the body and inquiring about the disease or casualty that had led to death.

Graunt had noticed that most of those who regularly bought the bills made little use of them. They merely scanned the total number of burials, looked for something unusual among the listed causes of death, and stored up the results for gossip at the next social occasion. Sure that better use could be made of this information, Graunt set out to collect as many of the past bills as were available. The statistics for the 57 years from 1604 to 1661 were reduced to a series of tables, which he published at some length in his *Natural and Political Observations.* Although Graunt claimed that his pamphlet required less than two hours' reading, it was studded with many conclusions of varying generality and validity, drawn from inspecting the data. Among the important statistical regularities Graunt observed: the number of male births was greater than the number of female births; women live longer than men; the number of people dying from most causes except epidemic disease was fairly constant from year to year; at the suitable age for marriage the numbers of women and men were about equal, so that "every woman may have a husband, without the allowance of polygamy." On the light side, contrasting burial statistics for men and women with physicians' claims that they had twice as many female as male patients, Graunt concluded that either physicians generally cured women's infirmities—or that compensating numbers of men died from their vices without recourse to medical aid.

In studying the Bills of Mortality, Graunt inquired "how many die usually before they can speak, or how many live past any assigned number of years." Because the

*Bills of Mortality recorded by London parish clerks. ("Bills of Mortality" [p. 183]
in* Devils, Drugs and Doctors *by Howard W. Haggard, M.D. Copyright 1929, 1957 by Howard W.
Haggard. Reprinted by permission of Harper & Row, Publishers, Inc.)*

bills lacked age data on childhood mortality, Graunt selected certain of the listed causes
of death—such as thrush, teeth and worms, abortives, convulsion, infant complaints,
and smothering—which would indicate a child up to the age of four or five. To the
numbers associated with these childhood causes of death, he added an estimated proportion of the deaths from more general causes such as smallpox, measles, and plague.
Deaths among the elderly were more readily counted, as "aged" was a category listed

by the searchers. Graunt assumed that this number represented persons over the age of 70, because "no man can be said to die properly of Age, who is much less. . . ."

To form some sort of mortality table, Graunt concluded from such conjectures that 36 out of every 100 people die by the age of 6, and at the other end of the life scale, hardly anyone lives to be 75. He then made an estimate of how many people die in each successive decade and summarized the results in a famous table, the London Life Table.

Age	Survivors
0	100
6	64
16	40
26	25
36	16
46	10
56	6
66	3
76	1

Graunt's importance both as statistician and empirical probabilist lay in this novel idea: A group of births followed through life and gradually (in those days it was not so gradual) reduced in number by death. Although based on scanty and uncertain data and containing the curiously arbitrary notion of a constant death rate after age 6, the London Life Table represented a tremendous advance from simple death totals to a new and graphical method of representing the age pattern of mortality. (The deaths in each decade really amount to about 3/8 of the survivors at the beginning of the decade.)

These researches secured for Graunt the honor of being one of the charter fellows of the Royal Society of London when it was founded in 1662. The 119 original fellows were scientists, doctors of medicine, and doctors of divinity, noblemen, lawyers, civil servants, and literary men; Graunt was the only shopkeeper. There is a story that the officers were uncertain whether to admit Graunt to their number, because he was only a tradesman in the city, and they sought the opinion of Charles II on the matter. The King replied that "they should certainly admit Mr. Graunt, and if they found any other such tradesmen, they should be sure to admit them also without delay." As it happened, the society was able to follow this advice while Charles was still alive and elect another distinguished shopkeeper a fellow in 1679. Anthony Leeuwenhoek, of Delft, was recognized for his pioneering microscopical observations.

Graunt's work drew the attention of others, notably the Huygens brothers in Holland, to the question of the length of life. Ludwig Huygens wrote to his older brother Christiaan (1629–1695) in 1669; "I have just been making a table showing how long people of a given age have to live. . . . Live well! According to my calculations you will live to about $56\frac{1}{2}$ and I to 55." From Graunt's figures, Christiaan then constructed his own mortality curve (the earliest known graph produced from statistical data) and used this to define such notions as the probability of death in a given time and the probability of surviving to a given age. These ideas were put on a sound mathematical footing in 1657, when Christiaan Huygens wrote *De Ratiociniis in Ludo Aleae* (On Reasoning in Games of Chance), the first text on probability theory to be published. It is only fair to point out that Huygen's treatise was prompted to a large extent by the work of Pascal and Fermat, with which he became acquainted on a visit to France in 1655.

The other source to which probability theory owes its birth and development is gambling. Gambling originated in the early stages of human history; it was not the invention of a single people but appeared at many places in the world. It is reasonable to guess that it evolved from some sort of divination rites, most likely divination by lot. (These rites are schemes whereby the deity consulted is given an opportunity of expressing its wishes.) The question was posed by the priest or suppliant, the dice were cast on the sacred ground, and the answer of the god deduced. Nothing was random, there was no chance; because the gods had influence over the course of earthly events, they could in particular interfere with the throwing of the dice. In this way, the ancient and universal practice of betting may have developed from casting lots in religious ceremonies.

Almost every primitive culture engaged in some form of dice play. The predecessor of the die, and the most common gambling device of early peoples, was a tarsal bone (the astragalus) from the hind foot of a hooved animal. Archeologists digging in pre-historic sites have uncovered many of these hard, solid bones among the debris. The astragalus, because of its peculiar oblong shape, can rest on only four sides, two of them broad and two narrow. The four sides differed sufficiently in appearance to make it possible to distinguish among them, so that the early astragali were not marked. One of the games in ancient Greece consisted in throwing four astragali together and noting which sides fell uppermost. The most widely used rule attributed the highest value to the throw that showed different faces for each of the four bones (the throw of Venus). The throw least valued among all others was called the dog, so that to throw the dog was to lose. This perhaps accounts for the phrase ''going to the dogs,'' that is, playing a losing game.

The six-sided die may have been obtained from the astragalus by grinding it down until a rough cube was formed. Whether the die evolved in this or in some other way, it was known well before the birth of Christ. The earliest known dice were excavated in northern Iraq and it is estimated they date from 3000 B.C. They were made of fired pottery, and the faces marked with from one to six dots (probably because there was no convenient symbolism for numbers). About 1400 B.C., the present arrangement of dots, with two partitions of seven imprinted on opposite faces, was introduced.

One question we shall never clearly resolve is when people first started to play games of chance. In the Egypt of 3500 B.C., a board game called ''Hounds and Jackals'' was played, in which counters were moved according to certain rules by throwing the astragali. In reference to a famine in Lydia (about 1500 B.C.), Herodotus wrote in his *History:*

> For some time the Lydians bore the affliction patiently, but finding that it did not pass away, they set to work to devise remedies for the evil. Various expedients were discovered by various persons; dice, hucklebones [astragali] and ball, and all such games were

invented, except table [an early form of backgammon?], the invention of which they did not claim as theirs. The plan adopted against the famine was to engage in games one day so entirely as not to feel any craving for food, and the next day to eat and abstain from games. In this way they passed eighteen years.

In the same vein, Gerolamo Cardan wrote in his *Liber de Ludo Aleae* (Book on Games of Chance) that various games were invented during the ten-year siege of Troy (circa 1200 B.C.) to bolster the morale of the surrounding Greek soldiers, who suffered from boredom. At some unknown time, dice finally supplanted astragali as instruments of play, and because cards were not introduced into Europe until the 1300s, gaming for amusement or profit must have been conducted mainly with dice for a thousand years.

By the time of the emergence of Rome, gambling with dice was a commonplace recreation, not only among the educated classes but also among the middle and lower classes. We are told that the Emperor Claudius (10 B.C.–A.D. 54) was so devoted to dicing that he wrote a book called *How to Win at Dice,* and that he used to play while driving, throwing on a board especially fitted into his carriage. Another regal player, the Emperor Marcus Aurelius (121–180), was accompanied everywhere by his personal croupier. Gaming reached such popularity with the Greeks and Romans that it was found necessary to issue laws forbidding it, except during certain seasons. Although gambling was in great favor among the Greeks and Romans, it was strictly forbidden by the Jews under penalty of death. The reason for the condemnation was that the gambler always expects to win, and therefore to get something for nothing; the rabbis considered this immoral and akin to robbery.

Throughout the Middle Ages, the Christian Church appears to have carried out a campaign against playing with dice and also cards. Their bans and prohibitions were not directed against the games as such but rather against the accompanying vices of drinking and swearing. For example, participants in the Third Crusade (A.D. 1190) were issued instructions that no person below the rank of knight was to gamble for money, whereas knights and clergy might play but could not lose more than 20 shillings in 24 hours. Medieval history is full of such attempts to prohibit gaming, with just such results as might be expected. In the thirteenth century, Louis IX of France (1214–1270) issued an edict forbidding not only the play of dice, but even their manufacture in his realm.

The exact origin of card playing is murky, with the invention variously credited to the Egyptians, the Chinese, and the Indians. It is known that card playing was introduced into Europe through the Crusades (the Crusaders also learned chess from the Arabs) and became firmly established in Western society during the 1300s. The first playing cards made in Europe were hand-painted, a luxury of great beauty available only to the upper classes. With the invention of printing, cards were manufactured in quantity; the presses of Johann Gutenberg turned out the 78-card tarot deck in the same year as his famous Bible (1440). About 1500, the French developed the present-day variety of spades, hearts, clubs, and diamonds. The now stylized face cards were originally portraits of actual persons; for instance, the four kings were the biblical David (spades), Charlemagne (hearts), Alexander the Great (clubs), and Julius Caesar (diamonds), representing the four monarchies or empires of the Jews, the Franks, the Greeks, and the Romans.

No sooner did playing cards become widespread than special rules were issued limiting their use—an effort as ineffective then as today. A series of English laws was passed during the reign of Edward III (1312–1377) that added dice and cards to the list of unlawful amusements in order to promote manly sports, such as archery. In 1397 the provost of Paris issued an edict prohibiting playing certain games on working days, among them games of cards. Nothing indicates more the persistence of gambling than the continued efforts on the part of civil and ecclesiastic authorities to control the evils associated with it.

One might assume that during the several thousand years of dice play preceding, say, the Renaissance, some rudiments of a probability theory would have appeared. Yet no link between mathematics and gambling was suspected. There is little evidence to suggest that anyone thought that calculating the frequency of dice falls was possible, or even that each face would turn up with equal probability. Several explanations for the late emergence of probabilistic notions have been offered. Possibly the dice used were imperfectly balanced, precluding noticeable regularity in their falls; or perhaps the absence of a suitable mathematical notation hampered a critical investigation into the laws of chance. A more powerful reason might be that randomness itself was an idea alien to the prevailing modes of thought. Not only according to the Christian Church's teachings on absolute truth but in accord also with far more ancient ideas of divination, God or the gods directed earthly events. Whatever the reasons might have been, it is undeniable that a doctrine of chance happenings took a remarkably long time to develop.

Cardan's *Liber de Ludo Aleae* contains the first reasoned considerations relating games of chance to a rudimentary theory of probability. Written around 1550, the manuscript was discovered posthumously among Cardan's papers in 1576. It was not printed until 1663; but Cardan's lectures were popular, so the contents were no doubt familiar to his pupils and friends. Cardan gave a rough definition of probability as a ratio of equally likely events: the probability of a particular outcome was the sum of the possible ways of achieving that outcome divided by "the whole circuit," the totality of possible outcomes of an event. (The "whole circuit" is 6 for one die.) Cardan investigated the probabilities of casting astragali and one or more dice, and he undertook to explain the occurrence of card combinations for the game of primero (similar to poker). A significant passage in a chapter entitled "On the Cast of One Die" reads:

> I am as able to throw 1, 3, or 5 as 2, 4 or 6. The wagers are therefore laid in accordance with this equality if the die is honest and, if not, they are made so much the larger or smaller in proportion to the departure from true equality.

For the first time, we find a transition from empiricism to the theoretical concept of a fair die. In making it, Cardan probably became the real father of modern probability theory.

One of the earliest problems that can be classified as a question in probability theory concerns the fair division of stakes between two players when the game is interrupted before its conclusion. This is first found in print in Fra Luca Pacioli's *Summa de Arithmetica, Geometrica, Proportioni, et Proportionalita* (1494), where among the "unusual questions," he proposed the following:

> A team plays ball so that a total of 60 points is required to win the game and the stakes are 22 ducats. By some incident, they cannot finish the game and one side has 50 points, and the other 30. What share of the prize money belongs to each side?

The form of the problem suggests that it is Arabic in origin, although it is not contained in Fibonacci's *Liber Abaci* (1202), which brought many Arabic puzzles to the West. The question is certainly not original with Fra Luca, having appeared in Italian mathematical manuscripts as early as 1380.

Fra Luca's response to the "problem of points" is that the stakes should be divided in the proportion 5:3, the ratio of the points already scored. In 1539, in a treatise entitled *Practica Arithmeticae Generalis* (Practical General Arithmetic), Cardan wrote with acid superiority about the errors in Fra Luca's work, "And there is an evident error in the determination of the shares in the game problem as even a child should recognize, while he [Fra Luca] criticizes others, and praises his own excellent opinion." Cardan pointed out that the solution did not take into account the number of points yet to be won by each team. He then propounded an alternative solution, equally incorrect: namely, if x and y are the number of points won by each of the teams when the total number required to win is z, the stakes should be divided in the ratio

$$1 + 2 + 3 + \cdots + (z - x):1 + 2 + 3 + \cdots + (z - y).$$

Cardan's arch enemy, Tartaglia, also dealt with fair stakes in his monumental *Generale Trattato* (General Treatise) of 1556. In a section entitled "Error di Fra Luca dal Borgo," Tartaglia pointed out, "His [Fra Luca's] rule seems neither agreeable nor good, since if one player has, by chance, ten points and the other no points, then by following this rule, the player who has the ten points would take all the stakes, which obviously does not make sense." As opposed to Fra Luca, who recommended that the stakes be divided in proportion to the number of points won by each player, Tartaglia recommended that the deviation from half of the stake should be proportional to the difference of the points won by the players. That is, Tartaglia's rule required that if z were the total number of points needed for completion of the game and x and y were the points in hand, then the player with the most points should take a portion

$$\frac{1}{2} + \frac{x - y}{2z}$$

of the stake. (Thus, the total stake in Fra Luca's problem would have been divided in the ratio 2:1.) Again we have an incorrect solution, because a fair division should be proportional to the probability of winning the whole stake. Toward the end of the section, Tartaglia seems to have given up, with the remark, "Therefore I say that the resolution of such a question is judicial rather than mathematical, so that in whatever way the division is made there will be cause for litigation." The "problem of points" remained in most arithmetic texts until well into the 1600s. The problem is so difficult that its solution (the correct ratio is 7:1) by Pascal in the spring of 1654 may well be considered a decisive breakthrough in the history of probability theory.

Blaise Pascal (1623–1662) is popularly linked with Fermat as one of the joint founders of probability theory. There is no doubt that Pascal was an exceptionally versatile man. It is unusual to find a person who is at the same time a gifted writer and a religious philosopher, as well as a creative mathematician and an experimental physicist. He is best known to the general reader as author of the first great prose classics in the modern French language, *Lettres Provinciales* and *Pensées*. As a mathematician, Pascal has been described as the greatest might-have-been in history. His mathematical reputation rests more on what he might have done than on what he

Blaise Pascal
(1623–1662)

(Smithsonian Institution.)

actually accomplished. During much of his life, his researches were impeded by poor health—he suffered from intolerable stomach pains accompanied by violent migraine headaches—and religious concerns. Pascal's three bursts of mathematical activity fell almost wholly in 1640, 1654, and 1658; hardly enough time to create the body of work of a great mathematician.

Blaise Pascal's mother died when he was but 3 years old, and he was brought up by his father, Etienne Pascal, a local judge who was himself a mathematician of some stature. Etienne Pascal was so desirous of supervising the education of his children, especially that of the precocious Blaise, that he sold his post and moved the family to Paris in 1630. Etienne's theories on education were somewhat unorthodox, and the young Pascal was reared without ever having gone to school or ever having studied at a university. He was taught exclusively by his father, who would permit no one to help or to interfere with his plan for teaching the boy. From an early age, Blaise appears to have been a sickly child, subject to nervous disorders. It is not clear whether Etienne Pascal feared that the charm of mathematics would distract the little boy from other interests or whether he was afraid of overtaxing his son's fragile health; but according to the father's planned course of education, the youngster was to be kept from studying mathematics until he was about 15 years old. To enforce this prohibition, all textbooks on mathematics were locked up, out of sight, and the subject was a forbidden topic in the daily instruction. But Blaise Pascal was one of those prodigies whose genius would have surmounted any method of schooling or restraints on his curiosity. Although the elder Pascal was an excellent tutor, he made a poor psychologist.

At the age of 12, the youngster's natural curiosity got the better of his father's educational scheme. Blaise insisted on knowing what geometry was, and when he was given evasive answers, he set to work on it himself. Etienne Pascal entered the play-room quietly one day and found his son drawing charcoal figures on the floor tiles. Calling lines and circles "bars" and "rounds," the boy had been working out for himself the principles of geometry, and was at the moment trying to prove his guess that the sum of the angles of a triangle is two right angles (Euclid's Proposition 32). With tears of joy the father made quick revision of his educational plans; Blaise was given a copy of Euclid's *Elements* to study.

Although the substance of this oft-told tale is no doubt true, it is frequently exaggerated to assert that the little Pascal rediscovered for himself all the propositions of Euclid up to the thirty-second and found them in the same order in which Euclid set them forth. But the order of Euclid's first 32 propositions is not so logically inevitable that any mathematical innocent, however gifted, would reproduce them as they stand. A more credible explanation is that with the help of the axioms that he had worked out, and through his own course of reasoning, Pascal reached Proposition 32. For a youngster of twelve to have discovered this on his own is already enough to remind us of a Mozart composing symphonies at the same age.

A feature of the time was the formation of sociable groups in which the arts and sciences could be discussed seriously, but informally, outside academic settings. When Father Marin Mersenne founded an "academy" to exchange information on mathematical and scientific topics of the day, Etienne Pascal was one of the first members. The evident genius of young Blaise, then 14 years old, prompted his father to bring the youth to the weekly gatherings and even encouraged him to contribute his own share of ideas. Having mastered Euclid's *Elements* without assistance, the boy was now making rapid progress in geometry. The bent of his mind, eager to know the reason for everything, could have found no more satisfying subject to study. At 16, the young Pascal put his mathematical abilities to the test before Mersenne's august circle by offering for discussion a handbill that bore the simple title *Essay pour les coniques*. Printed in 1640 on a single piece of paper, this broadsheet is, for all its brevity, one of the most fruitful pages in the history of mathematics.

The *Essay* contained the statements of a number of general theorems of a projective nature, including the equivalent of what has since become known as Pascal's mystic hexagon theorem: If a hexagon is inscribed in a circle, the three points of intersection of pairs of opposite sides lie on a straight line. To spell it out in more detail, Pascal had this to say: Inscribe any six-sided polygon in a circle (or more generally, in a conic) and label the vertices *A, B, C, D, E, F*. Extend a pair of opposite sides, say *AB* and *DE*, until they meet at a point *P*; extend another pair of opposite sides, say *AF* and *CD*, until they meet at a point *Q*; finally extend the third pair of sides so that they meet at a point *R*. The mystic hexagon theorem states that *P, Q,* and *R* will always lie on one straight line.

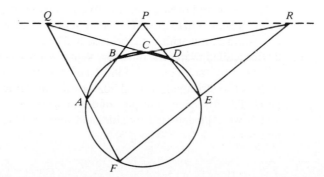

Pascal closed his essay by saying, "There are many other problems and theorems, and many deductions that can be made from what has been stated above. . . ." The *Essay pour les coniques* was an announcement of a treatise he was preparing. Mersenne reported that in this treatise Pascal had deduced no fewer than 400 propositions on conic sections from the mystic hexagon theorem (unfortunately, the work was never published). One of the propositions is the construction of the tangent to a conic at a given point; for this, it is enough to regard the tangent at a point *E* as a line through two coincident points and apply the mystic hexagon theorem to these and four other points of the conic.

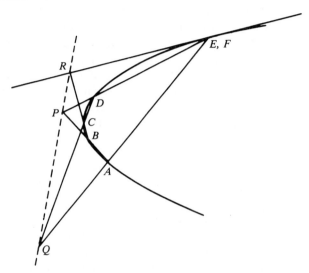

Amid the chorus of praise that greeted the *Essay,* the voice of Descartes was the exception. In reply to an enthusiastic letter from Mersenne, he could only grumble, "I cannot pretend to be interested in the work of a boy." The *Essay pour les coniques* was an auspicious beginning for a mathematical career, but it was not Blaise's habit to think long about any particular aspect of mathematics. He soon turned his attention, and limited energy, in another direction.

At the end of 1639, Etienne Pascal was appointed by the government to a high administrative post, where his principal task was to straighten out the chaotic tax records of the province of Normandy. The sight of his father wearily coping, night after night, with endless columns of figures put an idea into Blaise's head. Because these calculations required a machinelike precision, why not turn over this drudgery to its proper agent, a machine? The thought of using a mechanical calculator to speed up the execution of the arithmetic process had occurred to other scientists. Already, a clumsy device called Napier's rods, or Napier's bones, was used in some countinghouses. Invented by the Scotsman John Napier in 1617 (the same Napier who developed the tables of logarithms), this calculator was similar to a multiplication table with movable parts. The first nine multiples of a unit were inscribed on rods made of wood or bone, which were placed side by side, and then matched to produce the desired product of numbers.

For two years Blaise was completely absorbed in designing a clockwork apparatus that would reduce the two operations of elementary arithmetic—addition and multiplication—to the simple movement of gears and shafts. The new contrivance, which was called the Pascaline, resembled the desk calculator of the 1940s; a complete revolution of a numbered wheel communicated a movement of one unit to the wheel of the next higher order. A complex mechanism the size of a shoe box (14 inches long, 5 inches wide, and $3\frac{1}{2}$ inches high), it would add a column of eight figures.

Because the greatest shortcoming of the ''arithmetic machine'' was the excessive friction of the gears, Pascal spent the next ten years constructing some fifty models of various design and materials. The final version, for whose manufacture a royal monopoly had been obtained, was exhibited in Paris in 1652. Pascal had hopes of enriching himself by this invention; but its high price—a hundred livres was quite a sum—limited its sale and rendered the Pascaline more a curiosity rather than a useful device. The mechanism had no immediate successors, nor did it establish a vogue, so that it would be only fair to call Charles Babbage (1792–1871), rather than Pascal, the prophet who foresaw the future of the computer.

The story of calculating devices since the time of Pascal and Leibniz is highlighted by the invention of the so-called Difference Engine of Babbage, a Lucasian professor of mathematics at Cambridge. In 1822, Babbage built a small, hand-cranked machine that could generate logarithmic and astronomical tables to an accuracy of six decimal places. A successful demonstration convinced the British government to back the construction of a full-sized engine capable of twenty-place accuracy. Babbage's work took much longer than anticipated and official support was finally withdrawn in 1842, after the government had spent £17,000 and Babbage had contributed £6000 himself. Babbage, far from being discouraged, continued with the design and construction of a much more powerful tool—an Analytic Engine powered by steam—whose operating instructions were to be stored on punched cards. Although this machine uncannily foreshadowed modern equipment, it also was never completed, partly because of the enormous cost involved and partly because the contemporary techniques of precision engineering were unequal to the task.

In 1651, Etienne Pascal died and Blaise was left alone in Paris, unfettered by family associations; he is described as being, for the next two or three years, a man of the world leading a dissolute life. If he allowed the charms of society to capture him, Pascal had in no sense given up mathematics and science entirely. During this ''profane period'' of 1651–1654, he conducted various experiments on the pressure exerted by gases and liquids. (Earlier, in 1648, Pascal had demonstrated that the height of a column of mercury varied with altitude, by having measurements carried out simultaneously at the top and foot of a small mountain.) In this period, he also made numerous discoveries relating to the binomial coefficients, which were set forth in his *Traité du Triangle Arithmétique* (1654). And he laid the foundations of probability theory through his correspondence with Fermat. More than ever, he felt himself possessed by a zeal for scientific activity, which aside from its intrinsic interest, conferred on him a certain renown in the intellectual world and in society. But at the age of 31, Pascal experienced an intense religious ecstasy which was to transform his life completely.

November 23, 1654, was a dividing point in Pascal's life; on that day he resolved to abandon the study of mathematics to devote himself single-mindedly to religious activity. A commonly dismissed story is that his conversion was provoked by an

Pascal's Essay pour les Coniques *(1640). (From* A Concise History of Mathematics *by Dirk Struik, 1967, Dover Publications, Inc., N.Y.)*

accident on the bridge at Neuilly over the Seine. As the tale goes, Pascal was driving in a carriage with either four or six horses when the animals bolted. At a place on the bridge that had no protective barrier, the runaway horses leaped into the water and would have dragged the carriage after them had not the traces miraculously snapped. Pascal is supposed to have regarded this lucky escape from death as a symbol of God's will that he should thereafter consecrate his talents to the Christian faith. If we are to believe Pascal, his religious crisis was the result of a mystical experience in which God came to him and spoke for two hours. On emerging from his vision, Pascal seized a piece of paper and wrote at headlong speed an incoherent account of his revelation, copying it afterwards onto parchment. Both were found after his death, sewed into the lining of his clothing near his heart.

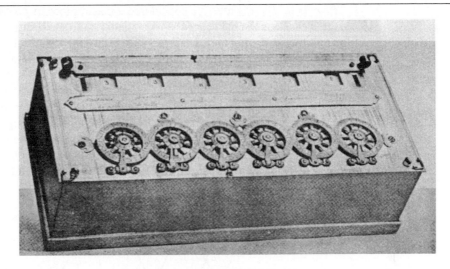

The calculating machine built by Pascal in 1642. *(Extract taken from* A History of Science, Technology and Philosophy in the 16th and 17th Centuries, *by A. Wolf. Reproduced by kind permission of Unwin Hyman Ltd.)*

During the years 1655–1658, Pascal composed those works that are at the center of his literary importance. At this time, the Catholic Church was involved in a doctrinal controversy over the matter of divine grace. The antagonists in the dispute were the Jesuits and another group within the Church known as the Jansenists, because of their adherence to the beliefs of Cornelius Jansen, Bishop of Ypres. When one of Pascal's friends, embracing the Jansenist cause, was condemned for heresy, Pascal undertook a vigorous defense. In a series of eighteen satirical pamphlets, under the title *A Letter Written to a Provincial by One of His Friends,* Pascal displayed the moral teachings of the Jesuits in the worst light. Written under the pseudonym Louis de Montalte and addressed to a nonexistent friend in the country, these *Provincial Letters* were published secretly at great risk to printer and author. (In seventeenth century France, Catholicism was the state religion and ridiculing sacred things was a political crime.) Yet the *Letters* caused a sensation, finding their way into every corner of France. The number of copies run off was enormous for the time (12,000 for the first letter printed in 1656) and the readership estimated at a million. In time, the controversy attending the *Provincial Letters* was forgotten, leaving only the clarity of Pascal's thoughts and the persuasiveness of his ideas.

At the height of the popularity of the *Provincial Letters,* Pascal suddenly abandoned the world of clandestine journalism to write an apology for the Christian religion. However, his always fragile health began to fail steadily in 1658; he died four years later at the age of 39, before the book was completed. The accumulated notes in which he had jotted down his thoughts in preparation for this monumental project were subsequently collected and published as the *Pensées de M. Pascal sur la Religion.* As the

Provincial Letters must be considered a masterpiece of polemical religious writing, so the *Pensées* must be regarded as a vindication and exaltation of faith. These works remain classics for their literary value in an age when their content has ceased to be important. Their influence on the literature of France and of the world was permanent.

For only one brief period, toward the middle of 1658, did Pascal's thoughts revert to mathematics. This was in connection with a curve that was enjoying a considerable vogue in its own special world, the cycloid (the curve traced out by a point on the circumference of a wheel, as the wheel rolls along a straight line). On account of its elegant properties and the endless quarrels that it has engendered among eminent mathematicians who have pretended to its favour, the cycloid has been called the Helen of Geometry. The accompanying figure shows a point P on a cycloid determined by three different positions of the circle.

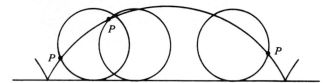

It happened that one night Pascal was unable to sleep owing to the tortures of a violent toothache. To take his mind off the pain, he began to reflect about the cycloid; after a time he realized that his tooth had ceased to ache. Taking this as a sign of divine permission to continue, Pascal worked furiously for eight days to solve anew many of the problems of the cycloid, such as the area under one arch or the volume of the solid obtained by revolving the curve about the base line. The solutions were later published in a brief *History of the Cycloid* and in four *Letters* written under the pseudonym Amos Dettonville (an anagram on the name Louis de Montalte, which Pascal had made famous through the *Provincial Letters*). Perhaps he did not want the world to know that he had lapsed from grace and engaged again in the science that he had once renounced.

Although not publicly available, the solution of the quadrature of the cycloid curve had been achieved some years before Pascal by Gilles Personne de Roberval (1602–1675), professor of mathematics at the Collège Royal in Paris and a member of the circle that gathered around Mersenne. The publication of his proof seems to have been delayed until 1693, when his *Traité des Indivisibles* appeared. Roberval's secrecy can perhaps be attributed to the customs governing the triennial renewal of the professorial chair he occupied. The renewal was made on the basis of an open mathematical competition, the questions for which were propounded by the incumbent. By not disclosing his methods, he could have provided himself with a favorable set of questions for the coming mathematical contests. (Roberval was indeed proclaimed winner of the 1634 competition, and held his office for the rest of his life.) But this practice also resulted in loss of credit to Roberval for priority of discovery.

To determine the area under half of one arch of the cycloid traced out by a circle of diameter $OC = 2a$, Roberval devised the following procedure: For each point P on the cycloidal arch, extend the chord DF to a point Q where $DF = PQ$. The locus of Q is a new curve OQP, which Roberval called the ''companion'' to the cycloid curve.

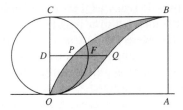

Roberval first proves that the companion curve divides the rectangle $OABC$ into two equal pieces, each of which has area πa^2. Indeed, the rectangle as a whole has altitude $2a$ and a base whose length is equal to the semiperimeter πa of the generating circle. Also all the horizontal lines that make up the region between the cycloid and its companion (shown shaded) correspond with, and are equal to, all the lines of the semicircle OFC. Roberval argues that the ''sum of these lines'' in this figure must therefore yield the area of the semicircle, namely, $\frac{1}{2}\pi a^2$. It follows that

$$OABP = OPBQ + OABQ = \pi a^2 + \tfrac{1}{2}\pi a^2 = \tfrac{3}{2}\pi a^2.$$

Upon doubling, the area under an arch of the cycloid is found to be three times that of the generating circle.

The work of Pascal on the cycloid had one by-product of surprising importance: It served as the immediate inspiration for Leibniz in his invention of the differential and integral calculus. One of Pascal's *Letters,* entitled *Traité des sinus du quart de circle* (1658), involved certain calculations that resemble the evaluation of the definite integral of the sine function. Pascal was, however, unable to hit upon the crucial point that would give a general theory of integration. Leibniz later wrote how when he was reading the *Traité* a light suddenly burst on him and then he realized what Pascal had not realized. (In 1703, in a letter to the mathematician James Bernoulli, he remarked that Pascal sometimes seemed to have bandages over his eyes.) Perhaps if Pascal had not died so young, or if he had not abandoned mathematics for the religious life, he would have had the honor accorded Newton and Leibniz for the discovery of the calculus.

The year 1654 is a landmark in the history of probability theory; indeed, it is usually called the date on which this science began. Bothered by certain problems in gambling, a member of the French nobility, the Chevalier de Méré, sent some questions to Pascal. His interest aroused, Pascal turned in correspondence to Pierre de Fermat, the leading mathematician of the day, and in the active interchange of letters that ensued, both men produced the basic results on which the subject now rests.

The initiator of this correspondence, Antoine Gombaud, Chevalier de Méré (1607–1684), was an interesting character. A soldier with a brilliant record, a linguist and classical scholar, he had traveled in England, Germany, Spain, and even America. The perfect courtier, de Méré became the self-appointed mentor of society, devoting his life to the vocation of teaching good manners. A favorite guest in the great chateaux, he was acquainted with most of the leading mathematicians of the period, including Pascal. Because de Méré was not a member of the continuance of what had been Mersenne's circle, they may have met in less academic surroundings—probably at dicing, which was a fashionable diversion in the drawing rooms of the seventeenth century. Pascal was then at the beginning of his so-called worldly period, and de Méré took some credit for having brought it on.

De Méré made a precarious living at cards and dice, wagering according to a private arithmetic of probabilities. Although not an accomplished mathematician, he knew enough about the subject to pose some interesting problems for which he got the ideas at his customary gambling resorts. That he had a good opinion of his own mathematical talent is clear from an excerpt of a letter written to Pascal sometime after 1656: ''You must realize that I have discovered in mathematics things so rare that the most learned among the ancients have never thought of them and which have surprised the best mathematicians in Europe.'' Leibniz, on the other hand, called him ''intelligent but half-learned and half-comprehending.''

One of de Méré's gaming problems gave Pascal the occasion to begin his now famous correspondence with Fermat. In a letter dated July 29, 1654, Pascal wrote:

> I haven't time to send you a solution of a difficulty which has puzzled M. de Méré. He has good intelligence, but he isn't a geometer and this, as you realise, is a bad fault.

A little later in the same letter Pascal reported on de Méré's views:

> The Chevalier de Méré said to me that he found a falsehood in the theory of numbers for the following reason. If one wants to throw a six with a single die, there is an advantage in undertaking to do it in four throws, as the odds are 671 to 625. If one throws two sixes with a pair of dice, there is a disadvantage in having only 24 throws. However, 24 to 36 (which is the number of cases for two dice) is as 4 to 6 (which is the number of cases on one die). This is the great ''scandal'' which makes him proclaim loftily that the theorems are not constant and Arithmetic is self-contradictory.

De Méré's reasoning in the passage just cited seems to be this. A perfect die has six faces, marked with from one to six dots; any one of these faces is equally likely to show up on a single cast, because nothing in the shape of the die or method of throwing it favors a particular face. Thus, the probability that a six will appear in a single throw is $\frac{1}{6}$. This might be restated by saying that the probability of not getting a six is $\frac{5}{6}$. If the die is thrown twice, there are $6 \cdot 6$ possible outcomes and $5 \cdot 5$ of them do not produce a six either time, so that the probability of not getting a six in two throws is $\left(\frac{5}{6}\right)^2$. Continuing the argument, it can be concluded that in n throws the probability of not getting a six is $\left(\frac{5}{6}\right)^n$. Hence, the opposite situation, having a six turn up at least once, occurs with probability

$$1 - \left(\frac{5}{6}\right)^n.$$

To have a better than even chance, one must have

$$1 - \left(\frac{5}{6}\right)^n > \frac{1}{2}.$$

It turns out that when $n = 4$,

$$1 - \left(\frac{5}{6}\right)^4 = 1 - \frac{625}{1296} = \frac{671}{1296},$$

which is greater than $\frac{1}{2}$. Thus de Méré could profitably bet on throwing a six at least once in four throws with a single die.

When throws with two dice are considered, the computations become more laborious. Suppose we paint one die red and the other green so that we can tell them apart. Let (a, b) denote the event in which the red die shows an a and the green die shows a b. Then all possible outcomes for a throw of these colored dice can be exhibited in a six-by-six array.

$$
\begin{array}{cccccc}
(1, 1) & (1, 2) & (1, 3) & (1, 4) & (1, 5) & (1, 6) \\
(2, 1) & (2, 2) & (2, 3) & (2, 4) & (2, 5) & (2, 6) \\
(3, 1) & (3, 2) & (3, 3) & (3, 4) & (3, 5) & (3, 6) \\
(4, 1) & (4, 2) & (4, 3) & (4, 4) & (4, 5) & (4, 6) \\
(5, 1) & (5, 2) & (5, 3) & (5, 4) & (5, 5) & (5, 6) \\
(6, 1) & (6, 2) & (6, 3) & (6, 4) & (6, 5) & (6, 6)
\end{array}
$$

Each of the combinations in which the faces have different values is counted twice, because they represent different throws. For instance, the $(1, 2)$ that corresponds to a one on the red die and a deuce on the green die is a different throw from the reverse order $(2, 1)$. A direct count shows that there are 36 possible outcomes in casting a pair of dice.

De Méré inquired into the number of rolls necessary to ensure a better than even chance of getting a double six. He thought that the smallest advantageous number of tosses should be 24, but patient observation had shown him that 25 throws were required. The answer 24 is gotten by an ancient but mistaken gambling rule that runs as follows. Consider a situation in which there is one chance in N of winning on a single trial, and let n be the number of trials required to have a better than even chance of success. De Méré's "rule" is that n/N is a constant. In the case in which one attempts to make a six with a single die, $N = 6$ and $n = 4$, so that $n/N = \frac{2}{3}$. Hence if $N = 36$, as for two dice in tosses aiming at a double six, n must be 24.

Pascal had little difficulty in disposing of de Méré's "great scandal." In the case of two dice, there are 36 possible throws and 35 of these do not give a double six; hence, the probability of not getting two sixes is $\frac{35}{36}$. If the dice are cast n times, then the probability of no throw of double six is $(\frac{35}{36})^n$. This means that the opposite event, getting double sixes at least once, occurs with probability

$$
1 - \left(\frac{35}{36}\right)^n.
$$

When $n = 24$ and 25, it is found that

$$
1 - \left(\frac{35}{36}\right)^{24} = 0.4914,
$$

$$
1 - \left(\frac{35}{36}\right)^{25} = 0.5055.
$$

Therefore, 25 is the smallest number of rolls of two dice that offers a probability greater than $\frac{1}{2}$ of achieving a double six.

9.2   Pascal's Arithmetic Triangle

The arithmetic triangle, now generally known as Pascal's triangle, is an infinite numerical table in "triangular form," where the nth row of the triangle lists the successive coefficients in the binomial expansion of $(x + y)^n$.

$(x + y)^0$					1					Row 0
$(x + y)^1$				1		1				Row 1
$(x + y)^2$			1		2		1			Row 2
$(x + y)^3$		1		3		3		1		Row 3
$(x + y)^4$	1		4		6		4		1	Row 4
$(x + y)^5$	1	5		10		10		5	1	Row 5

. . . .

Thus, for example, the numbers 1, 5, 10, 10, 5, 1 in our Row 5 are the coefficients in the expansion of $(x + y)^5$. The beginnings of the binomial theorem were found relatively early in the development of mathematics—the identity $(x + y)^2 = x^2 + 2xy + y^2$ occurred, in geometric language to be sure, in Book II of Euclid's *Elements*. Therefore, tables of binomial coefficients appeared in one form or another long before the publication of Pascal's famous *Triangle Arithmétique*. It is largely an accident of history that the arithmetic triangle should be named after Blaise Pascal rather than for one of his many anticipators in this arrangement of numbers.

A triangular arrangement of binomial coefficients through the eighth power, written in rod numerals and a round zero sign, is found in the treatise *The Precious Mirror of the Four Elements*, written by the Chinese mathematician Chu Shih-Chien in 1303. But a knowledge of the triangle is probably much older, for Chu spoke of the diagram as an ancient method that was not his. Chinese references indicate that the triangle appeared in a work of Chia Hsien about 1050. Indeed, a similar arrangement of binomial coefficients was known to the Arabs about the same time that it was being used in China. Omar Khayyam (circa 1050–1130), known in the West as the author of the *Rubaiyat*, referred in *On Demonstrations of Problems of Algebra and Almucabola* to a lost work in which he dealt with the arithmetic triangle. Another Persian mathematician, Al-Tusi (circa 1200–1275), in a work called *Collection on Arithmetic by Means of Board and Dust*, approximated the value of the square root $\sqrt{a^n + r}$ by $a + r/[(a + 1)^n - r]$, where the denominator was calculated by the binomial expansion. For this purpose, Al-Tusi furnished a table of binomial coefficients in triangular form up to the twelfth power. Thus, the so-called Pascal triangle, like the so-called Pythagorean theorem, is in reality the product of a much earlier Eastern culture.

The first triangular arrangement of the binomial coefficients to be printed in European books appeared on the title page of the *Rechnung* (1527) of Peter Apian

(1495–1552). Apian, a professor of astronomy at the University of Ingolstadt, is interesting if only because he taught in the Germanic tongue at a time when the prevailing custom was to use Latin. Although the arithmetic triangle had not previously been described in the West until depicted in Apian's text, it seems to have been discovered almost simultaneously by several authors of the 1500s. Michael Stifel (1486–1567) in his *Arithmetica Integra* (1544) calculated the device as far as the seventeenth line, though as he pointed out, there was no reason to stop there. Stifel's diagram took the form

1				
2				
3	3			
4	6			
5	10	10		
6	15	20		
7	21	35	35	
8	28	56	70	
9	36	84	126	126
10	45	120	210	252,

where each column after the first starts two places lower than the preceding one. Other schematic arrangements for the binomial coefficients were given by the old rivals Tartaglia and Cardan. Tartaglia in the *Generale Trattato* (1556) gave the numbers in the triangle through the eighth line, claiming the idea as his own invention. Cardan, in the work *Opus Novum de Proportionibus* (1570), presented a table to 17 lines, citing Stifel as the original discoverer.

Although Pascal was not the originator of the arithmetic triangle, being nearly the last of a long line of "discoverers," his name is forever linked with the triangle because he was the first to make any sort of systematic study of the relations it exhibited. The merits of Pascal's work in this regard are enough to justify the use of his name. The printing of Pascal's *Traité du Triangle Arithmétique* was finished towards the end of 1654 (Fermat received his copy sometime before September of 1654), but as Pascal had withdrawn from worldly matters, it was not generally distributed before 1665. The work is an exposition of the properties and relations between the binomial coefficients, and it includes a few general probability principles. In particular, in a section entitled *Utilization of the Arithmetic Triangle to Determine the Number of Games Required Between Two Players Who Play a Large Number of Games,* Pascal applied the arithmetic triangle to the problem of stakes in games of chance.

In the *Triangle Arithmétique,* Pascal did not write the arithmetic triangle from the top down as we now do, but instead expressed the table as shown:

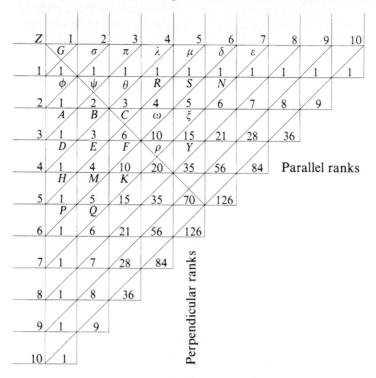

The numbers on the *n*th upward-sloping diagonal in this arrangement give the coefficients in the expansion of the binomial $(x + y)^n$. Thus, the figure that modern texts call Pascal's triangle differs from the triangle examined by Pascal himself by a rotation through 45°. In his scheme, Pascal called positions in the same vertical column cells of the same perpendicular rank, and those in the same horizontal row cells of the same parallel rank. Cells in the same northeast-running diagonal were said to be cells of the same base.

Pascal observed that his table could readily be enlarged by affixing further numbers, obtained without recourse to the binomial theorem. Once the 1s in the top line and left-hand column have been set down, any other entry is the sum of the number directly above the entry and the number immediately to the left of the entry. As he put it: "The number of each cell is equal to that of the cell which precedes it in its perpendicular rank, added to that of the cell which precedes it in its parallel rank." For instance, underneath the 5 in the second row and alongside the 10 of the third, a new number can be placed; this new entry is 15, the sum of 5 and 10. Pascal had no claim to originality here, for this generating rule had been accurately explained by several earlier European and Arabic "discoverers" of the arithmetic triangle.

When the arithmetic triangle is written in its present-day form,

$$
\begin{array}{ccccccccccc}
& & & & & 1 & & & & & \\
& & & & 1 & & 1 & & & & \\
& & & 1 & & 2 & & 1 & & & \\
& & 1 & & 3 & & 3 & & 1 & & \\
& 1 & & 4 & & 6 & & 4 & & 1 & \\
1 & & 5 & & 10 & & 10 & & 5 & & 1
\end{array}
$$

. . . .

it is equally easy to obtain the next horizontal row of numbers from any given row; each entry in the interior of the triangle is the sum of the numbers nearest (diagonally) to it in the row above. According to this rule of addition, the next horizontal row of the triangle, which has not been filled in, would consist of

$$1, 1 + 5 = 6, 5 + 10 = 15, 10 + 10 = 20, 10 + 5 = 15, 5 + 1 = 6, 1.$$

Knowing how to write down the arithmetic triangle permits quick computation of $(x + y)^n$ for small values of n. For instance, our newly added row furnishes the coefficients for $(x + y)^6$, so that

$$(x + y)^6 = x^6 + 6x^5y + 15x^4y^2 + 20x^3y^3 + 15x^2y^4 + 6xy^5 + y^6.$$

Straightforward expansions of $(x + y)^n$ had been known long before Pascal's time, but these were obtained by direct multiplication, not by applying any recursion formula for the coefficients.

Custom has settled on $\binom{n}{r}$ as the standard symbol for the coefficient of $x^{n-r}y^r$ in the expansion of $(x + y)^n$, so that we can write

$$
(x + y)^n = \binom{n}{0} x^n + \binom{n}{1} x^{n-1}y + \binom{n}{2} x^{n-2}y^2 + \cdots
$$
$$
+ \binom{n}{n-1} xy^{n-1} + \binom{n}{n} y^n.
$$

Euler designated the binomial coefficients by $\left(\dfrac{n}{r}\right)$ in a paper written in 1778, but not published until 1806, and used the device $\left[\dfrac{n}{r}\right]$ in a paper of 1781, published in 1784. The notation $\binom{n}{r}$, which was to become the more common convention, appeared about 1827.

Stated in this symbolism, the property of the numbers $\binom{n}{r}$ on which the arithmetic triangle is based is

$$\binom{n}{r-1} + \binom{n}{r} = \binom{n+1}{r}.$$

This rule, of course, does not give a formula for the binomial coefficients but tells us only how to build them from two numbers on the previous base. It would be desirable to have a computation scheme that is independent of such earlier knowledge. Contained within Pascal's treatise there is an explicit formula for $\binom{n}{r}$, to wit,

$$\binom{n}{r} = \frac{n(n-1)(n-2)\cdots(n-r+1)}{1\cdot 2\cdot 3\cdots r}, \qquad 1 \le r \le n.$$

If we use the space-saving notation $n!$ as an abbreviation for the product of the first n integers, then when $1 \le r \le n-1$, the foregoing formula can be expressed more simply as

$$\binom{n}{r} = \frac{n!}{r!(n-r)!}.$$

Of special interest to us is the symbol $0!$, which we shall define to have the value 1. With this convention, the expression $\dfrac{n!}{r!(n-r)!}$ represents $\binom{n}{r}$ even for $r = 0$ and $r = n$. For in these cases,

$$\binom{n}{0} = 1 = \frac{n!}{0!(n-0)!} \qquad \text{and} \qquad \binom{n}{n} = 1 = \frac{n!}{n!(n-n)!}.$$

The symbol $n!$, called ''n factorial,'' was first introduced by Christian Kramp of Strasbourg in his *Elements d'Arithmétique Universelle* (1808) as a convenience to the printer. It did not gain immediate acceptance, and the alternative symbols $\lfloor n$ and $\pi(n)$ were used until late in the nineteenth century. Some English textbooks suggested the reading ''n-admiration'' for $n!$, because the exclamation point ! is a note of admiration.

Pascal listed 19 properties, or consequences as he called them, of the binomial coefficients that could be discovered from the arithmetic triangle. Among them are the identities

$$\binom{n}{r} = \binom{n-1}{r-1} + \binom{n-2}{r-1} + \binom{n-3}{r-1} + \cdots + \binom{r-1}{r-1} \qquad (\textit{Consequence II})$$

$$\binom{n}{r} = \binom{n}{n-r} \qquad (\textit{Consequence V})$$

$$2^n = \binom{n}{0} + \binom{n}{1} + \binom{n}{2} + \cdots + \binom{n}{n} \qquad (\textit{Consequence VIII})$$

Loosely interpreted, these tell us that each number in the triangle is the sum of the entries in the preceding ''northeast-running'' diagonal, beginning with the entry above the given number; that the numbers in any base are symmetric with respect to the middle term or terms; and that the sum of the numbers in the nth base (that is, the sum of the coefficients in the expansion of $(x+y)^n$) equals 2^n. A less obvious relation developed by Pascal involves the ratio of successive terms in any base of the triangle. Consequence XII amounts to the statement

$$\frac{\dbinom{n}{r+1}}{\dbinom{n}{r}} = \frac{n-r}{r+1},$$

though Pascal had no symbols like $\dbinom{n}{r}$ available to him. He formulated this arithmetic fact in words, instead of symbols, as:

> In every arithmetic triangle, if two cells are contiguous in the same base, the upper is to the lower as the number of cells from the upper to the top of the base is to the number of those from the lower to the bottom, inclusive.

Consequence XII is of historical importance because its proof involves the first explicit formulation of the demonstrative procedure known as induction. All in all, it would perhaps be best for us to use the term ''mathematical induction,'' for the word ''induction'' by itself conveys an entirely erroneous idea of the method. It really has nothing to do with induction in the sense commonly understood in the experimental sciences, where it signifies drawing a general conclusion from particular cases. Mathematical induction is not a method of discovery but a technique of proving rigorously what has already been discovered.

Although the method is too well known to need an elaborate explanation here, let us briefly examine the principle of mathematical induction. It is perhaps the single most useful tool in the mathematician's kit. A theorem provable by mathematical induction involves a proposition—a formula, a statement of equality or of inequality, or the like—about an integer n that is to be shown true for all positive integers n, or at least for all integers greater than or equal to some one integer n_0. Such a proposition is usually denoted by the symbol $P(n)$, allowing us to speak of individual cases such as $P(1)$ or $P(10)$ corresponding to the integers 1 and 10. The proof involves two stages. The first part establishes that the proposition $P(n)$ is true for the lowest meaningful value of n, say $n = n_0$. The second part of the process consists in constructing a proof of the proposition $P(k + 1)$ based on the assumption that $P(k)$ is true. The rigorous verification that $P(k)$ implies $P(k + 1)$ is a purely deductive procedure using logic and known mathematical facts. Because of the ''hereditary nature'' of $P(n)$, we know that $P(n_0 + 1)$ is true because $P(n_0)$ is true. Then because $P(n_0 + 1)$ is true, $P(n_0 + 2)$ is also true, and so on, until all the positive integers $n \geq n_0$ have been exhausted.

To recapitulate, the proof of a proposition $P(n)$ by mathematical induction consists in showing that:

1. The statement $P(n_0)$ is true for some particular integer n_0.

2. The assumed truth of $P(k)$ implies the truth of $P(k + 1)$.

We can then conclude that the proposition $P(n)$ is true for all $n \geq n_0$. The proof of condition 1 is usually called the ''basis for the induction,'' and the proof of 2 is called the ''induction step.'' The assumptions made in carrying out the induction step are known as the ''induction hypotheses.'' Induction has been likened to an infinite row of dominoes, all standing on edge and arranged in such a way that when one falls it knocks down the next in line. If either no domino is pushed over (if there is no basis

Title page of the arithmetic of Peter Apian, showing the Pascal triangle. (From Rara Arithmetica *by David Eugene Smith, published by Chelsea Publishing Company in 1970.)*

for the induction) or the spacing between them is too large (the induction step fails), then the complete line will not fall.

In a typical illustration of mathematical induction, let us verify the assertion that the sum of the cubes of the first n positive integers is equal to the square of the nth triangular number. The appropriate statement $P(n)$ in this case is

$$1^3 + 2^3 + 3^3 + \cdots + n^3 = \left[\frac{n(n+1)}{2}\right]^2 = \frac{n^2(n+1)^2}{4}.$$

We shall show that $P(1)$ is true, and show how from $P(k)$ we can prove $P(k + 1)$. The verification of $P(1)$ presents no difficulty, since it says that

$$1^3 = 1 = \frac{1^2 \cdot 2^2}{4}.$$

We now go to the so-called induction step. Suppose that $P(k)$ is true, so that for this value of k,

$$1^3 + 2^3 + 3^3 + \cdots + k^3 = \frac{k^2(k + 1)^2}{4}.$$

To confirm that $P(k + 1)$ holds, we simply add the term $(k + 1)^3$ to both sides of the last equation, to obtain

$$1^3 + 2^3 + 3^3 + \cdots + k^3 + (k + 1)^3 = \frac{k^2(k + 1)^2}{4} + (k + 1)^3$$

$$= \frac{(k + 1)^2[k^2 + 4(k + 1)]}{4}$$

$$= \frac{(k + 1)^2(k + 2)^2}{4},$$

which is just the statement $P(k + 1)$. Notice that in this calculation we merely assumed $P(k)$ to be true in order to show that it would guarantee the truth of $P(k + 1)$. We did not concern ourselves with the actual validity of $P(k)$, but only with the fact that $P(k + 1)$ is a consequence of the assumed truth of $P(k)$. Having thus verified conditions 1 and 2, we see that $P(n)$ must hold for all positive integers according to the principle of mathematical induction.

Although Pascal gave an eminently satisfying explanation of mathematical induction, the gift of the idea of "reasoning by recurrence" can be found earlier, in the work of Francesco Maurolico (Franciscus Maurolycus). Still, Pascal was the first to recognize the value of this logical process and, through his *Triangle Arithmétique*, brought it into the common domain of the working mathematician. Fermat gave no evidence in his writings that he knew of the technique, although his method of infinite descent may be described, loosely, as "reverse induction."

Francesco Maurolico (1494–1575) is generally acknowledged to have been one of the foremost mathematicians of the sixteenth century. Born in Sicily of Greek parents, he was an ordained priest, at one time an abbot, and for many years a professor of mathematics at Messina. Except for short stays in Rome and in Naples, he lived his whole life in his native Sicily. Unfortunately, most of Maurolico's books were not published until after his death and hence came too late to have much influence on the sixteenth century renaissance of mathematical thought. For instance, of his treatises on the treasures of classical mathematics—a text on the *Conics* of Apollonius and a collection of writings of Archimedes—the works of Archimedes did not appear until 1685, the first edition of 1594 having been lost at sea. In Maurolico's time, only the first four books of Apollonius's *Conics* had survived in the original Greek. He tried to reconstruct the then missing Book V on maxima and minima from the brief references that the great Greek geometer had provided in the preface to the entire work. (The reconstruction of the lost works of classical antiquity, and of the last four books of the *Conics* in particular, was something of a vogue in Renaissance Italy.) Maurolico's reconstruction of the *Conics* was completed in 1547, but not published until 1654, by which time several other versions had appeared.

Maurolico, one of the great geometers of his time, also occupied himself with a geometric analysis of lenses, spherical mirrors, and the human eye. The chief record of his optical research, *Photismi de Lumine et Umbra* (1611) anticipated many of the findings in Kepler's own *Astronomiae Pars Optica* of 1604, although Kepler seems to have been unaware of his predecessor's work. Maurolico came remarkably close to the correct explanation of the origin of the rainbow.

The greatest number of Maurolico's mathematical writings are gathered together in his *Opuscula Mathematica* and *Arithmeticorum Libri Duo,* which were published as companion pieces at Venice in 1575. In the Pythagorean tradition, the *Arithmeticorum* contains a variety of the properties of integers noted by Nicomachus in the *Introductio Arithmeticae,* to which Maurolico added his own simple and ingenious proofs. For example, his Proposition 11 shows that the nth triangular number plus the preceding triangular number is equal to the nth square number. Expressed in modern notation, this is $t_n + t_{n-1} = s_n$, where t_n is represented by the formula

$$t_n = 1 + 2 + 3 + \cdots + n = \frac{n(n+1)}{2}.$$

Maurolico's simple proof is this: By the rule of formation of triangular numbers, $t_n = t_{n-1} + n$. Adding t_{n-1} to both sides of this equation gives

$$t_n + t_{n-1} = 2t_{n-1} + n$$

$$= (n-1)n + n = n^2 = s_n,$$

which was the relation to be established.

The first result in Maurolico's *Arithmeticorum* that is actually proved by mathematical induction is the proposition that the sum of the first n odd integers is equal to the nth square number; in present-day symbols, this is

$$1 + 3 + 5 + 7 + \cdots + (2n-1) = n^2.$$

Maurolico's proof is essentially translated as follows:

> By Proposition 13 [this proposition asserts that $n^2 + (2n+1) = (n+1)^2$] the first square number 1 added to the following odd number 3 makes the following square number 4; and this second square number 4 added to the third odd number 5 makes the third square number 9; and likewise the third square number 9 added to the fourth odd number 7 makes the fourth square number 16; and so successively to infinity the result is demonstrated by repeated application of Proposition 13.

This is a clear case of mathematical induction, with Maurolico's Proposition 13 providing the justification for the induction step.

Let us put the proof in the format that we have used earlier. Here

$$1 + 3 + 5 + 7 + \cdots + (2n-1) = n^2$$

is the statement $P(n)$. First observe that $P(1)$ is true, since $1 = 1^2$. Now, to obtain $P(k + 1)$ from $P(k)$, assume that

$$1 + 3 + 5 + 7 + \cdots + (2k - 1) = k^2.$$

Adding the next odd integer $2k + 1$ to both sides gives

$$1 + 3 + 5 + 7 + \cdots + (2k - 1) + (2k + 1) = k^2 + (2k + 1)$$
$$= (k + 1)^2,$$

which is just $P(k + 1)$. Therefore, the assumed truth of $P(k)$ implies the truth of $P(k + 1)$. It follows that $P(n)$ is true for all positive integers n.

Pascal was, as far as we know, the next person to use mathematical induction, and he did this repeatedly in the *Triangle Arithmétique* in connection with the arithmetic triangle and its applications. The induction proofs of Maurolico were presented in a somewhat sketchy style, but Pascal's followed more nearly modern lines. Although nowhere in the treatise was Maurolico's name mentioned, it is clear that Pascal was familiar with that part of the *Arithmeticorum* in which Maurolico used the new logical process. In a well-known letter from Pascal (written under the pseudonym Amos Dettonville) to the mathematician Carcavi, Pascal referred to Maurolico for a proof of the proposition that "twice the nth triangular number minus n equals n^2." He then said, "Cela est aise par Maurolico." ("That is easy, using Maurolico's [method].")

Contained in the proof of Pascal's Consequence XII of the *Triangle Arithmétique*, the result that we would write as

$$\frac{\binom{n}{r}}{\binom{n}{r + 1}} = \frac{r + 1}{n - r},$$

there is an explicit formulation of the principle of mathematical induction. The first part of Pascal's argument reads as follows:

> Although this proposition has an infinite number of cases, I will give a rather short demonstration by supposing two lemmas.
>
> The first one, which is self-evident, is that this proposition occurs in the second base; because it is apparent that ϕ is to σ as 1 is to 1[that is, for $n = 1$,
>
> $$\frac{\binom{1}{0}}{\binom{1}{1}} = \frac{0 + 1}{1 - 0}].$$
>
> The second one is that if this proposition is true for an arbitrary base, it will necessarily be true in the next base. From which it is clear that it will necessarily be true in all bases, because it is true in the second base by the first lemma; hence by means of the second lemma, it is true in the third base hence in the fourth base, and so on to infinity.
>
> It is therefore necessary only to prove the second lemma.

Pascal described here precisely the logical process now called mathematical induction.

It is interesting to translate Pascal's proof of his second lemma into modern notation, for as he said, "This proof is based only on the assumption that the proportion occurs in the preceding base, and that each cell is equal to its preceding plus the one above it."

As indicated in the statement of the lemma, we assume that the ratio in question holds in the kth case. That is, for all permissible values of r ($r = 0, 1, 2, \ldots, k$), suppose

$$\frac{\dbinom{k}{r}}{\dbinom{k}{r+1}} = \frac{r+1}{k-r}.$$

On the basis of this assumption, it is to be proved that

$$\frac{\dbinom{k+1}{r}}{\dbinom{k+1}{r+1}} = \frac{r+1}{(k+1)-r}.$$

We start by using the rule of formation of the numbers in the arithmetic triangle (namely that each entry is the sum of two numbers in the preceding base), to write

$$\frac{\dbinom{k+1}{r}}{\dbinom{k+1}{r+1}} = \frac{\dbinom{k}{r} + \dbinom{k}{r-1}}{\dbinom{k}{r+1} + \dbinom{k}{r}} = \frac{1 + \left[\dbinom{k}{r-1} \middle/ \dbinom{k}{r}\right]}{\left[\dbinom{k}{r+1} \middle/ \dbinom{k}{r}\right] + 1}.$$

If the induction hypothesis is now applied to the quotients of binomial coefficients that appear on the right-hand side of this equation, it follows that

$$\frac{\dbinom{k}{r-1}}{\dbinom{k}{r}} = \frac{r}{k-(r-1)} \quad \text{and} \quad \frac{\dbinom{k}{r+1}}{\dbinom{k}{r}} = \frac{k-r}{r+1}.$$

From this, we arrive at

$$\frac{\dbinom{k+1}{r}}{\dbinom{k+1}{r+1}} = \frac{1 + [r/(k+1-r)]}{[(k-r)/(r+1)] + 1} = \frac{(k+1)/(k+1-r)}{(k+1)/(r+1)} = \frac{r+1}{k+1-r},$$

which completes the induction step. Thus, the validity of the desired ratio for a certain value of n, namely $n = k$, implies its validity for $k + 1$. This is just what Pascal's second lemma asserts.

Maurolico and Pascal had no special designation for the technique of reasoning by recurrence. The earliest mathematician who appeared in the fixing of a name to this process of argumentation was John Wallis; in his *Arithmetica Infinitorum* (1656), he found *per modum inductionis* the ratio of the sum of the squares 0, 1, 4, 9, . . . , n^2 to the product $n^2(n + 1)$. Augustus De Morgan's article *Induction* (*Mathematics*) in the *Penny Cyclopedia* (1838) suggested the name *successive induction* for the method, but at the end of the article De Morgan incidentally referred to it as *mathematical induction*, which is the first use of the term. The expression *complete induction* attained popularity in Germany after Dedekind used it in a paper in 1887. In the present century, the name *mathematical induction* has gained complete ascendancy over other descriptive terms.

9.2 Problems

1. Prove by mathematical induction each of the following identities.

(a) $1 + 2 + 3 + \cdots + n = \dfrac{n(n + 1)}{2}$.

(b) $1^2 + 2^2 + 3^2 + \cdots + n^2$
$= \dfrac{n(n + 1)(2n + 1)}{6}$.

(c) $1 \cdot 2 + 2 \cdot 3 + 3 \cdot 4 + \cdots + n(n + 1)$
$= \dfrac{n(n + 1)(n + 2)}{3}$.

(d) $1^2 + 3^2 + 5^2 + \cdots + (2n - 1)^2$
$= \dfrac{n(4n^2 - 1)}{3}$.

(e) $1 + 2 \cdot 2 + 3 \cdot 2^2 + \cdots + n2^{n-1}$
$= (n - 1)2^n + 1$.

(f) $\dfrac{1}{1 \cdot 3} + \dfrac{1}{3 \cdot 5} + \dfrac{1}{5 \cdot 7} + \cdots$
$+ \dfrac{1}{(2n - 1)(2n + 1)} = \dfrac{n}{2n + 1}$.

2. Use mathematical induction to prove that for $n \geq 1$, the value $x^n - y^n$ is divisible by $x - y$. [*Hint:* Consider the identity

$x^{k+1} - y^{k+1} = x^k(x - y) + y(x^k - y^k).$]

3. Prove each of these divisibility statements by mathematical induction.

(a) $8 | 5^{2n} + 7$ for all $n \geq 1$.
[*Hint:* $5^{2(k+1)} + 7 = 5^2(5^{2k} + 7)$
$+ (7 - 5^2 \cdot 7).$]

(b) $3 | 2^n + (-1)^{n+1}$ for all $n \geq 1$.

(c) $21 | 4^{n+1} + 5^{2n-1}$ for all $n \geq 1$.

(d) $24 | 2 \cdot 7^n + 3 \cdot 5^n - 5$ for all $n \geq 1$.

4. Use mathematical induction to verify the formula

$1(1!) + 2(2!) + 3(3!) + \cdots + n(n!)$
$= (n + 1)! - 1$

for all $n \geq 1$.

5. Prove the following:

(a) $2^n < n!$ for all $n \geq 4$.
(b) $n^3 < 2^n$ for all $n \geq 10$.

6. Prove that for $n \geq 1$:

(a) $\dbinom{2n}{n} = \dfrac{1 \cdot 3 \cdot 5 \cdots (2n - 1)}{n!} 2^n.$

(b) $\dbinom{4n}{2n} = \dfrac{1 \cdot 3 \cdot 5 \cdots (4n - 1)}{[1 \cdot 3 \cdot 5 \cdots (2n - 1)]^2} \dbinom{2n}{n}.$

7. Show that the coefficient of x^r in the expansion of

$\left(x + \dfrac{1}{x} \right)^n$ is $\dbinom{n}{\frac{n - r}{2}}.$

8. Find the coefficient of the middle term in the expansion of $(x - 1/x)^{2n}$.

9. (a) For $n \geq 1$, use induction to prove that

$\dbinom{n}{0} + \dbinom{n}{1} + \dbinom{n}{2} + \cdots + \dbinom{n}{n} = 2^n.$

(b) From part (a), conclude that the sum of the numbers in any base of Pascal's triangle is twice the sum of the numbers in the preceding base.

10. For $n \geq 1$, derive the following identities involving the binomial coefficients:

(a) $\dbinom{n}{1} + 2\dbinom{n}{2} + 3\dbinom{n}{3} + \cdots + n\dbinom{n}{n}$
$= n2^{n-1}.$

[*Hint:* The left-hand side equals

$$n\left[\binom{n-1}{0} + \binom{n-1}{1} + \binom{n-1}{2}\right.$$
$$\left. + \cdots + \binom{n-1}{n-1}\right] \quad \text{for } n \geq 2.]$$

(b) $\binom{n}{0} - \binom{n}{1} + \binom{n}{2} - \cdots$

$+ (-1)^n \binom{n}{n} = 0.$

[*Hint:* Replace $\binom{n}{r}$ by

$$\binom{n-1}{r} + \binom{n-1}{r-1}.]$$

(c) $\binom{n}{0} + \binom{n}{2} + \binom{n}{4} + \binom{n}{6} + \cdots$

$= \binom{n}{1} + \binom{n}{3} + \binom{n}{5} + \cdots = 2^{n-1}.$

[*Hint:* Use part (b) and Problem 9(a).]

(d) $\binom{n}{0} - \frac{1}{2}\binom{n}{1} + \frac{1}{3}\binom{n}{2} - \cdots$

$+ \frac{(-1)^n}{n+1}\binom{n}{n} = \frac{1}{n+1}.$

[*Hint:* The left-hand side equals

$$\frac{1}{n+1}\left[\binom{n+1}{1} - \binom{n+1}{2}\right.$$
$$\left. + \binom{n+1}{3} - \cdots + (-1)^n \binom{n+1}{n+1}\right].]$$

11. Prove that for $n \geq 1$:

(a) $\binom{n}{r} < \binom{n}{r+1}$

if and only if $0 \leq r < \frac{1}{2}(n-1)$.

(b) $\binom{n}{r} > \binom{n}{r+1}$

if and only if $n - 1 \geq r > \frac{1}{2}(n-1)$.

(c) $\binom{n}{r} = \binom{n}{r+1}$

if and only if n is an odd integer and $r = \frac{1}{2}(n-1)$.

12. Creating an analogy with Pascal's arithmetic triangle, Leibniz formed the harmonic triangle:

$$\frac{1}{1}$$
$$\frac{1}{2} \quad \frac{1}{2}$$
$$\frac{1}{3} \quad \frac{1}{6} \quad \frac{1}{3}$$
$$\frac{1}{4} \quad \frac{1}{12} \quad \frac{1}{12} \quad \frac{1}{4}$$
$$\frac{1}{5} \quad \frac{1}{20} \quad \frac{1}{30} \quad \frac{1}{20} \quad \frac{1}{5}$$
$$\frac{1}{6} \quad \frac{1}{30} \quad \frac{1}{60} \quad \frac{1}{60} \quad \frac{1}{30} \quad \frac{1}{6}$$
$$\cdots \cdots \cdots$$

The right-hand edge consists of the reciprocals of the positive integers (these are the so-called harmonic numbers). Each number not on this edge is the difference of two entries, one diagonally above and to the right, the other immediately to the right of the number in question. Thus,

$$\tfrac{1}{20} = \tfrac{1}{12} - \tfrac{1}{30} = \tfrac{1}{4} - \tfrac{1}{5}.$$

(a) Write the next horizontal base of the harmonic triangle.

(b) Prove that the numbers in the nth horizontal base of this triangle are the reciprocals of the numbers in the corresponding base of the arithmetic triangle divided by $n + 1$, that is, if $\binom{n}{r}$ denotes the rth entry in the nth base of the harmonic triangle,

$$\left[\genfrac{}{}{0pt}{}{n}{r}\right] = \frac{1}{(n+1)\binom{n}{r}}, \qquad 0 \leq r \leq n.$$

[*Hint:* Use mathematical induction on n and the recursion formula

$$\left[\genfrac{}{}{0pt}{}{k+1}{r}\right] = \left[\genfrac{}{}{0pt}{}{k}{r}\right] - \left[\genfrac{}{}{0pt}{}{k+1}{r+1}\right].]$$

(c) Show that the harmonic triangle is symmetric by verifying that

$$\left[\genfrac{}{}{0pt}{}{n}{r}\right] = \left[\genfrac{}{}{0pt}{}{n}{n-r}\right] \qquad \text{for } 0 \leq r \leq n.$$

Christiaan Huygens
(1629–1695)

(Extract taken from A History of Science,
Technology and Philosophy in the 16th and 17th
Centuries, *by A. Wolf. Reproduced by kind
permission of Unwin Hyman Ltd.)*

9.3 The Bernoullis and Laplace

The earliest work on the mathematical treatment of probability is the *De Ratiociniis in Ludo Aleae* of the Dutch physicist Christiaan Huygens. It appeared first as an appendix to Frans van Schooten's *Exercitationes Mathematicae* (Mathematical Exercises), printed in Leyden in 1657. In the *Exercitationes Mathematicae,* van Schooten took up the most important individual results achieved by his pupils, one of whom was Huygens. Although Huygens wrote the original draft in Dutch, Latin was still thought to be the most appropriate language for expounding mathematics. When Huygens experienced some difficulty in coining suitable Latin terms for Dutch technical language, van Schooten provided the translation that was finally published—later the *De Ratiociniis in Ludo Aleae* came out in the vernacular. The most memorable of Huygen's discoveries in the tract was the important concept of mathematical expectation, or as he called it, "the value (price) of the chance" to win in a game. For half a century, this little pamphlet was essentially the only text available on probability theory. Then, in the early 1700s, it was superseded by elaborate treatises from the hands of James Bernoulli and Abraham De Moivre.

The Bernoulli family, Protestant in faith, fled Antwerp in 1583, after it had been captured by Catholic Spain. They first sought refuge in Frankfort, moving shortly thereafter to Switzerland, where the family settled in the free city of Basel. Basel had become a university town in 1460, a center for early printing that had attracted such Renaissance scholars as Erasmus, and a haven for the Huguenots during their long persecution. The founder of the Bernoulli dynasty married into one of the oldest families of the city and established himself as a prosperous merchant. During a century, this gifted family produced no fewer than eight mathematicians, several of whom brought fame to the country of their adoption. The first to attain prominence were the brothers James (Jacques, Jacob) Bernoulli and John (Jean, Johann) Bernoulli, grandsons of the fugitive from Antwerp. Their careers form the connecting link between the mathematics of the seventeenth and eighteenth centuries.

James Bernoulli
(1654–1705)

(Extract taken from A History of Science, Technology and Philosophy in the 18th Century, *by A. Wolf. Reproduced by kind permission of Unwin Hyman Ltd.)*

James Bernoulli (1654–1705), carrying out his father's wish for him to enter the ministry, took a degree in theology at the University of Basel in 1676. Meanwhile, he secured additional training in mathematics and astronomy, much against the wishes of his parents. Later in life, he chose as his motto *Invito patre sidera verso* ("I study the stars against my father's will"). During the years 1676–1682, Bernoulli traveled widely in France, England, and the Netherlands, familiarizing himself with the work of the leading mathematicians and scientists. There is no record that he encountered Christiaan Huygens at this time, but in England he met both Robert Boyle and Robert Hooke. In 1682, he opened a school for mathematics and science in his native city of Basel, presumably distilling the fruits of his new learning. Five years later, Bernoulli was appointed professor of mathematics at the University of Basel, holding the chair until his death.

The essentials of the differential calculus first appeared in print in a 6-page paper by Leibniz in the *Acta Eruditorum* of 1684, and the essentials of the integral calculus followed in 1686. Marred by misprints and poor exposition, these papers encountered an almost universal lack of comprehension. It is said that when the aging Huygens in 1690 wanted to master the newly proposed methods, there were not half a dozen men qualified to expound on the subject. One of the first of the continental mathematicians to achieve full understanding of Leibniz's abbreviated presentation was James Bernoulli, who taught the techniques to his younger brother John (John, in turn, instructed L'Hospital in the calculus, and the latter passed the knowledge on to Huygens). Unlike the reticent Newton, Leibniz corresponded extensively with fellow mathematicians on the subject of the infinitesimal analysis. He soon produced a group of enthusiastic adherents, who were convinced of its great potentialities and tried to expand the calculus in several different ways. Its rapid development into an instrument of great analytical power and flexibility was largely due to the ability of the Bernoulli brothers.

The *Acta Eruditorum,* in which Leibniz had published most of his papers, also opened its pages to the mathematical contributions of the Bernoullis. They achieved, with Leibniz, such a productivity that by the late 1600s almost all of what we now call elementary calculus had been created, along with the beginnings of ordinary differential equations.

It is curious that James Bernoulli should have been born in that same year in which the correspondence between Pascal and Fermat laid the foundations of probability theory; for of all Bernoulli's contributions to mathematics, the work for which he is best known is the *Ars Conjectandi* (The Art of Conjecturing). He had been writing the book off and on for 20 years, but was never satisfied with it. When he died in 1705, the all-but-complete manuscript was given to his 18-year-old nephew, Nicholas Bernoulli, with a view to editing it for publication. Nicholas, who was a former pupil of James, had already turned his attention to the theory of probability and so seemed the person best fitted to finish the project. Apparently the young man did not consider himself adequate to the task; and by his advice, the work was finally given to the printer in the state in which its author had left it. The *Ars Conjectandi* was published in Latin in 1713, eight years after James Bernoulli's death.

The *Ars Conjectandi* comes in four parts. The first is a reproduction of Huygen's *De Ratiociniis in Ludo Aleae,* accompanied by a commentary on all but one of Huygens's propositions. Bernoulli's commentary, which often offers alternative proofs of the fundamental propositions and in some cases extends them, is of more value than the original tract of Huygens. The second part of the *Ars Conjectandi* contains practically all the standard results on permutations and combinations in the form in which they are still expressed. Bernoulli said that others had treated these topics before him, especially van Schooten, Leibniz, and Wallis, and so intimated that his subject matter was not entirely new. In this part, he gave an array

$$
\begin{array}{cccccc}
1 & 0 & 0 & 0 & 0 & 0 \\
1 & 1 & 0 & 0 & 0 & 0 \\
1 & 2 & 1 & 0 & 0 & 0 \\
1 & 3 & 3 & 1 & 0 & 0 \\
1 & 4 & 6 & 4 & 1 & 0 \\
1 & 5 & 10 & 10 & 5 & 1 \\
& & \cdots\cdots\cdots
\end{array}
$$

that was substantially the same as Pascal's arithmetic triangle, and provided the first adequate proof of the binomial theorem for positive integral powers. Bernoulli's proof was by mathematical induction, a method he rediscovered from reading the *Arithmetica Infinitorum* (1659) of Wallis. Newton seems to have first stated the binomial theorem in general form for fractional exponents in a famous letter (1676) to Henry Oldenburg, the secretary of the Royal Society of London. But Newton gave no regular proof of his theorem, contenting himself with illustrating it by actual multiplication in some well-chosen examples. In 1774, Euler devised a proof for negative and fractional exponents, although the argument was faulty in that he failed to consider the convergence of the infinite series that arises in the case of negative exponents.

The third part of the *Ars Conjectandi* consists of 24 problems relating to the various games of chance that were popular in Bernoulli's day, and that were designed to furnish examples for the theory that had gone before in the book. The final part of the treatise is entitled "Applications of the Previous Study to Civil, Moral and Economic Problems." It was left incomplete by the author but nevertheless must be considered the most important part of the whole work. After a general discussion of philosophical problems connected with probability theory—probability as a degree of certainty, moral versus mathematical expectation, and so on—Bernoulli gave a proof of the celebrated theorem that bears his name. That he was proud of this result, the first limit theorem of probability theory, is apparent from his own words in the *Ars Conjectandi:*

> This is that problem which I am now about to impart to the public after having suppressed it for twenty years. It is difficult and it is novel, but is of such excellent use that it places a high value and dignity on every branch of this doctrine.

Bernoulli's theorem (which the French mathematician Poisson later called the "law of large numbers") is accurately stated as follows. If p is the probability of an event, if k is the actual number of times the event occurs in n trials, if $\epsilon > 0$ is an arbitrarily small number, and if P is the probability that the inequality $|k/n - p| < \epsilon$ is satisfied, then P increases to 1 as n grows without bound. Of course, it may happen that for some n, no matter how large, the difference $|k/n - p|$ will be greater than ϵ in a particular sequence of n trials. What Bernoulli's theorem asserts is that those sequences in which the inequality $|k/n - p| < \epsilon$ is satisfied are much more likely to occur.

With Bernoulli's theorem, the *Ars Conjectandi* ended abruptly. It is especially to be regretted that the promised applications to matters of economics and politics remained unfinished, because Bernoulli's theorem was to become the stepping-stone from insignificant problems with urns containing colored balls or dice tossing and card playing to valuable and scientifically justified applications of probability theory (such as mathematical statistics, demography, and the theory of random errors).

John Bernoulli (1667–1748), like his older brother James, ran counter to his father's plans regarding his work in life. Originally scheduled to take charge of the family's thriving trading interests, he escaped an unsuccessful apprenticeship as a salesman by getting permission to study medicine. Simultaneously, he was being privately tutored by his brother in the mathematical sciences. When Leibniz's papers began to appear in the *Acta Eruditorum,* John mastered the new methods and followed in James's footsteps as one of the leading exponents of the calculus. Through the intervention of Huygens, he was appointed professor of mathematics at Groningen in the Netherlands in 1695. At the death of James in 1705, he succeeded to his brother's vacated chair at Basel, where he remained for the rest of his life. Unfortunately, antagonism had developed and steadily increased between the two brothers, and their early collaboration eventually changed to rivalry.

For years there had been a certain friction between the brothers stemming from the arrogance of the older one in dealing with the younger, and from the desire of the younger brother for fame. What finally brought the quarrel into the open was the famous problem of quickest descent, or the brachistochrone problem (the name is derived from the Greek words meaning "shortest" and "time"). In a leaflet addressed

JACOBI BERNOULLI,
Profeſſ. Baſil. & utriuſque Societ. Reg. Scientiar.
Gall. & Pruſſ. Sodal.
MATHEMATICI CELEBERRIMI,

ARS CONJECTANDI,

OPUS POSTHUMUM.

Accedit

TRACTATUS

DE SERIEBUS INFINITIS,

Et EPISTOLA Gallicè ſcripta

DE LUDO PILÆ
RETICULARIS.

BASILEÆ,
Impenſis THURNISIORUM, Fratrum.

cIɔ Iɔcc xIII.

Title page of James Bernoulli's Ars Conjectandi *(1713). (By courtesy of Editions Culture et Civilisation.)*

(1696) to ''the shrewdest mathematicians of all the world,'' John Bernoulli posed the challenge of determining the curve down which a particle will slide in the shortest possible time from a given point to another not directly below. The search for the brachistochrome path had begun with Galileo, who believed incorrectly that the curve sought was an arc of a circle. Leibniz provided the answer on the day he received notice of the challenge, and accurately predicted a total of exactly five solutions—from Newton, Leibniz, L'Hospital, and the two Bernoullis. All these solutions were printed in the May issue of the *Acta Eruditorum* of 1697. John Bernoulli first found an incorrect solution, then tried to substitute his brother's proof for his own. The path of quickest descent turned out to be part of a cycloid, that fateful curve of the seventeenth century. When the Bernoullis found out that the solution involved an arc of an appropriate cycloid, they were amazed. As James Bernoulli pointed out in publishing his proof, a curve that had been investigated by so many mathematicians that there seemed to be

John Bernoulli
(1667–1748)

(Smithsonian Institution.)

nothing more to discover about it, now suddenly displayed an entirely new property. By itself the problem may not seem so very important, but its solution laid the foundations upon which Euler, Lagrange, and Legendre were later able to construct an entirely new branch of mathematics, called the calculus of variations.

John Bernoulli's bitterness increased when a French nobleman and amateur mathematician, the Marquis de L'Hospital (1661–1704), published under his own name various discoveries communicated to him by Bernoulli. Not quite convinced that he could master the newly invented calculus by himself, L'Hospital engaged the services of Bernoulli to spend several months in Paris in 1691 instructing him in the subject. L'Hospital even induced Bernoulli, in exchange for a yearly allowance, to continue these lessons after the latter's return to Basel. It was a curious financial arrangement, in which Bernoulli agreed to send L'Hospital and no one else his latest discoveries in mathematics, to be used as the marquis might please. This correspondence was subsequently organized by L'Hospital into *Analyse des Infiniment Petits* (1696), the first textbook on the differential calculus to appear in print. Supplying what many had awaited, namely an elementary introduction to the new infinitesimal analysis, its influence and popularity dominated the whole of the eighteenth century. In the preface, L'Hospital acknowledged his indebtedness to Leibniz and Bernoulli, especially "the young professor at Groningen," but only in very general terms: "I have made free of their discoveries, so that whatever they please to claim as their own I frankly return to them." John Bernoulli watched, with apparent jealousy, the growing success of his protégé's book. Following the author's death, he wrote letters claiming that much of the content of the *Analyse* was really his own property, practically accusing L'Hospital of plagiarism. Matters were clarified in 1921, when a manuscript of Bernoulli on the differential calculus, dating from 1691–1692, was discovered in the Basel University library. A comparison of Bernoulli's manuscript and the text of L'Hospital revealed a considerable overlapping, so that John Bernoulli seems to have been the true author of almost all of substance in the *Analyse*. One of the most noted contributions of Bernoulli that was incorporated in the work is the theorem, since known as L'Hospital's rule, for finding the limiting value of a fraction when numerator and denominator both tend to zero.

Leibniz recognized the importance of good notation in mathematics, writing to John Bernoulli, "As regards signs, I see clearly that it is to the interest of the Republic of Letters, and especially of students, that learned men should reach agreement on signs." In their correspondence, they discussed both the name and principal symbol of the integral. Leibniz favored the name *calculus summatorius* and ∫ as a symbol, bearing witness to the summation of a number of infinitely small areas. When Leibniz used the ∫ in his 1686 paper, it did not quite take its modern form. It was simply the small letter *s* as printed at that time, which resembled the modern type form for *f*. Bernoulli put forth the name *calculus integralis*, the idea being that some whole area was obtained by the summation of its parts, and the capital *I* as a sign of integration. In the end, Leibniz and Bernoulli reached a happy compromise, adopting Leibniz's elongated *S* for the symbol of the integral and Bernoulli's name *integral calculus* for the inverse of the *calculus differentalis*. The word "integral" itself was first used in print by James Bernoulli in the *Acta Eruditorum* for 1690, although John Bernoulli characteristically claimed the introduction of the term for himself.

Returning to the development of probability theory, we find that the next big milestone was the publication of De Moivre's *Doctrine of Chances: or, a Method of Calculating the Probability of Events in Play* (1718). This was the first English treatise on the subject. Abraham De Moivre (1667–1754) was a French Protestant who was forced to seek asylum in London after the revocation of the Edict of Nantes and the expulsion of the Huguenots (1685). There he took up his lifelong, unprofitable occupation as a private tutor in mathematics. He had hoped to obtain a university professorship, but he never succeeded in this, partly because of his non-British origin. A friend of Newton, he was appointed in 1710 to the partisan commission by means of which the Royal Society of London sought to review the evidence in the Newton–Leibniz dispute over the origin of the calculus. De Moivre supported himself for many years by solving problems proposed to him by wealthy patrons who wanted to know what stakes they should offer in games of chance. It is probable that the interest of these gentlemen led to De Moivre's memoir, the *Doctrine of Chances;* it contained numerous problems on throwing dice and drawing balls of various colors from a bag, as well as questions relating to life annuities. After both sight and hearing failed, he sank into a state of lethargy, sleeping 20 hours a day. Like Cardan, De Moivre predicted the day of his own death, or so the story goes. Finding that each day he was sleeping a quarter of an hour longer than on the preceding day, he calculated that he would die in his sleep on the very day in which he slept up to the limit of 24 hours. De Moivre died in 1754, at the age of 87; the cause of his death was recorded as "somnolence."

Because factorials increase with extreme rapidity, it is desirable to have an approximate expression for $n!$ that is convenient to manipulate both practically and theoretically. In the first edition of the *Doctrine of Chances*, De Moivre obtained a result (namely, for very large n, it is known that $n! \approx cn^{n + 1/2} e^{-n}$), which is complete except for the determination of an unknown constant factor c. In the second edition of 1738, he gave the final formula with due credit to Stirling: "I desisted in proceeding farther till my worthy friend Mr. James Stirling, who had applied after me to that inquiry, discovered that $c = \sqrt{2\pi}$." Hence, the celebrated approximation,

$$n! \approx \sqrt{2\pi n}\; n^n e^{-n},$$

which is usually known as Stirling's formula, is at least as much the work of De Moivre.

Before the 1770s, the calculus of probabilities had been largely restricted to the study of gambling and actuarial problems. The works of James Bernoulli and De Moivre had excited interest about the prospects for wider applications—to the estimation of errors in observation, to changes in the composition of population, or to the study of regularities that arise in political and social phenomena (for instance the measure of mean duration of life or marriage). But these had been little more than touched on. It was largely Laplace who went on to broaden the mathematical treatment of probability to areas of scientific research well beyond games of chance.

Pierre Simon Laplace (1749–1827) came from relatively humble origins in Normandy, a fact that in false shame he later found convenient to forget. Between the ages of 7 and 16, the young Laplace was educated in a school attached to a Benedictine priory in Beaumont; most of his teachers were members of the order. He entered the University of Caen at 16, intending to study theology, but soon decided that his true vocation lay in mathematics. Laplace seems to have remained at Caen for five years, where he wrote his first mathematical paper and was named a "provisional professor." When he decided to go to Paris, he was suitably furnished with several letters of recommendation to Jean d'Alembert, then the leading mathematician in France. The letters remained unnoticed, but Laplace caught d'Alembert's eye by addressing a long paper to him on the general principles of mechanics. This brought the enthusiastic reply, "You need no introduction; you have recommended yourself; my support is your due."

For men of science born without the advantages of rank or wealth, there were only a limited number of remunerative positions open in France in the 1700s. Apart from the opportunities to teach, one of the few salaried positions available was the coveted membership in the Académie Royale des Sciences; the small number of places (about 60 at a time) together with the large number of candidates provoked a high degree of competition for the honor. Mainly through D'Alembert's influence, Laplace was appointed professor of mathematics at the Paris Ecole Militaire in 1769 and elected to the select Académie in 1773. He was now launched on a career that was to become truly illustrious.

The great success of Newton's *Principia* made celestial mechanics an attractive field for research, and as soon as Laplace reached Paris, he developed a definite purpose for his life: to write a great treatise that would embrace all that was known about the mechanics of the heavens. He did not permit this activity to be greatly interrupted by the French Revolution or by the successive governmental changes through which he lived. During the most tumultuous period of French history, Laplace contrived to win appreciation and support for his talent from each ruling party in turn. Even though the stories about Laplace's political opportunism might be exaggerated, they are too much of a kind not to represent some core of truth.

Pierre Simon Laplace
(1749–1827)

(From A Concise History of Mathematics *by Dirk Struik, 1967, Dover Publications, Inc., N.Y.)*

he scholars of the royal artillery corps, Laplace happened
lieutenant named Napoleon Bonaparte (1769–1821); and
eers, the examinee passed. When a coup d'etat brought
Laplace set aside whatever republican principles he might
plete support to his old friend. Napoleon rewarded this
minister of interior, but dismissed him six weeks later in
Bonaparte. Looking back years later, Napoleon quipped
rywhere for subtleties, had only problematic ideas, and
tely small' into administration.'' Yet if Napoleon held a
administrator, it did not dim his respect for him as a
on for having removed him from office, Napoleon raised
le him a count. Laplace eventually became president of
ient of patronage is likely in turn to be generous with
sing that Laplace dedicated the next volume (Volume
e Céleste to ''Bonaparte, the Pacificator of Europe to
erity, her greatness and the most brilliant epoch of her
to depose Napoleon in 1814 (Laplace had had the fore-
aris during this awkward moment), Laplace shifted his
When the French monarchy was restored, Louis XVIII
lty by bestowing the title of marquis on Laplace and
Peers.
wn as the Newton of France died in March 1827 at age
quick to note that the month and year were exactly a
wton himself. When those around his deathbed were
veries he had made, Laplace replied, ''What we know
e ignorant of is immense.''

TRAITÉ

DE

MÉCANIQUE CÉLESTE,

PAR P. S. LAPLACE,

Membre de l'Institut national de France, et du Bureau
des Longitudes.

TOME PREMIER.

DE L'IMPRIMERIE DE CRAPELET.

A PARIS,

Chez J. B. M. DUPRAT, Libraire pour les Mathématiques,
quai des Augustins.

AN VII.

Title page of the first volume of Laplace's Traité de Mécanique Céleste *(1799).*
(By courtesy of Editions Culture et Civilisation.)

Although posterity may fault Laplace for his political maneuvering, his mathematical ability is beyond question. Laplace's greatest achievement was the *Traité de Mécanique Céleste,* published in five large volumes over 26 years (1799–1825). This modern *Almagest* was designed, in the author's own words, "to solve the great mechanical problems of the solar system and to bring theory to coincide so closely with observation that empirical equations should no longer be needed." The *Mécanique Céleste* completed the work Newton had begun, for it showed that all the movements of the known members of the planetary system were deducible from the law of gravitation. Within its 2000 pages were incorporated all the important discoveries of the previous century—of Newton, Clairaut, d'Alembert, Euler, and Lagrange—plus an immense amount of material that was original with Laplace himself. Even though it is granted that scientists of the time were far from conscientious in their citation practices, Laplace did not adequately acknowledge the contributions of others, but left the distinct impression that all the results contained in the *Mécanique Céleste* were his own.

The *Mécanique Céleste* was one of the great scientific works of the early 1800s. John Playfair in a review (1808) spoke of it as "the highest point to which man has yet ascended in the scale of intellectual attainment." The mathematical exposition in the *Mécanique Céleste* was extremely concise, and not easy reading. In preparing the manuscript for press, Laplace was frequently unable to reconstruct the details in his chains of reasoning. If he was satisfied that a conclusion was correct, he was content to insert the optimistic remark, "It is easy to see that . . . ," and give the result without further explanation. The American astronomer Nathaniel Bowditch (1773–1838), who translated four of the five volumes into English, observed, "I never came across one of Laplace's 'Thus it plainly appears' without feeling sure that I had hours of hard work before me to fill up the chasm and find out and show how it plainly appears."

There is a well-known anecdote of an encounter between Laplace and Napoleon over the *Mécanique Céleste*. When Laplace presented Napoleon with a copy of the monumental work, the latter teasingly chided him for an apparent oversight: "They tell me that you have written this huge book on the system of the universe without once mentioning its Creator." Whereupon Laplace drew himself up and bluntly replied, "I have no need for that hypothesis." Greatly amused by the comment, Napoleon mentioned the incident to the mathematician Lagrange, who is reported to have exclaimed, "Ah, but it is a beautiful hypothesis, it explains so many things."

Although many historians see Laplace's famous reply to Napoleon as an indication that he was not always ingratiating himself with the current holder of power, it is possible to read into the story something that does not appear at first glance. From an extensive comparison of earliest recorded and modern observations, Edmund Halley discovered in 1675 an irregularity in the motions of the planets Jupiter and Saturn. He noted that for many centuries the average angular velocity of Jupiter around the sun had been continually increasing, while for Saturn this measure had been continually decreasing. The implication was that the mean distance from the sun was always decreasing for Jupiter and increasing for Saturn. If the irregularities were to become progressively larger, the ultimate fate of Jupiter would be to fall into the sun, whereas Saturn would edge off course and be lost altogether in the solar system. Newton speculated that these seeming aberrations were somehow due to the mutual attraction of the planets but was unable to provide a mathematical theory that would account for it. In his opinion, the solar system was the deliberate design of the Creator ("This most beautiful system of the sun, planets and comets could only proceed from the counsel and dominion of an intelligent and powerful Being.") and at some moment God would intervene so as to redress the disorder introduced by the action of the planets on one another.

In a series of memoirs presented to the Académie des Sciences during the years 1784–1788, Laplace proved that the gradual changes in the elements of planetary orbits were periodic, during a very long period, about 900 years in the case of the mean motion of Jupiter and Saturn. This periodicity meant that an increase in the eccentricity of an orbit would be followed, not by a catastrophe, but by a decrease reestablishing the status quo. The need for divine intervention to correct erratic behavior now seemed less likely than Newton had assumed. This is what Laplace probably had in mind when he eliminated the action of God in celestial mechanics. Far from being an atheist, he thought of God as creating the solar system in the beginning in such a way that He did not subsequently need to intervene directly in its affairs.

If any important work of the period required an explanatory commentary, it was certainly Laplace's *Mécanique Céleste*. Mary Fairfax Somerville (1780–1872), an extraordinary, self-educated Scotswoman who had studied the treatise in Edinburgh, was persuaded by the Society for the Diffusion of Useful Knowledge to prepare a popular exposition of this work of Laplace for English readers. She was nearing 50 years of age, lacked formal training, and had never written for publication. Yet Laplace is said to have observed that Mary Somerville was the single woman who understood his great treatise. According to one story, when Laplace dined with the Somervilles he told Mary Somerville—not realizing that she had been the widow of Samuel Grieg—''Only two women have ever read the *Mécanique Céleste,* and both are Scotch, Mrs. Grieg and yourself.''

Although she was unsure of her qualifications, she accepted the assignment with the understanding that she would write in secret and that if the manuscript proved unsatisfactory it would be destroyed. Adding full mathematical explanations and diagrams to make Laplace's work comprehensible, her popularization was completed in 1830 under the title *The Mechanisms of the Heavens*. The first printing was small, 750 copies, because apparently the printer was dubious of the financial success of a book on such a subject. Nevertheless, it met with such great praise that the preface, containing the necessary mathematical background, was issued separately in 1832 as *A Preliminary Dissertation on the Mechanisms of the Heavens,* and both it and the previous volume went through many printings. They became required textbooks on higher mathematics and astronomy in British universities for a century, ''as essential works to those who aspire to the highest places in examinations.'' Hailed by the English press a ''Queen of Nineteenth-Century Science,'' Mary Somerville was the last of the great amateur expositors of science, producing compilations of the state of knowledge in a number of different areas. Chief among these were her *Connection of the Physical Sciences* (1834) and *Physical Geography* (1848). *Physical Sciences* ran through some ten editions during the next 40 years and was translated into French, German, and Italian. Her last work, the two-volume *Molecular and Microscopic Science,* was published when she was 89 years old. Accorded numerous honors, Mary Somerville was elected to the American Philosophical Society in 1869, at the same time as was Charles Darwin.

Laplace also wrote a popular work, *Exposition du Système du Monde* (1796), which was addressed to the nonprofessional reader. Serving in effect as a preface to the highly technical *Mécanique Céleste,* the *Exposition* put forth a speculative hypothesis about the origin of the solar system. Laplace was trying to account for the remarkable circumstance that the motions of all the members of the solar system were in the same direction. (The motion of the satellites of Uranus is in the opposite direction, but this was not known at the time.) Laplace suggested that the bodies of the solar system originated from a vast, rotating cloud of incandescent gas, a solar nebula. In cooling, such a cloud would condense and contract, forming a more concentrated fluid mass, whose rate of rotation would be increasing in the process. A stage would be reached in which the cohesive forces could no longer hold it together, and small pieces would then be thrown off; these would quickly cool and coalesce to form separate planets. This "nebula hypothesis" struck the imagination of the age and was in vogue for nearly a century. It has not survived as Laplace put it forth, but in modified form represents what is now thought to take place in the birth of stars.

Laplace's genius was also shown in his contributions to probability and to the emerging theory of mathematical statistics. From 1774 on, he published a series of memoirs that were eventually expanded into a mature final work, the 464-page *Théorie Analytique des Probabilités* (1812). Laplace seems to have left probability untouched between 1786, with a paper on the demography of France, and 1809, when his attention once again turned to the subject. In the interval, he was preoccupied with celestial mechanics, incorporating his own research and that of other astronomers in the *Mécanique Céleste;* and he also had such governmental responsibilities as serving on the Committee on Weights and Measures, which established the metric system. In 1809, Laplace moved back into probability theory by way of several papers on the analysis of probable error in observations. The result of greatest novelty here was his derivation of the so-called least-squares rule for minimizing the uncertainty in a number of in-dependent observations. Legendre had published the rule in 1805, and this may have been what turned Laplace back to probability theory after a lapse of a quarter-century.

The memoirs from his earlier years were revised mathematically, buttressed with new material, and published as the *Théorie Analytique des Probabilités.* By presenting the solution of almost every classical problem of probability theory, he traced the evolution of the subject, at the same time systematizing and extending the previously known but often uncoordinated results of many mathematicians. The remarks on the obscurity and lack of rigor in Laplace's exposition that the *Mécanique Céleste* called forth apply also to the *Théorie Analytique des Probabilités.* In a review (1837), De Morgan wrote:

> The *Théorie Analytique des Probabilités* is the Mont Blanc of mathematical analysis; but the mountain has this advantage over the book, that there are guides always ready near the former, whereas the student has been left to his own method of encountering the latter.

Although the reading may have been difficult, the work sustained three editions (1812, 1814, and 1820) and four supplements (1816, 1818, 1820, and 1825) during Laplace's lifetime.

Beginning with its second edition of 1814, the *Théorie Analytique des Probabilités* contained a lengthy preface of 153 pages, which was often republished as a separate booklet under the title *Essai Philosophique sur les Probabilités.* The *Essai* bears the same relation to its subject as the *Exposition du Système du Monde* does to the *Mécanique Céleste,* although this time the technical treatise preceded the popularization. It was almost entirely devoid of mathematical symbols or formulas, and its aim was to acquaint a broader circle of readers with the fundamentals of probability theory and its applications without resorting to higher mathematics. Indeed, large portions of the *Essai* are transcriptions into everyday language of certain parts of the *Théorie Analytique des Probabilités,* where the same material was developed more mathematically

ESSAI PHILOSOPHIQUE

SUR

LES PROBABILITÉS;

PAR M. LE COMTE LAPLACE,

Chancelier du Sénat-Conservateur, Grand-Officier de la Légion d'Honneur;
Grand'Croix de l'Ordre de la Réunion; Membre de l'Institut impérial et
du Bureau des Longitudes de France; des Sociétés royales de Londres
et de Gottingue; des Académies des Sciences de Russie, de Danemarck,
de Suède, de Prusse, d'Italie, etc.

PARIS,

Mᵐᵉ Vᵉ COURCIER, Imprimeur-Libraire pour les Mathématiques,
quai des Augustins, n° 57.

1814.

Title page of Laplace's Essai Philosophique sur les Probabilités *(1814). (By
courtesy of Editions Culture et Civilisation.)*

and more succinctly. Both the *Essai Philosophique sur les Probabilités* and the *Exposition du Système du Monde* originated in Laplace's responsibility for giving lectures on elementary mathematics during the year 1795–1796 at the Ecole Normale, one of the several new educational institutions within which the scientific life of France remained alive during the Revolution.

The *Théorie Analytique des Probabilités* consists of two books, the first of which, entitled *The Calculus of Generating Functions,* is devoted to analysis. In Book II, or *General Theory of Probability,* Laplace divided his subject into three sections: the theory of probability proper, limit theorems, and mathematical statistics (not yet distinguished as a separate discipline). On the first page of Book II, he gave what is regarded as the classical definition of the probability of an event:

The probability of an event is the ratio of the number of cases favorable to it, to the number of possible cases, when there is nothing to make us believe that one case should occur rather than any other, so that these cases are, for us, equally possible.

The general formula embodied in this statement, namely,

$$\Pr[\text{event}] = \frac{\text{number of favorable outcomes}}{\text{total number of outcomes}},$$

requires that all outcomes be equally possible. The interpretation of equipossibility as the insufficiency of grounds for preferring one outcome over another was not original with Laplace. In the *Ars Conjectandi,* Bernoulli wrote, ''All cases are equally possible, that is to say, each can come about as easily as any other.'' However, before the publication of *Ars Conjectandi,* no well-circulated work used the notion, and Bernoulli himself did not use it much.

Proceeding to the basic results of probability theory, Laplace derived, among other things, the widely used addition and multiplication theorems of probabilities. The modern formulations of these follow:

1. If A and B are mutually exclusive events (that is, they cannot both occur at the same time), then

$$\Pr[A \text{ or } B] = \Pr[A] + \Pr[B].$$

2. If A and B are two independent events (that is, the occurrence of one doesn't affect the probability of the other), then

$$\Pr[A \text{ and } B] = \Pr[A]\,\Pr[B].$$

For instance, if a card is drawn at random from a standard bridge deck and a second card drawn from a pinochle deck we might ask about the probability that both are aces. A bridge deck contains 52 cards of which 4 are aces, and a pinochle deck of 48 cards has 8 aces. Denoting the events of getting an ace from a bridge and a pinochle deck by A and B, respectively, we have

$$\Pr[A \text{ and } B] = \Pr[A]\,\Pr[B] = \frac{4}{52} \cdot \frac{8}{48} = \frac{1}{78}.$$

Laplace went on to consider a whole series of problems drawn from models offered by games: the probability of drawing all the numbers of a lottery at least once in a given series of selections; the probability of selecting a given number of balls of one color from an urn; the probability of selecting numbered balls from an urn in a specified order; and so on. Many of these involved what is now called a Bernoulli process or a Bernoulli trials experiment. Huygens's problems often were a point of departure for Bernoulli in developing new results in the *Ars Conjectandi.* His commentary on Proposition 12 of Huygens's *De Ratiociniis in Ludo Aleae* set forth what must have been the first mention in print of a series of independent trials with constant probability—hence the name, a Bernoulli trials experiment. In essence, a Bernoulli trials experiment is a sequence of independent, repeated trials in which each trial has only two possible outcomes, a success or a failure. If the probability of success on a single trial is denoted by p and the probability of failure by $q = 1 - p$, then p and q remain constant from trial to trial. Bernoulli showed that the probability of observing exactly r successes in n such trials was expressed by the rth term of the expansion for $(p + q)^n$; that is,

$$\text{Pr}[r \text{ successes and } n - r \text{ failures}] = \binom{n}{r} p^r q^{n - r}.$$

To see how Bernoulli derived this, let us consider one particular way in which exactly r successes can occur in n trials, namely, the instance $\text{SS} \cdots \text{SFF} \cdots \text{F}$ of r successes in the first r trials followed by $n - r$ failures in the remaining trials. Because the trials are independent, the probability of this particular event is the product of p^r for the r successes and q^{n-r} for the $n - r$ failures:

$$\text{Pr}[\text{S} \cdots \text{SF} \cdots \text{F}] = p^n q^{n - r}.$$

Now the total number of ways in which the n trials can be divided into exactly r successes and $n - r$ failures equals $\binom{n}{r}$. Because these individual ways are mutually exclusive events, each having the probability $p^r q^{n - r}$, the sum of their probabilities is

$$\binom{n}{r} p^r q^{n - r},$$

which is the probability sought.

Bernoulli's formula might best be understood by two examples, one in which the formula applies and another in which it does not. If six well-balanced dice are thrown, we might ask what is the probability that exactly two of the dice have top faces showing at least a five. This can be viewed as a Bernoulli trials experiment in which the probability p of success is $\frac{1}{3}$ (a five or a six showing) and failure $q = 1 - p = \frac{2}{3}$. The desired probability is therefore

$$\text{Pr}[2 \text{ successes and 4 failures}] = \binom{6}{2}\left(\frac{1}{3}\right)^2\left(\frac{2}{3}\right)^4 = 15\left(\frac{1}{9}\right)\left(\frac{16}{81}\right) = \frac{240}{729}.$$

Suppose, for a second illustration, that at a county fair a prize is offered to any person who in a sequence of four shots can hit the target three times in a row. What is the probability that a person will win the prize if his probability of hitting on any given shot is $\frac{3}{4}$? Considering the shots to be unrelated, we know that they constitute four Bernoulli trials with probability $p = \frac{3}{4}$ for success. If it were simply a matter of finding the probability of three and only three successes in four trials, Bernoulli's formula would tell us that this occurs with probability

$$\text{Pr}[3 \text{ successes and 1 failure}] = \binom{4}{3}\left(\frac{3}{4}\right)^3\left(\frac{1}{4}\right).$$

But the outcome SSFS, although it involves three successes, will not win any prize, because the three shots that hit the target are not three in a row. The prize will be won only in the three mutually exclusive cases SSSF, FSSS, and SSSS, whence the probability sought is

$$\text{Pr}[\text{SSSF}] + \text{Pr}[\text{FSSS}] + \text{Pr}[\text{SSSS}] = \left(\frac{3}{4}\right)^3\left(\frac{1}{4}\right) + \left(\frac{1}{4}\right)\left(\frac{3}{4}\right)^3 + \left(\frac{3}{4}\right)^4 = \frac{135}{256}.$$

Among the many problems Laplace solved in the *Théorie Analytique des Proba-bilités* is the celebrated "needle problem" due to Georges Louis Leclerc, Comte de Buffon (1707–1788). Buffon, who was the author of a 36-volume *Histoire Naturelle,* posed (1777) the following:

A large plane area is ruled with equidistant parallel lines, the distance between two con-secutive lines of the series being a. A thin needle of length $l < a$ is tossed randomly onto the plane. What is the probability that the needle will intersect one of the lines?

Using integral calculus, one can show that the formula $p = 2l/\pi a$ gives the probability p that the needle will fall across a line. Laplace was the first to point out that by throwing the needle many times and assuming the statistical estimate for p, one can use this formula to arrive at an approximate value for π. Sometime between 1849 and 1853, Johann Wolf, professor of astronomy at Bern, performed the actual experiment. In his demonstration, the width between the parallels was taken to be 45 mm and the needle used was 36 mm long. The needle was tossed 5000 times, and it cut a line 2532 times. Hence the relative frequency was

$$\frac{2532}{5000} = 0.5064.$$

Taking this value for the theoretical probability of intersections and setting

$$\frac{72}{45\pi} = 0.5064$$

produced $\pi = 3.1596$, which differs from the known value of π by less than 0.02. The best result of this kind was obtained in 1901 by Lazzerini, who by carrying out only 3408 trials, found π correct to six decimal places. Laplace was led to extend Buffon's problem to a grid of two mutually perpendicular sets of equidistant parallel lines. If the distances between members of the sets are a and b, the probability that a needle of length l (less than a and b) will intersect one of the lines is

$$p = \frac{2l(a + b) - l^2}{\pi ab}.$$

The latter part of the *Théorie Analytique des Probabilités* consists of the many applications that were radically to enlarge the realm of probabilistic analysis. In par-ticular, Laplace posed and solved numerous questions in demography, the one area for which assembling statistically significant information had begun by the end of the 1700s. Laplace considered such matters as the estimated increase in life span if a certain disease, say smallpox, were to be eliminated; the probability that the existing ratio of male to female births in Paris would continue for the next hundred years; and the evaluation of the correctness of witness testimonies and court decisions—Laplace ob-served that the probability of a correct unanimous verdict increases with the number of judges.

Estimating the population of a given nation had been considered an important problem in political arithmetic since the 1600s, for population was seen as an index to the prosperity of a nation. Demographers proposed to reach estimates of the popula-tion by determining a factor by which the average annual number of births should be multiplied to give the approximate total population. With data from a sample census, Laplace calculated the ratio of the population m in the sample districts to the annual number n of births in these districts. He therefore estimated the whole population of France from the formula

$$p = \left(\frac{m}{n}\right)N,$$

where N is the annual number of births for the nation as a whole. His sample, taken from 30 districts scattered over France, produced the figures 2,037,615 inhabitants and 215,599 births during a 3-year period (1800, 1801, 1802). Assuming the number of births in France each year to be 1 million, he concluded that the whole population contained people in the amount

$$P = 3\left(\frac{2,037,615}{215,599}\right)10^6 = 28,352,845.$$

Perhaps the more significant problem was his estimation of the degree of error in this figure (the first estimate of error inherent in sampling) and the computation of how large a sample would be needed to reduce the range of probable error to specified limits.

Although the *Théorie Analytique des Probabilités* is a monument to probability theory, it had a certain negative effect on the development of this discipline. The followers of Laplace promulgated the view that the political and social sciences held the same certainty as the natural sciences, and hence the analysis of social phenomena should fall within the scope of probability theory. This led to various unjustified and inappropriate applications of the theory, extending its methods to areas of no relevance to the subject, such as the correctness of court decisions. Many of these unsubstantiated applications had so little validity that they cast on probability theory a suspicion of quackery, holding it a kind of mathematical entertainment unworthy of serious consideration. The necessity of reexamining the logical foundations of the subject to secure its position as a full-fledged mathematical science became all too clear.

The concept of mathematical expectation, one of the most important notions of probability theory, was introduced by Huygens as "the value of a chance." We probably owe the modern terminology to van Schooten, because the word *expectatio* appears for the first time in his translation of Huygens's tract; for a long while *spes* ("hope") ran it a close second choice. To give a precise definition: If the possible outcomes of a game or experiment are given numerical values, $a_1, a_2, \ldots, a_n$ and these outcomes occur with probabilities $p_1, p_2, \ldots, p_n$, then the mathematical expectation E of the game is

$$E = a_1 p_1 + a_2 p_2 + \cdots + a_n p_n.$$

If the game has an expectation of zero, then it is said to be fair. The "expectation" is not to be interpreted as the value that will necessarily occur on a single play of the game but rather as the average payoff in a long run of similar gambles.

An example will illustrate this idea. Suppose a gambler rolls a fair die after which the casino pays him as many dollars as there are dots showing on the upper face. We may denote the individual outcomes by 1, 2, 3, 4, 5, 6 (the possible number of dots showing), each occurring with probability $\frac{1}{6}$. Thus, the gambler's expectation per game is

$$E = 1 \cdot \frac{1}{6} + 2 \cdot \frac{1}{6} + 3 \cdot \frac{1}{6} + 4 \cdot \frac{1}{6} + 5 \cdot \frac{1}{6} + 6 \cdot \frac{1}{6}$$

$$= \frac{1}{6}(1 + 2 + 3 + 4 + 5 + 6) = \frac{21}{6} = \frac{7}{2}.$$

This sum might be taken as a fair price to pay for the privilege of playing the game. In a long sequence of games, a gambler paying this price would have an average winning per game near zero.

For another illustration of mathematical expectation, consider the game of roulette. The wheel has 37 equally spaced slots, which are numbered from 0 to 36, inclusive. If a man playing roulette bets, say $1, on a given number and if the ball comes to rest in the slot bearing that number, he receives $36 from the croupier; that is, he recovers his bet and receives in addition 35 times the amount wagered. The player wins $35 with probability $\frac{1}{37}$, since all the slots are the same size, and loses $1 with probability $\frac{36}{37}$, so that his expectation is

$$E = 35 \cdot \frac{1}{37} + (-1)\frac{36}{37} = -\frac{1}{37} = -0.027.$$

This means that the player can expect to lose about 2.7 cents per bet in the long run.

Mathematical expectation need not be limited to games of chance but can be applied to any experiment that is repeated frequently and under identical conditions. Laplace, in his *Essai Philosophique sur les Probabilités,* calculated the expectation of a subsequent sunrise if it was given that the rising sun had been observed 1,826,623 times in succession, or each day for the previous 5000 years.

Probability theory abounds in paradoxes that wrench the common sense and trap the unwary. The most famous of these was first set forth in a letter of Nicholas Bernoulli, written in 1713 to Pierre Rémond de Montmort:

> Two players *A* and *B* agree to play a game in which *A* tosses a coin. The game is to continue until the first head appears. Player *B* will give player *A* a coin if a head appears on the first throw, two coins if a head appears for the first time on the second throw, four coins if on the third throw, 8 if on the fourth, and so on. What amount should *A* pay *B* before the start of the game to make it a fair game?

In 1738, Daniel Bernoulli (1700–1782) tackled the problem proposed by his older cousin and published his investigations in the *Proceedings* of the St. Petersburg Academy of Sciences. As a result, the problem has subsequently become famous under the title of the St. Petersburg problem, or the St. Petersburg paradox.

The question is: What sum should *A* pay *B* for the privilege of taking part in this game? For the game to be fair, *A* should agree to pay as an entrance fee the number of coins equal to A's mathematical expectation of gain. Let us therefore calculate this expectation. The number of outcomes is infinite, because there is no logical limit to the number of tails that might appear before the first head shows up. Thus, the expectation involves the sum of an infinite number of terms (an infinite series). If there is a

Daniel Bernoulli
(1700–1782)

(Smithsonian Institution.)

sequence of $n - 1$ tails before the first head appears, then B is to pay out 2^{n-1} coins for this outcome; and the probability that heads will show up for the first time on the nth toss is $(\frac{1}{2})^n$. The total expectation of A is therefore given by

$$1\left(\frac{1}{2}\right) + 2\left(\frac{1}{2}\right)^2 + 4\left(\frac{1}{2}\right)^3 + \cdots + 2^{n-1}\left(\frac{1}{2}\right)^n + \cdots$$

$$= \frac{1}{2} + \frac{1}{2} + \frac{1}{2} + \cdots,$$

which is infinite. The result seems to say that in order to play, A must first deposit with B an infinite sum of money; or put another way, no matter how much money A should offer B to enter the game, A would still come out the winner if the game were repeated on the same terms enough times. From a common-sense standpoint this seems palpably absurd. A person would not pay any considerable sum for the advantage that seems to be offered him in this game.

The French scientist Buffon conducted an experiment in tossing coins and found that in 2084 games, 1061 gave heads on the first toss, 494 on the second, 232 on the third, 137 on the fourth, 56 on the fifth, 29 on the sixth, 25 on the seventh, 8 on the eighth, and 6 on the ninth. Computing the various expectations, he concluded that B would have paid A a sum of 10,057 coins for the 2084 games, so that a modest entrance fee for each play would make the game fair. The "paradox" in the St. Petersburg paradox is that there is a discrepancy between the results of mathematical reasoning and the dictates of common sense and experience.

The St. Petersburg paradox puzzled mathematicians for generations and became one of the most hotly debated problems in the whole realm of probability theory. Numerous explanations were offered, some of them bordering on the ridiculous. Probably the first was given by Daniel Bernoulli, who distinguished between what he called "mathematical" explanation and "moral" expectation; if the mystical notion, or moral expectation, is applied to the problem in place of mathematical expectation, then a finite value is obtained for the expected gain. D'Alembert sought to resolve the paradox by introducing two forms of possibilities, "metaphysical" and "physical" (that a head would appear for the first time after 1000 throws was metaphysically possible, but quite impossible physically). Others preferred to reconcile the mathematical theory with common sense by pointing out that the fortune of the person whom we call A is necessarily finite, so that the problem is inherently impossible.

There is still no generally accepted verdict about how to treat the St. Petersburg paradox, except to note that mathematical expectation gives a meaningless result when applied to this problem. If one makes additional assumptions not warranted by the express conditions of the game, it is possible to arrive at a solution. For instance, suppose that instead of player B's paying out the amounts $1, 2, 4, \ldots, 2^{n-1}, \ldots$ to A, these are replaced by amounts $1, r, r^2, \ldots, r^{n-1}, \ldots$, where $0 < r < 2$. Then the mathematical expectation of A's gain is

$$
\begin{aligned}
E &= 1\left(\frac{1}{2}\right) + r\left(\frac{1}{2}\right)^2 + r^2\left(\frac{1}{2}\right)^3 + \cdots + r^{n-1}\left(\frac{1}{2}\right)^n + \cdots \\
&= \frac{1}{2} + \frac{1}{2}\left(\frac{r}{2}\right) + \frac{1}{2}\left(\frac{r}{2}\right)^2 + \cdots + \frac{1}{2}\left(\frac{r}{2}\right)^{n-1} + \cdots \\
&= \frac{1}{2}\left[1 + \left(\frac{r}{2}\right) + \left(\frac{r}{2}\right)^2 + \cdots + \left(\frac{r}{2}\right)^{n-1} + \cdots\right] \\
&= \frac{1}{2} \cdot \frac{1}{1 - (r/2)} = \frac{1}{2 - r},
\end{aligned}
$$

which is finite. For instance, if $r = \frac{3}{2}$, then A should pay the sum of 2 coins to play the game. However, as r approaches the value 2, the expectation grows larger without bounds; that is to say, it becomes infinite.

9.3 Problems

1. In the *Liber de Ludo Aleae*, Cardan held that since the probability of throwing a 6 with a single die is $\frac{1}{6}$, the probability that a 6 will appear when a pair of dice is rolled should be $2(\frac{1}{6}) = \frac{1}{3}$. What is the correct probability?

2. An Italian nobleman who was an amateur mathematician and an inveterate gambler had, by continued observation of a game in which three dice were simultaneously thrown, noticed that a sum of 10 was achieved more often than a sum of 9. He expressed his surprise at this and asked Galileo for an explanation. Calculate the probability, by enumerating the possible equally likely cases, that each sum will be rolled. (Assume that each of the three dice is painted a different color.)

3. The letters of the word "tailor" are written on different cards, and after the cards are thoroughly shuffled, four are drawn in order. What is the probability that the result spells the word "oral"?

4. (a) An urn contains 4 black and 5 white balls. A ball is drawn at random and then replaced, after which a second ball is drawn. Find the probability that the first is black and the second is white.

   (b) If a pair of fair dice is rolled, what is the probability that the sum shown is either exactly 9 or less than 4?

(c) A coin is tossed, then a card is drawn from an ordinary deck of 52 cards, and then a die is rolled. What is the probability of obtaining a head on the coin, an ace from the deck and a 5 on the die?

(d) If one letter is chosen at random from the word *boot* and one letter from the word *toot*, what is the probability that both letters match?

5. For any two events A and B, not necessarily mutually exclusive, the formula

$$\Pr[A \text{ or } B] = \Pr[A] + \Pr[B] - \Pr[A \text{ and } B]$$

holds. Use this to solve the following:

(a) A card is selected at random from an ordinary deck of 52 cards. What is the probability that it is either an ace or a spade?

(b) If an integer between 1 and 1000, inclusive, is selected at random, what is the probability that it begins with 1 or ends in 1?

6. Let A and B be events with $\Pr[B] \neq 0$. Then the *probability of A given B* written $\Pr[A|B]$, is

$$\Pr[A|B] = \frac{\Pr[A \text{ and } B]}{\Pr[B]}.$$

Use this formula to solve each of the following:

(a) Two fair dice are rolled and the numbers on the uppermost faces noted. Determine the probability that one die shows a 4, given that the sum of the numbers is 7.

(b) An urn contains red balls marked 1, 2, 3, a white ball marked 4, and blue balls marked 5, 6, 7, 8, 9, and a ball is selected at random. Determine the probability that it is red, given that the ball is known to have an even number.

(c) A single card is drawn from a standard deck of 52 cards and replaced by a joker, and then a second card is drawn. What is the probability that both cards are aces?

7. Find the probability that A will win each of the following two games proposed by Huygens in the *De Ratiociniis in Ludo Aleae.*

(a) A wagers B that given 40 cards of which 10 are of one color, 10 of another, 10 of another, and 10 of yet another, A will draw 4 so as to have one of each color.

(b) Twelve balls are taken, 8 of which are black and 4 white. A plays with B and undertakes, in drawing 7 balls blindfolded, to pick 3 white balls.

8. In the *Ars Conjectandi,* Bernoulli obtained the sum of the series

$$\frac{1}{1 \cdot 2} + \frac{1}{2 \cdot 3} + \frac{1}{3 \cdot 4} + \cdots + \frac{1}{n(n+1)}$$

by first writing

$$S = 1 + \frac{1}{2} + \frac{1}{3} + \frac{1}{4} + \cdots + \frac{1}{n+1}$$

$$S = \quad 1 + \frac{1}{2} + \frac{1}{3} + \cdots + \frac{1}{n} + \frac{1}{n+1}$$

and then subtracting. Use this technique to obtain the appropriate sum.

9. (a) A family has five children. If the probability of birth of either sex is assumed equally likely, what is the probability that exactly four of the children are boys?

(b) Suppose that a die is made by marking the faces of a regular dodecahedron with the numbers 1 through 12. What is the probability that on exactly three of six tosses, a number less than 4 turns up?

(c) Assume that the probability that a married couple will get a divorce within the next 10 years is $\frac{1}{5}$. Given eight randomly selected couples, find the probability that all eight will have the same partners after 10 years.

(d) An urn contains 3 red balls and 2 white balls. A ball is drawn, then replaced. Find the probability that a white ball will be drawn 4 times in a row.

10. Prove that the probability of obtaining at least r successes in n trials of a Bernoulli trials experiment, where the probability of success on a single trial is p, is given by

$\Pr[\text{at least } r \text{ successes in } n \text{ trials}]$

$$= \binom{n}{r} p^r (1-p)^{n-r} + \binom{n}{r+1} p^{r+1} \cdot$$

$$(1-p)^{n-r-1} + \cdots + \binom{n}{n} p^n (1-p)^0.$$

11. Using the formula in problem 10, solve the following:

   (a) If a fair die is tossed 5 times, what is the probability that at least three 6s are thrown?

   (b) A multiple-choice test has 4 "answers" listed for each of the 10 questions; only one answer is correct. If a student who knows nothing about the material guesses at the answers, what is the probability that the student will get at least 9 correct answers?

   (c) A weighted coin, when tossed, will land heads with probability $\frac{3}{7}$ and tails with probability $\frac{4}{7}$. If this coin is tossed a total of 6 times, what is the probability that it will land heads at most 4 times?
   [*Hint:* Pr[less than 5 successes]
   $= 1 -$ Pr[5 or more successes].]

12. A fair coin is tossed 6 times. Find the largest number n such that the probability of getting at least n heads is $\frac{1}{2}$ or more.

13. (a) Suppose that a gambler tosses 3 coins and receives \$8 if 3 heads appear, \$4 if 2 heads appear, \$2 if 1 head appears, and \$1 if no head appears. What is his mathematical expectation?

   (b) A game is devised in which a single die is thrown. A gambler agrees to pay out in dollars the amount shown on the die if the number is odd, and to collect the amount shown on the die if the number is even. What is his mathematical expectation?

14. (a) A player randomly draws a single card from an ordinary deck of 52 cards. If the card is an ace, the player receives \$6; and if the card is a king, a queen, or a jack, the player receives \$1. If any other card is drawn, the player loses \$1. Is this game fair?

   (b) Suppose that the rules of the game in part (a) are modified so that the player wins \$3 for an ace, \$2 for a king, and 50¢ for a queen or a jack, but loses 50¢ if the card has a face value of 10, 9, or 8 and loses 75¢ if it is a 7 or less. Is this game fair?

15. Nicholas Bernoulli discussed the following problem. Each of two players A and B has a deck with the same cards (or tickets). They draw the cards singly until each draws the same, in which case A wins. If such a match does not occur, then B

wins. The problem is to find the probability of a win for either player. Suppose that the players have 4 cards each, numbered 1, 2, 3, 4, and that player A draws them in their natural order—1, 2, 3, 4. Compute for each player the probability of winning [*Hint:* The number of permutations of the n integers 1 to n in which every number is in the wrong place is

$$N = n!\left(1 - 1 + \frac{1}{2!} - \frac{1}{3!} + \frac{1}{4!} + \cdots \right.$$
$$\left. + (-1)^n \frac{1}{n!}\right).]$$

16. The idea behind moral expectation was that it was necessary to represent the relative value of a given sum of money not by its face value, but in relation to the capital of the person under consideration. Gabriel Cramer (1704–1752) suggested that the value derived from a sum of money varied with the square root of the sum. Thus the moral expectation in his solution to the St. Petersburg paradox is

$$\sqrt{1}\left(\frac{1}{2}\right) + \sqrt{2}\left(\frac{1}{2}\right)^2 + \sqrt{4}\left(\frac{1}{2}\right)^3 + \cdots$$
$$+ \sqrt{2^{n-1}}\left(\frac{1}{2}\right)^n + \cdots.$$

What finite value does this series give?

Bibliography

Bacon, Harold M. "The Young Pascal." *Mathematics Teacher* 30(1937): 180–185.

Bell, A. E. *Christiaan Huygens, and the Development of Science in the Seventeenth Century.* London: Edward Arnold, 1947.

Bishop, Morris. *Pascal: The Life of Genius.* New York: Reynal & Hitchcock, 1936.

Boyer, Carl. "The First Calculus Textbooks." *Mathematics Teacher* 34(1946): 159–167.

———. "Cardan and the Pascal Triangle." *American Mathematical Monthly* 57(1950): 387–389.

Bussey, W. H. "The Origin of Mathematical Induction." *American Mathematical Monthly* 24(1917): 199–207.

Cajori, Florian. "The Origin of the Name 'Mathematical Induction'." *American Mathematical Monthly* 25(1918): 197–201.

Caulley, Maurice. *French Science and Its Principal Discoveries Since the Seventeenth Century.* Reprint ed. New York: Arno Press, 1975.

Crosland, Maurice. *The Society of Arcueil: A View of French Science at the Time of Napoleon I.* Cambridge, Mass.: Harvard University Press, 1967.

Daston, Lorraine. *Classical Probability in the Enlightenment.* Princeton, N.J.: Princeton University Press, 1988.

David, Florence N. "Studies in the History of Probability and Statistics I, Dicing and Gaming." *Biometrika* 42(1955): 1–15.

———. *Games, Gods and Gambling.* New York: Hafner, 1962.

Edwards, A. W. F. *Pascal's Arithmetic Triangle.* New York: Oxford University Press, 1987.

Eves, Howard. "The Bernoulli Family." *Mathematics Teacher* 59(1966): 276–278.

Gillispie, C. C. "Probability and Politics: Laplace, Condorcet and Turgot." *Proceedings of the American Philosophical Society* 116(1972): 1–20.

Glass, D. V. "John Graunt and his 'Natural and Political Observations'." *Proceedings of the Royal Society B* 159(1963–64): 2–37.

Goldstine, Herman. *A History of Numerical Analysis from the 16th through the 19th Century.* New York: Springer-Verlag, 1977.

Hacking, Ian. *The Emergence of Probability.* London: Cambridge University Press, 1975.

Heyde, C. C., and Seneta, E. *I. J. Bienaymé: Statistical Theory Anticipated.* New York: Springer-Verlag, 1977.

Kendell, M. G. "Studies in the History of Probability and Statistics. II. The Beginnings of a Probability Calculus." *Biometrika* 43(1956): 1–14.

———. "Where Shall the History of Statistics Begin?" *Biometrika* 47(1960): 447–449.

Kruger, Lornze; Daston, Lorraine; and Heidelberger, Michael, eds. *The Probabilistic Revolution.* 2 vols. Cambridge, Mass.: M.I.T. Press, 1987.

Laplace, Pierre Simon. *A Philosophical Essay on Probabilities.* Translated by F. Truscott and F. Emory. New York: Dover, 1951.

Lick, Dale. "The Remarkable Bernoulli Family." *Mathematics Teacher* 62(1969): 401–409.

Maistrov, L. E. *Probability: A Historical Sketch.* New York: Academic Press, 1974.

Mesnard, Jean. *Pascal, His Life and Works.* New York: Philosophical Library, 1952.

Ore, Oystein. "Pascal and the Invention of Probability." *American Mathematical Monthly* 67(1960): 409–419.

Patterson, Elizabeth. "The Case of Mary Somerville: An Aspect of Nineteenth-Century Science." *Proceedings of the American Philosophical Society* 118, No. 3(1974): 269–275.

Pearson, E. S., ed. *The History of Statistics in the 17th and 18th Centuries: Lectures by Karl Pearson.* New York: Macmillan, 1978.

Pearson, Karl. "James Bernoulli's Theorem." *Biometrika* 17(1925): 201–210.

———. "Laplace." *Biometrika* 21(1927): 202–216.

Pomeranz, Janet. "The Dice Problem—Then and Now." *College Mathematics Journal* 15(1984): 229–236.

Sheynin, O. B. "On the Early History of the Law of Large Numbers." *Biometrika* 55(1968): 459–467.

———. "P. S. Laplace's Work on Probability." *Archive for History of Exact Sciences* 16(1976): 137–187.

———. "Early History of the Theory of Probability." *Archive for History of Exact Sciences* 17(1977): 201–259.

Stigler, Stephen. "Studies in the History of Probability and Statistics XXXIV, Napoleonic Statistics: The Work of Laplace." *Biometrika* 62(1975): 503–517.

———. *The History of Statistics.* Cambridge, Mass.: Harvard University Press, 1986.

Struik, Dirk. "The Origin of L'Hospital's Rule." *Mathematics Teacher* 56(1963): 257–260.

Todhunter, Isaac. *History of the Mathematical Theory of Probability from the Time of Pascal to That of Laplace.* Reprint ed. New York: Chelsea, 1949.

Uspenski, J. V. *Introduction to Mathematical Probability.* New York: McGraw-Hill, 1937.

Wolf, A. *A History of Science, Technology and Philosophy in the 18th Century.* 2d ed. 2 vols. Gloucester, Mass.: Peter Smith, 1961.

CHAPTER **10**

The Renaissance of Number Theory: Fermat, Euler, and Gauss

Where there is number there is beauty.

PROCLUS

10.1 Marin Mersenne and the Search for Perfect Numbers

Very few people in the sixteenth century were seriously interested in mathematics or science. Probably there were only a few thousand of them—and most of those interested but not active. Many who worked in science or mathematics did so in isolation, for even within the university communities their interests were seen as only peripherally important. Science, still tainted with the practice of magic, was not yet a respectable intellectual occupation, so that recognition came to those who excelled in more central fields of scholarship. Outside of books, there was no regular way for learned men to keep in contact with one another and publication was at times difficult owing to ecclesiastical censorship and condemnation.

As more people engaged in scientific pursuits, however, it was increasingly seen that advancement in knowledge was powerfully affected by the extent and speed of the communication of discoveries. Continual testing of ideas keeps science sound as it leads to the abandonment of disproved hypotheses. In a remarkable development from Renaissance times, when almost all scientific discussion—what little there was—took place within the framework of the university curriculum, there emerged in the 1600s a new pattern of social organization, the learned societies and academies independent of the universities. Beginning with the establishment of small and relatively informal amateur circles, this period culminated in the founding of the Royal Society of London (1660) and L'Académie Royale des Sciences (1666). Both are still active and distinguished professional societies. These organizations worked chiefly for sharing scientific information within their own memberships and secondarily for making their collective findings generally available to other interested groups.

Contrasted with the groups that gathered about the famous philosophers of antiquity, this form of academy, organized and run by its members, was a product of sixteenth-century Italy. The numerous societies and academies that sprang up at this time (some claim more than 700) may be viewed as an attempt by groups of intellectuals to create for themselves a more congenial institution than the established university. Scholars sought to enrich their understanding by meeting in company to discuss subjects of common interest. The earliest of these modern academies tended to embrace almost the entire range of intellectual activity. After the mid-1500s, academies devoted to specific disciplines were founded; more than half were literary societies, and the rest were divided among theatrical, medical, legal, and (more rarely) scientific pursuits. Almost always they were looked on with disfavor both by the civil authorities, who saw subversive teaching and political intrigue in their often secret meetings, and by the Church, which feared the propagation of heresies and the power of uncontrolled associations.

The first academy devoted chiefly to science was probably the one Giambattista della Porta (1535–1615) established in Naples in 1560, called Academia Secretorum Naturae. To become a member, a man must have made some original discovery or communicated some previously unknown fact in natural science. The members of della Porta's academy called themselves the Otiosi, or Idlers, following the custom then prevalent in Italian societies of adopting humorously derogatory names. (Thus, one famous group of academicians was called the Umidi, or the damp ones (perhaps we could call them ''all wet'') and another bore the name Scomposti, which would mean ''the confused, disorganized ones.'') The title Academia Secretorum Naturae made the Church suspicious, and della Porta was denounced as a practitioner of the black arts. He surely encouraged the charge by writing a book called *Magia Naturalis* (Natural Magic). The secrecy of his society and its members' reputation for corresponding in cipher didn't help matters. Although della Porta was personally exonerated of all charges, he was nonetheless ordered to close his academy.

This short-lived attempt at formal scientific organization anticipated the more important Accademia dei Lincei (Academy of the Lynx-Eyed), which was founded in Rome in 1603 with the object of attempting new discoveries in physical science and publishing them to the world. Its emblem, a lynx clawing to death a Cerberus (a many-headed dog with a tail of snakes) was chosen to symbolize the struggle of scientific truth against ignorance. On any grounds of reasonable probability, the Accademia dei Lincei should have collapsed. For seven years, before it attracted any widely recognized scientist to membership, it consisted of exactly four members, all under 30 years of age. But in 1610 the Accademia reorganized on a larger scale, with Galileo and the aging della Porta among its new members. The revived organization published several books its members had written, including two by Galileo. The great man seemed proud to use the title *Lincean* in his correspondence and on the title pages of most of his subsequent books.

Galileo's new optical instrument was given the name *telescope* in 1611 at a meeting of the Accademia dei Lincei. From time to time Galileo had used a variety of descriptive words such as *eye-glass* or *perspective trunk;* but this device was too significant an instrument not to have its place in the language. A member of the Accademia coined the name by combining two Greek words meaning ''far'' and ''to look'' to describe Galileo's far-looking instrument. Galileo's ideas were so disturbing that some men refused even to look into his telescope, for fear of seeing things that might contradict traditional philosophy or religion. The study of physics and astronomy became, in fact, too dangerous to pursue; and in 1630, when the Holy Office condemned Galileo, the Accademia disbanded. It was reconstituted briefly in 1745, and permanently in 1801, so that the leading scientific society in Italy is still the Accademia dei Lincei.

The earliest-noted instance that has come down to us of a regular gathering of mathematicians is the circle held together by Father Marin Mersenne (1588–1648), an able mathematician and physicist, Jesuit-educated and a friar of the Order of Minims. A man who made little contribution himself in the way of new knowledge, Mersenne had the capacity for understanding and appreciating the work of others. Arriving in Paris in 1619, he found deplorable the lack of any sort of formal organization to which scholars might resort. He did his best to answer the need, partly by making his monastery a meeting place for those eager to discuss their results and to hear of similar work done elsewhere, and partly by acting as a link between scholars from different countries. At a time when publication of technical discoveries still lay in the future, Mersenne served as a personal and effective clearinghouse of scientific information. He made it his business to become acquainted with everyone of importance in the learned world through an elaborate network of correspondence by which he transmitted news of the advancement of science in return for more news. He stimulated scientific thought through the numerous questions he asked of anyone who might be able to contribute to finding an answer, and he communicated both questions and answers to others to elicit their reactions.

Mersenne's letter-writing fulfilled many functions of a modern scientific journal. After his death, letters were found in his cell from 78 correspondents scattered over Europe, among them Fermat in France, Huygens in Holland, Pell and Hobbes in England, and Galileo and Torricelli in Italy. During a visit to Italy in 1645, Mersenne had met Torricelli and discussed the latter's experimentation on the vacuum. Returning to Paris, Mersenne had the idea of taking a column of mercury to the top of a mountain to observe the effect of atmospheric pressure. But he missed his chance at scientific fame, because before Mersenne could carry out his plan, Pascal conducted the famous experiment demonstrating that the height of a mercury column varied with altitude.

Mersenne's most important contribution in the early days of the new science was maintaining more or less regular conferences for the exchange of ideas. At his monastery he gathered around him an intellectual circle, composed mainly of the scientific community of Paris but animated also by travelers passing through the city. These meetings seem to have taken place almost continually from 1635 until the end of Mersenne's life in 1648. An entirely informal gathering with no publications or sets of laws to bind its members, the circle could scarcely be considered an academy at all. Its significance lay in the fame of those who attended the meetings, among them Desargues, Roberval, Descartes, and the Pascals, father and son.

When any of the gathering had results to present, he would have them printed on loose sheets of paper, which would be distributed. At the age of 14, the precocious Blaise Pascal had been admitted to the circle of learned men with whom Mersenne surrounded himself, and at one of the weekly meetings, he presented a leaflet that contained the ''mystic hexagram'' theorem. The one-page *Essay pour les coniques* was in effect an announcement of a treatise that Pascal was preparing. It was never published and is now lost, though Leibniz saw it in Paris in 1676 and described its contents.

Pro-Copernican as early as 1629, Mersenne was virtually Galileo's representative in France. When he heard that Galileo was writing a book on the movement of the earth, he offered to get it published for him. However, Galileo was successful in having it published himself and copies were quickly sent to France. Not long after their arrival came word of Galileo's trial and condemnation and the burning of his book. Mersenne was affected by the Church's treatment of Galileo, but less so than one might imagine. He published, in 1634, under the title *Les Mécaniques de Galilée,* a version of Galileo's early lectures (around 1592) on mechanics, and in 1639, a year after the original publication, he translated the *Discorsi* into French. He did not translate Galileo's *Dialogo* when it appeared in 1632, possibly because he felt that Galileo had broken his promise to the pope not to come down heavily in favor of the Copernican theory; but he did bring out a summary account of parts of it. As Italian was little understood abroad, Mersenne was instrumental in popularizing Galileo's investigations on the Continent. It is worthy of notice that this was done by a faithful member of a Catholic religious order, at the height of the Church's hostility to science.

After Mersenne's death, the conferences continued to be held at regular intervals in various homes in and around Paris, including Pascal's. It is usual to regard the Académie Royale des Sciences, which was formed in 1666, as the more or less direct successor of these numerous private gatherings. The new academy was composed of two sections: a mathematical one including all "exact sciences," and a physical one concerned with the more "experimental sciences" such as physics, chemistry, botany, and anatomy. As its name indicates, this society was a royal institution as well as a scientific academy; its members acknowledged two masters, science and the crown. The state was willing to commit considerable funds to the support of the Académie so that it could reap the benefits of its scientific inquiry. Whereas the Royal Society in England was forever struggling with financial difficulties, the members of the French body drew fixed pensions and were to give all their time to the Académie. The resources of the royal treasury attracted scholars from all quarters, so that at times the Académie Royale des Sciences seemed more a European than a French gathering. At first its sessions were closed to the public, and even the minutes of the meetings remained the private property of the Académie until extracts were printed. After several years the Académie began to share its findings through the newly founded *Journal des savants.*

The *Journal des savants,* not originally affiliated with the Académie, was founded in 1665 to deal with arts, sciences, and theology. Its appearance met with such violent opposition from religious circles that its publication was temporarily suspended, its editor removed, and its sphere of activities seriously curtailed before it could resume publication. When it returned to publication, it did so as a quasi-official periodical of the Académie. Although in theory the Académie chose in favor of anonymous publication, seventeenth-century scientists were no less concerned with establishing their academic reputations than their modern counterparts are. Thus, after announcing a new idea or discovery in a closed meeting of the Académie, an individual would often send his findings to the *Journal des savants.*

The development of science in the 1600s was signally indebted to three associations: the Accademia dei Lincei in Rome, which existed for but a short time; the Académie Royale des Sciences in Paris; and the Royal Society of London. Paralleling events on the Continent, the founding of the Royal Society was an outgrowth of

A visit by Louis XIV in 1671 to the Académie des Sciences. (By courtesy of the University of California, Berkeley Library.)

informal meetings of friends, more or less learned but all deeply interested in experimental knowledge. Several English scientists had visited the academies in Italy as well as the conferences of Mersenne and others in Paris, and advocated forming a comparable organization in Great Britain. This led at first to two gatherings for amateurs in science, one centering in Oxford around Robert Boyle and the other in London at Gresham College, where the young Christopher Wren (remembered mainly for his architectural achievements) was the leading spirit in the discussion of matters of science.

It was to this group at London that the immediate rise of the Royal Society can be traced. The event that inaugurated the society as an organized body occurred on November 28, 1660, when twelve friends at Gresham College decided to initiate formal meetings each week—if possible, under the patronage of Charles II himself. Temporary officers were elected, rules drawn up, and a tentative list of members determined. An admission fee of one shilling and a further weekly fee were to be charged to defray the expenses of experimentation. The members agreed to refrain from any discussion of current news or questions of the day; science was to be the group's sole topic. In 1662 the Royal Society of London for the Promotion of Natural Knowledge received its formal charter and mace from Charles II, but unlike its French counterpart, it received little else. At first the number of members was limited to 55, but it was afterwards extended, and finally admission was left open to every proper candidate. The famous *Philosophical Transactions of the Royal Society,* which began to appear in 1665, and the *Journal des savants* were the first journals to include mathematical and scientific articles.

Mersenne's name is connected with one of the oldest problems in the theory of numbers, finding all perfect numbers. Early Greek mathematicians were especially intrigued by the relations between a number and its aliquot divisors (positive divisors less than the number itself). To an age for which numbers held mystical properties, it was considered remarkable that 6 is the sum of its aliquot divisors:

$$6 = 1 + 2 + 3.$$

The next number after 6 enjoying this feature is 28, since the positive divisors of 28 are 1, 2, 4, 7, 14, 28, and

$$28 = 1 + 2 + 4 + 7 + 14.$$

The Pythagoreans, in line with their philosophy of attributing certain social qualities, and later even ethical import, to numbers, called such numbers ''perfect.'' A precise definition is as follows.

Definition

A positive integer n is said to be perfect if n is equal to the sum of all its positive divisors, not including n itself.

If we let $\sigma(n)$ denote the sum of all the positive divisors of n, then the sum of the positive divisors less than n is given by $\sigma(n) - n$. Thus the condition ''n is perfect'' amounts to the requirement $\sigma(n) - n = n$, or equivalently,

$$\sigma(n) = 2n.$$

As examples, we have

$$\sigma(6) = 1 + 2 + 3 + 6 = 2 \cdot 6,$$

and

$$\sigma(28) = 1 + 2 + 4 + 7 + 14 + 28 = 2 \cdot 28.$$

For many centuries, philosophers were more concerned with the ethical and religious significance of perfect numbers than with their mathematical properties. In his *De Civitate Dei,* Saint Augustine (354–430) explained that although God could have created the world all at once, He preferred to take six days, because the perfection of the work is symbolized by the (perfect) number six. To quote Saint Augustine:

> Six is a number perfect in itself, and not because God created all things in six days; rather, the converse is true. God created all things in six days because this number is perfect, and it would have been perfect even if the work of six days did not exist.

Early commentators on the Old Testament argued that the perfection of the universe is represented by 28, the number of days the moon takes to circle the earth. In a similar vein, Alcuin of York (735–804), Charlemagne's educational advisor, observed that the whole human race is descended from the eight souls on Noah's ark and that this second creation was less perfect than the first, 8 being an imperfect number.

Only four perfect numbers were known to the ancient Greeks. Nicomachus of Gerasa, who summarized the existing knowledge of the theory of numbers in his *Introductio Arithmetica* (circa A.D. 100), lists

$$P_1 = 6, \qquad P_2 = 28, \qquad P_3 = 496, \qquad P_4 = 8128.$$

He said "the perfect numbers are easily enumerated and arranged in suitable order"— one found among the units, one among the tens, one among the hundreds, and one among the thousands (that is, below 10,000). Using this meager evidence, later writers conjectured that:

1. The nth perfect number P_n contains exactly n digits.

2. The even perfect numbers end alternately in the digits 6 or 8.

Both assertions are wrong. There is no perfect number with 5 digits; the next perfect number (first given correctly in an anonymous fifteenth-century manuscript) is

$$P_5 = 33,550,336.$$

Although the final digit of P_5 is 6, the succeeding perfect number

$$P_6 = 8,589,869,056$$

ends in 6 also, not 8 as conjectured. To salvage something in the positive direction, we shall show later that the even perfect numbers always do end in 6 or 8, but not alternately.

If nothing else, the magnitude of P_6 should convince the reader of the rarity of perfect numbers. It is not yet known whether there are finitely or infinitely many of them.

Determining the general form of all perfect numbers dates to ancient times. It was partially accomplished by Euclid when in Proposition 36 of Book IX of the *Elements,* he proved that if the sum

$$1 + 2 + 2^2 + 2^3 + \cdots + 2^{k-1} = p$$

is a prime number, then $2^{k-1}p$ is a perfect number (of necessity even). To quote Euclid's own words:

> If as many numbers as we please beginning from a unit be set out continuously in double proportion, until the sum of all becomes prime, and if the sum multiplied into the last makes some number, the product will be perfect.

For example, $1 + 2 + 4 = 7$ is a prime, hence $4 \cdot 7 = 28$ is a perfect number. Euclid's argument makes use of the formula for the sum of a geometric progression,

$$1 + 2 + 2^2 + 2^3 + \cdots + 2^{k-1} = 2^k - 1,$$

that is found in various Pythagorean texts. In this notation, the result reads as follows.

THEOREM   *If $2^k - 1$ is prime $(k > 1)$, then $n = 2^{k-1}(2^k - 1)$ is a perfect number.*

To demonstrate the theorem, it is necessary to know all the divisors of the integer $2^{k-1}(2^k - 1)$. Let us therefore take a detour and consider the problem of finding the divisors of an arbitrary positive integer. For this problem, we have a simple, concise answer.

LEMMA   *If $n = p_1^{k_1}p_2^{k_2} \ldots p_r^{k_r}$ is the prime factorization of the integer $n > 1$, then the positive divisors of n are precisely those integers d of the form*

$$d = p_1^{a_1}p_2^{a_2} \ldots p_r^{a_r}, \qquad \text{where } 0 \leq a_i \leq k_i \qquad (i = 1, 2, \ldots, r).$$

An example should clarify matters. For $n = 180$, we have the prime factorization

$$180 = 2^2 \cdot 3^2 \cdot 5.$$

The lemma asserts that the positive divisors of 180 are integers of the form

$$2^{a_1} \cdot 3^{a_2} \cdot 5^{a_3},$$

where $a_1 = 0, 1, 2$; $a_2 = 0, 1, 2$; $a_3 = 0, 1$. These are the integers

$$1 \cdot 1 \cdot 1, \quad 1 \cdot 1 \cdot 5, \quad 1 \cdot 3 \cdot 1, \quad 1 \cdot 3 \cdot 5, 1 \cdot 3^2 \cdot 1, \quad 1 \cdot 3^2 \cdot 5,$$

$$2 \cdot 1 \cdot 1, \quad 2 \cdot 1 \cdot 5, \quad 2 \cdot 3 \cdot 1, \quad 2 \cdot 3 \cdot 5, 2 \cdot 3^2 \cdot 1, \quad 2 \cdot 3^2 \cdot 5,$$

$$2^2 \cdot 1 \cdot 1, \quad 2^2 \cdot 1 \cdot 5, \quad 2^2 \cdot 3 \cdot 1, \quad 2^2 \cdot 3 \cdot 5, 2^2 \cdot 3^2 \cdot 1, \quad 2^2 \cdot 3^2 \cdot 5,$$

or put in sequential order,

$$1, 2, 3, 4, 5, 6, 9, 10, 12, 15, 18, 20, 30, 36, 45, 60, 90, 180.$$

This last lemma enables us to prove Euclid's theorem. It tells us that if $2^k - 1$ is prime, then each of

$$1, 2, 2^2, \ldots, 2^{k-1}, 2^k - 1, 2(2^k - 1), 2^2(2^k - 1), \ldots, 2^{k-1}(2^k - 1)$$

is a divisor of $n = 2^{k-1}(2^k - 1)$, and that these exhaust the possibilities for divisors. Thus there are two sums to be added:

$$S_1 = 1 + 2 + 2^2 + \cdots + 2^{k-1}$$

and

$$S_2 = (2^k - 1) + 2(2^k - 1) + 2^2(2^k - 1) + \cdots + 2^{k-1}(2^k - 1)$$

$$= (1 + 2 + 2^2 + \cdots + 2^{k-1})(2^k - 1)$$

$$= S_1(2^k - 1).$$

Added, these yield

$$\sigma(n) = S_1 + S_1(2^k - 1) = 2^k S_1.$$

But the formula for the sum of a geometric progression gives $S_1 = 2^k - 1$, so that our total is

$$\sigma(n) = 2^k(2^k - 1) = 2 \cdot 2^{k-1}(2^k - 1) = 2n,$$

the effect of which is to make n a perfect number.

In the eighteenth century, the great Swiss mathematician Leonhard Euler (1707–1783) showed that all even perfect numbers must be of the form expressed by Euclid's theorem. That is, an even number n is perfect if and only if it satisfies the formulation

$$n = 2^{k-1}(2^k - 1),$$

where $2^k - 1$ is prime. No odd perfect number has ever been found, yet it has never been proved that such a number cannot exist.

Although no one knows whether any exist, many interesting results have been obtained concerning the structure of odd perfect numbers. The bulk of present-day research concerns itself primarily with the determination of bounds on the number of prime factors of such a number, and on its size. What is currently known in this direction is that no odd perfect number can have fewer than eight different prime factors or be less than 10^{300}. Although all this lends support to the belief that there are no odd perfect numbers, only a proof of their nonexistence would be conclusive. "It must always stand to the credit of the Greek geometers," wrote the mathematician James Sylvester in 1888, "that they succeeded in discovering a class of perfect numbers which in all probability are the only numbers which are perfect."

Euclid's theorem immediately raises a new question: For what values of k is the integer $2^k - 1$ prime? One step toward a solution can be made at once; namely, if $2^k - 1$ is prime, then the exponent k must itself be a prime. Suppose to the contrary that k is a composite number, say $k = rs$, where $1 < r \leq s < k$. Using the formula

$$x^n - 1 = (x - 1)(x^{n-1} + x^{n-2} + \cdots + x^2 + x + 1),$$

we could then write

$$2^k - 1 = 2^{rs} - 1 = (2^r)^s - 1$$

$$= (2^r - 1)(2^{r(s-1)} + 2^{r(s-2)} + \cdots + 2^r + 1).$$

Because each factor on the right is plainly greater than 1, this violates the primality of $2^k - 1$; hence k must be a prime number.

For the primes $p = 2, 3, 5, 7$, the values, 3, 7, 31, 127 of $2^p - 1$ are prime numbers, so that by Euclid's formula

$$2(2^2 - 1) = 2 \cdot 3 = 6,$$

$$2^2(2^3 - 1) = 4 \cdot 7 = 28,$$

$$2^4(2^5 - 1) = 16 \cdot 31 = 496,$$

$$2^6(2^7 - 1) = 64 \cdot 127 = 8128$$

are all perfect numbers.

Many early writers erroneously believed that $2^p - 1$ is prime for every choice of the prime number p. But in 1536, Hudalrichus Regius in a work entitled *Utriusque Arithmetices* exhibited the correct factorization

$$2^{11} - 1 = 2047 = 23 \cdot 89.$$

If this seems a small accomplishment, it should be realized that his calculations were in all likelihood carried out in Roman numerals, with the aid of an abacus (not until the late 1500s did the Arabic numeral system win complete ascendancy over the Roman one). Regius also gave $p = 13$ as the next value of p for which the expression $2^p - 1$ is a prime. From this, one obtains the fifth perfect number

$$P_5 = 2^{12}(2^{13} - 1) = 33,350,336.$$

In 1603 Pietro Cataldi, who is chiefly remembered for his work on extracting roots by means of continued fractions, published a table of factors of all numbers up to 800, with a separate list of primes up to 750. By the laborious procedure of dividing by all primes not exceeding their respective square roots, Cataldi determined that $2^{17} - 1$ and $2^{19} - 1$ were both primes, and in consequence, that

$$P_6 = 2^{16}(2^{17} - 1) = 8,589,869,056$$

and

$$P_7 = 2^{18}(2^{19} - 1) = 137,438,691,328$$

were the sixth and seventh perfect numbers. He also stated that $2^n - 1$ was prime for $n = 2, 3, 5, 7, 13, 17, 19, 23, 29, 31$, and 37. In 1640, however, Fermat disproved Cataldi's claim by finding factors of $2^{23} - 1$ and $2^{37} - 1$, whereas Euler (1738) showed that $2^{29} - 1$ was composite.

One of the difficulties in discovering further perfect numbers was the lack of tables of primes and composites. The first fair-sized table, one giving the least prime factors of all numbers not divisible by 2 or 5, up to 24,000, was published by J. H. Rahn as an appendix to his *Teusche Algebra* (1659). In 1668, John Pell of England extended this table to include numbers up to 100,000. The table constructed by the Viennese schoolmaster Anton Felkel is interesting because of its singular fate. The first volume of Felkel's computations, giving factors of numbers not divisible by 2, 3, or 5 up to 408,000, was published in 1776 by the Imperial Treasury of Austria; when only a few subscriptions were received the remaining copies were confiscated, their pages used as paper in the manufacture of cartridges for a war against the Turks. Even more ill-starred was the table calculated by J. P. Kulik (1773–1863), which was deposited in the library of the Vienna Royal Academy in 1867 but was never published at all. Kulik, a professor of mathematics at the University of Prague, devoted 20 years of his life to preparing, unassisted and alone, the factors of numbers up to 100,000,000.

It has become traditional to call numbers of the form

$$M_n = 2^n - 1, \qquad n \geq 1,$$

Mersenne numbers, in honor of an incorrect but provocative assertion Marin Mersenne made about their primality. Those Mersenne numbers that happen to be prime are said to be Mersenne primes. As we have seen, the determination of Mersenne primes—and in turn, the quest for more even perfect numbers—can be narrowed to the case in which the exponent n is itself a prime. If it were known that there were infinitely many primes p for which M_p is prime, then there would exist an infinitude of (even) perfect numbers. Unfortunately, this is another one of those famous unresolved problems.

In the preface of his *Cogitata Physico-Mathematica* (1644), Mersenne stated that M_p was prime for $p = 2, 3, 5, 7, 13, 17, 19, 31, 67, 127$, and 257, and composite for all other primes $p \leq 257$. It was obvious to other mathematicians that Mersenne could not have tested for primality all the numbers he had announced; but neither could they. Whether Mersenne had available some theorem not yet rediscovered or was relying mainly on personal guessing is conjectural. He did admit that "to tell if a given number of 15 or 20 figures is prime or not, all time would not suffice for the test." The number M_{127} contains 39 digits, and M_{257} has 78.

For nearly 150 years, Cataldi's M_{19} remained the largest Mersenne prime. Then Euler (1732) verified that M_{31} was prime by examining all prime numbers up to 46,339 as possible divisors; but M_{67}, M_{127}, and M_{257} were beyond his technique. In any event, M_{31} was the largest known prime for the next century, and it gave rise to the ninth perfect number,

$$P_9 = 2^{30}(2^{31} - 1) = 2,305,843,008,139,952,128.$$

Euler, one of the greatest mathematicians of all ages, was not immune from making false claims about the occurrence of perfect numbers. He wrote:

> I venture to assert that aside from the cases noted [Euler earlier mentioned 11, 23, 29, 37, 43, 73, 83], every prime less than 50, and indeed than 100, makes $2^{n-1}(2^n - 1)$ a perfect number, whence the eleven values 1, 2, 3, 5, 7, 13, 17, 19, 31, 41, 47 of n yield perfect numbers.

Euler eventually (1753) detected his own error as to $n = 41$ and 47.

Not until 1947, after labor that was tremendous because of the unreliability of the desk calculators used, was examination completed on the prime or composite character of M_p for the last of the 55 primes in the range $p \leq 257$. We know now that Mersenne made five mistakes, although it is astonishing that it took over 300 years to set the good friar straight. The first error in Mersenne's statement was found when Pervusin (1883) and Seelhoff (1886) proved independently that M_{61} was prime. After Cole (1903) discovered factors for M_{67}, some of Mersenne's defenders suggested that 67 was merely a misprint for 61. Subsequent investigations revealed, however, three other errors in the list: M_{89} is prime (Powers in 1911), M_{107} is prime (Fauquembergue and Powers, independently, 1914–1917), and M_{257} is composite (Kraitchik in 1922).

Marin Mersenne
(1588–1648)

(By courtesy of Columbia University, David Eugene Smith Collection.)

A historical curiosity is that in 1876 Edouard Lucas worked a test whereby he was able to prove that the Mersenne number M_{67} was composite; but he could not produce the actual factors. At the October 1903 meeting of the American Mathematical Society, the American mathematician Frank Nelson Cole had a paper on the program with the modest title *On the Factorization of Large Numbers*. When called upon to speak, Cole walked to the blackboard, and saying nothing, raised the integer 2 to the 67th power; then he carefully subtracted 1 from the resulting number and let the answer stand. Silently he moved to a clean part of the board and multiplied, longhand, the product

$$193{,}707{,}721 \times 761{,}838{,}257{,}287.$$

The two calculations agreed. The story goes that for the first and only time on record, this august body gave the presenter of a paper a standing ovation. Cole took his seat without having uttered a word; no one asked him a question. (Later he confided to a friend that it took him ''three years of Sundays'' to find the factors of M_{67}.)

A question allied to the existence of Mersenne primes is whether there are infinitely many primes p such that the numbers M_p are composite. Even the answer to this is unresolved. However, much larger values of p are known for which M_p is composite than for which M_p is prime. For example,

$$M_{32376604223}$$

is composite. Consider the fantastic size of this number; if written in decimal form, its digits would more than fill the telephone books of all five boroughs of New York City.

After giving the eighth perfect number $2^{30}(2^{31} - 1)$, Peter Barlow in his book *Theory of Numbers* (published in 1811) concluded from its size that it ''is the greatest that ever will be discovered; for as they are merely curious, without being useful, it is not likely that any person will ever attempt to find one beyond it.'' The very least that can be said is that Barlow underestimated the obstinacy of human curiosity. Although the subsequent search for larger perfect numbers provides us with one of the fascinating chapters in the history of mathematics, an extended discussion would be out of place here.

It is worth remarking, however, that the first twelve Mersenne primes (hence, twelve perfect numbers) have been known since 1914. The twelfth in order of discovery, M_{89}, was the last Mersenne prime disclosed by hand calculation. Its primality was verified by both Powers and Cunningham in 1914, working independently and by different techniques. The prime M_{127} was found by Lucas in 1876 and for the next 75 years was the largest number actually known to be prime.

Calculations whose mere size and tedium repel the mathematician are grist for the mill of computers. Starting in 1952, twenty-one additional Mersenne primes, all huge, have come to light. The Mersenne prime M_{859433}, found in 1994, is the largest of the known prime numbers. It in its turn gives rise to the largest known even perfect number, the thirty-third one,

$$P_{33} = 2^{859432}(2^{859433} - 1),$$

an immense number of 517,431 digits.

Another type of number that has had a continuous history extending from the early Greeks to the present time constitutes the amicable numbers. Two numbers such as 220 and 284 are called amicable, or friendly, because they have the remarkable property that within each number is "contained" the other, in the sense that each number is equal to the sum of all the positive divisors of the other, not counting the number itself. Thus, as regards the divisors of 220,

$$1 + 2 + 4 + 5 + 10 + 11 + 20 + 22 + 44 + 55 + 110 = 284$$

while for 284,

$$1 + 2 + 4 + 71 + 142 = 220.$$

In terms of the σ function, amicable numbers m and n (or an amicable pair) are defined by the equations

$$\sigma(m) - m = n, \qquad \sigma(n) - n = m,$$

or what amounts to the same thing,

$$\sigma(m) = m + n = \sigma(n).$$

Down through their quaint history, amicable numbers have been important in magic and astrology, and in casting horoscopes, making talismans, and concocting love potions. The Greeks believed that these numbers had a particular influence in establishing friendship between two individuals. The philosopher Iamblichus of Chalcis (ca. A.D. 250–A.D. 330) ascribed a knowledge of the pair 220 and 284 to the Pythagoreans. He wrote:

> They [the Pythagoreans] call certain numbers amicable numbers, adopting virtues and social qualities to numbers, as 284 and 220; for the parts of each have the power to generate the other, according to the rule of friendship, as Pythagoras affirmed. When asked what is a friend, he replied, "Another I," which is shown in these numbers.

Biblical commentators spotted 220, the lesser of the classical pair, in Genesis 32:14 as numbering Jacob's present to Esau of 200 she-goats and 20 he-goats. According to one commentator, Jacob wisely counted out his gift (a "hidden secret arrangement") in order to secure the friendship of Esau. An Arab of the eleventh century, El Madschriti of Madrid, related that he had put to the test the erotic effect of these numbers by giving someone a confection in the shape of the smaller number, 220, to eat, while he himself ate the larger, 284. He failed, however, to describe what success that ceremony brought.

It is a mark of the slow development of number theory that until the 1630s no one had been able to add to the original pair of amicable numbers discovered by the Greeks. The first explicit rule described for finding certain types of amicable pairs is due to Thabit ibn Kurrah, an Arabian mathematician of the ninth century. In a manuscript composed at that time, he indicated:

> If the three numbers $p = 3 \cdot 2^{n-1} - 1$, $q = 3 \cdot 2^n - 1$, and $r = 9 \cdot 2^{2n-1} - 1$ are all prime and $n \geq 2$, then $2^n pq$ and $2^n r$ are amicable numbers.

It was not until its rediscovery centuries later by Fermat and Descartes that Thabit's rule produced the second and third pairs of amicable numbers. In a letter to Mersenne in 1636, Fermat announced that 17,296 and 18,416 were an amicable pair, and Descartes wrote to Mersenne in 1638 that he had found the pair 9,363,584 and 9,437,056. Fermat's pair resulted from taking $n = 4$ in Thabit's rule ($p = 23$, $q = 47$, $r = 1151$ are all prime) and Descartes's from $n = 7$ ($p = 191$, $q = 383$, $r = 73,727$ are all prime).

In the 1700s, Euler drew up at one clip a list of 64 amicable pairs; two of these new pairs were later found to be "unfriendly," one in 1909 and another in 1914. Adrien-Marie Legendre in 1830 found another pair, 2,172,649,216 and 8,520,191. Then in 1866 a 16-year-old Italian schoolboy, Niccolò Paganini, startled the mathematical world by announcing that 1184 and 1210 were amicable. This was the second lowest pair and had been completely overlooked by all previous investigators. Although the boy did not describe how he found the pair (probably by trial and error), the discovery links his name forever with the history of number theory.

Today more than 7500 amicable pairs are known, some of them running to 282 digits; these pairs include all those with values less than 10^{10}. It has not yet been established whether the number of amicable pairs is finite or infinite, nor has a pair been produced in which the numbers are relatively prime. Part of the problem is that in contrast with the single formula for generating perfect numbers (even), there is no known rule for finding all amicable pairs of numbers.

10.1 Problems

1. Show that the integer 496 is a perfect number by showing that $\sigma(496) = 2 \cdot 496$; in other words, sum up all the positive divisors of 496.

2. Verify that the integer $n = 2^{10}(2^{11} - 1)$ is not a perfect number, by showing that $\sigma(n) \neq 2n$. [*Hint:* $2^{11} - 1 = 23 \cdot 89$.]

3. Verify that:

   (a) No power of a prime can be a perfect number.
   (b) The product of two odd primes is never a perfect number. [*Hint:* Expand the inequality $(p - 1)(q - 1) > 2$ to get $pq > p + q + 1$.]

4. Prove that every even perfect number is also a triangular number. [*Hint:* Given the perfect number $p = 2^{k-1}(2^k - 1)$, show that $p = t_n$, where $n = 2^k - 1$.]

5. It has been observed that

$$6 = 2 + 2^2,$$

$$28 = 2^2 + 2^3 + 2^4,$$

$$496 = 2^4 + 2^5 + 2^6 + 2^7 + 2^8.$$

Can this procedure of representing even perfect numbers as sums of successive powers of 2 go on forever? [*Hint:* Establish the identity

$$2^{k-1}(2^k - 1) = 2^{k-1} + 2^k + 2^{k+1} + \cdots + 2^{2k-2}.]$$

6. Verify that when the reciprocals of all the positive divisors of a perfect number are added together, the total is 2. For example, the positive divisors of 6 are 1, 2, 3, and 6; and

$$\tfrac{1}{1} + \tfrac{1}{2} + \tfrac{1}{3} + \tfrac{1}{6} = 2.$$

[*Hint:* Note if $d \mid n$, then $n = dd'$ for some d'; hence, the sum

$$\Sigma_{d \mid n} 1/d = \Sigma_{d' \mid n}(d'/n) = \frac{1}{n} \Sigma_{d \mid n} d'.]$$

7. From the facts

$$28 = 1^3 + 3^3 \qquad \text{and}$$

$$496 = 1^3 + 3^3 + 5^3 + 7^3$$

conjecture a theorem and show that this guess holds for the next even perfect number.

8. For any even perfect number $n = 2^{k-1}(2^k - 1)$, prove that $2^k \mid \sigma(n^2) + 1$.

9. Prove that an integer n is one member of an amicable pair of numbers if and only if $s(s(n)) = n$, where $s(n)$ denotes the sum of the aliquot divisors of n.

10. Confirm that the pair of numbers found by the Italian schoolboy in 1866, namely $1184 = 2^5 \cdot 37$ and $1210 = 2 \cdot 5 \cdot 11^2$, actually form an amicable pair.

11. For a prime p, show that neither p nor p^2 can be one of an amicable pair.

12. Verify that the amicable pair 220 and 284 occurs on taking $n = 2$ in Thabit's rule; this rule yields amicable pairs for $n = 2, 4$, and 7, but for no other value $n < 200$.

10.2 From Fermat to Euler

Few periods were so fruitful for mathematics as the seventeenth century. Northern Europe alone produced as many men of outstanding ability as had appeared during the preceding millennium. At a time when such names as Desargues, Descartes, Pascal, Wallis, Bernoulli, Leibniz, and Newton were becoming famous, a certain French civil servant, Pierre de Fermat (1601–1665), stood as an equal among these brilliant scholars. Fermat, the "Prince of Amateurs," was the last great mathematician to pursue the subject as a sideline to a nonscientific career. By profession a lawyer and magistrate attached to the provincial parliament at Toulouse, he sought refuge from controversy in the abstraction of mathematics. Fermat evidently had no particular mathematical training and he evidenced no interest in its study until he was past 30; to him it was merely a hobby to be cultivated in leisure time. Yet no practitioner of his day made greater discoveries or contributed more to the advancement of the discipline. He was one of the inventors of analytic geometry (the actual term was coined in the early 1800s), he laid the technical foundations of differential and integral calculus, and with Pascal he established the conceptual guidelines of the theory of probability. Fermat's real love in mathematics was undoubtedly number theory, which he rescued from the realm of superstition and occultism, where it had long been imprisoned. His contributions there overshadow all else. It may well be said that the revival of interest in the

Pierre de Fermat
(1601–1665)

*(Reprinted with permission of Unwin Hyman Ltd.,
from* A History of Science, Technology and
Philosophy in the 16th and 17th Centuries, *by
A. Wolf.)*

abstract side of number theory began with Fermat; for by his refusal to accept rational solutions to diophantine problems, insisting rather on solutions in the integers, he represented a break with the classical tradition of the *Arithmetica*.

   With the fall of Constantinople to the Turks in 1453, the Byzantine scholars who had been the chief custodians of mathematics brought the ancient masterpieces of Greek learning to the West. It is reported that the remnants of the Greek text of Diophantus's *Arithmetica* were found in Padua about 1462 by Regionmontanus. Presumably, it too had been brought to Italy by the refugees from Byzantium. Regiomontanus observed, ''In these books the very flower of the whole of arithmetic lies hid,'' and he tried to interest others in translating it. Notwithstanding the attention that was called to the work, it remained practically a closed book until 1572, when the first translation and printed edition was brought out by the German professor Wilhelm Holzmann, who wrote under the Grecian form of his name, Xylander. (Bombelli had translated five of the seven books of the *Arithmetica* in 1570, using a Vatican manuscript, but this version of Diophantus has never come to light.) The *Arithmetica* became fully accessible to European mathematicians when Claude Bachet, borrowing liberally from Xylander, published (1621) the original Greek text, along with a Latin translation containing notes and comments. The Bachet edition probably has the distinction of being the work that first directed the attention of Fermat to the problems of number theory.

   Fermat, the son of a prosperous leather merchant, went from the University of Toulouse to the University of Orleans where he received a law degree in 1631. Because neither university boasted a mathematician or a scientist of particular note, his mathematical education probably did not extend beyond the first six books of Euclid's *Elements* and some symbolic algebra from Vièta's new treatise, *Introduction to the*

Analytic Art (1591). But by the spring of 1636, Fermat had set down a theory of what we should now call analytic geometry almost identical with what René Descartes had developed in his *La Géométrie* of 1637. Despite their simultaneous appearance on the mathematical scene, the two systems were entirely independent innovations. Descartes' *Géométrie* circulated widely in print; Fermat's *Introduction to Plane and Surface Loci* remained in manuscript form until 1679. The *Introduction* at least ended Fermat's isolation from the mathematical community at large, for it soon brought a letter from that connecting link between men of science, Father Marin Mersenne, inviting him to share his future findings with the Parisian circle.

Fermat preferred the pleasure he derived from mathematical research itself to any reputation that it might bring him. Indeed, he published only one important manuscript during his lifetime, and that just five years before his death, using the concealing initials M.P.E.A.S. Adamantly refusing to bring his work to the state of perfection that publication would demand, he thwarted the several efforts of others to make the results available in print under his name. Roberval, as early as 1637, offered to edit and publish some of Fermat's papers, but he was told, ''Whatever of my works is judged worthy of publication, I do not want my name to appear there.'' Yet the man who shunned formal publication still took pride in his achievements and did not want to remain entirely unknown. He carried on a voluminous correspondence with contemporary mathematicians in which he circulated his latest discoveries, offering to fill in gaps when leisure permitted. Most of what little we know about his investigations is found in the letters to friends with whom he exchanged problems and to whom he reported his successes. They did their best to publicize Fermat's talents by passing these letters from hand to hand or by making copies, which were dispatched over the Continent.

Fermat seems to have been one of those rare persons who was modest enough to believe that a few jottings of proof or even a statement of a theorem was enough for all the world to understand. This habit of communicating results piecemeal, usually as challenges, was particularly annoying to the Parisian mathematicians. At one point they angrily accused Fermat of posing impossible problems and threatened to break off correspondence unless more details were forthcoming.

Because his parliamentary duties demanded an ever greater portion of his time, Fermat was given to inserting notes in the margin of whatever book he happened to be using. Fermat's personal copy of the Bachet edition of Diophantus held in its margin many of his famous theorems in number theory. These were discovered five years after Fermat's death by his son Samuel, who brought out a new edition of the *Arithmetica* incorporating his father's celebrated marginalia. Because there was little space available, Fermat's habit had been to jot down some result and omit all steps leading to its conclusion. Posterity has often wished that Fermat had been a little less secretive about his methods, or at least that the margins of the *Arithmetica* had been wider.

Early in his career, Fermat paid much attention to classical divisibility questions concerning the function $\sigma(n)$, the sum of the divisors of n, such as finding the solutions of $\sigma(n) = 2n$ (the perfect numbers) or the solutions to $\sigma(n) = n + m = \sigma(m)$ (the amicable pairs). The crux of the problem of perfect numbers is to ascertain which numbers $2^p - 1$ are prime. While investigating this question, Fermat wrote to Mersenne, probably in June 1640, that he had discovered that if p were a prime, then $2p$ would divide the number $(2^p - 1) - 1$. Phrased differently, this says that p always divides $2^{p-1} - 1$. This was only a taste of more to come, for shortly thereafter Fermat announced that he was able to derive a far more general result. In a letter to his friend, Bernard Frénicle de Bessy, dated October 18, 1640, he stated essentially the following:

If p is a prime and a is any integer not divisible by p, then $a^{p-1} - 1$ is divisible by p.

For example, since 11 is prime, $3^{10} - 1$ must be divisible by 11; indeed, a simple calculation shows that

$$3^{10} - 1 = 59,048 = 11 \cdot 5368.$$

A variation in which there is no longer a divisibility restriction involving a and p, is:

For any number a and prime p, p divides $a^p - a$.

This result is now known as "Fermat's theorem," although some writers call it Fermat's little theorem, to distinguish it from the more famous last thorem.

As usual, Fermat did not say how he arrived at his theorem, writing to Frénicle merely, "I would send you the demonstration, if I did not fear its being too long." Nowhere among his papers did he leave a hint of a proof, a trace of his "demonstration." The first published proof was given by Euler in the *Proceedings* of the St. Petersburg Academy in 1736; however, manuscripts of Leibniz in the Hanover Library show that he had proved Fermat's theorem sometime before 1683. Euler's argument, like the one of Leibniz, is based on the use of the binomial expansion, and it is hardly possible to doubt that this was the proof that Fermat had in mind.

The proof rests on the almost evident point that if p is a prime, then the binomial coefficients $\binom{p}{k}$ are divisible by p for $k = 1, 2, \ldots, p - 1$, where

$$\binom{p}{k} = \frac{p(p - 1)(p - 2) \cdots (p - k + 1)}{1 \cdot 2 \cdot 3 \cdots k}.$$

From the definition of $\binom{p}{k}$, we have

$$1 \cdot 2 \cdot 3 \cdots k \binom{p}{k} = p(p - 1)(p - 2) \cdots (p - k + 1).$$

Here, p divides the right-hand side, so that it must also divide the left-hand side. But the product $1 \cdot 2 \cdot 3 \cdots k$, consisting of factors less than p, cannot be divisible by p; the implication is that p divides the coefficient $\binom{p}{k}$, as asserted.

We now turn to Fermat's theorem itself.

THEOREM

If p is a prime, then $a^p - a$ is divisible by p for any integer a.

Proof. When $p = 2$, the theorem can be checked easily. Here, we have $a^p - a = a^2 - a = a(a - 1)$, where the factors a and $a - 1$ are consecutive integers; thus one of them is even and their product is divisible by 2.

The argument showing that $a^p - a$ is divisible by any odd prime p involves induction on the integer a. If $a = 0$ or 1, the value of $a^p - a$ is zero, which is divisible by p; so let us consider $a > 1$. Assuming that the result holds for a, we must confirm its validity for the case $a + 1$. In light of the binomial theorem,

$$(a + 1)^p = a^p + \binom{p}{1}a^{p-1} + \binom{p}{2}a^{p-2} + \cdots + \binom{p}{p-1}a + 1.$$

Transposing gives

$$(a + 1)^p - a^p - 1 = \binom{p}{1}a^{p-1} + \binom{p}{2}a^{p-2} + \cdots + \binom{p}{p-1}a.$$

Because p divides every binomial coefficient on the right-hand side of this equation, it must divide the entire right-hand side, which is simply to say that p divides $(a + 1)^p - a^p - 1$. Combining this observation with the induction assumption that p divides $a^p - a$, one concludes that p divides the sum

$$[(a + 1)^p - a^p - 1] + (a^p - a) = (a + 1)^p - (a + 1).$$

Thus, the desired conclusion holds for $a + 1$ and consequently for all $a \geq 0$.

If a is a negative integer, there is no difficulty. For $a = -b$ where $b > 0$, so that

$$a^p - a = (-b)^p - (-b) = -(b^p - b).$$

Because b is positive, we know already that p divides $b^p - b$.   ∎

The main device by which Fermat proved difficult theorems was his method of infinite descent. He referred to this technique often in announcing to the world theorems "which I have shown by the method of infinite descent." Unfortunately, Fermat recorded most of his mathematics in his mind and not on paper, so he seldom explained how the descent was achieved. Perhaps his "proof" lay more in his faith in the applicability of the method than in actually having carried out the details in full. The method may be characterized in broad terms as follows. One assumes that there is a positive integer n having a certain property P. It may be possible to deduce from this assumption that there exists another positive integer $n_1 < n$ that also has the property P. Then, by a reapplication of the process, the property P will be found to hold for another, still smaller positive integer $n_2 < n_1$—and so on indefinitely. But this is impossible, for there are only a finite number of positive integers less than the given integer n. Hence there can be no integer at all possessing property P.

To clarify Fermat's method of infinite descent, let us use this technique to prove that $\sqrt{2}$ is irrational. Suppose, on the contrary, that $\sqrt{2} = a/b$, where a and b are positive integers. Now, the identity

$$\sqrt{2} + 1 = \frac{1}{\sqrt{2} - 1}$$

implies that

$$\frac{a}{b} + 1 = \frac{1}{(a/b) - 1} = \frac{b}{a - b}$$

and so

$$\sqrt{2} = \frac{a}{b} = \frac{b}{a-b} - 1 = \frac{2b-a}{a-b} = \frac{a_1}{b_1},$$

with $a_1 = 2b - a$ and $b_1 = a - b$. Because $1 < \sqrt{2} < 2$, or equivalently, $1 < a/b < 2$, we can multiply through by b to get the inequality $b < a < 2b$. The implication is that $0 < 2b - a = a_1$, while $2b < 2a$ gives

$$a_1 = 2b - a < a.$$

At first glance this may seem profitless, but much has been gained; starting with $\sqrt{2} = a/b$, we have ended with $\sqrt{2} = a_1/b_1$, where $0 < a_1 < a$. If the entire process is then repeated, it is found that $\sqrt{2} = a_2/b_2$, where a_2 is a positive integer less than a_1. Continuing in this way, we eventually obtain a sequence $a_1, a_2, a_3, \ldots$ such that

$$a > a_1 > a_2 > a_3 > \cdots > 0.$$

But this is impossible: Positive integers cannot be decreased in magnitude indefinitely. It follows that our original assumption that $\sqrt{2} = a/b$, where a and b are positive integers, is untenable and $\sqrt{2}$ must therefore be irrational.

Fermat published very little personally, preferring to communicate his discoveries in letters to friends (usually with no more than the terse statement that he possessed a proof) or to keep them to himself in notes in the margin of his *Arithmetica*. By far the most famous is the one, written in Latin about 1637, which states:

> It is impossible to write a cube as a sum of two cubes, a fourth power as a sum of two fourth powers, and, in general, any power beyond the second as a sum of two similar powers. For this, I have discovered a truly wonderful proof, but the margin is too small to contain it.

In this tantalizing aside, Fermat asserted that if $n > 2$, then the diophantine equation

$$x^n + y^n = z^n$$

had no solution in the integers except for the trivial solutions in which $z = x$ or $z = y$, with the remaining variable being zero.

The quotation just cited has come to be known as "Fermat's Last Theorem," or more accurately, Fermat's conjecture. All the results enunciated in the margin of his *Arithmetica* were later found to be true with the one exception of the last theorem, which still awaits proof or disproof. Fermat mentioned the cases of cubes and fourth powers repeatedly in his correspondence, which suggests that the success he had in proving these two instances led him to assume that the method of infinite descent would work for all exponents; but in no place is there any trace of this "truly wonderful proof" that he had purportedly found.

Most experts now feel that Fermat simply believed that he had a proof, when he did not. For 350 years, the last theorem resisted all attempts at solution by many great mathematicians and an even larger number of amateurs. Then, in the late decades of the twentieth century, rapid and substantial progress was made. Some mathematicians tried to subdue the conjecture by brute force, using computers to verify the conjecture for ever larger exponents. By 1976, it was shown to be true for all exponents up to

125,000. A major step forward took place in 1983 when a young West German, Gerd Faltings, proved that the Fermat equation $x^n + y^n = z^n$ has only a finite—as opposed to infinite—number of integral solutions for any given $n \geq 3$. If it could be subsequently demonstrated that the finite number of solutions was actually zero in each case, Fermat's claim would be settled once and for all. Another striking result, established in 1987, was that the last theorem is true for "almost all" values of the exponent n; that is, as n tends to infinity, the proportion of values for which the theorem is true approaches 100%. The first indication that the problem might have been finally solved came in the summer of 1993, when Andrew Wiles of Princeton University sketched a proof during a series of three lectures in Cambridge, England. Before his 200-page paper can be published, the details must be carefully scrutinized for any hidden flaws.

Resolving the problem of Fermat's last theorem would deprive it of a claim to attention but would not lessen its historical importance. The new techniques brought to bear on the claim and the new areas of mathematics spawned as a by-product may be more important to mathematics than the last theorem itself. In particular, Kummer was led to invent his ideal numbers out of which grew the theory of algebraic numbers. It will be of interest to trace this development.

When $n = pk$, where p is a prime, the Fermat equation is the same as

$$(x^k)^p + (y^k)^p = (z^k)^p.$$

If it could be shown that the equation $u^p + v^p = w^p$ cannot have any integral solutions different from zero, then in particular there would be no solution of the form $u = x^k$, $v = y^k$, $w = z^k$, and hence $x^n + y^n = z^n$ would not be solvable. Thus it suffices to prove Fermat's last theorem for $n = 4$, and for $n = p$ where p is an odd prime.

By 1839, the cases $n = 3, 4, 5$, and 7 had been eliminated. Fermat himself left a proof of the impossibility of satisfying the equation $x^4 + y^4 = z^2$ in integers x, y, z all different from zero; from this it follows immediately that the equation $x^4 + y^4 = z^4$ is equally impossible. Fermat's proof is important because it illustrates his method of infinite descent, and it represents the only instance of a detailed proof left by him. Leonhard Euler was the first to settle the case for cubes, also using an argument by infinite descent. The proof, which appears in his *Algebra* (1770), rests on nothing more than a good understanding of properties of numbers of the form $a^2 + 3b^2$, and there is no reason to doubt that Fermat had also found it. The next important advance came in 1825 when Lejeune Dirichlet read a paper at the Paris Academy in which he proved the theorem for the exponent $n = 5$. (The Revolutionary government had suppressed the Académie des Sciences in 1793, replacing it two years later with the Institut National.) His demonstration was incomplete, as Adrien-Marie Legendre pointed out. By techniques developed in Dirichlet's paper, Legendre provided an independent proof several weeks later. Each man acknowledged the contribution of the other, except that Legendre habitually referred to Dirichlet as M. Dieterich. Finally, in 1839, Gabriel Lamé was able to present a proof of the impossibility of solving the equation $x^7 + y^7 = z^7$ in nonzero integers.

With the increasing complexity of the arguments came a realization that successful resolution of the general case called for different techniques. The best hope seemed to lie in extending the meaning of "integer" to include a wider class of numbers. Then it would be possible, by attacking the problem within this enlarged system, to get more information than by using ordinary integers alone. The German mathematician Ernst Eduard Kummer (1810–1893) made the breakthrough.

At the March 1, 1847, meeting of the Paris Academy, Lamé, the author of the proof for the case $n = 7$, announced that he had discovered a proof of Fermat's last theorem for an arbitrary exponent and credited his colleague J. Liouville with having suggested the basic method. After Lamé had finished, however, Liouville addressed the meeting to decline the compliment. He pointed out several gaps in Lamé's proposed proof, the most serious of which was the tacit assumption that the usual rules of factorization of integers also would apply to "cyclotomic integers," those complex numbers of the form

$$a_0 + a_1r + a_2r^2 + \cdots + a_{n-1}r^{n-1},$$

where r is the complex nth root of unity ($r^n = 1$) and $a_0, a_1, \ldots, a_{n-1}$ are integers. Lamé admitted the shortcomings of his argument but believed (on the basis of extensive tables of factorizations) that the gap could be filled, affirming that the ordinary laws of integers held for cyclotomic integers when $n = 5$. So Lamé's proof was recognized as invalid, and the focus of activity shifted to answering Liouville's objection regarding the unique factorization of cyclotomic integers.

If the Parisians had been better acquainted with the published notices of the Berlin Academy, they would have realized that the problem had been settled already. In May of 1847, Liouville received a letter from Kummer in which he asserted:

> Concerning the elementary proposition for these complex numbers, that a composite complex number may be decomposed into prime factors in only one way, which you so correctly cite as lacking in this proof—a proof defective in other ways as well—I can assure you that it does not hold in general for complex numbers of the form
>
> $$a_0 + a_1r + a_2r^2 + \cdots + a_{n-1}r^{n-1},$$
>
> but it is possible to rescue it, by introducing a new kind of complex number, which I have called an ideal complex number.

Kummer enclosed a copy of the paper in which, three years earlier (1844), he had demonstrated the failure of unique factorization of cyclotomic integers; it had received little attention at the time, having been published in an obscure place (the University of Breslau, where Kummer was professor). His later attempts to find a way of restoring unique factorization had led him to his entirely new theory of ideal complex numbers. Kummer indicated that the outlines of his research were contained in a paper, "On the Theory of Complex Numbers," which had appeared in the *Proceedings* of the Berlin Academy in 1846; this work opened the way to the creation of the general theory of algebraic numbers. He went on to say, "I have considered already long ago the applications of this theory to the proof of Fermat's theorem and I succeeded in deriving the impossibility of the equation $x^n + y^n = z^n$ from two properties of the prime number n, so that it remains only to find out whether these properties are shared by all prime numbers."

A few weeks later, Liouville published both Kummer's letter and the little-known paper of 1844 in a privately established mathematical periodical called *Journal de mathématiques pures et appliquées.* Lamé allowed Liouville to print his ill-fated "proof" in the same issue, somewhat revised but with the disproved assumption of unique factorization of cyclotomic integers glaringly evident.

In the October 1847 issue of the *Proceedings* of the Berlin Academy, only a few months after Lamé's attempted proof opened the subject, there appeared Kummer's "Proof of Fermat's Theorem on the Impossibility of $x^p + y^p = z^p$ for an Infinite Number of Primes p." Kummer showed that Fermat's last theorem held for a certain class of prime exponents, which are too elaborate to be defined here, but which are now called regular primes. Moreover, he reduced the problem of determining whether a given prime is regular to a simple procedure involving ordinary arithmetic, and calculated that 37 is the smallest irregular prime. He had not yet noticed that 59 is irregular or in fact that the only irregular primes below 100 are 37, 59, and 67. Finally, in 1857, Kummer published new criteria for Fermat's last theorem to hold, conditions that would be satisfied when the prime exponent p is 37, 59, or 67, the values less than 100 for which he had not previously proved the theorem.

In 1850, the Paris Academy offered a prize of a gold medal valued at 3000 francs for a complete solution of Fermat's last theorem. When no proof was forthcoming, even on extension of the terminal date, the medal was awarded to Kummer as the author whose research most merited the prize, even though he had not submitted an entry in the competition. The use of the word *infinite* in the title of Kummer's monumental paper of 1847 is still unjustified; it is suspected that there exist infinitely many regular primes, but this has never been established. In the other direction, K. L. Jensen proved in 1915 that there exist infinitely many irregular primes of the form $4n + 3$, and in 1954 L. Carlitz gave a simpler proof of the weaker result that the number of irregular primes is infinite.

We might mention as a sidelight that in 1908 the mathematician P. Wolfskehl bequeathed 100,000 marks to the Academy of Science at Göttingen, to be paid for the first complete proof of Fermat's last theorem. The immediate result was a deluge of incorrect demonstrations by amateur mathematicians. Because only printed solutions were eligible, Fermat's conjecture is reputed to be the mathematical problem for which the greatest number of false proofs have been published; indeed, between 1908 and 1912 over 1000 alleged proofs appeared, mostly printed as private pamphlets. Suffice it to say, interest declined as the German inflation of the 1920s wiped out the monetary value of the prize.

The importance of Fermat's work resides not so much in any contribution to the mathematics of his own day as in its animating effect on later generations. Fermat complained that his interest in the new number theory found no echo among contemporary mathematicians. This perhaps explains some of his reluctance to write up his vaunted proofs. Whatever the reason, his refusal to publish robbed him of recognition for many striking discoveries, and his almost exclusive interest in number theory during the last fifteen years of his life led to a growing isolation from the mainstream of research. He tried to persuade Pascal to collaborate on the subject, but Pascal was not a number theorist by temperament, and after a certain time became much more interested in religion than in mathematics. Fermat's name slipped into relative obscurity and a century was to pass before a first-class mathematician, Leonhard Euler, either understood or appreciated the significance of what Fermat had been doing.

Because of the heritage of the scientific revolution of the 1600s one might have expected a whole new burst of discoveries to follow. But the hour had passed. Europe during the 1700s produced no figures of the epoch-making stature of Newton or Galileo. Instead there followed a long, slightly stunned period of assimilation during which a whole army of writers worked to make knowledge more widely available and intelligible.

The phenomenon characteristic of all levels of scientific activity was expansion. There was growth in the size of the profession, in the public it addressed, in the quantity of its publications, in the scope of its activities, and in the range of applications. Growth in the amount of printed matter available made possible the diffusion of scientific information with unheard-of speed and effectiveness. The production of books (some 2 million during the century), periodicals, and newspapers began at an unprecedented rate. Latin went out of use as the language of communication on weighty subjects. Thus, the intelligent layman had available, in his own language, a wide range of sources from which he might extract virtually unlimited amounts of information and even an occasional idea. At no other time in history have scientific principles and inventions held such a predominant share of public attention.

The eighteenth century has been variously described as the Age of Enlightenment, the Age of Reason, and the Age of Inquiry. The term "Enlightenment" is applied to a great movement of liberal and often iconoclastic ideas that sought to improve the practical conditions of human life through the power of reason. Although the same ideas were spread widely in Europe, they found their intellectual center in France in the works of Montesquieu, Diderot, Rousseau, and Voltaire. With few exceptions, these *philosophes* of the Enlightenment were not scientists themselves; nor were they philosophers, except in the sense that they made it their mission to refine, elaborate, and transmit the scientific ideas built up by the giants of the preceding century, especially Newton. The popularization of Newton's ideas on the Continent was largely the work of Voltaire. He spent nearly three years (1726–1729) in England, where he attended Newton's funeral, and from which he returned with a copy of the *Principia* in his bags. With the publication of Voltaire's popular *Elements de la philosophie de Newton* in 1738, the whole world of English empiricism and its implications became widespread knowledge among the literate public of Europe. Popular interest even called forth a special *Newtonianism for the Ladies* in Italian.

In the eyes of the eighteenth century, Newton's great achievement lay in showing that there was order to the universe and that it was to be described in the abstract terms of mathematics. Whereas Newton's laws of universal gravitation implied a certain mathematical regularity and uniformity of nature in the physical world, the *philosophes* set themselves the task of finding similar general laws throughout the whole range of human experience. Exaltation of science and an almost boundless faith in the powers of "reason" and in unfettered intelligence led them to reevaluate matters that had formerly been excluded from rational scrutiny: social conditions, religious dogma and authority, governmental policies. (To the *philosophes, reason* meant the inductive method of Newton—the collection and classification of facts, and the testing of conclusions by observation and experiment.) The underlying assumption was that there was a right answer to every question, that all problems could be answered with the same certain assurance as problems in mathematics.

The net effect of the Enlightenment was to produce a generation of people who were critical of existing institutions. This critical spirit was directed particularly at church and nobility, whose dogmatic claims to authority were alien to the new spirit of inquiry that demanded some sort of rational justification for the order of society. There arose an expectation, not so much of revolution but of a civilization of the future soon to be realized, more progressive and more equitable than the existing order. (The intellectual climate can, however, be numbered among the causes of the French Revolution of 1789 as well as the American Revolution of 1776.)

As science became fashionable at all literate levels of society, it grew in externals such as learned societies and periodicals. The beginnings of scientific journalism are generally placed at the year 1665 with the introduction of both the *Philosophical Transactions* and the *Journal des savants,* two publications that have persisted to this day. It has been estimated that in the period 1750–1789 close to 900 scientific periodicals were founded, many of them short-lived, as against a mere 35 before 1700. These early journals were predominantly vehicles for communicating scientific information— news, notes, and reports picked up from various sources—rather than repositories of scientific knowledge. They can hardly be said to have contained scientific contributions in the form of original papers as we know them today. Many of the communications were letters, either sent directly to the editor or conveyed to him by the recipient of the letter, who was then acknowledged in the publication as often as the writer himself. Despite their numbers, the journals did not immediately supplant books as the preferred mode of scientific publication; throughout the century the publication of short articles in the form of pamphlets continued.

Although the periodical literature of the 1600s consisted almost wholly of the journals of learned societies, a more popular type of periodical began to develop in the 1700s. The English took the lead with the publication of *The Tatler* (1709), *The Guardian* (1710), and *The Spectator* (1711). Essentially literary journals, they often printed articles of scientific and technical interest, culled from the publications of various scientific societies. The success of the *Spectator* (20,000 copies of some issues are said to have been sold) prompted the publication of a similar periodical in Paris, the *Spectateur Français* (1722). The end of the century saw the appearance of a superior kind of periodical, the *Philosophical Magazine* (1798), with its stated objective ''to diffuse philosophical [scientific] knowledge among every class of society, and to give the public as early account as possible of everything new or curious in the scientific world.''

Antecedents of the modern newspaper, brief, intermittent, and usually short-lived news sheets, had been published from quite early in the seventeenth century. Britain pioneered the first daily newspaper, the *Daily Courant,* in 1702. Three more were being published in London by 1724 and toward the end of the century some 42 papers were appearing. France had no daily paper until 1777, when the *Journal de Paris* was founded. Whereas the scholarly formed their academies, more ordinary people began to establish reading clubs. A group of people would rent a room or two, subscribe collectively to various periodicals and newspapers, and meet from time to time to discuss the contents of these publications.

Compendia of knowledge were known in European antiquity and medieval times, but most suffered from being the works of individual authors. The notion that

Jean D'Alembert
(1717–1783)

(From A Concise History of Mathematics *by Dirk Struik, 1967, Dover Publications, Inc., N.Y.)*

everything could still be ''discussed and analyzed, or at least mentioned'' led a team of French scholars of the eighteenth century to try to gather all existing knowledge into an orderly whole, alphabetically arranged and cross-referenced. The monumental work that bore the title *Encyclopédie ou dictionnaire raisonnée des sciences, des arts, et des métiers* began to appear in 1751 with the publication of the first of its 17 volumes of articles. The editors were Denis Diderot (1713–1784) and Jean d'Alembert (1717–1783), of whom the first wrote articles on particular arts and trades, and d'Alembert oversaw the mathematical subjects. Almost everybody who was anybody in France at this period took a hand in the gigantic publication, and distinguished contributors often wrote on unexpected subjects; Voltaire was the author of a piece on hemstitching, and Rousseau supplied the section on music.

More than a repository of knowledge, the *Encyclopédie* was the manifesto of the Enlightenment, a brilliant piece of propaganda in support of skepticism and exalting the scientific method as the supreme tool of reasoning. As such, it was subject to strict censorship and periodic suppression by the authorities, which helped to delay and maim the enterprise. After the first two volumes had appeared, any further sale or publication of the work was prohibited by decree of King's Council on the grounds that it tended to ''lay a foundation for error, corruption, irreligion, and incredulity.'' (After the historical and religious articles had left Diderot's desk, they were surreptitiously censored by the pious printer.) The 11 volumes of engraved illustrations, making a total of 28 in all, were finally completed in 1772, one year after the *Encyclopedia Britannica* made its debut with a mere 3 volumes. The effect of the 1000 or so engravings was to exhibit the state of manufacturing at that time by giving something like an anatomy of machines—overall views, cutaway drawings, and representations of individual parts, pieces, and tools—and industrial processes. There were, for instance, 55 plates on the textile industry, 11 on mining, and 3 on the manufacture of pins. The idea of presenting uniform industrial methods that any producer could adopt was neither easily achieved nor entirely welcomed. Many artisans, determined to guard their techniques in secret, resisted inquiries by Diderot and his agents or deliberately provided inaccurate descriptions of their procedures. For all the errors in fact, careless repetitions, and flagrant omissions, however, the *Encyclopédie* was the greatest achievement of its kind and the most influential work published in the eighteenth century.

The Enlightenment was certainly not a great period in the history of university education. Universities throughout Europe had with but few exceptions fallen from their high estate as centers of intellectual life. Political and ecclesiastical despotism had destroyed their freedom; Catholic governments required professors to take oaths of loyalty to the Church and its dogmas. The classical curriculum, to which established schools were limited by statute or tradition, had lost the power of inspiration that it had had in the days when people really spoke and wrote in Latin, and was hopelessly behind the times. Except for a few astronomical instruments, and terrestrial and celestial globes, most schools lacked the laboratory equipment for the study of science. Contempt for classical studies was not the rule in England, however, where the only universities, Oxford and Cambridge, enjoyed a certain prestige among the well-to-do classes. Because a vigorous parliamentary life put a premium on oratory, these universities continued to provide young men with brilliant instruction in ancient languages and literature.

Here and there on the Continent, there were universities that were genuine centers of research and at which first-rate teaching, at least in certain subjects, was to be found. In Germany, the most famous university of the 1700s was Leipzig (where Leibniz had been a student), and Leyden had replaced Padua as the leading European medical school. These institutions turned from Latin to the German language as the medium of communication in class and textbooks. The opening of a modern university in Göttingen (1737), with its emphasis on physics and astronomy, was a landmark in the history of German science. Within a generation, its library had grown to include 60,000 books and 100,000 pamphlets, making it the largest university library of the century. The absence of scientifically equipped French universities was partially remedied by the appearance of technical colleges and advanced military academies, such as the Ecole Militaire in Paris, during the second half of the 1700s.

On the whole, the Enlightenment in France went forward apart from that country's universities, and was even impeded in some ways by their rigidity and conservatism. Rousseau, obviously alluding to France, wrote in *Emile* (1762) of "those laughable places called Colleges." The wit and intellect of eighteenth-century France were to be found in her salons, country houses, and academies. A "salon" was simply the drawing room of a private house, with the company who habitually met there for social and cultural discussions. These "little courts" provided forums where nobles and men of letters were in some sense socially equal, and indeed where women conversed on the same footing as men. Salons differed completely from the clubs, composed exclusively of men, which were so prominent in English urban life. Great stress was laid on propriety of demeanor and diction, so that the activities of the salon have been somewhat uncharitably described as "a method of wasting time in perfect French."

The best French society showed as keen an interest in the sciences as in arts and letters. (John Locke, the great English philosopher, had said as early as 1693 that a gentleman must "look into" natural philosophy "to fit himself for conversation.") It was fashionable to attend the courses of public lectures on elementary science that were becoming increasingly common, and to perform scientific experiments of the most superficial kind. Although these drawing-room experiments in physics and electricity were more a social accomplishment than a serious exploration of nature, they at least indicated a widespread realization that the world was governed by laws that could be discerned by intelligent observation and careful measurement.

Academies and societies of learned men had existed from the time of the Renaissance. There were several reasons for scholars, or just interested amateurs of scholarship, to band together. The association conferred distinction and reputation on them, the publications of the organization presented an opportunity for placing the results of their studies before the learned world, and they could exercise their intellect and wit safely without the intervention of government censors. By the mid-1700s, each French provincial capital, like most of the larger cities of Western Europe, had its scientific, literary, artistic, or musical academy. Because these were often short-lived, unaffiliated, voluntary associations, only a few are still known, although they once were abundant. The members met regularly, read and discussed their dissertations, and encouraged merit by the offer of prizes. The Academy of Dijon, in 1749, made Rousseau famous by awarding him a prize for an essay in answer to the question: "Has the restoration of the arts and the sciences had a purifying effect on morals?" His *Discourse on the Arts and Sciences* was a tirade against all institutions of civilized life, and to that extent was in accord with the spirit of the *philosophes*. But he broke with them by attacking the cherished view that public enlightenment through the widespread use of human reason was the most fruitful of activities. According to Rousseau, the source of all social evils was to be found in the restless curiosity of which the arts and sciences were the final product. The eloquence and philosophy of the *Discourse* took the French intellectual community by storm, and established Rousseau's European reputation.

Whereas the 1600s had been an age of great amateurs, the success of scientific enterprise in the 1700s was continually providing more scientists with a livelihood, either as university professors or members of government-affiliated academies. Many of the reigning monarchs delighted in regarding themselves as patrons of learning, and the academies were the intellectual crown jewels of the royal courts. The motives of these rulers may not have been entirely philanthropic, but the fact remains that these bodies were in time to become important vehicles for raising the level of scientific learning. They provided salaries for distinguished scholars, published journals of research papers on a regular basis, and organized scientific expeditions. Beginning in 1721, the Académie des Sciences opened prize competitions in which monetary awards went to those offering the best solutions to problems the Académie posed. Academies of science sprang up in Germany (1700), Sweden (1710), Russia (1725), and Denmark (1742), but few of these institutions had an immediate effect on Europe's stock of scientific ideas. (The oldest American academy for the advancement of science is the American Philosophical Society Held at Philadelphia for Promoting Useful Knowledge, proposed by Benjamin Franklin in 1743 and finally organized in 1769.) The Royal Academy in Berlin, founded at the instigation of Leibniz, rapidly gained distinction through the generous patronage of Frederick the Great of Prussia. With its membership not limited to Germans, it had more the appearance of an association of expatriate *philosophes* than of an educational establishment.

Certainly there were no accomplishments in the eighteenth century comparable with the formulation of calculus in the seventeenth. Instead, the philosophical question of the logical structure of calculus seemed to absorb the energies of the eighteenth-century mathematicians. As the subject left the hands of Newton and Leibniz, it rested on no solid foundation; and few mathematicians, content with expanding its scope and effectiveness, paused to examine the basic contradictions in the theory. Such a state of affairs could not continue long unchallenged. A reaction set in against the total lack of clarity and agreement about the nature of the limit concept. Attacks came from all quarters; Voltaire called the calculus "the art of numbering and measuring exactly a thing whose existence cannot be conceived." The shrewdest attack of all, however, was delivered by a nonmathematician, George Berkeley (1685–1753), the bishop of the Irish diocese of Cloyne.

The most significant mathematical event in eighteenth-century England was the publication (1734) of a tract entitled *The Analyst: or a Discourse Addressed to an Infidel Mathematician,* in which Berkeley tried to show that the mysteries of calculus—although they led to true results—were no more securely grounded than the mysteries of religion. The "infidel mathematician" so addressed was supposed to be Newton's friend, the astronomer Edmund Halley. It appears that an acquaintance, when on his deathbed, refused Berkeley's spiritual consolation, because Halley had proved to the poor man the untenable nature of the doctrines of Christianity. This induced Berkeley to attack "the modern analysis" at its weakest spot: the foundations of the infinitesimal calculus. He taunted the proponents of calculus with "submitting to authority, taking things on trust, and believing points inconceivable"—precisely the charges against the followers of religious tenets. At one point in *The Analyst,* he derided the idea of instantaneous rates of change, using an expression that has become almost classic:

> And what are these fluxions? The velocities of evanescent increments. And what are these same evanescent increments? They are neither finite quantities, nor quantities infinitely small, nor yet nothing. May we not call them ghosts of departed quantities?

Without denying the utility of the new devices or the validity of the results, Berkeley developed an ingenious theory of compensating for errors that was meant to explain the correctness of the results of calculus; thus, "by virtue of a twofold mistake you arrive, though not at science, yet at truth."

It is difficult to understand why Berkeley thought that he could restore regard for religion by showing that mathematicians were as apt to rely on faith as theologians were. Be that as it may, Berkeley's criticism was well grounded and bore good fruit. Attention was thereby focused on some of the great weaknesses of the calculus, particularly the obvious need for a logical clarification. Almost every important mathematician of the period—Abraham De Moivre and Brook Taylor in England, Colin Maclaurin in Scotland, Joseph-Louis Lagrange in France, and the Bernoullis in Switzerland—made some effort to give the subject all the rigor of the demonstrations of the ancients. By far the ablest and most famous attempt to answer *The Analyst* was MacLaurin's *Treatise on Fluxions,* which appeared in 1742. MacLaurin took Greek demonstrative rigor as a model, and with a verbosity of which the ancients were guilty (the work consisted of 763 pages) tried to provide a geometric framework for Newton's doctrine of ''ultimate ratios.'' The effort was commendable, much praised, and much neglected; 59 years elapsed before a second edition appeared.

The influence of *Fluxions* on the progress of British mathematics in the 1700s was on the whole unfortunate. MacLaurin's skillful use of geometry persuaded Newton's countrymen to adhere rigidly to classical geometric methods at the moment when ''analysis'' (the study of infinite processes), not geometry, was on the rise in continental Europe. Thus, England dropped behind. With little regard for rigor, continental mathematicians pushed on with the purely formal development of several new branches of analysis: differential equations, calculus of variations, differential geometry, and the theory of functions of a complex variable. The center of activity shifted repeatedly across Europe, with no nation remaining the leader for any prolonged period.

The key figure in eighteenth century mathematics was Leonhard Euler (1707–1783), and the scene of his activity was chiefly Germany and Russia. Euler was the son of a Lutheran pastor who lived near Basel, Switzerland. His father earnestly wished him to enter the ministry and sent his son, at the age of 13, to the University of Basel to study theology. There the young Euler met John Bernoulli, then one of Europe's leading mathematicians, and befriended Bernoulli's two sons, Nicolaus and Daniel. Within a short time, Euler broke off the theological studies that had been chosen for him to apply himself exclusively to mathematics. He received his master's degree in 1723; and in 1727, when he was only 19, he won a prize from the Académie des Sciences for a treatise on the most efficient arrangement of ship masts.

For mathematicians beginning their careers in Switzerland in the early 1700s, there were few professorships and thus little chance of finding employment. When it was learned that the newly organized (1724) Academy of St. Petersburg was looking for personnel, many leading mathematicians from all over Europe, including Nicolaus and Daniel Bernoulli, went to Russia. On their recommendation, an appointment was also secured for Euler. Because of his youth (he was not yet 20) he had recently been denied a professorship in physics at the University of Basel and was only too ready to accept the invitation of the academy. On the very day in 1727 that Euler set foot on Russian soil, the liberal Empress Catherine I died, an event that nearly led to the dissolution of the academy. Euler gave up all hope of an academic career, became a ship's officer, and almost accepted a lieutenancy in the Russian navy. When another change of ruler brought a more favorable regime, conditions for scientific work improved, and Euler joined (1730) the academy's staff as professor of physics. Three years later when Daniel Bernoulli returned to Basel, Euler became his friend's successor as chief mathematician of the academy. Euler surprised the Russian mathematicians by computing in three days some astronomical tables whose construction was expected to take several months. Perhaps weakened by the exertion, however, he came down with a feverish illness that ultimately resulted in the loss of sight in one eye. In St. Petersburg, Euler met the versatile scholar Christian Goldbach, a man who subsequently rose from professor of mathematics to Russian minister of foreign affairs. With his interests, it seems likely that Goldbach was the one who first drew Euler's attention to the work of Fermat on the theory of numbers.

Euler eventually sickened of the political repression then prevalent in Russia and, in 1741, accepted the offer of Frederick the Great to direct the mathematical division

Leonhard Euler
(1707–1783)

(Smithsonian Institution.)

of the Berlin Academy, which was then rising in fame. The queen mother received him kindly at court, and asked why so distinguished a scholar should be so reticent and timid; he replied, "Madame, it is because I have just come from a country where, when one speaks, one is hanged." Despite a flattering letter of welcome, relations between Frederick and Euler soon spoiled. The youthful monarch felt it his duty to encourage mathematics, yet had little patience with a subject that seemed impractical to him. At court he preferred the company of polished philosophers such as Voltaire to the unsophisticated Euler, whom he cruelly called a "mathematical Cyclops." Euler's 25 years in Berlin were not altogether happy ones.

From his first years in Berlin, Euler kept in regular contact with the St. Petersburg Academy, so that he was in effect working simultaneously in two academies. His reputation among the Russians was such that they continued his salary for a long time after his departure; and when, during the Seven Years' War, Russian troops plundered Euler's estate, the commanding general compensated him handsomely for the loss, and the empress sent an additional sum of 4000 crowns. Heartened by the warmth of Russian feeling towards him, and at last unendurably offended by the contrasting coolness of Frederick, Euler returned to St. Petersburg in 1766 at the request of Catherine II, Catherine the Great.

This second Catherine was a keen patron of the intellectuals of the Enlightenment. In fact, it had been her financial support that helped to keep the *Encyclopédie* going during Diderot's darkest days. Diderot, short of funds, was about to sell his personal library to provide a dowry for his daughter when Catherine heard of his difficulties. She bought all Diderot's books, and not only allowed him to retain possession of them but also paid him an annual salary to act as librarian-custodian. Catherine tried unsuccessfully to persuade Diderot to transfer the publication of the *Encyclopédie* to Russia. Although Diderot refused to move to St. Petersburg permanently, he spent several months at Catherine's court in 1773, impudently arguing with her. Wearied by his insistent demands for reforms, she finally admonished him: ''You only work on paper, while I, poor Empress, work on human skin, which is more ticklish.''

There is a famous anecdote concerning Euler and Diderot that has circulated for so many years that historians tend to give it some credence. According to the tale told by Augustus De Morgan in his *Budget of Paradoxes* (1872), Catherine the Great persuaded Euler to join her in suppressing the atheistic views that the freethinking Diderot had been advancing among the younger members of her court. Diderot was informed that a learned mathematician possessed an algebraic proof of the existence of God and would give it to him before all the court, if he desired. When Diderot eagerly consented, Euler advanced toward him and said, gravely and in a tone of perfect conviction, ''Sir, $(a + b^n)/n = x;$ hence God exists. Can you answer that?'' It sounded like sense to the French philosopher, to whom, the story goes, all mathematics was inscrutable. Humiliated by the peals of laughter that greeted his embarrassed silence, the poor man asked Catherine's permission to return to France at once, permission that was graciously granted.

Although Diderot's alleged atheistic conversations with courtiers may possibly have become annoying to Catherine, the story runs counter to what we know of the characters of both men. There is no reason to think that Diderot, the author of five credible memoirs on mathematics, would have been nonplused by Euler's supposed formula. If the incident has any basis in fact, it is more likely that Diderot saw through the practical joke immediately, but was annoyed at the prospect of similar silly performances in the future and therefore went back to his country; and that not even ''at once.''

Shortly after Euler moved to St. Petersburg, he suffered the loss of sight in his good eye because of a cataract. He would not, however, permit blindness to diminish the rate of output of his scientific work. Euler's comment on his calamity was, ''I'll have fewer distractions.'' His memory was so phenomenal, and his speed of calculation so great, that he continued to produce almost as actively as before. Euler's faculty with figures was so prodigious that he could carry out in his head numerical computations that competent mathematicians had difficulty doing on paper. It is recorded that when two of his pupils, working the sum of a series to seventeen terms, disagreed in their results at the fiftieth significant figure, Euler arrived at the correct calculation in his own mind. For the next 17 years, Euler found selfless and devoted helpers in his sons and servants, to whom he dictated a vast store of mathematical papers. One of the first works dictated after the onset of his blindness, dictated to a valet who knew nothing of mathematics beyond simple computation, was his famous *Vollständige Anleitung zur Algebra* (Complete Instruction in Algebra; 1767). This book was subsequently published in many editions: English, Dutch, French, Italian, and Russian, and gave to algebra the form that it retains to our time. Euler continued his labors unabated until in 1783, while playing with his grandchildren and sipping tea, he suddenly suffered a brain hemorrhage. In the oft-quoted words of Condorcet, ''He ceased to calculate and to live.''

Without a doubt, Euler was the most versatile and prolific writer in the entire history of mathematics. Fifty pages were finally required in his eulogy merely to list the titles of his published works. He wrote or dictated over 700 books and papers in his lifetime, and left so much unpublished material that the St. Petersburg Academy did not finish printing all his manuscripts until 47 years after his death. The publication of Euler's collected works was begun by the Swiss Society of Natural Sciences in 1911, and it is estimated that more than 75 large volumes will ultimately be required for the completion of this monumental project. The best testament to the quality of these papers may be that on 12 occasions they won the coveted biennial prize of the Académie des Sciences. The first of these prizes, which Euler received at the age of 31, was for an essay on the theory of heat, and another involved a study of the tides, a subject that especially attracted scientists at that time. Euler also gained a share of the prize of £20,000 offered by the British parliament for a method of calculating longitude at sea to an accuracy of half a degree.

During his stay in Berlin, Euler fell into the habit of writing memoir after memoir, placing each when finished at the top of a pile of manuscript. Whenever material was needed to fill the academy's journal, the printers would help themselves to a few papers from the top of the stack. As the height of the pile increased more rapidly than the demand made on it, memoirs at the bottom would tend to remain in place a long time. This explains how it happened that when various papers of Euler were published, extension and improvements of the material contained in them had previously appeared in print under his name. It should be added that the manner in which Euler made his work public contrasted sharply with the secrecy customary in Fermat's time.

Besides the *Algebra,* another great and rich textbook was Euler's *Introductio in Analysin Infinitorum* (1748), a two-volume treatise covering a wide variety of subjects. The first of its two volumes was devoted to functions in general, and to logarithmic, exponential, and trigonometric functions in particular; it also dealt with power series, continued fractions, and various problems in number theory. The second volume contained analytic geometry of the plane and space, including a discussion of algebraic and transcendental curves. The *Introductio* did for analysis what Euclid's *Elements* had done for geometry and al-Khowârizmî's *Hisâb al-jabr w' al muqâbalah* for algebra. It was a classic text from which whole generations were inspired to learn their analysis, especially their knowledge of infinite series.

In a sense the *Introductio* created order in the still uncertain world of mathematical notation, for after its publication, Euler's conventions were used generally. He gave his approval for using the Greek letter π for the ratio between the circumference and diameter of a circle (π appeared in Oughtred's *Clavis Mathematicae* of 1647 to denote simply the circumference), and for e to represent the base of natural logarithms. (Euler had first used the symbol e in a letter to Goldbach in 1731.) One also meets here the modern abbreviations for the trigonometric functions, which he regarded as ratios, not lengths, and hence as dimensionless quantities. And there is also the famous "Euler identity" linking trigonometric and exponential functions:

$$e^{\sqrt{-1}x} = \cos x + \sqrt{-1} \sin x.$$

This formula was not entirely new. Roger Cotes (1682–1716) in the *Philosophical Transactions of London,* published in 1714, showed that

$$\sqrt{-1}x = \log(\cos x + \sqrt{-1} \sin x)$$

although not in this particular notation. (Cotes, who helped Newton prepare the second edition of the *Principia,* was the most mathematically gifted of all Newton's adherents. His death led Newton to lament, "If Cotes had lived, we might have learned something.") Near the end of Euler's life, in a memoir presented in 1777 to the Academy at St. Petersburg, he introduced the symbol i for $\sqrt{-1}$.

The *Introductio* was followed by *Institutiones Calculi Differentialis* (1755), and the three-volume *Institutiones Calculi Integralis* (1768–1770). These formed the most complete and systematic exposition of the calculus, including a theory of differential equations, to appear up to that time. The works that formed Euler's trilogy on analysis were exciting to read and skillfully presented, containing many new results of great value. They remained standard textbooks until the end of the century. Generations of young mathematicians followed Laplace's (1749–1827) advice: "Read Euler, he is our master in all."

Like most major mathematicians of the eighteenth century, Euler made significant contributions to the subject of infinite series. The *Introductio in Analysin Infinitorum,* for instance, contains a derivation of the now-familiar series expansion of e. As outlined in the next paragraph, Euler's argument is typical of the age in its reliance on formal manipulation and careless handling of limits and convergence.

Given $a > 1$, Euler first writes $a^w = 1 + kw$, where w is taken to be an infinitely small number ("so small that it is just not equal to zero") and k is a constant depending only upon a. For any real number x, he puts $j = x/w$; then

$$a^x = a^{jw} = (1 + kw)^j = (1 + kx/j)^j,$$

which can be expanded by Newton's binomial theorem into

$$a^x = 1 + \frac{j}{1!}\left(\frac{kx}{j}\right) + \frac{j(j-1)}{2!}\left(\frac{kx}{j}\right)^2 + \frac{j(j-1)(j-2)}{3!}\left(\frac{kx}{j}\right)^3 + \cdots.$$

Because w is infinitely small, j will be infinitely large; this allows Euler, in a loose notion of passing to the limit, to assume that

$$1 = \frac{j-1}{j} = \frac{j-2}{j} = \frac{j-3}{j} = \cdots.$$

Thus, he decides that

$$a^x = 1 + \frac{kx}{1!} + \frac{(kx)^2}{2!} + \frac{(kx)^3}{3!} + \cdots$$

so that, when $x = 1$,

$$a = 1 + \frac{k}{1!} + \frac{k^2}{2!} + \frac{k^3}{3!} + \cdots.$$

Because the number e is the value of a for which $k = 1$,

$$e = 1 + \frac{1}{1!} + \frac{1}{2!} + \frac{1}{3!} + \cdots.$$

(In effect, Euler has identified k with $\lim\limits_{w \to o} \dfrac{a^w - 1}{w}$, the latter limit being equal to $\log_e a$.)

The series obtained is well-suited for computing e, and Euler used it to obtain a numerical value to 23 decimal places:

$$e = 2.71828182845904523536028 \cdots .$$

Another achievement of Euler in this area deals with the summation of the reciprocals of the squares of the positive integers; that is, with the infinite series $1 + 1/2^2 + 1/3^2 + 1/4^2 + \cdots$. This series had provoked considerable discussion in England and France. Oldenburg, in a letter of 1673, had asked Leibniz for its sum and received no reply. In 1689, James Bernoulli showed that the series had a finite sum, but as to finding its actual value he "confessed that all his zeal had been mocked." Writing to Christian Goldbach in 1728, Daniel Bernoulli gave 8/5 as an approximate sum, whereas James Stirling later (1730) computed the sum to 16 decimal places, namely 1.6449340668482264 (the series converges slowly so that many terms have to be taken into account in order to achieve an accuracy of a few decimal digits). Neither mathematician appears to have recognized that their sums were about $\pi^2/6$, which Euler determined in 1735 to be the correct sum.

One approach taken by Euler, subsequently described in the *Introductio*, was to apply to infinite series the relations between roots and coefficients of polynomials. The starting point was a trigonometric series known to Newton:

$$\sin x = x - \frac{x^3}{3!} + \frac{x^5}{5!} - \frac{x^7}{7!} + \cdots .$$

He treated the right side as a polynomial of infinite degree, arguing that because $\sin x = 0$ has roots $0, \pm \pi, \pm 2\pi, \pm 3\pi, \cdots$, so will

$$0 = x - \frac{x^3}{3!} + \frac{x^5}{5!} - \frac{x^7}{7!} + \cdots .$$

Dividing this series by x in order to eliminate 0 as a root, and replacing x^2 by y, gave him a new polynomial equation

$$0 = 1 - \frac{y}{3!} + \frac{y^2}{5!} - \frac{y^3}{7!} + \cdots ,$$

with roots $\pi^2, (2\pi)^2, (3\pi)^2, \cdots$. Now it was known from the theory of equations that, for a polynomial with constant term equal to 1, the sum of the reciprocals of its roots is the negative of the coefficient of the linear term. Proceeding by analogy with the finite case, Euler therefore concluded that

$$\frac{1}{3!} = \frac{1}{\pi^2} + \frac{1}{(2\pi)^2} + \frac{1}{(3\pi)^2} + \cdots ,$$

or

$$\frac{\pi^2}{6} = \frac{1}{1^2} + \frac{1}{2^2} + \frac{1}{3^2} + \cdots .$$

He was led by this reasoning to sum the series that had defeated his predecessors.

Euler was much carried away with the theory of numbers, and his mathematical research supplied many of the proofs that Fermat claimed to possess—by techniques, to be sure, that were not those of Fermat. Having established Fermat's theorem [if p is a prime and gcd $(a, p) = 1$, then $a^{p-1} - 1$ is divisible by p], Euler demonstrated a more general result by introducing the phi function $\phi(n)$. It denotes the number of positive integers less than n and relatively prime to n, so that if n is prime, $\phi(n) = n - 1$. What Euler proved (1760) was:

If gcd$(a, n) = 1$, *then n divides the difference* $a^{\phi(n)} - 1$.

For a better understanding of Euler's generalization of Fermat's theorem, let us consider a numerical example. If, say, $n = 16$, then it is found that $\phi(n) = 8$; for among the integers between 1 and 15, there are eight that are relatively prime to 16, namely the integers 1, 3, 5, 7, 9, 11, 13, 15. Taking $a = 3$ [note that gcd $(3, 16) = 1$], Euler's result asserts that $3^8 - 1$ is divisible by 16. Indeed, one can easily check that

$$3^8 - 1 = 6561 - 1 = 6560 = 435 \cdot 16.$$

In the tradition of Diophantus and Fermat, another group of Euler's works involved the solution of indeterminate equations in integral values of the unknowns. In 1767, Euler gave a method for calculating the smallest integral solution of the equation $x^2 - ay^2 = 1$, where a is a positive nonsquare. In an earlier paper of 1732–1733, he had erroneously named this Pell's equation, and the misnomer has stuck. (John Pell, an English mathematician, did no more than copy the equation from Fermat's letters.) Euler extended Fermat's studies on the representation of integers by sums $x^2 + y^2$ by proving (1751) that every prime of the form $4k + 1$ can be uniquely expressed as the sum of two squares. Although Euler could not dispose of Fermat's last theorem, he did establish (1770) the impossibility of having nontrivial integral solutions of $x^n + y^n = z^n$ for the case $n = 3$.

Fermat, whose mathematical intuition was usually reliable, had observed that the integers

$$2^2 + 1 = 5, \qquad 2^{2^2} + 1 = 17, \qquad 2^{2^3} + 1 = 257, \qquad 2^{2^4} + 1 = 65,537$$

were all primes and expressed his belief that $2^{2^n} + 1$ would be prime for each value of n. In writing to Mersenne, he confidently announced, ''I have found that numbers of the form $2^{2^n} + 1$ are always prime numbers and have long since signified to analysts the truth of this theorem.'' However, Fermat bemoaned his inability to come up with a proof, and in subsequent letters, his growing tone of exasperation suggested that he was continually trying to do so. Through his uncanny ability at computation, Euler showed (1732) that $2^{2^5} + 1$ was not prime, one of its factors being 641:

$$2^{2^5} + 1 = 2^{32} + 1 = 4,294,967,297 = 641 \cdot 6,700,417.$$

Among Euler's accomplishments in number theory was a posthumously published paper (1862) giving the converse of Euclid's theorem on perfect numbers. The proof is so surprisingly elementary that we shall describe it here.

THEOREM

If n is an even perfect number, then n is of the form $2^{k-1}(2^k - 1)$, where $2^k - 1$ is a Mersenne prime.

Proof. To begin, we may write $n = 2^{k-1}m$, where m is an odd integer and $k \geq 2$. The divisors of $2^{k-1}m$ are $1, 2, 2^2, \ldots, 2^{k-1}$, each multiplied by the divisors $1, d_1, d_2, \ldots, m$ of m itself. Now these numbers are precisely the terms in the expansion of the product

$$(1 + 2 + 2^2 + \cdots + 2^{k-1})(1 + d_1 + d_2 + \cdots + m).$$

But this is the value of $\sigma(n)$, the sum of the divisors of n:

$$\sigma(n) = (1 + 2 + 2^2 + \cdots + 2^{k-1})(1 + d_1 + d_2 + \cdots + m)$$

$$= (1 + 2 + 2^2 + \cdots + 2^{k-1})\sigma(m).$$

Because the first factor on the right-hand side is a geometric progression, the last equation can also be written in the form

$$\sigma(n) = (2^k - 1)\sigma(m).$$

The requirement for n to be perfect entails that

$$\sigma(n) = 2n = 2^k m.$$

Together, these two expressions for $\sigma(n)$ yield

$$2^k m = (2^k - 1)\sigma(m),$$

which is simply to say that $2^k - 1$ divides the product $2^k m$. But $2^k - 1$ and 2^k are relatively prime integers, whence, as a consequence of Euclid's lemma, $2^k - 1$ must divide m; say $m = (2^k - 1)M$. The result of substituting this value of m into the last-displayed equation and canceling the common factor $2^k - 1$ is that $\sigma(m) = 2^k M$. Because m and M are both divisors of m, with $M < m$, we have

$$2^k M = \sigma(m) \geq m + M = (2^k - 1)M + M = 2^k M,$$

leading to $\sigma(m) = m + M$. The implication of this equality is that m has only two positive divisors, to wit, M and m itself. It must be that m is prime and the smaller divisor $M = 1$; in other words,

$$m = (2^k - 1)M = 2^k - 1$$

is a prime number. The original perfect even number n therefore has the form

$$n = 2^{k-1}m = 2^{k-1}(2^k - 1),$$

with $2^k - 1$ prime, completing the present proof. ∎

Many problems in number theory remain unsolved, but for combined simplicity of statement and difficulty of attack none surpasses the one posed by Christian Goldbach (1690–1764). Goldbach asserted in a letter to Euler dated June 7, 1742, that every even number was the sum of two odd primes. This proposition has generally been described as Goldbach's theorem, although a more appropriate designation would be

"Goldbach's conjecture." We should observe at the outset that if the term "prime" is taken in its modern sense, then the theorem is false. It fails in the case of the integer 2, which is even, but not the sum of two odd primes. We do not nowadays call 1 a prime, for if we did, the factorization of an integer into a product of primes would not be unique. The phrase "greater than 2" must therefore be inserted in the enunciation of the theorem: Every even number greater than 2 is the sum of two primes. A somewhat more general version is that every even integer greater than 4 is the sum of two odd primes. It is uncertain how much time Euler spent on the problem; but in replying to Goldbach, he wrote, "That every even number is a sum of two primes, I consider an entirely certain theorem in spite of the fact that I am not able to demonstrate it." He then countered with a conjecture of his own: Any even number greater than 6 that is of the form $4n + 2$ is the sum of two primes of the form $4n + 1$ (taking 1 as a prime of this latter type, where necessary). For instance, because 30 is an even number of the form $4n + 2$, we can write $30 = 13 + 17$ by Euler's conjecture.

There is no reasonable doubt that Goldbach's theorem is correct and that the number of representations is large when the even integer is large, but all attempts to obtain a proof have been completely unsuccessful. Probably the most important result obtained in modern times in regard to Goldbach's theorem is contained in a Russian paper by Schnirelman, published in 1930. Schnirelman proved that there exists a fixed number k such that every large number can be written as the sum of k or fewer primes. However, this was an existence theorem, which gave no clue about the actual magnitude of k. Since then various mathematicians have given values for k. The best known result is attributed to I. M. Vinogradov (1937):

> Every sufficiently large odd integer can be written as the sum of at most three prime numbers, and thus every sufficiently large integer is the sum of at most four primes.

One can use the Schnirelman method to determine that every even number is the sum of no more than 6 prime numbers.

10.2 Problems

1. Using the expansion

$$\frac{1}{1 + x^2} = 1 - x^2 + x^4 - x^6 + \cdots,$$

$$-1 < x < 1,$$

and termwise integration, obtain Leibniz's celebrated series for $\pi/4$:

$$\frac{\pi}{4} = \int_0^1 \frac{dx}{1 + x^2} = 1 - \frac{1}{3} + \frac{1}{5} - \frac{1}{7} + \cdots.$$

2. Carry out, as follows, the details in Pietro Mengoli's (1625–1687) proof of the divergence of the harmonic series

$$1 + \frac{1}{2} + \frac{1}{3} + \frac{1}{4} + \cdots.$$

(a) For $n > 1$, show that

$$\frac{1}{n - 1} + \frac{1}{n} + \frac{1}{n + 1} > \frac{3}{n}.$$

(b) By grouping the terms of the series by threes, verify the inequality

$$1 + \left(\frac{1}{2} + \frac{1}{3} + \frac{1}{4} \right) +$$

$$\left(\frac{1}{5} + \frac{1}{6} + \frac{1}{7} \right) + \cdots >$$

$$1 + \frac{3}{3} + \frac{9}{9} + \frac{27}{27} + \cdots .$$

3. Use Euler's identity $e^{iz} = \cos z + i \sin z$ to prove that $i^i = e^{-\pi/2}$.

4. Consider the polynomial equation

$$a_0 x^n + a_1 x^{n-1} + \cdots + a_{n-1} x + a_n = 0 \qquad a_0 \neq 0$$

with (complex) roots $r_1, r_2, \ldots, r_n$, so that

$$a_0 x^n + a_1 x^{n-1} + \cdots + a_{n-1} x + a_n$$
$$= a_0 (x - r_1)(x - r_2) \cdots (x - r_n).$$

By multiplying the linear factors on the right and comparing the resulting coefficients, prove that

$$r_1 + r_2 + r_3 + \cdots + r_n = -\frac{a_1}{a_0}$$

$$r_1 r_2 + r_1 r_3 + r_2 r_3 + \cdots + r_{n-1} r_n = \frac{a_2}{a_0}$$

$$\vdots$$

$$r_1 r_2 r_3 \cdots r_n = (-1)^n \frac{a_n}{a_0}$$

where the k'th equation asserts that the sum of all possible products of the roots, taken k at a time, equals $(-1)^k \frac{a_k}{a_0}$.

5. With the notation of Problem 4, let $a_n \neq 0$, so that the roots $r_1, r_2, \ldots, r_n$ are all nonzero. Establish the following:

(a) $\dfrac{1}{r_1} + \dfrac{1}{r_2} + \cdots + \dfrac{1}{r_n} = -\dfrac{a_{n-1}}{a_n}.$

(b) $a_0 x^n + \cdots + a_{n-1} x + a_n$

$$= a_n \left(1 - \frac{x}{r_1} \right)\left(1 - \frac{x}{r_2} \right) \cdots \left(1 - \frac{x}{r_n} \right).$$

6. (a) By Euler's methods, derive the summation

$$\frac{\pi^2}{8} = \frac{1}{1^2} + \frac{1}{3^2} + \frac{1}{5^2} + \frac{1}{7^2} + \cdots .$$

[*Hint:* Begin with the trigonometric series

$$\cos x = 1 - \frac{x^2}{2!} + \frac{x^4}{4!} - \frac{x^6}{6!} + \cdots ,$$

and use the fact that $\cos x = 0$ has roots

$$\pm \frac{\pi}{2}, \pm \frac{3\pi}{2}, \pm \frac{5\pi}{2}, \cdots .]$$

(b) From Euler's expansions of $\dfrac{\pi^2}{6}$ and $\dfrac{\pi^2}{8}$, formally obtain the equation

$$\frac{\pi^2}{12} = \frac{1}{1^2} - \frac{1}{2^2} + \frac{1}{3^2} - \frac{1}{4^2} + \cdots .$$

7. (a) Show, in the manner of Euler, that

$$\frac{\sin x}{x} =$$

$$\left(1 - \frac{x^2}{\pi^2} \right)\left(1 - \frac{x^2}{(2\pi)^2} \right)\left(1 - \frac{x^2}{(3\pi)^2} \right) \cdots .$$

[*Hint:* Interpret the right side of the expression

$$\frac{\sin x}{x} = 1 - \frac{x^2}{3!} + \frac{x^4}{5!} + \cdots$$

$$= 1 - \frac{y}{3!} + \frac{y^2}{5!} + \cdots$$

as a polynomial of infinite degree, and apply Problem 5(b).]

(b) Substitute $x = \frac{\pi}{2}$ in the formula of part (a) to obtain

$$\frac{2}{\pi} = \left(\frac{1 \cdot 3}{2 \cdot 2} \right)\left(\frac{3 \cdot 5}{4 \cdot 4} \right)\left(\frac{5 \cdot 7}{6 \cdot 6} \right)\left(\frac{7 \cdot 9}{8 \cdot 8} \right) \cdots .$$

This famous infinite product for π was published by John Wallis in his *Arithmetica Infinitorum* (1665).

(c) From Wallis's formula, conclude that

$$\sqrt{\pi} = \lim_{n \to \infty} \frac{(n!)^2 2^{2n}}{(2n)! \sqrt{n}}$$

[*Hint:* First prove that

$$\left(\frac{1 \cdot 3}{2 \cdot 2}\right)\left(\frac{3 \cdot 5}{4 \cdot 4}\right)\left(\frac{5 \cdot 7}{6 \cdot 6}\right) \cdots$$

$$\left(\frac{(2n-1)(2n+1)}{2n \cdot 2n}\right)$$

$$= \frac{[(2n)!]^2(2n+1)}{(n!)^4 \, 2^{4n}} .]$$

8. The following is an outline of a proof that e is irrational.

(a) Use the estimate $e < 3$ to show that

$$0 < e - \left(1 + \frac{1}{1!} + \frac{1}{2!} + \frac{1}{3!} + \cdots + \frac{1}{n!}\right)$$

$$< \frac{3}{(n+1)!}$$

for any positive integer n.

(b) Assume to the contrary that e is rational, say $e = a/b$, where a and b are positive integers. Pick $n > b$ and also $n > 3$. Substitute $e = a/b$ into the inequality of part (a) and multiply the inequality by $n!$. Prove that this leads to the existence of an integer N, which satisfies $0 < N < 3/4$.

10.3 The Prince of Mathematicians: Carl Friedrich Gauss

In number theory, Fermat was essentially alone for the whole of the seventeenth century, as Euler was for most of the century following, until he was joined by Lagrange. Joseph Louis Lagrange (1736–1813), Italian by birth, German by adoption, and French by choice, was next to Euler the foremost mathematician of the eighteenth century. When he entered the University of Turin, his great interest was in physics, but after chancing to read a tract by Halley on the merits of Newtonian calculus, he became excited about the new mathematics that was transforming celestial mechanics. He applied himself with such energy to mathematical studies that he was appointed, at age 18, professor of geometry at the Royal Artillery School in Turin. The French Académie des Sciences soon became accustomed to seeing Lagrange among the competitors for its biennial prizes. Between 1764 and 1788 he won five of the coveted prizes for his applications of mathematics to problems in astronomy.

In 1766, when Euler left Berlin for St. Petersburg, Frederick the Great arranged for Lagrange to fill the vacated post, accompanying his invitation with a modest message saying, "It is necessary that the greatest geometer of Europe should live near the greatest of Kings." (To d'Alembert, who had suggested Lagrange's name, the king wrote, "To your care and recommendation am I indebted for having replaced a half-blind mathematician with a mathematician with both eyes, which will especially please the anatomical members of my academy.") For the next 20 years, Lagrange was director of the mathematics section of the Berlin Academy. There he produced work of high distinction, which culminated in his monumental treatise, the *Mécanique analytique,* published in 1788 in four volumes. In this work he unified general mechanics and made of it, as the mathematician Hamilton was later to say, "a kind of scientific poem." Holding that mechanics was really a branch of pure mathematics, Lagrange so completely banished geometric ideas from the *Mécanique analytique* that he could boast in the preface that not a single diagram appeared in its pages.

Frederick died in 1787 and Lagrange, no longer finding a sympathetic atmosphere at the Prussian court, decided to accept the invitation of Louis XVI to settle in Paris, where he took French citizenship. But the years of constant activity had taken their toll. Lagrange fell into a deep mental depression that destroyed his interest in mathematics. So profound was his loathing for the subject that the first printed copy of the *Mécanique analytique,* the work of a quarter-century, lay unexamined on his desk for more than two years. Strange to say, it was the turmoil of the French Revolution that helped to awaken him from his lethargy.

Madame de Pompadour's cynical prophecy, ''After us, the deluge,'' had barely hinted at the floods that were in fact released at the storming of the Bastille. In the national upheaval that followed, anything that was more or less identified with the injustices practiced by the Old Regime came under suspicion and was destroyed. The universities, as well as the Académie des Sciences, were abolished by the National Convention in August of 1793. Because it was the ''academy'' as an aristocratic anachronism in a democratic society that was the target of the Revolution, and not learning itself, the solution was to start afresh. Two new institutions, bearing the humble titles Ecole Normale and Ecole Polytechnique, were to keep scientific education afloat in a sea of turmoil. The Ecole Polytechnique, created in 1794, was to produce the engineers so urgently required by a nation at war; technical instruction was to be firmly based on sound scientific training, and particularly on mathematics and physics.

The most famous embodiment of the revolutionary overhaul of French education was the Ecole Normale, originally intended to create, almost overnight, a cadre of educators for the nation. More than 1400 persons were selected by local officials to come to Paris and study at government expense. Candidates had to be at least 25 years old and of good character and unquestioned patriotism. In roughly four months' time, they were to learn the essential elements of a number of academic subjects, which on completing the course of training they in turn would teach to their own students. To guarantee the standards of excellence, the government found itself forced to recruit ex-academicians for the faculty. Lagrange, Laplace, Monge, and Vandermonde were in charge of lecturing on mathematics. Unfortunately missing were the victims of the Reign of Terror. Among those were the chemist Lavoisier, who had gone to the guillotine, the mathematician Condorcet, who had preferred to take poison in prison, and the astronomer de Saron, who before marching to the platform had spent his last hours calculating the orbit of a newly discovered planet. The outraged Lagrange, hearing that the Revolution had silenced yet another voice of progress, said, ''It took them only a moment to cause his head [Lavoisier's] to fall, and a hundred years will not suffice to produce its like.'' The fundamental weakness of the whole scheme was that in the absence of true universities, the Ecole Normale became the goal of all those who wanted to pursue scientific or literary studies at a higher level; professors gave lectures directed towards ''the heights of science rather than towards the art of teaching.'' The failure to perform the prescribed task brought increasing criticism, and after only three months and eleven days, the school closed its doors. In 1808, the Ecole Normale was reorganized by Napoleon as the Ecole Normale Superieure, with a view of securing a constant supply of instructors for secondary schools.

Lagrange, with Laplace as his assistant, taught elementary mathematics at the Ecole Normale, and when the short-lived institution had closed, lectured on analysis

Joseph Louis Lagrange
(1736–1813)

(Smithsonian Institution.)

at the Ecole Polytechnique. Although he had not lectured since his early days at Turin, having been under royal patronage in the interim, he seemed to welcome these appointments. The instructors, subject to constant surveillance, were pledged "neither to read nor repeat from memory," and transcripts of their lectures as delivered were inspected by the authorities. Despite the petty harassments, Lagrange gained a reputation as an inspiring teacher. His lecture notes on differential calculus formed the basis of another classic in mathematics, the *Théorie des fonctions analytiques* (1797).

Through his teaching at the Ecole Polytechnique, Lagrange became acquainted with Sophie Germain (1776–1831), a woman frequently called the Hypatia of the nineteenth century. Although the school did not accept women as students, she managed to secure Lagrange's lecture notes on analysis. Following the new revolutionary practice that allowed students to hand in written observations, she communicated her own thoughts to Lagrange, writing under the male pseudonym M. Leblanc (the name of a student enrolled at the school). Lagrange was so struck by the quality of her work that on learning the author's true identity, he praised Germain as one of the promising young mathematicians of the future. This accolade was vindicated when she proved (around the mid-1820s) that for each odd prime $p < 100$, Fermat's equation $x^p + y^p = z^p$ has no solution in integers not divisible by p; in other words, if x, y, z are nonzero integers satisfying $x^p + y^p = z^p$, then p divides xyz.

Lagrange's research covered an extraordinarily wide spectrum, but he possessed, much like Fermat and Euler before him, a special talent for the theory of numbers. Some of his work there included the first proof of a theorem stated by John Wilson in 1770 to the effect that n is prime if and only if n divides $(n - 1)! + 1$; the finding of all integral solutions of the diophantine equation $x^2 - ay^2 = 1$; and the solution of numerous problems posed by Fermat on how certain primes could be represented in particular ways. (Typical of these last-mentioned problems was the result that every prime p of the form $p = 8n + 1$ could be written as $p = a^2 + 2b^2$ for suitable integers a and b.) The discovery for which Lagrange acquired his greatest renown in number theory is that every positive integer can be expressed as the sum of four integral squares.

News of the fall of the Bastille was hailed everywhere on the Continent with satisfaction. But no other nation imitated France spontaneously; rather the French armies, with their goal of "natural frontiers" and their slogan, "War to the thrones, peace to the cottages," spread the principles of the Revolution. The conflict began in April 1792 and lasted, with brief interruptions and shifting alliances, until 1815. To repel the combined forces of all the old monarchies, it was necessary for France to return to absolutism—in this case, the dictatorship of a general, Napoleon Bonaparte. The statesmen of Europe held together a coalition of mutually hostile states just long enough to allow the weight of their numbers to ensure victory. On June 16, 1815, the British under Wellington and the Prussians under Blücher delivered the final blow to the Napoleonic epic at Waterloo, near Brussels. (Before the Revolution, war had been "the sport of kings" and armies were small; but at Waterloo, the massive armies of the allies numbered some 800,000 men—not all concentrated, to be sure—whereas Napoleon put together a force of 200,000 soldiers, of whom 130,000 were available.) Germany was more permanently transformed than most other parts of Europe during the Revolutionary-Napoleonic years, and under the domination of Prussia, it was to become the center of future development on the Continent.

With the end of the 1700s, Germany assumed the lead as the center of mathematical activity. The greatest mathematician of modern times, Carl Friedrich Gauss (1777–1855), was so very German that never in his life did he leave the country, not even for a visit. Unlike Fermat and Euler, who came from financially secure middle-class backgrounds, Gauss was born into a poor and unlettered family. His father managed to eke out a meager living in Brunswick, Germany, through hard work as stonecutter, gardener, canal worker, and finally foreman for a masonry firm. Had the father's views prevailed, the son would have followed one of the family trades; it was only by fortunate chance that Gauss did not become a bricklayer or a gardener. Much against his will, the father was persuaded to allow the gifted boy to acquire an appropriate education.

Gauss was one of those remarkable infant prodigies whose natural aptitude for mathematics soon becomes apparent. He used to say, laughingly, that without the help or knowledge of others he had learned to reckon before he could talk. As a child of three, according to a well-authenticated story, he corrected an error in his father's weekly payroll calculations. His arithmetical powers so overwhelmed his schoolmasters that by the time Gauss was 9 years old, they admitted that there was nothing more they could teach the boy. It is said that in his first arithmetic class, Gauss astonished his teacher by instantly solving what was intended as a long "busy work" problem, "Find the sum of all the numbers from 1 to 100." The young Gauss later confessed to having recognized the pattern

$$1 + 100 = 101, \quad 2 + 99 = 101, \quad 3 + 98 = 101, \ldots, \quad 50 + 51 = 101.$$

Because there are 50 pairs of numbers, each of which adds up to 101, the sum of all the numbers must be $50 \cdot 101 = 5050$. This technique provides another way of deriving the formula

$$1 + 2 + 3 + \cdots + n = \frac{n(n + 1)}{2}$$

for the sum of the first n positive integers. One need only display the consecutive integers 1 through n in two rows as follows:

$$
\begin{array}{cccccc}
1 & 2 & 3 & \cdots & n - 1 & n \\
n & n - 1 & n - 2 & \cdots & 2 & 1
\end{array}
$$

Adding the vertical columns produces n terms, each of which equals $n + 1$; when these terms are added, we get the value $n(n + 1)$. Because the same sum is obtained on adding the two rows horizontally, we arrive at the formula

$$n(n + 1) = 2(1 + 2 + 3 + \cdots + n).$$

This brilliance aroused the attention of influential people, particularly Duke Ferdinand of Brunswick, who became and remained Gauss's patron through many years. The duke's generosity made it possible for the young genius to attend Caroline College, a preparatory school in Brunswick, and later the University of Göttingen, where he remained only three years (1795–1798). During his student days at Caroline College, Gauss formulated the method of least squares, by which the most likely value of a variable quantity could be estimated from many random observations. He shared the honor of this discovery with Legendre, who first published the device independently in 1806 in his *Nouvelles methodes pour la determination des orbites des cometes*. When Gauss entered Göttingen, he was still undecided whether to become a mathematician or pursue a career in classical languages. March 30, 1796, marked the turning point in the choice of studies. On that day, when Gauss was still not 20 years old, he made a dramatic discovery that definitely decided him in favor of mathematics.

The problem of constructing regular polygons using only "Euclidean tools," that is to say, straightedge and compass, had long been laid aside in the belief that the ancients had exhausted all the possible constructions. What Gauss demonstrated in 1796 was that the regular polygon of 17 sides is so constructible—the first such advance since the time of Euclid. Gauss was so proud of this discovery that he expressed the wish to have a regular 17-sided polygon carved on his tombstone, just as Archimedes' grave was decorated with the figure of a sphere inscribed in a cylinder. Although this desire was not carried out, there is a 17-pointed star inscribed on the base of a monument erected in Gauss's memory in his native town of Brunswick. Apparently the stonemason refused to chisel a regular 17-sided polygon because he feared that the resulting figure would be indistinguishable from a circle.

Later, in the impressive *Disquisitiones Arithmeticae* of 1801, Gauss proved that a regular polygon of p sides, where p is an odd prime, is constructible by straightedge and compass if p is of the form $2^{2^k} + 1$. As we have already pointed out, the values of $2^{2^k} + 1$ for $k = 0, 1, 2, 3, 4$, namely

$$3, 5, 17, 257, \text{ and } 65{,}537$$

Carl Friedrich Gauss
(1777–1855)

(Smithsonian Institution.)

are all prime. Euler showed that $2^{2^5} + 1$ admits the factor 641; mathematicians now believe that $2^{2^k} + 1$ is composite for all $k \geq 5$, although there has yet been no proof of this.

At Göttingen, Gauss studied mathematics under Abraham Kastner; but since Kastner showed little understanding of Gauss's research, he tended to work quite independently of his teachers. There is a story that Gauss tried to interest Kastner in his construction for the regular 17-gon by pointing out that he had solved a seventeenth-degree algebraic equation to perform the construction. The professor, seeking to dismiss Gauss, replied that the solution was impossible. For this rebuff Gauss repaid Kastner, who prided himself on being something of a poet, by declaring him to be the best mathematician among poets and the best poet among mathematicians.

In 1798, Gauss returned to Brunswick, where he earned a somewhat precarious living by private tutoring. When he was unsuccessful in getting pupils, Duke Ferdinand granted him a fixed pension so that he could devote himself entirely to scientific work unhampered by financial worries. During this period, Gauss frequently consulted the mathematical library of the University of Helmstädt and there made the acquaintance of Johann Fredrich Pfaff, then the best-known mathematician in Germany. Gauss's first published scientific work was his famous dissertation, nominally written under Pfaff, on the basis of which he was granted his doctorate from Helmstädt without the usual examination. The doctoral thesis, bearing the title "New Proof of the Theorem That Every Integral Rational Function of One Variable Can Be Decomposed into Real Factors of the First or Second Degree," gave the first substantial proof (although not rigorous by modern standards) of the so-called fundamental theorem of algebra. This theorem states that every polynomial equation

$$x^n + a_1 x^{n-1} + \cdots + a_{n-1}x + a_n = 0$$

in which the coefficients are real or complex numbers has at least one root in the complex field, from which it follows that such a polynomial of degree n has precisely n roots (which may not be distinct). Gauss returned to the topic several times later on.

Although the name *fundamental theorem of algebra* appears to have been introduced by Gauss, the result itself was familiar, having resisted attempted demonstrations by d'Alembert (1746), Euler (1749), and Lagrange (1772). Many historians seem willing to give Albert Girard priority in the formulation of the fundamental theorem, for in his *L'Invention nouvelle en l'algèbre* of 1629 he asserted "Every equation of algebra has as many solutions as the exponent of the highest term indicates." Jean d'Alembert (who wrote many of the mathematical articles in the great French *Encyclopèdie*) made the first attempt to prove it. But he was able to show not even the existence of a root of the equation, only the form that the root takes. However, the "proof" was so widely accepted that the theorem came to be known, at least in France, as d'Alembert's theorem. Euler's contribution lay in showing that the complex roots occur in conjugate pairs; that is, when a root $a + b\sqrt{-1}$ occurs, there is also a conjugate root $a - b\sqrt{-1}$, so that the polynomial equation must contain a quadratic factor $x^2 + px + q$ with real coefficients. Gauss's original argument (1799) involved geometric considerations, but because his contemporaries wished an entirely algebraic proof, he gave three more demonstrations of the fundamental theorem, the last when he was 70 years old. The third proof (1816) necessitated using complex integrals and showed Gauss's early mastery of the new complex analysis. In the fourth proof (1849), which was really a worked-over version of the first, Gauss used complex numbers more freely, because as he said, they "are nowadays apparent to all."

From the fundamental theorem of algebra, Gauss went on to a succession of triumphs, each new discovery following on the heels of the previous one. The number-theoretic ideas that had been piling up since his days at Caroline College were finally brought together in *Disquisitiones Arithmeticae,* a work that instantly won Gauss recognition as a mathematical genius of the first order.

But the most extraordinary achievement of Gauss was more in the realm of theoretical astronomy than of mathematics. On the opening night of the nineteenth century, January 1, 1801, the Italian astronomer Piazzi discovered the first of the so-called minor planets (planetoids or asteroids), later called Ceres. After this newly found body, visible in its course only by telescope, passed the sun, neither Piazzi nor any other astronomer could find it again. Piazzi's observations extended over 41 days, during which the orbit swept out an angle of only 9 degrees. From the scanty data available, Gauss was able to calculate Ceres' orbit with amazing accuracy, and the elusive planet was rediscovered at the end of the year in almost exactly the position he had forecast. This success brought Gauss worldwide fame and led to the offer of an appointment at the St. Petersburg Academy, a post he declined.

Gauss's tour de force on Ceres has often been viewed as a disaster for mathematics. Just when he was experiencing an upsurge of ideas and publications in various fields of mathematics, he became absorbed in astronomy, a subject that was to remain his chief preoccupation for the next 20 years. He now developed a general theory of planetary and cometary orbits, taking into account not only the main gravitational attraction of the sun, but also the minor gravitational forces of the other planets. As the prying telescope picked up the first of a series of planetoids that lay between Mars and Jupiter—Pallas (1802), Juno (1804), and Vesta (1807)—Gauss immediately calculated the orbit of each new discovery. Today more than 1500 of these planetoids have been identified.

Using his superior methods, Gauss redid in an hour's time the calculations on which Euler had spent three days, and which sometimes are said to have led to Euler's loss of sight in one eye. Gauss remarked unkindly, ''I should also have gone blind if I had calculated in that fashion for three days.'' The method of orbital determination, including the use of least squares in minimizing the inevitable errors of observation, was published in 1809 as his second masterpiece, *Theoria Motus Corporum Coelestium in Sectionibus Conicus Solem Ambietium* (Theory of the Motion of the Heavenly Bodies About the Sun in Conic Sections). It was first written in German, but at the request of the publisher, who possibly believed that it would be read for centuries, was translated into Latin. The triumph of Gauss's methods was the calculation (in 1846 and so within Gauss's lifetime) of the orbit of the most distant planet, Neptune. Neptune was discovered as the result of observed disturbances in the motion of Uranus, and it was the first planet to be discovered on the basis of theoretical calculation.

When the patron of his student days, the Duke of Brunswick, was killed in the Battle of Jena (1806) leading the Prussian armies against Napoleon, Gauss was forced to seek some reliable livelihood to support his family. Reluctant to see Germany lose the most renowned mathematician in the world to St. Petersburg, his friends secured Gauss's appointment as director of the newly built observatory at the University of Göttingen, a position that he held until his death nearly a half-century later. It is said that except for a visit to Berlin in 1836 to attend a scientific meeting, Gauss never slept under any other roof than that of his own observatory.

Napoleon had a remarkable triumph at Jena, where he was victorious with only 26,000 men against Brunswick's 60,000. After this victory, Napoleon was free to reorganize northern Germany. By the Treaty of Tilsit in 1807, portions of Hanover, Hesse, and Brunswick west of the Elbe were given to the new Kingdom of Westphalia, to be ruled by Napoleon's youngest brother. In the following years, French war levies were severe, and Gauss, as a professor at Göttingen, was assessed an involuntary contribution of 2000 francs. The sum was exorbitant in that day's currency and quite beyond Gauss's ability to pay, particularly so because wage payments at Göttingen were frequently suspended. Laplace wrote to Gauss that he had himself paid the 2000-franc fine into the Treasury at Paris, considering it an honor to lift this undeserved burden from the shoulders of a friend. As soon as Gauss could, however, he repaid Laplace with interest at the current rate of exchange. For the rest of his life, Gauss resented the French, refusing to publish in their language and pretending ignorance of French when speaking to Frenchmen whom he did not know.

There was at least one other occasion when a French mathematician interceded with the authorities on Gauss's behalf. Sophie Germain (1776–1831), fascinated by Gauss's masterpiece on the theory of numbers, the *Disquisitiones Arithmeticae* (1801), had begun sending him some of the results of her own mathematical investigations. Because she feared that he might be prejudiced against a woman mathematician, she used the pen name *M. Leblanc;* and indeed Gauss did not suspect the true identity of M. Leblanc until 1806. The French forces had occupied Brunswick and were besieging

Hanover. Concerned over the great man's safety, Germain asked the general commanding the French troops, a family friend, to send an officer to see how Gauss was faring. It developed that Gauss was safe and well, but he denied any knowledge of Mlle. Germain because, after all, his only contact had been with M. Leblanc. The confusion was finally put right in an exchange of letters. The irony is that despite their extensive correspondence, Germain and Gauss never met; and before the University of Göttingen could award her the honorary doctor's degree that Gauss had recommended, Sophie Germain died.

Concurrently with his purely theoretical research in mathematics, Gauss pursued work in several related scientific fields, notably physics, mechanics, and theoretical astronomy. In the year 1831, Wilhelm Weber (1804–1891) was called to Göttingen at Gauss's suggestion to assume a chair in physics. There he became Gauss's collaborator in investigating the intensity of the earth's magnetic force. Gauss and Weber were the first to communicate by electromagnetic telegraph when in 1833, they connected the astronomical observatory and the physics laboratory with a mile-long double wire. Shortly thereafter, it was accidentally discovered (in Bavaria) that it was quite unnecessary to lay two wires, as the earth provided a return for the current. More efficient methods came in 1837 with the work of Samuel Morse.

By the mid-1800s, mathematics had grown into an enormous and unwieldy structure, divided into many fields in which only the specialist knew the way. Gauss was the last complete mathematician, and it is no exaggeration to say that he was in some degree connected with nearly every aspect of the subject. His contemporaries regarded him as *Princeps Mathematicorum* (Prince of Mathematicians), on a par with Archimedes and Isaac Newton. This is illustrated in a small incident: On being asked who was the greatest mathematician in Germany, Laplace answered, ''Why, Pfaff.'' When the questioner indicated that he would have thought Gauss was, Laplace replied, ''Pfaff is by far the greatest in Germany, but Gauss is the greatest in all Europe.''

Although Gauss adorned every branch of mathematics, he always held number theory in high esteem and affection. He is reputed to have insisted, ''Mathematics is the Queen of the sciences, and the theory of numbers is the Queen of mathematics.'' The publication of *Disquisitiones Arithmeticae* (Arithmetical Investigations), which appeared in 1801 when Gauss was 24 years old, laid the foundations of modern number theory. Legend has it that a large part of the *Disquisitiones Arithmeticae* had been submitted as a memoir to the French Academy the previous year and had been rejected in a manner that, even had the work been as worthless as the referees believed, would have been inexcusably rude. (In an attempt to lay this defamatory tale to rest, the officers of the academy made an exhaustive search of their permanent records in 1935 and concluded that the *Disquisitiones* was never submitted, much less rejected.) ''It is really astonishing,'' said Leopold Kronecker, ''to think that a single man of such young years was able to bring to light such a wealth of results, and above all to present such a profound and well-organized treatment of an entirely new discipline.''

In gratitude for all that the Duke of Brunswick had done for him, Gauss dedicated the *Disquisitiones* to his benefactor. ''Were it not for your unceasing benefits in support of my studies,'' said Gauss in the lengthy and flowery dedication, ''I would not have been able to devote myself totally to my passionate love, the study of mathematics.'' The words were not empty flattery, for if ever a patron deserved the homage of a protégé, Ferdinand deserved that of Gauss. The duke had financed the education of the young prodigy, granted him a yearly pension on completion of his studies, and even paid for the printing of his doctoral dissertation.

In the first chapter of *Disquisitiones Arithmeticae,* Gauss introduced the concept of congruence and the notation that made it such a powerful technique. He explained that he was induced to adopt the symbol ≡ because of the close analogy with algebraic equality. According to Gauss, "If a number n measures the difference between two numbers a and $b,$ then a and b are said to be congruent with respect to $n;$ if not, incongruent." In the form of a definition, this becomes the following.

Definition

Let n be a fixed positive integer. Two integers a and b are said to be congruent modulo n, symbolized by

$$a \equiv b \ (\text{mod } n)$$

if n divides the difference a − b, that is, provided that a − b = kn for some integer k.

To fix the idea let us consider the case $n = 7$. It is routine to check that

$$3 \equiv 24 \ (\text{mod } 7), \qquad -31 \equiv 11 \ (\text{mod } 7), \qquad -15 \equiv -64 \ (\text{mod } 7),$$

because $3 - 24 = (-3)7$, $-31 - 11 = (-6)7$, and $-15 - (-64) = 7 \cdot 7$. On the other hand, if $n \nmid (a - b)$, then, we say, a is *incongruent to b modulo n,* and we write $a \not\equiv b \ (\text{mod } n)$. For example, $25 \not\equiv 12 \ (\text{mod } 7)$, since 7 fails to divide $25 - 12 = 13$.

It is to be noted that any two integers are congruent modulo 1, whereas two integers are congruent modulo 2 when they are both even or both odd. Because congruence modulo 1 is not interesting, the usual practice is to assume that $n > 1$.

There is a useful characterization of congruence modulo n in terms of remainders on division by n.

THEOREM

For a and b arbitrary integers, a ≡ b (mod n) if and only if a and b leave the same nonnegative remainder when divided by n.

Proof. First, take $a \equiv b \ (\text{mod } n)$, so that $a = b + kn$ for some integer k. On division by n, we find that b leaves a certain remainder r; that is, $b = qn + r$, where $0 \leq r < n$. Therefore,

$$a = b + kn = (qn + r) + kn = (q + k)n + r,$$

which indicates that a has the same remainder as b.

On the other hand, suppose we can write $a = q_1 n + r$ and $b = q_2 n + r$, with the same remainder r $(0 \leq r < n)$. Then

$$a - b = (q_1 n + r) - (q_2 n + r) = (q_1 - q_2)n,$$

whence $n \mid (a - b)$. In the language of congruences, this says that $a \equiv b \pmod{n}$. ∎

Example. Because the integers -56 and -11 can be expressed in the form

$$-56 = (-7)9 + 7, \qquad -11 = (-2)9 + 7,$$

with the same remainder 7, by the foregoing theorem, $-56 \equiv -11 \pmod 9$. In the other direction, the congruence $-31 \equiv 11 \pmod 7$ implies that -31 and 11 have the same remainder when divided by 7. This is clear from the relations

$$-31 = (-5)7 + 4, \qquad 11 = 1 \cdot 7 + 4.$$

Congruence can be viewed as a generalized form of equality, in the sense that its behavior with respect to addition and multiplication is reminiscent of ordinary equality. Some of the elementary properties of equality that carry over to congruences appear in the next theorem.

THEOREM

Let $n > 0$ be fixed and a, b, c, d be arbitrary integers. Then the following properties hold.
(1) $a \equiv a \pmod n$.
(2) If $a \equiv b \pmod n$, then $b \equiv a \pmod n$.
(3) If $a \equiv b \pmod n$ and $b \equiv c \pmod n$, then $a \equiv c \pmod n$.
(4) If $a \equiv b \pmod n$ and $c \equiv d \pmod n$, then $a + c \equiv b + d \pmod n$, and $ac \equiv bd \pmod n$.
(5) If $a \equiv b \pmod n$, then $a + c \equiv b + c \pmod n$ and $ac \equiv bc \pmod n$.
(6) If $a \equiv b \pmod n$, then $a^k \equiv b^k \pmod n$ for any positive integer k.

Proof. For any integer a, we have $a - a = 0 \cdot n$, so that $a \equiv a \pmod n$. Now if $a \equiv b \pmod n$, then $a - b = kn$ for some integer k. Hence, $b - a = -(kn) = (-k)n$, and since $-k$ is an integer, this yields property (2).

Property (3) is slightly less obvious. Suppose that $a \equiv b \pmod n$ and also $b \equiv c \pmod n$. Then there exist integers h and k satisfying $a - b = hn$ and $b - c = kn$. It follows that

$$a - c = (a - b) + (b - c) = hn + kn = (h + k)n,$$

in consequence of which $a \equiv c \pmod n$.

In the same vein, if $a \equiv b \pmod n$ and $c \equiv d \pmod n$, then we are assured that $a - b = k_1 n$ and $c - d = k_2 n$ for some choice of k_1 and k_2. Adding these equations, one gets

$$(a + c) - (b + d) = (a - b) + (c - d) = k_1 n + k_2 n = (k_1 + k_2)n,$$

or as a congruence statement, $a + c \equiv b + d \pmod n$. As regards the second assertion of property (4), note that

$$ac = (b + k_1 n)(d + k_2 n) = bd + (bk_2 + dk_1 + k_1 k_2 n)n.$$

Because $bk_2 + dk_1 + k_1 k_2 n$ is an integer, this says that $ac - bd$ is divisible by n, whence $ac \equiv bd \pmod n$.

The proof of property (5) is covered by (4) and the fact that $c \equiv c \pmod{n}$. Finally, we obtain property (6) by making an induction argument. The statement certainly holds for $k = 1$, and we shall assume it is true for some fixed k. From (4), we know that $a \equiv b \pmod{n}$ and $a^k \equiv b^k \pmod{n}$ together imply that $aa^k \equiv bb^k \pmod{n}$, or equivalently, $a^{k+1} \equiv b^{k+1} \pmod{n}$. This is the form the statement should take for $k + 1$, so the induction step is complete. ∎

Before going further, you will want to know the great help that congruences can be in lightening labor when you are trying certain types of computations.

Example. Let us show that 641 divides $2^{2^5} + 1$, or in terms of congruence, that $2^{32} + 1 \equiv 0 \pmod{641}$. (This is the case that refutes Fermat's claim that all numbers of the form $2^{2^n} + 1$ are prime.) To prepare the ground, we observe that $5 \cdot 2^7 = 640 \equiv -1 \pmod{641}$. According to part (6) of the last theorem, both sides of this congruence can be raised to the fourth power, to give

$$5^4 \cdot 2^{28} \equiv (-1)^4 \equiv 1 \pmod{641}.$$

But $5^4 = 625 \equiv -16 = -(2^4) \pmod{641}$, which can be written

$$2^4 \equiv -(5^4) \pmod{641}.$$

These two congruences, together with parts (3) and (5) of the theorem, imply that

$$2^{32} + 1 = 2^4 \cdot 2^{28} + 1 \equiv -(5^4)2^{28} + 1 \equiv -1 + 1 = 0 \pmod{641}.$$

Thus, $641 \,|\, 2^{2^5} + 1$, as desired.

Example. For another illustration in the same spirit, we shall confirm that the Mersenne number $2^{83} - 1$ is divisible by 167, a fact Euler noticed in 1738. One approach to the calculation is through repeated squarings. We find that

$$2^8 = 256 \equiv 89 \pmod{167},$$

$$2^{16} \equiv 89^2 = 7921 \equiv 72 \pmod{167},$$

$$2^{32} \equiv 72^2 = 5184 \equiv 7 \pmod{167},$$

$$2^{64} \equiv 49 \pmod{167},$$

and so $2^{67} = 2^3 \cdot 2^{64} \equiv 8 \cdot 49 = 392 \equiv 58 \pmod{167}$. Combining these results, we obtain

$$2^{83} - 1 = 2^{16} \cdot 2^{67} - 1 \equiv 72 \cdot 58 - 1 = 4167 - 1 = 4166 \equiv 0 \pmod{167},$$

which shows that $167 \,|\, 2^{83} - 1$.

We shall now prove an earlier assertion regarding the final digits of even perfect numbers, namely, that all such numbers terminate in either 6 or 8. As a prelude, let us observe that any positive integer is congruent, modulo 10, to its last digit. The reason is that a positive integer n, expressed in decimal notation, takes the form

$$n = a_k 10^k + a_{k-1} 10^{k-1} + \cdots + a_1 10 + a_0,$$

where each a_i is an integer satisfying $0 \leq a_i \leq 9$, and the coefficients $a_0, a_1, \ldots, a_k$ are the digits of the number n. Because $10 \equiv 0$ (mod 10), we know by the theorem on the arithmetic of congruences that $10^i \equiv 0^i \equiv 0$ (mod 10). Therefore, by parts (4) and (5) of the same theorem, we conclude that

$$n \equiv a_k \cdot 0 + a_{k-1} \cdot 0 + \cdots + a_1 \cdot 0 + a_0 = a_0 \ (\text{mod } 10).$$

Thus to determine the final digit of a number, it suffices to find the number's congruence modulo 10.

THEOREM

An even perfect number n ends in the digit 6 or the digit 8; that is, $n \equiv 6$ (mod 10) or $n \equiv 8$ (mod 10).

> **Proof.** Being an even perfect number, n looks like $n = 2^{k-1}(2^k - 1)$, where the factor $2^k - 1$ is a prime. According to an earlier result, the exponent k must also be prime. If $k = 2$, then $n = 6$, and the asserted theorem holds. We can therefore confine our attention to the case $k > 2$. The proof falls into two parts, according as k takes the form $4m + 1$ or $4m + 3$.
>   If k is of the form $4m + 1$, then
>
> $$n = 2^{4m}(2^{4m+1} - 1)$$
> $$= 2^{8m+1} - 2^{4m} = 2 \cdot 16^{2m} - 16^m.$$

A straightforward induction argument will make it clear that $16^t \equiv 6$ (mod 10) for any positive integer t. Using this congruence, we get

$$n \equiv 2 \cdot 6 - 6 = 6 \ (\text{mod } 10).$$

Now in the case in which $k = 4m + 3$,

$$n = 2^{4m+2}(2^{4m+3} - 1)$$
$$= 2^{8m+5} - 2^{4m+2} = 2 \cdot 16^{2m+1} - 4 \cdot 16^m.$$

Recalling that $16^t \equiv 6$ (mod 10), we see that

$$n \equiv 2 \cdot 6 - 4 \cdot 6 = -12 \equiv 8 \ (\text{mod } 10).$$

Consequently, every even perfect number has a last digit equal to 6 or 8. ∎

We have seen that if $a \equiv b$ (mod n), then $ca \equiv cb$ (mod n) for any choice of the integer c. The converse, however, fails to hold. For an example as simple as any, note that $2 \cdot 4 \equiv 2 \cdot 1$ (mod 6), while $4 \not\equiv 1$ (mod 6). We emphasize that one cannot unrestrictedly cancel a common factor in the arithmetic of congruences.

With some suitable precautions, however, cancellation can be allowed. One setting is provided by the following lemma.

LEMMA

If $ca \equiv cb \pmod{n}$, and gcd $(c, n) = d$, then $a \equiv b \pmod{n/d}$.

Proof. By assumption, we can write

$$c(a - b) = ca - cb = kn$$

for some integer k. Because gcd $(c, n) = d$, there exist relatively prime integers r and s satisfying $c = dr$, $n = ds$. When these values are substituted in the displayed equation and the common factor d is cancelled, the net result is

$$r(a - b) = ks.$$

Hence, $s \mid r(a - b)$ and gcd $(r, s) = 1$. Euclid's lemma tells us that under these circumstances $s \mid a - b$, which can be recast as $a \equiv b \pmod{s}$. In other words, we have $a \equiv b \pmod{n/d}$. ∎

When the requirement that gcd $(c, n) = 1$ is added, the cancellation can be accomplished without a change in modulus.

COROLLARY

If $ca \equiv cb \pmod{n}$ with gcd $(c, n) = 1$, then $a \equiv b \pmod{n}$.

For an illustration of the lemma, consider the congruence $33 \equiv 15 \pmod{9}$, which is the same as $3 \cdot 11 \equiv 3 \cdot 5 \pmod{9}$. The lemma indicates that since gcd $(3, 9) = 3$, then $11 \equiv 5 \pmod{3}$.

Gauss, making an analogy with algebraic equations of the first degree, devoted the second section of the *Disquisitiones Arithmeticae* to the theory of linear congruences: An equation of the form $ax \equiv b \pmod{n}$ is called a *linear congruence,* and by a solution of such an equation we mean an integer x_0 for which $ax_0 \equiv b \pmod{n}$. By definition, $ax_0 \equiv b \pmod{n}$ if and only if $n \mid (ax_0 - b)$, or what amounts to the same thing, if and only if $ax_0 - b = ny_0$ for some integer y_0. Thus, the problem of finding all integers satisfying the linear congruence $ax \equiv b \pmod{n}$ is identical with obtaining all solutions of the linear diophantine equation $ax - ny = b$.

It is convenient to treat two solutions of $ax \equiv b \pmod{n}$ that are congruent modulo n as "equal" even though they are not equal in the usual sense. For instance, $x = 3$ and $x = -9$ both satisfy the congruence $3x \equiv 9 \pmod{12}$. Because $3 \equiv -9 \pmod{12}$, they are not counted as different solutions. In short, when we refer to the number of solutions of $ax \equiv b \pmod{n}$, we mean the number of incongruent integers satisfying this congruence.

With these remarks in mind, we can easily formulate the principal result on linear congruences.

THEOREM

The linear congruence $ax \equiv b$ (mod n) has a solution if and only if $d|b$, where $d = gcd\ (a, n)$. If $d|b$, then it has d mutually incongruent solutions modulo n.

Proof. We have already observed that the given congruence is equivalent to the linear diophantine equation $ax - ny = b$. From Chapter 5, it is known that the latter equation can be solved if and only if $d|b$; moreover, if it is solvable and x_0, y_0 is one specific solution, then any other solution has the form

$$x = x_0 + \frac{n}{d}t, \qquad y = y_0 + \frac{a}{d}t$$

for some choice of t.

Among the various integers satisfying the first of these formulas, consider those that occur when t takes on the successive values $t = 0, 1, 2, \ldots, d - 1$:

$$x_0, x_0 + \frac{n}{d}, x_0 + \frac{2n}{d}, \ldots, x_0 + \frac{(d-1)n}{d}.$$

We claim that these integers are incongruent modulo n, while any other such integer x is congruent to some one of them. If it happened that

$$x_0 + \frac{n}{d}t_1 \equiv x_0 + \frac{n}{d}t_2 \ (\text{mod } n),$$

where $0 \leq t_1 < t_2 \leq d - 1$, then one would have

$$\frac{n}{d}t_1 \equiv \frac{n}{d}t_2 \ (\text{mod } n).$$

Now gcd $(n/d, n) = n/d$, and so by the last lemma, the factor n/d could be cancelled to arrive at the congruence

$$t_1 \equiv t_2 \ (\text{mod } d),$$

which is to say that $d|t_2 - t_1$. But this is clearly impossible in view of the inequality $0 < t_2 - t_1 < d$.

It remains to argue that any other solution $x_0 + (n/d)t$ is congruent modulo n to one of the d integers listed. The division theorem permits us to write the number t as $t = qd + r$, where $0 \leq r \leq d - 1$. Hence

$$x_0 + \frac{n}{d}t = x_0 + \frac{n}{d}(qd + r)$$

$$= x_0 + nq + \frac{n}{d}r$$

$$\equiv x_0 + \frac{n}{d}r \ (\text{mod } n),$$

with $x_0 + (n/d)r$ as one of our d selected solutions. This ends the proof. ∎

The argument in this theorem brings out a point worth stating explicitly: If x_0 is any solution of $ax \equiv b \pmod{n}$, then the $d = \gcd(a, n)$ incongruent solutions are given by

$$x_0, \; x_0 + \frac{n}{d}, \; x_0 + \frac{2n}{d}, \; \ldots, \; x_0 + \frac{(d-1)n}{d}.$$

For your convenience, here is the form the theorem takes in the special case in which a and n are assumed to be relatively prime.

COROLLARY

If $\gcd(a, n) = 1$, *then the linear congruence* $ax \equiv b \pmod{n}$ has a unique solution modulo n.

We now pause to look at two concrete examples.

Example. Consider first the linear congruence $18x \equiv 30 \pmod{42}$. Because $\gcd(18, 42) = 6$ and 6 surely divides 30, the foregoing theorem guarantees the existence of exactly six solutions, which are incongruent modulo 42. By inspection, one solution is found to be $x \equiv 4 \pmod{42}$. Our theory tells us that the six solutions are as follows:

$$x \equiv 4 + \frac{42}{6}t \equiv 4 + 7t \pmod{42}, \qquad t = 0, 1, \ldots, 5,$$

or plainly enumerated,

$$x \equiv 4, 11, 18, 25, 32, 39 \pmod{42}.$$

Example. Let us solve the linear congruence $9x \equiv 21 \pmod{30}$. At the outset we know, since $\gcd(9, 30) = 3$ and $3 \mid 21$, that there must be three incongruent solutions.
Because the congruence $9x \equiv 21 \pmod{30}$ is equivalent to the diophantine equation

$$9x - 30y = 21,$$

we can begin by expressing $3 = \gcd(9, 30)$ as a linear combination of 9 and 30. Among the various possibilities, we see that $3 = 9(-3) + 30 \cdot 1$, and so if both sides are multiplied by 7,

$$21 = 7 \cdot 3 = 9(-21) - 30(-7).$$

Modulo 30, this becomes $9(-21) \equiv 21 \pmod{21}$. Thus, $x \equiv -21 \pmod{30}$ satisfies the congruence $9x \equiv 21 \pmod{30}$, or if one prefers to have positive numbers, $x \equiv 9 \pmod{30}$ is a solution. With this known, the three incongruent solutions can be obtained from the formula

$$x \equiv 9 + \tfrac{30}{3}t = 9 + 10t \pmod{30}, \qquad t = 0, 1, 2.$$

We end with $x \equiv 9, 19, 29 \pmod{30}$ as the three required solutions.

10.3   Problems

1. Verify each of the following assertions:

    (a) If $a \equiv b$ (mod n) and $m|n$, then $a \equiv b$ (mod m).
    (b) If $a \equiv b$ (mod n) and the integer $c > 0$, then $ca \equiv cb$ (mod cn).
    (c) If $a \equiv b$ (mod n) and the integers a, b, n are all divisible by $d > 0$, then $a/d \equiv b/d$ (mod n/d).
    (d) If $a \equiv b$ (mod n) and $a \equiv b$ (mod m) where gcd $(m, n) = 1$, then $a \equiv b$ (mod mn).

2. Give an example to show that $a^2 \equiv b^2$ (mod n) need not imply that $a \equiv b$ (mod n).

3. Prove that every integer is congruent modulo n to exactly one of the integers $0, 1, 2, \ldots , n - 1$.

4. Prove that if a is any odd integer, then $a^2 \equiv 1$ (mod 8). [*Hint:* Note that $a \equiv 1,3,5,$ or 7 (mod 8).]

5. For any integer a, prove the following statements:

    (a) $a^2 \equiv 0$ or 1 (mod 3);
    (b) $a^3 \equiv 0, 1$ or $- 1$ (mod 7);
    (c) $a^4 \equiv 0$ or 1 (mod 5).

6. Determine whether $5^{36} - 1$ is divisible by 13.

7. (a) Determine the remainder when $10^{49} + 5^3$ is divided by 7.
    (b) Find the remainder on dividing the sum

    $$1! + 2! + 3! + 4! + \cdots + 99! + 100!$$

    by 12. [*Hint:* Note that $4! \equiv 0$ (mod 12).]

8. Use congruence theory to show that 47 divides the Mersenne number $2^{23} - 1$, and that 223 divides $2^{37} - 1$.

9. Verify each of the following divisibility statements:

    (a) $3|4^{n + 1} + 5^{2n - 1}$ for all $n \geq 1$.
    [*Hint:* $4^{n + 1} + 5^{2n - 1} \equiv 1^{n + 1} + (- 1)^{2n - 1}$ mod (3).]
    (b) $5|7 \cdot 16^n + 3$ for all $n \geq 0$.
    (c) $7|5^{2n} + 3 \cdot 2^{5n - 2}$ for all $n \geq 1$.
    (d) $17|3^{4n + 2} + 2 \cdot 4^{3n + 1}$ for all $n \geq 0$.
    (e) $27|2^{5n + 1} + 5^{n + 2}$ for all $n \geq 0$.

10. Prove that if $n > 6$ is an even perfect number, then $n \equiv 4$ (mod 6). [*Hint:* $4^m \equiv 4$ (mod 6) for any $m \geq 1$.]

11. Prove that if p is a prime and $a^2 \equiv 1$ (mod p), then $a \equiv 1$ (mod p) or $a \equiv p - 1$ (mod p).

12. Suppose that the integer n is given in decimal form,

    $$n = a_k 10^k + a_{k - 1} 10^{k - 1} + \cdots + a_1 10 + a_0,$$

    where $0 \leq a_i \leq 9$ for all i. Prove that:

    (a) $n \equiv a_0 + a_1 + \cdots + a_k$ (mod 9); hence 9 divides n if and only if 9 divides the sum of its digits:
    (b) $n \equiv a_0 - a_1 + a_2 - \cdots + (- 1)^k a_k$ (mod 11); hence 11 divides n if and only if 11 divides the alternating sum of its digits.

13. Without actually performing the divisions, check the following numbers for divisibility by 9 or 11;

    (a) 113,058;   (b) 2,964,357;   (c) 176,521,221.

14. Using Problem 12, find the missing digit in the calculation:

    $$52,817 \cdot 3,212,146 = 169,655,x15,282$$

15. A palindrome is a number that reads the same backward as forward, such as 2662 or 9,351,539. Prove that every 6-digit palindrome is divisible by 11.

16. Solve the following linear congruences.

    (a) $6x \equiv 15$ (mod 21).
    (b) $36x \equiv 8$ (mod 102).
    (c) $13x \equiv 27$ (mod 52).

Bibliography

Archibald, Raymond C. "Gauss and the Regular Polygon of Seventeen Sides." *American Mathematical Monthly* 27(1920): 323–326.

Beiler, Albert. *Recreations in the Theory of Numbers.* 2d ed. New York: Dover, 1966.

Brown, Harcourt. *Scientific Organization in Seventeenth Century France 1620–1680.* New York: Russell & Russell, 1934.

Buhler, Walter. *Gauss: A Biographical Study.* New York: Springer-Verlag, 1981.

Burton, David M. *Elementary Number Theory.* 3d ed. Dubuque: Wm. C. Brown, 1993.

Bussey, W. H. "Fermat's Method of Infinite Descent." *American Mathematical Monthly* 25(1918): 333–337.

Dalmédico, Amy. "Sophie Germain." *Scientific American* 265(Dec. 1991): 117–122.

Dickson, Leonard E. "Fermat's Last Theorem and the Origin and Nature of the Theory of Algebraic Numbers." *Annals of Mathematics* 18, no. 4(1917): 161–187.

———. "Perfect and Amicable Numbers." *Scientific Monthly* 12(1921): 349–355.

———. *History of the Theory of Numbers.* 3 vols. New York: Chelsea, 1952.

Dunham, William. "Euler and the Fundamental Theorem of Algebra." *College Mathematics Journal* 16(1989): 249–268.

Dunnington, G. Waldo. *Carl Friedrich Gauss, Titan of Science.* New York: Hafner, 1955.

Edwards, H. M. "The Background to Kummer's Proof of Fermat's Last Theorem for Regular Primes." *Archive for History of Exact Sciences* 14(1975): 219–236.

Euler, Leonhard. *Introduction to Analysis of the Infinite, Book I.* Translated by John Blanton. New York: Springer-Verlag, 1988.

Eves, Howard. "Fermat's Method of Infinite Descent." *Mathematics Teacher* 53(1960): 195–196.

Freidberg, Richard. *An Adventurer's Guide to Number Theory.* New York: McGraw-Hill, 1968.

Gauss, Carl Friedrich. *Disquisitiones Arithmeticae.* Translated by Arthur Clarke. New Haven, Conn.: Yale University Press, 1966.

Hahn, Roger. *The Anatomy of a Scientific Institution: The Paris Academy of Sciences 1666–1803.* Berkeley: University of California Press, 1971.

Hall, Tord. *Carl Fredrich Gauss.* Translated by A. Froderberg. Cambridge, Mass.: M.I.T. Press, 1970.

Hardy, G. H., and Wright, E. M. *An Introduction to the Theory of Numbers* 10th ed. Oxford: Oxford University Press, 1992.

Houghton, Bernard. *Scientific Periodicals.* Hamden, Conn.: Linnet Books, 1975.

Kronick, David. *A History of Scientific and Technical Periodicals.* New York: Scarecrow Press, 1962.

Lee, Elvin, and Madachy, Joseph. "The History and Discovery of Amicable Numbers—Part 1." *Journal of Recreational Mathematics* 5(1972): 77–93.

Lutzen, Jesper. *Joseph Liouville 1809–1882.* New York: Springer-Verlag, 1990.

Mahoney, Michael. "Fermat's Mathematics: Proofs and Conjectures." *Science* 178(1972): 30–36.

———. *The Mathematical Career of Pierre de Fermat (1601–1655).* 2d ed. Princeton, N.J.: Princeton University Press, 1994.

Ogilvy, C. Stanley, and Anderson, John. *Excursions in Number Theory.* New York: Oxford University Press, 1966.

Ore, Oystein. *Number Theory and Its History.* New York: McGraw-Hill, 1948. (Dover reprint 1988.)

Ornstein, Martha. *The Role of Scientific Societies in the Seventeenth Century.* Repr. ed. London: Archon Books, 1963.

Plackett, R. L. "Studies in the History of Probability and Statistics XXIV. The Discovery of the Method of Least Squares." *Biometrika* 59(1972): 239–251.

Ribenboim, Paulo. *13 Lectures on Fermat's Last Theorem.* New York: Springer-Verlag, 1979.

Schaaf, William. *Carl Friedrich Gauss, Prince of Mathematicians.* New York: Franklin Watts, 1964.

Shields, Allen. "Lagrange and the Mécanique Analytique." *Mathematical Intelligencer* 10 no. 4(1988): 7–10.

Shoemaker, Richard. *Perfect Numbers.* Washington: National Council of Teachers of Mathematics, 1973.

Stephen, Sister Marie. "Monsieur Fermat." *Mathematics Teacher* 53(1960): 192–195.

Stigler, Stephen. "An Attack on Gauss, Published by Legendre in 1820." *Historia Mathematica* 4(1977): 31–35.

Uhler, Horace. "A Brief History of the Investigations on Mersenne Numbers and the Largest Immense Primes." *Scripta Mathematica* 18(1952): 122–131.

CHAPTER 11

Non-Euclidean Geometry: Bolyai and Lobachevsky

Geometry in every proposition speaks a language which experience never dares to utter; and indeed of which she but halfway comprehends the meaning.

WILLIAM WHEWELL

11.1   Attempts to Prove the Parallel Postulate

In writing the *Elements,* Euclid rested his imposing structure on certain statements, grounded in physical experience, which he regarded as self-evident truths. Distinct among these postulates is one whose tone differs greatly from the terseness and simple comprehensibility of the others. This is the famous parallel postulate, which Euclid phrased as, "If a straight line falling on two straight lines makes the interior angles on the same side less than two right angles, the two straight lines, if produced indefinitely, meet on that side on which are the angles less than two right angles." Not only did this statement lack the quality of self-evidence; it was actually open to doubt. The ancient Greeks were aware that certain curves may approach nearer and nearer and yet not meet, as the hyperbola approaches but does not meet its asymptotes. They saw no reason why the two straight lines in the parallel postulate might not exhibit the same behavior. Evidently Euclid himself did not quite trust the postulate, for he postponed using it in his proofs until he had reached Proposition 29 of Book I. The wish to remove this unwieldly—and suspect—thing from the set of postulates was quite natural. Starting in Euclid's own lifetime, attempts were made to change the definition of parallel lines, or to replace the troublesome postulate by a more acceptable but equivalent assertion, or to deduce it as a theorem from the other nine unquestioned axioms.

Many substitutes for Euclid's parallel postulate have been suggested through the ages. These alternatives are more intuitively appealing statements, but when they are carefully examined, they turn out to be logically equivalent to Euclid's axiom. The most frequent substitute is known as "Playfair's axiom," although it was stated as early as the fifth century by Proclus. John Playfair (1748–1819) held a chair in mathematics at the University of Edinburgh from 1785 until 1805. His *Elements of Geometry* was a detailed presentation of the first six books of Euclid, along with a supplement embracing the approximate calculation of π and some solid geometry. Finding Euclid's axiom on parallels unsatisfactory, Playfair proposed replacing it with the following statement: Through a given point, not on a given line, only one parallel can be drawn to the given line. The popularity of Playfair's *Elements of Geometry,* of which ten editions were published between 1795 and 1846, doubtless led to the appearance of his version of the postulate in most modern geometry textbooks.

The parallel postulate, an assertion about intersecting lines, and Playfair's axiom, an assertion about parallel lines, are equivalent statements. This means that they mutually imply each other (the other, unquestioned axioms holding). That the parallel postulate implies its modern counterpart is evident from our earlier study of Euclidean geometry. Thus it remains for us to derive the parallel postulate, accepting as data the remaining nine axioms of Euclid's geometry and Playfair's axiom.

THEOREM
Playfair's axiom implies the parallel postulate.

Proof. Let the lines l and l' be cut by a transversal t in points P and Q, forming a pair of interior angles $\angle 1$ and $\angle 2$ on one side of the transversal. Assume that the sum of $\angle 1$ and $\angle 2$ is less than two right angles.

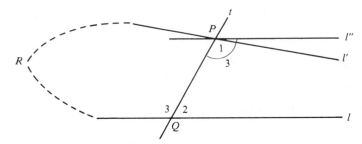

If $\angle 3$ is the supplement of $\angle 2$ on the side of t opposite $\angle 1$ and $\angle 2$, then

$$\angle 3 + \angle 2 = 180° > \angle 1 + \angle 2,$$

which implies that $\angle 3 > \angle 1$. Through P, construct a line l'' making with t an angle equal to and alternately interior to $\angle 3$. (Proposition 23 of Euclid I permits this.) Then by Proposition 27, l'' is parallel to l. Because $\angle 3 > \angle 1$, we have l' and l'' as distinct lines. From Playfair's axiom, which tells us that only one line can be drawn through P parallel to l, we conclude that l and l' must meet at some point R. If R were on the side of t opposite $\angle 1$ and $\angle 2$, then $\angle 1$ would be an exterior angle of triangle PQR. The exterior angle theorem (Proposition 16) would then guarantee that $\angle 1$ is greater than $\angle 3$, which is impossible. Thus l and l' intersect on the side of t containing $\angle 1$ and $\angle 2$, proving the parallel postulate. ∎

Mathematicians from the Greeks onward have compiled a long list of statements equivalent to the parallel postulate. Among them are:

- A line that intersects one of two parallel lines intersects the other also.
- There exist lines that are everywhere equidistant from one another.
- The sum of the angles of a triangle is equal to two right angles.
- For any triangle, there exists a similar noncongruent triangle.
- Any two parallel lines have a common perpendicular.
- There exists a circle passing through any three noncollinear points.
- Two lines parallel to the same line are parallel to each other.

Those geometers who were not content to accept Euclid's postulate on parallels did not wish merely to know whether some simpler equivalent assumption could be substituted for it (although this question is not without interest). Their real aim was to learn whether Euclid's form of the postulate was deducible from the other 9 axioms and the first 28 propositions of the *Elements*, which did not depend on the parallel postulate.

One of the earliest efforts to prove the parallel postulate was made by Proclus (410–485) in his *Commentary on the First Book of Euclid's Elements*. After remarking that ''others before us have classed it among the theorems and demanded a proof of this which was taken as a postulate by the author of the *Elements*,'' he went on to point out a fallacy contained in a demonstration offered by the noted astronomer Ptolemy, and then to submit a ''proof'' of his own. Proclus derived what we call Playfair's axiom, hence by implication, the parallel postulate. In substance, the argument ran as follows.

Let l be a given line, with P any point not on it. Assume that Q is the foot of the perpendicular from P to l and let l' be the line perpendicular to PQ at P. The lines l and l', forming equal alternate interior angles with PQ, are parallel. Thus it suffices to show that any other line l'' through P meets l. Pick any point R on l'' in the region between l and l', and take S to be the foot of the perpendicular from R to l'. As the

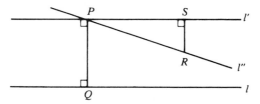

point R moves along l'', away from P, the length of the segment RS increases. Eventually, this length becomes greater than the distance between the parallels, namely, the length of the segment PQ. Then R will cross over to the other side of l, so that l'' cuts l. This shows that l' is the only line through P parallel to l.

It is easy enough to pick out the flaw in Proclus's reasoning. He made free with the assumption that two parallel lines are everywhere the same distance apart. (His argument would suffice if one merely granted that the perpendicular distance between them is bounded.) Yet there is no justification in Postulates 1–4 for this. Indeed, the supposition that parallel lines are a constant distance from one another can be shown to imply the parallel postulate, so that Proclus was tacitly assuming a property of parallel lines that is equivalent to Postulate 5. Despite his ambitious efforts, he succeeded only in begging the question.

All the numerous and varied attempts to prove the postulate on parallels failed, although for a time some mathematicians thought they had succeeded. Like Proclus, many failed because their arguments were marred by the use, open or hidden, conscious or unconscious, of some assumption equivalent to the fifth postulate. Some failed because their reasoning was otherwise fallacious. Some of these alleged "proofs," such as that presented by John Wallis (1616–1703), had a certain seductive appeal. While lecturing at Oxford University in 1663, Wallis proposed replacing the parallel postulate by a new axiom he felt was more plausible and should therefore be given precedence. He suggested: To each triangle, there exists a similar triangle of arbitrary magnitude. Using this assumption and the other axioms of Euclid, Wallis was able to demonstrate that Playfair's axiom holds, hence so does its equivalent, the parallel postulate. He proceeded thus:

Given a point P not on the line l, construct a line l' through P parallel to l in the familiar way, that is, by dropping a perpendicular PQ to l and erecting a perpendicular l' to PQ at P. The crux of the argument is to show that if l'' is any other line through P, then l'' necessarily meets l. Pursuing this aim, pick R to be any point on l'' in the region between l and l' and drop a perpendicular RS to PQ.

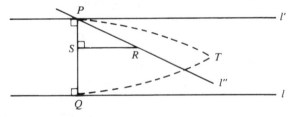

Because we are taking Wallis's axiom as part of our hypothesis, there exists a triangle PQT similar to triangle PSR and such that T is on the same side of PQ as R is. Then, in compliance with the definition of similar triangles, $\angle TPQ = \angle RPS$. The implication is that since these angles have the segments PS and PQ as a common side, PR and PT coincide. This puts the point T on the line l''. Again, by similarity, $\angle PQT = \angle PSR = 90°$, which makes TQ a perpendicular to PQ at Q. But l is the unique perpendicular to PQ at Q, so that T lies on l. Thus l and l'' meet at the point T, allowing us to conclude that the only line through P parallel to l is l'.

Actually Wallis fared no better than Proclus. Although the above "proof" seems perfectly reasonable, it rests on the assumption of the existence of two similar but noncongruent triangles. This turns out to be another of the equivalents of the parallel postulate. Like the many others who tried to demonstrate "the enemy of geometers," Wallis was guilty of the circular reasoning of assuming what he set out to prove.

This is not the place to review the entire history of the vain efforts to establish the truth of Euclid's fifth postulate as a matter of proof; it is enough to note that all inevitably failed. Although the results of these investigations were on the whole negative, some writers, such as Saccheri, Lambert, and Legendre, made important contributions to what we now call non-Euclidean geometry, though each was unaware of the true meaning at the time. The work of Girolamo Saccheri (1667–1733), a professor at the University of Pavia, deserves attention, because he seems to have been the first to study the logical consequences of an actual denial of the famous postulate. Saccheri was a Jesuit priest who taught in a succession of colleges of his order in Italy. He was considered a brilliant teacher and a man of such remarkable memory, we are told, that he could play three games of chess at a time without seeing any of the boards. Before going to Pavia in 1697 to occupy the chair in mathematics, Saccheri taught philosophy for three years in Turin. The result of this experience was the publication of a work on logic, *Logica Demonstrativa* (1697), concerning the compatibility of definitions and postulates. As a logician, Saccheri became impressed with the deductive power of *reductio ad absurdum*, or the indirect method of reasoning, used early in the *Elements*. According to this method, one shows that if the desired conclusion were not true, then a contradiction (absurdity) would follow. The question of Euclid's fifth postulate interested Saccheri throughout his lifetime and it is natural that in searching for material to which his logical principles could be applied, he would turn to the problem of parallels.

In the year of his death, Saccheri published a little treatise of some 101 pages, bearing the intriguing title *Euclides ab omni naevo vindicatus,* commonly translated *Euclid Vindicated of Every Blemish.* This was an attempt to clear Euclid of the criticisms arising from the doubtful status of the parallel postulate; and the title was perhaps intended in response to Sir Henry Savile, who in his *Praelectiones on Euclid* called the theory of parallels a blemish (*naevus*) on geometry. Saccheri's novel procedure for vindicating Euclid was to assume the parallel axiom to be false and to develop the resulting consequences, hoping to reach a contradiction; this would entitle him to affirm the parallel postulate by reductio ad absurdum. Denying the parallel postulate gave Saccheri a new premise for his reasonings and led him to expect more success than his predecessors had found.

The fundamental figure of Saccheri's investigations was a quadrilateral *ABCD* in which the sides *AD* and *BC* were equal and perpendicular to the base *AB*. These quadrilaterals have subsequently become known as Saccheri quadrilaterals. Without using the parallel postulate or any of its consequences, Saccheri was able to prove that the angles at *C* and *D,* the summit angles of the quadrilateral, were equal. For triangles *BAD* and *ABC* are congruent by the side-angle-side proposition (Proposition 4 of Euclid I), hence *BD* = *AC*. But this in its turn makes triangles *ADC* and *BCD* congruent by

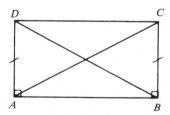

the side-side-side proposition (Proposition 8 of Euclid I). As corresponding parts of congruent triangles, it follows that $\angle C = \angle D$. Now there are three possible alternatives, giving rise to three hypotheses:

1. $\angle C = \angle D > 90°$ (hypothesis of the obtuse angle).
2. $\angle C = \angle D < 90°$ (hypothesis of the acute angle).
3. $\angle C = \angle D = 90°$ (hypothesis of the right angle).

Euclid's parallel postulate is equivalent to the third of these possibilities, so that to assume that the summit angles are either obtuse or acute is an implicit denial of the parallel axiom. This is exactly what Saccheri's proof by reductio ad absurdum demanded, to show that the hypothesis of the obtuse angle and the hypothesis of the acute angle both led to contradictions. The hypothesis of the right angle, and in consequence the parallel postulate, should then hold.

Saccheri proved that if any one of the three hypotheses were true for one of his quadrilaterals, then it would be true for every such quadrilateral. He went on to show that according to the hypothesis of the obtuse angle, the hypothesis of the acute angle, or the hypothesis of the right angle held, the sum of the angles of a triangle would be, respectively, greater than, less than, or equal to two right angles. The reasoning was as follows. Given a triangle ABC, let D and E be the midpoints of sides AC and BC, respectively. Drop perpendiculars AF, BG, and CH from the vertices A, B, and C to the line through D and E. The triangles AFD and CHD are congruent; and also the

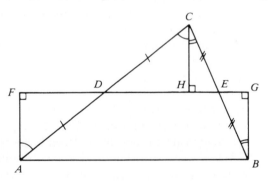

triangles BGE and CHE are congruent by the side-angle-angle proposition (Proposition 26 of Euclid I). This means that $AF = CH = BG$, making $ABGF$ a Saccheri quadrilateral with right angles at F and G. Then

$$\angle FAB = \angle FAD + \angle DAB = \angle HCD + \angle DAB$$

and

$$\angle GBA = \angle GBE + \angle EBA = \angle HCE + \angle EBA.$$

The result of adding these equalities is that

$$\angle FAB + \angle GBA = \angle DAB + (\angle HCD + \angle HCE) + \angle EBA$$

$$= \angle BAC + \angle ACB + \angle ABC,$$

or in words, the sum of the summit angles of the Saccheri quadrilateral *ABGF* equals the sum of the angles of the triangle *ABC*. Knowing this, one can easily see that if Saccheri's hypothesis of the obtuse angle holds (so that in particular, each of ∠*FAB* and ∠*GBA* is obtuse), then the sum of the angles of any triangle is greater than two right angles. If the hypothesis of the acute angle holds, the same sum is less than two right angles. If the hypothesis of the right angle holds, this sum equals two right angles.

Saccheri had no trouble in eliminating the hypothesis of the obtuse angle. After a carefully reasoned chain of thirteen propositions, he found that the hypothesis of the obtuse angle implied the parallel postulate, which in its turn implied that the sum of the angles of a triangle would have to equal two right angles. But this second implication contradicts the just-demonstrated result that on the supposition of the hypothesis of the obtuse angle, the sum of the angles of a triangle is greater than two right angles. In Saccheri's own colorful words from *Euclides Vindicatus:* ''The Hypothesis of the Obtuse Angle is absolutely false because it destroys itself.'' We might remark that the unconscious assumption that a straight line is infinite played a part in reaching this conclusion, since Euclid's Proposition 16 (the exterior angle theorem) was used.

The task of disposing of the ''hostile Acute Angle Hypothesis,'' as Saccheri called it, was much more elusive. Though he proved one theorem after another, he never found the sought-after contradiction. Without realizing it, Saccheri was on the threshold of discovering the first non-Euclidean geometry. Although the theorems he was obtaining seemed at variance with experience, they formed a geometry that in internal coherence was not inferior to Euclid's. But Saccheri was so convinced by what he was trying to do that he permitted his faith in Euclidean geometry to interfere with his logical perseverance. When no contradiction presented itself, the devoted geometer forced his mathematical development into an unsatisfactory ending by treating a point at infinity (a limit point of the plane) as if it were a point of the plane. Lamely, he concluded that two distinct lines that meet at an infinitely distant point can both be perpendicular at that point to the same straight line. Saccheri viewed this as a contradiction to Proposition 12 of Euclid I, according to which there is a unique perpendicular to a line at each point of the line. Triumphantly he announced, ''The Hypothesis of the Acute Angle is absolutely false, being repugnant to the nature of a straight line.'' Saccheri was satisfied that he had finally removed the uncertainty surrounding Euclid's axiom system, declaring at the close of his work, ''The foregoing considerations seem to me sufficient to clear Euclid of the faults with which he has been reproached.'' But a violation of intuitive ideas does not make a logical contradiction, and the problem of the necessity of the parallel postulate remained.

Had Saccheri unflinchingly admitted that his reasoning led to no contradiction but merely results that seemed paradoxical in that they clashed sharply with the familiar geometry, he would have anticipated the discovery of non-Euclidean geometry by at least one hundred years. As it was, his treatise was not read widely enough to have much influence and was virtually forgotten until the Italian mathematician Eugenio Beltrami rescued it from oblivion in 1889. Beltrami hailed Saccheri as the ''Italian precursor of the Hungarian Bolyai and the Russian Lobachevsky,'' the two individuals who are usually credited with the concept of a logically consistent geometry based on a denial of the parallel postulate. But the final discovery of non-Euclidean geometry did not depend on the pioneering work of Saccheri, because neither Bolyai nor Lobachevsky seemed to have heard of the book or its author.

EUCLIDES

AB OMNI NÆVO VINDICATUS:

SIVE

CONATUS GEOMETRICUS

QUO STABILIUNTUR

Prima ipſa univerſæ Geometriæ Principia.

AUCTORE

HIERONYMO SACCHERIO

SOCIETATIS JESU

In Ticinenſi Univerſitate Matheſeos Profeſſore.

OPUSCULUM

EX.ᴹᴼ SENATUI

MEDIOLANENSI

Ab Auctore Dicatum.

MEDIOLANI, MDCCXXXIII.

Ex Typographia Pauli Antonii Montani.   *Superiorum permiſſu*

Title page of Saccheri's Euclides ab omni naevo vindicatus *(1733). (Reprinted by permission of Open Court Publishing Company, La Salle, Illinois, from* Euclides Vindicatus *by Girolamo Saccheri, edited and translated by G. B. Halsted.)*

Indeed, had Saccheri actually accomplished his purpose and proved the parallel postulate from the remaining axioms of Euclidean geometry, he would not have vindicated Euclid. Quite to the contrary, he would have dealt a terrible blow to Euclid. Euclid was vindicated by the discovery of non-Euclidean geometry, for its existence demonstrates that the parallel postulate is independent of Euclid's other axioms, so that it truly widens the axiomatic base on which Euclid's geometry stands. We must admire the Great Geometer all the more; the introduction of the fifth postulate, so decidedly unaxiomatic in appearance, yet an independent postulate, was a stroke of pure genius.

There is a notable resemblance between Saccheri's *Euclides Vindicatus* and the *Theorie der Parallellinien* (written in 1766) of the German mathematician Johann Heinrich Lambert (1728–1777), a colleague of Euler and Lagrange at the Berlin

Academy of Sciences. For the fundamental figure of his investigations, Lambert adopted a quadrilateral having three right angles and then examined the hypotheses in which the remaining angle was in turn obtuse, acute, and right. Lambert disposed of the obtuse-angle hypothesis in much the manner of Saccheri by establishing the contradictory result that the parallel postulate could be proved as a theorem from it. He too derived many non-Euclidean facts from the acute angle hypothesis, but unlike Saccheri, he was well aware that he had reached no contradiction. Lambert showed what no one had noticed before—that in this new geometry the angle sum of a triangle increases when the area decreases. (By the angle sum of a triangle, we mean the sum of its three interior angles.) He did this by proving that the area of a triangle is proportional to its angular defect, that is, the amount by which its angle sum falls short of two right angles. In modern symbols, for a triangle with angles a, b, c, he demonstrated that the area $A = k^2[180° - (a + b + c)]$, where k is a constant of proportionality.

Although the range of Lambert's research interests was enormous, he is remembered today mainly for having given the first rigorous proof that the number π is irrational. In a paper presented to the Berlin Academy in 1768, he showed that if x is a nonzero rational number, then neither e^x nor $\tan x$ can be rational (earlier, in 1737, Euler had established that e and e^2 are irrational). Because $\tan \pi/4 = 1$, a rational number, it can be inferred that $\pi/4$, hence π itself, is irrational. Lambert also suggested that e and π are transcendental numbers; but proof of this would have to wait a hundred years.

Perhaps the most tireless pursuer of a proof of the parallel postulate was the third member of the ''great trio'' Lagrange, Laplace, and Legendre. Adrien-Marie Legendre (1752–1833) came from a well-to-do family in the south of France but spent the greater part of his life in Paris. He was educated there at the College Mazarin and from 1775 until 1780 was a professor at the Ecole Militaire, where Laplace was also teaching. Two years after resigning his position to reserve more time for research, Legendre won a prize offered by the Berlin Academy with an essay on the path of a projectile in a resisting medium (*Recherches sur la trajectoire des projectiles dans les milieux résistants*). This brought him to the attention of the scientific community and secured his appointment to the Académie des Sciences, first as an adjunct member in 1783 and then as an associate member in 1785. Following the closing of the Académie in 1793, Legendre was not invited to cooperate in the reorganization of public education. This may have been because he was not in favor with the revolutionary government or because of some other reason. His name does not figure among those of the professors at either the Ecole Normale or the Ecole Polytechnique. Nor was he on the list of 48 scholars whom the government selected (1795) to form the nucleus of the new Institut National des Sciences et des Arts as Lagrange and Laplace were. His colleagues redressed this oversight by electing him a resident member of the mathematics section. During this period of turmoil in France, Legendre was, however, made a member of several public commissions. He served, in particular, as one of the three commissioners who were to oversee the triangulations necessary for determining the standard meter. (This new unit of measure was originally intended to represent the 10-millionth part of the distance from the North Pole to the equator, calculated from the measured length of the meridian arc between Dunkirk and Barcelona.) He later was appointed to succeed

Laplace as examiner in mathematics of the graduates of the Ecole Polytechnique destined for the artillery. It is said that Legendre was not so well regarded by his countrymen as Lagrange and Laplace were, for, according to Lao Simons, "He sought recognition which apparently came unsought to both Lagrange and Laplace." Whatever the faults of his personality, Legendre was a man of integrity who always spoke his mind, even when his own interests were adversely affected. Refusing to vote in favor of the government's candidate for the Institut National, he was deprived (1824) of his pension and died in poverty.

Legendre's achievements in mathematics, although they were considerable, did not approach the achievements of his two compeers, Lagrange and Laplace. That his work was nonetheless held in high regard is indicated by a comment of the secretary of the French Academy, Jean-Baptiste Elie de Beaumont, who wrote, "Laplace . . . has earned a right to be styled the Newton of France; Legendre, more profound than popular, was our Euler." Like Euler, Legendre had interests covering the breadth of mathematics. He seems to have been most pleased with his research in number theory, celestial mechanics, and the theory of elliptic functions, because although he took up other problems in the course of his lifetime, he always returned to these.

In 1785, Legendre read a memoir to the Académie with the title *Recherches d'analyse indetermine*. This contained the celebrated quadratic reciprocity law, which is concerned with the solvability of the pair of quadratic congruences $x^2 \equiv q \pmod{p}$ and $x^2 \equiv p \pmod{q}$, where p and q are distinct odd primes. The following holds.

Either both congruences $x^2 \equiv q \pmod{p}$ and $x^2 \equiv p \pmod{q}$ are solvable or both unsolvable; except in the case in which p and q are each of the form $4n + 3$, in which event one of the congruences is solvable whereas the other is not.

Legendre tried a long and imperfect demonstration of this, but slipped in assuming an obvious theorem that was as difficult to prove as the law itself. Undaunted, he tried another proof in his *Essai sur la théorie des nombres* (1798). This too contained a gap, because Legendre took for granted that there were an infinite number of primes in certain arithmetic progressions. At 18, Gauss (in 1795) had independently discovered this reciprocity law, and after a year's unremitting labor, he obtained the first complete proof. "It tortured me," said Gauss, "for a whole year and eluded my most strenuous efforts before, finally, I got the proof explained in the fourth section of the *Disquisitiones Arithmeticae*." In the *Disquisitiones Arithmeticae* (published in 1801, though finished in 1798) Gauss attributed the quadratic reciprocity law to himself. Two pages after its proof, he vaguely alluded to Legendre's contribution, saying, "Legendre in his excellent tract in *Mem. Acad. des Sci.* 1785 arrived at a theorem which is basically the same as the fundamental theorem." If Legendre was nettled by this passing reference, it would not be the last time that he would feel that Gauss had given him insufficient recognition. His *Essai sur la théorie des nombres,* which represented the first "modern" treatise devoted exclusively to number theory, passed through two editions, one appearing in 1798 and the other in 1808. In the second edition, Legendre adopted the proof of the quadratic reciprocity law given by the young Gauss. Legendre's *Essai* was later expanded into his *Théorie des nombres*. The two-volume third edition (1830), together with Gauss's *Disquisitiones Arithmeticae,* became the dominant work on the subject for the remainder of the century.

Although the primes occur among the positive integers in a most irregular manner, there are ways in which their overall distribution seems quite regular. By inspecting tables of prime numbers, both Legendre and Gauss—the latter still in his early teens— hoped to find a simple function whose values approximated those of $\pi(x)$, the function which gives the number of primes less than or equal to x. Legendre, in his *Essai sur la théorie des nombres,* stated that the function

$$\frac{x}{\log x - 1.08366}$$

agrees very well with $\pi(x)$ as long as x is not greater than 1 million. Although Gauss discovered that the integral

$$\int_2^x \frac{dt}{\log t}$$

produces a better approximation than Legendre's guess, he never published the results of his investigation. The first major progress concerning $\pi(x)$ is attributed to the Russian mathematician P. L. Chebyshev (1821–1894). Around 1850, he showed that the inequalities

$$(0.92)\,\frac{x}{\log x} < \pi(x) < (1.11)\,\frac{x}{\log x}$$

are valid for all x sufficiently large; and if the limit

$$\lim_{x \to \infty} \frac{\pi(x)}{x/\log x}$$

exists, then its value must be 1. However, the existence of the limit turned out to be very difficult to prove. It was finally established in 1896, a hundred years after Legendre's and Gauss's conjectures, by means of new and powerful methods primarily from analysis. This celebrated achievement has since become known as the prime number theorem.

The usual procedure in astronomy is for different observers, using different instruments, to make numerous observations at a variety of locations. Because the results of these observations are subject to errors arising from the reactions of the observers or the precision of their instruments, determining the most probable value of an observed quantity becomes a question. The method of least squares is so called because the sum of the squares of the differences between the observed values and the true value of the observed quantity must be minimized. This method was discovered independently, and almost simultaneously, by Legendre and Gauss. Gauss had been using the basic idea since 1794 or 1795, when he was a student at Caroline College in Brunswick preparing for his university studies. But the first explicit account of the method was published by Legendre in 1805 in his paper *Nouvelles méthodes pour la détermination des orbites des comètes.* He gave it the name *méthode des moindres quarrés,* showed that the rule of the arithmetic mean is a particular case of the general principle, and presented examples of its application to the determination of the orbit of a comet. He offered no probabilistic justification, however, that the results thus obtained were the "best" or the most plausible results the observations were capable of affording. It remained for Gauss in his famous astronomical text *Theoria Motus Corporum Coelestium* (Theory of the Motion of the Heavenly Bodies) of 1809 to derive the law of probability of error that is the basis for the method of least squares. This was the work in which Gauss explained his solution of one of the most interesting and difficult problems in astronomy, finding the orbit of the lost planet Ceres. At one point he stated, "Our principle, which we have made use of since the year 1795, has lately been published by Legendre. . . ." Legendre was disturbed by Gauss's use of the phrase "our principle" and wrote to him in censure:

> There is no discovery that one cannot claim for himself by saying that one had found the same thing some years previously; but if one does not supply the evidence by citing the place where one has published it, this assertion becomes pointless and serves only to do a disservice to the true author of the discovery.

It was not that Gauss treated casually what others did, only that he considered a result to be his if he gave the first rigorous demonstration of it. Legendre, on the other hand, felt that publication established priority, even if an argument that was merely plausible took the place of a complete proof. All discussion of priority rights between the two was futile. Because each clung to the correctness of his position, neither took heed of the other. Gauss could only express regret about the competition between them. Writing to Heinrich Olbers, his associate and a noted astronomer, Gauss was led to remark:

> It seems to be my fate to compete with Legendre in almost all my theoretical works. So it is in the higher arithmetic, in the researches on transcendental functions connected with the rectification of the ellipse, in the fundamentals of geometry, and now here again.

A large portion of Legendre's research was devoted to elliptic functions. (The name *elliptic* arises because these functions exist in the expression for finding the length of an arc of an ellipse.) A systematic account of the theory of elliptic functions is contained in Legendre's three-volume *Exercises du calcul intégral* (1811, 1817, 1819), a great and rich textbook on the integral calculus that rivalled Euler's *Institutiones Calculi Integralis* in its comprehensiveness. The third volume includes long tables of elliptic integrals calculated by Legendre himself at immense labor. This material was further developed in another three-volume treatise, *Traité des fonctions elliptiques* (1825, 1826, 1830), the most significant of Legendre's works in higher mathematics. The third volume, which although composed in 1830 did not appear until a few weeks before his death, presents an account of Abel's contemporaneous research and also Jacobi's in the same area. Notwithstanding the disparity in their ages, Legendre had developed an active correspondence with Carl Gustav Jacobi (1804–1851) about elliptic functions. On hearing that many of Jacobi's results had been obtained independently by Gauss but never published, Legendre wrote (1827) to the young man to share his indignation:

> How can M. Gauss have dared to tell you that the greater part of your theorems on elliptic functions were known to him and that he discovered them as early as 1808? . . . This extreme impertinence is incredible on the part of a man who has sufficient personal merit to have no need of appropriating the discoveries of others. . . . But this is the same man

who, in 1801, wished to attribute to himself the discovery of the law of quadratic reciprocity published in 1785 and who wanted to appropriate in 1809 the method of least squares published in 1805.

Apparently, the passage of time did not soften Legendre's sense of bitterness over having to share his honors with Gauss.

The work of Legendre that mainly concerns us is his *Eléments de géométrie* (1794). As the teaching of geometry began to be stressed in the eighteenth-century universities, many geometry textbooks were published, some introducing novel ideas. Legendre's *Eléments de géométrie* presents an interesting twofold contrast when compared with contemporary English and French geometries. In the universities of Great Britain, Euclid met little competition, being practically the only geometry textbook used. To English authors of versions of the *Elements,* the order of the propositions in Euclid was absolutely essential to rigorous demonstration. Rather than depart from the original, they spent their efforts purging the text of faults that had crept in over time. The French, on the other hand, maintained a critical attitude toward Euclid as a textbook for beginners. Many elementary geometries appeared in France in the 1700s in which rigor and formalism were sacrificed for the sake of a new sequence of propositions better fitted to the needs of mathematical novices. These endeavors to make geometry palatable were frequently criticized for lightening work for the examiner as well as for the student. In a move away from the loose, intuitive presentations, Legendre's *Eléments de géométrie* undertook to revive a taste for rigorous demonstration in France. Legendre simplified and rearranged many of the proofs of Euclid's *Elements,* but on the whole approached the severity of the ancient treatment more closely than his predecessors did. The clarity of Legendre's exposition and the attractiveness of his style made the *Eléments de géométrie* one of the most successful textbooks ever written. Twenty editions of this book, comprising some 100,000 copies, appeared in France alone before the author's death in 1833. Perhaps the only complaint was that Legendre, prepossessed by the methods of Euclid, relied excessively on reductio ad absurdum, a method that "convinces but does not satisfy the mind."

It was natural that shortly after the War of Independence began there should be a distinct lessening of British influence and a corresponding increase in French influence on the scientific life of those colonies that were to become the United States. When the American Academy of Arts and Sciences was formed in Boston in 1780 at the suggestion of John Adams, its founders publicly stated that it was their intention "to give it the air of France rather than that of England and to follow the Royal Academy rather than the Royal Society." Little wonder that in mathematics, as in the physical sciences, translations of French textbooks came to dominate teaching in the colleges of the United States. Legendre's name became known to a great army of students, young people who had never heard of Lagrange or Laplace, when the *Eléments de géométrie* completely supplanted Euclid in American schools.

We shall not speak here at any great length about geometry in the educational scheme of the United States, but a few words might prove helpful. Harvard, which was the only North American college in the 1600s, was founded in 1636. In its early days, geometry played a minor role in Harvard's curriculum, taught only one day a week in the last year of college. As soon as the subject began to receive serious attention, Euclid's *Elements* became the accepted text. The earliest known use of Euclid on the

Adrien Marie Legendre
(1752–1833)

(From A Concise History of Mathematics *by Dirk Struik, 1967, Dover Publications, Inc., N.Y.)*

college level was at Yale (opened in 1701) in 1733, and then at Harvard in 1737. In the latter part of the 1700s, when the average age of freshmen in the American colleges was gradually increasing, geometry constantly crept downward until it was finally taught as a first-year subject. According to a graduate of the Harvard class of 1798, ''The sophomore year gave us Euclid to measure our strength.'' By 1818, geometry had been placed in the first year at Harvard, and in 1844, it was made a mathematical requirement for entrance to the college, though only the preliminary notions were required.

The first translation into English of Legendre's *Eléments de géométrie* was made in 1819 by John Farrar, a professor of mathematics at Harvard from 1807 to 1836; it appeared in a total of ten editions, the last in 1841. The next translation of Legendre was brought out in 1822 by the famous Scottish essayist and historian Thomas Carlyle, who in his early life was a teacher of mathematics. Carlyle's translation—with Charles Davies, a professor at West Point, as editor—was the most popular textbook in the United States during the nineteenth century. It ran through about thirty American editions, the last not appearing until 1890. Indeed, ''Davies-Legendre'' was adopted at Yale as late as 1885, when Euclid was finally discarded as a text. American teachers of the first half of the nineteenth century seemed willing to turn to the elementary textbooks of France, not only for geometrical instruction, but for a variety of subjects. The French influence was so powerfully felt that a writer to an early periodical devoted to mathematics, *The Mathematical Diary,* was led to complain:

> Our elementary works on the pure and physico-mathematical sciences have savored too generally of a foreign mint. It would certainly be more creditable to us as a nation, and becoming us as an independent people, to rely less on bald translations and compilations of and from transatlantic publications, and more upon our own exertions.

But not until William Chauvenet prepared his *Treatise on Elementary Geometry* in 1870 was there any successful American rival to Legendre.

Legendre's investigations of the provability of the parallel postulate extended over 40 years and appeared mostly in appendixes of successive editions of the *Eléments de géométrie.* All his attempts to derive the postulate from the other Euclidean axioms were deficient in that each one rested on some hypothesis that was logically equivalent to the desired statement. One unsuccessful attack was impaired by the assumption that there existed similar triangles of different sizes, another by the assumption of the existence of a circle passing through any three noncollinear points. Legendre never gave up trying, and in a final monograph, *Réflexions sur différentes manières de démontrer de la théorie de parallèles,* which appeared in the year of his death, he brought together some half-dozen alleged proofs of the immortal postulate.

Legendre, much like Saccheri and Lambert before him, approached the thorny question of the parallel axiom from the side of the angle sum of a triangle. He correctly perceived that Euclid's fifth postulate was equivalent to the theorem that the angle sum of a triangle was equal to two right angles. Here he argued as follows.

THEOREM

If the sum of the angles in a triangle is equal to two right angles, then Euclid's parallel postulate holds.

Proof. We need only show how to obtain Playfair's axiom from the hypothesis. Starting with a point P not on a given line l, construct a parallel to l through P by first dropping a perpendicular PQ to l and then taking l' perpendicular to PQ at P. As a step to establishing that l' is the only parallel to l through P, choose points $Q_0 = Q, Q_1, Q_2, \ldots, Q_n, \ldots$ on l, all on the same side of PQ, such that

$$Q_0Q_1 = QP, \qquad\qquad Q_1Q_2 = Q_1P,$$

$$Q_2Q_3 = Q_2P, \ldots, Q_nQ_{n+1} = Q_nP, \ldots .$$

This produces a sequence of isosceles triangles,

$$\Delta Q_0PQ_1, \quad \Delta Q_1PQ_2, \quad \Delta Q_2PQ_3, \ldots, \Delta Q_nPQ_{n+1}, \ldots .$$

To simplify notation, set $\alpha_n = \angle Q_{n-1}PQ_n = \angle Q_{n-1}Q_nP$ for $n = 1, 2, 3, \ldots .$

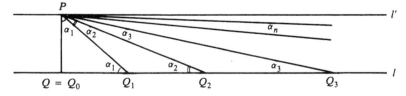

Because $\angle Q_{n-1}Q_nP$ and $\angle PQ_nQ_{n+1}$ are supplementary angles for all n, we have

$$\alpha_n + \angle PQ_nQ_{n+1} = 180°, \qquad n = 1, 2, 3, \ldots .$$

The assumption that the angle sum of all triangles is 180° also tells us, when it is applied to triangle $Q_nPQ_{n+1,}$ that

$$\angle PQ_nQ_{n+1} + 2\alpha_{n+1} = 180°, \qquad n = 1, 2, 3, \ldots$$

When we tie together these last two equalities, we see this situation:

$$\alpha_n = 180° - \angle PQ_nQ_{n+1} = 2\alpha_{n+1}, \qquad n = 1,2,3, \ldots ,$$

whence

$$\alpha_2 = \frac{\alpha_1}{2}, \qquad \alpha_3 = \frac{\alpha_2}{2} = \frac{\alpha_1}{2^2}, \ldots , \alpha_{n+1} = \frac{\alpha_n}{2} = \frac{\alpha_1}{2^n}, \ldots .$$

Now let l'' be any line through the point P, and distinct from l'. If α is the angle between l'' and l', then for n sufficiently large, $\alpha_1/2^n$ can be made less than α; that is, $\alpha_{n+1} < \alpha$.

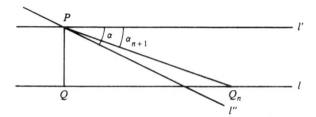

This means that l'' will lie in the interior of $\angle QPQ_n$, hence will intersect the side QQ_n of triangle PQQ_n. Because l'' must cut line l, l' is the only line through P parallel to l, which gives us the conclusion we wanted. ∎

Keeping this last theorem in view, Legendre tried to show that if one assumed the angle sum of a triangle to be either greater than 180° or less than 180°, then a contradiction would be reached. Therefore, the sum of the angles of a triangle would necessarily equal 180°, from which the truth of the parallel postulate would follow. Legendre's ''proof'' was widely circulated because of its appearance in the twelfth and successive editions of the *Eléments de géométrie*. It began with Legendre's replacing a triangle *ABC* with another triangle in which one angle was at most half as large as *A*. (The result is more profound than it at first seems, for in the absence of the parallel postulate there is no reason to think that all triangles have the same angle sum.) Because the forthcoming lemma has something of the format of Euclid's Proposition 16, it is not surprising that Legendre, like all his predecessors, took for granted that straight lines were infinite. This quiet assumption did not bother people at the time, but ultimately became as vital as the parallel postulate itself.

LEMMA

Let $\angle A$ be any angle of a triangle ABC. Then there exists a triangle $A_1B_1C_1$ with the same angle sum as triangle ABC, and such that $\angle A_1 \le \frac{1}{2} \angle A$.

Proof. Let D be the midpoint of side BC; extend AD to a point E in such a manner that $AD = DE$. Because $BD = CD$, and $AD = DE$, and $\angle BDA = \angle CDE$ (vertical angles are equal by Euclid's Proposition 15 of Book I), the triangles

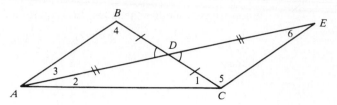

BDA and *CDE* are congruent from the side-angle-side proposition. Following the notation of the angles in the diagram, the implication is that $\angle 3 = \angle 6$ and $\angle 4 = \angle 5$. Hence, by substitution,

$$\angle 1 + (\angle 2 + \angle 3) + \angle 4 = \angle 1 + \angle 2 + \angle 6 + \angle 5.$$

Thus the triangle *ACE* has the same angle sum as triangle *ABC*. To complete the argument, note that $\angle 2 + \angle 6 = \angle 2 + \angle 3 = \angle A$. This means that either $\angle 2$ or $\angle 3$ must be less than or equal to $\frac{1}{2}\angle A$, for if both were greater than $\frac{1}{2}\angle A$, then their sum would be greater than $\angle A$. If $\angle 2 \leq \frac{1}{2}\angle A$, relabel *A* as A_1 and the other two vertices of triangle *ACE* as B_1 and C_1; for the case in which $\angle 3 \leq \frac{1}{2}\angle A$, take *E* as A_1. Then triangle $A_1B_1C_1$ satisfies the requirements of the lemma. ∎

With the aid of this lemma, Legendre had little trouble in disposing of the assumption that the sum of the angles of a triangle could be greater than two right angles. The resulting theorem is usually, but mistakenly, called Legendre's first theorem—mistakenly, because Saccheri had already established the theorem almost a century earlier when he showed that his hypothesis of the obtuse angle could not hold.

THEOREM   *The angle sum of a triangle is always less than or equal to 180°.*

Proof. Assume to the contrary that there exists a triangle *ABC* whose angle sum is $180° + \alpha$, where α is a positive number of degrees. By a direct appeal to the preceding lemma, it is possible to find a triangle $A_1B_1C_1$ with the same angle sum as triangle *ABC*, namely $180° + \alpha$, in which $\angle A_1 \leq \frac{1}{2}\angle A$. Applying the lemma to triangle $A_1B_1C_1$, we obtain a third triangle $A_2B_2C_2$, whose angle sum is also $180° + \alpha$ and one of whose angles, $\angle A_2$, is such that

$$\angle A_2 \leq \frac{1}{2}\angle A_1 \leq \frac{1}{2^2}(\angle A).$$

It is obvious that this procedure leads by mathematical induction to a sequence of triangles,

$$\Delta A_1B_1C_1, \qquad \Delta A_2B_2C_2, \qquad \Delta A_3B_3C_3, \ldots,$$

each with angle sum $180° + \alpha$ and each containing an angle, $\angle A_n$, for which

$$\angle A_n \leq \frac{1}{2^n}(\angle A).$$

By taking n sufficiently large, $(1/2^n)(\angle A)$ can be made less than α, whence $\angle A_n$ is less than α. When this is done, one sees that

$$180° + \alpha = \angle A_n + \angle B_n + \angle C_n < \alpha + \angle B_n + \angle C_n,$$

which entails that

$$180° < \angle B_n + \angle C_n.$$

But this violates the fact that the sum of any two angles of a triangle is less than two right angles, or 180° (Proposition 17 of Euclid I). Having obtained the required contradiction, one can conclude that the angle sum of a triangle is not greater than 180°, and so all is proved. ■

Having correctly established that the sum of the angles in a triangle cannot be more than two right angles, Legendre next tried to rule out the possibility that this sum is less than two right angles. In the proof we are quoting here, he reasoned as follows. Assume that the angle sum of triangle ABC is less than 180°, say 180° − α, where α is a positive number of degrees, and that $\angle A$ is the smallest angle of the triangle. On side BC, construct a triangle BCD congruent with triangle ABC, by drawing $\angle CBD$ equal to $\angle ACB$ and taking $BD = AC$. Through D draw a line that meets the extensions of sides AB and AC at points E and F, respectively, thereby forming new triangles BED and CDF.

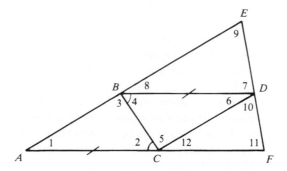

Triangles ABC and BCD, being congruent, must have the same angle sum, to wit, 180° − α:

$$\angle 1 + \angle 2 + \angle 3 = 180° - \alpha, \qquad \angle 4 + \angle 5 + \angle 6 = 180° - \alpha.$$

From Legendre's first theorem, it is also known that the angle sum of triangles BED and CDF cannot be greater than 180°:

$$\angle 7 + \angle 8 + \angle 9 \leq 180°, \qquad \angle 10 + \angle 11 + \angle 12 \leq 180°.$$

Thus, the sum of all the angles of the four triangles cannot exceed 4(180°) − 2α. Of these angles, each of the groups with vertices at B, C, and D adds up to 180°:

$$\angle 3 + \angle 4 + \angle 8 = \angle 6 + \angle 7 + \angle 10 = \angle 2 + \angle 5 + \angle 12 = 180°,$$

from which it is evident that

$$\angle 1 + \angle 9 + \angle 11 + 3(180°) \leq 4(180°) - 2\alpha$$

or

$$\angle 1 + \angle 9 + \angle 11 \leq 180° - 2\alpha.$$

Whereas the original triangle *ABC* had an angle sum of not more than $180° - \alpha$, that of the triangle *AEF* is not more than $180° - 2\alpha$. An application of the same construction to triangle *AEF* yields another triangle having an angle sum not greater than $180° - 2^2\alpha$. If the argument is repeated *n* times, a triangle is produced whose angle sum does not exceed $180° - 2^n\alpha$. For *n* sufficiently large, $2^n\alpha$ would be greater than 180°. Thus the process leads to a triangle whose angles have a negative sum—an absurd situation. The proof having come to a contradiction (said Legendre), the angle sum of a triangle cannot be less than 180°.

Legendre concluded that because the sum of the angles of a triangle could neither be greater nor less than 180°, it would have to equal 180°; if this equality were granted, the parallel postulate would follow. Unfortunately, there is a defect in the proof he proposed—the angle sum of a triangle cannot be less than 180°. To construct his sequence of triangles, he assumed that through any point in the interior of an angle it is always possible to draw a line that meets both sides of the angle. Although it is not immediately apparent, this assumption turns out to be another equivalent of the parallel postulate. Legendre's demonstration failed, although he himself never saw through the matter and thought that he had finally settled the question.

One can hardly blame Legendre for being convinced that he had cleared up the uncertainty surrounding the parallel postulate. Good and bad mathematicians alike had fallen on this slippery ground. In fact, so lengthy and so persistent were their efforts that in 1767 d'Alembert called the state of the theory of parallels ''the scandal of elementary geometry.'' The great analyst Lagrange, according to a story told by Augustus De Morgan in the *Budget of Paradoxes,* presented a paper on parallel lines to the French Academy, but broke off his reading part way through with the exclamation, ''I must meditate further on this.'' With this he put the paper in his pocket and never afterwards spoke of it publicly.

Through all of this it must be remembered that for centuries Euclidean geometry had been the most firmly established branch of mathematics, reputedly the most complete, and one whose authority was derived mainly from the clarity of its axioms. Euclid's choice of these ''self-evident truths'' had been dictated by his intuitive observation of the world about him. But because the fifth postulate involved an infinite concept and therefore could not be verified experimentally, it was seen as a shocking flaw in geometry. A recurring challenge to geometers was to discover whether this postulate could be deduced from the others. If it could, this would transform it into a theorem, not an axiom, of the system and put it beyond question. Yet by the beginning of the 1800s the problem of the necessity of the parallel postulate was not in much better shape than in antiquity. Twenty centuries of fruitless effort, and particularly the latest unsuccessful investigations of Legendre, did strike a spark of doubt in the minds of some mathematicians. The conviction had begun to grow that a proof of the parallel postulate was not possible within Euclid's own system.

11.1   Problems

1. Find the flaw in the following attempted proof of
   the parallel postulate by Wolfgang Bolyai (1775–
   1856). Given any point P not on a line l, construct
   one parallel l' to l through P in the usual way, by
   dropping a perpendicular PQ to l and erecting l'
   perpendicular to PQ. Let l'' be any line through P
   distinct from l'. To see that l'' meets l, pick a point
   A on PQ between P and Q. Extend PQ beyond Q to
   a point B so that $AQ = QB$. Now let R be the foot
   of the perpendicular from A to l'' and extend AR
   beyond R to a point C so that $AR = RC$. Then A, B,
   C are not collinear; hence there exists a unique
   circle passing through them. Because l is the
   perpendicular bisector of the chord AB of this circle
   and l'' is the perpendicular bisector of the chord
   AC, l and l'' must intersect in the center of the
   circle.

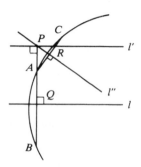

2. Find the flaw in the following attempted proof of
   the parallel postulate by A. M. Legendre (1752–
   1833). Start with a point P not on a line l, with PQ
   perpendicular to l at Q and line l' perpendicular to
   PQ at P; then l' is parallel to l. Let l'' be any line
   through P different from l'. If we assume that l'' is
   not perpendicular to l, then it makes an acute angle
   with l. Pick any point R on l'' such that $\angle QPR$ is
   acute. Construct $\angle QPR'$, with R' on the opposite
   side of PQ from R, so that $\angle QPR' = \angle QPR$. Then
   Q is inside $\angle R'PR$, and because the line l passes
   through Q, it will intersect one of the sides of
   $\angle R'PR$. If l meets the side of $\angle R'PR$ containing
   PR, then l meets l''. If l meets the side of $\angle R'PR$
   containing PR', say at A, then choose a point B on
   the opposite side with $PB = PA$. Then triangles
   PBQ and PAQ are congruent by the side-angle-side
   proposition; hence $\angle PQB = \angle PQA = 90°$. This
   puts the point B on l, so that l'' and l meet at B.

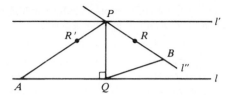

3. Prove the following sequence of results attributed
   to Legendre.

   (a) If the angle sum of a triangle is equal to two
       right angles, then the same is true of any
       triangle obtained from it by drawing a line
       through a vertex to a point on the opposite
       side.

   (b) If there exists one triangle with angle sum
       equal to two right angles, then there exists an
       isosceles right triangle whose angle sum is two
       right angles and whose legs are longer than
       any given segment. [*Hint:* Within the existing
       triangle construct an isosceles right triangle
       with angle sum equal to two right angles.
       (Begin by dropping a perpendicular from a
       vertex to a side.) Now join eight such
       congruent triangles to form a square whose
       sides are twice the length of one of the
       congruent triangles' (perpendicular) sides.
       Draw the diagonal of the square and repeat the
       process with one of the triangles you get.]

   (c) Legendre's second theorem: If there exists one
       triangle whose angle sum is equal to two right
       angles, then every triangle has an angle sum
       equal to two right angles. [*Hint:* Because any
       triangle can be divided into two right triangles,
       it suffices to consider an arbitrary right
       triangle ABC. By part (b), there exists an
       isosceles right triangle DEF whose angle sum
       is two right angles and with sides greater than
       those of ABC. Split DEF into triangles so that
       one of them, $A'B'C'$, is congruent with
       triangle ABC.]

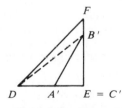

   (d) If there exists one triangle whose angle sum is
       less than two right angles, then every triangle
       has an angle sum less than two right angles.

In Problems 4–10, you may assume Euclid's first four postulates, and hence any result that is derived from them.

4. A Saccheri quadrilateral is a quadrilateral $ABCD$ in which $\angle A$ and $\angle D$ are right angles and the sides AB and DC are equal. Side AD is known as the base, the opposite side BC as the summit, and $\angle B$ and $\angle C$ as the summit angles. Verify each of the following.

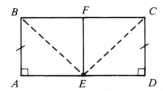

(a) A line that passes through the midpoints of the base and the summit of a Saccheri quadrilateral is perpendicular to both of them. [*Hint:* First show that $\triangle BAE$ is congruent to $\triangle CDE$, and then that $\triangle BFE$ is congruent to $\triangle CFE$.]

(b) If a line is the perpendicular bisector of the base of a Saccheri quadrilateral, then it is also the perpendicular bisector of the summit.

(c) The lines containing the base and summit of a Saccheri quadrilateral are parallel.

5. Consider a quadrilateral $ABCD$ whose base angles $\angle A$ and $\angle B$ are right angles. Prove the following assertions.

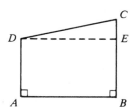

(a) If $AD < BC$, then $\angle C < \angle D$. [*Hint:* There exists a point E on BC with $BE = AD$; apply the exterior angle theorem to $\triangle DEC$.]

(b) If $\angle C < \angle D$, then $AD < BC$. [*Hint:* Suppose to the contrary that $AD = BC$ or $AD > BC$.]

(c) If $\angle C = \angle D$, then $ABCD$ is a Saccheri quadrilateral.

6. Let $ABCD$ and $XYZW$ be Saccheri quadrilaterals with bases AD and XW. Show that if $AD = XW$ and $AB = XY$, then $BC = YZ$, and $\angle B = \angle Y$, and $\angle C = \angle Z$.

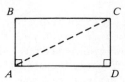 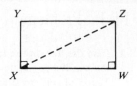

7. Fill in any missing details in the following proof that the base of a Saccheri quadrilateral is not longer than the summit. Given a Saccheri quadrilateral $A_1B_1B_2A_2$, with base A_1A_2, let l be the line containing A_1 and A_2. Pick distinct points $A_3, A_4, \ldots, A_n, A_{n+1}$ on l, appearing in this order, so that $A_kA_{k+1} = A_1A_2$ for $k = 2, 3, \ldots, n$. Also let B_k be the point on the same side of l as B_1 such that B_kA_k is perpendicular to l at A_k and $B_kA_k = B_1A_1$, where $k = 3, 4, \ldots, n+1$. In this way a sequence of n Saccheri quadrilaterals is set up, end to end. (We do not know whether the points $B_1, B_2, \ldots, B_{n+1}$ are collinear.)

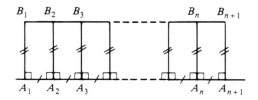

By Problem 6, we have $B_1B_2 = B_2B_3 = \cdots = B_{n-1}B_n = B_nB_{n+1}$. By the polygon inequality, it also follows that

$$A_1A_{n+1} \le A_1B_1 + B_1B_2 + B_2B_3 + \cdots + B_nB_{n+1} + B_{n+1}A_{n+1},$$

or $nA_1A_2 \le 2A_1B_1 + nB_1B_2$. But then

$$n(A_1A_2 - B_1B_2) \le 2A_1B_1$$

for any positive integer n. If $A_1A_2 > B_1B_2$, the displayed inequality leads to a contradiction.

8. Prove these statements.

(a) The angle sum of a right triangle is always less than or equal to 180°. [*Hint:* Given a right triangle ABD, with right angle at A, construct

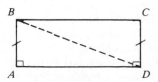

a Saccheri quadrilateral *ABCD* having base
AD. According to Problem 7, we find *BC* ≥
AD; whence in triangles *BAD* and *DCB*,
∠*BDC* is greater than or equal to ∠*ABD*.]

   (b) The angle sum of an arbitrary triangle is
always less than or equal to 180°.

9. Prove that if Saccheri's hypothesis of the acute
angle holds, then the summit of a Saccheri
quadrilateral is longer than its base; from this it
follows by Problem 8 that the angle sum of a
triangle is less than 180°. [*Hint:* Consider a
Saccheri quadrilateral *ABCD*, with base *AD*. Let *E*
and *F* be the midpoints of the base and summit,
respectively. If *AE* ≥ *BF*, pick a point *G* on *BF*
extended, so that *GF* = *AE*; then *AGFE* is a
Saccheri quadrilateral, with base *EF*. By
hypothesis, the summit angle *EAG* is acute, which
leads to a contradiction. Thus *AE* < *BF* and
similarly *ED* < *FC*.]

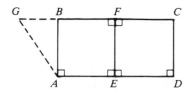

10. Prove that if Saccheri's hypothesis of the acute
angle holds, then the segment joining the midpoints
of the base and summit of a Saccheri quadrilateral
is shorter than each side; hence, there exist parallel
lines that are not equidistant from one another.
[*Hint:* See Problems 4 and 5.]

11. Explain the fallacy in the following "proof" that
all triangles are isosceles. Given a triangle *ABC*, the
angle bisector from vertex *A* will meet the
perpendicular bisector of the side *BC* at a point *P*,
which is either inside, on, or outside triangle *ABC*.
From *P* drop perpendiculars to *AB* and *AC*, meeting
these sides, or their extensions, at *D* and *E*,
respectively. Several diagrams are shown herewith
and the discussion applies equally to each.

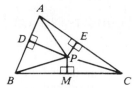

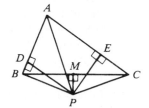

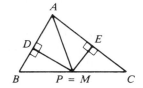

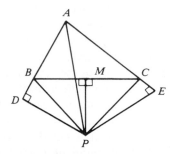

The right triangles *APD* and *APE* have the common
hypotenuse *AP*, and the sides *PD* and *PE* are equal.
(Points on the bisector of an angle are equidistant
from the sides of the angle.) These triangles are
therefore congruent, whence, as corresponding
sides, *AD* and *AE* are equal. Also, the right triangles
PMB and *PMC* are congruent by the side-angle-side
theorem. Because *PD* = *PE* and *PB* = *PC*, it
follows that the right triangles *PDB* and *PEC* are
congruent, and so *DB* = *EC*. In the first three
diagrams, lengths are added to give *AB* = *AD* +
DB = *AE* + *EC* = *AC*; in the fourth diagram,
lengths are subtracted to give *AB* = *AD* − *DB* =
AE − *EC* = *AC*. In either event, *AB* = *AC*, and the
triangle *ABC* is isosceles.

11.2   The Founders of Non-Euclidean Geometry

By the beginning of the 1800s, the question whether Euclid's parallel postulate could be demonstrated as a logical consequence of the others remained unresolved. At some point, mathematicians were bound to realize that the unending record of failure in the search for a proof of the troublesome postulate was not due to any lack of ingenuity on their part, but rather to the actual independence of the fifth postulate from the other axioms. This meant that it would be possible, by substituting a contrary axiom about parallels, to develop an equally valid companion geometry to Euclid's. When this idea finally dawned, it came not to one but to three mathematicians, more or less simultaneously and in widely separated parts of Europe. Cases of simultaneity or near simultaneity of discovery had happened before in the history of mathematics—as in the seventeenth century with the invention of calculus by Newton in England and Leibniz in Germany. And it would happen again. When the spirit of the time is ripe for a new result to come forth, it will not long be delayed. Thus, it was in the first third of the nineteenth century that Gauss in Germany, Bolyai in Hungary, and Lobachevsky in Russia were discriminating enough to reject a postulate that for 2000 years had been the main cornerstone of geometry. In Einstein's striking phrase, they ''challenged an axiom.'' Although each was unaware of the thoughts of the other two, each conceived the notion of replacing the parallel postulate with the counterintuitive axiom that ''through a point not on a line there exists more than one parallel to the line,'' while retaining all the other axioms. This was a daring innovation in its day, for it went against the time-honored tradition that axioms must be self-evident truths. Through the customary rules of deduction, the three men developed a sequence of theorems resembling those Saccheri obtained from his hypothesis of the acute angle. Notwithstanding the breach of one's intuitive picture of space (the angle sum of a triangle, for instance, does not add up to two right angles), their geometric system is free from internal contradiction. Whereas Saccheri had been too awed by a conviction of the absolute necessity of Euclidean geometry to recognize what he had done, Gauss, Bolyai, and Lobachevsky appreciated the revolutionary implication of their discovery—the geometry of Euclid, being not the only logically consistent geometry, might not even be the correct system for describing the physical world.

From the detailed publication of much of Gauss's correspondence, we now know that Gauss was the first to reach any advanced conclusions concerning a consistent geometry different from that of Euclid. Because he failed to make public his thoughts on the subject, the honor for discovering non-Euclidean geometry is usually divided between John Bolyai and Nicholas Lobachevsky. We can at least credit Gauss with having realized that along the path Bolyai and Lobachevsky traveled, complete success was bound to be achieved. Scholars have long remarked that Gauss's publications are not an adequate reflection of his full greatness, for he published relatively little of his work—only about half of the innovative ideas with which he is credited. In his writings, Gauss was as exact and elegant as Lagrange, but more difficult to follow than even Laplace. Not only did he remove every trace of the reasoning by which he reached his conclusions, but also he made a concerted effort to give proofs that although strictly logical were as concise as possible. The result was always a finished work of art, for until Gauss had polished a paper over and over, he refused to allow it to appear in print. These demands for excellence found expression in his unofficial coat of arms, a tree bearing only seven fruits. The motto beneath reads *Pauca Sed Matura* (''Few, but ripe'').

Ideas came so quickly from Gauss's fertile imagination that each one inhibited the development of the preceding. Because of the great length of time needed to polish a paper to the point of perfection on which he insisted, he developed the habit of leaving his rough notes in a desk drawer, often not even telling anyone about them. This reluctance to make known many of his original ideas provided considerable frustration for Gauss's contemporaries, who often found, laboring for years on some important new development, that Gauss had anticipated them. One instance is the theory of elliptic functions, about which Gauss discovered many of the results Abel and Jacobi subsequently published. There is a story that Jacobi, who was considered the ''second-best'' mathematician in Europe, came to Gauss to relate a new creation only to have the latter pull from his desk drawer some papers that contained the same discovery. In exasperation, Jacobi remarked, ''It is a pity that you did not publish this result, since you have published so many poorer papers.''

Gauss began to meditate about parallel lines as early as 1792. He was only 15 at the time, but had already told his friend Schumacher that if the Euclidean system were not the only ''true'' geometry (that is, the correct idealization of the properties of actual physical space), then he could work out another logically consistent geometry. Unfortunately, he refrained from publishing any statement of his theory, large or small, throughout his lifetime. What we know of Gauss's successive ventures into the subject can be gleaned only from letters to interested colleagues and some notes left behind after his death. Through this correspondence, one can trace the development of Gauss's thought, from skepticism concerning the purported proofs of the parallel postulate to confidence in the validity of his new geometry.

At first, Gauss blundered in good company by trying to deduce the truth of the fifth postulate from the other axioms and the 28 propositions proved without it. He soon recognized the profound difficulties involved, and in a letter written to William Bolyai in 1804, he spoke of a ''group of rocks'' on which his attempts had always been wrecked, adding: ''I have still ever the hope that these rocks sometime and, indeed, before my death will permit a passage.'' As late as 1813, there was still no light, for we read: ''In the theory of parallels we are even now not further than Euclid. This is the shameful part of mathematics, which sooner or later must receive a wholly new form.''

Over the years, Gauss cautiously and reluctantly concluded that the parallel postulate could not be proved on the basis of the remaining nine axioms. His nearest-to-public utterance on the matter was a book review (1816) of a certain tract on parallels, in which he spoke of ''vain attempts to conceal the gulf, which one cannot conceal, with an unsound web of sham proof.'' Although this mere hint that the parallel axiom cannot be proved was ''besmirched with mud'' by Gauss's critics, his doubts were not dispelled. A letter written to Heinrich Olbers in 1817 marked a turning point in Gauss's thinking:

> I am becoming more and more convinced that the necessary truth of our [Euclidean] geometry cannot be demonstrated, at least not by the human intellect to the human understanding. Perhaps in another life, we shall obtain insights into the nature of space which are now beyond our reach.

From that time onward, he seemed firmly convinced that the parallel postulate was independent of the other Euclidean axioms, so that it would be possible to adopt a contradictory axiom and build an entirely new geometry logically as legitimate as Euclid's. In rejecting the fifth postulate, he chose to assume (as Bolyai and Lobachevsky would after him) that through a given point there could be drawn more than one parallel to a given line. As Gauss developed the consequences of this assumption, he mentioned something of his work to a few trusted friends but concealed his investigations from the world at large.

Like Newton, Gauss had an intense dislike of controversy and seemed unwilling to venture publicly into any area that might subject him to vulgar attack. He was certain that his discoveries in this alternative system of geometry would shock layman and mathematician alike. Indeed, his results ran so counter to the accepted view, expressed by Kant, that our inner consciousness allows us to imagine geometric figures with Euclidean properties only, that he felt by divulging them he might subject himself to ridicule. Thus he expressly begged his friends to keep silent about the information he imparted to them. The fullest indication of Gauss's feelings is contained in a letter written in 1824 to Franz Taurinus. After pointing out an error in an attempted proof of the parallel postulate by Taurinus, Gauss went on to say:

> The assumption that the sum of the three angles of a triangle is less than 180° leads to a curious geometry, quite different from our own [the Euclidean], but thoroughly consistent, which I have developed to my satisfaction. . . . The theorems of this geometry appear to be paradoxical and, to the uninitiated, absurd; but calm, steady reflection reveals that they contain nothing impossible. . . . In any case, consider this a private communication, of which no public use or use leading to publicity is to be made. Perhaps I shall myself, if I have at some future time more leisure than in my present circumstances, make public my investigations.

Gauss continued with his development of the fundamental results of a new geometry and was again considering writing them up, possibly to be published after his death. He took up the matter in a reply (1829) to the astronomer Friedrich Bessel:

> It may take a very long time before I make public my investigations on the issue. In fact, it may not happen during my lifetime, since I fear the scream of the Boeotians [a figurative reference to the dullards, for the Boeotians were reputed to have been one of the more simple-minded Greek tribes] were I to completely express my views.

It may seem strange to the modern observer that Gauss, who was recognized in his own lifetime as one of the greatest mathematicians, would choose to withhold publication of his non-Euclidean geometry lest its unpopularity damage his reputation. But in the early 1800s, the immense authority of the German philosopher Immanuel Kant (1724–1804) still dominated the intellectual world. Kant's reputation as a thinker of the most abstract kind has tended to overshadow his reputation as a scientist, but until 1770 he was primarily attracted to the study of the natural sciences rather than philosophy. Newtonian physics and its astronomical ramifications especially aroused his interest. Kant had neither the mathematical background nor the necessary means of observation and experiment to be a strict scientist. He did, however, contribute to scientific knowledge to the extent of anticipating the nebula hypothesis of the origin of the solar system some 40 years before the publication of Laplace's *Système du monde*.

Kant's speculations first appeared in 1755 in a small, anonymous volume entitled *Universal Natural History and Theory of the Heavens.* (It may have been anonymous for fear of the reaction of the Protestant clergy.) The main object of the book was to establish the existence of God by showing that the Newtonian universe was not the product of mere chance but instead designed to evolve in an orderly fashion. Unfortunately, the publisher went bankrupt just at the time the work was to appear, so that Kant failed to get an adequate hearing for his theory. In 1763, he tried again to put his views before the public, this time in a tract called *The Only Possible Argument for a Demonstration of the Existence of God.* Such studies made Kant's name widely known in Germany, and several times he was offered chairs at universities. Wishing to remain in his native Königsberg, he continued in the laborious role of *privatdozent* (a private teacher, recompensed only by such fees as the students are willing to pay). As such, he gave an enormous number of lectures on a bewildering variety of subjects: physics, mathematics, logic, ethics, and anthropology. In 1764, he was offered, but wisely refused, the professorship in poetry at Königsberg, and in 1765, he was made assistant librarian at the university. It was only in 1770, at the age of 46, that Kant was appointed professor of logic and metaphysics at Königsberg, the post he held until his death in 1804. The twelve years from 1769 to 1781 were spent in the most profound speculation of the philosophical issues he ultimately presented in the *Critique of Pure Reason* (1781).

Kant's problem was to reconcile his belief in the preestablished harmony of the universe with the confusion of data collected by experimental observation. In this, he succeeds quite deftly in the *Critique of Pure Reason,* his treatise on the theory of knowledge, a subject that he created. Kant sought to bridge the gulf between the British empiricists, who held that all knowledge arises from experience, and the equally dogmatic continental rationalists, who professed that all knowledge arises from unaided thought. He agreed that knowledge begins with external experience, so far as some sensation must precede and arouse the operations of thought; but, he maintained, the mind can act on sense impressions only because it is already endowed with ''intuitions'' of space and time that are independent of experience and that mold it. As Kant declared in the *Critique of Pure Reason,* ''the concept of [Euclidean] space is by no means of empirical origin, but is the inevitable necessity of thought,'' and as a consequence the sole spatial relations acceptable to the mind were those of Euclid. No other system of geometry could possibly exist, for no other geometry was thinkable. To dispute the idea that the parallel postulate was inherent in the structure of the mind itself as a divinely implanted intuition was to challenge Kant's theory of knowledge, which Kant called ''the Copernican revolution in philosophy.''

For himself, Gauss doubted that the mind compels us to view the world in only one way. He refrained from public controversy with its attendant criticism, yet thought that determining the geometry of space was an empirical question, to be verified like any other physical law by actual measurement. The theorem concerning the sum of the angles of a triangle seemed to provide an experimental means for deciding once and for all which geometry, his or Euclid's, best described the space of our physical experience. In the traditional geometry, the Euclidean, the angle sum of a triangle is always 180°, but Gauss's new geometry predicted a result less than 180°. With this in mind, Gauss laid out a triangle in the neighborhood of Göttingen whose sides were 40 or so miles long and whose vertices were carefully surveyed points on three mountain summits. Had he detected a significant deviation from 180°, the experiment would have been conclusive. When the data was collected, the result was an angle sum within 2″ of 180°, a difference that could be charged to the unavoidable errors in observation. Gauss concluded that with the instruments at his disposal, the measurement of physically defined triangles would not discriminate between the two geometries.

It is remarkable how closely the trains of thought of the mathematicians involved in developing non-Euclidean geometry were related. Lobachevsky conducted a similar experiment with angle sums, the results of which were published in 1829 in his treatise *On the Foundations of Geometry*. The only difference was that he analyzed some existing astronomical measurements to test which geometry best accounted for the data. Lobachevsky calculated that the angle sum of the triangle determined by the earth, the sun, and the fixed star Sirius differed from 180° by an amount less than 0.000004″. Although this led him to infer that "the exactitude of traditional geometry is far-reaching," the matter was far from settled in favor of Euclidean geometry, because Lobachevsky had no way of knowing whether he was measuring a significantly large area of space. Later mathematicians were subsequently to argue that no amount of evidence could be obtained to threaten the claim that space was Euclidean, hence the question should not be seriously proposed, even for testing; in that case, the Euclidean theory was preferable on the grounds of elegance and simplicity.

Sometime toward 1830, Gauss took alarm over his long delay in committing his non-Euclidean geometry to paper and thought that if he did not get his ideas out at once he might never do so. In a letter dated May 17, 1831, Gauss wrote to the astronomer H. C. Schumacher:

> In the past few weeks I have begun to write down some of my meditations [on the theory of parallels], a part of which I have never previously put in writing, so that already I have had to think it all through anew three or four times. But I wished that this should not perish with me.

Found among the papers after his death, there was a brief synopsis of the new theory of parallels. Because Gauss was not resolute enough to risk coming out in the open with his findings, he did not produce the revolution in mathematics that Bolyai and Lobachevsky were to bring about; his lack of moral courage in facing "the scream of the Boeotians" (that is, the followers of Kant) made it inevitable that posterity withhold a portion of the honor that could have been entirely his.

Perhaps the reason that Gauss did not go further in recording his earlier investigations was that on February 14, 1832, he received a copy of the famous *Appendix* of John Bolyai. John's father, Wolfgang Bolyai (1775–1856) had studied at Jena, then afterwards at Göttingen, from 1796 to 1799. Gauss was among Wolfgang Bolyai's close friends at Göttingen. Bolyai, after returning to Hungary, maintained a correspondence with Gauss that lasted, sometimes with years-long interruptions, for the remainder of their lives. During their student days, the two men had frequently discussed problems related to the theory of parallels. In 1804 Bolyai, convinced that he had succeeded in proving the parallel postulate, sent a little tract on this subject, *Theoria Parallelarum,* to Gauss. The reasoning was incorrect, and Gauss in replying tactfully pointed out the error of assuming that through any three noncollinear points there passes a circle. Bolyai mailed a supplementary paper to Gauss four years later, but when there was no answer, he became discouraged with his "theory of the Göttingen parallels" and turned to other matters. His ideas on elementary mathematics were collected in a large two-volume *Tentamen Juventutem Studiosam in Elementa Matheseos Purae* (An Attempt to Introduce Studious Youth to the Elements of Pure Mathematics), published in 1832–1833 although it bears the date 1829. It contained an *Appendix* composed by his son, John.

The chief claim to fame of Wolfgang Bolyai must doubtless be that he was the father of John Bolyai (1802–1860). The father gave the young boy early instruction in mathematics and his progress was so quick that when Wolfgang was sick on one occasion, he did not hesitate to send his 13-year-old son to teach the college classes in his stead. The elder Bolyai hoped that John would go to Göttingen and study with his old friend Gauss; in 1816, Bolyai wrote to ask whether Gauss would take John into his household as an apprentice mathematician and to inquire what the youngster should study in the meantime. Gauss never answered this letter and never wrote again for 16 years. Instead, John entered the Imperial Engineering Academy in Vienna in 1817, to receive a military education. He finished his courses at the academy in 1823, and as a sublieutenant, entered on an army career. During the next 10 years, he built up a reputation as a profound mathematician, an impassioned violin player, and an expert fencer who dueled regularly. (He once accepted the challenge of 13 cavalry officers on the condition that he be allowed to play a violin piece after every two duels; he vanquished all 13.) Plagued with intermittent fever, Bolyai was pensioned off by the service in 1833 as a semi-invalid. Under Gauss's care and direction, the younger Bolyai might have enriched the great learned journals with discovery after discovery. As it was, the sum of his publications was two dozen pages.

John Bolyai had from his father some of the inspiration to original research that the latter had received from Gauss. While still at the academy, John was giving his attention to the parallel postulate, though first on the side of proving it. Recalling his own unsuccessful efforts, Wolfgang urged the boy not to become preoccupied with the theory of parallels:

> Do not waste one hour's time on that problem. It does not lead to any result; instead it will come to poison all your life. . . . I believe that I myself have investigated all conceivable ideas in this connection.

His father's advice was founded on bitter experience, but far from being dissuaded, John was moved to greater efforts.

After several vain attempts to prove the parallel postulate, the young man began to think of it as an independent assertion and succeeded in constructing an entirely new geometry based on the denial of this axiom. He did not know that the same idea had already occurred to Gauss and to Lobachevsky. His sense of joy and triumph at solving what had baffled the world for 2000 years was expressed in a letter to his father in 1823:

> I am resolved to publish a work on parallels as soon as I can complete and arrange the material, and the opportunity arises. At the moment I still do not clearly see my way through, but the path which I have followed is almost certain to lead me to my goal, provided it is at all possible. I have not quite reached it, but I have discovered things so

wonderful that I was astounded and it would be an everlasting pity if these things were lost. When you, my dear father, see them, you will understand. All I can say at present is that out of nothing I have created a strange new world. All that I have sent you previously is like a house of cards in comparison with a tower.

The expansion and arrangement of the ideas proceeded more slowly than John had expected, and not until 1825 did he commit to writing a treatise on the subject. His father suggested that his be translated into Latin and issued as an appendix to the planned *Tentamen;* moreover, he urged that this be done without delay:

> It seems to me advisable, if you have actually succeeded in obtaining a solution of the problem, that, for a two-fold reason its publication be hastened: first, because ideas easily pass from one to another who, in that case, can publish them; secondly, because it seems to be true that many things have, as it were, an epoch in which they are discovered in several places simultaneously, just as the violets appear on all sides in the springtime.

Wolfgang Bolyai's vivid presentiment was fulfilled by the coincident discovery of non-Euclidean geometry by his son, and by Gauss and Lobachevsky. John Bolyai's treatise was printed separately under the title *Appendix Scientiam Spatii Absolute Veram Exhibens* (Appendix Explaining the Absolutely True Science of Space) and appeared before the volume of the *Tentamen* with which it was afterwards bound up. Excluding the title page and one that gave an index of notation, the work itself contained only 24 pages—which have been described by George B. Halsted (1853–1922) as "the most extraordinary two dozen pages in the whole history of thought."

Knowing Gauss's interest in the subject of non-Euclidean geometry, Wolfgang sent an advance copy of his son's work to Gauss in June 1831. The elder Bolyai no doubt felt that his belief in his son had been vindicated, and John must have expected that public praise from Gauss would make him famous. The copy did not reach its destination, and a second copy of the *Appendix* was mailed in January 1832. Gauss's reaction was typical—sincere approval, but lack of support in print. In February, Gauss wrote in great enthusiasm to his former student C. L. Gerling, who was professor of astronomy at Marberg, saying:

> Let me add further that I have this day received from Hungary a little work on non-Euclidean geometry in which I find all my own ideas and results developed with great elegance, although in a form so concise as to offer great difficulty to anyone not familiar with the subject. . . . I regard this young geometer Bolyai as a genius of first order.

Then in March, he replied to his "old, unforgettable friend" Wolfgang Bolyai. It is easy to see how this well-intentioned letter had a devastating effect on the young Bolyai:

> If I begin by saying that I dare not praise this work, you will of course be surprised for a moment; but I cannot do otherwise. To praise it would amount to praising myself. For the entire content of the work, the approach which your son has taken, and the results to which he is led, coincide almost exactly with my own meditations which have occupied my mind for the past thirty or thirty-five years. . . . It was my plan to put it all down on paper eventually, so that at least it would not perish with me. So I am greatly surprised to be spared this effort, and am overjoyed that it happens to be the son of my old friend who outstrips me in such a remarkable way.

Wolfgang may have been pleased because he understood that his son was being praised, but John was sadly distressed by the great mathematician's reply. He even imagined that before the *Appendix* was published, his father had secretly confided some of his results to Gauss, who was now trying to claim them as his own. John eventually set aside these suspicions, but like many others who found that Gauss had anticipated them, he still felt cheated of his earned honors. Certainly Gauss, who was quite sensitive to all forms of suffering, had no desire to hurt him; but Bolyai's mental depression deepened when the *Appendix* met with complete indifference from other mathematicians, and for long periods he did almost no creative work. The unhappy fact remains that he never published anything further, although he did leave behind some thousand pages of manuscript. The final disappointment came in 1848, when Lobachevsky's *Geometrische Untersuchungen zur Theorie der Parallellinien* (1840) reached him through his father, and he learned that the young Russian had obtained priority by publishing the radically new geometry in 1829.

It was long after Bolyai's death that recognition as one of the founders of non-Euclidean geometry finally came to him. The view that Euclidean geometry was the only possible geometry was so firmly rooted that his *Appendix* was practically forgotten for 35 years. In fact, the Hungarian version of volume one of the *Tentamen,* which was published in 1834, did not contain the *Appendix* at all. It was rescued from oblivion when Richard Baltzer drew public attention to the work of Bolyai and Lobachevsky in his *Elemente der Mathematik* (1867). Precipitated by Baltzer, translations of the *Appendix* into French (1868), Italian (1868), German (1872), and English (1891) made the masterpiece accessible to the international scientific community.

Although we shall not make any lengthy analysis of the cultural milieu in which Lobachevsky worked, we should say a few words about the institutional setting of mathematics in Russia in the early 1800s. Russian scientific and mathematical contact with the outside world was always scanty and intermittent. Toward the end of the 1700s, an intense suspicion of Western ideas had led to a perilous state of decline in science. The display piece of Russia's contribution to modern scientific thought, the St. Petersburg Academy, had lapsed into inactivity, its staff reduced to 14 full members. Censorship of the journals of the scientific bodies of the West deprived the academy even of news of developments abroad. Aside from the Academy, the empire's only other center of scientific learning was Moscow University.

When Alexander I became tsar in 1801, he responded to the mounting demands for extensive educational reform by reopening Dorpat University and founding four new universities: Vilna (1802), Kazan (1804), Karkov (1804), and St. Petersburg (1819). Recognizing that scientific knowledge can be translated into economic and military strength, he insisted that the new schools give great weight to the various natural and mathematical sciences in their curricula. Professors were directed to incorporate the latest scientific ideas in their lectures, with the implication that they were expected to keep abreast of developments in their fields. They were required to hold monthly meetings devoted to discussions of scientific papers presented by the teaching staff. These directions were unduly optimistic. Because Russia never really had a system of lower schools that would provide the training for those willing to pursue advanced studies, the new universities were hard-pressed to fill their classrooms. Kazan had only 40 students in 1809, and the more established Moscow University listed an enrollment of 135. What students were available were mostly drawn from the theological seminaries, and although they were well versed in Greek and Latin, they knew little science or mathematics and no other modern European language.

The problems of the fledgling universities were compounded by an acute shortage of candidates for teaching positions. The authorities tried to overcome this deficiency by hiring some established foreign scholars, including several distinguished members of Western learned societies. (The scientific fortunes of the St. Petersburg Academy would have been immensely improved if only Gauss had chosen to accept the appointment offered him.) Even though these foreign instructors usually could not lecture in the Russian language, they managed to reestablish the avenues of mathematical contact with the West. By bringing to Russia the newest ideas of Lagrange, Laplace, Gauss, and other stalwarts of the day, they nurtured a generation of native mathematicians who were to make important strides during the second quarter of the nineteenth century.

Unfortunately for the development of the universities, the first decade of Alexander's reign marked the last time in the nineteenth century that the Russian government regarded higher education as an unqualified good. Greatly shocked by Napoleon's invasion of Russia in the latter half of 1812, the Russian people reacted in two ways. First, the ancient Russian distrust of outsiders was revived by the pillage of the country by foreigners; secondly, there was a great surge of religious mysticism in giving thanks to God, who once more had flung out the invaders and saved Mother Russia. With the awakening of national pride and the religious reaction against rationalism, it was inevitable that the Russian state began turning against the existing university system, based as it was on Western, particularly German, models. In 1815, the government decreed that henceforth all university lectures must be delivered in the Russian language. To save Russia from the misfortunes of Germany, where "the professors of godless universities imbue hapless youth with the subtle poison of skepticism and hatred of authority," Russian scholars were forbidden to study in Germany and then the universities were prohibited from employing any Russian citizen who had attended a German university. There were dismissals at Karkov and Dorpat, but most of the foreign professors departed voluntarily, with the result that Russia's universities once again withered into intellectual wastelands.

The first mathematician to express his results on non-Euclidean geometry openly, both orally in 1826 and in print in 1829, was a product of Russia's new university system. Nicolai Ivanovitch Lobachevsky (1793–1856) was the son of a poor government clerk, who died when the boy was only seven. The widow, finding herself in extreme poverty, moved the family to remote Kazan on the threshold of Siberia, where she succeeded in getting her three sons admitted to the secondary schools on public scholarships. In 1807, the young Lobachevsky entered Kazan University as a free student, planning to prepare for a medical career. He was destined to spend the next 40 years of his life there, as student, teacher, and administrator, an isolated scholar in Europe's easternmost university.

In the hope of transforming Kazan from an intellectual outpost on the periphery of the West into the equal of any European university, the authorities had recently acquired the services of four distinguished German professors. Among these was the mathematician Johann Bartels (1769–1836), one of Gauss's early teachers at Caroline College in Brunswick. Bartels came with the reputation of being more skilled in the art of teaching than in independent scientific investigation, and under his tutelage, Lobachevsky soon found himself deeply involved in the study of mathematics. In 1811,

Nicolai Lobachevsky
(1793–1856)

(From A Concise History of Mathematics *by Dirk Struik, 1967, Dover Publications, Inc., N.Y.)*

he graduated from Kazan with a master's degree in physics and mathematics; and the following year, he began his career by holding geometry classes for minor government officials, who by law needed a college degree to advance to higher civil service positions. Two years later, Lobachevsky received a regular appointment at the university as an assistant professor, lecturing on various aspects of mathematics, physics, and astronomy. He was promoted to full professor in 1816, at the early age of 23.

It is not clear just when Lobachevsky shook himself free from Euclid's 2000 years of authority. No doubt Bartels was familiar with Gauss's special interest in non-Euclidean geometry, and Lobachevsky probably first heard about it from him. But Bartels never saw Gauss after 1807 and received (1808) only one letter from his former pupil during his entire stay at Kazan, so that it is unlikely that he had any inkling how the whole problem should be treated. We know that until the early 1820s (long after Bartels had returned to Germany), Lobachevsky was still working along conventional lines and not searching for a new geometry. A manuscript for a geometry textbook that he drew up in 1823, presumably for classroom use, contained three attempted "proofs" of Euclid's parallel postulate. Despite the proofs that he put forth, Lobachevsky made the significant statement that all such proofs "no matter of what kind, can only be regarded as clarifications, but do not deserve to be called mathematical proofs in the full sense." Although beginning to appreciate the difficulties encountered in the attempts to prove the postulate, he apparently had not then excluded the possibility that a demonstration might yet be found. It might well have been that Lobachevsky, in trying to bring his textbook to satisfactory completion, turned from the attempt to perfect the structure of Euclidean geometry and entered an entirely new path that would lead him to success. The manuscript itself was never published (that is, until 1910), rejected by the government press on the grounds that it used meters as units of measure—innovations of the French Revolution.

The next three years saw the evolution of an alternative geometry that did not involve the dubious postulate. At a meeting of the mathematics and physics faculty of

Kazan University held on February 23, 1826, Lobachevsky delivered a paper called *Exposition succinte des principes de la géométrie avec une démonstration rigoureuse du théorème des parallèles*. In spite of the ominous-sounding "rigorous proof of theorem of parallels" in the latter half of the title, it is likely that the lecture contained the basic outline of his historic attempt to establish a new geometric system independent of the fifth postulate. No part of this French manuscript has ever been found, but its essence was incorporated into his memoir *On the Foundations of Geometry*, printed in installments in the Kazan *Messenger*, a monthly journal published by the school, in 1829–1830. This work, the first account of non-Euclidean geometry to appear in print, marked the official birth of the subject. It was in Russian, so it remained unknown to most foreigners, and even at home it attracted little attention. The indifference was partly due to a lapse of literary judgment, which let Lobachevsky slight his creation with the deprecating name *imaginary geometry*.

In the hopes of acquainting the learned world with his grand departure from Euclidean geometry, Lobachevsky submitted the manuscript of *On the Foundations of Geometry* to the St. Petersburg Academy of Sciences to be considered for publication in one of its scholarly journals. The verdict was that the work was not precise enough and was unworthy of being printed under the auspices of the academy. The reviewer, Mikhail Ostrogradski, failing to notice the startling theme of the paper, concentrated his remarks instead on two definite integrals that entered into certain of Lobachevsky's calculations. Thinking that these integrals were themselves some "new method," the reviewer complained that one of them was false and the second "easily deducible from the elementary principles of the calculus of integration." A St. Petersburg journal followed the lead by printing an uncomplimentary review of *On the Foundations of Geometry*, with the apparent intent of making Lobachevsky and his work objects of public ridicule.

A lesser man might have given up the struggle at this point, but Lobachevsky responded by writing a series of papers in an effort to convince the mathematicians of the world of the merits of his research. The rejected manuscript was expanded into *New Elements of Geometry, with a Complete Theory of Parallels*, which appeared in the recently founded scientific journal of Kazan University during the years 1835–1838. This time Lobachevsky made no effort to communicate with the Academy. Then he published (1835) a long article that bore the title "Imaginary Geometry" in Moscow University's *Messenger of Europe*, the first time one of his studies was printed outside of Kazan. He also translated it into French to be carried by *Crelle's Journal* (1837). Perhaps the best summary of his new geometry was a little book of 61 pages, *Geometrische Untersuchungen zur Theorie der Parallellinien* (Geometrical Investigations on the Theory of Parallels), published in Berlin in 1840. Gauss first learned of Lobachevsky's contributions to non-Euclidean geometry when he received a copy of this last work from the author, and he replied to him in congratulatory fashion. That the great German mathematician was deeply impressed is indicated by a letter to Schumacher in 1846:

> I have recently had occasion to look through again that little volume by Lobachevsky. . . . You know that for fifty-four years now (even since 1792) I have held the same conviction. I have found in Lobachevsky's work nothing that is new to me, but the development is made in a way different from that which I have followed, and certainly by Lobachevsky in a skillful way and a truly geometrical spirit.

Gauss studied Russian in his old age, so as to be able to consult Lobachevsky's other works in the original language. He also arranged (1842) for Lobachevsky to become a corresponding member of the Göttingen Academy of Sciences. But he stubbornly withheld the public support that would have made the new ideas mathematically acceptable.

At age 34, Lobachevsky was elected rector of Kazan University, a post he held for 19 years. The honor, coming as it did in 1827, was more likely a reward for his dedicated service as chairman of the university's library and construction committees than a recognition of his attempt to erect an entire geometry on new foundations. As he neared old age, a bureaucratic revamping of the administrative structure of the Kazan school district brought Lobachevsky's abrupt dismissal from the university. Sorely hurt, he lingered on in the neighborhood of Kazan, trying to be of some use by occasionally showing up for doctoral examinations. Finally, the premature death of a son coupled with rapid failure of his eyesight took its toll on his health and spirit.

The two heretic geometers—Lobachevsky, who misnamed his creation *imaginary geometry,* and Bolyai who used the nobler designation *absolute science of space*— were denied the welcome that so many centuries of anticipation seemed to promise. No leading mathematician of the time gave a word of public approval to what Hilbert later called "the most suggestive and notable achievement of the last century." Both men were aware of the revolutionary magnitude of their geometrical theories and expected to be rightfully heralded; both were bitterly disappointed with the response but reacted in entirely different ways. Bolyai withdrew from all mathematical activity in apparent disgust with the world and himself. To the last days of his life, Lobachevsky waged a resolute and uncompromising struggle to gain wider circulation for his ideas. In 1855, on the occasion of the fiftieth anniversary of the founding of Kazan University, he made a final effort by publishing a comprehensive exposition of his system of geometry. Under the title *Pangeometry or a Summary of the Geometric Foundations of a General and Rigorous Theory of Parallels,* it appeared simultaneously in Russian and French in the scientific journal of Kazan University. The treatise was not written in his own hand, but dictated, for Lobachevsky had become blind. Although he died without seeing his ideas become a recognized part of mathematical thought, they achieved gradual acceptance during the remainder of the century. In a lecture delivered in 1873, the English mathematician William Kingdon Clifford went so far as to compare the effect of Lobachevsky's new doctrine with the scientific revolution wrought by Copernicus's heliocentric system saying, "What Vesalius was to Galen, what Copernicus was to Ptolemy, that was Lobachevsky to Euclid." As if to atone for the initial neglect, much of the mathematical world now refers to the non-Euclidean geometry that he developed as Lobachevskian geometry.

As we have remarked several times, the chief difference between the axiomatic structure of classical Euclidean geometry and the new geometry of Lobachevsky resides in their respective parallel axioms. Lobachevsky replaced Euclid's fifth postulate with a contrary one and kept all the remaining postulates unaltered. From the time of Proclus, most geometers accepted Playfair's axiom as the appropriate wording for a postulate on parallels. If this version is chosen, then a contrary statement says that through any point P not on a line l there exists either no parallel or more than one parallel to l. Because the assumption of no parallel would contradict the infinite extent of a straight line (a tacit assumption Euclid and others used), Lobachevsky took the following supposition as a starting point:

LOBACHEVSKIAN
PARALLEL
POSTULATE

There exist two lines parallel to a given line through a given point not on the line.

By parallels are here meant lines that do not intersect *l*. From this postulate, together with the remaining axioms of Euclid, plus some additional axioms required for the rigorous foundation of Euclidean geometry, Lobachevsky developed his new geometry to the point of working out its trigonometry. Since the technical content is neither simple nor short, we shall content ourselves with a few remarks concerning parallel lines.

All the Euclidean propositions that do not depend on Euclid's parallel postulate, in particular the first 28, were available to Lobachevsky. Thus, he knew that given a point *P* not on a line *l* one could construct a line *l'* through *P* parallel to *l*. The procedure would be to drop a perpendicular *PQ* to *l* and take *l'* perpendicular to *PQ* at *P*; having a common perpendicular segment, *l* and *l'* would then be parallel, by Proposition 27 of Euclid I. What additional information could Lobachevsky glean from his new parallel postulate?

Consider one of the half-lines *PS* of *l'*. Of the various lines through *P* that lie in the interior of *SPQ*, some (such as the line through *P* and *R*) will intersect *l*; others will not. This situation allows us to divide these lines into two sets $\mathcal{S}_1$ and $\mathcal{S}_2$: in $\mathcal{S}_1$ are just those lines that intersect *l*, whereas in $\mathcal{S}_2$ are just those lines that do not meet *l*. Each line in $\mathcal{S}_1$ "precedes" each line in $\mathcal{S}_2$ in the sense of making a smaller angle with the segment *PQ*. Now there exists a line *l** through *P* that brings about the division of the lines into two sets. Because *l** either meets *l* or does not meet it, *l** must be either the "last" line in $\mathcal{S}_1$ (it is preceded by every member of $\mathcal{S}_1$) or the "first" line in $\mathcal{S}_2$ (it precedes every member of $\mathcal{S}_2$). To which set does this boundary line between the members of $\mathcal{S}_1$ and $\mathcal{S}_2$ belong?

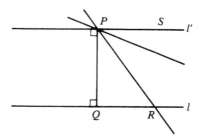

To see that there is no last line in $\mathcal{S}_1$, let us assume to the contrary that *l** is in $\mathcal{S}_1$. Then *l** intersects *l* at a point *E*. Choose any point *F* on *l* such that *E* is between *Q* and *F*. Then the line through *P* and *F* is in $\mathcal{S}_1$. Because *l** precedes this line, *l** is not the last line in $\mathcal{S}_1$, and a contradiction is reached. This means that the dividing line of the two sets is the first of those that do not intersect *l*. We shall call *l** the right limiting parallel to *l* through *P*. We choose this terminology because as a point *R* on *l* recedes endlessly from *Q* on *l*, the line through *P* and *R* approaches the limiting position *l**. An analogous discussion, dealing with lines through *P* that lie in the interior of *TPQ*, gives rise to a left limiting parallel l_*. That is, there are two lines *l** and l_* through *P* that do not intersect *l*, and are such that any line through *P* lying within the angle formed by the two lines and containing *PQ* intersects *l*.

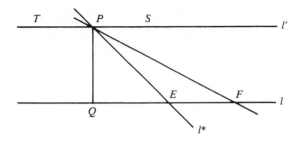

These limiting parallels are situated symmetrically about PQ in the sense that the angles α and β that l^* and l_* make with PQ are equal. Suppose one of the angles, say α, is greater than the other. Then there exists a line $\bar{l}$ between PQ and l^* that intersects l at a point R such that $\angle RPQ = \beta$. Let R' be the point on the opposite side of PQ from R, and satisfying $R'Q = RQ$. Then triangles $R'PQ$ and RPQ are congruent, so that the corresponding angles $\angle QPR'$ and $\angle QPR$ are equal. Hence, $\angle QPR' = \beta$. The implication is that the line through P and R' coincides with l_*. But this is impossible, since l_* does not intersect l. This contradiction allows us to discard the situation in which $\alpha > \beta$; similarly, $\alpha < \beta$ cannot hold and so $\alpha = \beta$.

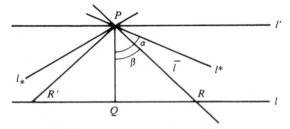

The angles α and β that l^* and l_* make with PQ are called the angles of parallelism at P with respect to l, and are important to Lobachevsky's development (for instance, he showed that the size of α would depend on the length of the perpendicular PQ). It is already known that $\alpha = \beta \leq 90°$. If $\alpha = \beta = 90°$, then l^* and l_* would be the same straight line; namely, both would coincide with the perpendicular l' through P to PQ. Any other line through P would enter the interior of one of the angles α or β and would therefore have to intersect l. It follows that l' is the only line through P parallel to l, a denial of Lobachevsky's parallel axiom. Thus, the equal angles α and β must be acute. We have proved one of the first theorems of Lobachevsky's geometry.

THEOREM

If P is any point not on a line l, there exist two distinct lines l^* and l_* that do not intersect l and that form equal acute angles with the perpendicular PQ to l.

Actually, there are infinitely many lines through P that do not intersect l, whence it is true that an infinite number of parallels to l through P exist. Take any line $\bar{l}$ through P lying in one of the pair of vertical angles bounded by l^* and l_* and not containing PQ. If $\bar{l}$ meets l, say at the point R, then because l^* passes through the vertex P of triangle PQR, it must intersect the opposite side QR. From this contradiction, it can be concluded that any such line $\bar{l}$ is parallel to l. This concludes our brief exposition of Lobachevsky's geometry.

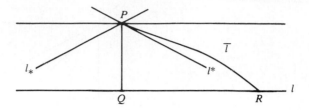

11.3   The Age of Rigor

For 35 years after the appearance of the pioneering work of Bolyai and Lobachevsky, the subject of non-Euclidean geometry was essentially ignored. The relative obscurity of the two young founders contributed to this delay, and so did the slow passage of scientific knowledge from one part of Europe to another. Few mathematicians could read Russian, the language in which Lobachevsky usually published. Further, Kant's views on the Euclidean nature of space were still powerfully dominant. Kant had held that experience could not contradict these axioms, because intuitions of a Euclidean space are part of the human manner of perceiving reality. Not until the 1870s was the revolutionary significance of the new geometric ideas really understood, following the belated (1868) appearance in print of the famous probationary lecture Bernhard Riemann gave in 1854.

Bernhard Riemann (1826–1866), who was born about the time non-Euclidean geometry was discovered, was the son of a Lutheran pastor. At age 19, he enrolled at the University of Göttingen as a student of theology and philosophy, a choice dictated by his wish to please his father. He also attended lectures on mathematics and finally received the elder Riemann's permission to devote himself entirely to the subject. After a year at Göttingen, where he found the instruction antiquated (even Gauss taught only elementary courses), Riemann was drawn to Berlin to study under such great mathematicians as Jacobi, Dirichlet, and Eisenstein. (During the Revolution of 1848, as a member of the loyal student corps, he spent two days on guard duty protecting the king at the royal palace.) In 1849 he returned to Göttingen to complete his training for a doctorate. His thesis of 1851, written under Gauss, dealt with surfaces over a complex domain; these are now called Riemann surfaces. This work at once established his reputation as a mathematician of first rank. Perhaps because the thesis involved many original ideas that had never gotten beyond Gauss's desk drawer, it excited Gauss to an unusual degree. His official report to the faculty stated: ''The dissertation submitted by Riemann offers convincing evidence of the author's thorough and penetrating investigations in those parts of the subject treated in the dissertation, of a creative, active, truly mathematical mind, and of a gloriously fertile originality.'' On Gauss's recommendation, Riemann became a privatdozent at Göttingen in 1854, subsisting on fees paid to him directly by those students who chose to attend his lectures. In 1857 he was promoted to the paying position of assistant professor. When Gauss died in 1855, his chair as professor of mathematics had gone to Dirichlet; when Dirichlet passed away four years later, Riemann himself succeeded to the position. But he had already contracted tuberculosis and was a dying man. In an attempt to cure the illness in a warmer climate, Riemann spent his last years in Italy, where he died in 1866 at the age of 39.

Bernhard Riemann
(1826–1866)

(The Bettmann Archive.)

On admission as a privatdozent at Göttingen, Riemann was called on to prove his mettle as a lecturer by delivering a *Habilitationsvortrag* (probationary lecture) before the faculty. For this trial, he submitted a list of three possible topics to the faculty; he felt well prepared to discuss either of the first two topics. Riemann rashly listed as his third offering a subject on which Gauss had reflected some 60 years, the foundations of geometry. Contrary to Riemann's expectations, Gauss selected this last topic for the test lecture. The strain of carrying out this difficult assignment, while also working as an assistant to Wilhelm Weber in a mathematical physics course brought on a temporary nervous breakdown.

In comfortable retrospect, the reading of *Über die Hypothesen welche der Geometrie zu Grunde liegen* (On the Hypotheses that Underlie the Foundation of Geometry) on June 10, 1854, is seen as one of the highlights of modern mathematical history. Because Riemann adapted his lecture to the intended audience, the entire philosophical faculty at Göttingen, it contained no specific examples and practically no formulas. Yet despite its intuitive character, it was extraordinarily powerful in generalities and suggestive in nature. It is said that nobody present understood Riemann's approach to geometry except the aged, legendary Gauss; and Weber said that even Gauss was perplexed.

Although Riemann's inaugural lecture did not immediately affect the intellectual world, its publication two years after his death caused a stir among those mathematicians who sought to fill in the details. With the discovery of competing geometries, no one geometry could be regarded as a collection of truths about physical space. Riemann, in assessing just what facts we can hold as certain, had the marvelous perception that the space of our experience may yet be finite. As he affirmed in the lecture:

> In the extension of space constructions to the immeasurably great, we must distinguish between unboundedness and infinite extent; the first pertains to the extent relations, the latter to the measure relations.

The surface of a sphere, with great circles understood as "lines," furnishes a good illustration of what Riemann meant; a "line" is not infinite in extent, because after a finite prolongation it returns on itself, yet it is unbounded in the sense that it can be traversed endlessly about the sphere. Riemann also argued in favor of geometry as an empirical science. He held that because observation on physical space had not confirmed the existence of parallel lines, one could, with as much reason as not, assume that every pair of lines would meet at some finitely distant point.

Mainly as a result of Riemann's thinking, attention was directed to a different type of non-Euclidean geometry, one that takes as its alternative to the parallel postulate the statement that there are no parallels to a line through a point not on the line, in short, that any pair of lines must meet somewhere. The geometry stemming from this assumption requires an even more drastic break with tradition than that made by Gauss, Bolyai, and Lobachevsky, for the Riemannian parallel axiom affects the import of the remaining axioms. In fact, it conflicts with a combination of Postulates 1 and 2 of Euclid.

When considering Euclidean geometry, we observed that Proposition 16 of Book I, the so-called exterior angle theorem, implies the existence of parallel lines. Thus, the supposition that there are no parallel lines would be inconsistent with the axioms of Euclid that led to this proposition. (A set of axioms is said to be consistent if they and the consequent theorems, taken together, produce no logical contradiction.) A brief examination of Euclid's proof of Proposition 16 reveals that in it he unconsciously assumed the infinite extent of straight lines; if lines are considered finite in length yet boundless, then his argument fails. In order to avoid a contradiction in the geometry suggested by Riemann, it is necessary to modify Euclid's axiom scheme further, by replacing Postulate 2 by a Postulate 2': A straight line is boundless. Another disturbing factor is that in assuming that two straight lines always meet, we allow the possibility that they will intersect at more than one point. This conflicts with Postulate 1 of Euclid, which requires that two points determine one and only one line. As if two serious departures from Euclid are not enough, Riemannian theory necessitates discarding the traditional form of Postulate 1 and adopting instead Postulate 1': Two distinct points determine at least one line.

The first of the non-Euclidean geometries involved nothing more than replacing the parallel postulate by the contrary statement that through a point not on a given line there passes more than one parallel to the line, but the new geometry of Riemann was achieved by making three changes in the axiom system framed by Euclid. At first sight this may seem an exorbitant price to pay, but the geometry developed from these hypotheses turned out to be as consistent as Euclid's. In fact, the resulting system corresponded to what Saccheri obtained under his hypothesis of the obtuse angle. The theorems are, to be sure, radically different from those to which we are accustomed and cannot fail to be a little disconcerting at first. For instance, one of Riemann's results is that all straight lines have the same finite length, and another asserts that all perpendiculars to a straight line must meet in a point. You can probably almost anticipate the theorem on angle sums of triangles: The sum of the angles of any triangle is greater than 180°. This may appear to contradict Legendre's first theorem. In the demonstration of that theorem, however, Legendre used the property that a straight line is infinite, a fact unavailable to Riemann.

This second period in the development of non-Euclidean geometry focused on the problem of its logical consistency, a problem whose solution brought to a close the many efforts to deduce the parallel postulate as a theorem from Euclid's other assumptions. As far as the creators of non-Euclidean geometry went in their investigation of the subject, no evident contradictions were encountered. They left open, however, the possibility that further effort might yet produce a contradictory statement as a theorem of the system. If this were to happen, not only would the approach of Saccheri be vindicated, but all the work of Gauss, Bolyai, Lobachevsky, and Riemann would prove to be, quite exactly, much ado about nothing. Consequently, doubt hung over their enterprise like a threatening cloud. Although they felt confident that their new geometries were as valid as Euclid's, this feeling was still an act of faith.

In the development of any axiom system—one defining a geometry or some other structure—there is the task of proving that the system is consistent by showing that no contradictions can occur. The consistency of any set of axioms is ordinarily established by finding a model whose elements and relations are specific interpretations of the undefined technical terms of the axiom system, and in which the axioms are realized as true statements about the model. Because the model itself will have been constructed within the framework of some other set of postulates, the most we can say about the axiom system under investigation is that it is relatively consistent with the system of the model. That is, the system of axioms that interests us will be seen to be consistent, provided that the axioms in the system that furnished the model are themselves consistent. This sort of consistency proof, of course, merely shifts the burden of proof of consistency from one system to another; and yet it is the best we can provide. Because no absolute method for testing the consistency of a set of axioms has ever been derived, one has to be content with relative consistency.

The basic idea of a proof of the relative consistency of non-Euclidean geometry was given in 1868 by the Italian geometer Eugenio Beltrami (1835–1900) in a paper entitled *Saggio di interpretazione della geometria non-Euclidea* (Essay on the Interpretation of Non-Euclidean Geometry). His method consisted of finding a model within Euclidean geometry that with suitable interpretations had the same postulation structure as the non-Euclidean geometry of Bolyai and Lobachevsky. The Euclidean surface on which Beltrami gave a partial representation of non-Euclidean geometry is called a pseudosphere, because it has certain properties in common with a sphere. The "pseudosphere" is the surface of revolution obtained by revolving a curve known as a tractrix about its asymptote. The resulting figure, which resembles an infinitely long horn, has constant negative curvature. Provided that the role of "straight line" between two points on the pseudosphere is taken by the geodesic connecting the points (the path of shortest length, like a great circle on a sphere), the geometry on this surface satisfies the postulates of Bolyai and Lobachevsky. The axioms of classical non-Euclidean geometry fit the geodesics on the pseudosphere, so every theorem, as a logical consequence of the axioms, can be interpreted as a fact on the pseudosphere. (Actually, Beltrami succeeded in representing only a limited portion of the Lobachevsky plane on the pseudosphere, and the task of interpreting all this non-Euclidean geometry fell to Felix Klein in 1871.) Beltrami's model made the "imaginary geometry" of Lobachevsky "real," in the mathematical sense, by giving it a readily visualizable interpretation on a familiar Euclidean surface. More important, the geometry on the pseudosphere provided an interpretation in which Euclid's parallel postulate could not possibly hold, yet in which all his other axioms were true.

Beltrami's success in realizing the geometry of Bolyai and Lobachevsky on the surface of the pseudosphere showed that non-Euclidean geometry is as logically consistent as the geometry of Euclid. Every theorem of their geometry, when interpreted in the model at hand, gets translated into an ordinary Euclidean result; any possible inconsistency within classical non-Euclidean geometry would reveal itself in a corresponding inconsistency in the Euclidean geometry of geodesics on the pseudosphere. Thus, if one is willing to grant that Euclidean geometry is consistent (and no satisfactory proof of this has ever been given), the non-Euclidean geometry of Bolyai and Lobachevsky is equally correct.

The discovery of a model for non-Euclidean geometry also revealed the inherent impossibility of proving the parallel postulate on the basis of the other axioms of Euclid. If the parallel postulate could be deduced from the remaining axioms, then it could be converted into a theorem in non-Euclidean geometry. (Aside from the fifth postulate, the axioms of Euclid are the same as the axioms of Bolyai and Lobachevsky.) But this theorem would contradict the form of the parallel hypothesis Bolyai and Lobachevsky used, making their non-Euclidean geometry inconsistent; and this is not the case. It is now clear why 2000 years of efforts to prove Euclid's parallel postulate from the rest of the postulates were in vain. If Euclidean geometry is assumed consistent, no proof can ever be found.

The non-Euclidean geometries had not yet become familiar topics among mathematicians when in 1871 and 1873 Felix Klein (1849–1925) published two monographs entitled *Über die Sogenannte Nicht-Euklidische Geometrie* (On the So-Called Non-Euclidean Geometry). Klein pointed out a clear distinction between two geometrical theories that assume Riemann's parallel postulate. In one of the systems, two straight lines meet at a single point, and in the other, two straight lines intersect at two points. His essential work in these publications was to furnish models for Lobachevskian geometry and the two kinds of Riemannian geometry. Perhaps the simplest model for the first type of Riemannian geometry is gotten by taking only half of a sphere with its boundary circle (one must identify as equal any two points on the boundary that are diametrically opposite). The great circle arcs on the hemisphere are the ''lines'' of this geometry. The sphere itself provides a model for the other type of geometry. We cannot be certain that Riemann recognized the possibility of both geometries; his celebrated memoir did not deal in specifics. For the two new non-Euclidean geometries, the case for consistency can be reduced to that for Euclidean geometry; that is, if one is willing to believe that Euclidean geometry is consistent, then so also are the two types of Riemannian geometry. The unanswered question whether Euclidean geometry can ever be shown to be internally consistent will be deferred for the time being.

Klein's most important achievement in geometry was the creation of the *Erlanger Programm,* a bold proposal to use the group concept to classify and unify the principal geometries then existing. When Klein was 23 years old, he had been made a full professor at the University of Erlangen. By the custom in German universities, he gave (1872) an inaugural lecture for presentation to the philosophical faculty and the senate. The views expressed in this address, ''A Comparative Review of Recent Research in Geometry,'' are known as the *Erlanger Programm* for geometry. In essence, Klein described geometry as the study of those properties of figures that remain invariant

Felix Klein
(1849–1925)

(*From* A Concise History of Mathematics *by Dirk Struik, 1967, Dover Publications, Inc., N.Y.*)

under a particular group of transformations. Not only did this provide a neat way of codifying all the seemingly unrelated geometries that were known in the late 1800s, but it indicated that new geometries could be defined by starting with different transformation groups.

In 1886, Klein accepted a call to the University of Göttingen to fill the role of leading research mathematician. His extensive activity rescued the university from the mathematical doldrums into which it had fallen after the premature death of Riemann and helped raise Göttingen to greater renown than even in the days of Gauss. Klein's stimulating influence can be seen from the number (48) of dissertations that he personally supervised. His "favorite pupil" in these Göttingen days was the Englishwoman Grace Chisholm Young.

Grace Chisholm was a pioneer in mathematics, one of the few women of her generation to achieve an international reputation in the subject. After graduating from Girton College, Cambridge, she had been advised to continue her studies under Klein. Permission was not entirely his to give, resting instead with the Ministry of Culture in Berlin; at that time, women could not be officially admitted to German universities. Through Klein's committed support and influence approval was secured, as long as the only exceptional cases were foreign graduate students. In the Fall of 1893, Grace Chisholm was one of three women allowed to attend regular lectures and seminars at Göttingen. (Also arriving on the same errand was Mary Winston, who had just completed a year of graduate study at the newly opened University of Chicago; with the awarding of her degree under Klein's direction in 1897, she became the first American woman to obtain a Ph.D. in mathematics from a foreign university.) Grace Chisholm's doctorate in 1895 was the first awarded in Germany, through the normal examination process, to a woman in any field whatever.

Returning to England upon completion of her doctoral research studies, she married William Henry Young (1863–1942), one of her former tutors. Young, between the ages of 25 and 35, had published practically no mathematics, but deliberately set himself to earning quite large sums through the private teaching of undergraduates. At the end of their first year together, the couple agreed to "throw up lucre," go abroad, and devote themselves energetically to research in the theory of functions of a real variable. Living mostly in Göttingen until 1908 and then in Switzerland, they carried out investigations that helped make England part of the modern mathematical world. Grace Chisholm Young, either alone or in collaboration with her husband, wrote 31 papers and several books.

The most conspicuous and distinctive feature of nineteenth-century mathematics was the introduction of a strictly logical approach; indeed, historians have frequently labeled this period the Age of Rigor in mathematics. A critical spirit soon permeated the whole of analysis, beginning with the calculus, and toward the end of the century this led to an overhaul of the foundations of geometry. In the extremely rapid development of the calculus in the 150 years after Newton and Leibniz discovered it, mathematicians plunged ahead without logical support. They were aware of the requirement of proof, but this tended to take the form of a search for an explanation that would be intuitively plausible instead of mathematically exact. Rules of procedure—differentiation and integration, but also rearrangement of series—were normally used in a formalistic way without regard for their validity. The success in applying the conclusions to physical problems gave confidence that the mathematics itself must be correct. "Go forward and faith will come to you," d'Alembert is supposed to have said. The highly unsatisfactory state of the logical underpinnings of the calculus could not long continue unchallenged, and in the second half of the 1700s, an effort was made to supply the missing rigor in its fundamental conceptions.

For the calculus to be put on a sound footing, the fetters of geometrical constructions and mechanical analogies had to be removed; and the subject had to be built anew solely on considerations derived from arithmetic and algebra. Many mathematicians of the time, among them d'Alembert, insisted that the logical basis of the calculus was to be found in the method of limits. Although the leaders of this movement understood the idea of a limit a little better than most of their contemporaries, they failed to give the concept a clear and precise definition that would make it logically tenable. Augustin Louis Cauchy (1789–1857) finally developed an acceptable theory of limits, and in doing so removed much of the doubt about the soundness of the notion.

Cauchy acquired his early education from his father, a barrister and master of classical studies. In 1805 he entered the Ecole Polytechnique in Paris, proceeding two years later to the civil engineering school Ecole des Ponts et Chaussées (School of Bridges and Roads). Thereafter he worked briefly (1810–1813) as a military engineer at the harbor of Cherbourg, where Napoleon had started to build a naval base for his intended invasion of England. Cauchy first attracted the attention of the leading mathematicians of France, among them the aging Lagrange, by a series of brilliant memoirs that date from 1811 when he was but 22 years old. For reasons of health, Cauchy was persuaded by Lagrange to abandon the profession of engineering and to devote himself to pure science and mathematics. Cauchy returned to Paris in 1813 to accept an instructorship at the Ecole Polytechnique, rising to full professor in 1816. As we have observed, it was the practice of scientific academies to offer prizes for solving major outstanding problems. Cauchy won a grand prize from the French Académie des Sciences in 1816 for a 300-page paper on the propagation of waves at the surface of a liquid.

Although he was worthy of admission to the Académie, Cauchy lost the respect of many scientists at this time by being appointed to the Académie by royal decree instead of being elected by its members. Moreover, the seat that became Cauchy's had become vacant because of the political expulsion of the noted geometer, Gaspard Monge. An active partisan of the Revolution, and subsequently an enthusiastic supporter of Napoleon, Monge was not forgiven by the returned rulers of France during the Bourbon Restoration of 1814.

The changing political situation in France had unexpected effects on Cauchy's career. He was an ardent royalist, supporting the Bourbon monarchy almost to the point of perversity. When Louis-Philippe came to power following the Revolution of 1830, Cauchy refused, as a loyal supporter of the deposed King Charles X, to swear allegiance to the new government. Abandoning his academic post at the Ecole Polytechnique, Cauchy went into a self-imposed exile. He did not regain public employment in France for eighteen years. Cauchy spent part of this time teaching at the University of Turin (1831–1833) and part in Prague (1833–1838), where the ex-monarch had summoned Cauchy to tutor his grandson. When Cauchy returned to France in 1838, he was permitted to resume his post at the Académie but not his professorship at the Ecole Polytechnique, because he was still adamant in refusing to take an oath of allegiance. He had to be content to support himself by teaching at a Jesuit college. When Louis-Philippe was himself ousted in the Revolution of 1848 and oaths were no longer required, Cauchy was appointed professor of celestial mechanics at the Sorbonne, where he remained for the rest of his career.

Like Euler's, Cauchy's mathematical output was prodigious and embraced almost all branches of the subject. In total he wrote eight full-length books as well as 789 papers, which in a modern edition fill 26 large volumes. Cauchy was so active that he had to establish a private journal, the *Exercices de mathématiques* (1826–1830), continued in a second series as *Exercices d'analyse mathématique et de physique,* whose issues were composed entirely of Cauchy's own work. In 1835, the Académie des Sciences began to publish a weekly bulletin, the *Comptes rendus*. Cauchy inundated this journal with a succession of his papers to such an extent that the Académie became alarmed at the rapidly mounting printing bill; to cope with Cauchy's staggering productivity, it passed a rule (which is still in force today) limiting each contribution to a length of four pages. In all, the *Comptes rendus* printed 589 articles by Cauchy.

Cauchy was not the first mathematician to attempt a logical underpinning for the calculus. In his influential textbook *Théorie des fonctions analytiques* (1797), Lagrange tried to base the whole subject on a series expansion suggested by Brook Taylor. Lagrange implicitly assumed that any function f has an infinite Taylor series of the form

$$f(x + h) = f(x) + ph + q\frac{h^2}{2!} + r\frac{h^3}{3!} + \cdots,$$

where $p, q, r, \ldots$ are new functions of x different from but somehow derived from f. In the expression

$$\frac{f(x + h) - f(x)}{h} = p + V,$$

V goes to zero as h does, so Lagrange's idea was to define the derivative of f with respect to x as the coefficient of h in the linear term of the Taylor series expansion for $f(x + h)$. He introduced an accent mark as the notation to indicate the derivative, writing $p = f'(x)$, and called this the first derived function of f (thus providing the source of our term "derivative"). Higher order derivatives are defined to be the coefficients of $h^2/2!$, $h^3/3!$, . . . and denoted by $f''(x)$, $f'''(x)$, ... Through this new approach Lagrange concluded that the calculus could be "redeemed from all considerations of the infinitely small, from vanishing quantities, from limits and from fluxions, and reduced to algebraic analysis of finite quantities."

The *Théorie des fonctions analytiques* was a work of grand design, systematically developing the major properties of the calculus based on this definition of the derivative. By freeing the differential calculus from "every illicit supposition" and relying on the method of power series, Lagrange declared that he could bring to the subject "the rigor of the ancient demonstrations." Although the *Théorie des fonctions analytiques* was greatly applauded at first, mathematicians soon perceived certain flaws in Lagrange's scheme. Expanding a function in a Taylor series is not an orthodox algebraic process, as Lagrange erroneously thought. It is still necessary to consider the matter of convergence, which must be discussed in terms of limits; his claim to have circumvented the limit concept was therefore illusory. Nor does every function allow a Taylor series expansion. For a function to have this series representation, all its derivatives must first exist: the series is constructed from the derivatives, and not the other way around. A further difficulty, which Cauchy was the first to point out, is that a given Taylor series does not determine a function uniquely. For example, the functions $f(x) = e^{-x^2}$, with $f(0) = 0$; and $g(x) = e^{-x^2} + e^{-x^2}$, with $g(0) = 0$, both have the same Taylor series at $x = 0$ (all the derivatives of e^{-1/x^2} vanish at $x = 0$).

Even though the Taylor series process proved inadequate for his primary purpose, Lagrange nonetheless made a start in the direction of setting the known results of the calculus within a rigorous framework. Inspired by Lagrange's investigations, Cauchy took up the task of providing a logically consistent foundation for the subject, one which did not rest on any appeal to geometric notions such as slope or tangent. His program of rigorization rested on "purely analytic and direct proof," with an inequality-based translation of the limit concept serving as the cornerstone. Many of the details of Cauchy's new approach to analysis were developed in the classrooms of the Paris colleges in which he taught. In an age that made little distinction between research work and teaching material, he was not above putting his notes into book form "for the greater use of the students." Cauchy's elementary textbooks, combining clear exposition with strict reasoning, were far ahead of the loose and slack works of the previous generation, and set the style for future mathematical writing.

The formulation of elementary calculus current today in most textbooks is essentially what Cauchy expounded in three great treatises: *Cours d'analyse de l'Ecole Royale Polytechnique* (1821), *Résumé des lecons sur le calcul infinitesimal* (1823), and *Lecons sur le calcul différentiel* (1829). The first and most important of these works is based, as the title indicates, on the lectures in analysis that he had been presenting at the Ecole Polytechnique. In the introduction to the *Cours d'analyse*, Cauchy was explicit in his aspiration to remove the immense obscurity that reigned in analysis: "As to methods, I tried to fill them with all the rigor one requires in geometry, so as never to resort to the reasoning taken from the generality of algebra.'' In the ''generality of algebra'' he included the habits of assuming for complex numbers all the properties of real numbers, and for infinite series the properties of finite sums.

The *Cours d'analyse* contains the definition of limit that was used until the 1870s, when the modern ε-δ version was first put forth. Banishing the rough, geometrically intuitional concepts of the early analysts, Cauchy appealed to the notions of number, variable, and function, saying:

> When successive values attributed to a variable approach indefinitely to a fixed value so as to end by differing from it by as little as one wishes, this last is called the limit of all the others.

From this purely arithmetical definition of limit, Cauchy then went on to give the central concepts of the calculus—continuity, differentiability, and the definite integral—a formal precision that had been lacking in the work of his predecessors. It is remarkable how closely his definition of a continuous function, for instance, parallels what we use today:

> The function $f(x)$ is continuous within given limits if between these limits an infinitely small increment i in the variable x always produces an infinitely small increment, $f(x + i) - f(x)$, in the function itself.

Because he swept away the old foundations based largely on faith and formal techniques, Cauchy has come to be regarded as the creator of calculus in the modern sense. Calculus after Cauchy was no longer a set of problem-solving operations but a set of theorems based on rigorous definitions.

Indiscriminate manipulation of infinite series, with the question of convergence blithely ignored, had gone on for over a century before the time of Cauchy. Although better mathematicians were generally wary of divergent series, lesser talents were often led to completely absurd results when working with them. A well-known example involves the series expansion

$$\frac{1}{1 + x} = 1 - x + x^2 - x^3 + \cdots.$$

Guido Grandi (1671–1762), a professor at Pisa, in his *Quadratura Circuli et Hyperbolae* of 1703, set $x = 1$ in this expression and grouped the terms on the right-hand side in pairs to obtain sums of both 0 and 1/2:

$$\frac{1}{1 + 1} = (1 - 1) + (1 - 1) + \cdots = 0 + 0 + \cdots = 0.$$

This, he argued, proved mathematically that the world could be created out of nothing. Cauchy showed that the convergence, and thus the meaning, of the series in question is assured only when $|x| < 1$.

To bring order to these logically questionable practices, Cauchy devotes much of the *Cours d'analyse* to a careful treatment of series. He rejects the attribution of a sum to a nonconvergent series, because no numerical verification is possible; and stresses that the convergence of an infinite series $u_0 + u_1 + u_2 + \cdots$ is determined by the tendency of the finite, partial sums $S_n = u_0 + u_1 + \cdots + u_n$ to a limiting value:

If, for ever increasing values of n, the sum S_n indefinitely approaches a certain limit S, the series will be called convergent, and the limit in question will be called the sum of the series. If, on the contrary, while n increases indefinitely, the sum S_n does not approach a fixed limit, the series will be divergent and will no longer have a sum.

A page later Cauchy indicates the profound result that today is called the "Cauchy convergence criterion." Specifically: A sequence S_n converges to a limit if and only if the difference $S_m - S_n$ can be made less in absolute value than any assignable quantity, provided m and n are sufficiently large. The great merit in this criterion is that it provides an internal condition that can be checked by looking at the sequence itself, not at a proposed limit. Cauchy has little difficulty in proving that his convergence principle is necessary, but he avoids the difficulty of showing its sufficiency and simply claims that when the condition is fulfilled, the convergence of the sequence is assured. A more satisfactory proof requires a finer analysis of the structure of the real line (more precisely, the completeness of the real numbers), something that was unavailable at that time.

The topics that Cauchy treated in the *Cours d'analyse* were subtle, and by modern standards even he made mistakes at the lowest level of his analysis. For instance, when handling infinite series of functions he did not distinguish between what we now call pointwise convergence and uniform convergence over an interval. The most famous "false theorem" contained in the *Cours d'analyse* asserts that the sum-function $S(x) = u_0(x) + u_1(x) + u_2(x) + \cdots$ of a convergent series of continuous functions $u_n(x)$ is itself a continuous function. In an 1826 paper on the binomial series, Abel pointed out that this assertion is false. He remarked, "It seems to me that this theorem of Cauchy admits exceptions. For example the series

$$\sin(x) - \frac{1}{2} \sin(2x) + \frac{1}{3} \sin(3x) - \cdots$$

is not continuous at $x = (2n + 1)\pi$, n being an integer."

Cauchy's standard of rigor was immeasurably in advance of his contemporaries—except for Gauss, whose insistence on careful deduction was the admiration of all. This standard also served as a model for a generation of mathematicians. "He is at present," Abel said of Cauchy, "the only one who knows how mathematics should be treated." Gradually, mathematicians came to realize that the last word in rigor had not been said and such verbal phrases of Cauchy's as "approaches indefinitely," "infinitely small increase," and "as little as one pleases," lacked meaningful exactness. The figure who towers above all others in the movement to make these ideas still more precise, whose name has become a synonym with rigor, is Weierstrass.

Many great mathematicians give evidence of their unusual ability at an early age, but there are those who abandon other lines of work to turn to mathematics at a later period. Weierstrass is in this latter class; he showed his special aptitude along this line at an age greater than that ever reached by Galois, and at which Gauss had already completed the *Disquisitiones Arithmeticae*. Weierstrass was such a striking exception to the common notion that a mathematician of first rank must make his mark early in life that it is worth taking a closer look at his career.

Karl Weierstrass (1815–1897) was hindered in his early years by a domineering father, a customs officer in a salt works, who tried to force his son into an uncongenial

Karl Weierstrass
(1815–1897)

(From A Concise History of Mathematics *by Dirk Struik, 1967, Dover Publications, Inc., N.Y.)*

occupation. The young boy had made a brilliant record in secondary school, winning several prizes before graduation. At his father's insistence, Weierstrass spent the four years from 1834 to 1838 at the University of Bonn, trying to master law and public finance; this was the usual course of study for one seeking an administrative post in the Prussian civil service. More attracted to a wild life of drinking and fencing, he came to shun all lectures and left the university without receiving a diploma. Weierstrass next enrolled (1839) in an accelerated program at the Academy of Münster, which was near his home, to prepare himself for teaching mathematics at the secondary level. He was fortunate to attend the lectures of Christoph Gudermann (1798–1852), his only significant mathematics teacher, who aroused his interest in the newly created theory of elliptic functions of Abel and Jacobi. We are told that at the opening lecture of Gudermann's course on elliptic functions there were 13 students, but for the rest of the semester the only one in attendance was Weierstrass. As part of his state examination, he was given a mathematical problem to be written and completed in six months' time. Weierstrass, on his own request, was assigned a difficult task by Gudermann, namely, to represent an elliptic function as the quotient of two convergent power series. The response, Weierstrass's first mathematical paper, was praised by Gudermann as an important advance. The article might have changed the course of Weierstrass's life if it had appeared in print, but because Münster awarded only teaching certificates and not doctorates, publication of such papers was not customary there. (The paper remained in manuscript until 1894, when it was printed in Weierstrass's collected works.)

Weierstrass finally secured his teaching license at the age of 26, and from 1841 until 1854 presented such subjects as science, writing, and gymnastics to youngsters in obscure secondary schools in Prussia. In the "unending dreariness and boredom of these years," with no access to a mathematical library nor any colleague for mathematical discussions, he carried on a remarkable study of abelian functions. It is reported that when Weierstrass once missed the start of morning classes, the school principal went to his home only to find that Weierstrass had worked the whole night through on his research, completely unaware of the approach of dawn. He asked the principal to excuse his tardiness, for he was on the verge of a discovery that would surprise the mathematical world. These ideas eventually appeared in the program of studies for 1848–1849 of the Catholic Gymnasium in Braunsberg, East Prussia, perhaps the last place that one would look for an epochmaking memoir. Had Weierstrass's first publication caught the eye of any professional mathematician, it would indeed have created a furor. As it was, the article was totally unintelligible to the students, parents, and teachers who read the program.

Fortunately for mathematics, Weierstrass's next memoir, *Zur Theorie der Abelschen Functionen* fared better, appearing in *Crelle's Journal* for 1854. This research paper elicited enormous interest, with the mathematician Liouville calling it ''one of those works that mark an epoch in science.'' By 1855, the year of Gauss's death, Weierstrass was just gaining the recognition his talent deserved. The University of Königsberg bestowed an honorary doctorate on the once obscure teacher, and the ministry of education granted him a year's leave with pay to pursue his research.

Germany at this time was not a unified nation but a collection of many small, independent states. It had no educational center such as Paris or London; yet the various states took pride in their institutions of higher learning and eagerly competed for great scholars. It was therefore not uncommon to find German professors transferring from one chair to another with a higher endowment. Gauss's death brought about several such shifts, thereby opening the way for Weierstrass. When Dirichlet was called to succeed Gauss at Göttingen, Dirichlet's place in Berlin was occupied by his friend Kummer, who had been at Breslau. Determined not to return to the provinces of Prussia, Weierstrass applied for Kummer's vacant post at Breslau. Kummer recommended against Weierstrass's application, apparently feeling that the publication of one paper, however noteworthy, was not a sufficient guarantee of future productivity. However, when Weierstrass was offered a special professorship at any Austrian university of his choice, Kummer changed his mind. He secured (1856) a position for Weierstrass at the Industrial Institute at Berlin (the forerunner of Berlin Technical University) teaching technical subjects, and at the same time an assistant professorship at the University of Berlin. Another 8 years passed before, in 1864, he was able to withdraw completely from the Industrial Institute and assume a full professorship at the University of Berlin, where he spent the last 40 years of his life. What is noteworthy is that Weierstrass did not begin his professional tenure at Berlin until an age when most mathematicians cease their creative work.

Weierstrass, despite his late start, was the world's greatest analyst during the last third of the nineteenth century—the ''father of modern analysis.'' It is difficult to date his contributions, because he was slow in publication, preferring instead to make his ideas first known to the general mathematical public through his lectures at the University of Berlin. No one else offered the same subject matter, so these lectures enjoyed an increasing reputation and attracted the most gifted students from around the world. Weierstrass's rare mathematical gifts were matched by his resourceful teaching. Each of his meticulously prepared lectures was a logically intricate exercise in creative work for himself and his students alike. He often took his listeners to the frontiers of research, for the content of his course was usually new mathematics in the making. Indifferent to fame, he endeared himself to students with the liberality with which he allowed them to develop results that he would not take time to write up. The lecture notes and personal research of such students as Georg Cantor (1845–1918), Sonya Kovalevsky (1850–1891), Gösta Mittag-Leffler (1846–1927), Carl Runge (1856–1927), Max Planck (1858–1947), Otto Holder (1859–1937), and David Hilbert (1862–1943) spread the new analysis created by Weierstrass. One of those who attended his lectures in 1872 on the calculus of variations, Oskar Bolza (1857–1942), went on to teach at the newly founded University of Chicago for 17 years; there he was influential in building up a mathematics faculty of unusual strength.

Any discussion of Weierstrass's life would be incomplete without mentioning his close personal relationship with the most talented of all his students, Sonya Kovalevsky. Unlike many of her female contemporaries, who were prohibited from developing their mathematical talents, Sonya Kovalevsky was fortunate in having the opportunity to attend a German university. But it cannot be said that Germany had any real sympathy for the higher education of women. Until the turn of the century, German universities had admitted no women students except those who came from abroad, and even when this barrier had fallen, women in Prussia still did not have the right to qualify for higher degrees.

Sonya Korvin-Krukovsky (1850–1891) later known as Sonya Kovalevsky, was born in Moscow to a family of the minor nobility, her father being an artillery general. Although regarded as the greatest woman mathematician before 1900, she did not receive a formal mathematical education in her own homeland. In her early childhood she was exposed to mathematics in a curious way. During the renovation of the family country estate, one of the children's rooms was temporarily wallpapered with some lithographed notes of Ostrogradski's lectures on differential and integral calculus. In her published recollections, *A Russian Childhood,* Kovalevsky recalled that she had "passed hours before that mysterious wall, trying to decipher even a single phrase and to discover the order in which the sheets ought to follow each other." In St. Petersburg, at the age of 15, when she took more rigorous private lessons under the tutelage of a professor at the naval academy, the teacher was astounded with the quickness with which she grasped the ideas "just as if I had known them before."

Russia's dismal showing in the Crimean War (1853–1856) had been widely attributed to its scientific and technological backwardness, a backwardness for which a disproportionate emphasis on the classical university education was partly responsible. The immediate reaction was another reform of the university system; purchase of foreign textbooks was allowed without interference from the censors, students regained their traditional right to attend Western universities, and the decree against hiring foreign professors was dropped. Whereas a new university statute of 1863 opened the institutions of higher learning to all social classes, women were still barred by law; the lower urban classes and peasantry were still kept out by tradition and economics. The very idea of higher education for women seemed ludicrous, and there was strong resistance to any suggestion that they deserved the same secondary-school training as men.

One of the few ways for a Russian woman to continue her education was to study in a foreign university. It was not unusual at that time for a young Russian woman to enter into a marriage of convenience in order to gain the freedom to travel abroad, with

Sonya Kovalevsky
(1850–1891)

(The Bettmann Archive.)

or without the attendance of her husband. Sonya Korvin-Krukovsky followed this pattern; in 1868 she married Vladimir Kovalevsky, a geology student at Moscow University. In the spring of 1869, the Kovalevskys left for Heidelberg University, one of Germany's oldest institutions and then famous for such figures of modern science as Bunsen, Helmholtz, and Kirchhoff. Madame Kovalevsky obtained special permission to attend several courses, but the teaching of the mathematician Leo Konigsberger, one of Weierstrass's first pupils, made the strongest impression on her. At the end of 1870, she transferred to Berlin to learn from the master himself. (Her husband had already left for Jena to get a doctorate.)

At the age of 55, the lifelong bachelor Weierstrass met the 20-year-old Russian student. Soon impressed by her ability, Weierstrass claimed that Kovalevsky had ''the gift of intuitive genius'' to a degree he had seldom found even among his older and more advanced students. But the University of Berlin was far more conservative than Heidelberg, and as a woman, Kovalevsky was not allowed to take courses. Weierstrass first tried to get the university senate to admit her to his lectures, and when this was refused, he offered to work with her privately. For the next four years, he shared his lecture notes, his unpublished manuscripts, and his latest theories on analysis with a ''refreshingly enthusiastic participant.'' By 1874, Sonya Kovalevsky had completed three outstanding research papers: *On the Theory of Partial Differential Equations, On the Reduction of a Definite Class of Abelian Integrals,* and *Observations on Laplace's Research on the Form of Saturn's Rings.* The strength of these works, along with Weierstrass's strong recommendation, qualified her (1874) for a doctorate without examination, and in absentia, from the University of Göttingen (which she never actually attended). She had petitioned that the oral examination be waived as her German was inadequate. This was the first time that a woman had applied for a higher degree in mathematics, so Weierstrass requested that the university be particularly careful to uphold high standards in judging her. The degree was awarded summa cum laude.

In spite of an advanced degree and strong letters of recommendation from Weierstrass, Kovalevsky was rebuffed in her attempts to get an academic position anywhere in Europe. For this reason, she returned to St. Petersburg. During her absence from Russia, the advocates of higher education for women had managed to institute a program known as Higher Courses for Women in several universities, Moscow in 1872 and St. Petersburg in 1878. Kovalevsky's mathematical training qualified her to teach in the program, but Russian law made it impossible for a woman to occupy a university position, even if she met the highest standards of scholarship. For nine years, she occupied herself in various nonscientific pursuits: writing newspaper articles, theater reviews, poetry, and even a small novel. Although Weierstrass's letters kept her in touch with recent developments, she seemed on the verge of abandoning mathematics. At his suggestion, she read a paper on abelian integrals at a scientific congress in St. Petersburg in 1879 and another on the refraction of light through a crystalline medium at Odessa in 1883. Then in 1883, Kovalevsky was invited by Gösta Mittag-Leffler, another of Weierstrass's distinguished students, to teach at the newly founded University of Stockholm. One newspaper greeted her arrival:

> Today we do not herald the arrival of some vulgar insignificant prince of noble blood. No, the Princess of Science, Madame Kovalevsky has honored our city with her arrival. She is the first woman to lecture in all of Sweden.

Though the "princess" was first appointed a privatdozent, the lowest possible rank on the university scale, she was soon given a post that carried a salary with it. Mittag-Leffler managed to raise private funds to match the small sum the university offered. Over the succeeding four years, Kovalevsky taught (in imperfect German) twelve courses on the newest and most advanced topics in analysis.

Sonya Kovalevsky's prominence as a professional mathematician reached its peak in 1888, when she received the famous Prix Bordin from the French Académie des Sciences for her paper *On the Rotation of a Solid Body About a Fixed Point*. Voting unanimously in favor of the winning memoir, the selection committee, unaware that the winner was a woman since each entry was submitted anonymously, "recognized in this work not only the power of an expansive and profound mind, but also a great spirit of invention." Because of the exceptional merit of the work, the monetary value of the prize was raised from 3000 to 5000 francs, a considerable sum of money at that time. Weierstrass rejoiced at this, writing to Kovalevsky, "Competent judges have now given their verdict that my favorite pupil, my 'weakness,' is not a frivolous marionette." When word of her great success reached Russia, Kovalevsky finally received recognition in her own country. The St. Petersburg Academy of Sciences elected her a corresponding member, a purely honorific title, to be sure, but she was the first woman so honored. In 1889, Stockholm appointed her a full professor of mathematics, the office she held until her untimely death. Less than two years later, at the height of her career, Sonya Kovalevsky died of influenza, which at the time was epidemic. Mittag-Leffler wrote of her:

> She came to us from the center of modern science full of faith and enthusiasm for the ideas of her great master in Berlin, the venerable old man who has outlived his favorite pupil. Her works, all of which belong to the same order of ideas, have shown by new discoveries the power of Weierstrass' system.

Weierstrass often demonstrated the need for a strictly logical approach to mathematics by constructing counterexamples to plausible and widely held notions. The prime instance of this is the case of the continuous, but nowhere differentiable, function. Characteristic of the first half of the nineteenth century was the loose way in which geometrical and other intuitive ideas were used in analytic proofs. Although Cauchy had clarified the foundations of calculus, nearly all the mathematicians of his era believed erroneously that a continuous function, pictured by their spatial intuition as a ''smooth'' curve, must be differentiable at most points. André-Marie Ampère, who is best known for his work in electricity, wrote a paper *Recherches sur quelques points de la théorie des fonctions dérivées* in 1806. In this work, he tried to prove (or so it seems, because the paper is confusing to read) the false proposition that every continuous function has a derivative except possibly for a few isolated places. His contemporaries realized that the demonstration was wrong, but there was little doubt about the correctness of the assertion. The mathematical world was profoundly surprised when Weierstrass read a paper to the Berlin Academy of Sciences in 1872, presenting an example of a continuous function that has a derivative at no point of the real line—or what is the same, a continuous curve having no tangent at any point. Weierstrass's example,

$$\sum_{n=0}^{\infty} a^n \cos (b^n x), \qquad 0 < a < 1, \, ab > 1 + \tfrac{3}{2},$$

is supposed to have been given in his classroom lectures as far back as 1861, although it was published, with his assent, for the first time in 1874 by one of his pupils, Paul du Bois-Reymond. An equally rude shock to the geometric preconceptions of what a continuous curve ought to be was Guiseppe Peano's discovery (1890) of a continuous curve that passes through every point of a square. Those mathematicians who considered the class of continuous functions coextensive with the functions pictured by their geometric intuition treated the examples of Weierstrass and Peano as pathological cases, quite outside the field of orthodox mathematics. ''I turn away with fear and horror,'' wrote Hermite, ''from this lamentable sore of continuous functions without derivatives.'' But the real significance of the varieties of abnormal behavior that continuous functions can show is that it taught mathematicians caution about the deceptive evidence of quasi-geometric proofs.

The critical reorganization of the foundations of analysis, fostered largely by Cauchy in the early 1800s, was continued with notable success in the second half of the century by Weierstrass and his followers. It is customary to say that Weierstrass ''arithmetized analysis'' (a phrase Felix Klein coined in 1895), by which is meant that he freed analysis from the geometrical reasonings and intuitive understandings so prevalent at the time. Among other things, this called for giving a more precise formulation of the idea of a limit. Cauchy's definition had used the expression ''approaches indefinitely,'' which suggests vague intuitions of continuous motion. With Weierstrass, the now-accepted ε-δ terminology became part of the language of rigorous analysis:

A function $f(x)$ has a limit L at $x = x_0$ if for any positive number ε, there exists a δ such that $|f(x) - L| < ε$ for all x in the deleted interval $0 < |x - x_0| < δ$.

The unequivocal language and symbolism—the fearsome ''epsilontics'' of contemporary undergraduates—banished the older vague ideas and made really convincing proofs possible. Along with the new definition, Weierstrass also gave us the standard notation $\lim_{x \to x_0}$ to express "the limit as x approaches x_0." In a paper of 1841, he wrote "lim," and in 1854, "$\lim_{n = \infty} a_n = \infty$."

Another cause for concern was the lack of precise meaning in the phrase "real number." Cauchy in his *Cours d'analyse* had regarded a real number as the limit of an infinite sequence of rational numbers. But this involves a serious error in reasoning. The limit, if irrational, does not logically exist until irrational numbers are defined; that is, use of the definition of limit requires the demonstrated existence of the very quantity whose definition is being attempted. Aware of the shortcomings in Cauchy's approach, Weierstrass advocated developing an algebraically self-contained notion of a real number that in no way presupposed their existence. The ultimate aim was to rest the majestic edifice of analysis on the arithmetic of the positive integers alone, to "arithmetize analysis." Weierstrass, Cantor, and Dedekind are credited with having made the basic advance in building rigorous theories of the real-number system. However, no account given by Weierstrass himself survives except in the lecture notes of his students. The Cantor-Dedekind theory of irrational numbers is perhaps a story best avoided, for as one recent writer said, "The irrational number, logically defined, is an intellectual monster."

The interest in postulational methods generated by Weierstrass's arithmetization of analysis carried over into attempts to give Euclidean geometry a strict axiomatic foundation. Euclid's *Elements* contain the earliest attempt to axiomatize geometry, but it was far, very far, from being a perfect work. Definitions were given that did not define, axioms were implied but not stated, and certain proofs were needlessly complicated. One should not, of course, be too hard on Euclid. All writers on elementary geometry to almost 1900 used their visual imaginations freely, though usually unconsciously, in assuming further facts not guaranteed by Euclid's axioms. Recognition of these deficiencies finally forced mathematicians to reexamine the foundations of Euclidean geometry. The aim of geometers in the late 1800s was to give a set of independent axioms from which it would be possible to prove, with the rigor the times demanded, all the familiar theorems of Euclid's geometry. Unlike Euclid, who tried to give a description of the basic objects with which the axioms deal, these geometers understood that not all terms can be defined; the properties of the undefined terms are specified solely by the axioms.

The first rigorous attempt to remove the hidden assumptions that marred Euclid's work was made by Moritz Pasch (1843–1930) in his *Vorlesungen Über Neuere Geometrie* (1882). Pasch's philosophy for choosing the axioms of his treatment of geometry was that the initial notions, the so-called nuclear concepts, should have a common basis of acceptance in human experience. He emphasized, however, that after the basic axioms had been introduced, the logical deduction of the remaining propositions should proceed without regard to empirical significance. The term *point* was allowed, but not *line*, because no one had ever observed a complete straight line; instead, the notion of *line segment* was taken as undefined. Likewise, *planar surface*, but not *plane*, was a primitive term. Pasch axiomatized the order of points on a line (or the relation of "betweenness") for the first time; Euclid never mentioned this notion explicitly but used it frequently. Pasch's groundbreaking ideas were modified and elaborated by a flourishing Italian school of geometers whose leaders were Peano, Veronese, Pieri,

Padoa, Burali-Forti, and Levi-Civita. Guiseppe Peano's (1858–1932) *I Principii di geometria, logicamente espositi* (1889) is largely a translation of Pasch's treatise into the notation of a symbolic logic that he had invented. The highly complicated symbolism made the work difficult to follow, even the subject of ridicule by some mathematicians. Mario Pieri (1860–1904), a pupil of Peano, proposed a novel system of axioms for Euclidean geometry based on the two undefined concepts of *point* and *motion*. However, the postulational development of the subject that clings closest to the Euclidean tradition, with its customary primitive notions of *point*, *line*, and *plane*, was attributable to the great German mathematician David Hilbert. Because of the clear simplicity of its axioms, stated with a minimum of unfamiliar and often unnecessary symbolism, his treatment came to receive the widest acceptance. Hilbert succeeded, as neither Pasch nor Peano had been able, in convincing mathematicians of the abstract and purely formal nature of geometry.

David Hilbert (1862–1943), the son of a district judge, was born near Königsberg in East Prussia. In 1880, he enrolled at the local university, the University of Königsberg, where Immanuel Kant had studied and taught. Although Königsberg was far removed from the center of things in Göttingen and Berlin, it had developed a distinguished mathematical tradition in the 1800s, beginning with the activity of Carl Gustav Jacobi in the early decades and maintained subsequently by Ferdinand Lindemann. Lindemann had proved the long-suspected transcendence of the number π, but accomplished little of note thereafter. Under his direction, Hilbert completed his doctoral thesis in 1885. Over the next few years, he rose in the academic ranks at Königsberg. He qualified as a privatdozent in 1886, was appointed to an assistant professorship in 1892, and in the following year advanced to full professor, succeeding Lindemann. Then, on the initiative of Felix Klein, he was offered a chair in mathematics at Göttingen. Klein later recalled, ''My colleagues had suggested at the time that I should call a comfortable young colleague for me, but I told them, 'I shall call the most uncomfortable one.' '' Hilbert arrived at Göttingen in 1895, almost exactly one hundred years after the arrival of Gauss, to remain there until his official retirement in 1930. The world had looked on Gauss as the greatest mathematician living in the first decades of the nineteenth century, the man who kept Göttingen alive with creative impulse. Hilbert was to play a similar role in the new century.

During the winter term of 1898–1899, Hilbert broke with his current interest in the theory of algebraic number fields to present a series of lectures on the postulational development of Euclidean geometry. These were edited in a slim volume *Grundlagen der Geometrie* (Foundations of Geometry) published as a memorial address in connection with the celebration at the unveiling of the Gauss-Weber monument at Göttingen in June 1899, the same year in which Pieri's studies were made available. Within a few months after its printing, Hilbert's 92-page book attracted attention all over the mathematical world. A French edition appeared soon after the German publication; and an English rendition was brought out in 1902 by E. J. Townsend, who received his doctorate under Hilbert. The work was gradually modernized and expanded through a sequence of appendixes that appeared in seven German editions in the author's lifetime. An eleventh edition (1972), prepared by Paul Bernays, is twice as long as the original. Legendre's *Eléments de géométrie* passed through more editions than Hilbert's *Grundlagen der Geometrie,* but the two books are not comparable. Legendre's work was a textbook addressed to schoolboys, whereas Hilbert's, though widely read by mathematicians, exerted no immediate influence on elementary teaching. Yet next to the *Elements* itself, the *Grundlagen der Geometrie* has become the second most influential work written on elementary geometry. The axiomatic method as it is understood today was initiated by Pasch, but it was the *Grundlagen der Geometrie,* backed by Hilbert's great mathematical authority, that was most responsible for spreading the axiomatic viewpoint.

Hilbert began his classic *Grundlagen der Geometrie* with the words, ''Let us consider three distinct systems of things.'' The ''things,'' his undefined terms, are called *points, lines,* and *planes;* and these are connected by three undefined relations, which he chose to indicate by the familiar names *incidence* (on), *order* (betweenness), and *congruence.* The 21 axioms of his treatment are given in 5 groups dealing respectively with incidence, order, congruence, parallelism, and continuity. This arrangement illuminates how the logical difficulties in Euclid were overcome. For instance, his axioms of congruence remedy the defect of superposition—it was postulated that two triangles are congruent if two sides and the included angle of one are congruent with two sides and the included angle of the other. Hilbert's second group of axioms built on the earlier work of Pasch by including what is today called Pasch's postulate: A line that intersects one side of a triangle, not at a vertex, must intersect another side of the triangle. The fourth group consists of just one axiom, Playfair's form of the parallel postulate. The number of axioms may appear unusually large when compared with certain other treatments, such as that given by Oswald Veblen (in 1904), which contains the two primitive notions ''point'' and ''between'' and twelve axioms; but it is no worse than the axiomatization Pieri devised, with its 20 postulates. (Two of Hilbert's axioms were later discovered to be implied by the others, so that in the original formulation, the axioms did not form an independent set.)

Hilbert was chiefly responsible for acceptance of the view that a system of axioms can have an intrinsic integrity independent of any physical reality, that is, by being a consistent system. He is often quoted as having urged: ''It must be possible to replace in all geometric statements the words point, line, plane by table, chair, mug.'' In other words, because one is not concerned with the nature of points, lines, and planes except so far as their properties are exhibited in the axioms themselves, one could as well call these undefined things by any convenient names. Euclid believed the axioms to be self-evident truths derived from experience with actual space, but to Hilbert they were merely the ground rules from which one set out to develop the logical consequences. Physical reality had no place in his order of ideas; the discovery of non-Euclidean geometry had, after all, revealed the folly of relying too heavily on spatial intuition. Hilbert became the champion of the formalistic school of thought in which the whole of mathematics is conceived in the form of theorems, meticulously symbolized, and deduced from uninterpreted axioms. This attitude was summed up by Hilbert's remark: ''Mathematics is a game played according to certain rules with meaningless marks on paper.'' Of course, not everyone subscribed to this position. When Hilbert first came to the notice of the mathematical world in 1888 through his great work on the theory of invariants, Paul Gordan (1837–1912) proclaimed, ''This is no longer mathematics, it is theology.'' Later he conceded that ''theology has its merits.''

As long as Euclid's geometry was considered a catalog of truths about the physical world, the question of its consistency never arose. But once it had been reduced to an exercise in formal logic, the absolute necessity arose for a proof of freedom of contradiction. That the consistency of the non-Euclidean geometries had been shown to depend on the consistency of Euclidean geometry added further impetus to the problem. By interpreting a point as an ordered pair of real numbers and a line as a linear equation, Hilbert constructed (1904) a model of Euclidean geometry within the domain of arithmetic; thus, any contradiction drawn from the axioms of the former would be recognizable in the latter. The logical admissibility of Euclidean geometry came down to granting the consistency of the arithmetic of the real numbers, which still left one with the basic problem of demonstrating the latter. The questions had been simply shifted from one branch of mathematics to another, but even this was a substantial achievement.

With continuing optimism, Hilbert and his disciples set about trying to prove the consistency of arithmetic itself. Efforts to solve this problem revealed the astonishingly complex logical structure of mathematics. In 1931, a 25-year-old mathematician at the University of Vienna, Kurt Gödel, published a paper that left little hope that a system of axioms wide enough to encompass arithmetic could be supported by a proof of consistency. Bearing the forbidding title *Über Formal Unentscheidbare Satze der Principia Mathematica und Verwandter Systeme* (On Formally Undecidable Propositions of Principia Mathematica and Related Systems), Gödel's paper is a milestone in the history of modern logic, and led to an invitation for him to join the Institute for Advanced Study in Princeton.

Gödel's main conclusions were twofold. In the first place, he showed that the formalization of logic contained in the famous *Principia Mathematica* (3 volumes, 1910–1913) of Whitehead and Russell had inherent limitations when applied to sufficiently rich deductive systems. Specifically, in any system with "enough" axioms—such as Hilbert's treatment of Euclidean geometry or Peano's axiomatization of arithmetic—there are certain statements that can never be proved or refuted by using only the methods and concepts of that system; that is, undecidable propositions exist. It is natural to ask whether this defect could be remedied by adding further assumptions that would allow one to derive previously undecidable propositions. The answer is no, for Gödel also proved that new undecidable propositions could always be constructed in the enlarged system. In short, there is no system in which all mathematical questions expressible in the system can be decided on the basis of its axioms. This brings us to Gödel's second surprising conclusion, namely, the statement that the system is consistent is one of the undecidable propositions. Consistency can never be established by operating within the system in question; it must involve essentially new methods that are incapable of formalization in the system. As Hermann Weyl (1885–1955), a leading mathematician of the recent era, so aptly put it, "If the game of mathematics is actually consistent, then the formula of consistency cannot be proved within this game." Gödel's incompleteness theorems dealt a mortal blow to the final objective of Hilbert's career, a formalized version of all classical mathematics.

Solving specific problems has always been important in determining the lines of development in mathematics. The calculus of variations, for instance, owes its origin to John Bernoulli's determination of the "curve of quickest descent," and the attempts to solve Fermat's last theorem extended the boundaries of number theory. Hilbert's conviction that great individual problems were the lifeblood of mathematics ("If one seeks methods without any definite problems in mind, his search is then mostly in vain.") led him to put forth a list of problems on which he believed mathematicians should concentrate their efforts. The first International Congress of Mathematicians met at Zurich in 1897, and the second was held at Paris in 1900. At the Paris congress, Hilbert was invited to give one of the leading addresses. With the new century stretching before him, Hilbert felt it appropriate to review the basic trends of mathematical research at the end of the nineteenth century and to indicate what he thought the future might hold. He did this by formulating 23 unsolved problems, extending over all fields, that he felt should occupy the attention of mathematicians in the next era. (The second of these was to establish that the axioms of arithmetic are consistent.) Hilbert's prestige induced many followers to consider the problems he had listed, and less than a year after the Paris congress two were solved by his own pupils. The list became something of a chart by which mathematicians measured their progress in the new century. "A mathematician," wrote Hermann Weyl forty years later, "who had solved one of them thereby passed on to the honors class of the mathematical community."

11.4   Arithmetic Generalized

As we saw in Chapter 7, Italian mathematicians of the 16th century introduced expressions for the square roots of negative numbers to satisfy the demand that all quadratic and cubic equations have solutions. But they displayed a skeptical attitude toward such expressions, referring to the mysterious quantities $a \pm \sqrt{-b}$ as "impossible" or "nonexistent" numbers. In 1637, when Descartes published his *Géométrie,* he contributed the term "imaginary" as a name. We read there:

> Neither the true nor false [negative] roots are always real, sometimes they are imaginary.

The realization that the use of these new numbers enabled a polynomial of degree n to have n roots led to their reluctant acceptance. During the 18th century imaginary numbers were used extensively in higher analysis, especially by Euler, because they produced concrete results. It was assumed not only that imaginaries exist for the ordinary purposes of analysis, but that they obey the same rules of algebraic operation as the familiar real numbers. Euler was the first to employ the now-standard notation i for the imaginary unit $\sqrt{-1}$; in a memoir, *De Formulis Differentialibus Angularibus,* presented to the St. Petersburg Academy in 1777, he wrote:

> In the following, I shall denote the expression $\sqrt{-1}$ by i so that $ii = -1$.

Before Euler, the symbol $\sqrt{-1}$, as distinct from $\sqrt{-a}$, seldom if ever occurred.

Although complex numbers (that is, numbers of the form $a + bi$ with a,b real) were being admitted in formal calculations on an equal footing with real numbers, doubts concerning their precise meaning and nature continued to plague mathematicians. Evidence of Euler's concern with their vague status appears in his celebrated *Algebra* (1770), where he remarks that "such numbers, which by their nature are impossible, are ordinarily called imaginary or fanciful numbers because they exist only in the imagination." It would be almost three centuries after their introduction before an adequate theory would be available to interpret them properly. In the absence of a full understanding of these new numbers, progress in justifying their logical foundation moved along two lines: one approach sought to anchor the complex numbers in a geometric interpretation, whereas the other called for expanding the number concept to a wider ''field of arithmetic'' that would embrace them.

John Wallis seems to have been the first to attempt, although unsuccessfully, any graphic representation of the complex numbers. In his *Algebra* of 1673, he suggested that, because $\sqrt{-bc}$ is the mean proportion between $+b$ and $-c$, the geometric interpretation of $\sqrt{-bc}$ could be obtained by applying the Euclidean mean-construction to two directed line segments representing $+b$ and $-c$. But after touching on this notion, he did nothing of consequence with it.

The geometric interpretation of a complex number as a point in the plane is a simple idea, but it took a long time to break through. When it finally came, it occurred at nearly the same time to three persons who had no connection with or knowledge of each other: a Norwegian surveyor and cartographer Caspar Wessel (1745–1818), a French-Swiss bookkeeper Jean Robert Argand (1768–1822), and that greatest of all German mathematicians, Carl Friedrich Gauss (1777–1855).

Caspar Wessel's fame is based on his only mathematical paper, one in which he established his priority in publication of the geometric representation of complex numbers. In 1797, he presented his ideas on the subject before the Royal Academy of Sciences of Denmark; the paper, written in Danish, was published two years later in the *Philosophical Transactions* of the Academy. It speaks well for the Academy that the members received Wessel's work sympathetically, for he was neither one of its members nor was he considered a mathematician. Unfortunately, his account appeared in a journal that few European scholars were likely to read. It passed unnoticed for a century until, on the 100th anniversary of its publication, the Academy issued a French translation, *Essai sur la représentation analytique de la direction.*

In the opening paragraph of this paper, Wessel indicated the objective of his study:

> This present attempt deals with the question, how we may represent direction analytically; that is, how shall we express right lines [line segments] so that in a single equation involving one unknown line and others known, both the length and direction of the unknown line may be expressed.

His first step was to give a definition of addition of line segments (vectors), placing the initial point of one at the terminal point of another: ''Two right lines are added if we write them in such a way that the second line begins where the first one ends, and then pass a right line from the first to the last point of the united lines; this line is the sum of the united lines.'' He observed that for this addition, ''the order in which these lines are taken is immaterial;'' that is, the commutative law holds.

Next, Wessel turned to the multiplication of line segments. His approach involved setting up two mutually perpendicular coordinate axes, with $+1$ denoting a unit in one direction and $+\epsilon$ a unit perpendicular to it:

> Let $+1$ designate the positive rectilinear unit and $+\epsilon$ a certain other unit perpendicular to the positive unit and having the same origin; then the direction angle of $+1$ will be equal to 0°, that of -1 to 180°, that of $+\epsilon$ to 90°, and that of $-\epsilon$ to $-90°$ or 270°. By the rule that the direction angle of the product shall equal the sum of the angles of the factors,

we have: $(+1)(+1) = +1; (+1)(-1) = -1; (+1)(+\epsilon) = +\epsilon; (+1)(-\epsilon) = -\epsilon; (-1)(-\epsilon)$ $= +\epsilon; (+\epsilon)(+\epsilon) = -1; (+\epsilon)(-\epsilon) = +1; (-\epsilon)(-\epsilon) = -1$. From this it is seen that ϵ is equal to $\sqrt{-1}$, and the divergence of the product is determined such that not any of the common rules of operation are contravened.

Thus, only in the course of computing the multiplication table for the four different units $+1$, -1, $+\epsilon$, $-\epsilon$ does Wessel indicate that $\epsilon = \sqrt{-1}$. He stated that any line segment (Wessel called these "indirect lines") can be represented by the expression $a + b\epsilon$, and derived the rule

$$(a + b\epsilon)(c + d\epsilon) = (ac - bd) + (ad + bc)\epsilon$$

for their multiplication.

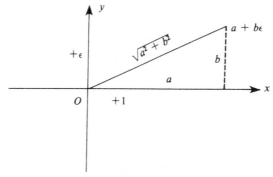

Because Wessel's treatment anticipated not only the notion of a vector space but also an algebra, it was unfortunate for mathematics that it lay buried for a century.

Jean Robert Argand's contribution, *Essai sur une manière de représenter les quantités imaginaire dans les constructions géométriques,* fared somewhat better than the paper of Wessel. It was privately printed in 1806, in a small edition that lacked the author's name on the title page. The work might soon have been forgotten except for a peculiar chain of events in 1813 that rescued it from oblivion. Argand had shown his treatise to Adrien-Marie Legendre before its publication, and Legendre discussed the interesting deliberations of this unknown mathematician in a letter to the brother of J. F. Français. Français saw the letter, and was so taken with the notions in it that he developed them further in a publication (1813) in the journal *Annales de mathématiques.* In the final paragraph he pointed out that he had taken some of his concepts from Legendre's letter and expressed the hope that the anonymous "first author of these ideas" would identify himself and publish his results. Hearing of Français's paper, Argand responded with an article in the next issue of the *Annales* in which he summarized the main points of his original work. But, despite its great merit, this publication was without significant influence.

Wessel and Argand were little known; thus it was the authority of Carl Gauss that brought about the general acceptance of the interpretation that his two predecessors had failed to put across. Gauss seems to have been in possession of a geometric theory of complex numbers around the turn of the century. In his (1799) doctoral dissertation on the fundamental theorem of algebra, he employed the idea without explicit mention; in a letter of 1811 to Bessel, it is clearly outlined; and finally in 1831, in a commentary on his paper *Theoria Residuorum Biquadraticorum* it is publicly described. The novelty is that Gauss gives the representation of complex numbers as points in the plane, rather than as directed line segments as did Wessel and Argand. That is, he replaced the number $a + bi$ by the point (a,b). With this approach, nothing remained of the mystic flavor that was so long attached to these numbers. Gauss affirms:

> That this subject [of imaginary magnitudes] has hitherto been considered from the wrong point of view and surrounded by a mysterious obscurity, is to be attributed largely to an ill-adapted notation. If for instance, $+1$, -1, $\sqrt{-1}$ had been called direct, inverse, and lateral units, instead of positive, negative, and imaginary (or even impossible) such an obscurity would have been out of the question.

He added the opinion that his presentation "puts the true metaphysics of imaginary numbers in a new light," and that all the difficulties would now disappear. Just as the real numbers can be interpreted as representing a line, so the complex numbers can represent a plane. Gauss also introduced the technical term "complex number" for the quantity $a + bi$, as opposed to the phrase "imaginary number," which he thought imputed some dark mystery to these numbers.

As a result of these endeavors, we find French mathematicians referring to the "Argand diagram" in texts, whereas Germans speak of the "Gaussian plane." The Norwegians, with becoming modesty, refrain from such patriotics.

The climax of the attempt to establish the theory of complex numbers on a firm mathematical footing can be found in the work of the Irish scientist, William Rowan Hamilton (1805–1865).

Hamilton was a child prodigy whose maturity fulfilled all that his early precocity promised. The fourth of nine children of a practicing attorney, he was born in Dublin, Ireland, and except for short visits elsewhere spent his whole life in that same city. Hamilton's preliminary education, which was carried on at home, was mainly in languages, classics, and mathematics. The young Hamilton was proficient in Oriental as well as European tongues. By the time he was 7, he could read Latin, Greek, Hebrew, French, and Italian; at the age of 13 he had a working knowledge of as many languages as the years he had lived. Hamilton's artistic inclination was just as strong; often he said that, although he made his living as a mathematician, he was a poet at heart.

The young Hamilton's thoughts were turned to mathematics by a meeting in 1818 with Zerah Colburn, an American youngster who gave a demonstration in Dublin as a "lightning calculator." The meeting was in the nature of a public test of arithmetic skills, and Hamilton came out second best, which only induced this competitive youth to begin a furious study of the original texts of Euclid (in Greek), Newton (in Latin), and Laplace (in French). At 17 he detected an error in Laplace's *Mécanique céleste,* which was communicated to the president of the Royal Irish Academy. After reading Hamilton's article on the subject the president is supposed to have remarked, "This young man, I do not say will be, but is, the first mathematician of his age."

In 1823, Hamilton took the entrance examination for Trinity College, Dublin, and came out first in a field of 100 candidates. Once there, his progress continued to be brilliant. He achieved the previously unheard-of distinction of winning the highest possible marks in both mathematics and English verse. In 1827, he presented to the Royal Irish Academy his "Theory of Systems of Rays," a long and technical treatise

William Rowan Hamilton
(1805–1865)

(From A Concise History of Mathematics *by Dirk Struik, 1967, Dover Publications, Inc., N.Y.)*

on geometric optics. This paper led to his appointment (while still an undergraduate) as professor of astronomy at Trinity College; it was understood that, except for twelve annual lectures, Hamilton would be left free to pursue his own lines of interest. His professorship automatically included the titles of Astronomer Royal of Ireland and Director of the Dunsink Observatory, positions that he retained until his death. Hamilton devoted himself entirely to theoretical studies and did almost nothing as a practical astronomer, perhaps because the research equipment was poor and inadequate and he himself lacked technical training. He was knighted in 1835 as an honor for his work in optics, particularly the result that light refracts in certain biaxial crystals in a conical configuration of rays.

Although such ideas as negative and imaginary numbers appeared essential for algebra, Hamilton could not reconcile himself to the interpretations set forth in his day. Until these could be adequately defined, algebra remained for him "obscure and doubtful." Thus, in the early 1830s, Hamilton struggled to clarify the shaky logical foundations of the subject, hoping to create "a SCIENCE properly so called; strict, pure and independent; deduced by valid reasonings from its own intuitive principles." Influenced largely by Kant's *Critique of Pure Reason,* he concluded that the mental intuition of time is the rudiment from which such a science may be constructed. Hamilton maintained that since geometry is a science of space, and since time and space are "pure sensuous forms of intuition," algebra must be a science about time.

Hamilton's *Theory of Conjugate Functions, or Algebraic Couples: With a Preliminary Essay on Algebra as the Science of Pure Time* is his attempt to found algebra on a set of axioms based on "order and continuous progression, or, as it might be called, PURE TIME." Although portions of this work had been read to the Royal Irish Academy as early as 1833, the entire treatise was first published in the volume of the Academy's transactions for 1837. The paper is divided into three parts: five pages of *General Introductory Remarks,* the 95-page *Essay on Algebra as the Science of Pure Time,* and the 29-page *Theory of Conjugate Functions, or Algebraic Couples.*

The middle section, the *Essay,* is one of the earliest efforts to list systematically the properties of the real number system. Starting with the intuitive notion of order in time, Hamilton builds up the natural numbers through a sequence of equal time steps taken from an arbitrarily chosen zero moment. This approach allows him to define negative numbers by a temporal opposite direction, steps backward in time. After obtaining the laws of rational numbers, his assumption of ''continuous progression'' from moment to moment in time permits a natural extension to properties of irrationals. The metaphysical effluent accompanying Hamilton's presentation discouraged many from reading it, and it was to have little influence on the arithmetization of analysis done in Germany during the last half of the century.

In the *Algebraic Couples* portion of his paper, Hamilton develops the complex numbers in terms of ordered pairs of real numbers in almost the same way it is done in modern texts. He begins by taking time steps to form, as he called it, a number couple. Considering possible ways of combining these couples, he arrives at the rules

$$(b_1,b_2) + (a_1,a_2) = (b_1 + a_1, b_2 + a_2)$$

$$(b_1,b_2) \cdot (a_1,a_2) = (b_1 a_1 - b_2 a_2, b_2 a_1 + b_1 a_2)$$

along with the inverse operations of subtraction and division. In introducing these definitions, Hamilton felt it necessary to protect himself from the possible objection that they are wholly arbitrary, pointing out:

> Were these definitions even altogether arbitrary they would at least not contradict each other, nor the earlier principles of Algebra. It would be possible to draw legitimate conclusions by rigorous mathematical reasoning from premises thus arbitrarily assumed. But persons who have read with attention the foregoing remarks of this theory, and have compared them with the Preliminary Essay, will see that these definitions are really not arbitrarily chosen. Though others might have been assumed, no others would be equally proper.

Hamilton then goes on to demonstrate that addition and multiplication are commutative, and that multiplication distributes over addition. He misses the associative law, probably because it did not occur to him that there might be an algebraic system in which it would not hold.

Hamilton believed that imaginary numbers had no real meaning. They could not be properly defined and therefore had to be excluded from ordinary algebra. The advantage of his number couples was that they provided a means of expressing complex numbers that avoided any reference to imaginaries. To obtain the traditional form of a complex number, we note that

$$(a_1,a_2) = (a_1,0) + (a_2,0)(0,1) = a_1 + a_2\sqrt{-1}.$$

Hamilton expanded on this idea by writing:

> In the THEORY OF SINGLE NUMBERS, the symbol $\sqrt{-1}$ is <u>absurd</u>, and denotes an IMPOSSIBLE EXTRACTION, or a merely <u>IMAGINARY NUMBER</u>; but in the THEORY OF COUPLES, the same symbol $\sqrt{-1}$ is <u>significant</u>, and denotes a POSSIBLE EXTRACTION, or a REAL COUPLE, namely (as we have just now seen) the <u>principal square-root of the couple</u> $(-1,0)$. In the latter theory, therefore, though not in the former, this sign $\sqrt{-1}$ may be properly employed; and we may write, if we choose, for any couple (a_1,a_2) whatever,
>
> $$11(a_1,a_2) = a_1 + a_2\sqrt{-1}. \ldots$$

In this way, ''the metaphysical stumbling-blocks'' that beset algebra were cleared away. For some historians of mathematics, the conception of complex numbers as number couples is Hamilton's greatest achievement in algebra, even more significant than his later discovery of quaternions.

Hamilton closed his paper with the statement that "the author hopes to publish hereafter many other applications of this view; especially to Equations and Integrals, and to a Theory of Triplets. . . ." The triplets that he sought were hypercomplex numbers that were related to three-dimensional space just as the usual complex numbers are related to two-dimensional space. His hopes were finally fulfilled in 1843 with his derivation of "quaternions" of four numbers, after a long and fruitless search for triplets.

By analogy with the complex numbers, Hamilton wrote his triplets as $a + bi + cj$, where a, b, c are real numbers and $i^2 = j^2 = -1$. Addition of such expressions presented no difficulty; they are added by adding their real or "scalar" parts and adding the coefficients of each of the units i and j to form new coefficients for these units:

$$(a + bi + cj) + (a' + b'\,i + c'\,j) = (a + a') + (b + b')i + (c + c')j.$$

But Hamilton was frustrated by his repeated failure to be able to define multiplication of triplets in a way that would preserve the properties of ordinary complex numbers.

Now the modulus of a complex number $a + bi$ is $a^2 + b^2$; and the "law of the moduli," as Hamilton called it, states that the product of the moduli of two complex numbers equals the modulus of the product of the two numbers. Hamilton was particularly concerned that this law also hold for triplets. Consider the simplest case of the product of two triplets, the square of $a + bi + cj$. Assuming that the multiplication can be carried out and the terms collected, one gets

$$(a + bi + cj)^2 = a^2 - b^2 - c^2 + 2abi + 2acj + 2bcij,$$

which has modulus

$$(a^2 - b^2 - c^2)^2 + (2ab)^2 + (2ac)^2 + (2bc)^2 = (a^2 + b^2 + c^2)^2 + (2bc)^2.$$

The law of the moduli is not fulfilled unless the ij-term is removed from the expansion of $(a + bi + cj)^2$. The term must be either suppressed by setting $ij = 0$, or somehow included in one of the other three terms. Taking ij to be zero did not appear to Hamilton to be quite right:

> Behold me therefore tempted for a moment to fancy that $ij = 0$. But this seemed odd and uncomfortable, and I perceived that the same suppression of the term which was de trop might be attained by assuming what seemed to me less harsh, namely that $ji = -ij$. I made therefore $ij = k, ji = -k$, reserving to myself to inquire whether k was 0 or not.

His thought here is that, if the order of multiplication is scrupulously respected, there are actually two terms involving the product of i and j; that is, $2bcij$ should be instead written as $bc(ij + ji)$. The law of the moduli could be satisfied simply by assuming that $ij + ji = 0$, without taking either ij or ji to be zero separately.

The next step was to "try boldly then the general product of triplets." Under the supposition that $ij = -ji = k$, Hamilton calculated

$$(a + bi + cj)(x + yi + zj)$$
$$= (ax - by - cz) + i(ay + bx) + j(az + cx) + k(bz - cy).$$

Setting $k = 0$, he once again asked whether the law of the moduli is satisfied. In other words, does the equation

$$(a^2 + b^2 + c^2)(x^2 + y^2 + z^2) = (ax - by - cz)^2 + (ay + bx)^2 + (az + cx)^2$$

hold? Clearly the answer is no; for the left-hand side of the proposed equation exceeds the right-hand side by $(bz - cy)^2$, which is the square of the coefficient of k in the expansion of the product. Thus, the assumption that $k = 0$ is untenable. However, letting $k \neq 0$ is not entirely satisfactory either, because the product of two triplets should again be a triplet; and the product as indicated above contains four terms rather than three. For nearly ten years Hamilton was unable to get beyond this impasse. Every morning on coming downstairs to breakfast, his eldest son would ask him, ''Well, Papa, can you multiply triplets?'' And each time he had to confess ruefully, ''No, I can only add and subtract them.''

In a moment of insight, which occurred while he was strolling with his wife by the Royal Canal in Dublin in 1843, Hamilton suddenly realized that his difficulties would vanish if he multiplied expressions with four terms rather than three; that is, if he took k to be a third distinct imaginary unit in addition to i and j. He described this dramatic event in a letter to one of his sons as follows:

> An electric current seemed to close; and a spark flashed forth, the herald (as I foresaw immediately) of many long years to come of definitely directed thought and work, by myself if spared, and at all events on the part of others, if I should ever be allowed to live long enough distinctly to communicate the discovery. I pulled out on the spot a pocket-book, which still exists, and made an entry then and there. Nor could I resist the impulse—unphilosophical as it may have been—to cut with a knife on a stone of Brougham Bridge, as we passed it, the fundamental formula with the symbols i, j, k;
>
> $$i^2 = j^2 = k^2 = ijk = -1,$$
>
> which contains the solution of the Problem, but of course as an inscription, has long since mouldered away.

Hamilton called these new expressions ''quaternions,'' or four-fold numbers. They are hypercomplex numbers of the form $q = a + bi + cj + dk$, where a, b, c, d, are real numbers and i, j, k satisfy $i^2 = j^2 = k^2 = -1$. Having already assumed that $i^2 = j^2 = -1$ and $k = ij = -ji$, it seemed clear to Hamilton that he should have

$$k^2 = (ij)(ij) = -(ji)(ij) = -ji^2j = j^2 = -1.$$

In order to test the law of the moduli, he still needed values for ik and kj. Not yet sure that the associative law held for quaternions, Hamilton tentatively concluded:

> . . . that we had probably $ik = -j$, because $ik = iij$, and $i^2 = -1$; and that in like manner we might expect to find $kj = ijj = -i$.

The associative law would also have provided the value for ki, since $ki = (-ji)i = -ji^2 = (-j)(-1) = j$. But Hamilton preferred to argue by analogy:

> . . . from which I thought it likely that $ki = j$, $jk = i$, because it seemed likely that if $ji = -ij$, we should have also $kj = -jk$, $ik = -ki$.

Summarizing the multiplication "assumptions" (as Hamilton called them) for the quaternion units, we have

$$i^2 = j^2 = k^2 = -1, ij = -ji = k, jk = -kj = i, ki = -ik = j.$$

These are the fundamental rules of calculation that Hamilton scratched on the nearby bridge with his penknife; they would preserve his priority until he could announce the discovery to the Academy. They have long since faded, but today there is a commemorative tablet on Brougham Bridge that reads: "Here as he walked by on the 16th of October 1843, Sir William Rowan Hamilton in a flash of genius discovered the fundamental formula for quaternion multiplication $i^2 = j^2 = k^2 = ijk = -1$ and cut it in a stone of this bridge."

Note that for two quaternions q and q', the product qq' does not in general equal $q'q$. Hamilton's quaternion algebra obeys all the fundamental laws of traditional arithmetic except for the commutative law. That he was willing to abandon the commutative law while retaining the associative law is often regarded as a stroke of genius. To depart from long-established tradition and accept strange innovations was every bit as difficult for the algebraist as it had been for the founders of non-Euclidean geometry. Just as the break with the parallel postulate had paved the way for all sorts of new geometries, this bold sacrifice of the commutative law was to bring forth a host of new algebras in which the fundamental laws did not necessarily all apply. Indeed, within three months of Hamilton's creation of the quaternions, John Graves arrived at the "octonions," a system with eight unit elements; multiplication in this algebra was not only noncommutative but not even associative.

Hamilton was convinced that in the quaternions he had found the right instrument to provide a mathematical description of our world of time and space. Time is a scalar and points in space are specified by three real coordinates; together four components are required, just as in a quaternion. For the remaining 22 years of his life, Hamilton's scientific career was devoted almost exclusively to an elaboration of the theory of quaternions, applying them to dynamics, astronomy, and the theory of light. By the end of 1865, there had been 150 papers published on the subject, with 109 of them written by Hamilton. Ten years after his initial discovery, he brought out the *Lectures on Quaternions,* a massive work running to 736 pages plus an additional 64-page preface. The cumbersome *Lectures* proved to be unreadable to all save the most intrepid mathematicians. A colleague wrote Hamilton that the book would "take any man a twelvemonth to read, and a near lifetime to digest." At the urging of friends Hamilton began to write an introductory manual on quaternions, complete with examples and problems; but again his excessive verbosity carried him away, and this too expanded beyond reasonable bounds. The *Elements of Quaternions,* posthumously edited by Hamilton's son in 1866, was even longer than the *Lectures* (762 closely printed pages).

The quaternions never fulfilled Hamilton's expectation of becoming the mathematical language applicable to the physical world; few important physical discoveries were made by quaternion methods. As a basic tool for scientists, they proved to be simply too complicated for quick mastery and easy application. The first and most profitable departure from Hamilton's creation was made by the American Josiah Willard Gibbs (1839–1925). Using just the vector portion $v = bi + cj + dk$ of a quaternion to represent physical quantities, Gibbs, in the early 1880s, built up a new system called

Arthur Cayley
(1821–1895)

(From A Concise History of Mathematics *by Dirk Struik, 1967, Dover Publications, Inc., N.Y.)*

three-dimensional "vector analysis." In place of Hamilton's single quaternion product, he introduced two different types of multiplication: namely, the scalar or dot product of v and $v' = b'i + c'j + d'k$ defined by

$$v \cdot v' = bb' + cc' + dd',$$

and the vector or cross product given by

$$v \times v' = (cd' - c'd)i + (db' - b'd)j + (bc' - b'c)k.$$

Despite the spirited advocacy of Hamilton's devoted followers, the vector analysis of Gibbs eventually prevailed, replacing the quaternions for the practical purposes of physics and engineering.

Although the failure of the quaternion cause tended to diminish Hamilton's historical stature, the long view justifies his tremendous labor. Quaternions, with their abandonment of commutativity, were a key step in the development of modern algebra. They showed that it was possible consciously to construct new elements of algebra rather than finding them from elements of existing algebras. Once this possibility opened up, many people seized the opportunity—including Graves. Had Graves not left the matter of announcing the discovery of the octonions up to Hamilton, these numbers might be known today as the Graves numbers rather than the Cayley numbers. Unfortunately, Hamilton did not act right away, and in the meantime Arthur Cayley published a paper describing an algebra essentially identical to Graves's octonions.

Arthur Cayley was born in 1821 at Richmond, in Surrey, during a short visit by his parents to England. His earliest years were spent in St. Petersburg, where his father was a partner in a firm of Russian merchants. In 1829 the family took up residence in England where Arthur, at the age of 14, was sent to King's College School in London. There the young Cayley gave such indications of mathematical genius that school officials persuaded his father to abandon his intentions of bringing up the boy in the family business and to send him instead to Cambridge. Accordingly, Cayley began his university career at Trinity College, at the age of 17. In 1841, while still an undergraduate of 20, he published a work in the *Cambridge Mathematical Journal,* thus beginning an astounding series of papers that were to enrich various scientific periodicals. Cayley finished his student days in the following year by winning Cambridge's two highest honors, Senior Wrangler in the Mathematical Tripos and the more difficult Smith's Prize for the essay of greatest merit on any subject in mathematics.

At Cambridge from the 1780s onward, bachelor's degrees were awarded on the basis of a single examination called the Senate-House Examination. This consisted almost entirely of mathematics (including "arithmetic, algebra, fluxions, the doctrine of infinitesimals and increments, geometry, trigonometry, optics, and astronomy") with some philosophy. Apparently the college tutors were interested in having a subject of instruction that lent itself to examination purposes. The heavy emphasis on mathematics so dominated the Cambridge curriculum "to the neglect of the classics" that in 1824 a second, classical, examination was instituted; and from that date the old Senate-House Examination became known as the Mathematical Tripos. But the Classical Tripos was a voluntary examination open only to those candidates who had already secured mathematical honors, a rule that must have excluded many. (The word "tripos" itself had a medieval origin; in the 15th century the "ould bachelor" who conducted oral examinations at Cambridge sat upon a three-legged stool and was therefore called "Mr. Tripos.") It was customary to publish the results of the Mathematical Tripos in strict order of merit, with the individual who earned the year's highest score being given the title Senior Wrangler; after him came the other Wranglers, Senior Optimes, and Junior Optimes. After them were listed all the other students down to the last man, the Wooden Spoon. Nothing for a student in any university was deemed comparable to the distinction of being Senior Wrangler ("the first of the firsts"), for in its prime, the Mathematical Tripos was the most severe mathematical challenge ever devised: four days of tests, up to ten hours a day.

Stories were current in Cambridge of the equanimity with which Cayley treated his success in the Tripos examination. A concluding feature of the examination was a three-hour paper consisting of problems representing the utmost range of the examiner's fancy. On the evening after the grand competition, a friend rushed into Cayley's room with disconcerting news of Cayley's chief rival. "I've just seen Smith," the friend announced, "and he told me that he did all the questions within two hours." Cayley responded, "Likely enough he did: I cleaned up that paper in 45 minutes."

Upon completing his degree, Cayley was immediately elected a Fellow of Trinity College, at the youngest age for any student of that century. According to university statutes, a Fellow was required to take Holy Orders in the Church of England or to vacate the position after seven years. (Before the Oxford and Cambridge Act of 1877, it was contrary to the statutes to have a wife and a fellowship at the same time.) Many scholars held their fellowships for a brief period only, and then sought careers in which they would be free to marry; in such careers they would usually earn more than the modest stipends offered by their colleges. So it was with Cayley, who left the academic world at the end of three years to enter the legal profession. He practiced in London for fourteen years as a conveyancer, drawing up deeds for the transference of property. But his real occupation was pure mathematics, and he wrote something approaching 300 mathematical papers, incorporating in them some of his most significant discoveries. Determined to preserve a portion of his time for research, Cayley rejected much of the legal work that came his way, making a comfortable living but turning his back upon the opportunity to become wealthy.

It was during this phase of his life that Cayley met James Joseph Sylvester, who was likewise dividing his allegiance between the law and mathematics. Thus began a lifelong friendship which would produce the theory of invariants (the two came to be known as "the invariant twins") as well as important contributions to higher geometry, combinatorics, and the theory of matrices. Sylvester gratefully acknowledged that Cayley was the person ". . . to whom I am indebted for my restoration to the enjoyment of mathematical life. . . ." Between 1851 and 1855, when Sylvester left to accept a teaching appointment at the Royal Military Academy at Woolwich, both men established solid mathematical reputations. Often they would walk around the Courts of Lincoln's Inn, with Cayley, in Sylvester's words, "habitually discoursing pearls and rubies."

Around 1861, the Lucasian professorship of mathematics at Cambridge—the chair made illustrious by Barrow, Newton, and Cotes—fell vacant. Marking the growing acceptance of science as part of a liberal education, the chair was filled by the physicist George Stokes, who was Senior Wrangler and first Smith's Prizeman the year before Cayley. However, Cambridge wished to have Cayley also, and in 1863 created a new Sadlerian Professorship of Pure Mathematics and promptly offered it to Cayley. Sacrificing any lucrative future in the legal profession for a modest provision in his truer vocation, he held this chair until his death in 1895. It is said that Cayley brought mathematical glory to Cambridge second only to that of Newton.

In the Spring of 1882, Cayley delivered an extended series of lectures at Johns Hopkins University, where his friend and fellow worker, Sylvester, was then in charge of the mathematics department. The presence, at the same time, of two of the world's most prominent mathematicians was said to have made Baltimore "the stronghold of mathematics in America." Although he wrote but one extensive work, the *Treatise on Elliptic Functions* (1876), Cayley's output of papers and memoirs was prodigious. His most productive period, the years between 1863 and 1883, saw the publication of 430 papers. Many of these appeared in the *Quarterly Journal of Mathematics,* a periodical that he, Sylvester, and Stokes helped to found in 1855. As Cayley continued in creative activity until the end of his life, Cayley's grand total of 966 articles ranks him third, after Euler and Cauchy, as the most prolific writer of mathematics. His *Collected Mathematical Papers* (1889–1898) fill 13 large volumes.

It is customary to view Cayley as the creator of an algebra of matrices that did not require repeated reference to the equations from which their entries were taken. Cayley's interpretation grew out of his interest in linear transformations of the form

$$T_1 \qquad \begin{aligned} x' &= ax + by \\ y' &= cx + dy \end{aligned}$$

that may be viewed as transforming an ordered pair (x,y) into a pair (x',y'). In searching for "an abbreviated notation for a set of linear equations," Cayley symbolized the foregoing transformation by the square array

$$\begin{bmatrix} a & b \\ c & d \end{bmatrix}$$

of its coefficients or "elements," and called this a (square) matrix of order 2. Two such matrices are equal provided their corresponding elements are equal. If the transformation just given is followed by a second linear transformation,

$$T_2 \quad \begin{aligned} x'' &= ex' + fy' \\ y'' &= gx' + hy', \end{aligned}$$

then it is a simple matter to show that (x'',y'') can be obtained directly from (x,y) by using the single transformation

$$T_2T_1 \quad \begin{aligned} x'' &= (ea + fc)x + (eb + fd)y \\ y'' &= (ga + hc)x + (gb + hd)y. \end{aligned}$$

This led Cayley to define an operation of matrix multiplication by transferring the rule for computing the product of linear transformations to the matrices that represent them:

$$\begin{bmatrix} e & f \\ g & h \end{bmatrix} \begin{bmatrix} a & b \\ c & d \end{bmatrix} = \begin{bmatrix} ea + fc & eb + fd \\ ga + hc & gb + hd \end{bmatrix}.$$

Cayley was the first to realize that square arrays themselves could be treated as algebraic quantities. He had a sufficiently clear idea of their various properties in the mid-1840s, but it was not until 1858 that he put forth his results in a paper called *A Memoir on the Theory of Matrices*. Defining addition of matrices by

$$\begin{bmatrix} a & b \\ c & d \end{bmatrix} + \begin{bmatrix} e & f \\ g & h \end{bmatrix} = \begin{bmatrix} a + e & b + f \\ c + g & d + h \end{bmatrix},$$

Cayley showed that in the resulting system addition is both associative and commutative; and that multiplication is associative and distributive over addition. The matrix

$$I = \begin{bmatrix} 1 & 0 \\ 0 & 1 \end{bmatrix}$$

is called the identity matrix (of order 2), because it leaves each second-order matrix fixed under multiplication:

$$\begin{bmatrix} a & b \\ c & d \end{bmatrix} I = \begin{bmatrix} a & b \\ c & d \end{bmatrix} = I \begin{bmatrix} a & b \\ c & d \end{bmatrix}.$$

As with Hamilton's quaternions, matrix multiplication is not commutative. This is seen by the following example:

$$\begin{bmatrix} 1 & 1 \\ 0 & 1 \end{bmatrix} \begin{bmatrix} 0 & 1 \\ 1 & 0 \end{bmatrix} = \begin{bmatrix} 1 & 1 \\ 1 & 0 \end{bmatrix} \neq \begin{bmatrix} 0 & 1 \\ 1 & 1 \end{bmatrix} = \begin{bmatrix} 0 & 1 \\ 1 & 0 \end{bmatrix} \begin{bmatrix} 1 & 1 \\ 0 & 1 \end{bmatrix}.$$

Also, the product of two matrices may be zero (that is, equal to the matrix all of whose elements are zero) even though neither factor is zero.

The sole theorem contained in the 1858 *Memoir* ("a remarkable theorem," said Cayley) is the famous Cayley-Hamilton theorem, which asserts that any square matrix "satisfies its own characteristic polynomial equation." To illustrate this situation, consider an arbitrary second-order matrix $A = \begin{bmatrix} a & b \\ c & d \end{bmatrix}$. If vertical lines represent the determinant of a matrix, then the polynomial

$$p(x) = \mid A - xI \mid = \begin{vmatrix} a - x & b \\ c & d - x \end{vmatrix}$$

is the characteristic polynomial of A. The developed expression of this determinant is

$$p(x) = x^2 - (a + d)x + (ad - bc).$$

When A is substituted into the polynomial, we get

$$p(A) = A^2 - (a + d)A + (ad - bc)I = \begin{bmatrix} 0 & 0 \\ 0 & 0 \end{bmatrix}.$$

It is in this sense that A satisfies its characteristic polynomial.

Cayley did not prove the Cayley-Hamilton theorem in general. He gave a computational verification for matrices of order 2, noted that he had confirmed the same result for matrices of order 3, and concluded that it was unnecessary " . . . to undertake a formal proof of the theorem in the general case of a matrix of arbitrary degree." This reflects Cayley's lack of interest in proofs where inductive evidence seemed convincing. Sylvester subsequently carried these investigations much further, and may be viewed as the main developer of the theory of matrices. Indeed, the introduction of the term "matrix" into the mathematical literature seems to be due to Sylvester (1848). Cayley, on the other hand, came up with the notion of a square matrix between a pair of vertical lines to denote a determinant and used (1843) pairs of double vertical lines to indicate matrices.

We have seen that multiplication of quaternions, and also of matrices, broke the laws of ordinary algebra by being noncommutative operations. Once Hamilton had showed that the study of algebra need not treat just the real or complex numbers, new types of systems were rapidly created. A radically different system, one that differs fundamentally from traditional algebra even though addition and multiplication are both commutative, was developed by George Boole. He called the system an "algebra of logic," whose general symbols could be interpreted either as sets or as propositions.

George Boole (1815–1869) was born in humble circumstances in Lincoln, England, a shoemaker's son. He went to elementary school and for a short time he attended a school for commercial subjects, but beyond this meager education he was entirely self-taught. Acutely conscious of England's class distinction, the young Boole learned Greek and Latin, without outside help, in the hope that such knowledge would improve his social standing. This project led to his first published work, a translation into verse of an ancient Greek ode; his father sent it to the local newspaper along with a note indicating that the author was 14 years of age. (The note's benefit was mixed, for the town schoolmaster insisted that the translation could not be the product of an untutored boy.) Faced with the necessity of supporting his poverty-stricken parents, Boole took up teaching in elementary schools when he was 16 years old. Four years later, he opened his own day school in his hometown.

The need to prepare his pupils in mathematics aroused Boole's interest in the subject. He mastered, again by his own unaided efforts, the works of the great mathematicians: Newton's *Principia,* Lagrange's *Mécanique analytique,* and Laplace's *Mécanique céleste.* As Boole's mathematical investigations proceeded he began contributing a stream of original papers to the recently established (1837) *Cambridge Mathematical Journal* and also to the *Philosophical Transactions of the Royal Society.* Boole's reputation was established by his essay on the calculus of operations, "On a General Method in Analysis"; it was published in the *Philosophical Transactions* in 1844 and awarded the Mathematical Medal of the Royal Society for the best paper in the most recent three years. A sideline pertains here: Boole's manuscript had nearly been rejected by the Council of the Royal Society, but one member steadfastly argued that the author's poverty and obscurity was no reason that his paper should be summarily dismissed.

This award-winning paper was followed in 1847 by a slim 82-page pamphlet entitled *The Mathematical Analysis of Logic, Being an Essay Towards a Calculus of Deductive Reasoning,* Boole's first work on the subject in which he was to earn enduring fame. By a curious coincidence it appeared almost simultaneously—at least in the same month, and some say on the same day—as his friend Augustus De Morgan's book *Formal Logic.* Early in the following year Boole penned *The Calculus of Logic,* giving some further developments of his system.

In 1849, at the urging of De Morgan and others, Boole applied for the professorship in mathematics at the newly formed Queen's College in Cork, Ireland. Although he was without a university education or degree, he secured the position on the basis of his research publication. While there he married Mary Everest, a niece of Sir George Everest, who surveyed the highest peak of the Himalayas. Boole remained in Cork until his premature death in 1864 from pneumonia, which he contracted after walking two miles in a drenching rain and, soaked to the skin, dutifully lecturing to his class.

Boole was a member of the growing body of English mathematicians who liberated algebra from "common arithmetic" by suggesting that the rules that symbols obey are the important thing in algebra, and not so much the meaning that one may attach to the symbols. In particular, the symbols of algebra need not stand for numbers. In the opening section of *The Mathematical Analysis of Logic,* Boole writes:

> Those who are acquainted with the present state of the theory of Symbolic Algebra are aware that the validity of the process of analysis does not depend on the interpretation of the symbols which are employed, but solely upon the laws of their combination.

This aspect of his work made Boole a pioneer in the evolution of modern abstract algebra.

The idea of using algebraic symbolism not only to expedite reasoning about numerical quantities but to impart precision to the logical methods of reasoning can be traced to Leibniz in the seventeenth century. But it was not until the mid-1800s that symbolic logic began to emerge fully as a special branch of mathematics, and its early growth was primarily a consequence of the efforts of Boole and De Morgan. The signally important contribution was Boole's *An Investigation of the Laws of Thought, on Which are Founded the Mathematical Theories of Logic and Probabilities* (1854), which expanded and clarified the content of his earlier pamphlet. The philosopher Bertrand Russell was later (1901) to assert that "Pure Mathematics was discovered by Boole in a work which he called *The Laws of Thought.* . . . His work was concerned with formal logic and this is the same thing as mathematics.'' As Boole says in the opening sentences of *The Laws of Thought,* his object is to show that the reasoning processes that are studied in logic can be formalized and carried out in an algebra of logic:

> The design of the following treatise is to investigate the fundamental laws of those operations of the mind by which reasoning is performed; to give expression to them in the language of a Calculus, and upon this foundation to establish the science of Logic and construct its method.

He goes on to construct an algebra of classes, now known as Boolean algebra, whereby logical problems can be solved by a process of formal calculation.

The logical calculus that Boole described hewed closely to arithmetic algebra, in that he chose to use the symbols of arithmetic, and had only one property that departed from its usual laws (the novel algebraic law being $x^2 = x$, which is only true arithmetically if x is 0 or 1). The letters $x, y, z, \ldots$ represented various classes (sets) of things and the equality sign $=$ between two classes indicated that they had the same members. Boole employed the notation $x \cdot y$, or simply xy, to stand for the intersection of the two classes, that is, the class consisting of all those things common to x and y. In the same vein, $x + y$ represented the union of x and y, but only if they were disjoint. The use of $x + y$ to denote the exclusive sense of ''or'' led to difficulties; most of Boole's successors in mathematical logic took $x + y$ to mean the class of elements in x or y, or both.

The symbols 0 and 1 had special meanings, with 1 designating the entire universe of things—the ''universe of discourse'' as Boole called it—whereas 0 stood for the class that had no members, the empty or null class. These interpretations accorded with the behavior of 0 and 1 in ordinary arithmetic:

$$x \cdot 1 = x, \quad x \cdot 0 = 0.$$

The notation $1 - x$, or for brevity $\bar{x}$, indicated the class complementary to x; that is, the class of all elements in the universe that do not belong to x. Its use allowed Boole to represent the inclusive union of x and y (the inclusive ''or'') as $x + \bar{x} \cdot y$. More generally, $x - y$ stood for the class of things in x but not in y and was assumed to have meaning only if y is contained in x.

Logical relations were built up from such symbols, so that $x \cdot \bar{x} = 1$ was interpreted, for instance, as asserting that the universe is made up of elements that belong to the class x and those that do not. The resulting system was similar in many respects to traditional arithmetic. Boole assumed, either explicitly or implicitly, that the following familiar laws held:

$$x + y = y + x, \quad x \cdot y = y \cdot x$$

$$x + (y + z) = (x + y) + z, \quad x \cdot (y \cdot z) = (x \cdot y) \cdot z$$

$$x \cdot (y + z) = x \cdot y + x \cdot z, \quad x \cdot (y - z) = x \cdot y - x \cdot z.$$

But his new algebra differed essentially from the algebra of numbers in the properties expressed by the equations $x \cdot x = x$ and $x + x = x$. Another algebraic rule that had no counterpart in ordinary algebra was $x \cdot (1 - x) = 0$, derived by Boole from $x^2 = x$.

After Boole's death, his pioneering work was developed much further by De Morgan and the American logician Charles Sanders Peirce (1809–1890). They independently enunciated the principle of duality: To every relation involving logical addition and multiplication there is a corresponding dual relation, which is obtained by an interchange of the signs $+$ and $\cdot$, as well as the symbols 0 and 1. (For example, dual to $x \cdot \bar{x} = 0$ is the relation $x + \bar{x} = 1$.) De Morgan and Peirce also enriched the young science of mathematical logic with the discovery of what are still called De Morgan's Rules; in Boole's notation these read

$$\overline{x \cdot y} = \bar{x} + \bar{y}, \quad \overline{x + y} = \bar{x} \cdot \bar{y}.$$

11.4 Problems

1. Verify that quaternion multiplication is noncommutative by computing the product of the quaternions $q = 1 + i + k$ and $q' = 2 + j - k$ in both orders.

2. In the algebra of quaternions, if

$$q = a + bi + cj + dk \quad \text{and}$$
$$\bar{q} = a - bi - cj - dk,$$

show each of the following:

(a) $q = (a + bi) + (c + di)j$;
(b) $qi = iq$ if and only if $c = d = 0$;
(c) $qi = i\bar{q}$ if and only if $b = 0$;
(d) If $q \neq 0$, then q has an inverse q^{-1} under multiplication; that is, $qq^{-1} = 1 = q^{-1}q$.
  [*Hint:* Take $q^{-1} = \bar{q}/(a^2 + b^2 + c^2 + d^2)$.]
(e) q and $\bar{q}$ are both roots of the quadratic polynomial
  $p(t) = t^2 - 2at + (a^2 + b^2 + c^2 + d^2)$.
  [*Hint:* Note that $q^2 = (a^2 - b^2 - c^2 - d^2) + 2abi + 2acj + 2adk$.]

3. Establish that whenever b, c, d are real numbers satisfying $b^2 + c^2 + d^2 = 1$, then the quaternion $q = bi + cj + dk$ has the property that $q^2 = -1$; hence, there are infinitely many quaternions whose squares equal -1.

4. Using Hamilton's interpretation of complex numbers as "number couples" (that is, ordered pairs of real numbers), confirm the following:

(a) multiplication is both commutative and associative;
(b) the couple $(1,0)$ satisfies $(a,b) \cdot (1,0) = (a,b)$ for any (a,b);

(c) if $(a,b) \neq (0,0)$, then (a,b) has an inverse under multiplication, in the sense that there exists a couple (x,y) satisfying $(a,b) \cdot (x,y) = (1,0)$. [*Hint:* Consider $(x,y) = (a/a^2 + b^2, -b/a^2 + b^2)$.]

5. Given two ordered triples of real numbers, say $x = (x_1,x_2,x_3)$ and $y = (y_1,y_2,y_3)$, define their cross product $x \times y$ by

$$x \times y =$$
$$(x_2 y_3 - y_2 x_3, -(x_1 y_3 - y_1 x_3), x_1 y_2 - y_1 x_2).$$

For ordered triples x, y, z, establish that

(a) $x \times 0 = 0 \times x = 0$, where 0 is the triple having all entries zero;
(b) $x \times y = -y \times x = -(y \times x)$;
(c) $(x + y) \times z = (x \times z) + (y \times z)$ and $x \times (y + z) = (x \times y) + (x \times z)$;
(d) if $e_1 = (1,0,0)$, $e_2 = (0,1,0)$, $e_3 = (0,0,1)$, then

$$e_1 \times e_2 = e_3, \ e_2 \times e_3 = e_1, \ e_3 \times e_1 = -e_2.$$

6. Show that in general the associative law does not hold for the cross product.

7. In the algebra of matrices of order 2, show the following:

(a) $\begin{bmatrix} 1 & -1 \\ 1 & -1 \end{bmatrix}^2 = \begin{bmatrix} 0 & 0 \\ 0 & 0 \end{bmatrix}$, hence there exist nonzero matrices whose squares are zero.

(b) There is no matrix A for which $A^2 = \begin{bmatrix} 0 & 1 \\ 0 & 0. \end{bmatrix}$.

(c) $\begin{bmatrix} 1 & 1 \\ 0 & 0 \end{bmatrix}^2 = \begin{bmatrix} 1 & 1 \\ 0 & 0 \end{bmatrix}$, and

$\begin{bmatrix} 1 & 1 \\ 0 & 0 \end{bmatrix}\begin{bmatrix} 1 & 0 \\ -1 & 0 \end{bmatrix} = \begin{bmatrix} 0 & 0 \\ 0 & 0 \end{bmatrix}$.

(d) If n is a positive integer, then

$$\begin{bmatrix} 1 & 1 \\ 0 & 1 \end{bmatrix}^n = \begin{bmatrix} 1 & n \\ 0 & 1 \end{bmatrix}.$$

8. The transpose of a matrix A, denoted by A^t, is the matrix obtained from A by interchanging its successive rows and columns; for example,

$$\begin{bmatrix} a & b \\ c & d \end{bmatrix}^t = \begin{bmatrix} a & c \\ b & d \end{bmatrix}.$$

For matrices of order 2, prove the following facts from Cayley's *Memoir:*

   a. $(A^t)^t = A$.
   b. $(AB)^t = B^t A^t$.
   c. If A has an inverse A^{-1} under multiplication, then $(A^{-1})^t = (A^t)^{-1}$.

9. Establish the Cayley-Hamilton theorem for matrices of order 2: if A is a matrix of order 2 and $p(t)$ is its characteristic polynomial, then $p(A) = 0$ (the zero matrix of order 2).

10. At the end of his *Memoir,* Cayley writes, ''If L, M are skew convertible matrices, that is, $LM = -ML$ of order 2, and if these matrices are such that $L^2 = -I, M^2 = -I$; then putting $N = LM = -ML$, we obtain

$$L^2 = -I, M^2 = -I, N^2 = -I,$$
$$L = MN = -NM, M = NL = -LN,$$
$$N = LM = -ML.''$$

Illustrate this situation using the matrices

$$L = \begin{bmatrix} 0 & i \\ i & 0 \end{bmatrix} \text{ and}$$

$$M = \begin{bmatrix} i & 0 \\ 0 & -i \end{bmatrix} (i^2 = -1).$$

11. In abstract algebra, a group is defined to be a set G of elements on which there is an operation $*$ defined. The operation satisfies the following four properties:

   (a) For all a,b, in G, $a*b$ is in G.
   (b) For all a,b,c, in G, $(a*b)*c = a*(b*c)$.
   (c) There exists an element e in G such that $a*e = a = e*a$ for all a in G.
   (d) For each element a in G, there exists an element a^{-1} in G such that $a*a^{-1} = e = a^{-1}*a$.

Let G be the set of all matrices $A = \begin{bmatrix} a & b \\ c & d \end{bmatrix}$ of order 2 for which $|A| = ad - bc \neq 0$. Prove that G is a group in the sense of abstract algebra, where matrix multiplication plays the role of the operation $*$.

[*Hint:* For A in G, take

$$A^{-1} = \begin{bmatrix} d/|A| & -b/|A| \\ -c/|A| & a/|A| \end{bmatrix}.]$$

12. A Boolean algebra is a set B of elements on which there are operations $\vee$ and $\wedge$ satisfying the following properties for all a,b,c in B:

   (1) $a \vee b$ and $a \wedge b$ are in B.
   (2) $a \vee b = b \vee a$ and $a \wedge b = b \wedge a$.
   (3) $a \vee (b \vee c) = (a \vee b) \vee c$ and $a \wedge (b \wedge c) = (a \wedge b) \wedge c$.
   (4) $a \wedge (b \vee c) = (a \wedge b) \vee (a \wedge c)$ and $a \vee (b \wedge c) = (a \vee b) \wedge (a \vee c)$.
   (5) There exist elements 0 and 1 in B such that for all a in B,
   $$a \vee 0 = a \text{ and } a \wedge 1 = a.$$
   (6) For each a in B, there exists an element a' in B such that
   $$a \vee a' = 1 \text{ and } a \wedge a' = 0.$$
In a Boolean algebra B, prove that for all a,b in B

   (a) $a \vee 1 = 1$ and $a \wedge 0 = 0$.
   [*Hint:* $1 = a \vee a' = a \vee (a' \wedge 1)$.]
   (b) $a \vee a = a$ and $a \wedge a = a$.
   [*Hint:* $a \vee a = (a \vee a) \wedge 1 = (a \vee a) \wedge (a \vee a')$.]
   (c) $a \wedge (a \vee b) = a$ and $a \vee (a \wedge b) = a$.
   [*Hint:* $a = a \wedge 1 = a \wedge (b \vee 1)$.]
   (d) a' is unique; that is, if $a \vee b = 1$ and $a \wedge b = 0$, then $b = a'$.
   [*Hint:* $b = b \vee 0 = b \vee (a \wedge a') = (b \vee a) \wedge (b \vee a') = 1 \wedge (b \vee a')$.]
   (e) $(a \vee b)' = a' \wedge b'$ and $(a \wedge b)' = a' \vee b'$.

13. For any a,b,c in a Boolean algebra B, establish the following:

   (a) $a \vee c = b \vee c$ and $a \vee c' = b \vee c'$ together imply that $a = b$.
   (b) $a \wedge c = b \wedge c$ and $a \wedge c' = b \wedge c'$ together imply that $a = b$.

14. Let B be the set consisting of the integers $1,3,7$ and 21. For a,b in B, define

$$a \vee b = \text{lcm } (a,b) \text{ and } a \wedge b = \text{gcd } (a,b).$$

Also put $a' = 21/a$. Show that B with these operations is a Boolean algebra. [*Hint*: Notice that if $a = 3^r 7^s$ and $b = 3^u 7^v$, where $r,s,u,v = 0,1$, then gcd $(a,b) = 3^k 7^j$, where $k = \min \{r,u\}$ and $j = \min \{s,v\}$; also lcm $(a,b) = 3^n 7^m$ where $n = \max \{r,u\}$ and $m = \max \{s,v\}$].

15. Let B be the set of all logical propositions, that is, declarative sentences that are either true or false but not both.

(a) If p,q are in B, define $p \vee q$ to be the proposition "*p* or *q*." Assume $p \vee q$ is true if at least one of p and q is true.

(b) If p,q are in B, define $p \wedge q$ to be the proposition "*p* and *q*." Assume $p \wedge q$ is true if both p and q are true.

(c) If p is in B, define p' to be the proposition "not *p*." Assume p' is true when p is false, and false when p is true.

(d) If p and q are propositions formed by combining propositions $r,s,t \ldots$ with the connectives $\vee, \wedge, '$, write $p = q$ provided p and q have the same truth value for every assignment of truth or falsity to any of $r,s,t, \ldots$

(e) Let 1 represent a proposition that is always true, and 0 represent a proposition which is always false.

Establish that B forms a Boolean algebra, the algebra of logical propositions.

16. Suppose that B is the Boolean algebra of logical propositions.

(a) By assigning truth values to the propositions p and q, verify the following:
$(p \wedge p')' = 1$, $(p \wedge q') \vee (p' \wedge q) = 0$, and $(p \vee q) \vee (p' \wedge q') = 1$.

(b) Show that the cancellation law does not hold in B; that is, $p \vee r = q \vee r$ (or $p \wedge r = q \wedge r$) does not necessarily imply $p = q$.

(c) Verify the results in part (a) by using the laws of Boolean algebra as given in Problem 12.

Bibliography

Adler, Claire. *Modern Geometry.* 2d ed. New York: McGraw-Hill, 1967.

Archibald, Raymond C. "Remarks on Klein's 'Famous Problems of Elementary Geometry'." *American Mathematical Monthly* 21(1914): 247–259.

Belhoste, B. *Augustin-Louis Cauchy: A Biography.* Translated by F. Ragland. New York: Springer-Verlag, 1990.

Birkhoff, Garrett, ed. *A Source Book in Classical Analysis.* Cambridge, Mass.: Harvard University Press, 1973.

Blumenthal, Leonard. *A Modern View of Geometry.* San Francisco: W. H. Freeman, 1961.

Bonola, Roberto. *Non-Euclidean Geometry.* Translated by H. S. Carslaw. New York: Dover, 1955.

Bottazzini, Umberto. *The Higher Calculus: A History of Real and Complex Analysis from Euler to Weierstrass.* Translated by Warren Van Egmond. New York: Springer-Verlag, 1986.

Bunt, Lucas. "Equivalent Forms of the Parallel Axiom." *Mathematics Teacher* 60(1967): 641–652.

Cajori, Florian. "Attempts Made During the Eighteenth and Nineteenth Centuries to Reform the Teaching of Geometry." *American Mathematical Monthly* 17(1910): 181–201.

Cartwright, Mary. "Grace Chisholm Young." *Journal of the London Mathematical Society* 19(1944): 185–192.

Cooke, Roger. *The Mathematics of Sonya Kovalevskaya.* New York: Springer-Verlag, 1984.

Coolidge, John. "Six Female Mathematicians." *Scripta Mathematica* 17(1951): 20–31.

Coxeter, H. S. M. *Non-Euclidean Geometry.* Toronto: University of Toronto Press, 1947.

———. "The Erlangen Program." *Mathematical Intelligencer* 0(1977): 22–30.

———. "Gauss as a Geometer." *Historia Mathematica* 4(1977): 379–396.

Crowe, Michael. *A History of Vector Analysis.* New York: Dover Publications, 1985.

Dauben, Joseph. "Cauchy and Bolzano: Tradition and Transformation in the History of Mathematics" in *Transformation and Tradition in the Sciences.* Edited by Everett Mendelsohn. Cambridge: Cambridge University Press, 1984.

Eves, Howard. *A Survey of Geometry.* Vol. 2. Boston: Allyn and Bacon, 1965.

Feldmann, Arthur. "Arthur Cayley—Founder of Matrix Theory." *Mathematics Teacher* 55(1962): 482–484.

Feur, Lewis. "Sylvester in Virginia." *Mathematical Intelligencer* 9, no. 2(1987): 13–19.

Fitzpatrick, Sister Mary of Mercy. "Saccheri, Forerunner of Non-Euclidean Geometry." *Mathematics Teacher* 57(1964): 323–331.

Freudenthal, H. "Did Cauchy Plagiarize Bolzano?" *Archive for History of Exact Sciences* 7(1971): 375–392.

Galos, Ellery B. *Foundations of Euclidean and Non-Euclidean Geometry,* New York: Holt, Rinehart & Winston, 1968.

Gans, David. *An Introduction to Non-Euclidean Geometry.* New York: Academic Press, 1973.

Gratton-Guinnes, I. *The Development of the Foundations of Mathematical Analysis from Euler to Riemann.* Cambridge, Mass.: M.I.T. Press, 1970.

———. "Bolzano, Cauchy and the 'New Analysis' of the Early Nineteenth Century." *Archive for History of Exact Sciences* 6(1970): 372–400.

Gray, Jeremy. *Ideas of Space: Euclidean, Non-Euclidean and Relativistic.* Oxford: Clarendon Press, 1979.

Greenberg, Marvin. *Euclidean and Non-Euclidean Geometry: Development and History.* 3/E. San Francisco: W. H. Freeman, 1993.

Halstead, G. B. "Lobachevsky." *American Mathematical Monthly* 2(1895): 137–139.

———. "Bolyai Janos (John Bolyai)." *American Mathematical Monthly* 5(1898): 35–41.

———. "Gauss and Non-Euclidean Geometry." *American Mathematical Monthly* 7(1900): 247–252.

Hankens, Thomas. *Jean d'Alembert: Science and the Enlightenment.* London: Oxford University Press, 1970.

———. "Algebra of Pure Time: William Rowan Hamilton and the Foundations of Algebra" in *Motion and Time, Space and Matter: Interrelations in the History and Philosophy of Science,* ed. by P. J. Mackamer and R. G. Turnbull. Columbus, Ohio: Ohio State University Press, 1976: 327–359.

———. "Triplets and Triads: Sir William Rowan Hamilton on the Metaphysics of Mathematics." *Isis* 68(1977): 175–193.

———. *Sir William Rowan Hamilton.* Baltimore: Johns Hopkins Press, 1980.

Hans, Nicholas. *The Russian Tradition in Education.* Westport, Conn.: Greenwood Press, 1973.

Harley, Robert. "George Boole, F.R.S." in George Boole's Collected Works. La Salle, Ill.: Open Court, 1952.

Hendry, John. "The Evolution of William Rowan Hamilton's View of Algebra as the Science of Pure Time." *Studies in History and Philosophy of Science* 15(1984): 63–81.

Hilbert, David. *The Foundations of Geometry.* Translated by E. J. Townsend. Chicago: Open Court, 1902.

Jones, Phillip. "Early American Geometry." *Mathematics Teacher* 37(1944): 3–11.

Kattsoff, Louis. "The Saccheri Quadrilateral." *Mathematics Teacher* 55(1962): 630–636.

Kennedy, Don. *Little Sparrow: A Portrait of Sophia Kovalevsky.* Athens, Ohio: Ohio University Press, 1983.

Kennedy, Hubert. "Giuseppe Peano at the University of Turin." *Mathematics Teacher* 61(1968): 703–706.

———. "The Origins of Modern Axiomatics: Pasch to Peano." *American Mathematical Monthly* 79(1972): 133–136.

———. *Peano: Life and Works of Giuseppe Peano.* Dordrecht, Holland: D. Reidel, 1980.

Kennedy, Hubert, ed. *Selected Works of Giuseppe Peano.* Toronto: University of Toronto Press, 1973.

Kleiner, Israel. "Thinking the Unthinkable: The Story of Complex Numbers." *Mathematics Teacher* 81(1988): 583–592.

Koblitz, Ann Hibner. *A Convergence of Lives. Sofia Kovalevskaia: Scientist, Writer, Revolutionary.* Boston: Birkhauser, 1983.

———. "Sofia Kovalevskaia and the Mathematical Community." *Mathematical Intelligencer* 6(no. 1, 1984): 20–29.

Kovalevskaya, S. *A Russian Childhood.* Translated by B. Stillman. Berlin: Springer-Verlag, 1978.

Kulczycki, C. E. *Non-Euclidean Geometry.* New York: Pergamon, 1961.

Lanczos, C. "William Rowan Hamilton—An Appreciation." *American Scientist* 55(1967): 129–143.

Lewis, Florence. "History of the Parallel Postulate." *American Mathematical Monthly* 27(1920): 16–23.

Lipski, Alexander. "The Foundation of the Russian Academy of Sciences." *Isis* 44(1953):349–354.

Maiers, Wesley. "Introduction to Non-Euclidean Geometry." *Mathematics Teacher* 57(1964): 457–461.

Manning, K. B. "The Emergence of the Weierstrassian Approach to Complex Analysis." *Archive for History of Exact Sciences* 12(1975): 297–383.

Martin, George. *The Foundations of Geometry and the Non-Euclidean Plane.* New York: Intext Educational Publishers, 1975.

Miller, Arthur. "The Myth of Gauss' Experiment on the Euclidean Nature of Physical Space." *Isis* 70(1972): 345–348.

Monastyrsky, Michael. *Reimann, Topology and Physics.* Translated by James King and Victoria King. Boston: Birkhauser, 1987.

Neuenschwander, Edwin. "Riemann's Example of a Continuous Nondifferentiable Function." *Mathematical Intelligencer* 1 no. 1(1978): 40–44.

Novy, Lubos. *Origins of Modern Algebra.* Leyden, The Netherlands: Noordhoff, 1973.

O'Neill, John. "Formalism, Hamilton and Complex Numbers." *Studies in History and Philosophy of Science* 17(1986): 351–372.

Polubarinova-Kochina, P. Ya. "Karl Theodor Wilhelm Weierstrass (on the 150th anniversary of his birthday)." Translated by R. Davis. *Russian Mathematical Surveys* 21(May–June 1966): 195–206.

Prenowitz, Walter, and Jordan, Meyer. *Basic Concepts of Geometry*. New York: Blaisdell, 1965.

Rootsebaar, B. van. "Bolzano's Theory of Real Numbers." *Archive for History of Exact Sciences* 2(1964): 168–180.

Rosenfeld, B. A. *A History of Non-Euclidean Geometry: Evolution of the Concept of Geometric Space*. Translated by Abe Shenitzer. New York: Springer-Verlag, 1988.

Rowling, Raymond, and Levine, Maita. "The Parallel Postulate." *Mathematics Teacher* 62(1969): 665–669.

Simons, Lao G. "The Influence of French Mathematicians at the End of the Eighteenth Century upon the Teaching of Mathematics in American Colleges." *Isis* 15(1931): 104–123.

Sjöstedt, C. E. *Le Axiome de Paralleles: De Euclides a Hilbert*. Uppsala, Sweden: Interlingue-Fundation, 1968.

Toepell, Michael. "Origins of David Hilbert's Grundlagen der Geometrie." *Archive for History of Exact Sciences* 35(1986): 329–344.

Trudeau, Richard. *The Non-Euclidean Revolution*. Boston: Birkhauser, 1987.

Van der Waerden, B. L. "Hamilton's Discovery of Quaternions." *Mathematics Magazine* 49(1976): 227–234.

Vucinich, Alexander. "Nikolai Ivanovich Lobachevskii: The Man Behind the First Non-Euclidean Geometry." *Isis* 53(1962): 465–481.

———. *Science in Russian Culture: A History to 1860*. Stanford, Calif.: Stanford University Press, 1963.

Wolfe, Harold. *Introduction to Non-Euclidean Geometry*. New York: Holt, Rinehart and Winston, 1945.

Wylie, C. R., Jr. *Foundations of Geometry*. New York: McGraw-Hill, 1964.

CHAPTER 12

The Theory of Sets: Georg Cantor

The solution of the difficulties which formerly surrounded the mathematical infinite is probably the greatest achievement of which our age has to boast.

BERTRAND RUSSELL

12.1 Cantor and Kronecker

Unlike England, France, and Russia, where the academies more than compensated for the meager development of scientific thought in the universities, Germany had universities that assumed responsibility for scientific and mathematical research. "The education of German universities," said one French writer, "begins where that of most nations in Europe ends." The German university in its characteristic form was precipitated by the utter defeat of Prussia by Napoleon at Jena in 1806. The humiliating Treaty of Tilsit (1807) stripped the country of all its territories west of the Elbe—in all, about half the territory and population—laid upon it a heavy indemnity of 120,000,000 francs, and compelled it to support an army of occupation of 150,000 soldiers. One result of the treaty was that Prussia lost all its universities except for three along the Baltic coast; the loss of those at Göttingen and Halle was the most severe blow. The throne and people of Prussia turned to education, the only area in which the French left them free to act as the means to the moral and physical regeneration of their country. Said the king, Frederick William III:

> We have indeed lost territory, and it is true that the state has declined in outward splendor and power, and for that very reason it is my earnest desire that the greatest attention be paid to the education of the people. . . . The state must regain in intellectual force what it lost in physical force.

To carry out this aim, a series of laws was passed establishing a universal, compulsory system of state education that was to inculcate patriotism in the oncoming generation. Prussia became a nation of pupils and schoolmasters. Then with dramatic suddenness came a renaissance of spirit, out of which rose the Prussia that was later to unify the German states. When its carefully trained and completely equipped armies achieved the military defeat of France in 1871, the schoolmaster of Prussia was held to have triumphed at last.

At the time of Prussia's deepest national despair, a new university was founded in Berlin (1810), given an annual money grant, and assigned a royal palace for a home. The University of Berlin was not intended to be a mere addition to the existing universities, but rather the embodiment of a new conception of higher education. The new university was intended primarily to develop knowledge; secondarily and perhaps as a concession, it was to train the professional classes. The main emphasis for both students and professors was on original research, not teaching skill and examining; therefore, positions were offered only to those who had proved themselves capable of advancing knowledge. The lecture hall took the place of classroom recitation, and the seminar, in which a small group of advanced students investigated a problem under the direction of a professor, became a prominent feature of every department. (Weierstrass, in a joint undertaking with Kummer in 1861, introduced the first seminar in Germany devoted entirely to mathematics.) Although the appointment of all professors lay in the hands of the minister of education, the university was given full liberty to manage its own affairs in regard to studies and administration. With it came the modern academic freedom to pursue the truth in whatever way seemed best; the sober search for truth, without reference to where the truth led, was the watchword of the University of Berlin.

As the 1800s advanced, other universities were founded on the new model of Berlin: Breslau (1811), Bonn (1818), and Munich (1826). The unquestioned superiority of their libraries, laboratories, and scholars explained the scientific renown that Germany enjoyed abroad. The German universities, with their combination of lecture, seminar, and laboratory, were felt to be about the only institutions in the world where a student could obtain training in how to do scientific and scholarly research. As might be expected, this reputation for academic leadership brought clients from all over the world. In particular, a steady stream of American students sought specialized training in Germany.

Until the mid-1800s, there was hardly any advanced work in mathematics in the United States worth mentioning. A modest beginning had been made by Nathaniel Bowditch (1773–1838), a largely self-taught mathematician, who translated four of the five volumes of Laplace's *Mécanique céleste,* the first great scientific work issued in the United States. Bowditch also wrote a classic on navigation, *New American Practical Navigator* (1802). These writings earned him membership in the Royal Society of London. Benjamin Peirce (1809–1880), a professor of mathematics at Harvard University and one of the fifty original incorporators of the National Academy of Sciences (1863), was generally regarded as the leading American mathematician of his day. His most outstanding work was a memoir *Linear Associative Algebra,* read to the National Academy in 1866–1870. It was printed only in lithograph form for private circulation in 1870, and at long last published in the *American Journal of Mathematics* (1881). The memorable opening sentence states, ''Mathematics is the science which draws necessary conclusions.'' On page five, there is reference to the ''mysterious formula'' connecting π, $e,$ and $i,$ namely $i^{-i} = e^{\pi/2}$, which evidently had a strong hold on Peirce's imagination. After proving it in one of his classes on analysis, he said to his students, ''Gentlemen, this is surely true, it is absolutely paradoxical, we can't understand it, and we haven't the slightest idea what the equation means, but we may be sure that it means something very important.''

A Yale University physical chemist, Josiah Willard Gibbs (1839–1903), wrote for use by his students a pamphlet, *Elements of Vector Analysis,* which was a great simplification and improvement of Hamilton's vector calculus. The value of Gibbs's work was so little appreciated by the authorities at Yale that for ten years he served without pay, living on his inherited income. Only when Gibbs was invited to join the faculty of the new Johns Hopkins University was Yale persuaded to provide him with a salary; this was still, however, only two-thirds of what had been offered him in Baltimore. There is a story that the German mathematician and physicist Hermann von Helmholtz on his visit to Yale in 1893 expressed regret at missing an opportunity to talk with Gibbs; the university officials, perplexed, looked at one another and said, ''Who?''

For most of the nineteenth century, there was no good understanding of mathematical research in the United States, much less the capacity for training research mathematicians. It was considered impossible for an American to enter the field of advanced mathematical study without going abroad to seek fulfillment at first hand under a European professor. Germany proved to be particularly attractive, partly because of the brilliance of its teachers and partly because of the relatively inexpensive German doctorate, which could be obtained fairly quickly by a well-prepared student. Then there was the matter of prestige; to have studied at a German university placed one in a superior class, and the actual attainment of a German degree was looked on as an infallible passport into American academic circles. One estimate about 1900 indicated that 20 percent of the mathematical faculty of American universities had at some time studied in Germany. Clearly there was a significant German influence on mathematics in the United States.

For study in Germany, Göttingen was the most frequent choice, with Berlin and Leipzig the next most popular. The Americans at the University of Göttingen were so numerous as to have their own letterhead, ''The American Colony in Göttingen.'' These American scholars returned from Europe with a new zeal for research in fields they had not been aware of. They also brought back, for better or worse, a taste for abstract mathematics rather than for its applications. Thus, the early 1890s witnessed a great awakening in American mathematics, when many able and enthusiastic young men, largely trained in Germany, set about raising the subject to the same level as that pursued in Europe.

At the same time, a notable number of German mathematicians joined American university faculties, where they rendered distinguished service. When the University of Chicago opened its doors in 1892, Eliakim Hastings Moore, an American who had studied in Berlin during 1855–1886, was appointed professor and acting head of the mathematics department. Moore persuaded the trustees to employ for him two unusually fine scholars, Oskar Bolza and Heinrich Maschke, both former students at Berlin and Ph.D.'s of Göttingen. The young and vigorous department at Chicago with its trio of leaders soon became a major American research center. Bolza's lectures in particular were responsible for a highly productive American school in the modern calculus of variations.

The real credit for the sudden outburst of mathematical energy in the United States in the late 1800s must go to Johns Hopkins University, which was founded by a merchant of the same name in 1876. Johns Hopkins and Chicago started primarily as graduate schools, and although they maintained undergraduate colleges, instruction at the lower level was regarded as a subordinate function from the start. Under the powerful influence of Johns Hopkins, the already existing graduate schools began to awaken, to modernize their hidebound curriculums, and to undertake research as part of their academic responsibility. Old faculty members with poor qualifications gave way to younger and better-trained mathematicians.

In 1877, the distinguished British mathematician James Joseph Sylvester (1814–1897) was called to Johns Hopkins to take charge of the mathematics department of America's newest seat of learning. Sylvester was admitted to Cambridge University in 1831, becoming one of the best scholars in his class. Cambridge was controlled by the Church of England, and the laws required signing articles of faith before a diploma could be conferred. Sylvester was Jewish, and he was unwilling to submit to this religious test and so was barred from receiving a degree. What was far worse, he was deprived of the fellowship to which his standing in class entitled him. For three years following 1838, he taught at University College, London, where his old teacher De Morgan was one of his colleagues. At the age of 37, Sylvester accepted a chair in mathematics at the University of Virginia. He resigned, after a few months' service, however, because he was dissatisfied about the faculty's failure to sustain him in exercising his authority over an insulting student in his course. (The young man had been reading a newspaper while the lecture was under way, and when reprimanded, refused to leave.) After this, Sylvester left the academic world for a time, serving as an actuary and a lawyer from 1845 to 1855. He seems to have given some private instruction in mathematics, counting Florence Nightingale among his pupils. In 1855, Sylvester commenced new duties as professor of mathematics at the Royal Military Academy in Woolwich, a position he held until he was 55, the age set by English military law for granting pensions. He was thus free to accept a position at Johns Hopkins at its organization. The annual salary, $6,000, was generous for those days. With British conservatism, Sylvester stipulated that it should be paid in gold.

Sylvester's reputation and scholarship drew to him a small body of earnest students seeking his guidance. A graduate of Vassar College, Christine Ladd, obtained special permission to hear Sylvester's lectures—but only his—although Johns Hopkins did not officially admit women at that time. She was later allowed to attend other mathematics classes, including those of the renowned logician and philosopher Charles Sanders Peirce. By 1882, as Peirce's student, she had written a dissertation on ''The Algebra of Logic''; but not until 1926, as part of its fiftieth anniversary exercise, did the University finally award Ladd the Ph.D. degree that she had earned years earlier.

Doctoral degrees in mathematics had been given by Yale since 1862, and by Harvard since 1873. The first such degree actually conferred on an American woman was that granted by Columbia University to Winifred Edgerton in 1886.

Edgerton's counterpart in England was Charlotte Angas Scott of Girton College, Cambridge. Girton was one of the new women's colleges whose students, beginning in 1878, were allowed to attend most Cambridge University lectures—at first, on the condition that they be accompanied by chaperones. After 1881, Girton College undergraduates were formally admitted to Cambridge degree examinations (though they could not receive the resulting degree) on the same terms as male students. Just one year earlier Scott had been allowed to participate in Cambridge's highly-prized Mathematical Tripos, an informal arrangement at the discretion of the individual examiners.

Though she was judged as standing ''equal in proficiency to the eighth Wrangler'' she could neither attend the award ceremony nor have her name officially read out; but when the eighth place was announced the students chanted ''Scott of Girton, Scott of Girton!'' (Ten years later the brilliant individual performance of Philippa Fawcett in the Tripos placed her ''above the Senior Wrangler.'') Although Scott spent nine years at Cambridge, Cambridge gave no advanced degrees to women and indeed did not do so until 1948. Thus she actually had to earn her doctorate in mathematics (1885) from the University of London, by external examination. Immediately after obtaining her degree, Scott assumed the chairmanship of the mathematics department at the newly founded Bryn Mawr College. She was the first woman in the United States to hold a Ph.D. in mathematics and directed seven doctorates granted by Bryn Mawr.

It is due to the enthusiasm and ability of Sylvester more than any other one man that mathematical science in America received its remarkable impetus in the late 1800s. During his few years at Johns Hopkins, he founded the first mathematical research journal in the United States, the *American Journal of Mathematics* (1878). The purpose of the journal was to make accessible the papers written by Sylvester, his pupils, and other mathematicians trained in America. European contributors added to its prestige. Of the first 90 writers submitting articles, 30 were from foreign countries, and a third of the rest were students of Sylvester. In the period 1878–1900, Johns Hopkins awarded 32 doctorates in mathematics, compared with 15 from Yale and 9 from Harvard.

The University of Chicago had an immediate effect on American mathematics by granting 10 doctoral degrees between 1896 and 1900. Its doctoral graduates were to make up a large part of the faculties of the universities throughout the Middle West and Mississippi Valley.

As a result of the sudden increase in the tempo of mathematical research, the number of journals grew, and meetings for the reading of papers became more numerous and active. The New York Mathematical Society, which was organized in 1888, was at first little more than a mathematical club meeting at Columbia University. As its membership widened, it adopted the name *American Mathematical Society* in 1894. Simultaneously, its periodical, the *Bulletin of the New York Mathematical Society,* whose publication began in 1891, became the *Bulletin of the American Mathematical Society.* The activity of the members of the society and the quality of the papers prepared led to the publication of its most prestigious research journal, *The Transactions of the American Mathematical Society,* beginning in 1900.

Sylvester resigned his position at Johns Hopkins in 1883 to become a professor of geometry at Oxford University. The school's administrator, ''not wishing to lose the impulse already given to mathematical studies among us,'' invited the young German mathematician Felix Klein to succeed Sylvester. After some hesitation, Klein refused the offer and went instead to Göttingen, where he succeeded in making it as prestigious and as great as Berlin, if not greater. Klein attended the International Scientific Congress at the World's Fair in Chicago in 1893 (commemorating the four-hundredth anniversary of the discovery of America). And afterward, he delivered a series of lectures at Northwestern University, the first colloquium of American mathematicians. By the end of the century, a well-known French mathematician, C. A. Laisant, was able to write of the situation in the United States:

Mathematics in all its forms and in all its parts is taught in numerous universities, treated in a multitude of publications, and cultivated by scholars who are in no respect inferior to their fellow mathematicians in Europe. It is no longer an object of import from the old world, but it has become an essential article of national production, and this production increases each day both in importance and in quantity.

This is not the place to recount in detail the events whereby Prussia reached ascendancy in a united Germany. Suffice it to say that the long-desired unification of the German states, with the exception of Austria, was accomplished as a result of the Franco-Prussian War in 1871. The new German empire was a union of the governments of 26 states of various sizes and one administrative territory, the conquered provinces of Alsace and Lorraine. The rights enjoyed by the member states were not equal, because they were all in one way or another subject to Prussia. The constitution gave Prussia a degree of prominence and power that was consistent with its territory, population, and military prowess in bringing about the stunning victory over France. Not only was the hereditary leadership of the empire vested in the king of Prussia with the title *German emperor,* but the minister-president was nearly always the imperial chancellor, the head of the federal government. Thus the new empire did not reflect the submergence of Prussia in Germany but represented the extension of Prussian influence to the whole nation. Although Germany was to become the most powerful of the continental countries, it was, like the old Prussian kingdom, an autocratic military state.

People have come to look back on the last third of the nineteenth century in Germany as a golden age of mathematical scholarship; and they are not unjustified in doing so, as even a short list of its university professors will indicate. Although mathematical generations inevitably overlap—as their ideas do also—the great names on the scene after 1870 are Georg Cantor, Richard Dedekind, Paul Gordan, Eduard Heine, David Hilbert, Otto Holder, Adolf Hurwitz, Felix Klein, Leopold Kronecker, Ernst Kummer, Ferdinand Lindemann, Rudolph Lipschitz, Hermann Minkowski, Moritz Pasch, and Karl Weierstrass. One consequence of this galaxy of brilliance was that a state of intense rivalry and sometimes of bitter enmity existed continually in German mathematical circles. This was particularly manifest in the loudly voiced doubts over one of the most disturbingly original contributions to mathematics in 2500 years, Cantor's theory of infinite sets. Whether the violent opposition was brought on more by the strangeness of the idea of the "actually infinite" or more by the forceful personalities of the individuals involved is hard to say. The result was the same.

Georg Ferdinand Cantor (1845–1918), although he was born in St. Petersburg and lived there until 1856, should properly be ranked among the German mathematicians, because he was educated and employed in German universities. His stockbroker father had urged him to study engineering, a more profitable pursuit than mathematics, and with this intention Cantor began his university studies at Zurich in 1862. The elder Cantor finally agreed to allow his son to follow a career in mathematics, so that after a semester at Zurich he moved to the University of Berlin. There he attended the lectures of the great triumvirate, Weierstrass, Kummer, and Kronecker. In 1867, he received his Ph.D. from Berlin, having submitted a thesis on problems in number theory, a thesis that in no way foreshadowed his future work. Two years later, Cantor accepted an appointment as privatdozent at Halle University, where he remained until his retirement in 1913.

Georg Cantor
(1845–1918)

(By courtesy of Columbia University, David Eugene Smith Collection.)

Influenced by Weierstrass's teaching on analysis, Cantor's initial research dealt with trigonometric series. A sequence of five articles issued between 1870 and 1872 culminated in showing that the uniqueness of the representation of a function by a trigonometric series holds even if convergence is renounced for an infinite set of points in the interval [0, 2π]. Because Cantor's uniqueness proof depended heavily on the nature of certain point sets in the real line, and only to a lesser extent on trigonometric series, it was only natural for him to explore the consequences of the former. The birth of set theory can be marked by Cantor's next published paper, *Über eine Eigenshaft des Inbegriffes aller reellen algebraischen Zahlen* (On a Property of the System of all the Real Algebraic Numbers), which is found in *Crelle's Journal* for 1874. Over the next two decades, the need for comparing the magnitudes of infinite sets of numbers led Cantor, almost against his will, to his notion of transfinite numbers, and to immortality. Growing out of specific problems posed by trigonometric representation, and reaching full articulation in Cantor's lengthy survey *Beitrage zur Begründung der Transfiniten Mengenlehre* of 1895 (translated into English in 1915 under the title *Contributions to the Founding of the Theory of Transfinite Numbers*), set theory gained an autonomy as a mathematical discipline.

The year 1872 was significant for mathematics in many ways. Cantor laid the outline of an entirely new field of research. There was also Klein's celebrated inaugural address when he became professor at Erlangen. The year also saw Weierstrass's presentation to the Berlin Academy of an example of a continuous nondifferentiable function. And Dedekind published *Stetigheit und irrationale Zahlen,* in which he constructed the irrational numbers in terms of his famous ''cuts.'' The problem of irrational numbers had existed from the time of the Pythagoreans, but until 1872 no successful attempt had been made to give them a precise mathematical meaning. They ''existed'' as decimal approximations, and the logical basis of, say, π was no more sure than its approximation to 707 places by William Shanks in 1853. (It took him 15 years of calculation, and later an error was found in the 528th place.) Cantor's attention was directed toward these matters when he realized that an understanding of the nature of the elusive irrationals lay at the root of his proof of the uniqueness of the trigonometric representation. In his paper of 1872, the year of Dedekind's construction, Cantor devised a rigorous formulation of the irrational numbers by means of what we should today call Cauchy sequences. Thus, during the 1870s, Weierstrass, Dedekind, and Cantor all succeeded in developing algebraically self-contained theories of the irrational numbers; but they substituted an appeal to set-theoretic intuition for the limit concept.

Cantor, in the first sentence of his great synoptic work of 1895, tried to define what he meant by a set (*Menge*, in the German). The words are not novel now, although they were then:

> By a set we are to understand any collection into a whole M of definite and distinguishable objects of our intuition or our thought. These objects are called the elements of M.

Although "collection into a whole" is at best a paraphrase of the notion of set, the terms *definite* and *distinguishable* had a clear meaning to Cantor. The intended meaning of the former was that given a set M, one should be able to decide whether any particular element would belong to M; the attribute *distinguishable* is interpreted as meaning that any two elements of the same set are different. The implication is that a set is determined solely by what is in it, that is, by its elements.

Cantor conceived of the notion of set in as general a way as possible. There was no restriction whatever on the nature of the considered objects nor on the way they were collected into a whole. Because his definition was not precise enough to prohibit him from considering such things as the "set of all sets," it ultimately led to some famous paradoxes concerning the infinite. (Paradoxes are apparently contradictory results obtained by apparently impeccable logic.) These paradoxes, which threatened the very foundations of logic and mathematics, necessitated the refinement of Cantor's naive concept of "set." The attempted improvements in the definition were so unsuccessful in identifying the notion that today we find it convenient to take *set* and *element* as undefined terms.

It should be emphasized that Cantor was not the only one, or even the first, to be interested in the properties of infinite sets. Galileo noticed the curious circumstance that part of an infinite set could, in a certain sense, contain as many elements as the whole set. In his *Dialogue Concerning the Two Chief World Systems* (1632), he made the telling observation: "There are as many squares as there are numbers because they are just as numerous as their roots." He asked which of the two sets, squares or natural numbers, could be the larger one. Seeing in this discovery only a puzzle, he abandoned the subject because it was not amenable to reason.

Cantor gave a precise meaning to "as many" by interpreting the phrase to require that there exist a one-to-one correspondence between the two sets in question:

> Two sets M and M' are equivalent (equipotent, equinumerous), written $M \sim M'$, if there exists a one-to-one correspondence between their elements.

It is clear that two finite sets are equivalent provided that they have the same number of elements. But Cantor's definition of equivalence does not use the notion of finiteness in any essential way. It depends only on the idea of one-to-one correspondence, which can be applied to all sets, finite or not. In Galileo's example, the set of natural numbers is equivalent to the set of perfect squares via the mapping that sends a natural number n to its square n^2. This shows that a set may be equivalent to a subset of itself.

Up to now the terms *finite set* and *infinite set* have been used in an informal way, but they can be given a precise meaning through the notion of equivalence. Because everyday experience involves encounters with finite sets only, the usual custom is first to define a finite set in the positive sense, and then to take an infinite set as one that is not finite:

> A set M is finite if either it is empty or there exists a natural number n such that $M \sim \{1, 2, 3, \ldots, n\}$; otherwise, M is infinite.

The first positive steps toward a theory of sets were taken in the mid-nineteenth century by Bernhard Bolzano (1781–1848), a Bohemian priest who was dismissed from his post as professor of religion at the University of Prague for heresy. Although Bolzano was concerned mainly with social, ethical, and religious questions, he was attracted by logic and mathematics, especially analysis. Unfortunately, most of Bolzano's mathematical writings remained in manuscript form and did not attract the attention of his contemporaries or directly influence the development of the subject. (Many were published for the first time in 1962.) Bolzano's small tract *Paradoxien des Unendlichen* (Paradoxes of the Infinite), which was published three years after his death by a student he had befriended, contains many interesting fragments of set theory; in fact, the term *set* made its initial appearance here. Familiar with Galileo's paradox on the one-to-one correspondence between natural numbers and perfect squares, Bolzano expanded the theme by giving more examples of correspondences between the elements of an infinite set and a proper subset. What had perplexed Galileo and what Bolzano had regarded as a curious property of infinite sets was elevated by Dedekind—who earned his doctor's degree under Gauss—to the status of a definition of the infinite. In 1888, Dedekind published a small pamphlet, *Was sind und was sollen die Zahlen* (The Nature and Meaning of Numbers), in which he proposed a definition of *infinite* that had no explicit reference to the concept of natural number:

> A set M is infinite if it is equivalent to a proper subset of itself; in the contrary case, M is finite.

This was adopted by Cantor, whose work developed along a direction parallel to that of his personal friend, Dedekind.

Cantor spent considerable effort defending himself against the opposition of many mathematicians who regarded the *infinite* more as a description of unbounded growth, expressed by some symbol like ∞, than of an attained quantity. According to the traditional conception, the infinite was something "increasing above all bounds, but always remaining finite." As it appeared in the work of Cantor, it was "fixed mathematically by numbers in the definite form of a completed whole." What disturbed the critics most was that an "actually infinite" set was an abstraction to which there could be no corresponding physical reality—there was no evidence that infinite collections of physical objects existed. Whose views carried more authority than those of the Prince of Mathematicians, Carl Friedrich Gauss? The influence of this monumental figure most surely set the tone of the mathematical world up to almost the end of the century. In a famous letter to Schumacher, written in 1831, Gauss posed his horror of the infinite:

> As to your proof, I must protest most vehemently against your use of the infinite as something consummated, as this is never permitted in mathematics. The infinite is but a figure of speech; an abridged form for the statement that limits exist which certain ratios may approach as closely as we desire, while other magnitudes may be permitted to grow beyond all bounds. . . . No contradictions will arise as long as Finite Man does not mistake the infinite for something fixed, as long as he is not led by an acquired habit of mind to regard the infinite as something bounded.

Not satisfied with merely defining infinite sets, Cantor proposed something even more shocking and impious—endowing each set with a number representing its plurality. This would allow him to distinguish infinite sets by "size," and to show, for example, that there are "more" real numbers than there are integers. Some mathematicians of the day could accept, albeit reluctantly, Cantor's infinite sets, taking an attitude that has been compared with that of a gentleman toward adultery: better to commit the act than utter the word in the presence of a lady. It was an actually infinite *number* that was forbidden, and its use forced Cantor to live the rest of his life within a storm.

Cantor's attempt to measure sets led him to introduce the notion of cardinal numbers. In his earliest papers, he found it prudent to adopt a neutral attitude toward cardinal numbers, saying what they are supposed to do and not what they actually are:

> Two sets have the same cardinal number or have the same power if they are equivalent.

Thus, for Cantor, a cardinal number is "something" attached to a set in such a way that two sets are assigned the same cardinal if and only if they are equivalent. In his final (1895) exposition of their theory, he tried to remove this vagueness by means of a definition "by abstraction":

> If we abstract both from the nature of the elements and from the order in which they are given, we get the cardinal number or power of the set.

The cardinal number of the set M was thus taken to be the general concept common to all sets equivalent to M. The process of double abstraction, or disregarding both the special properties of the elements and any ordering within the set, is the origin of the double bar in Cantor's symbol $\overline{\overline{M}}$ for the cardinal number of the set M. The modern notation $o(A)$ for the cardinal number of A will serve our purposes quite well.

Cantor's "definition" of cardinal number can hardly be regarded as satisfactory, and various attempts were made to formalize the concept. The logician Gottlob Frege in his *Grundlagen der Arithmetik* of 1884 suggested a definition that did not become widely known until Bertrand Russell, who had arrived at the same idea independently, gave prominence to it in his *Principles of Arithmetic* (1903). This so-called Frege-Russell definition is beautiful in its simplicity: The cardinal number of a set A is the set of all sets equivalent to A. (Unless logical precautions are taken, of course, there may not exist a set that contains all sets with a given property.) On the other hand, John Von Neumann (1928) selected a fixed set from the set of all sets equivalent to A

Leopold Kronecker
(1823–1891)

(By courtesy of Columbia University, David Eugene Smith Collection.)

to serve as the cardinal of *A*. Whatever a cardinal number is is perhaps immaterial; all we need assert is that it is an object shared by just those sets that are equivalent to each other. The cardinal number of a finite set is said to be a finite cardinal, whereas the cardinal number of an infinite set is called a "transfinite cardinal."

In building up an arithmetic of transfinite numbers analogous to ordinary arithmetic, Cantor became a mathematical heretic. The outcry was immediate, furious, and extended. Cantor was accused of encroaching on the domain of philosophers and of violating the principles of religion. Yet, in this bitter controversy, he had the support of certain colleagues, most notably Dedekind, Weierstrass, and Hilbert. Hilbert was later to refer to Cantor's work as "the finest product of mathematical genius and one of the supreme achievements of purely intellectual human activity."

Cantor's former professor, Leopold Kronecker (1823–1891) became the focus of Cantor's troubles, a sort of personal devil. Kronecker had made important contributions to higher algebra, but in matters pertaining to the foundations of mathematics he did little more than openly criticize the efforts of his contemporaries. The son of a wealthy businessman in Liegnitz, Prussia, Kronecker was provided private tutoring at home until he entered the local gymnasium, where Ernst Kummer happened to be teaching. (Because no university position was open at the time Kummer was awarded his Ph.D., he taught for ten years in his old gymnasium.) While still at Liegnitz, Kronecker became interested in mathematics through Kummer's stimulation and encouragement. In 1841, he enrolled at the University of Berlin, then the mathematical capital of the world, where he studied with Dirichlet, Jacobi, and Eisenstein.

The German student of that day was free to attend the lectures of his choice or even to move from one university to another, restricted by no formal curriculum and responsible only, in the end, to his examiners. Following this custom, Kronecker spent the summer of 1843 at the University of Bonn, then migrated to Breslau for two semesters. There his former teacher Kummer was then a professor. Subsequently Kronecker returned to Berlin to write a thesis on algebraic number theory under Dirichlet. Temporarily obliged to leave the academic world in order to manage the prosperous family business, Kronecker was for eleven years unable to return to Berlin and to his hobby, mathematics. By this time the University of Berlin was beginning to experience a new flowering in mathematics, brought on by the arrival of both Kummer and Weierstrass. On Kummer's nomination, Kronecker was elected a member of the Berlin Academy of Sciences in 1860; this position entitled him to deliver lectures at the university and he regularly availed himself of the prerogative, beginning in 1861. Because the wealthy Kronecker could afford to teach without holding a chair, he refused the professorship in mathematics at Göttingen held successively by Gauss, Dirichlet, and Riemann. Feeling the onset of a decline in productivity, Kummer suddenly decided to retire in 1882. Kronecker was then called upon to succeed his old mentor, thus becoming the first person to hold a position at Berlin who had also earned a doctorate there.

Kronecker was a tiny man, who was increasingly self-conscious of his size with age. He took any reference to his height as a slur on his intellectual powers. Making loud voice of his opinions, he was venomous and personal in his attacks on those whose mathematics he disapproved; and his opinions relative to the new theory of infinite sets were ones of ire and indignation. As Cantor's bold advance into the realm of the infinite was based largely on nonconstructive reasoning, Kronecker categorically rejected the ideas from the start. He asserted dogmatically, "Definitions must contain the means of reaching a decision in a finite number of steps, and existence proofs must be conducted so that the quantity in question can be calculated with any required degree of accuracy." Any discussion of infinite sets was, according to Kronecker, illegitimate since it began with the assumption that infinite sets exist in mathematics.

Kronecker not only objected strenuously to Cantor's uninhibited use of infinite sets, but to most of contemporary analysis. His principal concern was with the new formulations of irrational numbers, Dedekind's by his device of "cuts," Weierstrass's by classes of rational numbers. Kronecker felt that these produced numbers that could have no existence. Returning to the ancient Pythagorean vision, Kronecker gave loud voice to the view that all mathematics must be built up by finite processes from the natural numbers. This counterrevolutionary program is revealed in his oft-quoted dictum, "God created the natural numbers, and all the rest is the work of man." It is not too surprising that Kronecker found Weierstrass's analysis unacceptable, lacking as it did constructive procedures for determining quantities whose being was merely established by the free use of "theological existence proofs." One day he reduced the distinguished old man to tears with an abrasive remark about "the incorrectness of all those conclusions used in the so-called present method of analysis." Seeing in these words an attempt by Kronecker to tear down a whole life's work, Weierstrass severed all ties with his erstwhile colleague.

Although Kronecker's notion of mathematical existence angered and embittered Weierstrass, it was the high-strung Cantor who was wounded most seriously by such uncompromising skepticism. Cantor had hoped to obtain a professorship at the University of Berlin, possibly the highest German distinction that could be secured during the period of his productivity. But the opposition to his work was growing, especially for its use of the "actually infinite." In Berlin, the almost omnipotent Kronecker blocked Cantor's every attempt to improve his position; and when a professorship at Göttingen was to be made, Cantor was passed over in favor of lesser candidates. All Cantor's professional career, some 44 years, was spent at Halle University, a small school without particular reputation. The temperamental Cantor suffered deeply under what he considered Kronecker's malicious persecution, with the tragic outcome that he sustained a complete nervous breakdown in 1884. Although he recovered from this crisis within a year, mental illness was to plague him through the remainder of his life.

In the hostile intellectual world, Cantor found an influential friend in Gösta Mittag-Leffler, who had studied under Weierstrass in Berlin. Mittag-Leffler's wife was a millionaire, so that he was financially able to establish a new mathematical journal, *Acta Mathematica*. Hoping to make a noteworthy start, he proposed issuing French translations of the most important papers that Cantor had so far published. No doubt he had in mind the good fortune of Crelle, who began his journal with a plentiful supply of work by Abel. These translations, which appeared in volume two (1883) of *Acta Mathematica,* contributed to the spread of Cantor's ideas on set theory.

Even so sympathetic a supporter as Mittag-Leffler failed to appreciate the revolutionary character of Cantor's research. He asked Cantor to withdraw a comprehensive account of the properties of ordered sets that was intended for the seventh volume of *Acta Mathematica* (1885–1886). Mittag-Leffler suggested that because the paper did not contain the solution of some important problem, it would be better not to publish it but to allow the results to be rediscovered—in say one hundred years' time—when it would be found that Cantor had possessed them much earlier. Aggrieved at being told that his paper was "one hundred years premature," Cantor nevertheless complied with this unfortunate request. As events fell out, the corrected page proofs of the rejected article were indeed rediscovered among Cantor's surviving papers. The paper was published, 85 years later, in the 1970 *Acta Mathematica.*

Cantor had become exhausted in the hard struggle to gain recognition for his work. Beginning with the first of his attacks of depression in 1884, the rest of Cantor's life was punctuated by bouts of mental illness which would force him to spend time in various sanitoria. His intervening periods of clarity were more often devoted to Elizabethan scholarship and religious writings than to mathematical activity. Apart from a short article in 1892 setting forth the "diagonal argument" for the uncountability of the real numbers, Cantor published little on set theory until his comprehensive *Contributions to the Founding of the Theory of Transfinite Numbers* (in the *Mathematische Annalen* for 1895 and 1897). This two-part memoir was less a collection of new ideas than a final statement of many of the most important results going back to 1870.

Only by the 1900s, when Cantor had ceased his research, did his ideas at last begin to receive some recognition. Initial distrust by the mathematical world turned into appreciation and even admiration. Of the various awards and honorary degrees belatedly bestowed on Cantor, the Sylvester Medal of the Royal Society of London (1904) is worth particular mention because it is so rarely given. The first comprehensive textbook on set theory and its applications to the general theory of functions, *The Theory of Sets of Points,* was published in 1906 by William Henry Young and his wife Grace Chisholm Young. Initial homage came mainly from abroad, and as late as 1908, Cantor complained to the Youngs of the lack of importance attached to his work by the Germans "who do not seem to know me, although I have lived among them for fifty-two years.'' In 1915, an event of international importance was planned at Halle to celebrate his seventieth birthday; but because of the war, only a few close German friends were able to gather to pay him honor. Cantor died in a psychiatric clinic in Halle in 1918.

From our present vantage point, we see that Cantor won from the next generation of mathematicians the recognition that most of his contemporaries denied him. Although the discovery of the paradoxes of the infinite were to force the modification of many of his ideas, the main concepts of set theory survived to become cornerstones in the foundations of many other branches of mathematics. Kronecker, on the other hand, despite his great authority, failed to gain supporters for his ''mathematical nihilism.'' Faithful adherence to the position that existence statements are meaningless unless they contain a construction for the asserted object would result in the abandonment of much of modern mathematics. Kronecker was contending against the unquestionable fact that proofs of pure existence often produce the most general results with the least effort. An inflexible advocate of his convictions, Kronecker in his violent opposition of Cantor's work succeeded only in curbing its early development for two decades.

12.2   Counting the Infinite

As Cantor turned away from the established traditions of mid-nineteenth-century analysis and focused on linear points sets, a new era in mathematics opened. In 1874, he published his first purely set theoretic work, *On a Property of the Collection of All Real Algebraic Numbers,* in which he made a distinction between two types of infinite sets on the real line. Dedekind, writing in his memoir *Stetigkeit und die Irrational-zahlen* (1872), had already perceived this distinction: ''The line L is infinitely richer in point-individuals than is the domain R of rational numbers in number-individuals.'' The term ''countable'' was later used by Cantor to describe the simplest kind of infinite, one that has the power of the natural numbers.

Definition

A set A is said to be countable (denumerable, enumerable) if there is a one-to-one correspondence between it and the set N of natural numbers. Infinite sets which are not countable are called uncountable (nondenumerable).

This definition affords us a certain convenience. If a set A is countable, then for a particular one-to-one correspondence between A and N that element of A associated with the natural number n may be labeled a_n. This allows us to write $A = \{a_1, a_2, a_3, \ldots, a_n, \ldots\}$, with the elements of A listed in the form of a sequence. The converse is also true: A set that can be designated $\{a_1, a_2, a_3, \ldots, a_n, \ldots\}$ is countable.

The first thing to notice about countability is that it is the smallest kind of infinity.

THEOREM

Every infinite set contains a countable subset.

Proof. Let an infinite set A be given. Choose an element of A and call it a_1. Because A is an infinite set, $A - \{a_1\}$ is infinite also; choose one of its elements and call it a_2. Then $A - \{a_1, a_2\}$ is an infinite set; indeed, after k repetitions of this selection process, $A - \{a_1, a_2, \ldots, a_k\}$ is still infinite, and we can choose from it a next element a_{k+1}. Because there is no stage of this process at which we lack a successor to the elements already chosen, this selection scheme produces a countable subset $\{a_1, a_2, \ldots, a_k, \ldots\}$ of A. ∎

The next theorem uses in its proof the well-ordering principle for positive integers: Any set of positive integers has a smallest element.

THEOREM   *A subset of a countable set is either finite or countable.*

> **Proof.** Let B be a subset of the countable set $A = \{a_1, a_2, \ldots\}$. We shall list the elements of B in the order in which they occur in A, calling b_1 the element of B that has the least subscript when viewed as an element of A. If $B - \{b_1\}$ is not empty, it too has an element with the least subscript when viewed as an element of A; call this element b_2.
>
> If after k repetitions of this process, it is found that $B - \{b_1, b_2, \ldots, b_k\}$ is empty, then clearly $B = \{b_1, b_2, \ldots, b_k\}$ is a finite subset of A.
>
> If on the contrary, $B - \{b_1, b_2, \ldots, b_k\}$ remains nonempty for each positive integer k, then we can always choose b_{k+1} as its element with the least subscript in A. In this way it is possible to construct a countable subset $\{b_1, b_2, \ldots\}$ of B.
>
> To see that each element b of the set B is a member of the countable subset $\{b_1, b_2, \ldots\}$, recall that since $b \in A$, then $b = a_n$ for some n. At some stage in the construction of $\{b_1, b_2, \ldots\}$, certainly by the nth step, a_n must have been that element of B with the least subscript when viewed as an element of A. Then $b = a_n$ belongs to the countable set $\{b_1, b_2, \ldots\}$; indeed it is one of its first n elements. Hence $B = \{b_1, b_2, \ldots\}$ is a countable subset of A. ∎

The springboard from which to prove many results on countable sets is the following theorem.

THEOREM   *The union of a countable number of countable sets is a countable set.*

> **Proof.** We consider a countable collection $\{A_1, A_2, \ldots, A_n, \ldots\}$ of sets A_i each of which is itself countable. Thus, for each i, the set A_i can be displayed in sequence form as $A_i = \{a_{i1}, a_{i2}, \ldots, a_{in}, \ldots\}$. The theorem will become evident when a sequence is constructed in which all the elements of all the set A_i appear. Because

$$
\begin{aligned}
A_1 &= \{a_{11}, a_{12}, a_{13}, \ldots, a_{1n}, \ldots\}, \\
A_2 &= \{a_{21}, a_{22}, a_{23}, \ldots, a_{2n}, \ldots\}, \\
A_3 &= \{a_{31}, a_{32}, a_{33}, \ldots, a_{3n}, \ldots\}, \\
&\ \ \vdots \\
A_i &= \{a_{i1}, a_{i2}, a_{i3}, \ldots, a_{in}, \ldots\}, \\
&\ \ \vdots
\end{aligned}
$$

we may use the back-and-forth diagonal path indicated to list all the a_{ij} systematically as a sequence:

$$a_{11}, a_{12}, a_{21}, a_{31}, a_{22}, a_{13}, a_{14}, a_{23}, a_{32}, a_{41}, \ldots$$

Every element of each set A_i will eventually be encountered somewhere on the path; hence a one-to-one correspondence between the set of a_{ij} and the set N of natural numbers is implied (any element that is repeated can be deleted from the list when we come to it a second time).

It is not hard to write down the actual formula giving the one-to-one correspondence from $\cup A_i$ to N being used in this process. In fact, one can easily show that

$$f(a_{ij}) = \frac{(i + j - 1)^2 + i - j + 1}{2},$$

but the description of our listing is so simple that there is no need to use this explicit correspondence. ∎

With this theorem we acquire several useful corollaries.

COROLLARY 1   *The union of a finite number of countable sets is countable.*

Proof. If countable sets $A_1, A_2, \ldots, A_n$ are to be considered, we may use the theorem on taking $A_{n+1}, A_{n+2}, \ldots$ to be the empty set. ∎

COROLLARY 2   *The set Z of integers is countable.*

Proof. Certainly the set N of natural numbers is countable. We display the set of all the integers

$$Z = N \cup \{0, -1, -2, -3, \ldots, -n, \ldots\}$$

to see that Z is a union of two countable sets. ∎

COROLLARY 3   *The set Q of rational numbers is countable.*

Proof. First consider the positive rational numbers, classifying them by denominator:

$$A_1 = \left\{ \frac{1}{1}, \frac{2}{1}, \frac{3}{1}, \ldots, \frac{n}{1}, \ldots \right\},$$

$$A_2 = \left\{ \frac{1}{2}, \frac{2}{2}, \frac{3}{2}, \ldots, \frac{n}{2}, \ldots \right\},$$

$$A_3 = \left\{ \frac{1}{3}, \frac{2}{3}, \frac{3}{3}, \ldots, \frac{n}{3}, \ldots \right\},$$
$$\vdots$$
$$A_i = \left\{ \frac{1}{i}, \frac{2}{i}, \frac{3}{i}, \ldots, \frac{n}{i}, \ldots \right\},$$
$$\vdots$$

That is, all positive rational numbers with denominator 1 appear in A_1, those with denominator 2 appear in A_2, and so forth. The union $\cup A_i$ is the set Q_+ of positive rational numbers, and just as in the proof of the theorem, we see that it is countable. With the obvious modifications in the sets A_i, it can also be inferred that the set Q_- of negative rationals is countable. Because

$$Q = Q_+ \cup \{0\} \cup Q_-,$$

an appeal to Corollary 1 completes the proof that Q is a countable set. ∎

A more "visual" proof of the countability of Q_+ is obtained by writing out the positive rationals as the array

All that is left to do is to start counting down the diagonals in the manner of the last theorem; that is, we traverse the diagram as shown by the arrows, discarding rational numbers like $\frac{2}{2}$, $\frac{2}{4}$, $\frac{3}{3}$, and $\frac{4}{2}$, which are equal to numbers that have been previously passed. The enumeration according to our procedure begins with

$$1, 2, \frac{1}{2}, \frac{1}{3}, 3, 4, \frac{3}{2}, \frac{2}{3}, \frac{1}{4}, \frac{1}{5}, 5, \ldots.$$

In this way, we get an infinite sequence in which every positive rational number occurs exactly once.

By what we have just proved, we must conclude that the set Q of rational numbers and the set Z of integers are of "equal size," despite the inclusion of the second set in the first. You will probably want to object vigorously. Doesn't this contradict the famous principle formulated by the Greeks that the whole is greater than any of its parts? Cantor realized that in dealing with infinite sets this principle, which holds for finite sets, must be abandoned. The sets Q and Z have the same number of elements because they are equivalent; their elements can be matched against each other. If we accept "equivalent" as the criterion for comparing the sizes of infinite sets, then we must put aside the traditional way of thinking and agree that "the part has the power of the whole" in this instance. This is the essence of Galileo's paradox.

Because all the infinite sets considered so far have had the same power, one might get the impression that all infinite sets are equivalent. Cantor's theory would be trivial if there were no uncountable sets, no kinds of infinity beyond the countably infinite. Cantor gave two proofs that the set of real numbers cannot be arranged in one-to-one correspondence with the natural numbers. The first, which appeared in the pathbreaking paper *On a Property of the Collection of All Real Algebraic Numbers* (1874), involved taking a nested sequence of closed intervals and claiming the existence of a limiting number contained within all these intervals. We shall describe Cantor's second proof (1891), this being both simpler in form and more general in application. It uses what is today known as Cantor's diagonal argument.

THEOREM   *The set R of real numbers is uncountable.*

Proof. Let us assume that the theorem is false, so that the real numbers form a countable set. Each infinite subset of the real numbers must then be countable, by an earlier theorem. The subset we want to consider is the set of all real numbers x satisfying $0 < x < 1$. Every such real number has a nonterminating decimal expansion $0.x_1x_2x_3 \ldots x_n \ldots$, where each x_i represents a digit in the expansion; that is, $0 \le x_i \le 9$. Not only is there an expansion of this form for each real number between 0 and 1, but some numbers have two such expansions. The confusion is caused by decimals in which all the digits assume the value 9 after a certain point, numbers like $0.36999 \ldots$, which is no different from $0.37000 \ldots = 0.37$. To eliminate this ambiguity, let us rule out those expansions ending in an infinite string of zeros. In other words, we shall not identify the real number $\frac{37}{100}$ by $0.37000 \ldots$, but rather by $0.36999 \ldots$

The set of real numbers between 0 and 1, being countable, can be displayed as $\{a_1, a_2, a_3, \ldots, a_n, \ldots\}$, and each a_i has an infinite decimal expansion. Let us say that $0.a_{i1}a_{i2}a_{i3} \ldots a_{in} \ldots$ is the decimal expansion of a_i. We can then write the elements of the set under consideration as the array

$$a_1 = 0.a_{11}a_{12}a_{13} \ldots a_{1n} \ldots,$$

$$a_2 = 0.a_{21}a_{22}a_{23} \ldots a_{2n} \ldots,$$

$$a_3 = 0.a_{31}a_{32}a_{33} \ldots a_{3n} \ldots,$$

$$\vdots$$

$$a_i = 0.a_{i1}a_{i2}a_{i3} \ldots a_{in} \ldots,$$

$$\vdots$$

where $0 \le a_{ij} \le 9$ for all i and j.

We intend to construct a real number b, with $0 < b < 1$, which appears nowhere on our list. The desired contradiction is established and the theorem proved when such a number b is produced. Our assumption that the set of real numbers is countable has led to a claim that any real number between 0 and 1 must be one of the listed a_i.

Looking down the diagonal of the preceding array, we shall form b as an infinite decimal, digit by digit, so that it disagrees at some decimal place with each of the a_i. For a given value of i, if the "diagonal digit" $a_{ii} = 1$, then we put $b_i = 2$, and if $a_{ii} \neq 1$, then we put $b_i = 1$. To illustrate: If $a_1 = 0.31429 \ldots$, $a_2 = 0.81621 \ldots$, $a_3 = 0.58207 \ldots$ happened to be the first three numbers in our list, then the decimal expansion of b would start off as $b = 0.121. \ldots$ (We have avoided the possible ambiguity that might arise from an infinite sequence of zeros by making sure that the decimal expansion of b doesn't have any.)

This procedure defines a new number b that equals none of the a_i just listed, since $b_i \neq a_{ii}$; that is, the decimal expansion of b differs from the decimal expansion of a_i at least in the ith place. But b is surely a real number between 0 and 1, which contradicts our assertion that the set of all such numbers is countable. ∎

Although this last theorem confounded hopes of a certain tidiness of the infinite sets, it also meant that the varieties of infinity are richer than first expected. The distinction between countable sets and uncountable sets is not an empty one. Cantor, somewhat at a loss for new symbols for transfinite cardinals, called the power of the natural numbers $\aleph_0$, where $\aleph$ (*aleph*) is the first letter of the Hebrew alphabet. The countable cardinal $\aleph_0$ is the first of the transfinite cardinals. By showing that the set of real numbers was uncountable, Cantor demonstrated that their cardinal number c (for "continuum") strictly exceeded $\aleph_0$.

Although any countable set has cardinality $\aleph_0$, it does not follow that any uncountable set must have cardinality c. The test for whether a given set has cardinality c is to see whether it can be placed in one-to-one correspondence with the set of all real numbers. For instance, the set of real numbers can be regarded as the union of the rational numbers and the irrational numbers, where the rationals have been shown to be countable. Were the irrational numbers to form a countable set also, a reference to an earlier theorem would convince us that the reals are countable also. We have just proved that this is not the case, so we must conclude that the set of irrational numbers is not countable. Nothing that has been said so far, however, convinces us that the set of irrational numbers has the same power as the continuum does. And from another point of view, were we to find a set whose power exceeds c, it would be a noncountable set.

For the moment we have been considering a division of the real numbers into two subsets: those that are rational and those that are not. It is also possible to think of another partition of the real numbers, into those that are roots of equations and those that are not.

Definition

A number, real or complex, is said to be an algebraic number *if and only if it is the root of the algebraic equation*

$$a_0 x^n + a_1 x^{n-1} + \cdots + a_{n-1}x + a_n = 0, \qquad a_0 \neq 0,$$

where n is a natural number and each a_i is an integer.

A moment's pause will show that any rational number is algebraic; if r is a rational number, say $r = s/t$ where s and t are integers, then r is the solution of the equation $tx - s = 0$. On the other hand, not all algebraic numbers are rational; the irrational number $\sqrt{2}$ is algebraic, because it is a solution of $x^2 - 2 = 0$. Similarly, $\frac{1}{2}\sqrt[3]{3}$, as a root of the equation $8x^3 - 3 = 0$, is an algebraic number.

Let us use the letter A to denote the set of real algebraic numbers. Then in the hierarchy of subsets of real numbers, we have an ascending chain of sets:

$$Z \subseteq Q \subseteq A \subseteq R.$$

The question naturally arises whether the set A constitutes the totality of all real numbers, or whether there are real numbers that are not algebraic. Such numbers are called "transcendental," for as Euler said, "They transcend the power of algebraic methods." Just as it was by no means obvious that irrational numbers exist, so it was by no means obvious that transcendental numbers exist. No satisfactory answer to the question of their existence was found until 1844, when the great French analyst Joseph Liouville (1809–1882) obtained a whole class of such numbers, namely, those defined by the infinite series

$$\sum_{n=1}^{\infty} a_n 10^{-n!},$$

where each a_n is an arbitrary integer between 1 and 9. These so-called Liouville numbers are characterized by increasingly long blocks of zeros interrupted by a single nonzero digit, as with

$$\alpha = 10^{-1} + 10^{-2} + 10^{-6} + 10^{-24} + 10^{-120} + \cdots$$

$$= 0.110001000000000000000000010. \ldots$$

The proof of their transcendence was eventually set forth (1851) by Liouville in his own periodical, *Journal des mathématiques,* in a memoir under the title: "On a very extensive class of quantities which are neither algebraic nor reducible to algebraic irrationals."

For quite a time, Liouville's transcendental numbers were the only ones known. The situation changed radically when Cantor, in his 1874 paper, used relatively simple set-theoretic methods to show that "almost all" real numbers are transcendental, not algebraic.

To develop the proof of the countability of the set of algebraic numbers, we begin by considering an arbitrary algebraic equation

$$a_0 x^n + a_1 x^{n-1} + \cdots + a_{n-1} x + a_n = 0,$$

where $a_0, a_1, \ldots, a_n$ are integers and $a_0 > 0$. The *height* of this equation is defined to be the integer

$$h = n + a_0 + |a_1| + \cdots + |a_{n-1}| + |a_n|.$$

Because the integers a_0 and n are at least one, the height $h \geq 2$. Thus, for instance, $3x^2 - 2x + 1 = 0$ has height $h = 2 + 3 + 2 + 1 = 8$. For any fixed height h, the integers $n, a_0, a_1, \ldots, a_{n-1}, a_n$ can be specified in only a finite number of ways, thereby leading to a finite number of equations; and each such equation can have at most as many different roots as its degree. Thus, there are just a finite number of algebraic numbers arising from equations of a given height. By grouping the algebraic equations according to height, starting with those of height 2, then taking those of height 3, and so on, one can write down the set of algebraic numbers in a sequence. This is what Cantor actually did.

When an equation's height is 2, we must have $n = a_0 = 1$ and $a_1 = 0$, so that the equation must be $x = 0$. Then 0 is the sole root of this equation, hence the first algebraic number.

Let us arrange equations of height 3 (and also equations of larger heights) first according to the increasing degrees of the equations; then let us arrange all equations of the same degree according to the size of their initial coefficients, those with the same first coefficient according to the second, and so on. This enumeration scheme yields:

Equation	Roots
$x \pm 1 = 0$	± 1
$2x = 0$	0
$x^2 = 0$	0

We shall be explicit about one more step by considering equations that satisfy the condition $n + a_0 + |a_1| + \cdots + |a_n| = 4$. There are various possibilities, as listed here:

Equation	Roots
$x \pm 2 = 0$	± 2
$2x \pm 1 = 0$	$\pm \frac{1}{2}$
$3x = 0$	0
$x^2 \pm x = 0$	$0, \pm 1$
$x^2 \pm 1 = 0$	$\pm 1, \pm i$
$2x^2 = 0$	0
$x^3 = 0$	0

We are trying to list distinct, real algebraic numbers, so let us at each stage discard imaginary numbers and any numbers that are roots of an equation of lower height. Where there are several real roots of the same equation, we shall order the numbers according to their increasing magnitude. With these conventions, we get the following sequence of algebraic numbers:

$$0, -1, 1, -2, 2, -\frac{1}{2}, \frac{1}{2}, -3, 3, -\frac{1}{3}, \frac{1}{3}, -\sqrt{2}, \sqrt{2}, -\frac{\sqrt{2}}{2}, \frac{\sqrt{2}}{2}, \ldots$$

An algebraic number is a root of an algebraic equation and every such equation has a height; we cannot "miss" any algebraic number with this scheme. Thus, our argument proves a theorem.

THEOREM | *The set A of (real) algebraic numbers is countable.*

The initial half of Cantor's 1874 paper established the countability of certain sets, such as the algebraic numbers, which scarcely seemed at first glance to possess this property. But it was the second half that contained the more profound result, and the first great triumph of set theory: The set T of transcendental numbers is uncountable. The set of real numbers is the disjoint union of all real algebraic numbers (A) and all real transcendental numbers, $R = A \cup T$. If it happened that T were countable, then R as the union of two countable sets would itself be a countable set. This contradicts an earlier theorem, so the transcendentals must be uncountable. The remarkable thing about this argument is that it demonstrates the existence of an uncountable set of real numbers (to wit, the transcendental numbers), no member of which had been constructed or exhibited in any way. To Kronecker, such nonconstructive existence proofs were sheer nonsense, without any hope of redemption.

Cantor's proof of the uncountability of the transcendentals has only a theoretical character and is not of much use in determining whether certain specific numbers are actually transcendental. The problem of the transcendence of the classical constants e and π attracted mathematicians as soon as Liouville had justified the distinction between algebraic and transcendental numbers. Because the numbers e and π are closely connected by the Euler equation $e^{\pi i} + 1 = 0$, the investigation of their nature was carried on at much the same time. (Felix Klein once observed that Euler's celebrated formula "is certainly one of the most remarkable in mathematics," relating five important symbols, each with its own history.) The irrationality of e had been demonstrated earlier by Euler (in 1737, published in 1744), and Liouville had shown in 1840 that neither e nor e^2 could be a root of a quadratic equation with integral coefficients. This was the first step forward in verifying that e cannot be classed among the algebraic numbers. But more than 30 years passed before Charles Hermite (1822–1901), in 1873, published a memoir entitled *Sur la fonction exponentielle,* in which he succeeded in establishing the transcendental character of e. Although Hermite's paper marked the beginning of a prosperous period in the recognition of specific transcendental numbers, he himself turned his attention elsewhere. He expressed his view in a letter to a former pupil, Carl Wilhelm Borchardt:

> I shall risk nothing on an attempt to show the transcendence of π. If others undertake it, no one will be happier than I at their success, but believe me, my dear friend, this cannot fail to cost them some effort.

Within a decade, Ferdinand Lindemann was able to confirm in an article *Über die Zahl π* (1882) that π is transcendental, modeling his proof on Hermite's. Lindemann's argument required the theorem that for any distinct algebraic numbers $\alpha_1, \alpha_2, \ldots, \alpha_n$, real or complex, and any nonzero algebraic numbers $\beta_1, \beta_2, \ldots, \beta_n$, the expression

$$\beta_1 e^{\alpha_1} + \beta_2 e^{\alpha_2} + \cdots + \beta_n e^{\alpha_n}$$

must always be nonzero. Because the complex number i, as a root of the equation $x^2 + 1 = 0$, is algebraic, and since $e^{\pi i} + e^0 = 0$, it follows that the number πi and therefore π cannot be algebraic.

Lindemann's victory over the obstinate π left the doubters still skeptical, and loud among them rose the voice of Kronecker. In a conversation with Lindemann, he complained, "Of what use is your beautiful investigation regarding π? Why study such problems when irrational [hence, transcendental] numbers do not exist?"

The proof of the transcendence of π was far more exciting than the proof for e, because it put an end to the ancient dream of "squaring the circle," that is, constructing with straightedge and compass alone a square that equaled a given circle in area. This requires the construction of a line segment whose length is $\sqrt{\pi}$, which can be accomplished if a line segment of length π is constructible. The construction of a segment of specified length is possible only if that length is a root of a special algebraic equation. But Lindemann showed that π is not a root of any algebraic equation, whence a segment of length π is not constructible by Euclidean tools (nor is any transcendental length). A man distinguished more by industry and determination than by mathematical brilliance, Lindemann achieved greater fame than Hermite for this discovery based on Hermite's work.

To prove that some specific real number is transcendental is usually difficult, and only recently have such numbers as $2^{\sqrt{2}}$ and $e^{\pi} = (-1)^{-i}$ been disposed of. At the International Congress of Mathematicians held in Paris in 1900, Hilbert asked, as the seventh of his 23 outstanding unsolved problems, whether α^{β} is transcendental for any algebraic number $\alpha \neq 0, 1$ and any algebraic irrational β. Later in a number theory lecture at Göttingen (1919), he speculated that the resolution of the problem lay further in the future than a proof of Fermat's last theorem, and that no one present in the lecture hall would live to see it successfully concluded. Modern progress has been more rapid than Hilbert anticipated; for the desired transcendence was established, independently and by different methods, in 1934 by A. O. Gelfond in Russia and T. Schneider in Germany. Still the best efforts of mathematicians have succeeded in proving the transcendence of only a relatively limited class of numbers. Such numbers as e^e, π^e, 2^e, π^{π}, and 2^{π} have not yet been classified as algebraic or transcendental.

Before ending this digression, we should observe that although the matter of its transcendence was put to rest, there was still a concern with obtaining an accurate numerical value for π. In 1853, the Englishman William Shanks (1812–1882) used the infinite series for the arctangent function

$$\arctan x = x - x^3/3 + x^5/5 - \cdots, \qquad |x| \leq 1$$

together with the formula

$$\pi = 16 \arctan(1/5) - 4 \arctan(1/293)$$

to hand-calculate the first 607 purported digits in the decimal expansion of π. He later returned to his computations and by 1874 had worked out a total of 707 digits. For the next 70 years this approximation stood as the accepted decimal value of π. The error in Shanks's evaluation was not caught until 1945, when it was observed that his expansion for π seemed to disfavor the digit 7; that is, 7 appeared noticeably less frequently than did any of the other nine digits. A check on his accuracy revealed erroneous figures from the 528th decimal place onward (the 528th digit should be 4, but Shanks called it 5). Within the next four years, the decimal representation had been extended to 1120 correct digits, using a desk calculator.

With the advent of computers, the evaluation of the digits of π proceeded with a frenzy. The first such determination in 1949 produced 2037 digits in 70 hours elapsed time. By 1961 at least 100,000 decimal places were available. The number of known digits increased rapidly as larger and faster machines came on the scene. A million decimal figures were reached in 1973 with a calculation that required only 24 hours running time; and, by 1987, the 100 million digit mark was passed in roughly a day and a half of work.

Certainly such accuracy far exceeds utilitarian concerns; a value for π correct to 20 decimal places is sufficient for any imaginable application. But in recent years the computation of its decimal expansion has become a popular check on the reliability of new supercomputers and their software. The latest feat carried the approximation of π beyond the billion-digit barrier, when 1,011,196,691 decimal places were calculated; this number, if printed out, would stretch nearly halfway across the United States. Because π-records are made to be broken, many hundreds of millions of digits more will no doubt soon be aroused from their deep slumber. Probably the most significant mathematical motivation for these large-scale calculations is to investigate whether the digits in the decimal expansion of π are "statistically random"; that is, whether the expansion shows no preponderance of any one of the ten digits 0 through 9. The hundreds of millions of digits now known suggest that this is the case, but questions concerning the distribution of decimal digits of particular numbers such as π, e and even $\sqrt{2}$ appear beyond the scope of current mathematical techniques.

Having wandered down a lesser lane, let us return to the main highway, the development of set theory.

Cantor, having succeeded in proving the existence of infinite sets with the same "power" and with different "powers," went on to attack new and bolder problems. In a paper *A Contribution to Manifold Theory* submitted to *Crelle's Journal* in 1877 and published the following year, he showed that the points in a square, "clearly two-dimensional," can be put in one-to-one correspondence with the points of a straight line segment, "clearly one-dimensional." Quite unprepared for this paradoxical result, which seemed to cloud the concept of dimension, Cantor tried to discuss it with fellow mathematicians, but they treated the whole idea as absurd, even with contempt. Cantor himself found the result so odd that he wrote to Dedekind, "I see it, but I do not believe it," and asked his friend to check the details of the proof. (Because Dedekind recognized immediately that Cantor's one-to-one mapping was not continuous, he did not read the same significance into Cantor's counterintuitive discovery that Cantor himself had.) Publication of the paper was postponed time and time again in favor of manuscripts submitted at a later date. The presence of the skeptical Kronecker on the editorial board seemed to impede its progress. After Dedekind intervened, the difficulties were eventually resolved, but Cantor never again permitted his work to appear in *Crelle's Journal*.

To establish that the unit square S, defined by $0 \leq x \leq 1$ and $0 \leq y \leq 1$, and the interval [0, 1] are equivalent, it is sufficient to match their points in a one-to-one fashion. Give a point (x, y) of S, let us represent x and y as infinite decimals,

$$x = 0.x_1x_2x_3 \ldots , \qquad y = 0.y_1y_2y_3 \ldots ,$$

with the usual proviso that in any decimal terminating in an infinite string of 0s, the 0s are replaced by 9s (with the single exception of the number 0 itself). By taking digits alternately from x and y, it is possible to form the decimal representation

$$z = 0.x_1 y_1 x_2 y_2 x_3 y_3 \ldots$$

of a number z in the interval $0 \leq z \leq 1$. Conversely, a knowledge of the decimal expansion of z permits us to reconstruct x and y by "unlacing" the digits of z. The obvious approach then is to pair the point (x, y) of the square with the point z of the interval. Indeed, this was the strategy that Cantor used when he sent his first "proof" to Dedekind.

There is one drawback to all this, which Dedekind quickly pointed out to Cantor; although to each (x, y) there corresponds a single z, there exist values of z that arise from no (x, y) by the previous procedure. If, for example, $z = 0.3404040404 \ldots ,$ then unlacing the digits leads to

$$x = 0.300000 \ldots , \qquad y = 0.44444. \ldots$$

And because x consists exclusively of zeros after the first digit, it is not written in admissible decimal form. Moreover, if the trailing 0s are replaced by 9s, then the infinite decimal form for this number x does not correspond to the specified number z. The difficulty can be remedied by breaking z into blocks of digits, instead of single digits, each block ending with the first nonzero digit encountered. The transition from z back to (x, y) can be accomplished by alternating the blocks when forming the decimal expansions of x and y. Thus, for instance,

$$z = 0.2|7|03|009|4|06| \ldots$$

would be paired with the point of the square having

$$x = 0.2034 \ldots , \qquad y = 0.700906. \ldots$$

By treating blocks as single digits for purposes of interlacing, we can also go from (x, y) to z in this manner. Thus, the one-to-one correspondence between geometric figures of different dimensions is established:

THEOREM

The set of points in the unit square has cardinal number c.

The publication of Cantor's 1878 paper destroyed the feeling that the plane is richer in points than the line and forced mathematicians to take a fresh look at the concept of dimension. Because Cantor's argument did not involve continuous mappings from one dimension to another, there was a flurry of activity to show that the result failed under the additional assumption that the mapping between the spaces should be continuous. Not until the appearance of an article by L. E. J. Brouwer in 1911, however, would there be established a rigorous proof of the invariance of dimension under continuous one-to-one mappings.

Cantor's most important investigations into set theory are spread across a series of six papers entitled *Über unendliche lineare Punktmannichfaltigkeiten* (On Infinite Linear Point Sets) published in the German journal *Mathematische Annalen* in the period 1879–1884. These brilliant papers, some of whose French translations appeared in Mittag-Leffler's *Acta Mathematica,* constitute the acme of Cantor's lifework. The underlying concept is that of an accumulation point (a point p is an accumulation point of a set P if every neighborhood of p contains points of P). The fundamental theorem is the so-called Bolzano-Weierstrass theorem, first proved by Weierstrass in his lectures at Berlin in the 1860s and known to Cantor from these; this result states that every bounded infinite set of points in Euclidean n-space possesses at least one accumulation point. From this basic idea and theorem flowed a host of new types of sets that today lie at the foundations of the theory: closed sets, perfect sets, sets of first and second category, dense sets, and so forth.

Cantor's use of transfinite cardinals to compare sizes of infinite sets created a storm in the camp of the orthodox. Cantor defined an order relation in which the cardinal number of a set A is "smaller" than the cardinal number of a set B provided that A is equivalent to a subset of B. A precise definition is given.

Definition

Let α and β be two given cardinals and A and B sets with $\alpha = o(A)$ and $\beta = o(B)$. We write $\alpha \leq \beta$ if there exists a one-to-one mapping from A into B. We also write $\alpha < \beta$ if $\alpha \leq \beta$, but not $\alpha = \beta$.

This definition confirms our intention that $o(A) \leq o(B)$ whenever $A \subseteq B$, because if A is a subset of B, then the inclusion mapping $i: A \to B$ defined by $i(a) = a$ is one-to-one, so that we have $o(A) \leq o(B)$. In general, to show that $o(A) < o(B)$, it must be demonstrated that there exists a one-to-one mapping from A into B, but no one-to-one mapping of A onto B. A note of caution—this is not equivalent to the statement "There is a one-to-one mapping of A that is into B, but not onto B."

We are now in a position to tie up two loose ends. We have observed that the set of irrational numbers and the set of transcendental numbers are both uncountable, but we have assigned a cardinal number to neither. Cantor showed, in fact, that whereas the algebraic numbers are countable, the transcendental numbers have cardinality c, the power of the continuum. Because every transcendental number is irrational, the irrational numbers must then also have cardinality c; for using T and I to denote the sets of transcendental numbers and irrational numbers, we have a chain of set inclusions

$$T \subseteq I \subseteq R.$$

This leads to a corresponding inequality involving cardinals:

$$c = o(T) \leq o(I) \leq o(R) = c.$$

There is no alternative but to conclude that $o(I) = c$.

The inequality $\aleph_0, < c$ raises the question whether there are any sets with cardinality between $\aleph_0$ and c. On the face of the matter, there seems no reason why some uncountable set should not have cardinality less than c; yet all attempts to discover a set of real numbers that is infinite and uncountable, but whose power is less than that of the continuum, have been unsuccessful. The conjecture that there is no cardinal number α satisfying $\aleph_0 < \alpha < c$ is customarily known as the continuum hypothesis. This hypothesis may also be stated: Every infinite subset of R has cardinal number either $\aleph_0$ or c. It is frequently alleged that Cantor's emotional breakdown in 1884 was caused by the strain of his prolonged, futile efforts to find a set of the desired intermediate cardinality. First he thought that he had succeeded in proving that the continuum hypothesis was true; the next day, he asserted that he could demonstrate its falsehood. Then he withdrew this claim and announced a new proof of the conjecture. In the end, an embarrassed Cantor must have seen that all his purported proofs of the intransigent hypothesis were invalid.

When Hilbert delivered his famous address of 1900 on outstanding problems awaiting solution by future mathematicians, the continuum hypothesis headed the list. Yet for decades the question eluded all efforts at resolution, and it is today "settled" in an unlooked-for sense. Kurt Gödel announced in 1938–1939 (and published in 1940) that the continuum hypothesis was consistent with the current axioms of set theory and hence could not be disproved. Twenty-four years later, Paul Cohen (1963) demonstrated that the continuum hypothesis was independent of the other axioms of set theory, thereby showing that it could not be proved within the framework of these axioms. Thus, on the basis of our present understanding of sets, Cantor's conjecture remains in a sort of limbo, as an undecidable statement. It is ironic that so specific a problem as the first on Hilbert's list—that ambitious program put forward in such a spirit of optimism—should have its status changed from "unknown" to "unknowable."

12.2   Problems

1. Prove that the following sets are countable:

   (a) $\{2, 2^2, 2^3, \ldots, 2^n, \ldots\}$.

   (b) $\{1, \frac{1}{2}, \frac{1}{3}, \ldots, 1/n, \ldots\}$.

   (c) $\{5, 10, 15, \ldots, 5n, \ldots\}$.

   (d) $\left\{\dfrac{1}{2}, \dfrac{2}{3}, \dfrac{3}{4}, \ldots, \dfrac{n}{n+1}, \ldots\right\}$.

2. Verify that the set N_e of all even natural numbers and the set N_o of all odd natural numbers are countable; do the same for the sets Z_e and Z_o of all even and odd integers.

3. Prove, by confirming that the function $f : Q_+ \to N$ defined by $f(m/n) = 2^m 3^n$ is one-to-one, that the set Q_+ of positive rational numbers is countable. [*Hint*: Notice that $Q_+ \sim f(Q_+) \subseteq N$.]

4. Use the theorems in this section to show that the set of prime numbers is a countable set.

5. Establish that the Cartesian product $A \times B$ (that is, the set of all ordered pairs (a, b) with a in A and b in B) of two countable sets A and B is countable; in particular, conclude that $N \times Z$, $Z \times Z$, and $Q \times Q$

are countable sets. [*Hint*: Show that $A \times B = \cup (A \times \{b\})$, where $A \times \{b\} \sim A$ for any b in B.]

6. If S is the set of all right triangles whose sides have integral lengths, then S is a countable set. Prove this statement.

7. Let $\mathscr{C}$ be the set of all circles in the Cartesian plane that have rational radii and centers at points whose coordinates are both rational. Show that $\mathscr{C}$ forms a countable set. [*Hint*: Consider the mapping $f : \mathscr{C} \to Q \times Q \times Q$ defined by $f(C) = (x, y, z)$, where (x, y) is the center and z the radius of a circle C in $\mathscr{C}$.]

8. Prove the following.

   (a) If $Z_n[x]$ is the set of all polynomials of degree n with integral coefficients, then $Z_n[x]$ is countable. [*Hint*: Consider the function $f : Z_n[x] \to Z \times Z \times \cdots \times Z$ defined by

   $$f(a_n x^n + \cdots + a_1 x + a_0)$$
   $$= (a_n, \ldots, a_1, a_0).$$

   Note that $Z \times Z \times \cdots \times Z \sim N \times N \times \cdots \times N \sim N$.]

   (b) The set $Z[x]$ of all polynomials with integral coefficients is countable. [*Hint*: Show that $Z[x] = \cup Z_n[x]$.]

9. Use Cantor's diagonal argument to show that the set of all infinite sequences of 0s and 1s (that is, of all expressions such as 11010001 . . .) is uncountable.

10. Determine whether each of the following sets is countable or uncountable.

   (a) The set of all numbers of the form $m/2^n$, where m is an integer and n is a natural number.
   (b) The set of all straight lines in the Cartesian plane, each of which passes through the origin.
   (c) The set of all straight lines in the Cartesian plane, each of which passes through the origin and a point having both coordinates rational.
   (d) The set of all intervals on the real line having both endpoints rational.
   (e) Any infinite set of nonoverlapping intervals on the real line.

11. Prove that the set L of Liouville numbers, and hence the set of transcendental numbers, has cardinality c. [*Hint*: Consider the function $f : L \rightarrow [0,1]$, which is defined by sending $\sum_{n=1}^{\infty} a_n 10^{-n!}$ to $0.a_1 a_2 a_3. \ . \ . \ .$]

12. Verify that the function $f : R \rightarrow [0,1]$ defined by

$$f(x) = \frac{1}{2}\left(1 + \frac{x}{1 + |x|}\right)$$

is one-to-one, so that $R \sim [0,1]$.

13. Establish that any nondegenerate interval in R has cardinality c. [*Hint*: Show that $[0,1] \sim [a,b]$ via the function $f(x) = a + (b - a)x$ and $[0,\infty] \sim [0,1]$ via

$$f(x) = \frac{x}{1 + x} .]$$

12.3   The Paradoxes of Set Theory

It has been said that the relation between Cantor's theory of sets and mathematics was like the course of true love, never running smooth. About 1900, just when Cantor's ideas were beginning to gain acceptance, a series of entirely unexpected logical contradictions were discovered in the fringes of the theory of sets. Curiously, these were at first called ''paradoxes,'' rather than flat contradictions, and regarded as little more than mathematical oddities. The feeling was that the conceptual apparatus of set theory was not yet quite satisfactorily constituted and that some slight alteration of the basic definitions would set things right. Then, in 1902, the British philosopher, mathematician, and social reformer Bertrand Russell (1872–1970) offered a paradox in which Cantor's very definition of set seemed to lead to the contradiction. The simplicity and directness of Russell's paradox shook the very foundations of logic and mathematics; and the tremors are still being felt today.

This most notorious of the modern paradoxes appeared in Russell's *Principles of Mathematics,* published in 1903. Before we examine Russell's paradox, let us observe that some sets are members of themselves and some are not. The set of all abstract ideas, for example, is an abstract idea, but the set of all stars is not a star. Most sets are not elements of themselves; those that are tend to be far-fetched in description.

With this in mind, we can formulate Russell's paradox quite simply, using the bare notions of set and element. If one naively accepts the Cantorian view that every condition determines a set, then it is obviously possible to consider the set of all sets that have the property of not being elements of themselves, that is, the set

$$S = \{A \,|\, A \text{ is a set and } A \notin A\}.$$

Here, we have used the symbol $\notin$, standing for the phrase ''does not belong,'' for the first time. The use of $\in$ to denote set membership was initiated by Peano in his *Arithmetices Principia* (1889); it is an abbreviation of the Greek word εστι, meaning ''is.'')

Now S itself is a set, so that by the law of the excluded middle, which says that every proposition is either true or false, $S \in S$ or $S \notin S$ is a true assertion; but, it is easily seen, each of the two cases leads to a contradiction. For if $S \in S$, it follows that S must be one of the sets A described in the condition; hence, $S \notin S$ is an impossible situation. On the other hand, if $S \notin S$, then S satisfies the property by which one determines which sets are elements of S; thus $S \in S$, which is equally impossible. Because either case leads to contradiction, the paradox is apparent.

Russell's paradox jolted the mathematical world and proved to be a personal disaster for the logician Gottlob Frege. Frege had labored for more than ten years in the production of the second volume of his treatise on the logical foundations of arithmetic, *Grundgesetze der Arithmetik*. In the preface, Frege had written as follows: "The whole of the second part is really a test of my logical convictions; it is impossible that such an edifice could be erected on an unsound basis. . . ." At the very time the volume was coming near the end of its printing before being offered to the public, Frege received a personal letter from Russell announcing his discovery of the paradox mentioned earlier. Frege barely had time to compose an appendix to the *Grundgesetze,* which said in part:

> A scientist can hardly meet with anything more undesirable than to have the foundation give way just as the work is finished. I was placed in this position by a letter from Mr. Bertrand Russell as the printing of the present volume was nearing completion.

Frege's immediate reaction, as shown in his prompt reply to the young British logician, was one of consternation: "Arithmetic has begun to totter." On the other hand, many distinguished mathematicians rejoiced in the paradox. The elder statesman of French mathematics, Henri Poincaré, who had little faith in mathematical logic, was overjoyed that the carefully constructed logical foundation was insufficient to bear the weight of arithmetic. He exclaimed, "It is no longer sterile, it begets contradictions." This may also have been a play on the saying, "Logic is barren, where mathematics is the most prolific of mothers."

Russell's paradox was not, to be sure, the first paradox noted in set theory. The earliest of the paradoxes—one based on the consideration of the "set of all ordinal numbers"—was published in 1897 by the Italian mathematician Cesare Burali-Forti, and known to Cantor at least two years earlier. Then, in 1899, Cantor discovered a paradox that had to do with his theory of cardinal numbers. This was based on another far-reaching theorem of his (1883), which said that for any set A, its power set had a larger cardinal number than that of A. By the power set of A was meant the set of all subsets of A; our notation for the power set of A is $P(A)$.

The result that needs to be proved first in deducing Cantor's theorem may be stated as a theorem.

THEOREM   *For any set A, there does not exist a function mapping A onto its power set P(A).*

Proof. The general plan of attack is to show, by an indirect argument, that no such function exists. Therefore we start with the assumption, to be refuted in the

Henri Poincaré
(1854–1912)

(From A Concise History of Mathematics *by Dirk
Struik, 1967, Dover Publications, Inc., N.Y.)*

end, that there is a function $f:A \to P(A)$ that is onto $P(A)$. This means that to each
element $a \in A$, there is assigned a subset $f(a)$ of A. Let us consider the set B
defined by

$$B = \{a \in A \mid a \notin f(a)\}.$$

The set B is a subset of A, so that $B = f(b)$ for some $b \in A$ (recall that f maps
onto $P(A)$). We now ask an innocent question: Is the element b a member of the
set B?

If $b \in B$, then by the very definition of the set B, we must have $b \notin f(b) = B$,
which is impossible. On the other hand, $b \notin B = f(b)$ implies that b satisfies the
defining property of B and so $b \in B$. This is also nonsense. It follows that no
function f as described here can exist, and the proof of the theorem is thereby
concluded. ∎

With this accomplished, we are ready to give a proof of Cantor's theorem:

CANTOR'S THEOREM

For any set A, $o(A) < o(P(A))$.

Proof. We must show first that $o(A) \le o(P(A))$ and then that $o(A) \ne o(P(A))$. The
meaning of the statement $o(A) \le o(P(A))$ is that A is equivalent to a subset of
$P(A)$; that is, there exists a one-to-one correspondence between A and a subset of
$P(A)$. Quite obviously the function that takes each $a \in A$ into the single-element
set $\{a\}$ in $P(A)$ defines such a correspondence.

For the task of proving that $o(A) \ne o(P(A))$, it is enough to show that there is
no one-to-one correspondence between A and $P(A)$. Because such a
correspondence would certainly be a mapping of A onto $P(A)$, it cannot exist, by
the previous theorem. This gives the desired conclusion. ∎

Cantor's theorem not only answers the question whether to every cardinal number there exists a still larger cardinal (in particular, whether there are cardinals larger than c) but also furnishes a way of constructing a strictly increasing sequence of transfinite cardinal numbers. For in the special case of the cardinal number c, the theorem indicates that $c = o(R) < o(P(R))$. By iterating the operation of forming the power set, an unending hierarchy of infinite cardinals can be obtained:

$$c < o[P(R)] < o\{P[P(R)]\} < o(P\{P[P(R)]\}) < \cdots.$$

Here we see infinity on infinity, each incomparably larger than the last, in a process that never ends. The imagination is beggared, but the cardinal numbers are not.

A paradox arises when we consider the most comprehensive of all sets, the set U that contains all sets. By Cantor's theorem, $o[P(U)] > o(U)$. Because U is the set of all sets and $P(U)$ is a set (the elements of which happen to be subsets of U), then $P(U)$ is contained in U; hence, $o[P(U)] \leq o(U)$, and we have a contradiction. Although this paradox was published only posthumously with Cantor's correspondence in 1932, rumor of it reached Russell in 1901, and he then devised a paradox of his own. The Cantor paradox requires a good deal of mathematical machinery, enough so as to make it suspect on various grounds; but not so the Russell paradox, which uses only simple and well-established principles.

It is natural that after the shock of the paradoxes, the foundations upon which mathematics—and in particular, set theory—had been built were scrutinized as never before. For ages, the reasoning used in mathematics had been regarded as a model of logical perfection. Mathematicians prided themselves that theirs was the one science so irrefutably established that in its long history it had never had to take a backward step. But now many mathematicians turned away from Cantor's ideas and ceased to work on aspects of their discipline that depended on an unqualified acceptance of set theory. Doubts voiced in France reached such proportions that at the International Congress of Mathematicians held in Rome in 1908, the eminent Henri Poincaré, who was regarded as something of an oracle on mathematics, went so far as to say, "Later mathematicians will regard set theory as a disease from which one has recovered." Other mathematicians believed that Cantor's basic tenets were essentially correct, but set theory in its existing form was too naive. They felt that set theory must be built on logical and consistent foundations if Cantor's innovations were to be secured for posterity.

The fatal flaw in early set theory proved to be Cantor's broad approach, which permitted any conceivable property to give rise to a set, namely, the set of all elements that possessed the property. Russell's paradox showed, without any considerations involving cardinal numbers, that one cannot allow arbitrary conditions to determine sets and then indiscriminately permit the sets so formed to be members of other sets. Because the difficulty appeared to originate in the liberality with which Cantor's theory allowed the formation of sets, it seemed that the very concept of "set" as it then stood was inherently faulty. The immediate aim was to restrict the definition of set in such a way as to forestall the emergence of those sets that entered into the paradoxes, and yet to allow mathematics the greatest possible latitude for development. Each of the various solutions proposed that proved to be successful lay the blame on the introduction of sets that were "excessively large." Thus, an essential ingredient in any formal theory of sets was to be an axiom guaranteeing a "limitation of size."

In his original development of set theory, Cantor relied on intuition, rather than any set of axioms, in deciding which objects were to be sets. But common sense turned out not to be a good enough lighthouse to keep set theory from being wrecked on the shoals of the paradoxes. The first successful axiomatic treatment of set theory was published by the German mathematician Ernst Zermelo (1871–1953) in 1908. Zermelo, who received a doctorate from the University of Berlin in 1894 with a dissertation on the calculus of variations, began his career not in set theory, but in mathematical physics. He became curious about Cantor's ideas, and as an assistant professor at Göttingen, he lectured on the subject during the winter semester 1900–1901. The following year (1901), he published his first relevant paper consisting of several results on the arithmetic of cardinal numbers. In a famous memoir, *Foundations of a General Manifold Theory* (1883), Cantor stated that "every well-defined set can be brought into the form of a well-ordered set." (An ordered set is well-ordered if every nonempty subset has a "first" element.) And Cantor promised to return in a future publication to this "law of thought which seems to be both fundamental, rich in consequences, and particularly remarkable for its generality." This promise was never fully carried out. When Hilbert presented his famous 23 problems at the 1900 International Congress of Mathematicians, he indicated that the discovery of such a proof was one of the tasks challenging mathematicians the world over. Zermelo supplied the critical proof of the well-ordering theorem in 1904, a proof that unleashed much spirited controversy. Zermelo based his argument on a powerful new, and suspect, device, the axiom of choice. The axiom of choice (a name given by Zermelo) asserted that from any given collection of disjoint nonempty sets, it would be possible to choose exactly one element from each set and thereby form a new set. Intuitively, this axiom allowed for a simultaneous but independent selection from each of an infinite number of sets. The idea of making infinitely many choices was not entirely new, having been an important part of many mathematical arguments around 1900. As early as 1883, Cantor himself had unconsciously applied the choice axiom; and in an 1890 article on differential equations, Peano had incidentally alluded to and rejected it ("as one cannot apply infinitely many times an arbitrary law by which one assigns to a class A an individual of that class. . . ."). The first explicit mention of the statement of the axiom was by Beppo Levi in 1902, in considering a proof of a theorem on cardinal numbers.

The controversy touched off by the axiom of choice reminds one of another famous axiom, Euclid's parallel postulate. This time the dispute centered on the question of what are admissible methods in mathematics; for the essence of the axiom of choice is that it is an existential statement giving no constructive definition of the representative elements involved in its use. One of those who resolutely opposed nonconstructive methods in set theory, Emile Borel (1871–1956), insisted: "Any argument where one supposes an arbitrary choice to be made an uncountably infinite number of times . . . [is] outside the domain of mathematics." Another such mathematician, Jacques Hadamard (1865–1963), crystallized the whole controversy into the question, "Can the existence of a mathematical entity be proved without defining it?" In a paper written in 1908, Zermelo furnished a second proof of the well-ordering theorem, in which the objections to the first were discussed at length. He concluded his spirited defense with the statement "No mathematical error can be demonstrated in my [earlier] proof." Once again, the axiom of choice was used in the same way, the only difference from the 1904 article being in the remaining set-theoretic axioms. Although Zermelo's two papers raised as many questions as they professed to settle, their great merit lay in the formal recognition of the principle of arbitrary choice as an independent method of proof.

Zermelo took a decisive step in the attempt to rehabilitate the heretofore haphazard formulation of Cantor's set theory. In the same issue of the *Mathematische Annalen* that contained his second proof of the well-ordering theorem, he also published *Investigations into the Foundations of Set Theory*. He hoped that this paper, which presented a strictly axiomatic theory of sets, would in turn serve as a basis for all mathematics. Zermelo did not say what sets are but simply postulated them together with their basic properties. The conciseness of his system of axioms is surprising. Only seven axioms, involving just two undefined technical terms (*set* and *membership,* the latter relation denoted by $\in$) sufficed to build up the set theory required for practically all mathematics. The original axioms of Zermelo, as amended by Fraenkel, von Neumann, and others, are now called the Zermelo-Fraenkel axioms for set theory.

Zermelo, to avoid paradoxes that would render his system useless, refused to admit into the club of decent sets those collections that were ''too big.'' He observed that mathematicians would not normally think of using such sets as ''the set of all sets'' or ''the set of all sets which are not elements of themselves.'' He held that the sets needed in practice were always built up by means of given operations from certain simple sets (like the sets of natural numbers or real numbers) that are known about to begin with. Thus, Zermelo formulated a principle by which he guaranteed the existence of certain subsets.

AXIOM OF SPECIFICATION

To every set A and every definite property P(x) there corresponds a set whose elements are exactly those elements x in A for which the property P(x) holds.

The essential difference in Zermelo's system was that both a property $P(x)$ and a preexisting set A were needed to form a new set, whereas in Cantor's original scheme only the property $P(x)$ was required. Rather than proclaim the existence of sets, the specification axiom posits the existence of certain subsets of a given set; a set is admitted into the theory only by being related to a known set. (The vague notion of ''definite property'' gave rise to misgivings almost from the outset; the idea was made more precise by Fraenkel in 1922, and somewhat differently, by Skolem in 1922–1923.) Zermelo also recognized that because the axiom of choice did not follow from any previously known principle of mathematics or logic, he must make it one of his seven axioms.

How do the restrictions implied by Zermelo's axiomatization avert the disaster of the paradoxes? The critical set that appears in Russell's paradox, ''the set of all sets which are not elements of themselves,'' cannot be formed in Zermelo's formal set theory; the best one can do is to produce the set

$$S = \{A \in \mathcal{A} \mid A \text{ is a set and } A \notin A\},$$

where $\mathcal{A}$ is a set (of sets) known to exist. Notice that $S \in S$ is impossible; for then $S \in \mathcal{A}$ and $S \notin S$, a contradiction. Thus, $S \notin S$. The implication is that $S \notin \mathcal{A}$. Indeed, if $S \in \mathcal{A}$, then S would (because $S \notin S$) satisfy the condition determining the members of S and we should have $S \in S$, which would be a contradiction. The outcome of Russell's argument has changed completely. All it shows is that if $\mathcal{A}$ is any set that exists, then the set S cannot be an element of $\mathcal{A}$. This does not lead to a contradiction, so Russell's paradox cannot be reproduced in axiomatic set theory.

Let us now consider the paradox of Cantor, which derives its origin from the possibility of constructing the set of all sets. Suppose for the moment that there is a set $\mathcal{U}$ that contains every set. Then because $\mathcal{U}$ contains every set, $S \in \mathcal{U}$; this violates the conclusion of the last paragraph. The set of all sets, no matter how curiously natural it seems, does not exist within Zermelo's system; and Cantor's paradox falls away. As Zermelo hoped to show, axiomatization was a successful antidote to the paradoxes.

Declaring that "I have not yet even been able to prove my axioms are consistent, though this is certainly essential," Zermelo left the difficult questions of consistency and independence to his successors. Absolute consistency turned out to be a blind alley. According to Gödel's famous incompleteness theorem, it is impossible in certain logical systems—Zermelo-Fraenkel set theory, for example—to demonstrate the internal consistency of the system by methods formalizable within the system itself. In a series of lectures at Princeton in 1938, published afterward as *The Consistency of the Axiom of Choice and of the Generalized Continuum Hypothesis with the Axioms of Set Theory*, Gödel proved that if the other axioms present in set theory are consistent with one another, then the system obtained by adjoining the axiom of choice will not give rise to any contradictions. This was, in other words, a proof of the relative consistency of the axiom of choice. Cohen (1963) succeeded in showing that the negation of both the axiom of choice and the continuum hypothesis is also consistent with the rest of the Zermelo-Fraenkel axioms (provided they are consistent themselves).

The combined results of Gödel and Cohen imply that the choice axiom is an independent axiom of set theory; its use or rejection in an axiom system is a matter of personal inclination. Because so many profound theorems have been obtained using the axiom of choice, without visible alternative, the average mathematician would probably speak in favor of its retention. On the other hand, because it has troubled the consciences of so many, it is important to know which theorems have been proved with the aid of the axiom of choice, and the extent to which the axiom is needed in the proof. An alternative proof without the axiom of choice would then be desirable.

The paradoxes of the infinite laid the stage for a modern "crisis in the foundations," not unlike the profoundly disturbing situation that arose when the Pythagoreans unexpectedly discovered incommensurable quantities. It appeared to many that the entire structure of mathematics was weak or at least built on weak foundations. In the early 1900s, widely different diagnoses of the ills of mathematics divided mathematicians into various enemy camps. The three main schools of thought concerning the origin and nature of mathematics are usually distinguished as the logistic (also called logicistic), formalistic, and intuitionistic schools, and their best-known proponents identified as Bertrand Russell, David Hilbert, and L. E. J. Brouwer, respectively. Although the adherents of these factions had the common purpose of coming to grips with the destructive paradoxes, the radically different ways they chose to accomplish this led to some sharp conflicts. It was almost as though mathematics had become a kind of religious fanaticism instead of a labor of love.

The chief characteristic of the logistic school was its uncompromising insistence that logic and mathematics were related as earlier and later parts of the same subject, that mathematics was ultimately derivable from logic alone. Logic was not simply an instrument in the construction of mathematical theories; mathematics now became the offspring of pure logic. According to the logistic philosophy, all mathematical concepts must be given definitions only from ideas that are a part of logic; and all mathematical statements must be deducible from universally recognized assumptions of logic. Of course, to say mathematics is logic is merely to replace one undefined term by another; logic is as much in need of definition as mathematics is.

The first determined effort to rewrite the established body of mathematics in logical symbolism was made by the German logician and philosopher Gottlob Frege (1848–1925). Frege received his doctorate in mathematics from Göttingen in 1873. The following year he began his teaching career at the University of Jena, where he remained for 45 years. Frege's mathematical work was almost wholly confined to mathematical logic and foundations. His small but weighty treatise, entitled *Begriffsschrift, einer der arithmetischen nachgebildete Formelsprache des reinen Denkens* (Conceptual Notation, a Symbolic Language of Pure Thought Modeled on the Language of Arithmetic), published in 1879, was a natural milestone in the history of modern logic. More than twenty years passed, however, before Bertrand Russell sensed the greatness of the achievement. In the *Begriffsschrift*, the whole calculus of propositions and the device known as quantification theory was presented for the first time. Frege framed a formal language of symbols, which was intended to be adequate for the exposition of any mathematical statement, together with certain rules of inference for expressing any train of proof.

During the succeeding years, Frege turned to the actual formalization of a particular branch of mathematics; and in two principal works, he chose arithmetic as his subject: *Die Grundlagen der Arithmetik* (The Foundations of Arithmetic), which appeared in 1884, and *Grundgesetze der Arithmetik* (Fundamental Laws of Arithmetic), of which volume 1 was published in 1893 and volume 2 in 1903. Owing to the complexity of the symbolism—"a monstrous waste of space," said one critic—and the novelty of the approach, these writings passed almost wholly without recognition until Russell devoted an appendix to them in *Principles of Mathematics* (1903). Frege's system of axioms, as we observed earlier, introduced sets in a way that led directly to one of the newly discovered paradoxes. When the *Grundgesetze* was at the printshop, Russell found the fatal flaw and communicated it to Frege in a letter, asking, "Is the set of all sets which are not members of themselves a member of itself?" After acknowledging the contradiction contained in his system, it seems that Frege was never able to regain his former faith in the possibility of a purely formal presentation of arithmetic. He effectively abandoned his creative research and died a bitter man, convinced that his life's work had been for the most part a failure. Frege's death went virtually unmarked by the scholarly world, a tragic fate for a man who singlehandedly created a revolution in logic.

Another great pioneer in the study of the foundations of mathematics, and a master of the art of reasoning by formal logic, was the Italian mathematician Giuseppe Peano (1858–1932). Peano's first attempt at deducing the truths of mathematics from pure logic was his *Arithmetices principia, nova methodo exposita* of 1889, a small tract of

Gottlob Frege
(1848–1925)

(By courtesy of John R. Parsons.)

29 pages written almost entirely in the symbols of what he called mathematical logic. Among the body of signs Peano invented for this were ∈ (belongs to), ∪ (logical sum or union), ∩ (logical product or intersection), and ⊃ (logical *implies* or *contains*). The "horseshoe" symbol ⊃ became the standard notation for material implication after Whitehead and Russell used it in their *Principia Mathematica*. The *Arithmetices principia* is noteworthy for containing the first statement of the famous postulates for the natural numbers, perhaps the best known of Peano's achievements.

Peano's goal was to take mathematics as he found it and translate it completely into his designed language of signs, thereby making the principles of logic the vehicle of demonstration. This was carried out in the *Formulaire de mathématiques,* which expressed in symbols all the known definitions, proofs, and theorems of extensive tracts of mathematics. Essentially a series of reports by Peano and his collaborators, the *Formulaire* was published in five successive editions or volumes. The first edition appeared in 1895; the last was completed in 1908 and contained some 4200 theorems. The final version came out under the title *Formulario mathematico* and was written in a new international language of Peano's invention, which he called *latino sine flexione* (Latin without grammar), afterward named *Interlingua*. Peano spent almost the whole of his life in Turin. He entered the University of Turin as a student in 1876 and held a position there from 1880 to 1932. He also held a position at the military academy, which was next door, from 1886 to 1901. When Peano adopted the *Formulaire* as a textbook, the pre-engineering students at the academy rebelled, complaining that the lectures contained an excess of queer new symbolism. His attempt to regain popularity by passing everyone who registered for his course proved ineffectual, and Peano was forced to resign his professorship. Another example of Peano's infatuation with formalism is his famous paper (1890) in the *Mathematische Annalen* concerning the existence of solutions of real differential equations. The article was so clothed in the language of symbolic logic that it could be read only by a few of the initiated and did not generally become available to the working mathematician. It became something of the fashion to refer scornfully to Peano's symbolic language as Peanese.

At the first International Congress of Philosophy at Paris in 1900, the Italian phalanx of Peano, Burali-Forte, Padoa, and Pieri dominated the discussion. Bertrand Russell's meeting with Peano brought about, in the Englishman's own words, ''a turning point in my intellectual life.'' Russell asked Peano for copies of his published works, quickly mastered the techniques of mathematical logic, and returned home to write *Principles of Mathematics*. In the *Principles of Mathematics,* which contains the first explanation of the paradox that bears his name, Russell put forward the opinion that logic is the progenitor of mathematics:

> The present work has to fulfill two objects, first, to show that all mathematics follows from symbolic logic, and secondly, to discover, as far as possible, what are the principles of symbolic logic itself.

The book was to appear in two volumes, the first of which was to confine itself to a popularly understandable explanation, avoiding symbolism. The second volume was to provide the logical proofs that would support the view that mathematics and logic are identical. The proposed second volume never came out, because the revision of material necessitated writing a wholly new work, the *Principia Mathematica.*

The apex of the logistic conception of mathematics is *Principia Mathematica,* written by Russell in stages in collaboration with his friend and older colleague Alfred North Whitehead (1861–1947). A massive work of bewildering complexity, its three volumes compose more than 2000 pages. The *Principia* must thus be regarded as one of the classics of mathematical literature, more often quoted than read. Frege and Peano were, so to speak, the godfathers of the *Principia.* Russell was unacquainted with Frege's writings during most of the preparation of the *Principles of Mathematics* (''Professor Frege's work, which largely anticipates my own, was for the most part unknown to me when the printing of the present work began''). In the preface of the *Principia,* however, the authors acknowledged their obligation: ''In all questions of logical analysis, our chief debt is to Frege.'' On the formal side, the new symbolism that Whitehead and Russell adopted had its roots in Peano's work. They explained that they were obliged to renounce ordinary language and write almost entirely in symbols, because no words had the exact value of the symbols; they held that without the symbolic form of the work, they would have been unable to perform the requisite reasoning. Taken as a whole, the *Principia* constituted a formidable effort to prove the logistic thesis that mathematics was indistinguishable from logic. Russell later wrote:

> If there are still those who do not admit the identity of logic and mathematics, we may challenge them to indicate at what point, in the successive definitions and deductions of *Principia Mathematica,* they consider that logic ends and mathematics begins.

After analyzing the paradoxes, Russell arrived at the view that they all resulted from the same circular kind of reasoning, a misuse of self-referential expressions. To quote Russell's words in the *Principia*:

> The principle which enables us to avoid illegitimate totalities may be stated as follows: ''Whatever involves all of a collection must not be one of the collection.'' . . . We shall call this the ''vicious-circle principle,'' because it enables us to avoid the vicious circles involved in the assumption of illegitimate totalities.

To ensure strict observance of the vicious-circle principle, he introduced his theory of logical types. The theory set up a hierarchy of levels of elements. On the lowest level there are "individuals"; on the next level, sets of individuals; on the next level, sets of sets of individuals, and so on—a different type of object at each level. In applying the theory, one follows the rule that sets have as members only those things from the next lower level. This idea made set theory secure against the "illegitimate totalities," but it led to other difficulties in the body of mathematics itself—many important theorems not only could not be proved but could not even be expressed. (Among them was Cantor's theorem that there are more real numbers than positive integers.) Adherence to Russell's rule of hierarchies introduced complications in Dedekind's theory of the real line. For instance, the least upper bound of a bounded set, because it is defined in terms of a set of real numbers, must be of a higher type than the real numbers, and so itself is not a real number.

Russell overcame this weakness by positing a new axiom, which he called the axiom of reducibility. (The axiom is less a part of the deductive system of logic than a rule for the manipulation of symbols.) The sole justification for the axiom of reducibility was that there seemed to be no other way out of the particular difficulty engendered by the theory of types. It was the artificial, *ad hoc* character of this device that brought forth an overwhelmingly negative response from mathematicians. It became the main bone of contention for the critics of the system of the *Principia*. They argued that the axiom was incompatible with the integral program of logicism, that the claim could no longer be made that the axioms were purely logical or that the resulting system was founded exclusively on logic. One critic stated that the introduction of the axiom of reducibility into the *Principia,* without any supporting reason, was an act of hara-kiri. Russell felt that this was not the sort of axiom with which one could rest content, and he conjectured that some less objectionable axiom might give the desired results. But he could find no satisfactory alternative in his attempt to drop it from the second edition (1925) of the *Principia.* As the authors admitted, "This axiom has a purely pragmatic justification: It leads to the desired results, and to no others." The impetus of the logistic movement seemed to falter with the publication of the third great tome of the *Principia,* for beyond the distrust inspired by the axiom of reducibility was another difficulty. Russell had shown merely that all the known paradoxes were circumvented in his system; he had not shown that his system would remain free of contradictions.

Whitehead, when he began his collaboration with Russell, was already the author of two tracts on geometry, and also a book on the symbolic structure of algebra, *A Treatise on Universal Algebra* (1898). The preparation of a fourth volume of the *Principia,* on the logical foundations of geometry, which was to be written largely by Whitehead, was interrupted by the First World War. At a time when pacifism aroused bitter emotions, Russell's opposition to the war caused him to be denounced as unpatriotic. In 1916, he was prosecuted and fined £100 for writing a pamphlet containing "statements likely to prejudice the recruiting and discipline of His Majesty's Forces." The conviction, and Russell's general unpopularity, led to his dismissal from Trinity College, Cambridge. Then, in 1918, he was sentenced to six months' imprisonment for another article libeling the American army. In prison, Russell found the leisure to write his *Introduction to the Philosophy of Mathematics,* which was published in 1919. His subsequent works on moral and social issues were more popular.

Whitehead was also inclined to philosophy, going to the United States in 1924 as professor of philosophy at Harvard. As the authors of the *Principia* moved into this new territory, the leadership in research into the foundations of mathematics passed to the great German authority, David Hilbert. The younger generation was probably thankful to be delivered from such a dry and desolate undertaking as the Russell-Whitehead enterprise; and besides, Hilbert's axiomatic approach appeared more in keeping with the traditional character of mathematics.

Hilbert's interest in foundations dated to his investigations of geometry in the 1890s. His masterful *Grundlagen der Geometrie* (Foundations of Geometry) was an attempt to rewrite Euclid in accordance with the principles of Peano, although without Peano's unfamiliar logical symbolism. That is, the *Grundlagen* laid out a rigorous axiomatic treatment of elementary geometry that avoided any illegal appeal to intuition. The question that Hilbert explicitly formulated for the first time was the consistency of his set of axioms: that there should be no paradox or contradiction consisting of two theorems, one of which would be the negative of the other. As a natural evolution of this work, at the Second International Congress of Mathematicians (1900), he offered as one of his list of the most challenging current problems that of the consistency of the axioms of arithmetic. By the time of the Third International Congress, which was held at Heidelberg in 1904, the emergence of the paradoxes had brought mathematicians into uncertainty. Hilbert, speaking before those gathered at Heidelberg, volunteered his service in the reconstruction of mathematics. Having achieved initial success with the axiomatization of geometry, he saw no reason why the same approach could not be applied to other areas—indeed, to all mathematics. After considering the problems that beset a rigorous development of the number system, he offered the outline of a concrete plan:

> I believe that all the difficulties that I have touched upon may be overcome, and an entirely satisfactory foundation of the number concept can be reached, by a method which I call the axiomatic method, and whose leading idea I wish now to develop.

Hilbert did not act on his Heidelberg proposal for many years; instead, he became absorbed first in integral equations and then later in mathematical physics. He next returned to the old problem, publicly at least, when he delivered an address in 1917, *Axiomatisches Denken,* before the Swiss Mathematical Society. In the interim, the nagging questions concerning the logical foundations of mathematics had reached a critical stage. To make matters worse, L. E. J. Brouwer of the University of Amsterdam was winning converts to his distinctly personal philosophy of mathematics known as "intuitionism." One of the fundamental canons of this thinking was that the analysis of Weierstrass and the concept of the infinite as it appeared in the work of Cantor were "built on sand." Fearing that his cherished theorems, his paradise, would be among the sacrifices required, Hilbert returned in earnest to the task of providing mathematics with a secure foundation. In a lecture delivered at Munster in 1925 to honor the memory of Weierstrass he declared stoutly, "No one will expel us from this paradise Cantor has created for us."

It was a fundamental thesis of Hilbert that once the consistency of any axiom system had been established, then its use was "legitimate"; and that mathematical existence was nothing less than consistency:

David Hilbert
(1862–1943)

*(By courtesy of Columbia University, David
Eugene Smith Collection.)*

> If the arbitrarily given axioms do not contradict each other through their consequences
> . . . then the objects defined through the axioms exist. That, for me, is the criterion of
> truth and existence.

Thus, to salvage traditional mathematics in the face of Brouwer's attacks, Hilbert proposed a bold new program. It required first that the whole of existing mathematics, including logic, should be axiomatized; and second, that this axiomatic theory should then be proved consistent by simple finitary arguments. This approach to foundations became known as formalism.

Before Hilbert's proposal, the method used in consistency proofs for an axiomatic theory was merely to shift the burden to another area of mathematics. In the *Grundlagen der Geometrie,* for instance, consistency of geometry had been established only in relative terms, assuming the consistency of arithmetic; but doubt about the consistency of arithmetic cast a shadow over the whole enterprise. Hilbert proposed a more thoroughgoing cure for those ills that the set paradoxes had generated in mathematics. His aim was to furnish an absolute consistency proof for some axiomatic system within which all mathematics could be deduced. For if a system strong enough to embrace the notion of infinite set and the operations on such sets could be shown to be mathematically incapable of producing an inconsistency, then no contradiction would ever be forthcoming as a result in set theory. The establishment of the consistency of classical mathematics was Hilbert's golden dream.

As a first step toward carrying out such a program, Hilbert introduced the notion of a formal theory, a completely symbolic axiomatic theory in which there was explicitly incorporated a system of logic. (The formal theory is usually a formalization of some more intuitively conceived theory of the ordinary mathematical kind.) Hilbert's idea was the axiomatic development pushed to its extreme. Even the logical methods used uncritically in carrying out a mathematical proof had to be themselves subjected to formalization. Hilbert denied, however, that logic was more than mathematics and should precede it. He was concerned with extracting from the whole of logic only so much as was needed to reason in his formalism. Particularly helpful in symbolizing the statements of mathematics was the *Principia* of Russell and White-head, which Hilbert called "the crowning achievement of the work of axiomatization"; for by actually exhibiting the details, the *Principia* showed that all mathematical state-ments could be translated into a small body of symbols and that the laws of reasoning reduced to several simple rules for combining these symbols. Hence, an appropriate formal theory could be considered equivalent to the whole of mathematics.

A formal theory, as set up by Hilbert, starts with a stock of symbols and the rules governing them. The rules consist of a certain set of initial formulas involving the symbols, the axioms, and certain explicitly stated rules of inference determining how further assertible formulas are to be constructed. A proof in such a theory is a finite sequence of formulas, each of which either is an axiom or is obtainable from one of the earlier formulas in the sequence by applying the rules of inference. The last formula in the sequence is, by definition, a theorem. The aim is to make certain that no one of these formalized reasonings will ever lead to a contradiction. Hilbert insisted that proofs must have a finite character, arguing that man is capable of only a finite number of logical deductions, so that any contradiction that might arise must do so after a finite number of operations. In Hilbert's formalism, proofs themselves become the objects of a mathematical study, which he called proof theory, or metamathematics. In this way, investigation into the nature of mathematical proofs becomes another branch of mathematics.

Mathematics, for Hilbert, became a purely formal calculus, an almost mechanical manipulation of symbols devoid of concrete content. John von Neumann, a collaborator with Hilbert on axiomatic foundations, said at the time, "We must regard classical mathematics as a combinatorial game played with symbols." The comment is perhaps unfortunate in that it suggests that formalistic mathematics is a trivial game played with meaningless marks on paper. Pure axiomatics presupposes an already existing intuitive theory, which it represents in idealized form: "The axioms and demonstrable theorems which arise in our formalistic game are the images of the ideas which form the subject matter of ordinary mathematics." The formalist position is that the inherent structure of intuitive proofs is fully reflected in the combinatorial relations between formal expressions—the outward and visible signs of mathematical reasoning. Hilbert argued, "My theory of proof actually is nothing more than the description of the innermost processes of our understanding and it is a protocol of the rules according to which our thought actually proceeds."

During the decade 1920–1930, Hilbert and his two young assistants, Wilhelm Ackermann and Paul Bernays, worked diligently at carrying through the objectives of the formalist program. After some partial success, there was every reason to believe that a few years' sustained effort would succeed in establishing the hoped-for consis-tency of the formal equivalent of classical mathematics, that it was now merely a matter of finding the correct technique. But such optimistic expectations were dealt a stag-gering setback when Kurt Gödel, a 25-year-old mathematician at the University of Vienna, announced an important and dismaying discovery. Gödel proved his result in 1930, an abstract was published at the end of the year, and the full details appeared early in 1931. Hilbert originally intended to prove consistency of a formal system by such means as could be formalized in the system itself. Gödel showed this program to be incapable of fulfillment. According to Gödel, the consistency of a formal system strong enough to be considered a foundation of mathematics could not be established from within the system by strictly finitary methods. Thus, either the system was inconsistent to begin with, or if consistent, the limitations prescribed by Hilbert were inadequate to formalize a proof of its consistency. Each formal system needed a wider and more inclusive system to demonstrate its consistency. This melancholy revelation effectively brought to an end the initial phase of the formalistic movement.

While Hilbert proposed to save and safeguard the customary formulation of mathematics by a consistency proof, the Dutch mathematician L. E. J. Brouwer was ready to sacrifice those parts of the subject in which he felt that language had outrun clear meaning. Better to be rid of these offensive embellishments, beautiful in form but hollow in content. It was Brouwer's intention to develop an intuitionistic mathematics, using only those constructions that had a clear intuitive justification. Intuitionism, though in some way anticipated by Kant, Kronecker, and Poincaré, was shaped as a definite philosophy of mathematics by Brouwer and his Amsterdam school. Brouwer had a rare insight into the defects of classical mathematics (the familiar mathematics using the logic of Aristotle), and he made a valiant attempt to set things right, beginning in 1907 with his doctoral dissertation. The dissertation, entitled *On the Foundations of Mathematics,* was a penetrating criticism directed against Cantor's transfinite numbers, against the logicism of Russell, and against Hilbert's axiomatic method. Some of the ideas he put forth in his dissertation were sketchy and had to be revised and expanded in later papers. A modest six-page criticism written in 1908, *On the Unreliability of Logical Principles,* questioned the logic used in traditional reasoning. This was followed in 1912 by *Intuitionism and Formalism,* a memoir that consolidated Brouwer's program. Two articles of faith emerged: the reduction of mathematics to the ultimate intuition of the natural numbers and the rejection of the unrestricted application of the law of the excluded middle.

Another no less important feature of the intuitionistic position was its challenge to the logicist presumption of the priority of logic over mathematics. Where Frege, Peano, and Russell thought a logical symbolism was a prerequisite for mathematical knowledge, the intuitionists repudiated any such requirement. Brouwer, in his dissertation, argued that mathematics was a mental construction, a free creation of the mind, completely independent of language and logic. Although mathematical language, whether ordinary or symbolic, may be unavoidable from a practical standpoint, it is nothing more than an imperfect tool used by mathematicians to assist them in memorizing their results and communicating them to one another. Symbolic logic does not represent an essential feature of mathematical reasoning and by itself can never create new mathematics. The symbolic language must not be confused with the mathematics it conveys; it reflects, but does not contain, mathematical reality. ''These paradoxes arise . . . ,'' in the words of Brouwer, ''when the language which accompanies mathematics is extended to a language of mathematical words which is not connected with mathematics.'' He upbraided the formalists for building self-subsistent verbal edifices that could be studied apart from any intuitive interpretation; to that extent, they reduced mathematics to a meaningless ''game of formulas'':

What it [the formalist school] seems to have overlooked is that between the perfection of mathematical language and the perfection of mathematics proper, no clear connection can be seen.

Brouwer's conception of mathematics had some affinity to the conception of Immanuel Kant, both men finding the source of mathematical truth in intuition. Kant, in his *Critique of Pure Reason,* argued the case for believing that a substantial part of theoretical knowledge has an inescapably a priori nature. It was his firm conviction that the axioms of arithmetic and geometry had this a priori character; that is, they were judgments independent of experience and not capable of analytic demonstration. For Kant, the possibility of disproving arithmetical and geometrical laws was entirely unthinkable. Although Brouwer rejected Kant's discredited notion that our geometry was based on an a priori intuition of space, he retained the complementary thesis that a temporal intuition would allow us to conceive one object, then one more, then another, and so on indefinitely. Thus, the natural numbers were accepted by Brouwer, not on the basis of any axiom system such as Peano's, but as arising from some primordial instinct of the passage of time; and all further mathematical objects would have to be constructed out of these numbers by intuitively clear finite methods. This constructive tendency in mathematics had been espoused apart from and before intuitionism in the work of Kronecker and Poincaré. Kronecker, you will remember, objected to introducing into mathematics objects that could not be produced by any kind of finite construction. This was the view that led to his notorious feud with Cantor. As Kronecker himself said in a striking phrase, ''God made the natural numbers, and all the rest is the work of man.'' Another early forerunner of the spirit of intuitionism was Poincaré, who contended that mathematical induction was a pure intuition of mathematical reasoning, not just an axiom that was useful in some systems. Brouwer agreed with Poincaré on this point, but rejected his opinion that mathematical existence coincides with any demonstration of noncontradiction.

In his famous paper, *On the Unreliability of Logical Principles,* Brouwer challenged the view that the classical logic founded on the authority of Aristotle has a universal validity independent of the subject matter to which it is applied. Brouwer maintained that the common belief in the applicability of classical logic to mathematics could only be considered a phenomenon of the history of civilization, of the same order as the oldtime belief in the rationality of π. According to his interpretation, traditional logic arose from the mathematics of finite sets and their subsets; then, forgetful of this limited origin, people mistook that logic for something that had an a priori existence independent of mathematics; and finally, they unjustly applied it, by virtue of its a priori character, to the mathematics of infinite sets. Hermann Weyl said, ''This is the Fall and original sin of set theory, for which it is justly punished by the antinomies.'' That paradoxes showed up was not surprising to Brouwer, only that they showed up so late in the day.

A specific example of a logical principle, valid in reasoning with finite sets, that Brouwer rejected for infinite sets was the law of the excluded middle. The supposition underlying this principle is that each mathematical statement is determinably true or false, independent of the means available to us of recognizing its truth value. For Brouwer, purely hypothetical truth values were an illusion. A given mathematical statement was true only when a certain self-evident construction had been effected in a finite number of steps. Because it could not be guaranteed beforehand that such a construction could be found, we should have no right to assume of each statement that it is either true or false. For instance, Brouwer asked whether it is true or false ''that in the decimal expansion of π there occur ten successive digits forming a sequence of 0123456789.'' This would evidently require that we either indicate a sequence of 0123456789 in π or demonstrate that no such sequence could appear, and no method exists for making the decision, so one could not apply the law of the excluded middle here to conclude that the statement would have to be either true or false. On the other hand, it is legitimate from the intuitionists' standpoint to assert that the number $10^{10^{10}} + 1$ is either prime or composite, without being able to say which alternative actually holds. There is a method, if one were to take the trouble to apply it, that is in principle effective for deciding which alternative is correct.

Brouwer's criticism of the logic used in classical reasoning arose from his refusal to accept nonconstructive existence proofs; that is, proofs that establish the existence of something without providing an effective means for finding it. Giving a nonconstructive existence proof, observed Weyl, is like informing the world that somewhere there is a buried treasure, but not saying where it is. These proofs tend to take the form of an indirect demonstration and as a rule rely on the law of the excluded middle. (Euclid's proof of the infinitude of primes is a perfect illustration.) Brouwer would have none of this; for him, ''mathematical existence'' was synonymous with actual constructibility. It is not enough, in showing that an infinite set has members, to demonstrate that the assumption that the set is empty leads to a contradiction; one must exhibit a process that at least in principle will enable an element of the set to be found or constructed. The intuitionists regarded the idea that there might exist a set that couldn't be constructed as a piece of hopelessly confused metaphysics. By refusing to define a set by means of an attribute characteristic of its elements, and limiting themselves to constructive methods, Brouwer's followers did away with the problems raised by the paradoxes of set theory.

The abandonment of the law of the excluded middle, and with it the nonconstructive existence proofs, was too radical a step for Hilbert to accept. ''Taking the law of the excluded middle from mathematicians,'' he exclaimed, ''is the same as prohibiting the astronomer his telescope or the boxer the use of his fists.'' For his part, Brouwer would not go along with the proposition that proving classical mathematics to be consistent would restore its meaning. Thus, he wrote, ''Nothing of mathematical value will be attained in this manner; a false theory which is not halted by a contradiction is none the less false, just as a criminal act not forbidden by a reprimanding court is none the less criminal.''

At the height of this noisy battle between Göttingen and Amsterdam, the most gifted of Hilbert's pupils, Hermann Weyl (1885–1955), accepted Brouwer's main conclusions and began to fight actively on his side. Weyl had entered Göttingen in 1903 and remained there, first as a student and then as a privatdozent, until his call to the Polytechnicum in Zürich in 1913. Weyl concluded that despite the introduction of Weierstrassian rigor, analysis was still not well-founded, and that part of classical mathematics should be swept away. In his monograph *Das Kontinuum* (1918), he wrote:

> In this little book, I am not concerned to disguise the ''solid rock'' on which the house of analysis is built with a wooden platform of formalism, in order to talk the reader into believing at the end that this platform is the true foundation. What will be proposed is rather the view that this house is largely built on sand.

Strong words were answered in kind. Hilbert refused to accept the mutilation of mathematics that Brouwer's standpoint demanded, comparing these attacks with the earlier negativism of Kronecker. Speaking in Hamburg in 1922:

> What Weyl and Brouwer are doing is mainly following in the path of Kronecker; they are trying to establish mathematics by throwing everything overboard that does not suit them and dictatorially promulgating an embargo. The effect of this is to dismember and cripple our science and to run the risk of losing a large part of our most valuable posses- sions . . . Brouwer's program is not, as Weyl believes it to be, the Revolution, but only the repetition of a vain Putsch, which then was undertaken with greater dash, yet failed utterly.

The anger and determination with which Hilbert made the above declaration is most readily understandable when one remembers that Hilbert made his early reputa- tion by using nonconstructive methods. His existential proof of Gordan's theorem in invariant theory had perplexed his contemporaries and provoked Gordan's outraged cry of ''theology.'' For Hilbert's intransigent adversary, Brouwer, it was proofs like this that had to be jettisoned. Because Hilbert would not follow in the treasonous rejection of the greater part of mathematics and set theory, the battle was joined. Those determined not to be driven from a ''paradise'' must continually struggle to protect it.

We have now examined three rival philosophies concerning mathematical activity; it is pure logic for the logicist, the manipulation of abstract symbols for the formalist, constructions in the medium of temporal intuition for the intuitionist. Each of these schools of thought contained serious defects, and none achieved its objective of pro- viding a universally acceptable approach to mathematics. In the last analysis, it seems that we are no closer to understanding the ultimate meaning of mathematics than the founders of these movements some 75 years ago. Today, foundations of mathematics is not a field in which the same intense activity takes place as occurred in the early part of the century. Modern investigators are less inclined to dogmatism than Brouwer or Hilbert was, and no particular doctrine any longer pretends to represent the true mathematics. Despite the indecisive outcome of the original undertaking, the work done by the three schools had one great value; it revealed the extreme subtlety and complexity of the interplay between logic and mathematics.

12.3 Problems

1. The argument involved in establishing Russell's paradox can be used to show that $P(N)$ is an uncountable set. That is, suppose to the contrary that $N \sim P(N)$ via the function $f{:}N \to P(N)$. Put $B = \{n \in N \,|\, n \notin f(n)\}$. Because $B \in P(N)$, it follows that $B = f(b)$ for some $b \in N$. Complete the argument by reasoning until a contradiction is obtained.

2. Verify that $P(R^{\#})$ is uncountable, where $R^{\#}$ is the set of real numbers. [*Hint*: Apply Cantor's theorem.]

3. For any sets A and B, either prove or give a counterexample for each of the following assertions.

   (a) If $A \subseteq B$, then $P(A) \subseteq P(B)$.
   (b) $P(A \cup B) = P(A) \cup P(B)$.
   (c) $P(A \cap B) = P(A) \cap P(B)$.
   (d) $P(A \times B) = P(A) \times P(B)$.

4. Prove that if A and B are sets for which $A \sim B$, then $P(A) \sim P(B)$. [*Hint*: A function $f : A \to B$ induces a function $f^{*}{:}P(A) \to P(B)$, defined by taking $f^{*}(S) = \{f(s) \,|\, s \in S\}$.]

5. If N denotes the set of natural numbers, establish that:

   (a) The set of all infinite subsets of N is uncountable.

   (b) The set of all finite subsets of N is countable. [*Hint*: With each finite subset $\{n_1, n_2, \ldots, n_k\}$ of N, with $n_1 < n_2 < \cdots < n_k$, associate the number $2^{n_1}3^{n_2} \cdots p_k^{n_k}$, where p_k is the kth prime.]

6. Let a and b be cardinal numbers and A and B be sets such that $a = o(A)$ and $b = o(B)$. Then the sum and product of a and b can be defined as follows:

$$a + b = o(A \cup B), \quad \text{provided } A \cap B = \emptyset;$$
$$a \cdot b = o(A \times B).$$

Using these definitions, prove that:

   (a) $x + 0 = x, x \cdot 0 = 0, x \cdot 1 = x$. [*Hint*: Recall that $0 = o(\emptyset)$.]

   (b) $\aleph_0 + \aleph_0 = \aleph_0$. [*Hint*: $\aleph_0 + \aleph_0 = o(N_e \cup N_o)$, where N_e and N_o are the even and odd natural numbers, respectively.]

   (c) $\aleph_0 \cdot \aleph_0 = \aleph_0$.

   (d) $c + \aleph_0 = c$. [*Hint*: $c + \aleph_0 = o(I \cup Q)$, where I and Q are the irrational and rational numbers, respectively.]

   (e) $c + c = c$. [*Hint*: $c + c = o([0,\infty) \cup (-\infty,0))$.]

   (f) $c \cdot \aleph_0 = c$. [*Hint*: $[0,1) \times N \sim (0,\infty)$ using $f(r,n) = r + n$.]

   (g) $c \cdot c = c$.

7. (a) Could there exist a town in which the barber shaves all men who do not shave themselves? [*Hint*: If such a town exists, who shaves the barber? In this traditional conundrum, it is presumed that barbers are male.]

   (b) Could there exist a book that lists in its bibliography exactly those books that do not list themselves in their bibliographies?

8. Suppose that a lexicon is drawn up containing all the words that occur in the text of this book; names and punctuation marks and also mathematical symbols are counted as words. Let S be the set of natural numbers that are defined by sentences containing at most 50 words (a word being counted each time it occurs), all of them chosen from our lexicon; then the set S is finite. Consider the natural number defined as follows:

Let n be the smallest natural number, in accordance with the usual ordering of the natural numbers, that cannot be defined by means of a sentence containing at most 50 words, all taken from our lexicon.

Show that n is defined in no more than 50 words, but is not a member of S (this is Berry's paradox).

9. Let S be the set of all decimals that are defined by sentences containing a finite number of words, all taken from our lexicon; then S is a countable set of real numbers $r_1, r_2, r_3, \ldots$. Consider the real number r defined as follows:

If the digit in the nth decimal place of r_n is denoted by r_{nn}, then construct the real number $r = 0 \cdot a_1 a_1 a_3 \ldots$ so that in nth digit $a_n = 1$ if $r_{nn} \neq 1$, and $a_n = 2$ if $r_{nn} = 1$.

Show that r is defined in a finite number of words but is not a member of S (this is Richard's paradox).

10. Show, as an example of a nonconstructive existence proof, that there exists a solution of the equation $x^y = z$ with x, y irrational and z rational. [*Hint*: Consider $\sqrt{2}^{\sqrt{2}}$.]

11. (a) Consider the number $n = (-1)^k$, where k is the number of the first decimal place in the decimal expansion of π where the sequence of consecutive digits 0123456789 begins; or if no such number k exists, then $n = 0$. Would the intuitionist accept the statement that n is either positive, negative, or zero?

   (b) The intuitionist views the following as a situation in which the statement p is not the same as "not not p." Write $r = 0.3333 \ldots$, breaking this off as soon as a sequence of consecutive digits 0123456789 has appeared in the decimal expansion of π. Thus, if the 9 of the first sequence 0123456789 in π is the kth digit after the decimal point, then

$$r = 0.33333 \ldots 3 \ (k \text{ decimals})$$
$$= \frac{10^k - 1}{3 \cdot 10^k} .$$

Let p be the statement "r is a rational number." Show that "not p" leads to a contradiction, so that "not not p" must hold. On the other hand, the intuitionist would not conclude that r is rational, because there is no effective way of calculating a and b for which $r = a/b$.

Bibliography

Bernays, Paul, and Fraenkel, Abraham. *Axiomatic Set Theory.* Amsterdam, Holland: North-Holland, 1968.

Beth, Evert W. *The Foundations of Mathematics.* 2d rev. ed. Amsterdam, Holland: North-Holland, 1965.

———. *Mathematical Thought: An Introduction to the Philosophy of Mathematics.* Dordrecht, Holland: D. Reidel, 1965.

———. *Aspects of Modern Logic.* Dordrecht, Holland: D. Reidel, 1971.

Birkhoff, George D. "Fifty Years of American Mathematics 1888–1938." In *Semicentennial Addresses of the American Mathematical Society.* New York: American Mathematical Society, 1938.

Bishop, Errett. "The Crisis in Contemporary Mathematics." *Historia Mathematica* 2(1975): 507–517.

Black, Max. *The Nature of Mathematics: A Critical Survey.* London: Routledge & Kegan Paul, 1965.

Bochner, Maxime. "The Fundamental Conceptions and Methods of Mathematics." *Bulletin of the American Mathematical Society* 11(1904): 115–135.

Breuer, Joseph. *The Theory of Sets.* Translated by H. Fehr. Englewood Cliffs, N.J.: Prentice-Hall, 1958.

Brouwer, L. E. J. "Intuitionism and Formalism." *Bulletin of the American Mathematical Society* 20(1913): 81–96.

Browder, Felix, ed. *Mathematical Developments Arising from Hilbert's Problems (Proceedings of Symposia in Pure Mathematics,* vol. 28). Providence, R.I.: American Mathematical Society, 1976.

Bynum, Terrell Ward. *Gottlob Frege: Conceptual Notations and Related Articles.* Oxford: Oxford University Press, 1972.

Cantor, Georg. *Contributions to the Founding of the Theory of Transfinite Sets.* Translated by P. E. B. Jordain. La Salle, Ill.: Open Court, 1952.

Cohen, Paul. *Set Theory and the Continuum Hypothesis.* New York: W. A. Benjamin, 1966.

Dauben, Joseph W. "The Trigonometric Background to Georg Cantor's Theory of Sets." *Archive for History of Exact Sciences* 7(1971): 142–170.

———. "The Invariance of Dimension: Problems in the Early Development of Set Theory and Topology." *Historia Mathematica* 2(1975): 273–288.

———. "Georg Cantor: The Personal Matrix of His Mathematics" *Isis* 69(1978): 534–550.

Dawson, John. "The Gödel Incompleteness Theorem from a Length-of-Proof Perspective." *American Mathematical Monthly* 86(1979): 740–747.

Dedekind, Richard. *Essays on the Theory of Numbers.* Translated by W. W. Beman. Chicago: Open Court, 1901.

De Sua, Frank. "Consistency and Completeness—A Resume." *American Mathematical Monthly* 63(1956): 295–305.

Dresden, Arnold. "Brouwer's Contributions to the Foundations of Mathematics." *Bulletin of the American Mathematical Society* 30(1924): 31–40.

Dummett, Michael. *Frege: Philosophy of Language.* New York: Harper & Row, 1973.

———. *Elements of Intuitionism.* Oxford: Oxford University Press, 1977.

———. *Frege: Philosophy of Mathematics.* Cambridge: Harvard University Press, 1991.

Edwards, Harold. "An Appreciation of Kronecker." *Mathematical Intelligencer* 9, no. 1(1987): 28–35.

Enderton, Herbert. *Elements of Set Theory.* New York: Academic Press, 1977.

Fang, J. *Hilbert: Towards a Philosophy of Modern Mathematics.* Hauppauge, N.Y.: Paideia Press, 1970.

Fraenkel, Abraham. *Abstract Set Theory.* 3d rev. ed. Amsterdam, Holland: North-Holland, 1966.

———. *Set Theory and Logic.* Reading, Mass.: Addison-Wesley, 1966.

Frege, G. *The Foundations of Arithmetic.* Translated by J. L. Austing. Evanston, Ill.: Northwestern University Press, 1968.

———. *Posthumous Writings.* Edited by Hans Hermes. Chicago: University of Chicago Press, 1979.

Garciadiego, Alejandro. *Bertrand Russell and the Origins of the Set-Theoretic Paradoxes.* Boston: Birkhauser, 1992.

Gödel, Kurt. *The Consistency of the Axiom of Choice and the Generalized Continuum Hypothesis with the Axioms of Set Theory.* Princeton, N.J.: Princeton University Press, 1940.

———. *On Formally Undecidable Propositions of Principia Mathematica and Related Systems.* New York: Basic Books, 1962.

Goldfarb, Warren. "Logic in the Twenties." *Journal of Symbolic Logic* 44(1979): 351–368.

Grattan-Guinness, I. "An Unpublished Paper of Georg Cantor: Principien einer Theorie der Ordnungstypen." *Acta Mathematica* 124(1970): 65–170.

———. "Towards a Biography of Georg Cantor." *Annals of Science* 27(1971): 345–391.

———. "How Bertrand Russell Discovered His Paradox." *Historia Mathematica* 5(1978): 127–137.

Green, Judy and LaDuke, Jeanne. "Women in the American Mathematical Community: The Pre-1940 Ph.D.s." *Mathematical Intelligencer* 9(no. 1, 1987): 11–23.

Halmos, Paul. *Naive Set Theory*. Princeton, N.J.: D. Van Nostrand, 1960.

Hardy, G. H. "William Henry Young." *Journal of the London Mathematical Society* 17(1942): 218–237.

———. *Bertrand Russell and Trinity*. Cambridge: Cambridge University Press, 1970.

Hatcher, William. *Foundations of Mathematics*. Philadelphia: W. B. Saunders, 1968.

Heyting, A. *Intuitionism: An Introduction*. Amsterdam, Holland: North-Holland, 1966.

Heyting, A., ed. *L. E. J. Brouwer: Collected Works*. Amsterdam, Holland: North-Holland, 1975.

Hilbert, David. "Mathematical Problems: Lecture Delivered Before the International Congress of Mathematicians at Paris in 1900." *Bulletin of the American Mathematical Society* 8(1902): 437–479.

Hilbert, David, and Ackermann, Wilhelm. *Principles of Mathematical Logic*. Translated by L. M. Hammond. New York: Chelsea, 1950.

Johnson, Phillip. "The Early Beginnings of Set Theory." *Mathematics Teacher* 63(1970): 690–692.

———. *A History of Set Theory*. Boston: Prindle, Weber & Schmidt, 1972.

Kamke, E. *Theory of Sets*. Translated by F. Bagemihl. New York: Dover, 1950.

Kattsoff, Louis. *A Philosophy of Mathematics*. Freeport, N.Y.: Books for Libraries Press, 1969.

Katz, Kaila and Kenschaft, Patricia. "Sylvester and Scott." *Mathematics Teacher* 75(1982): 490–494.

Kennedy, Hubert. *Peano: Life and Works of Giuseppe Peano*. Dordrecht, Holland: D. Reidel, 1980.

Kenschaft, Patricia. "Charlotte Angas Scott, 1858–1931." *College Mathematics Journal* 18(1987): 98–110.

Kilmister, C. W. *Language, Logic and Mathematics*. New York: Barnes & Noble, 1967.

Kleene, Stephen. *Introduction to Metamathematics*. Princeton, N.J.: D. Van Nostrand, 1952.

———. *Mathematical Logic*. New York: John Wiley, 1967.

Kneebone, G. T. *Mathematical Logic and the Foundations of Mathematics*. Princeton, N.J.: D. Van Nostrand, 1963.

Korner, Stephan. *The Philosophy of Mathematics*. London: Hutchison, 1960.

Kuratowski, K., and Fraenkel, A. *Axiomatic Set Theory*. Amsterdam, Holland: North-Holland, 1968.

Levy, Azriel. *Basic Set Theory*. Berlin: Springer-Verlag, 1979.

Maziarz, Edward. *The Philosophy of Mathematics*. New York: Philosophical Library, 1950.

Middleton, Karen. "The Role of Russell's Paradox in the Development of Twentieth Century Mathematics." *Pi Mu Epsilon Journal* 8(1986): 234–241.

Moore, Gregory. "The Origins of Zermelo's Axiomatization of Set Theory." *Journal of Philosophical Logic* 7(1978): 307–329.

———. *Zermelo's Axiom of Choice: Its Origins, Development and Influence*. New York: Springer-Verlag, 1982.

———. "Sixty Years After Godel." *Mathematical Intelligencer* 13, no. 3(1991): 6–11.

Mostowski, Andrzej. *Thirty Years of Foundational Studies*. New York: Barnes & Noble, 1958.

Nagel, Ernest, and Newman, James. "Goedel's Proof." *Scientific American* 194(June 1956): 71–86.

Nagel, Ernest; Suppes, Patrick; and Tarski, Alfred, eds. *Logic, Methodology and Philosophy of Science*. Stanford, Calif.: Stanford University Press, 1962.

Newman, H. H. "Hermann Weyl." *Journal of the London Mathematical Society* 33(1958): 500–511.

Nidditch, P. H. *The Development of Mathematical Logic*. New York: Dover, 1962.

Parshall, Karen. "Eliakim Hastings Moore and the Founding of a Mathematical Community in America." *Annals of Science* 41 (1984): 313–333.

———. "America's First School of Mathematical Research: James Joseph Sylvester at the Johns Hopkins University." *Archive for the History of Exact Sciences* 38(1988): 153–196.

———. "A Century-Old Snapshot of American Mathematics." *Mathematical Intelligencer* 12, no. 3(1990): 7–11.

———. "The 100th Anniversary of Mathematics at the University of Chicago." *Mathematical Intelligencer* 14, no. 2(1992): 39–44.

Pierpont, James. "The History of Mathematics in the Nineteenth Century." *Bulletin of the American Mathematical Society* 11(1904): 136–159.

———. "Mathematical Rigor, Past and Present." *Bulletin of the American Mathematical Society* 34(1928): 23–53.

Reid, Constance. *Hilbert*. New York: Springer-Verlag, 1970.

———. *Courant in Göttingen and New York*. New York: Springer-Verlag, 1976.

Ross, Steve. "Bolzano's Analytic Programme." *Mathematical Intelligencer* 14, no. 3(1992): 45–53.

Russell, Bertrand. *Introduction to Mathematical Philosophy*. 2/E. New York: Macmillan, 1924.

———. *Principles of Mathematics*. 2/E. New York: W. W. Norton, 1938.

———. *The Autobiography of Bertrand Russell*. 3 vols. London: George Allen and Unwin Ltd., 1967–1969.

Sierpinski, Waclaw. *Cardinal and Ordinal Numbers*. 2d ed. Warsaw: Polish Scientific Publications, 1965.

Snapper, Ernst. "The Three Crises in Mathematics: Logicism, Intuitionism and Formalism." *Mathematics Magazine* 52(1979): 207–216.

———. "The Russell Paradox." *Pi Mu Epsilon Journal* 8(1986): 281–291.

Stabler, E. Russell. "An Interpretation and Comparison of Three Schools of Thought in the Foundations of Mathematics." *Mathematics Teacher* 28(1935): 5–35.

———. *An Introduction to Mathematical Thought*. Reading, Mass.: Addison-Wesley, 1953.

Stoll, Robert. *Set Theory and Logic.* San Francisco: W. H. Freeman, 1961.

———. *Sets, Logic and Axiomatic Theories.* 2d ed. San Francisco: W. H. Freeman, 1961.

Styazhkin, N. I. *History of Mathematical Logic from Leibniz to Peano.* Cambridge, Mass.: M.I.T. Press, 1969.

Suppes, Patrick. *Axiomatic Set Theory.* Princeton, N.J.: D. Van Nostrand, 1960.

Van Heijenoort, Jan, ed. *From Frege to Gödel: A Source Book in Mathematical Logic, 1879–1931.* Cambridge, Mass.: Harvard University Press, 1967.

Van Stigt, W. P. *Brouwer's Intuitionism.* New York: North-Holland/Elsevier, 1990.

Von Helmholtz, H. *Counting and Measuring.* New York: D. Van Nostrand, 1930.

Wang, Hoa. *A Survey of Mathematical Logic.* Amsterdam, Holland: North-Holland, 1963.

Wavre, Rolin. "Is There a Crisis in Mathematics?" *American Mathematical Monthly* 41(1934): 488–499.

Weyl, Hermann. "David Hilbert and His Mathematical Work." *Bulletin of the American Mathematical Society* 50(1944): 612–654.

———. *Philosophy of Mathematics and Natural Science.* Rev. ed. Princeton, N.J.: Princeton University Press, 1949.

———. "A Half Century of Mathematics." *American Mathematical Monthly* 58(1951): 523–553.

Wheeler, Lynde. *Josiah Willard Gibbs: The History of a Great Mind.* New Haven, Conn.: Yale University Press, 1951.

Whitehead, A. N., and Russell, B. *Principia Mathematica.* 2/E. 3 vols. Cambridge: Cambridge University Press, 1965.

Wilder, Raymond. *The Foundations of Mathematics.* New York: John Wiley, 1965.

———. *The Evolution of Mathematical Concepts: An Elementary Study.* New York: John Wiley, 1968.

Young, W. H., and Young, G. C. *The Theory of Sets of Points.* Cambridge: Cambridge University Press, 1906.

Zuckerman, Martin. *Sets and Transfinite Numbers.* New York: Macmillan, 1974.

CHAPTER 13

Extensions and Generalizations: Hardy, Hausdorff, and Noether

" . . . A mathematician, like a painter or a poet, is a maker of patterns. If his patterns are more permanent than theirs, it is because they are made of ideas."

G. H. HARDY

13.1 Hardy and Ramanujan

In the late decades of the nineteenth century, the study of higher mathematics in England was centered at Cambridge University. And at Cambridge, any thought of a serious mathematical education was badly distorted by the local obsession with the Mathematical Tripos, the most notorious, most challenging test that any university has known.

For a student taking a degree at this time, the Tripos was a series of examinations, spread over more than a week, and given in the middle of the fourth year of residence. Preparation for the Tripos became the student's chief mathematical activity. Few undergraduates paid any attention to the professors: Cayley, for instance, lectured to classes of only two or three. Instead, most students entrusted their mathematical education almost entirely to private tutors or "coaches." These coaches had all the past Tripos papers at their fingertips and knew all the tricks of the examiners. For handsome fees, they would prepare their pupils for the venerable contest. The most famous coach, Edward Routh, achieved remarkable results. Between 1858 and 1888 he produced 27 Senior Wranglers (those scoring highest on the examination) and 41 winners of the Smith's Prize.

Because of the exaggerated importance attached to the Tripos, candidates were more concerned with their place in the examination's Order of Merit list than in learning mathematics. To attain high ranking, the Cambridge coaches drove their pupils ferociously. They were drilled for up to seven hours a day in solving difficult problems against time: a good memory and facile manipulative skills were critical, whereas rigor in argument was generally derided. Curiosity about mathematical topics that were not apt to appear on the examination was firmly discouraged. Coaches forbade the reading of great mathematical works such as Laplace's *Mécanique céleste*. Many insisted that their pupils not attend professorial lectures. The process was a terrible ordeal for students. Bertrand Russell, who stood as Seventh Wrangler in 1893, later remarked, ''When I finished my Tripos, I sold all my mathematical books and made a vow that I would never look at a mathematical book again.''

The Mathematical Tripos generated the sort of interest as would a horse race: serious contenders for Senior Wrangler were the subject of betting by all and sundry. Sometimes there were surprises. William Thompson, later to be made Lord Kelvin, was easily the best mathematician of his year, an odds-on favorite to be first. On the day the Tripos results were read, Kelvin stayed in bed late and sent his college servant out to hear the news. When the servant returned and Kelvin demanded to know who was second, he was greeted with the reply, ''You, sir.'' Someone in the examination hall had evidently memorized better, or written faster, than Kelvin.

During the nineteenth century, classical applied mathematics had become an English specialty, typified by the work of Green, Kelvin, Stokes, Rayleigh, and Maxwell. Cambridge paid homage to their achievements by making the Tripos primarily a test of mathematical physics. Only rarely were its problems related to physical questions not contained in Newton's *Principia*. When it came to the *Principia*, students were ''expected to know any lemma in that great work by its number alone, as if it were one of the Commandments or the 100th Psalm.'' There was little on the examination that could be called contemporary analysis, which is to say the advanced developments depended on infinite processes—limits, differentiation, integration, summation of series, and the like—and in particular the theories of functions of a real or a complex variable.

The effect of the Tripos was to train some outstanding applied mathematicians but to stifle pure mathematics. Although Cayley and Sylvester were notable exceptions, the area was largely ignored by students aspiring chiefly to a high ranking in the Order of Merit. In the Cambridge of the 1880s, when the Tripos stood at the zenith of its reputation, English mathematics was somewhere near its lowest ebb: self-supporting, self-satisfied, and indifferent to the developments that were taking place on the Continent.

England's isolation from the current thinking at Paris, Berlin, and Göttingen was broken by Andrew Forsyth (1858–1942), Cayley's successor as Sadlerian Professor at Cambridge. Forsyth's *Theory of Functions of a Complex Variable,* which was published in 1893, ''burst with the splendour of a revelation.'' Some hailed it as having a greater influence on English mathematics than any work since Newton's *Principia.* The faults of the *Theory of Functions* are much more obvious today than its merits. Because Forsyth had little faculty for displaying the crucial points of a delicate argument, his exposition is often obscure and illogical. Nevertheless, for all its shortcomings *Theory of Functions* was the work that brought the methods of modern analysis to Cambridge.

The principal architect of an English school of mathematical analysis was Godfrey Harold Hardy (1877–1947). For more than a quarter of a century Hardy dominated English mathematics, through both the significance of his work and the force of his personality.

A product of England's finest private schools, the young Hardy went to Cambridge in 1896 on an Entrance Scholarship. Starting something of a trend among the better students, he took Part I of the Mathematical Tripos in the third instead of the normal fourth year, and he finished Fourth Wrangler. (By Hardy's day, the Tripos had been divided into two parts, with the advanced Part II being taken at a later stage by those seeking fellowships.) Relieved of two years of tedium, he was ready to begin learning genuine mathematics. Hardy was later to write of the astonishment with which he read Camille Jordan's *Cours d'analyse de L'Ecole Polytechnique* and to realize for the first time " . . . what mathematics really meant." In 1900, he placed highest in Part II of the Tripos and succeeded automatically to a fellowship. Three years later Hardy received an M.A. degree, which was the highest academic degree awarded by Cambridge. (Cambridge did not offer the doctorate, a German innovation, until after the Great War of 1914–1918 and in hopes of attracting American students who would otherwise have gone to Germany.) At the expiration of his fellowship in 1906, Hardy joined the Cambridge faculty as a lecturer in mathematics.

Hardy's research specialty was "analytic number theory," an area of mathematics that uses analysis to answer questions about number theory. Developments stemming from Bernhard Riemann's epoch-making eight-page paper "On the Number of Primes Less than a Given Magnitude" (1859) had made the field sufficiently promising. Riemann took as his starting point a remarkable formula discovered by Euler over a century earlier,

$$\sum_{n=1}^{\infty} 1/n^s = \prod_{p} \frac{1}{1 - 1/p^s},$$

where the infinite product is taken over all the prime numbers p, and the sum is over all positive integers n. This formula results from expanding the factor involving p as a geometric series

$$\frac{1}{1 - 1/p^s} = 1 + 1/p^s + 1/p^{2s} + 1/p^{3s} + \cdots .$$

On multiplying these series for all primes p, we get a sum of terms of the form

$$1/(p_1^{k_1} p_2^{k_2} \ldots p_r^{k_r})^s$$

where $p_1, p_2, \ldots, p_r$ are distinct primes and $k_1, k_2, \ldots, k_r$ are positive integers. By the Fundamental Theorem of Arithmetic, the products $p_1^{k_1} p_2^{k_2} \ldots p_r^{k_r}$ so obtained yield precisely the positive integers, allowing us to conclude that the sum in question is simply $\sum_{n=1}^{\infty} 1/n^s$. This sum, which is a function of a real variable s, is called the (Riemann) zeta function and denoted by $\zeta(s)$:

$$\zeta(s) = \sum_{n=1}^{\infty} 1/n^s.$$

Because the series for $\zeta(1)$ diverges, Euler's formula implies the existence of an infinitude of prime numbers; for if there were only finitely many primes, then the product on the right-hand side of the formula

$$\sum_{n=1}^{\infty} 1/n = \prod_{p} \frac{1}{1 - 1/p}$$

would be a finite product and hence would have a finite value.

Godfrey Harold Hardy
(1877–1947)

(Courtesy of the Master and Fellows of Trinity College, Cambridge.)

Using the zeta function, Euler proved that the sum of the reciprocals of the prime numbers diverge. It is known that

$$\zeta(2) = \sum_{n=1}^{\infty} 1/n^2 = \pi/6,$$

a result that Euler also obtained, and that $\zeta(4) = \pi^4/90$. One immediate consequence of this is that $\zeta(2)$ and $\zeta(4)$ are both transcendental numbers. A more recent (1979) gain along these lines was to establish that $\zeta(3)$ is irrational; but whether it is transcendental remains unknown.

Riemann's key idea was to extend the zeta function $\zeta(s)$ so that, instead of being restricted to a real variable, s is allowed to be a complex number $s = a + bi$. Among the many questions that can be asked about the complex function $\zeta(s)$, a paramount one concerns the location of its zeros. Riemann stated that $\zeta(s)$ vanishes when $s = -2n$, the "trivial zeros," and that all other complex zeros must lie in the so-called critical strip $0 < a < 1$. He went on to conjecture that these zeros are on the vertical axis of symmetry, the critical line $a = 1/2$; that is, if $\zeta(s) = 0$ for a complex number $s = a + bi$, with $0 < a < 1$, then s is of the form $1/2 + bi$. This is the famous Riemann Hypothesis, still open to proof or disproof—by universal agreement the outstanding unsolved problem in mathematics. Hilbert listed it as the eighth problem in his address before the Paris Congress of 1900. He obviously thought it incredibly difficult, once remarking that if he awakened after having slept for 1000 years his first question would be, "Has the Riemann Hypothesis been proved?"

Hardy was the first to give any sort of answer to the Riemann Hypothesis when, in 1914, he established that infinitely many zeros of $\zeta(s)$ are located on the critical line (this does not preclude the existence of infinitely many that are not). Current opinion is predominantly in favor of Riemann's celebrated conjecture, because there is considerable numerical evidence in support of it. Recent computer calculations have verified that the first 1,500,000,000 nontrivial zeros of the zeta function, ordered by the size of their imaginary part, lie on the critical line.

There is a pleasant anecdote in this connection. Hardy was returning from Denmark to England on a day when the weather conditions in the North Sea channel were unusually rough. With Fermat's famous marginal note in mind, he sent a message to a friend: "Have proved the Riemann Hypothesis." He was confident that God would not let him die with such undeserved glory, so that his safe return was assured.

Perhaps Hardy's greatest service to mathematics in this period was his well-known book *A Course in Pure Mathematics,* published in 1908. It was designed to give the undergraduate student a rigorous exposition of the basic ideas of analysis—limits, continuity, convergence of series, and the like. Writing in the preface, Hardy decried the neglect of analysis in England: "I have indeed in examination asked a dozen candidates, including several future Senior Wranglers, to sum the series $1 + x + x^2 + \cdots$ and not received a single answer that was not practically worthless."

Running through numerous editions and translated into several languages, *A Course in Pure Mathematics* transformed the state of university teaching. The remarkably modern second edition, which was brought out in 1914, acknowledges the changing curriculum. Hardy now introduces the Bolzano-Weierstrass Theorem, Eduard Heine's result asserting that a continuous function on a closed interval is uniformly continuous, and the Borel Covering Theorem (which concerns the question: if a set S is covered by any collection of open intervals, under what conditions is it possible to choose a finite number of sets from the collection and still cover $S?$). Subsequent editions of the text were but minor variations of the 1914 edition.

Hardy despised the old Tripos system and was a leading advocate of its reform. In his view, the excessive concentration on examination topics drew the students' attention away from modern mathematics, contributing to England's backwardness. He would have preferred to do away with the Tripos altogether, but settled for the abolition of the strict Order of Merit: degree candidates still took Part I of the examination but were ranked only by broad classes—Wrangler, Senior Optime, or Junior Optime—in which their names appeared in alphabetical order. With the adoption of the new regulations in 1910, the practice of "coaching" disappeared almost at once.

Hardy's name is inevitably linked with that of John Endsor Littlewood (1885–1977). A Cambridge graduate, Littlewood had been Senior Wrangler in 1905, but the fellowship that normally would have been his went to someone else. After spending three years as a lecturer at Manchester University, he joined the Cambridge faculty to succeed Alfred North Whitehead. Hardy found in Littlewood a partner to help strengthen and build on the foundations of analysis. Together they carried on the most prolonged (35 years), extensive, and fruitful collaboration in the history of mathematics. They wrote nearly 100 papers together, the last one published a year after Hardy's death. It was often joked that there were only three great English mathematicians in those days: Hardy, Littlewood, and Hardy-Littlewood. One mathematician, upon meeting Littlewood for the first time, exclaimed, "I thought that you were merely a name used by Hardy for those papers which he did not think were quite good enough to publish under his own name."

Hardy's sympathy with Bertrand Russell's pacifism put him in an unpleasant position during and immediately after the war; some of his Cambridge colleagues scarcely spoke to him. In 1919, he was only too ready to accept the Savilian Chair in geometry at Oxford. Although Littlewood—who served as an artillery officer from 1914 until 1918—remained at Cambridge, their mathematical partnership suffered no interruption. Because they seldom, if ever, met together to discuss or write mathematics, the only noticeable change was a collaboration by mail instead of by college messenger. Contemplating this, they drew up a set of "axioms" expressing the personal freedom of their cooperation. One of their principles was: if one wrote a letter to the other, the recipient was under no obligation to reply to it, or even to read it. Another, designed to prevent quarrels, was: it made no difference if one of them had not contributed the least bit to the contents of a paper appearing under both their names. With these guidelines, the two mathematicians entered the most productive decade of their far-reaching and intensive joint work. Hardy established a school of analysis at Oxford, gathering about him a team of colleagues and research students; but Cambridge, far more than Oxford, was still the center of English mathematics. After a lapse of eleven years Hardy returned to Cambridge to assume its senior mathematical chair, the Sadlerian Professorship. He held the position until his retirement in 1942.

There are few major problems in analysis or the theory of numbers to which Hardy and Littlewood did not make significant contributions. A primary interest of theirs was Waring's Problem. Much like Fermat's Last Theorem, it is a simply stated assertion about the positive integers, which gives no suggestion of the difficulty or mathematical depth of a correct solution. The problem began in 1770 when Edward Waring conjectured, on the basis of empirical evidence, that every number is the sum of at most 4 squares, 9 cubes, 19 fourth powers, and so on. It has become traditional to let $g(k)$ denote the least integer such that every positive integer can be expressed as the sum of at most $g(k)$ positive k'th powers.

At about the same time that Waring recorded his conjecture, the value $g(2) = 4$ was confirmed by Lagrange in his classic "four-square theorem." The general result that $g(k)$ is finite for all k is attributable to Hilbert (1909), who gave an existence proof that shed no light on how many k'th powers are needed. Shortly thereafter, it was shown that $g(3)$ does indeed equal 9. Using the powerful techniques of analysis, Hardy and Littlewood established (1921) that all sufficiently large integers can be written as the sum of 19 or fewer fourth powers. Because $79 = 4 \cdot 2^4 + 15 \cdot 1^4$ requires 19 fourth powers, $g(4) \geq 19$. This observation, together with the Hardy-Littlewood result, suggested that $g(4) = 19$ as Waring had guessed and raised the possibility that its actual value could be settled by direct, exhaustive computation. In 1986, it was finally verified that $g(4) = 19$.

Another topic that drew the attention of the two collaborators was a variation of the Goldbach Conjecture called the "three-primes problem": can every odd integer $n \geq 7$ be written as the sum of three prime numbers? In 1922, Hardy and his younger colleague showed that, assuming the Riemann Hypothesis holds, there exists a positive integer N such that every odd integer $n \geq N$ is the sum of three primes. They also conjectured that every large integer is a sum of a prime and two squares, an assertion that was subsequently verified.

Srinivasa Ramanujan
(1887–1920)

(Courtesy of the Master and Fellows of Trinity College, Cambridge.)

In 1913, Hardy "discovered" Srinivasa Ramanujan (1887–1920)—by which he meant that he was the first really competent mathematician to see and judge Ramanujan's work. The discovery led to an association that Hardy was to call "the most romantic incident in my life." India has from time to time produced mathematicians of remarkable power, but Ramanujan is universally considered to have been its greatest genius.

Ramanujan was born in the southern Indian town of Erode, near Madras, the son of a bookkeeper in the shop of a cloth merchant. He began his single-minded pursuit of mathematics when, at the age of 15 or 16, he borrowed a copy of Carr's *Synopsis of Pure Mathematics*. This unusual book contained the statements of more than 6000 theorems, very few with proofs. Ramanujan undertook the task of establishing, without help, all the formulas in the book. In 1903 he won a scholarship to the University of Madras, only to lose it a year later for neglecting other subjects in favor of mathematics. He dropped out of college in disappointment and wandered the countryside for the next several years, impoverished and unemployed. Compelled to seek a regular livelihood after marrying, Ramanujan secured (1912) a clerical position with the Madras Port Trust Office, a job that left him enough time to continue his work in mathematics. After publishing his first paper in 1911, and two more the next year, he gradually gained recognition.

At the urging of influential friends, Ramanujan began a correspondence with Hardy, who was by then recognized as the leading British pure mathematician. Appended to his letters to Hardy were lists of new theorems, 120 in all, some definitely proved and others only conjectured. Hardy took the time and, with Littlewood's help, made the considerable effort to analyze Ramanujan's findings. Their conclusion was that "they could only have been written down by a mathematician of the highest class;

they must be true because if they were not true, no one would have the imagination to invent them.'' Hardy immediately invited Ramanujan to come to Cambridge University to develop his already great, but untrained, mathematical talent; for up to that time Ramanujan had worked in almost total isolation from modern European mathematics.

Supported by a special scholarship, Ramanujan arrived in England in April of 1914. There, under the guidance of Hardy and Littlewood, he had three years of un-interrupted activity. Some 32 of Ramanujan's 37 published papers took shape during the period 1914–1917, 7 written in collaboration with Hardy. Hardy wrote to Madras University saying, ''He will return to India with a scientific standing and reputation such as no Indian has enjoyed before.'' England gave Ramanujan such honors as were possible. In 1916, he received a B.A. from Cambridge ''by research,'' his only college degree. The Royal Society—England's preeminent scientific body—made Ramanujan a Fellow in 1918, and Trinity College, Cambridge, elected him a Fellow later in the same year. He was the first Indian to gain either of these high distinctions.

Even as Ramanujan's prominence grew, his health deteriorated disastrously. In 1917 he became incurably ill with a disease that was then believed to be tuberculosis. The exact nature of his illness is not known, but the decline in his health was doubtless accelerated by the difficulty Ramanujan had in maintaining an adequate vegetarian diet in war-rationed England. Early in 1919 when the seas were finally considered safe for travel, he returned to India. In extreme pain, Ramanujan continued to do mathematics while lying in bed. He died the following April, at the age of 32.

The theory of partitions is one of the outstanding examples of the success of the Hardy-Ramanujan collaboration. A ''partition'' of a positive integer n is a way of writing n as a sum of positive integers, the order of the summands being irrelevant. The integer 5, for example, may be partitioned in seven ways: 5, 4 + 1, 3 + 2, 3 + 1 + 1, 2 + 2 + 1, 2 + 1 + 1 + 1, 1 + 1 + 1 + 1 + 1. If $p(n)$ denotes the total number of partitions of n, then the values of $p(n)$ for the first six positive integers are $p(1) = 1$, $p(2) = 2$, $p(3) = 3$, $p(4) = 5$, $p(5) = 7$ and $p(6) = 11$. Actual computation shows that the partition function $p(n)$ increases very rapidly with n; for instance, $p(200)$ has the enormous value

$$p(200) = 3,972,999,029,388.$$

Although no simple formula for $p(n)$ exists, one can look for an approximate formula giving its general order of magnitude. In 1918, Hardy and Ramanujan proved what is considered one of the masterpieces in number theory: namely, that for large n the partition function satisfies the relation

$$p(n) \approx \frac{e^{c\sqrt{n}}}{4n\sqrt{3}},$$

where the constant $c = \pi (2/3)^{1/2}$. For $n = 200$, the right-hand side of the relation is approximately $4 \cdot 10^{12}$, which is remarkably close to the actual value of $p(200)$.

According to Hardy, Ramanujan could remember the idiosyncrasies of numbers in almost uncanny ways: Littlewood is said to have remarked, ''Every positive integer was one of his personal friends.'' There is a well-known story that Hardy once visited Ramanujan as he lay ill in a hospital and observed incidentally that he had arrived in a taxi with a rather dull number, 1729. ''No,'' Ramanujan replied without hesitation, "it is a very interesting number; it is the smallest number expressible as the sum of two cubes in two different ways." Ramanujan had immediately recognized that $1729 = 1^3 + 12^3 = 9^3 + 10^3$. Hardy then asked for the smallest number that is a sum of two fourth powers in two different ways. After a moment's reflection, Ramanujan responded that there was no obvious example and that the first such number must be very large. (In fact, the number is $635{,}318{,}657 = 59^4 + 158^4 = 133^4 + 134^4$.)

As a further example of Ramanujan's creativity, we mention his unparalleled ability to come up with infinite series representations for π. Computer scientists have exploited his series

$$\frac{1}{\pi} = \frac{\sqrt{8}}{9801} \sum_{n=0}^{\infty} \frac{(4n)!}{(n!)^4} \frac{[1103 + 26390n]}{396^{4n}}$$

to calculate the value of π to millions of decimal digits; each successive term in the series adds roughly eight more correct digits. Ramanujan discovered fourteen other series for $1/\pi$, but he gave almost no explanation as to their origin. The most remarkable of these is

$$\frac{1}{\pi} = \sum_{n=0}^{\infty} \binom{2n}{n}^3 \frac{42n + 3}{2^{12n+4}} .$$

This series has the property that it can be used to compute the second block of k (binary) digits in the decimal expansion of π without calculating the first k digits.

Despite the brevity of Ramanujan's life, his influence is still evident in many parts of mathematics and its allied fields. He left behind three notebooks—composed between 1903 and 1914—recording the statements of approximately 3000 results, with scarcely any indication of proof. The task of editing and deciphering these notebooks is only now nearing completion; most of the incorporated material has been substantiated, but there remain many asserted theorems and identities so startling that no one knows how to derive them. In 1976, a "lost notebook" of Ramanujan, with more than 100 pages listing 600 formulas one after another, was found tucked away in the Cambridge University Library. Apparently written after his return to India, the notebook's discovery caused roughly as much stir in the mathematical world as a previously unknown symphony of Beethoven might generate in the musical world. Ramanujan has bequeathed an unexpected legacy that will keep mathematicians busy for many more decades.

13.2 The Beginnings of Point-Set Topology

While English mathematics was waking from its slumber, pure mathematicians on the Continent had not been entirely idle. Convergence problems connected with trigonometric series led Georg Cantor, during the years 1872 to 1890, to investigate properties of certain infinite subsets of the real line. For the sake of such investigations he introduced the basic concept of limit point of a set and the associated ideas of closed set, derived set, and dense set:

A real number x is a limit point of a set X of real numbers provided that for every positive number ε there is an element y of the set X such that $0 < |x - y| < \varepsilon$.

Cantor described a set as being closed if it contained its limit points, while the collection of limit points of a given set was called its derived set. Through these notions, he inaugurated the study of what is known today as point-set or general topology, that part of topology most relevant to analysis. Loosely speaking, point-set topology may be considered as an abstract investigation of the limit point concept in more generalized spaces of unspecified elements.

The term ''topology'' appears to have been coined by Joseph Listing—a pupil of Gauss and later professor of physics at Göttingen—for the title of his 1847 textbook *Verstudeien zur Topologie;* it is derived from two Greek words: *topos* meaning ''place'' or ''surface,'' and *logos* meaning ''study.'' A rival name, now obsolete, was ''analysis situs'' (*situs,* for ''site'').

The development of point-set topology did not go beyond the real line, plane, and higher-dimensional Euclidean spaces for several decades, although the definitions introduced there have a more general validity. This was recognized by Maurice Fréchet (1878–1973), who may be regarded as the creator of the first systematic point-set theory in ''abstract spaces.'' An active researcher, Fréchet had more than twenty papers in print before his thesis for the doctorate at Paris was published in 1906. It is likely that this thesis had more impact on the mathematical world than anything else he ever wrote.

The outlines of Fréchet's thesis took shape in a series of five notes that appeared in the *Comptes Rendus* of the Academie des Sciences in 1904–1905. In these articles, he was intent on building up an abstract theory that closely resembles the point-set theory of classical analysis. Because the most important and best known results of analysis were those based on limits, Fréchet considered abstract spaces in which limit points of sets or sequences are definable.

To ask whether x is a limit point of a set X, it is essential to be able to say when x is close enough to X. Fréchet suggested and explored several ways of generalizing the intuitive notion of ''closeness,'' but his most influential proposal was the concept of what is now called a ''metric space.'' (The term was introduced by Felix Hausdorff, using the German name *metriche Raum* in his book of 1914.) Such spaces are equipped with a notion of distance between two points.

Fréchet recognized that some suitable assumptions about distance are needed; and, for the concept to be useful, these assumptions should be broad enough to include the most familiar spaces of nineteenth-century mathematics. Abstracting from the real line, where the standard method of measuring the distance between points x and y is by the nonnegative number $|x - y|$, he defined his generalized distance $d(x,y)$ to be a nonnegative real number satisfying the three conditions

1. $d(x,y) = 0$ if and only if $x = y$.

2. $d(x,y) = d(y,x)$.

3. $d(x,y) \leq d(x,z) + d(z,y)$.

Condition 3, whose antecedents go back at least to Euclid's *Elements* (Proposition 20 of Book I: In any triangle, two sides taken together in any manner are greater than the remaining one) is called the ''triangle inequality.''

By a metric space, we mean nothing more than a set X having a distance function governed by the conditions above. The elements of a metric space are usually referred to by the generic name "points," with the number $d(x,y)$ being the distance from the point x to the point y; Fréchet used the French word *ecart*, meaning "difference," for $d(x,y)$.

An obvious example of a metric space is Euclidean n-space, whose points are ordered n-tuples of real numbers. In the case $n = 2$, this gives the coordinate plane, and in the case $n = 3$ we get the usual three-dimensional space. For arbitrary $x = (x_1,x_2, \ldots , x_n)$ and $y = (y_1,y_2, \ldots , y_n)$, define $d(x,y)$ by

$$d(x,y) = \left(\sum_{i=1}^{n} (x_i - y_i)^2 \right)^{1/2},$$

the standard Euclidean distance between points. The set of all real-valued continuous functions on the closed interval [0,1] make for a more interesting example of a metric space. For two such functions f and g, we let

$$d(f,g) = \max\{ |f(x) - g(x)| \; |x \in [0,1]\}.$$

That is, $d(f,g)$ is the largest of the values of $|f(x) - g(x)|$ as x varies within the interval [0,1], the "largest separation" of the functions.

The existence of an idea of distance lets us formalize what we mean by saying that x is a limit point of a subset A of a metric space X. The essential idea is that the points of A different from x can be "arbitrarily close to" x: x is a limit point of A provided that for any $\varepsilon > 0$ there is some point y in A, different from x, such that $d(x,y) < \varepsilon$. Closed sets can be defined exactly as in the set theory of the real line; that is, if a subset $A \subseteq X$ contains each of its limit points, then we declare that it is closed.

Using this generalized theory of limits, Fréchet carried over to metric spaces the familiar concepts of classical analysis that depended mainly on distance. For example, convergence of sequences can be defined in this way: the sequence x_1, x_2, x_3, . . . of points on X converges to x if the sequence of real numbers $d(x_1,x)$, $d(x_2,x)$, $d(x_3,x)$, . . . converges to 0; that is, for each positive number ε there is a positive integer n_0 such that $d(x_n,x) < \varepsilon$ for all $n \geq n_0$. Another central theme of analysis, that of a continuous function, makes sense in a metric space setting. Specifically, if X and X' are both metric spaces with respective distance functions d and d', then a function $f:X \rightarrow X'$ is continuous at a point $x \varepsilon X$ provided that $d'(f(x_n), f(x)) \rightarrow 0$ whenever $d(x_n,x) \rightarrow 0$. Though the definition may sound a bit intricate, it is a straightforward adaptation of the familiar definition of sequential continuity that is found in standard calculus texts.

Fréchet's thesis laid the groundwork, but it was Hausdorff's *Grundzüge der Mengenlehre* (*Foundations of Set Theory*) that marks the emergence of set-theoretical topology as a cohesive mathematical discipline. Until the late 1920s, the *Grundzüge* was the most convenient single source from which the succeeding generation of young mathematicians could learn the elements of set theory and point-set topology. Eminently readable, the text exerted a greater influence on the development of these subjects during their formative years than any other work. What is perhaps remarkable is that Hausdorff was not primarily a topologist, nor had he published anything at all on the theory of topology or metric spaces prior to the appearance of the *Grundzüge*.

The son of a wealthy merchant, Felix Hausdorff (1868–1942) earned his doctorate in astronomy from Leipzig University in 1891. From 1902, when he was appointed an associated professor at Leipzig, his attention seems to have focused on Cantor's theory of sets. Hausdorff's lectures on the subject in the summer of 1902 were most likely the first course in set theory anywhere in Germany; oddly enough, in his 44 years at Halle, Cantor himself never lectured on set theory. Hausdorff opted to leave Leipzig in 1910 for an associate professorship at Bonn, where he wrote his classic textbook; the principal features of his theory of topological spaces based on neighborhoods were presented in the summer semester of 1912. He subsequently (1913) accepted a professorship at Greifswald, but later (1921) returned to Bonn until forcibly retired by the Nazis in 1935.

Hausdorff's approach to point-set topology was to let the notion of "neighborhood of a point" play the fundamental role. Neighborhoods as sets of some kind already appeared in Fréchet's work, where distances were used to define them: In a metric space X, a spherical neighborhood $S_\varepsilon(x)$ of the point x is the set of those points y in X satisfying $d(x,y) < \varepsilon$, the number $\varepsilon > 0$ being the radius of the neighborhood. Hausdorff wanted to retain the concept of Fréchet's neighborhood but rid himself of any dependence upon distance. In the seventh chapter of the *Grundzüge*, "Point Sets in General Spaces," he defines what he calls a topological space. It is a set X of points x, to each of which there corresponds a family of subsets U_x, called the neighborhoods of x, which satisfy the conditions:

1. To each point x there corresponds at least one neighborhood U_x, and U_x contains x.

2. If U_x and V_x are neighborhoods of the same point x, then there exists a neighborhood W_x of x such that $W_x \subseteq U_x \cap V_x$.

3. If y is a point in U_x, then there exists a neighborhood U_y of y such that $U_y \subseteq U_x$.

4. For distinct points x and y, there exist two disjoint neighborhoods U_x and U_y.

Because spherical neighborhoods in a metric space satisfy these "neighborhood axioms," Hausdorff's topological space (or a Hausdorff space, as it is generally called today) is the more general concept.

Hausdorff developed the fundamental topological concepts from his theory of neighborhoods. The idea of a limit point of a set carries over to the setting of a Hausdorff space X as follows: x is a limit point of a subset $A \subseteq X$ provided that every neighborhood U_x of x contains a point of A different from x. Convergent sequences can be similarly defined in terms of neighborhoods, by saying that the sequence $x_1,x_2,x_3,$ . . . converges to x if for each neighborhood U_x of x there is an integer n_0 such that $x_n \in U_x$ for all $n \geq n_0$. This is just a direct extension of what occurred with metric spaces, a generalization of a generalization.

The *Grundzüge* was the source of the vigorous growth in point-set topology in the 1920s and 1930s. Many mathematicians added new ideas and results to the field. In the period right after the war a pair of young Russians, Paul Alexandroff (1896–1983) and Paul Urysohn (1898–1924), introduced the definition of compactness in the presently accepted sense using open coverings. They also proved that a metric space is compact if and only if each infinite subset possesses a limit point, the so-called Bolzano-Weierstrass property. (Urysohn's premature death at the age of 26 occurred when, swimming with Alexandroff on the coast of Brittany, he was dashed against the rocks.) The two also produced the first major work on what is known as the "metrization problem." Although every metric space must be a topological space, there do exist topological spaces whose neighborhoods cannot be specified by any distance function. The metrization problem was to find topological conditions under which a topological space can be considered a metric space; in other words, distance definitions under which the limit points for all subsets in the resulting metric space will be the same as those in the topology already associated with the space. The search for necessary and sufficient conditions for the metrizability of a given topological space was not satisfactorily concluded until the early 1950s.

The theory of abstract spaces is a typical example of the internationalism of mathematics. The originators of the subject were mainly French and German, but its most active workers in the 1920s and 1930s were Russian and Polish. After independence was restored to Poland in 1918, many distinguished Polish mathematicians returned to the country from emigration or from exile. They decided to rejuvenate the Polish mathematical tradition by concentrating research on one or two branches of the subject at first. One specialization chosen for this daring and novel approach was the budding theory of topological spaces. This led to what would soon be called the "Polish School" of mathematics: a small group of scholars with common interests, working on similar problems in close contact with each other. In the first rank of the Polish School were such luminaries as Casimir Kuratowski (1896–1980), Waclaw Sierpinski (1882–1969), and Stefan Banach (1892–1945). Their first success came in 1920 with the publication of *Fundamenta Mathematicae,* a journal devoted not to mathematics as a whole but to set theory and its related questions. The initial issue, consisting of 24 articles, was designed "to introduce all the Polish mathematicians who are interested in the theory of sets." *Fundamenta Mathematicae* was immediately developed into a unique periodical that attracted international recognition and co-workers from abroad. The appearance of Banach's doctoral thesis in the journal in 1922 is sometimes said to have marked the birth of functional analysis. *Fundamenta Mathematicae* was joined by an equally famous periodical, *Studia Mathematica* (commencing in 1929), which was primarily concerned with problems in functional analysis.

A second world war intervened in 1939, virtually halting mathematical activity in Europe. The war was a particular tragedy for mathematical progress in Poland; many of the finest Polish mathematicians were murdered or died in the concentration camps. When *Fundamenta Mathematicae* resumed publication (1945), the editors dedicated the volume to their colleagues, contributors to the journal, who had perished in the war.

Hausdorff was also a casualty of those perilous years. With the rise of Nazism in Germany, the liberal professions such as law, medicine, and teaching were rigidly controlled and regimented. Increasingly repressive legislation led to the dismissal of university professors deemed to be political or racial enemies of the state. Ernst Zermelo, for instance, had his (honorary) professorship at the University of Freiburg rescinded for refusing to give the Hitler salute. The climax came with the sweeping Nuremburg Laws of 1935, which deprived all Jews of citizenship. As a consequence of this "purification of education," Hausdorff was forced to leave his position at Bonn. Although he remained active mathematically for several more years, his research could only be published outside of Germany—most notably in *Fundamenta Mathematicae.* In January 1942, when internment in a concentration camp became imminent, Hausdorff, his wife, and her sister committed suicide together.

13.2   Problems

1. Indicate why each of the following functions fails to be a distance function for R:
   - (a) $d(x,y) = |x + y|$
   - (b) $d(x,y) = x^2 + y^2$
   - (c) $d(x,y) = |x^2 - y^2|$
   - (d) $d(x,y) = ||x| - |y||$

2. Suppose that X is any nonempty set and for $x,y \in X$ define $d(x,y)$ by

$$d(x,y) = \begin{cases} 0 & \text{if } x = y \\ 1 & \text{if } x \neq y. \end{cases}$$

   Prove that (X,d) is a metric space, called the discrete space.

3. Let $X = R \times R$, the usual Euclidean plane. For points $x = (x_1,x_2)$, $y = (y_1,y_2)$, verify that both functions below are distance functions for X:
   - (a) $d(x,y) = |x_1 - y_1| + |x_2 - y_2|$;
   - (b) $d(x,y) = \max\{|x_1 - y_1|, |x_2 - y_2|\}$, that is, the larger of $|x_1 - y_1|$ and $|x_2 - y_2|.$

4. Let X be the set of all continuous functions $f: [a,b] \to R$. For two functions $f,g \in X$, define $d(f,g)$ by

$$d(f,g) = \int_a^b |f(x) - g(x)| \, dx.$$

   Show that (X,d) is a metric space. [*Hint:* Recall that if $h(x) \geq 0$ on $[a,b]$, then $\int_a^b h(x)dx \geq 0$; and if $\int_a^b h(x)dx = 0$ for $h(x) \geq 0$, then $h(x) = 0$ for all $x \in [a,b]$.]

5. Let $d: X \times X \to R$ be a function that satisfies the following: For all x,y,z in X,

$d(x,y) = 0$ if and only if $x = y$,
$d(x,y) \leq d(z,x) + d(z,y).$

   Show that $d(x,y) \geq 0$ and $d(x,y) = d(y,x)$ for all $x,y \in X$. Hence, a function satisfying the two given properties is a distance function for X.

6. For the metric spaces in Problems 2 and 3, sketch the spherical neighborhoods $S_1(x)$.

7. In a metric space (X,d), prove that the limit of a convergent sequence $x_1,x_2,x_3, \ldots$ is unique. [*Hint:* Suppose that $x,y \in X$ are both limits of the sequence and show that $d(x,y) = 0$.]

8. For R, with the standard Euclidean distance function $d(x,y) = |x - y|$, determine which of the following subsets are closed sets:
   - (a) $(0,1] \cup \{2\}$
   - (b) $[0,\infty)$
   - (c) $\{1,1/2,1/3,1/4, \ldots \}$
   - (d) Q, the set of rational numbers

9. Consider the metric space $(R \times R, d)$, where the distance function is given by
$$d(x,y) = \sqrt{(x_1 - y_1)^2 + (x_2 - y_2)^2}$$
   for $x = (x_1, x_2)$, $y = (y_1,y_2)$. Which of the subsets of $R \times R$ are closed?
   - (a) $\{(x,y) \mid x = 1\}$
   - (b) $\{(x,y) \mid x < 0\}$
   - (c) $\{(x,y) \mid x^2 + y^2 \leq 1\}$

10. Let x and y be points of a metric space (X,d), with $x \neq y$. If $\varepsilon = d(x,y)/2$, establish that the spherical neighborhoods $S_\varepsilon(x)$ and $S_\varepsilon(y)$ are disjoint; that is, Condition 4 of Hausdorff's definition of a topological space is satisfied.

11. For a Hausdorff topological space, express the notion of a closed set in terms of neighborhoods.

13.3   Toward the Computer Age

No discussion of twentieth-century mathematics would be complete without mentioning Emmy Noether, considered the greatest female mathematician up to her time. Amalie Emmy Noether (1882–1935) was born in the small South German town of Erlangen, where her father, Max Noether, was a professor at the university. Much like the Bernoullis in Switzerland, the Noethers provide a striking example of a mathematically talented family. Max Noether (1844–1921) was a distinguished mathematician who played a considerable part in the development of the theory of algebraic functions, and Emmy Noether's younger brother Fritz later

Amalie Emmy Noether
(1882–1935)

(Bryn Mawr College Archives.)

became a professor of applied mathematics at Breslau. However, nothing in the woman's early years seemed to foreshadow the unmistakable mathematical genius that she would later show. Somewhat reminiscent of Gauss, Emmy Noether seemed to favor the study of languages at first; indeed, after graduating from secondary school she passed the tests that would qualify her to teach French and English. From 1900 to 1902, she studied mathematics and languages at the University of Erlangen, one of two women among nearly a thousand students enrolled. Conditions had changed little during the thirty years since Sonya Kovalevsky went to Heidelberg; female students, unable to enroll in the usual sense, could merely audit lectures on an unofficial basis and then only with the permission of the professor giving the course—a permission frequently denied. The one noteworthy difference was that a woman, having passed through the required courses or not, was allowed to take a final university examination leading to a degree; and Emmy Noether did so in the summer of 1903.

Having decided to specialize in mathematics, Emmy Noether attended classes at Göttingen during the winter of 1903. Mathematics at the University of Berlin had reached its peak during the "heroic period" 1855–1891, when the immense talents of Kummer, Weierstrass, and Kronecker provided leadership. Following Kummer's retirement and Kronecker's sudden death in 1891, with less distinguished men filling the principal positions, Berlin relinquished its primacy in mathematics; and Göttingen quickly regained preeminence in Germany. The great tradition of the university was currently being carried on by a quartet of full professors: Felix Klein, David Hilbert, Hermann Minkowski, and Carl Runge. The legendary Klein "ruled Göttingen like a god," but as he began to devote more time to administrative matters, Hilbert took over the role of leading mathematician. Had Emmy Noether remained at Göttingen, she no doubt would have been attracted—as she later was—to Hilbert's axiomatic approach to mathematics. After only one term, however, she returned to Erlangen, where educational opportunities had improved to the point that women could now be registered and tested in the manner formerly reserved for men.

By 1907, Emmy Noether had completed her doctoral thesis, *On Complete Systems of Invariants for Ternary Biquadratic Forms,* under the tutelage of one of the most prominent mathematicians of the day and an old family friend, Paul Gordan. The dissertation itself, which was not an epoch-making enterprise, ended with a list of 331 forms written in symbolic notation; she was later to dismiss it as ''a jungle of formulas.'' As an extreme example of formal computation, the dissertation was entirely in line with the spirit of the earlier work of Gordan, whom his admiring colleagues called the King of the Invariants. The theory of algebraic invariants was one of the branches of mathematics much in vogue in the early 1900s, and when Gordan was once asked about its value, he said, ''Oh, it is very useful indeed; one can write many theses about it.''

Emmy Noether spent her next few years in Erlangen, publishing half a dozen papers and occasionally substituting for her father at the university when he was ill. During this time, Hilbert was working on the mathematical aspects of a general theory of relativity. Because he ran into problems that required a knowledge of algebraic invariants, he invited Emmy Noether to come to Göttingen in 1916 and assist him. Although Göttingen had been the first university in Germany to grant a doctoral degree to a woman—to Sonya Kovalevsky in absentia and to Grace Chisholm Young through the regular examination process—it was still reluctant to offer a teaching position to a woman, no matter how great her ability and learning. Resistance was particularly high among the classicists and historians of the philosophical faculty, who had to vote on Noether's ''habilitation,'' which carried with it the license to deliver lectures as a privatdozent. In a well-known rejoinder, Hilbert supported her application by declaring during a university senate meeting: ''I do not see that the sex of the candidate is an argument against her admission as a privatdozent; after all, we are a university and not a bathhouse.'' When the appointment failed to win approval, Hilbert bridged the matter by letting her deliver lectures in courses that were announced under his name. She continued in that insecure status until 1919, when she was at last obtained the desired position of privatdozent. Three years later she was appointed *nichtbeamteter ausserordentlicher Professor* (unofficial professor-extraordinary), a merely honorary title that carried neither obligation nor remuneration. Subsequently, Emmy Noether was entrusted with a lectureship in algebra, which carried with it a very modest salary, the first and only salary she was ever to be paid at Göttingen.

Not long after Germany's defeat in the Great War, foreign students once again thronged to Göttingen because of the reputation of its great scholars. Although Emmy Noether never reached the academic standing due her in her own country, she nonetheless became the center of the most fertile group of young algebraists in Europe. According to Norbert Wiener (1894–1964), for many years a professor of mathematics at Massachusetts Institute of Technology, ''her many students flocked around her like a clutch of ducklings about a kind of motherly hen.'' She was particularly popular with the Russian visitors; when they began going around the university in their shirtsleeves, some of the more reserved Göttingen professors dubbed the informal style the Noether-guard uniform.

The mathematics that grew out of her papers following 1920 and the lectures she gave at Göttingen to the ''Noether boys'' made Emmy Noether one of the pioneers of modern algebra. Whereas classical algebra was concerned chiefly with the theory of algebraic equations, modern algebra tends to concentrate on the study of the formal properties of sets on which certain abstract operations are defined. Under the influence of Hilbert's axiomatic thinking, Emmy Noether sought a system of axioms for "rings" (we are indebted to Dedekind for the term itself) that would allow her to subsume a number of earlier results under a general theory. These axioms appeared in 1921 in her now famous paper *The Theory of Ideals in Rings.*

Historically, several of the fundamental notions in Emmy Noether's abstract theory of ideals can be traced to the work of Dedekind, Kronecker, and Lasker. The use of ideal numbers in algebraic number theory was initiated by Kummer (1844), who found that he needed unique factorization in order to help to prove certain cases of Fermat's Last Theorem. Kummer's ideal numbers, though ordinary numbers, belonged to a more extensive field than the one in which the factorization was attempted. Dedekind took a different approach; rather than suitably expand the field at hand, he sought to restore the desired unique factorization by introducing certain subsets (in the same field), which he called "ideals" in honor of Kummer's vision of ideal numbers. Dedekind, by substituting relations among sets for relations among numbers, was able to state and prove theorems analogous to the theorems on the factorization of the integers into primes. The principal result was that each nonzero ideal in the ring of algebraic integers in fixed algebraic number field could be uniquely represented as a product of a finite number of prime ideals. This was published in a supplement to a second edition of Dirichlet's *Lectures on Number Theory* (1871), which Dedekind edited. The idea of a decomposition theory for ideals in rings where a unique decomposition into prime factors does not exist seems to have originated with Kronecker.

It is often said that Hilbert's 1890 proof of Gordan's theorem slew invariant theory; and it is a fact that publication in the subject diminished rapidly. But if Hilbert ended the old ways of doing invariant theory, he opened a new chapter in the development of ideal theory. For framed in the language of modern algebra, his proof showed that any ideal in a polynomial ring is finitely generated. A complete theory of ideals for polynomial rings was obtained by Emanuel Lasker (1868–1941), better known to non-mathematicians as world chess champion for many years. Lasker, who took his doctoral degree under Hilbert's guidance in 1905, established that every polynomial ideal is a finite intersection of primary ideals. Emmy Noether, in her 1921 paper, *The Theory of Ideals in Rings,* generalized Lasker's primary decomposition theorem to arbitrary commutative rings satisfying an ascending chain condition for ideals—that is, to rings in which any strictly ascending chain of ideals in finite. The ascending-chain condition is a weak restriction; it holds in all polynomial domains over any field and in many other cases. In recognition of Noether's inauguration of the use of chain conditions in algebra, rings in which the ascending-chain condition for ideals hold are today called Noetherian. In a second important article, *Abstract Construction of Ideal Theory in the Domain of Algebraic Number Fields* (1927), Noether did for abstract rings what Dedekind had done for rings of algebraic numbers; namely, she formulated five axioms that ensure the possibility of factoring every ideal into a finite product of prime ideals (rings satisfying these axioms are known as Dedekind rings). This pioneer work of Emmy Noether is a cornerstone of the modern algebra course now presented to every mathematics graduate student.

Being relatively unknown both inside and outside of Germany, Emmy Noether required someone capable of popularizing the abstract theory of ideals that she had developed. B. L. van der Waerden, who came to Göttingen in the fall of 1924, eventually served in this way. Van der Waerden spent a year studying with Noether, before returning to the University of Amsterdam to complete his doctorate. He was, at 22 years of age, already regarded as one of the most gifted mathematical talents in Europe. Quickly mastering Noether's ideas, he later gave them brilliant exposition in his two-volume *Moderne Algebra* (1930); reprinted in numerous editions and translated into many different languages, it became the standard work in the field. A large part of what is contained in the second volume of *Moderne Algebra* must be regarded as Noether's property.

While Emmy Noether and her school were making the abstract side of algebra the fashion, Göttingen was clouded over by the threat of coming political events. Then, during the spring of 1933, the storm of the Nazi revolution, that modern Black Plague, swept over Germany. On January 30, the aged and confused President von Hindenburg resolved a parliamentary impasse by appointing Adolf Hitler to the post of chancellor. Although the decision was applauded by many, Hindenburg's former comrade-in-arms, Ludendorff, saw the future more realistically. Two days later he wrote to the president, "Because of what you have done, coming generations will curse you in your grave." In the March elections Hitler won 44 percent of the popular vote, the most he ever received under free conditions. Although the election did not bring him a majority of seats in the Reichstag, he achieved a majority by banning the Communist deputies and arresting a number of Socialists. Hitler asked the Reichstag for sweeping powers that would allow his government to dispense with constitutional procedures and limitations as it carried through "the political and moral disinfection of public life." With the passage of the Enabling Act on March 23, Germany's fate was sealed for the next twelve years.

The total elimination of Jewish influence in Germany had been a Nazi obsession from the outset. Premonitions of the horrors to come were found in the law for the restoration of the professional civil service (April 7) and its supplementary decrees, which deprived Jews of their positions in the state bureaucracy, the judiciary, the professions, and the universities. The law against the overcrowding of German schools and institutions of higher learning (April 25) deprived their children of the right of higher education.

Among the many victims of these invidious measures was Emmy Noether. Summarily placing her on leave until further notice, the new rulers of Germany deprived her of even the modest position she had in Göttingen. Despite the efforts of Hilbert to have her reinstated, Emmy Noether, as well as other Jewish professors, was in a hopeless situation. Forced to emigrate from her native land, she accepted a visiting professorship at Bryn Mawr College, close to Philadelphia, beginning in the fall of 1933. This convenient location, close to the Institute for Advanced Study, allowed her to give weekly lectures at the newly founded Institute. These activities were cut short, however, by her sudden death in 1935 from complications following an operation that seemed to have been completely successful.

Like Sonya Kovalevsky, the greatest woman mathematician before the twentieth century, Emmy Noether died at the height of her career. Beyond that the two female scholars had little in common. Whereas Kovalevsky was able to enthrall the middle-aged Weierstrass as much with her beauty as with the depth of her mind, "no one would contend,'' wrote Hermann Weyl of Noether, ''that the Graces stood by her cradle.'' Heavy of appearance and loud of voice, she ''looked like an energetic and nearsighted washerwoman.'' Kovalevsky was as fully gifted in her literary talents as in mathematics; she wrote poetry, a novel, popular articles on literary and scientific themes, and an autobiography, and even shared in writing a play. But Noether's only true and lasting love was mathematics. A simpler, less tormented personality, she was no doubt the happier of the two women. She had been described as ''warm like a loaf of bread. . . . There irradiated from her a broad, comforting, vital warmth.'' In delivering her eulogy, her old friend Hermann Weyl best summed it up:

> Two traits determined above all her nature: First, the native productive power of her mathematical genius. She was not clay, pressed by the artistic hands of God into a harmonious form, but rather a chunk of human primary rock into which he had blown his creative breath of life. Second, her heart knew no malice; she did not believe in evil—indeed it never entered her mind that it could play a role among men. This was never more forcefully apparent to me than in the last stormy summer, that of 1933, which we spent together in Göttingen. The memory of her work in science and of her personality among her fellows will not soon pass away. She was a great mathematician, the greatest, I firmly believe, that her sex has ever produced, and a great woman.

Developments in the German universities following the spring of 1933 are well known. Mass dismissal of ''racially undesirable'' professors took place, accompanied by political appointments and promotions for those who conformed to the ideas of the Nazi regime. Books were burned and boycott lists drawn up, and manuscripts had to be submitted for censorship. Academic self-government was lost. The universities never recovered from the expulsion of the Jewish professors and the voluntary resignations of those scholars who realized that serious study would be impossible under the totalitarian government. Within the first year of Hitler's regime, faculties showed an average numerical decline of 16 percent, with 45 percent of the established positions changing hands over the next five years. The most promising of the displaced scholars were to give other lands the benefit of their intellectual energy and imagination, while the chairs that they might have filled with distinction in their own country fell to lesser talents. Robbed of all independence, respected academic institutions were changed into ''brown universities.''

Hilbert was left practically alone in an ''empty'' Göttingen—the honorable mathematics tradition first kindled by Gauss, Dirichlet, and Riemann now broken. Among the scholars of whom the university had once been proud, Emmy Noether, Richard Courant, Hermann Weyl, Otto Neugebauer, Felix Bernstein, Hans Lewy, and Paul Bernays had all taken refuge outside of Germany. In the phrase of Weyl, ''Göttingen scattered into the four winds.'' When Hilbert was asked by the Nazi minister of education how mathematics was progressing at Göttingen now that it was freed of Jewish influence, he could only reply, ''Mathematics at Göttingen? There is really none any more.'' The era of mathematics on which Hilbert had impressed the seal of his spirit had drawn to a close.

Many an American university reaped the benefit of Hitler's insane racial policies by adding one or more German mathematicians to its faculty. During the first wave of emigration that began in 1933, Princeton chose Salomon Bochner (Munich); Yale, Max Zorn (Halle); Pennsylvania, Hans Rademacher (Breslau); New York University,

Richard Courant (Göttingen); University of Kentucky, Richard Brauer (Königsberg); and the list could be extended. In Princeton, the presence of Hermann Weyl and Albert Einstein made the Institute for Advanced Study something of a reception center for refugee mathematicians and physicists. (When the Institute for Advanced Study was founded in 1933, the original six professors in its School of Mathematics were J. W. Alexander, A. Einstein, M. Morse, O. Veblen, J. von Neumann, and H. Weyl; Kurt Gödel accepted an offer of permanent membership in 1938 after the German annexation of Austria.) As Nazism continued to spread over Europe, more and more of the mathematicians who had been driven from their homelands made their way to the United States. Emil Artin, Paul Erdos, Richard von Mises, Georg Polya, Stanislaw Ulam, Andre Weil, Antoni Zygmund—names that are familiar to American mathematics—all joined the exodus to more friendly surroundings. The enrichment of American mathematics by this massive injection of European talent helped raise it to new heights.

Another significant influx of creative people took place a half-century after the flight from the Nazis. The dramatic political events that led to the breakup of the Soviet Union produced a fresh exodus of highly qualified mathematicians. Prompted by disastrous economic conditions and the fear that their nation might return to a new form of totalitarianism, practically all the leading mathematicians under the age of 40 emigrated. In the late 1980s a gifted young Russian's situation was desperate: during increasingly hard times, a university professor's monthly salary was not enough to nourish a person for a week in Moscow. Many of these displaced scholars found refuge in American universities. The "great migration" of European mathematicians to the United States in the late 1930s and early 1940s was an unqualified success from the American point of view, transforming the country into the center of world mathematics. It is too early to report, but not too early to predict, an equal enrichment of American mathematics by this second wave of newcomers.

In the years between these two migrations the branches of mathematical investigation were enormously widened, and its methods profoundly deepened. Perhaps the most far-reaching development has been the effect of the "computer revolution," which can be regarded as a continuation of the Scientific-Industrial Revolution, on the subject. Rapid advances in computer technology have led to an intriguing interplay between mathematician and machine. The computer has been an invaluable research tool in furnishing counterexamples to outstanding conjectures or in verifying conjectures up to specific numerical bounds. For example: Goldbach's conjecture has been confirmed for the first $2 \cdot 10^{10}$ even integers; Fermat's Last Theorem is known to be true for exponents up to 1,000,000; the initial billion and a half zeros of the zeta function have been calculated; and the expansion of π has been carried out to just more than 1 billion decimal places. In such instances of large-scale calculation, the computer serves to generate new data. As for the theorems, its use in assisting numerically to prove new results is as yet rather rare. There have been some remarkable successes with previously intractable problems; but perhaps the computer's most impressive contribution to mathematics was the verification in 1976 of the famous Four-Color Conjecture.

For more than a hundred years, the Four-Color Conjecture was one of the most popular and challenging problems in mathematics. In nontechnical terms the conjecture is usually stated as follows: any conceivable map drawn on a plane or on the surface of a sphere can be colored, using only four colors, in such a way that adjacent countries have different colors. Adjacent countries are those that border one another along a line, rather than having just a finite set of points as a common boundary. Moreover, the territory of each country must consist of a single connected region. In the case of a map of the United States, for example, Arizona and Colorado may be colored the same because they meet only at a point, but the two physically separated pieces of Michigan must be colored differently.

If practical mapmakers were aware of the Four-Color Conjecture, they certainly kept the secret well. The first known document indicating the problem is a letter dated October 23, 1852, from Augustus De Morgan to his friend William Rowan Hamilton, the inventor of quaternions. Earlier in the month, Francis Guthrie (1831–1899) had noticed that four colors suffice to distinguish the various counties on a map of England. Francis asked his younger brother Frederick, still a student of De Morgan at University College London, if it could be shown mathematically that coloring any map would require only four colors. Unable to answer his brother's question, Frederick brought the problem to the attention of De Morgan, who could not find any method for determining its truth or falsity. For his part, Hamilton failed to recognize the conjecture's importance, replying merely that he did not wish to work on this ''quaternion of colors'' soon. In fact, he never tried it at all.

The coloring problem was entirely neglected for a quarter of a century. Other English mathematicians learned of it in 1878, when Arthur Cayley presented the conjecture at a meeting of the London Mathematical Society. The first printed reference to the problem is a four-line report of Cayley's remarks, ''On the Colouring of Maps,'' which appeared in the Society's *Proceedings*. Interest was immediate. Arthur Kemp (1849–1922), a barrister and member of the Society (and the author of a short, celebrated book with the provocative title *How to Draw a Straight Line*) published a paper in 1879 in the newly founded *American Journal of Mathematics*. In it he claimed to prove that four colors suffice for coloring any map on a sphere. For more than a decade Kemp's extremely clever argument was accepted; but in 1890, Percy Heawood pointed out a fatal flaw in the reasoning. Heawood's modest paper of six pages, ''Map-Colour Theorems,'' was not entirely destructive: it included a simplification of Kemp's proof showing that each map drawn on the plane or sphere can be colored by at most five colors, the Five-Color Theorem.

Heawood's analysis of Kemp's purported proof showed that the problem is more subtle than had first been believed. During the subsequent years it attracted the attention of dedicated amateurs and distinguished mathematicians, inspiring progress in the development of mathematical methods yet always denying the final step of proof. One significant advance occurred in 1922 when it was shown that an arbitrarily drawn map of 25 or fewer countries is four-colorable; thus, any counterexample to the conjecture would have to be a map of at least 26 countries. This lower bound was gradually raised, finally reaching 96 countries before all such results were rendered superfluous. This is because in the summer of 1976, the Four-Color Conjecture was finally confirmed by Kenneth Appel and Wolfgang Haken of the University of Illinois. The two colleagues presented their proof at a meeting of mathematicians in Toronto—to be rewarded with only polite applause for the solution to such a longstanding problem. Shortly later, a full account was published. The question seemed to be whether they had actually provided a "rigorous demonstration" of the Four-Color Conjecture. Their argument contains several hundred pages of complex detail, requiring more than 1200 hours of time on a large computer. The coloring of certain complex configurations is deduced from the coloring of others involving fewer regions, thereby "reducing" the type of map that needs to be considered. Appel and Haken found 1936 reducible cases, each of which involved a computer search of up to 500,000 logical options to confirm its reducibility.

Even in its solution the Four-Color Conjecture remains an enigma, for the fundamental novelty in the Appel-Haken proof is the unprecedented use of a computer to establish a mathematical theorem. In the years since the conjecture was proven, other important results have been obtained with computer aid. But as the first instance of such a proof, the coloring-proof provoked considerable controversy within the mathematical community. Because it is currently impossible to verify the correctness of the argument without a computer-facilitated analysis, there is a tendency on the part of many mathematicians to mistrust the whole thing. It cannot be ruled out that a short and convincing proof of the conjecture may yet be found, but it is just as conceivable that the only valid proofs will involve massive computations requiring computer assistance. If this is the case, we must acknowledge that a new and interesting type of theorem has emerged, one having no verification by traditionally accepted methods. Admitting these theorems will mean that the apparently secure notion of a mathematical proof is open to revision.

Bibliography

Appel, Kenneth, and Haken, Wolfgang. "Every Planar Map Is Four Colorable." *Journal of Recreational Mathematics* 9(1976–1977): 161–169.

———. "The Solution of the Four-Color Map Problem." *Scientific American* 237(Oct. 1977): 108–121.

Berndt, Bruce. "Ramanujan—100 Years Old (Fashioned) or 100 Years New (Fangled)?" *Mathematical Intelligencer* 10, no. 3(1988): 24–29.

———. "Srinivasa Ramanujan." *The American Scholar* 58(1989): 234–244.

Bollabás, Béla, ed. *Littlewood's Miscellany.* Cambridge: Cambridge University Press, 1986.

Borwein, Jonathan, and Borwein, Peter. "Ramanujan and Pi." *Scientific American* 258(Feb. 1988): 112–117.

Brewer, James W., and Smith, Martha. *Emmy Noether, A Tribute to Her Life and Work.* New York: Marcel Decker, 1981.

Burkill, J. C. "John Edensor Littlewood." *Bulletin of the London Mathematical Society* 11(1979): 59–70.

Coxeter, H. S. M. "The Four-Color Map Problem." *Mathematics Teacher* 52(1959): 283–289.

———. "The Mathematics of Map Coloring." *Journal of Recreational Mathematics* 2(1969): 3–12.

Dick, Auguste. *Emmy Noether 1882–1935.* Basel, Switzerland: Birkhauser-Verlag, 1970.

Dresden, Arnold. "The Migration of Mathematicians." *American Mathematical Monthly* 49(1942): 415–429.

Edwards, Harold. *Riemann's Zeta Function.* New York: Academic Press, 1974.

Hardy, Godfrey H. "The Indian Mathematician Ramanujan." *American Mathematical Monthly* 44(1937): 137–155.

———. *A Mathematician's Apology.* Cambridge: Cambridge University Press, 1940. Reprint, with a forward by C. P. Snow, 1967.

Kanigel, Robert. *The Man Who Knew Infinity: A Life of the Genius Ramanujan.* New York: Charles Scribner's Sons, 1991.

Kimberling, Charles. "Emmy Noether." *American Mathematical Monthly* 79(1972): 136–149.

Kuzawa, Sister Mary Grace. *Modern Mathematics: The Genesis of a School in Poland.* New Haven, Conn.: College and University Press, 1968.

———. "*Fundamenta Mathematicae:* An Examination of Its Founding and Significance." *American Mathematical Monthly* 77(1970): 485–492.

May, Kenneth. "The Origin of the Four-Color Conjecture." *Isis* 56(1965): 346–348.

Neville, E. H. "Andrew Russell Forsythe." *Journal of the London Mathematical Society* 17(1942): 237–256.

Ranganathan, S. *Ramanujan: The Man and the Mathematician.* Bombay: Asia Publishing House, 1967.

Reingold, Nathan. "Refugee Mathematicians in the United States of America 1933–1941: Reception and Reaction." *Annals of Science* 38(1981): 313–338.

Roth, Leonard. "Old Cambridge Days." *American Mathematical Monthly* 78(1971): 223–226.

Rowe, David. "Klein, Hilbert, and the Göttingen Mathematics Tradition." *Osiris Second Series* 5(1989): 186–213.

Saaty, Thomas, and Kainen, Paul. *The Four-Color Problem: Assaults and Conquest.* New York: McGraw Hill, 1977.

Shields, Allen. "Felix Hausdorff: Grundzüge der Mengenlehre." *Mathematical Intelligencer* 11, no. 1(1989): 6–8.

Steen, Lynn. "Solution of the Four Color Problem." *Mathematics Magazine* 49(1976): 219–222.

Stewart, Ian. "The Formula Man." *New Scientist* 1591(Dec. 1987): 24–28.

Taylor, Angus. "A Study of Maurice Fréchet: I. His Early Work on Point Set Theory and the Theory of Functionals." *Archive for History of Exact Sciences* 27(1982): 233–295.

————. "A Study of Maurice Fréchet: II. Mainly His Work on General Topology." *Archive for History of Exact Sciences* 34(1985): 279–380.

Titchmarsh, E. C. "Godfrey Harold Hardy." *Journal of the London Mathematical Society* 25, Part 2(1950): 81–101.

Tymoczko, Thomas. "Computers, Proofs and Mathematics: A Philosophical Investigation of the Four-Color Proof." *Mathematics Magazine* 53(1980): 131–138.

Van der Waerden, B. L. "On the Source of My Book, 'Moderne Algebra'." *Historia Mathematica* 2(1975): 507–511.

————. "The School of Hilbert and Emmy Noether." *Bulletin of the London Mathematical Society* 15(1983): 1–7.

Weyl, Hermann. "Emmy Noether." *Scripta Mathematica* 3(1935): 201–220.

Wiener, Norbert. "Godfrey Harold Hardy, 1877–1947." *Bulletin of the American Mathematical Association* 55(1949): 72–77.

General Bibliography

Abetti, Giorgio. *The History of Astronomy*. Translated by B. Abetti. London: Abelard-Schuman, 1952.

Alic, Margaret. *Hypatia's Heritage: A History of Women in Science from Antiquity Through the Nineteenth Century*. Boston: Beacon Press, 1986.

Archibald, R. C. *Outline of the History of Mathematics*. Herbert Elsworth Slaught Memorial Paper No. 2. Buffalo, N.Y.: Mathematical Association of America, 1949.

Ball, W. W. Rouse. *A Short Account of the History of Mathematics*. 5th ed. New York: Macmillan, 1912. (Dover reprint, 1960).

Bell, Eric T. *The Development of Mathematics*. 2d ed. New York: McGraw-Hill, 1945. (Dover reprint, 1992).

———. *Men of Mathematics*. New York: Simon & Schuster, 1965.

———. *Mathematics: Queen and Servant of Science*. Washington, D.C.: Mathematical Association of America, 1987.

Bernal, J. D. *Science in History*. 4 vols. Cambridge, Mass.: M.I.T. Press, 1971.

Berry, Arthur. *A History of Astronomy from Earliest Times Through the Nineteenth Century*. New York: Dover Publications, 1961.

Bourbaki, Nicolas. *Elements of the History of Mathematics*. New York: Springer-Verlag, 1993.

Boyer, Carl B. *A History of Mathematics*. 2d ed. New York: John Wiley, 1991.

Cajori, Florian. *A History of Elementary Mathematics*. Rev. ed. New York: Macmillan, 1930.

———. *A History of Mathematics*. 5th ed. New York: Chelsea, 1991.

———. *A History of Mathematical Notations*. 2 vols. Chicago: Open Court, 1929. (Dover reprint, 1993).

Calinger, Ronald, ed. *Classics of Mathematics*. Oak Park, Ill.: Moore Publishing Co., 1982.

Coolidge, Julian. *The Mathematics of Great Amateurs*, 2d ed. Oxford: Clarendon Press, 1990.

Dedron, P., and Itard, J. *Mathematics and Mathematicians*. Translated by J. V. Field. 2 vols. London: Transworld Publications, 1973.

Devlin, Keith. *Mathematics: The New Golden Age*. Harmondsworth, England: Penguin Books, 1988.

Dörrie, Heinrich. *100 Great Problems of Elementary Mathematics: Their History and Solution*. Translated by D. Antin. New York: Dover Publications, 1965.

Dreyer, J. L. E. *A History of Astronomy from Thales to Kepler*. New York: Dover Publications, 1953.

Dunham, William. *Journey Through Genius: The Great Theorems of Mathematics*. New York: John Wiley, 1990.

Durbin, John R. *Mathematics: Its Spirit and Evolution*. Boston: Allyn and Bacon, 1973.

Eves, Howard. *In Mathematical Circles*. 2 vols. Boston: Prindle, Weber & Schmidt, 1969.

———. *The Other Side of the Equation: A Selection of Mathematical Stories and Anecdotes*. Boston: Prindle, Weber & Schmidt, 1971.

———. *Great Moments in Mathematics*. 2 vols. Washington, D.C.: Mathematical Association of America, 1980.

———. *An Introduction to the History of Mathematics*. 6th ed. Philadelphia: Saunders, 1990.

Eves, Howard, and Newsom, Carroll. *An Introduction to the Foundations and Fundamental Concepts of Mathematics*. Rev. ed. New York: Holt, Rinehart and Winston, 1965.

Fink, Karl. *A Brief History of Mathematics*. Translated by W. W. Beman and D. E. Smith. Chicago: Open Court, 1900.

Gillispie, C. C., ed. *Dictionary of Scientific Biography*. 18 vols. New York: Charles Scribner, 1970–1990.

Gindikin, Semyon. *Tales of Physicists and Mathematicians*. Translated by Alan Shuchat. Boston: Birkhauser, 1988.

Grattan-Guinness, I., ed. *From the Calculus to Set Theory 1630–1910*. London: Duckworth, 1980.

Hart, Ivor. *Makers of Science. Mathematics, Physics and Astronomy*. Freeport, N.Y.: Books for Libraries Press, 1968.

Hofmann, Joseph E. *A History of Mathematics to Eighteen Hundred*. New York: Philosophical Library, 1957.

———. *Classical Mathematics, A Concise History of the Classical Era in Mathematics*. New York: Philosophical Library, 1959.

Hollingdale, Stuart. *Makers of Mathematics*. Harmondsworth, England: Penguin Books, 1989.

Hooper, Alfred. *Makers of Mathematics*. New York: Random House, 1948.

Howson, Geoffrey. *A History of Mathematics Education in England*. Cambridge: Cambridge University Press, 1982.

Karpinski, Louis. *The History of Arithmetic*. Chicago: Rand McNally, 1925.

Katz, Victor. *A History of Mathematics, An Introduction*. New York: HarperCollins, 1993.

Kemble, Edwin. *Physical Science, Its Structure and Development*. Cambridge, Mass.: M.I.T. Press, 1966.

Klein, Felix. *Development of Mathematics in the 19th Century*. Translated by M. Ackerman. Brookline, Mass.: Math Sci Press, 1979.

Kline, Morris. *Mathematics in Western Culture*. New York: Oxford University Press, 1953.

————. *Mathematics in the Physical World*. New York: Thomas Y. Crowell, 1959.

————. *Mathematics, a Cultural Approach*. Reading, Mass.: Addison-Wesley, 1962.

————. *Mathematical Thought from Ancient to Modern Times*. New York: Oxford University Press, 1972.

————. *Mathematics: The Loss of Certainty*. New York: Oxford University Press, 1979.

————. *Mathematics and the Search for Knowledge*. New York: Oxford University Press, 1985.

Kramer, Edna E. *The Main Stream of Mathematics*. Greenwich, Conn.: Fawcett Publications, 1964.

————. *The Nature and Growth of Modern Mathematics*. New York: Hawthorn Books, 1970.

LeLionnais, F., ed. *Great Currents of Mathematical Thought*. Translated by R. Hall. 2 vols. New York: Dover Publications, 1971.

Lenard, Philipp. *Great Men of Science*. Translated by S. Hatfield. Freeport, N.Y.: Books for Libraries Press, 1970.

Mason, S. F. *Main Currents of Scientific Thought*. New York: Henry Schuman, 1953.

Meschkowski, Herbert. *The Ways of Thought of Great Mathematicians*. San Francisco: Holden-Day, 1964.

Morgan, Bryan. *Men and Discoveries in Mathematics*. London: John Murry, 1972.

Newman, James, ed. *The World of Mathematics*. 4 vols. New York: Simon & Schuster, 1956.

Newsom, Carroll. *Mathematical Discourses: The Heart of Mathematical Science*. Englewood-Cliffs, N.J.: Prentice-Hall, 1964.

Osen, Lynn M. *Women in Mathematics*. Cambridge, Mass.: M.I.T. Press, 1974.

Pannekoch, A. *A History of Astronomy*. New York: Interscience, 1961.

Perl, Teri. *Math Equals: Biographies of Women Mathematicians*. Menlo Park, California: Addison-Wesley, 1978.

Pledge, H. T. *Science Since 1500: A Short History of Mathematics, Physics, Chemistry and Biology*. New York: Harper Brothers, 1959.

Rowe, David, and McCleary, John, eds. *The History of Modern Mathematics*. 2 vols. San Diego, California: Academic Press, 1989.

Sanford, Vera. *A Short History of Mathematics*. Boston: Houghton Mifflin, 1930.

Sarton, George. *A History of Science*. 2 vols. New York: W. W. Norton, 1952.

Scott, Joseph F. *A History of Mathematics from Antiquity to the Beginning of the Nineteenth Century*. 2d ed. London: Taylor and Francis, 1969.

Sedgwick, W. T., and Tyler, H. W. *A Short History of Science*. New York: Macmillan, 1921.

Simmons, George. *Calculus Gems: Brief Lives and Memorable Mathematics*. New York: McGraw-Hill, 1992.

Singer, Charles. *A Short History of Scientific Ideas to 1900*. New York: Oxford University Press, 1959.

Smith, David E. *Mathematics*. Boston: Marshall Jones, 1923.

————. *History of Mathematics*. 2 vols. Boston: Ginn & Company, 1923–1925. (Dover reprint, 1958).

————. *A Source Book in Mathematics*. New York: McGraw-Hill, 1929. (Dover reprint, 2 vols., 1959).

Smith, David, and Ginsburg, Jekuthiel. *A History of Mathematics in America Before 1900*. Carus Mathematical Monograph No. 5. Chicago: Open Court, 1934.

Stillwell, John. *Mathematics and Its History*. New York: Springer-Verlag, 1989.

Struik, Dirk J. *A Concise History of Mathematics*. Rev. ed. New York: Dover Publications, 1967.

————. *A Source Book in Mathematics, 1200–1800*. Cambridge, Mass.: Harvard University Press, 1969.

Tarwater, Dalton, ed. *The Bicentennial Tribute to American Mathematics, 1776–1976*. Washington, D.C.: Mathematical Association of America, 1977.

Taton, Rene, ed. *History of Science*. Translated by A. J. Pomerans. 4 vols. New York: Basic Books, 1963–1966.

Tee, G. J. ''The Pioneering Women Mathematicians.'' *Mathematical Intelligencer* 5 (No. 4, 1983): 27–36.

Temple, George. *100 Years of Mathematics*. New York: Springer-Verlag, 1981.

Tietze, Heinrich. *Famous Problems of Mathematics*. New York: Graylock, 1965.

Turnbull, Herbert W. *The Great Mathematicians*. New York: New York University Press, 1969.

Van der Waerden, B. L. *Geometry and Algebra in Ancient Civilizations*. New York: Springer-Verlag, 1983.

————. *A History of Algebra: From al-Khowarizmi to Emmy Noether*. New York: Springer-Verlag, 1985.

Weil, Andre. *Number Theory: An Approach Through History, from Hammurapi to Legendre*. Boston: Birkhauser, 1984.

Wilder, Raymond L. *Mathematics as a Cultural System*. London: Pergamon, 1981.

Willerding, Margaret. *Mathematical Concepts, an Historical Approach*. Boston: Prindle, Weber & Schmidt, 1967.

Woodruff, L., ed. *The Development of the Sciences*. New Haven, Conn.: Yale University Press, 1941.

Young, Laurence. *Mathematicians and Their Times*. Amsterdam, Holland: North Holland, 1981.

Additional Reading

Barrows, John. *Pi in the Sky: Counting, Thinking and Being.* Oxford: Clarendon Press, 1992.

Bucciarelli, Louis, and Dworsky, Nancy. *Sophie Germain: An Essay in the History of the Theory of Elasticity.* Dordrecht, Holland: D. Reidel, 1980.

Davis, Donald. *The Nature and Power of Mathematics.* Princeton, N.J.: Princeton University Press, 1993.

Davis, Philip, and Hersh, Reuben. *The Mathematical Experience.* Boston: Birkhauser, 1981.

———. *Descartes' Dream: The World According to Mathematics.* San Diego: Harcourt Brace Jovanovich, 1986.

Gilles, Donald, ed. *Revolutions in Mathematics.* New York: Oxford University Press, 1992.

Gowing, Ronald. *Roger Cotes—Natural Philosopher.* Cambridge: Cambridge University Press, 1983.

Guicciardini, Niccolo. *The Development of Newtonian Calculus in Great Britain 1700–1800.* Cambridge: Cambridge University Press, 1989.

Guillen, Michael. *Bridges to Infinity: The Human Side of Mathematics.* Los Angeles: Jeremy P. Tarcher Inc., 1983.

Hoffman, Paul. *Archimedes' Revenge: The Joys and Perils of Mathematics.* New York: Fawcett Crest, 1988.

Hyman, Anthony. *Charles Babbage, Pioneer of the Computer.* Princeton, N.J.: Princeton University Press, 1982.

King, John. *The Art of Mathematics.* New York: Plenum Press, 1992.

Kitcher, Philip. *The Nature of Mathematical Knowledge.* New York: Oxford University Press, 1983.

Maor, Eli. *To Infinity and Beyond: A Cultural History of the Infinite.* Boston: Birkhauser, 1987.

———. *e: The Story of a Number.* Princeton, N.J.: Princeton University Press, 1993.

Peterson, Ivars. *The Mathematical Tourist: Snapshots of Modern Mathematics.* New York: Freeman, 1988.

Richards, Joan. *Mathematical Visions: The Pursuit of Geometry in Victorian England.* San Diego: Academic Press, 1989.

Stewart, Ian. *The Problems of Mathematics.* 2d ed. Oxford: Oxford University Press, 1992.

Tiles, Mary. *The Philosophy of Set Theory: An Introduction to Cantor's Paradise.* Oxford: Basil Blackwell, 1989.

Yaglon, I. M. *Felix Klein and Sophus Lie: The Evolution of the Idea of Symmetry in the 19th Century.* Boston: Birkhauser, 1988.

The Greek Alphabet

The Greek Alphabet								
Letters		*Names*	*Letters*		*Names*	*Letters*		*Names*
A	α	alpha	I	ι	iota	P	ρ	rho
B	β	beta	K	κ	kappa	Σ	σ ς	sigma
Γ	γ	gamma	Λ	λ	lambda	T	τ	tau
Δ	δ	delta	M	μ	mu	Υ	υ	upsilon
E	ε	epsilon	N	ν	nu	Φ	φ	phi
Z	ζ	zeta	Ξ	ξ	xi	X	χ	chi
H	η	eta	O	ο	omicron	Ψ	ψ	psi
Θ	θ	theta	Π	π	pi	Ω	ω	omega

Solutions
to Selected
Problems

Section 1.2, p. 16

1. (a) 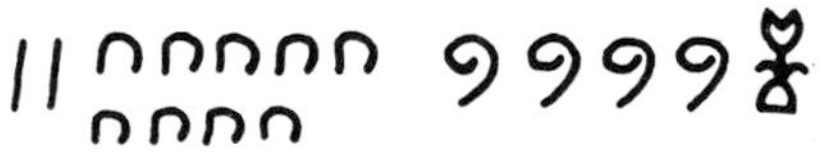 ‖ ∩∩∩∩∩ ∩∩∩∩ 𝟫𝟫𝟫𝟫 ♥

   (c) 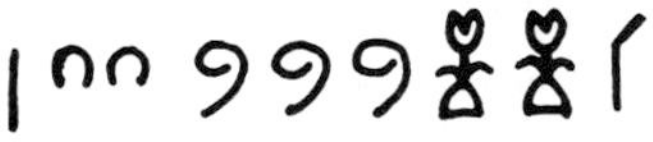 | ∩∩ 𝟫𝟫𝟫 ♥♥ (

   (e) 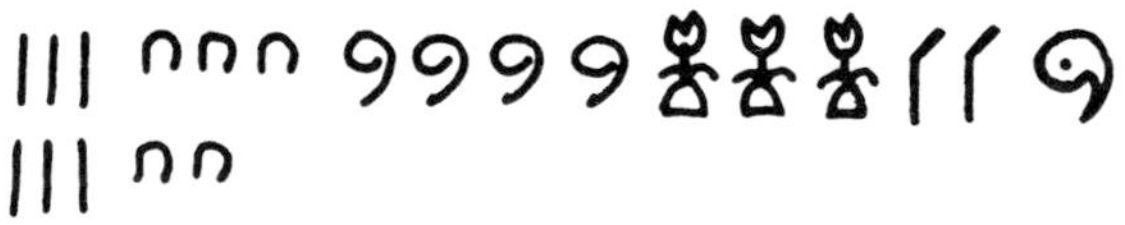 ‖‖ ∩∩∩ ‖‖ ∩∩ 𝟫𝟫𝟫𝟫 ♥♥♥ ((⌐⌐

2. (a) 648.   (b) 140,060.

3. (a)  ‖‖ 𝟫𝟫𝟫 ‖ 𝟫𝟫𝟫 ♥ ♥♥♥ (

   (c) 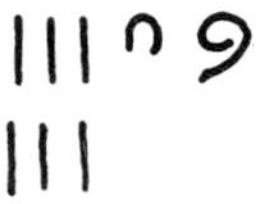 ‖‖ ∩ 𝟫 ‖‖

5. (a)  ⛢ 𝟫 ⟆

   (c)  ⋈ ⅄ 𝟫 θ

   (e) 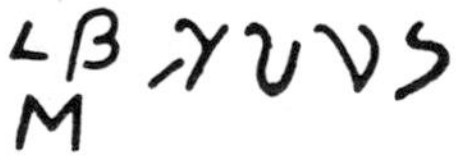 ∟β ⅄ 𝜐 ⋁⟆ M

6. (a) 1234.   (c) 55,555.
7. (a) χλ.   (c) βωξγ.

8. (a)  ΗΗΗ ⌐ ΔΔΔ ⌐Ι

   (c)  Χ ⌐ ΗΗΗΗ ⌐ ΔΔΔΔ ⌐ΙΙΙΙ

   (e)  ⌐ ΜΜΧΧΧΧ ⌐ ΗΗΗ ΙΙ

9. (a) 2756.   (c) 2977.

10. (a)  Χ ΗΗΗ ⌐

    (c) 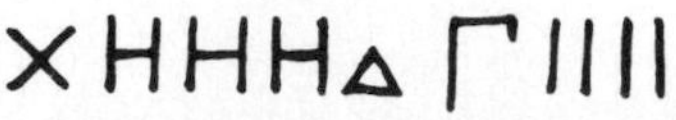 Χ ΗΗΗ Δ ⌐ΙΙΙΙ

11. (a) MCDXCII. (c) MCMXCIX. (e) $\overline{\text{CXX}}$MMMCDLVI.
12. (a) 124. (c) 1748. (e) 19000.
13. (a) CMXIX. (c) LXX. (e) XCI.

Section 1.3, p. 27

1. (d) $1234 = 20,34 =$

   (e) $12,345 = 3,25,45 =$

   (f) $123,456 = 34,17,36 =$

2. (a) Among other possibilities, $\qquad = 886$.

3. $\dfrac{1}{6} = 0;10 \qquad \dfrac{1}{9} = 0;6,40 \qquad \dfrac{1}{5} = 0;12 \qquad \dfrac{1}{40} = 0;1,30 \qquad \dfrac{5}{12} = 0;25$

4. (a) 5025. (b) $12\dfrac{1}{16}$. (c) $\dfrac{193}{960}$. (d) $83\dfrac{3}{4}$.

5. $12,3,45;6$

6. (c) $1066 =$ (d) $57,942 =$

   (e) $123,456 =$

7. (a) 666,666. (b) 7725. (c) 123,321. (d) 9,623,088.

9. (a) (b) (c)

10. (a) $236 =$ (d) $1066 =$

11. (a) 83. (b) 470. (c) 29,005. (d) 5634.

12.

13. (c) 1066 =   (d) 57,942 =   (e) 123,456 =

14. (a) 93,707.   (b) 1,086,220.   (c) 5,832,244.

15. (a)   (b)   (c)

17. (a) 54   (b) 36   (c) 7   (d) 12   (e) 6   (f) 11   (g) 5

Section 2.3, p. 49

2. (a) 23. (b) $2 + \dfrac{1}{4} + \dfrac{1}{8}$. (c) $5 + \dfrac{1}{6} + \dfrac{1}{18}$. (d) $88 + \dfrac{1}{3}$. (e) $7 + \dfrac{1}{2} + \dfrac{1}{8}$.

3. (a) $430 + \dfrac{1}{8}$. (b) $17 + \dfrac{1}{16}$. (c) $3 + \dfrac{1}{2} + \dfrac{1}{4} + \dfrac{1}{8} + \dfrac{1}{16}$.

6. (b) $\dfrac{2}{25} = \dfrac{1}{15} + \dfrac{1}{75}; \dfrac{2}{65} = \dfrac{1}{39} + \dfrac{1}{195}; \dfrac{2}{85} = \dfrac{1}{51} + \dfrac{1}{255}$.

7. (b) $\dfrac{2}{21} = \dfrac{1}{14} + \dfrac{1}{42}; \dfrac{2}{75} = \dfrac{1}{50} + \dfrac{1}{150}; \dfrac{2}{99} = \dfrac{1}{66} + \dfrac{1}{198}$.

10. $2/n = \dfrac{1}{4} \cdot 1/n + \dfrac{7}{4} \cdot 1/n.$

11. $\dfrac{13}{15} = \dfrac{1}{15} + 6\left(\dfrac{2}{15}\right) = \dfrac{1}{2} + \dfrac{1}{5} + \dfrac{1}{6};$

$$\frac{9}{49} = \frac{1}{49} + 4\left(\frac{2}{49}\right) = \frac{1}{7} + \frac{1}{28} + \frac{1}{196};$$

$$\frac{19}{35} = \frac{1}{35} + 9\left(\frac{2}{35}\right) = \frac{1}{5} + \frac{1}{7} + \frac{1}{10} + \frac{1}{14} + \frac{1}{35}.$$

12. (a) Possible answers are: $\frac{3}{7} = \frac{1}{4} + \frac{1}{7} + \frac{1}{28}; \frac{4}{15} = \frac{1}{4} + \frac{1}{60}; \frac{7}{29} = \frac{1}{5} + \frac{1}{29} + \frac{1}{145}.$

(b) $\frac{3}{7} = \frac{1}{3} + \frac{1}{11} + \frac{1}{231}; \frac{4}{15} = \frac{1}{4} + \frac{1}{60}; \frac{7}{29} = \frac{1}{5} + \frac{1}{25} + \frac{1}{725}.$

13. If $m + 1 = nk$, then $n/m = 1/k + 1/km; \frac{2}{5} = \frac{1}{3} + \frac{1}{15}.$

14. $\frac{9}{13} = \frac{2}{13} + 7\left(\frac{1}{13}\right) = \frac{1}{2} + \frac{1}{7} + \frac{1}{26} + \frac{1}{91}.$

15. $\frac{2}{5} = \frac{1}{5} + \left(\frac{1}{6} + \frac{1}{30}\right) = \frac{1}{5} + \left(\frac{1}{6} + \frac{1}{30}\right) + \left(\frac{1}{7} + \frac{1}{42}\right) + \left(\frac{1}{31} + \frac{1}{930}\right).$

16. $\frac{2}{11} = \frac{1}{6} + \frac{1}{66}; \frac{2}{17} = \frac{1}{9} + \frac{1}{153}.$

17. The sums $2 + 6 = 3 + 5 = 1 + 3 + 4 = 8$ yield $\frac{2}{15} = \frac{1}{10} + \frac{1}{30} = \frac{1}{12} + \frac{1}{20} =$

$\frac{1}{15} + \frac{1}{20} + \frac{1}{60}$, whereas $2 + 4 + 6 + 12 = 24$ gives $\frac{2}{43} = \frac{1}{43} + \frac{1}{86} + \frac{1}{129} + \frac{1}{258}.$

18. The sum $15 + (6 + 3) = 24 = 2 \cdot 12$ yields $\frac{2}{15} = \frac{1}{12} + \frac{1}{30} + \frac{1}{60}$, whereas

$43 + (18 + 9 + 2) = 72 = 2 \cdot 36$ gives $\frac{2}{43} = \frac{1}{36} + \frac{1}{86} + \frac{1}{172} + \frac{1}{774}.$

19. $\frac{1}{2} + \frac{1}{10}; \frac{1}{2} + \frac{1}{5}; \frac{1}{2} + \frac{1}{5} + \frac{1}{10}; \frac{1}{2} + \frac{1}{3} + \frac{1}{15}.$

20. $10 + \frac{2}{3} = 10 + \frac{1}{2} + \frac{1}{6}; 12; 17 + \frac{1}{2}.$

22. $1 + \frac{1}{4} + \frac{1}{76}.$

23. $12 + \frac{2}{3} + \frac{1}{42} + \frac{1}{126}.$

24. (a) Solve the equations

$$x + (x + d) + (x + 2d) + (x + 3d) + (x + 4d) = 100$$

$$x + (x + d) = \frac{1}{7}((x + 2d) + (x + 3d) + (x + 4d))$$

to get $x = \frac{10}{6}, d = \frac{55}{6}.$

(b) Because $1 + \left(6 + \frac{1}{2}\right) + 12 + \left(17 + \frac{1}{2}\right) + 23 = 60$, and $60\left(1 + \frac{2}{3}\right) = 100,$

multiply each of $1, 6 + \frac{1}{2}, 12, 17 + \frac{1}{2}, 23$ by $1 + \frac{2}{3}.$

Section 2.4, p. 58

1. (a) 640 cubic cubits. (b) 20 square khets. (c) $\dfrac{1}{2} + \dfrac{1}{4}$.

2. (b) $3\dfrac{1}{8}$.

3. $\pi r^2/(2r)^2 = \dfrac{11}{14}$ implies $\pi = \dfrac{22}{7}$.

4. Aryabhata's rule gives the correct area.
6. The Babylonian formula gives $V = 180$, as compared with the correct value of $V = 56\pi \sim 176$.

7. $V = \dfrac{1188}{7}$.

Section 2.5, p. 69

1. (a) $\dfrac{19}{15} = 19(0;4) = 1;16$.

   $\dfrac{5}{3} = 5(0;20) = 1;40$.

   $\dfrac{10}{9} = 10(0;6,40) = 1;6,40$.

   (b) $\dfrac{10}{9} = \dfrac{10 \cdot 60}{9 \cdot 60} = \dfrac{10 \cdot 20}{3 \cdot 60} = \dfrac{10 \cdot 20 \cdot 20}{60^2} = 1;6,40$.

3. $x = \dfrac{35}{8} = 4;22,30, y = \dfrac{33}{8} = 4;7,30$.

4. $x = 8, y = 2$.
5. $x = 18, y = 6$.
6. (a) $x = 8, y = 2$. (b) $x = 7, y = 3$. (c) $x = 5, y = 3$.
7. $x = 15, y = 12$.
8. $x = 0;30, y = 0;20, z = 6$.
9. $x = 0;30$.
10. $x = 14$.
11. The sides are 30 and 25.

12. $x = \dfrac{2}{3} = 0;40$.

13. All parts have $x = 30, y = 20$ as a solution.

Section 2.6, p. 78

1. 12.   2. 768.
3. $x = 20, y = 12$.
4. $x = 18, y = 60, z = 40$.
5. $b_1 = 10, b_2 = 5, s_1 = s_2 = 20$.   6. $d^2 \sim 1782.7, d^2 \sim 1701.6$.

7. $\sqrt{2} \sim \dfrac{17}{12}, \sqrt{5} \sim \dfrac{9}{4}, \sqrt{17} \sim \dfrac{33}{8}$.

8. $\sqrt{720} \sim \dfrac{51841}{1932} \sim 26.83, \sqrt{63} \sim \dfrac{32257}{4064} \sim 7.93$.

Section 3.2, p. 101

1. $8t_n + 1 = s_{2n+1}$.

3. $9t_n + 1 = t_{3n+1}$.

4. (a) $56 = 55 + 1$.   (c) $185 = 91 + 66 + 28$.

6. $1225 = t_{49} = s_{35}$, $41,616 = t_{288} = s_{204}$.

7. (a) $o_n = n(n + 1) = 2(1 + 2 + \cdots + n)$.

   (c) $o_n + n^2 = n(n + 1) + n^2 = 2n^2 + n = \dfrac{2n(2n + 1)}{2} = t_{2n}$.

   (e) $n^2 + 2o_n + (n + 1)^2 = n^2 + 2n(n + 1) + (n + 1)^2$
   $$= 4n^2 + 4n + 1 = (2n + 1)^2.$$

8. $9 = 0^2 + 3 + 6 = 2^2 + 2^2 + 1$,
   $81 = 0^2 + 3 + 78 = 5^2 + 1^2 + 55$.

10. $[n(n - 1) + 1] + [n(n - 1) + 3] + \cdots + [n(n - 1) + (2n - 1)]$
   $$= n[n(n - 1)] + [1 + 3 + \cdots + (2n - 1)]$$
   $$= n^2(n - 1) + n^2$$
   $$= n^3.$$

11. (a) $[1 + 2 + 3 + \cdots + (n - 1) + n] + [(n - 1) + \cdots + 3 + 2 + 1]$
   $$= \frac{n(n + 1)}{2} + \frac{(n - 1)n}{2}$$
   $$= n^2.$$

   (c) $1 \cdot 2 + 2 \cdot 3 + 3 \cdot 4 + \cdots + n(n + 1)$
   $$= (1^2 + 1) + (2^2 + 2) + (3^2 + 3) + \cdots + (n^2 + n)$$
   $$= (1^2 + 2^2 + 3^2 + \cdots + n^2) + (1 + 2 + 3 + \cdots + n)$$
   $$= \frac{n(n + 1)(2n + 1)}{6} + \frac{n(n + 1)}{2}$$
   $$= \frac{n(n + 1)(n + 2)}{3}.$$

   (e) $1^3 + 3^3 + 5^3 + \cdots + (2n - 1)^3 = \left[\dfrac{(2n)(2n + 1)}{2}\right]^2 - [2^3 + 4^3 + 6^3 + \cdots + (2n)^3]$
   $$= n^2(2n + 1)^2 - 8(1^3 + 2^3 + 3^3 + \cdots + n^3)$$
   $$= n^2(2n + 1)^2 - 8\left[\frac{n(n + 1)}{2}\right]^2$$
   $$= n^2(2n^2 - 1).$$

12. (a) $(a + d) + (a + 2d) + (a + 3d) + \cdots + (a + nd)$
   $$= na + (1 + 2 + 3 + \cdots + n)d$$
   $$= na + \frac{n(n + 1)d}{2}$$
   $$= n\left[\frac{(a + d) + (a + nd)}{2}\right].$$

14. If n is odd, say $n = 2m + 1$, then
   $(t_1 + t_2) + (t_3 + t_4) + \cdots + (t_{n-2} + t_{n-1}) + t_n$
   $$= 2^2 + 4^2 + \cdots + (2m)^2 + \frac{(2m + 1)(2m + 2)}{2}$$
   $$= 4(1^2 + 2^2 + \cdots + m^2) + (2m + 1)(m + 1)$$
   $$= 4\frac{m(m + 1)(2m + 1)}{6} + (2m + 1)(m + 1)$$
   $$= \frac{n(n + 1)(n + 2)}{6}.$$

16. $T_n = \dfrac{n(n+1)(n+2)}{6} = \dfrac{n+1}{6}[n(n+2)] = \dfrac{n+1}{6}[n(n+1)+n]$

$$= \dfrac{n+1}{6}(2t_n + n).$$

17. $t_n + 2T_{n-1} = \dfrac{n(n+1)}{2} + 2\left[\dfrac{(n-1)n(n+1)}{6}\right]$

$$= n(n+1)\left[\dfrac{1}{2} + \dfrac{n-1}{3}\right]$$

$$= \dfrac{n(n+1)(2n+1)}{6}$$

$$= 1^2 + 2^2 + 3^2 + \cdots + n^2.$$

Section 3.3, p. 115

2. $(12,5,13)$, $(8,6,10)$.

4. $x^2 + (x+1)^2 = (x+2)^2$ implies that $(x-3)(x+1) = 0$.

5. (b) $(3,4,5)$, $(20,21,29)$, $(119,120,169)$, $(696,697,985)$, $(4059,4060,5741)$.

6. (a) $1 < 7/5 < 41/29$, $3/2 > 17/12 > 99/70$,

(b) $2 - (3/2)^2 = -1/4$, $2 - (7/5)^2 = 1/25$.

$2 - (17/12)^2 = -1/144$, $2 - (41/29)^2 = 1/841$.

7. (a) $x_1 = 2$, $x_2 = 12$, $x_3 = 70$, $x_4 = 408$, $x_5 = 2378$,

$y_1 = 3$, $y_2 = 17$, $y_3 = 99$, $y_4 = 577$, $y_5 = 3363$.

(c) $y_n^2 - 2x_n^2 = 1$ implies that $y_n/x_n = \sqrt{2 + (1/x_n)^2}$.

8. (a) 2, 1.5, $1.41666\ldots$, $1.41422\ldots$.

10. (a) $x_1 = 3$, $x_2 = 8$, $x_3 = 22$, $x_4 = 60$, $x_5 = 164$, $x_6 = 448$,

$y_1 = 5$, $y_2 = 14$, $y_3 = 38$, $y_4 = 104$, $y_5 = 284$, $y_6 = 776$.

(b) $5^2 - 3 \cdot 3^2 = -2$, $7^2 - 3 \cdot 4^2 = 1$.

(d) $\dfrac{5}{3}, \dfrac{7}{4}, \dfrac{19}{11}, \dfrac{26}{15}, \dfrac{71}{41}, \dfrac{97}{56}$.

11. (b) $\dfrac{265}{153} = \dfrac{1}{15}\left(26 - \dfrac{1}{51}\right) = \dfrac{1}{15}\sqrt{\left(26 - \dfrac{1}{51}\right)^2} < \dfrac{1}{15}\sqrt{26^2 - 1} = \sqrt{3}$.

12. (a) $\dfrac{26}{15}$. (b) $\dfrac{1351}{780}$.

13. (b) $\dfrac{105}{15} < \sqrt{50} < \dfrac{99}{14}$, $\dfrac{119}{15} < \sqrt{63} < \dfrac{127}{16}$, $\dfrac{147}{17} < \sqrt{75} < \dfrac{156}{18}$.

15. (b) $\dfrac{a}{AD} = \dfrac{c}{a}$ and $\dfrac{b}{BD} = \dfrac{c}{b}$ gives $a^2 = c \cdot AD$ and $b^2 = c \cdot BD$.

16. (b) $\dfrac{1}{2}a\left(\dfrac{a^2}{b}\right) + \dfrac{1}{2}ab = \dfrac{1}{2}c\left(\dfrac{ac}{b}\right)$ implies that $a^2 + b^2 = c^2$.

17. $\dfrac{1}{2}(a+b)^2 = \dfrac{1}{2}ab + \dfrac{1}{2}ab + \dfrac{1}{2}ac$ implies that $a^2 + b^2 = c^2$.

19. $\dfrac{a}{b} = \dfrac{AE}{BC} = \dfrac{AE}{BD} = \dfrac{AH}{HB} = \dfrac{OH - a}{b - OH}$.

22. (a) $(\sqrt{n})^2 = \left(\dfrac{n+1}{2}\right)^2 - \left(\dfrac{n-1}{2}\right)^2$. (b) $(2\sqrt{n})^2 = (n+1)^2 - (n-1)^2$.

23. If h is the hypotenuse of the n'th triangle, then $h^2 = (\sqrt{n})^2 + 1^2$.

Section 3.4, p. 125

1. Because $\dfrac{\text{area I}}{\text{area II}} = \dfrac{AB^2}{AC^2} = \dfrac{AB^2}{2AB^2} = \dfrac{1}{2}$,

    it follows that

    $$\text{area lune} = \text{area semicircle on } AC - \text{area II}$$
    $$= \text{area semicircle on } AC - 2 \text{ area I}$$
    $$= \text{area } \triangle ABC.$$

4. The equation $x^4 = (ay)^2 = a^2(2ax)$ gives $x^3 = 2a^3$.

Section 3.5, p. 135

1. If $AK < AG$, then with A as center and AK as radius draw a quarter circle KPL. Let FK perpendicular to AD intersect the quadratrix at F; join AF and extend it to meet the circumference BED at E and the circumference

    KPL at P. Reasoning as before, it follows that $FK = \text{arc } PK$; hence $\dfrac{1}{2}AK \cdot FK = \dfrac{1}{2}AK \cdot \text{arc } PK$, or area triangle

    $AKF = \text{area sector } AKP$, a contradiction.

2. (a) In polar coordinates, the defining property $\dfrac{\angle BAD}{\angle EAD} = \dfrac{AB}{FH}$ of the quadratrix becomes

    $$\dfrac{\pi/2}{\theta} = \dfrac{a}{r \sin \theta}.$$

    (b) $\displaystyle\lim_{\theta \to 0} r = \lim_{\theta \to 0} \left(\dfrac{2a}{\pi} \cdot \dfrac{\dfrac{1}{\sin \theta}}{\theta} \right) = \dfrac{2a}{\pi} \cdot 1.$

3. The similarity of triangles FBA and FBE implies that $\dfrac{x}{b} = \dfrac{a}{x}$.

4. First obtain a right triangle whose area is equal to that of the circle; then construct a rectangle equal in area to the triangle. Now, use the previous problem to construct a square whose area is that of the rectangle.

6. In polar coordinates, the defining property of the limaçon is $r - 2 \cos \theta = 1$.

    Since $r = \sqrt{x^2 + y^2}$ and $\cos \theta = \dfrac{x}{r}$, this becomes $r^2 - 2x = r$ or $(r^2 - 2x)^2 = r^2$.

Section 4.2, p. 158

1. Triangles DAB and CBA are congruent by the side-angle-side theorem; hence, $\angle DBA = \angle CAB = \angle CBA$, which contradicts Common Notion 5.

2. $\alpha + \beta = 180° = \beta + \gamma$ implies that $\alpha = \gamma$.

3. Because $\angle ABC < \angle ACD$ by the exterior angle theorem, it follows that $\angle ABC + \angle ACB < \angle ACD + \angle ACB = 180°$.

4. Triangle ABD is isosceles, hence $\angle ABD = \angle ADB$. Applying the exterior angle theorem, $\angle ABC > \angle ABD = \angle ADB > \angle ACB$.

5. Triangles GBC and DEF are congruent by the side-angle-side theorem; hence, $\angle C = \angle F = \angle BCG$, which contradicts Common Notion 5.

8. Because, in area, $ABE = DCF$, it follows that
$$\begin{aligned} ABCD = ABGD + BCG &= (ABE - DEG) + BCG \\ &= (DCF - DEG) + BCG \\ &= EGCF + BCG = EBCF. \end{aligned}$$

11. By Problem 8, it follows that, in area, $ABDE = ABKH = BLSR$, and $ACFG = ACJH = RSMC$. Hence, $ABDE + ACFG = BLSR + RSMC = BLMC$.

13. (a) 666 2/3 paces.　　(b) 250 paces.

14. (a) 9 people, 70 wen.

　　(b) 35 3/4 ounces gold, 29 1/4 ounces silver.

Section 4.3, p. 175

1. (a) $x^2 + 8x = 9$ implies that $x^2 + 8x + 16 = 25$; hence $(x + 4)^2 = 25$ or $x + 4 = 5$, yielding $x = 1$.

　　(d) If $3x^2 + 10x = 32$, then $9x^2 + 30x = 96$ and so $y^2 + 10y = 96$, where $y = 3x$. It follows that $y^2 + 10y + 25 = 121$ or $(y + 5)^2 = 121$. Hence $y + 5 = 11$, yielding $y = 6$ and $x = 2$.

2. If $x = y - 4$, then $x^2 + 8x = 9$ becomes $y^2 = 25$; hence $y = 5$ and $x = 1$.

3. $(x + a/2)^2 = (a/2)^2 + b^2$ implies that $x^2 + ax = b^2$.

5. 2.

6. 16.

7. 4 and 6.

8. 3 or 49/3.

9. 2 and 8.

Section 4.4, p. 188

1. (a) If $a|b$, then $b = ar$ for some r; hence, $bc = a(rc)$ or $a|bc$.

　　(c) If $ac|bc$, then $bc = arc$ for some r; hence, $b = ar$ or $a|b$.

　　(e) If $a|b$ and $c|d$, then $b = ar$ and $d = cs$ for some r,s; hence, $bd = (ac)(rs)$ or $ac|bd$.

3. $66 = 5 + 61 = 7 + 59 = 13 + 53 = 19 + 47 = 23 + 43 = 29 + 37$.

　　$96 = 7 + 89 = 13 + 83 = 17 + 79 = 23 + 73 = 29 + 67 = 37 + 59 = 43 + 53$.

4. $51 = 47 + 2 \cdot 2, 53 = 47 + 2 \cdot 3, 55 = 41 + 2 \cdot 7, 57 = 53 + 2 \cdot 2, 59 = 53 + 2 \cdot 3, \ldots$

5. 85.

6. If $n^3 - 1 = (n - 1)(n^2 + n + 1)$ is prime, then $n - 1 = 1$.

7. 11, 13, 15, 17; 101, 103, 105, 107, 109.

8. $6 = 17 - 11, 12 = 23 - 11, 18 = 29 - 11, 24 = 31 - 7, 30 = 37 - 7, 36 = 43 - 7, \ldots$

9. $29 = 23 + 19 + 17 - 13 - 11 - 7 - 5 + 3 + 2 + 1$,

　　$37 = 31 - 29 + 23 - 19 + 17 - 13 + 11 + 7 + 5 + 3 + 2 - 1$.

10. (a) If $m = 2k$, then $m^2 = 4k^2$; while if $m = 2k - 1$, then $m^2 = 4(k^2 + k) + 1$.

　　(c) If $m = 6k + 1$, then $m^2 = 12(3k^2 + k) + 1$; while if $m = 6k + 5$, then $m^2 = 12(3k^2 + 5k + 2) + 1$.

11. If $a = 3k$, then $3|a$; if $a = 3k + 1$, then $3|(a + 2)$; if $a = 3k + 2$, then $3|(a + 1)$. In any case, $3|a(a + 1)(a + 2)$.

12. If $a = 3k + 1$, then $a^2 - 1 = 3(3k^2 + 2k)$; while if $a = 3k + 2$, then $a^2 - 1 = 3(3k^2 + 4k + 1)$.

13. If $2|(a + 1)^2 - a^2$, then $2|(2a + 1)$; hence, $2|(2a + 1) - 2a$, or $2|1$, a contradiction.

15. $\gcd(143,227) = 1$, $\gcd(136,232) = 8$, $\gcd(272,1479) = 17$.

16. (a) $\gcd(56,72) = 8 = 4 \cdot 56 + (-3)72$, (b) $\gcd(119,272) = 17 = 7 \cdot 119 + (-3)272$.

17. If $d|a$ and $d|(a + 1)$, then $d|(a + 1) - a$, so $d|1$; hence, $d = \pm 1$.

18. Because both 3 and 8 divide $a(a + 1)(a + 1)(a + 2)$, with $\gcd(3,8) = 1$, it follows that $3 \cdot 8$ divides this product.

19. If $p \mid a^n$, with p prime, then $p \mid a$; hence $p^n \mid a^n$.

20. (b) $1234 = 2 \cdot 617$, $10140 = 2^2 \cdot 3 \cdot 5 \cdot 13^2$, $36,000 = 2^5 \cdot 3^2 \cdot 5^3$.

21. (a) 17 and 257.

    (b) 7, 31, 127, and 8191.

22. Because $3p = (a + 1)(a - 1)$, with p prime, either $3 \mid a + 1$ or $3 \mid a - 1$. But $3 \mid a + 1$ leads to a contradiction. If $a - 1 = 3k$ for some k, then $3p = (3k + 2)(3k)$ or $p = (3k + 2)k$; it follows that $k = 1$ and $p = 5$.

23. $6 = 13 - 7 = 19 - 13 = 29 - 23 = 37 - 31 = \ldots$

25. 71 and 13,859.

Section 4.5, p. 199

2. Because $AB = AD \sin \beta$, $AC = AD \sin \alpha$, $BD = AD \sin(90° - \beta)$ and $CD = AD \sin(90° - \alpha)$, Ptolemy's theorem becomes
$$AD \sin \alpha \cdot AD \sin(90° - \beta) = AD \sin \beta \cdot AD \sin(90° - \alpha) + BC \cdot AD.$$

3. Ptolemy's theorem implies that
$$AB \cdot PC = PA \cdot BC + PB \cdot AC,$$
where $AB = BC = AC$.

4. (b) $\pi \sim 3\dfrac{17}{120} = 3.14166. \ldots$ (c) $\sqrt{3} \sim 1.73205. \ldots$

6. Take $d = 0$ to get $k = \sqrt{s(s - a)(s - b)(s - c)}$.

8. $AX = XP = XD$.

Section 4.6, p. 214

1. (a) If r is the radius of the base of the cylinder and h is its height, then its surface area equals
$$s = 2\pi rh = \pi x^2, \text{ where}$$
$$\frac{h}{x} = \frac{x}{2r}.$$

   (c) If r is the radius of the base of the cone and s is its slant height, then
$$\frac{\frac{1}{2}(2\pi r)s}{\pi r^2} = \frac{s}{r}.$$

   (e) If r is the radius of the sphere, then its volume equals
$$V_s = \frac{4}{3}\pi r^3 = 4\frac{(\pi r^2)r}{3} = 4V_c.$$

2. If the sphere has a radius r, then

   (a) $V_c = (\pi r^2)(2r) = \dfrac{3}{2}\left(\dfrac{4}{3}\pi r^3\right) = \dfrac{3}{2}V_s.$

   (b) $A_c = (2\pi r)(2r) + 2(\pi r^2) = \dfrac{3}{2}(4\pi r^2) = \dfrac{3}{2}A_s.$

3. $AB + BF = DB + BF = DF = FC.$

6. The area of the "shoemaker's knife" is
$$A = \frac{\pi}{8}AB^2 - \frac{\pi}{8}AC^2 - \frac{\pi}{8}CB^2 = \frac{\pi}{8}(AB^2 - AC^2 - CB^2) = \frac{\pi}{4}PC^2.$$

7. Because $PRCS$ is a parallelogram, its diagonals PC and RS bisect each other.

8. Because CD is bisected at O, it follows that
$$AD^2 + AC^2 = 2(CO^2 + AO^2) \text{ and } PQ = AO + OD = AD.$$
Therefore
$$AB^2 + CD^2 = 4(AO^2 + CO^2) = 2(AD^2 + CO^2) = 2(PQ^2 + AC^2).$$
This implies that the area of the "salt cellar" is
$$A = \frac{\pi}{8}AB^2 + \frac{\pi}{8}CD^2 - 2\left(\frac{\pi}{8}AC\right)^2 = \frac{\pi}{4}PQ^2.$$

9. Note that $a = \dfrac{1}{2}\displaystyle\int_0^{2\pi} r^2 d\theta = \dfrac{1}{2}\displaystyle\int_0^{2\pi} (a\theta)^2 d\theta = \dfrac{4a^2\pi^3}{3}$.

10. Let $\angle POA = \beta$, so that $P = (a\beta, \beta)$. Then $Q = \left(\dfrac{a\beta}{3}, \beta\right)$ and $R = \left(\dfrac{2a\beta}{3}, \beta\right)$.

The circles with center O and radii OQ and OR meet the spiral in the points
$$V = \left(\frac{a\beta}{3}, \frac{\beta}{3}\right) \text{ and } U = \left(\frac{2a\beta}{3}, \frac{2\beta}{3}\right). \text{ Hence, } \angle VOA = \frac{\beta}{3}.$$

11. The point $P = \left(\dfrac{\pi a}{2}, \dfrac{\pi}{2}\right)$, while $O = (0,0)$, hence $OP = \dfrac{\pi a}{2}$.

12. In polar coordinates, the equation of the tangent to the spiral $r = a\theta$ at the point $A = (2\pi a, 2\pi)$ is
$$r(2\pi\cos\theta - \sin\theta) = 4\pi^2 a.$$
The tangent intersects the line $\theta = \dfrac{3\pi}{2}$ at the point $B = \left(4\pi^2 a, \dfrac{3\pi}{2}\right)$.

Section 5.3, p. 236

1. 15, 5 and 25.

2. 30, 25 and 35.

5. If $a = 3$, then the squares are $\left(\dfrac{17}{2}\right)^2$ and $\left(\dfrac{23}{2}\right)^2$.

6. If $12 + x = (x - 4)^2$, then the number is $\dfrac{35}{4}$.

7. If $x^2 + 4x + 2 = (x - 2)^2$, then the numbers are $\dfrac{1}{4}$ and $\dfrac{5}{4}$.

8. If $10x + 54 = 64$, then the numbers are 1, 7, 9.

9. If $x + (4x + 4)^2 = (4x - 5)^2$, then the numbers are 9, 328, and 73.

10. 8 and 2.

11. If $8x(x^2 - 1) + (x^2 - 1) = (2x - 1)^3$, then the numbers are $\dfrac{112}{13}$ and $\dfrac{27}{169}$.

12. If $(x^2 + 4) - 4x = 64 = 4^3$, then the triangle has sides 40, 96, and 104.

14. (a) $x = 20 + 9t$, $y = -15 - 7t$.
    (c) $x = 176 + 35t$, $y = -1111 - 221t$.

15. (a) $x = 1$, $y = 6$.
    (c) No positive solutions.

17. 28 pieces is one answer.

19. One answer is 1 man, 5 women, and 14 children.

20. 56 and 44.

21. 59 is one answer.

22. 119 is one answer.

23. 1103 is one answer.

Section 5.4, p. 242

1. (a) Because $r(ar^2 + brs + cs^2) = -ds^3$, it follows that $r \mid ds^3$; hence, $r \mid d$.

2. (a) $-\dfrac{3}{2}$.

   (c) $-\dfrac{1}{3}$, $-\dfrac{1}{2}$, 1.

   (e) No rational solutions.

5. (a) The equation $8x^3 - 6x - 1 = 0$ has no rational roots, hence $\cos 20°$ is not a constructible real number; it follows that an angle of $20°$ is not constructible.

Section 6.2, p. 260

1. 18 and 32.

2. $10\dfrac{1}{2}$.

3. The system $x + 23 = 2y$
   $$y + 23 = 3z$$
   $$z + 23 = 4x$$
   has $x = 9$, $y = 16$, $z = 13$ as a solution.

4. $(2n + 5)^2 = (2n + 3)^2 + ((2n + 3)^2 - (2n + 1)^2) + 8$.

5. (a) Let $x = 5$ and y be any square such that $x + y$ and $x - y$ are also squares.

6. (a) If a^2, b^2, c^2 are three squares in arithmetic progression, with common difference d, then
   $$x = \frac{2b^2}{d} \text{ is a solution.}$$

   (b) $3^2 + 4^2 + 12^2 = 13^2$.

8. (b) $481 = 15^2 + 16^2 = 20^2 + 9^2$.

9. (a) If $c^2 + d^2 = k^2$, then $(a^2 + b^2)(c^2 + d^2) = u^2 + v^2$ implies that
   $$a^2 + b^2 = \left(\frac{u}{k}\right)^2 + \left(\frac{v}{k}\right)^2.$$

   (b) $61 = \left(\dfrac{39}{5}\right)^2 + \left(\dfrac{2}{5}\right)^2$.

10. If $u = 1$, then $t = 47$; hence, the amounts would be 33, 13, and 1.

14. $x = \frac{1}{2}\left(a + \sqrt{a^2 - 4\,(ca)/b}\right) = 8$, $y = \frac{1}{2}(a - \sqrt{a^2 - 4\,(ca)/b}) = 2$.

15. $x = \frac{1}{2}\left[(a - 1) + \sqrt{(2b + 1) - a^2}\right] = 7$,
    $y = \frac{1}{2}\left[(a + 1) - \sqrt{(2b + 1) - a^2}\right] = 3$.

16. $x = (ab - c)/(b + 1) = 4$, $y = (a + c)/(b + 1) = 8$.

17. $x = \sqrt{abc} = 12$, $y = \sqrt{(ab)/c} = 3$.

Section 6.3, p. 268

1. $50 = F_4 + F_7 + F_9$, $\quad 75 = F_3 + F_5 + F_7 + F_{10}$.

2. (b) $F_2 + F_4 + \cdots + F_{2n} = (F_1 + F_2 + F_3 + \cdots + F_{2n}) - (F_1 + F_3 + \cdots + F_{2n-1})$
   $$= (F_{2n+2} - F_{2n}) - 1 = F_{2n+1} - 1.$$

3. $7|F_8$, $11|F_{10}$.

4. If $d = \gcd(F_n, F_{n-1})$, then $d|(F_{n+1}F_{n-1} - F_n^2)$ or $d|(-1)^n$.

5. $\gcd(F_{15}, F_{20}) = 5$.

6. If $F_n|F_m$, then $\gcd(F_n, F_m) = F_n$; by problem 5, $n = \gcd(n, m)$, hence $n|m$.

7. (a) If $2|F_n$, then $F_3|F_n$ and so $3|n$.

   (c) If $4|F_n$, then $3|n$; since $2|n$ also it follows that $6|n$.

Section 7.3, p. 299

1. (a) 1, 2, 3.   (c) $-2, -2 + 4\sqrt{2}, -2 - 4\sqrt{2}$.

3. (a) $2\sqrt[3]{4} - 2\sqrt[3]{2}$ (c) $2 + \sqrt[3]{25} - \sqrt[3]{5}$.   (e) 8.

4. $-2, -2 + \sqrt{11}, -2 - \sqrt{11}$.

5. 6, 4.

6. $-48, 52$.

7. $4 + \sqrt{16 - \sqrt[3]{16}}, 4 - \sqrt{16 - \sqrt[3]{16}}$.

8. $14 - 2\sqrt{11} - 2\sqrt{13 - 2\sqrt{11}}$.

9. $x = -4 + \sqrt[3]{190 + \sqrt{35757}} + \sqrt[3]{190 - \sqrt{35757}}$.

10. $x = \dfrac{1}{3} + \sqrt[3]{\dfrac{\sqrt{2107}}{3} + \dfrac{1}{27}} - \sqrt[3]{\dfrac{\sqrt{2107}}{3} - \dfrac{1}{27}}$.

11. $x = -\dfrac{7}{3} + \sqrt[3]{\dfrac{413}{27} + \dfrac{\sqrt{1960}}{27}} + \sqrt[3]{\dfrac{413}{27} - \dfrac{\sqrt{1960}}{27}}$.

12. If $x = \dfrac{27}{y} - y$, then $x^3 + 81x = 702$ becomes $y^6 + 702y^3 = 27^3$, with $y = 3$ as one solution.

   If $y = \dfrac{2}{z} - z$, then $y^3 + 6y = 7$ becomes $z^6 + 7z^3 = 8$, with $z = 1$ as one solution.

13. 6.

14. (a) -6.

   (c) $y^3 = \dfrac{19}{13}y - \dfrac{56}{27}$ has $y = -\dfrac{1}{3}$ as one solution.

Section 7.4, p. 307

1. (a) $\sqrt{\dfrac{3}{2}} \pm \sqrt{\sqrt{6} - \dfrac{3}{2}}, -\sqrt{\dfrac{3}{2}} \pm \sqrt{-\sqrt{6} - \dfrac{3}{2}}$.   (c) $\dfrac{-3 \pm \sqrt{5}}{2}, \dfrac{3 \pm \sqrt{-7}}{2}$.

2. $\dfrac{-1 \pm \sqrt{-3}}{2}, \dfrac{-3 \pm \sqrt{-7}}{2}$.   3. $1, -1, -4, -4$.

5. $10\left(\sqrt{\sqrt{5} + \dfrac{5}{4}} - \sqrt{\dfrac{5}{4}}\right)$.   6. 2.

Section 8.2, p. 343

1. The triangles ABC and DBE are similar, so that $\dfrac{BA}{BC} = \dfrac{BC}{BE}$ or $\dfrac{1}{a} = \dfrac{b}{BE}$;

   hence, $BE = ab$.

4. The equation $x^4 + x^2 + 3x + 1 = (x + 1)(x^3 - x^2 + 2x + 1) = 0$ has no positive roots, which implies that $x^3 - x^2 + 2x + 1$ has no positive roots.

5. The equation $x^6 + x^5 + 2x^4 + x^3 - 1 = (x + 1)(x^5 + 2x^3 - x^2 + x - 1) = 0$ has just one positive root.

6. Because $f(x) = x^{2n} - 1$ has one variation in sign and $f(-x) = x^{2n} - 1$ also has one sign variation, the equation $f(x) = 0$ cannot have more than one positive or more than one negative root. But 1 and -1 are clearly roots, so that there are $2n - 2$ complex roots.

7. (a) $f(x) = x^3 + 3x + 7$ has no variations in sign and therefore $f(x) = 0$ has no positive root. Now $-f(-x) = x^3 + 3x - 7$ has just one sign variation, hence $f(x) = 0$ may have one negative root. It must therefore have two complex roots.

8. (a) Because there are no variations of sign in either $f(x)$ or $f(-x)$, there are no real roots of $f(x) = 0$.

10. (a) $x^4 - 3x^2 + 6x - 2 = (x^2 + 2x - 1)(x^2 - 2x + 2)$, hence the roots of the quartic are $-1 \pm \sqrt{2}, 1 \pm \sqrt{-1}$.
    (b) $x^4 - 2x^2 - 8x - 3 = (x^2 - 2x - 1)(x^2 + 2x + 3)$, hence the roots of the quartic are $1 \pm \sqrt{2}, -1 \pm \sqrt{-2}$.

Section 8.4, p. 389

1. $\dfrac{\pi}{4} = \left(1 - \dfrac{1}{3}\right) + \left(\dfrac{1}{5} - \dfrac{1}{7}\right) + \cdots + \left(\dfrac{1}{2n - 1} - \dfrac{1}{2n + 1}\right) + \cdots$

   $= \dfrac{2}{1\cdot3} + \dfrac{2}{5\cdot7} + \cdots + \dfrac{2}{(2n - 1)(2n + 1)} + \cdots$

   $= \dfrac{2}{(2 - 1)(2 + 1)} + \dfrac{2}{(6 - 1)(6 + 1)} + \cdots + \dfrac{2}{(4n - 3)(4n - 1)} + \cdots$

   $= 2\left[\dfrac{1}{2^2 - 1} + \dfrac{1}{6^2 - 1} + \cdots + \dfrac{1}{(2(2n - 1)^2 - 1)} + \cdots\right].$

2. $2P_n - (n - 1)P_{n-1} = 2n! - (n - 1)(n - 1)!$

   $= (n - 1)![2n - (n - 1)]$

   $= (n - 1)![n + 1] = n! + (n - 1)! = P_n + P_{n-1}.$

3. $6 = 1 + 2 \cdot 2 + 1 = (1 + \sqrt{-3}) + 2\sqrt{(1 + \sqrt{-3})(1 - \sqrt{-3})} + (1 - \sqrt{-3})$

   $= (\sqrt{1 + \sqrt{-3}} + \sqrt{1 - \sqrt{-3}})^2.$

5. $\log\left(\dfrac{1 + x}{1 - x}\right) = \log(1 + x) - \log(1 - x)$

   $= \left(x - \dfrac{x^2}{2} + \dfrac{x^3}{3} - \dfrac{x^4}{4} + \cdots\right) - \left(-x - \dfrac{x^2}{2} - \dfrac{x^3}{3} - \dfrac{x^4}{4} - \cdots\right)$

   $= 2\left(x + \dfrac{x^3}{3} + \dfrac{x^5}{5} + \cdots\right).$

8. (a) $(1 + x)^{-1} = 1 + (-1)x + \dfrac{(-1)(-2)}{2!}x^2 + \dfrac{(-1)(-2)(-3)}{3!}x^3 + \cdots$

   $= 1 - x + \dfrac{(-1)^2 2!}{2!}x^2 + \dfrac{(-1)^3 3!}{3!}x^3 + \cdots$

   $= 1 - x + x^2 - x^3 + \cdots.$

7. Any perfect number can be expressed as the sum of consecutive odd cubes; in fact,
$$2^{2k}(2^{2k+1} - 1) = 1^3 + 3^3 + 5^3 + \cdots + (2^{k+1} - 1)^3 \text{ for all } k \geq 0.$$

8. Because $n^2 = 2^{2k-2}p^2$, where $p = 2^k - 1$ is prime,
$$\sigma(n^2) = (1 + 2 + 2^2 + \cdots + 2^{2k-2})(1 + p + p^2)$$
$$= (2^{2k-1} - 1)(2^{2k} - 2^{k+1} + 2^k + 1).$$
Thus, $\sigma(n^2) + 1 = 2^k N$.

9. If n and m form an amicable pair, then $\sigma(n) = n + m = \sigma(m)$.
The relation $m = \sigma(n) - n$ yields $\sigma(\sigma(n) - n) = \sigma(n)$. It follows that $\sigma(s(n)) = s(n) + n$, or $s(s(n)) = n$. Conversely, if this condition holds, then n and $s(n) = \sigma(n) - n$ form an amicable pair.

10. $\sigma(1184) = (1 + 2 + 2^2 + 2^3 + 2^4 + 2^5)(1 + 37)$
$$= 63 \cdot 38 = 2394 = 1184 + 1210.$$

11. Assume to the contrary that p and n form an amicable pair. Then the condition
$$1 + p = \sigma(p) = p + n$$
implies that $n = 1$. But $\sigma(1) = 1 \neq p + 1 = \sigma(p)$.

Section 10.2, p. 486

2. (a) Because $\dfrac{2n^2}{n^2 - 1} > 2$, it follows that $\dfrac{n}{n - 1} + \dfrac{n}{n + 1} > 2$, whence

$$\dfrac{n}{n - 1} + 1 + \dfrac{n}{n + 1} > 3.$$

3. $e^{-\pi/2} = e^{\pi i^2/2} = (e^{\pi i/2})^i$.

5. (b) $a_n(x - r_1)(x - r_2) \cdots (x - r_n) =$

$$a_n r_1 r_2 \cdots r_n (-1)^n \left(1 - \dfrac{x}{r_1}\right)\left(1 - \dfrac{x}{r_2}\right) \cdots \left(1 - \dfrac{x}{r_n}\right).$$

7. (c) $\left(\dfrac{1 \cdot 3}{2 \cdot 2}\right)\left(\dfrac{3 \cdot 5}{4 \cdot 4}\right)\left(\dfrac{5 \cdot 7}{6 \cdot 6}\right) \cdots \left(\dfrac{(2n - 1)(2n + 1)}{2n \cdot 2n}\right)$

$$= \dfrac{1}{(2^n n!)^2} \left(\dfrac{1 \cdot 2^2 \cdot 3}{2^2 \cdot 1 \cdot 1}\right)\left(\dfrac{3 \cdot 4^2 \cdot 5}{2^2 \cdot 2 \cdot 2}\right) \cdots \left(\dfrac{(2n - 1)(2n)^2(2n + 1)}{2^n \cdot n \cdot n}\right)$$

$$= \dfrac{1}{(2^n n!)^2} \dfrac{(2n)!(2n)!(2n + 1)}{(2^n n!)^2}.$$

Section 10.3, p. 504

1. (a) Because $a \equiv b \pmod{n}$, it follows that $a - b = kn$ for some k.
If $m | n$, then $n = rm$ for some r. Hence $a - b = k(rm) = (kr)m$, and so $a \equiv b \pmod{m}$.

(c) Suppose $a \equiv b \pmod{n}$, so that $a - b = kn$ for some k. If $a = rd$, $b = sd$ and $n = td$ for some r, s, t, then $(rd) - (sd) = k(td)$ or $r - s = kt$. Hence, $r \equiv s \pmod{t}$; that is, $a/d \equiv b/d \pmod{n/d}$.

2. $5^2 \equiv 1^2 \pmod 8$, but $5 \not\equiv 1 \pmod 8$.

4. If $a \equiv 1 \pmod 8$, then $a^2 \equiv 1^2 \equiv 1 \pmod 8$; if $a \equiv 3 \pmod 8$, then $a^2 \equiv 9 \equiv 1 \pmod 8$; if $a \equiv 5 \pmod 8$, then $a^2 \equiv 25 \equiv 1 \pmod 8$; if $a \equiv 7 \pmod 8$, then $a^2 \equiv 49 \equiv 1 \pmod 8$.

5. (a) If $a \equiv 0 \pmod 3$, then $a^2 \equiv 0 \pmod 3$; if $a \equiv 1 \pmod 3$, then $a^2 \equiv 1 \pmod 3$; if $a \equiv 2 \pmod 3$, then $a^2 \equiv 1 \pmod 3$.

6. $5^{36} - 1 = (25)^{18} - 1 \equiv (-1)^{18} - 1 = 1 - 1 = 0 \pmod{13}$, hence $13 | 5^{36} - 1$.

7. (a) $10^{49} + 5^3 \equiv 10(100)^{24} + (-2)^3 \equiv 3(2)^{24} - 1 \equiv 3(8)^8 - 1 \equiv 3 - 1 \equiv 2 \pmod 7$.

8. $2^{23} - 1 = 2^5(2^9)^2 - 1 \equiv (-15)5^2 - 1 \equiv (-3)5^3 - 1 \equiv (-3)31 - 1 \equiv 0 \pmod{47}$.

9. (c) $5^{2n} + 3 \cdot 2^{5n-2} \equiv 4^n + 3 \cdot 2^{5(n-1)+3} \equiv 4^n + 3 \cdot 4^{n-1} \equiv 7 \cdot 4^{n-1} \equiv 0 \pmod 7$.

   (e) $2^{5n+1} + 5^{n+2} = 2(32)^n + 25(5)^n \equiv 2 \cdot 5^n + (-2)5^n \equiv 0 \pmod{27}$.

10. If k is of the form $4n + 1$, then
$$2^{k-1}(2^k - 1) = 2^{4n}(2 \cdot 2^{4n} - 1) = 4^{2n}(2 \cdot 4^{2n} - 1) \equiv 4(2 \cdot 4 - 1) \equiv 4 \pmod 6.$$
A similar argument holds if k is of the form $4n + 3$.

13. (a) $9 | 113{,}058$ since $9 | (1 + 1 + 3 + 0 + 5 + 8)$.

14. $x = 9$.

16. (a) $x \equiv 6, 13$, and $20 \pmod{21}$.

    (c) No solutions, since gcd $(13,52) \nmid 27$.

Section 11.1, p. 525

1. Bolyai's assumption that there exists a circle passing through any three noncollinear points is equivalent to assuming Euclid's parallel postulate.

2. Legendre's assumption that every line through a point in the interior of an angle must meet one of the sides is equivalent to assuming Euclid's parallel postulate.

4. (a) Because triangles BAE and CDE are congruent, $BE = CE$. This implies that triangles BFE and CFE are congruent, whence $\angle BFE = \angle CFE = 90°$.

   (c) The base and summit have a common perpendicular, hence are parallel by Euclid's Proposition 27 on alternate interior angles.

5. (a) Because $ABED$ is a Saccheri quadrilateral, it follows that $\angle ADC > \angle ADE = \angle BED > \angle BCD$.

   (c) Either $AD < BC$, $AD > BC$ or $AD = BC$. If $AD < BC$, then $\angle C < \angle D$ by part (a), a contradiction; if $AD > BC$, then $\angle C > \angle D$, a contradiction. Thus, $AD = BC$.

6. Triangles ACD and XZW are congruent, so that $\angle CAD = \angle ZXW$; hence, $\angle BAC = \angle YXZ$. This implies that triangles ABC and XYZ are congruent by the side-angle-side theorem.

8. (b) Consider a triangle PQR which contains no right angle. Because it cannot have more than one obtuse angle, it contains at least two acute angles, say at vertices P and Q. Drop a perpendicular RS from R to the side PQ. Apply part (a) to the resulting right triangles PRS and QRS.

10. Use the figure in Problem 4. Since $\angle B < \angle BFE = 90°$, Problem 5(b) implies that in quadrilateral $AEFB$, one has $AB > EF$; similarly, in quadrilateral $DEFC$, it follows that $CD > EF$.

11. If P lies outside of triangle ADC, point D lies on the extension of AB beyond B, and point E lies between A and C, then one cannot obtain $AB = AC$ from the earlier equations by either addition or subtraction.

Section 11.4, p. 579

2. (c) If $iq = -b + ai - dj + ck = -b + ai + dj - ck = qi$, then $-d = d$ and $c = -c$, so that $c = d = 0$.

7. (b) Suppose $\begin{bmatrix} a & b \\ c & d \end{bmatrix}^2 = \begin{bmatrix} a^2 + bc & ab + bd \\ ac + dc & cd + d^2 \end{bmatrix} = \begin{bmatrix} 0 & 1 \\ 0 & 0 \end{bmatrix}$.

   Then $(a + d)c = 0$. Either $c = 0$, whence $a = d = 0$, which implies that $0 = 1$; or else $a = -d$, again yielding $0 = 1$.

   (d) Use induction on n. If the result holds for $n = k$, then
$$\begin{bmatrix} 1 & 1 \\ 0 & 1 \end{bmatrix}^{k+1} = \begin{bmatrix} 1 & 1 \\ 0 & 1 \end{bmatrix}^k \begin{bmatrix} 1 & 1 \\ 0 & 1 \end{bmatrix} = \begin{bmatrix} 1 & k \\ 0 & 1 \end{bmatrix} \begin{bmatrix} 1 & 1 \\ 0 & 1 \end{bmatrix} = \begin{bmatrix} 1 & k+1 \\ 0 & 1 \end{bmatrix}.$$

8. (c) Because $AA^{-1} = I = A^{-1}A$, taking transposes gives
$$(A^{-1})^t A^t = I = A^t (A^{-1})^t$$
and so $(A^{-1})^t$ is the inverse of A^t.

Section 12.2, p. 610

1. (a) $\{2, 2^2, 2^3, \ldots\} \sim N$ using the function f defined by $f(2^n) = n$.
   (b) $\{5, 10, 15, \ldots\} \sim N$ using the function f defined by $f(5n) = n$.
2. If Z_o^+ denotes the set of positive odd integers, then $Z_o^+ \sim N$ using the function $f(2n - 1) = n$; similarly, if Z_o^- denotes the set of negative odd integers, then $Z_o^- \sim N$ using the function $g(-(2n - 1)) = n$. Now, $Z_o = Z_o^+ \cup Z_o^-$, hence Z_0 is countable.
4. The prime numbers form an infinite subset of the countable set N and so are countable themselves.
5. Kf A is countable, then $A \times \{b\}$ is countable for each $b \in B$. Hence, $A \times B = \cup_{b \in B}(A \times \{b\})$ is a countable union of countable sets, which makes $A \times B$ countable.
6. If t is a right triangle having sides of integral lengths a,b,c, define the function f by $f(t) = (a,b,c)$. Then $S \sim f(S) \subseteq Z \times Z \times Z$. Since $Z \times Z \times Z$ is countable, so is the set $f(S)$ and, in its turn, the set S.
10. (a) The set of numbers of the form $m/2^n$ is an infinite subset of the countable set Q, hence is a countable set.
    (c) Let ℓ be the line through the origin and the point (r,s) where $r, s \in Q$. Then ℓ has a rational slope t. Identify ℓ with the ordered triple $(r,s,t) \in Q \times Q \times Q$.
    (e) Consider any infinite set S of nonoverlapping intervals. For each interval $I \in S$, select a single rational number (in lowest terms) r_i in I. A one-to-one function $f : S \to Q$ is defined by letting $f(I) = r_i$. Then $S \sim f(S) \subseteq Q$.
11. Note that $L \sim f(L) = (0,1]$, where the interval $(0,1]$ is uncountable.

Section 12.3, p. 628

2. By Cantor's theorem, $o(P(R)) > o(R) = c > \aleph_0$.
3. (a) Suppose that $A \subseteq B$. If $C \in P(A)$, then $C \subseteq A \subseteq B$; hence, $C \subseteq P(B)$. This shows that $P(A) \subseteq P(B)$.
   (c) Let $C \in P(A \cap B)$, so that $C \subseteq A \cap B$. Then $C \subseteq A$, which means $C \in P(A)$; and $C \subseteq B$, which means $C \in P(B)$. Thus $C \in P(A) \cap P(B)$, implying that $P(A \cap B) \subseteq P(A) \cap P(B)$.
4. If A and B are in one-to-one correspondence via the mapping $f : A \to B$, then $f^* : P(A) \to P(B)$ will also be one-to-one. For suppose that $C \neq C'$, where $C, C' \in P(A)$; say, there is some element $x \in C$ with $x \notin C'$. Then $f(x) \in f(C)$, but $f(x) \notin f(C')$; for if $f(x) = f(x')$ where $x' \in C'$, then the one-to-one nature of f would imply that $x = x' \in C'$, a contradiction. Thus, $f^*(C) \neq f^*(C')$.
5. (a) Suppose to the contrary that the set of countable subsets of N can be arranged in a sequence $A_1, A_2, A_3, \cdots$. By a diagonal argument, construct a subset A of N which is different from each A_n.
6. (a) If $x = o(A)$, then $x + 0 = o(A \cup \varnothing) = o(A) = x$. Also, $x \cdot 1 = o(A \times \{1\}) = o(A) = x$.
   (c) $\aleph_0 \cdot \aleph_0 = o(N \times N) = \aleph_0$.
   (e) $c + c = o([0,\infty) \cup (-\infty,0)) = o(R) = c$.
   (g) $c \cdot c = o(R \times R) = o(R) = c$.
8. Because the natural number n is defined in no more than 36 words, all of them taken from our lexicon, n is contained in the set S. On the other hand, on account of its definition, n cannot be contained in S. This leads to a formal contradiction.
9. Because the sentence contains a finite number of words, the real number r which it describes will be in the set S. But, owing to its very definition, r differs from each member of S in at least one decimal place; hence this number is not in S.

10. The number $\sqrt{2}^{\sqrt{2}}$ is either rational or irrational. If it is rational, then we have our example. If it is irrational, put $x = \sqrt{2}^{\sqrt{2}}$ and $y = \sqrt{2}$, so that $x^y = (\sqrt{2}^{\sqrt{2}})^{\sqrt{2}} = (\sqrt{2})^2 = 2$, which is certainly rational.

11. (a) To decide on the value of n, we must either find a sequence of digits 0123456789 in π or demonstrate that no such sequence can exist. At present, there is no method that would enable us to do either. From the intuitionist point of view, only constructible entities exist, so they would accept no statement regarding n.

Section 13.2, p. 646

1. (a) $d(x,y) = 0$ does not necessarily imply that $x = y$.
   (b) $x = y$ does not necessarily imply that $d(x,y) = 0$.
2. The only way in which $d(x,y) \le d(x,z) + d(z,y)$ could fail to hold is if $d(x,y) = 1$ and $d(x,z) = d(z,y) = 0$. This is impossible, because it would imply that $x = z = y$, whence $d(x,y) = 0$.
5. Taking $x = y$ gives $0 = d(y,y) \le d(z,y) + d(z,y)$; hence, $d(z,y) \ge 0$ for all z,y in X. If $z = y$, then $d(x,y) \le d(y,x) + d(y,y) = d(y,x)$. Thus, $d(x,y) \le d(y,x)$ for all x,y in X. Interchanging x and y yields $d(y,x) \le d(x,y)$, implying that $d(x,y) = d(y,x)$.
7. Suppose that $\epsilon = d(x,y) > 0$. Now $d(x_n,x) < \epsilon/2$ and $d(x_n,y) < \epsilon/2$ when $n \ge n_0$ for some n_0. Thus $d(x,y) \le d(x,x_n) + d(x_n,y) < \epsilon/2 + \epsilon/2 = \epsilon$, which is a contradiction. Hence $d(x,y) = 0$ and so $x = y$.
8. $[0, \infty)$ is a closed set.
9. The sets $\{(x,y) \mid x = 1\}$ and $\{(x,y) \mid x^2 + y^2 \le 1\}$ are both closed.

Index

Early Modern Period (Seventeenth and Eighteenth Centuries)

MATHEMATICAL		GENERAL	
1588–1648	Marin Mersenne	1607	Jamestown founded
1596–1650	René Descartes	1608	Telescope invented
1601–1665	Pierre de Fermat	1611	King James Bible
1616–1703	John Wallis	1618–1648	Thirty Years' War
1623–1662	Blaise Pascal	1619	Savilian Professorship (Oxford)
1629–1695	Christiaan Huygens	1620	Landing of Pilgrims
1630–1677	Isaac Barrow	1636	Harvard College founded
1635–1703	Robert Hooke	1642–1649	English Civil War
1642–1727	Isaac Newton	1658	Death of Cromwell
1646–1716	Gottfried Leibniz	1662	Royal Society of London
1654–1705	James Bernoulli	1663	Lucasian Professorship (Cambridge)
1656–1742	Edmond Halley	1666	Académie des Sciences
1661–1704	Marquis de l'Hospital	1682	*Acta Eruditorum*
1667–1733	Girolamo Saccheri	1683	Turks defeated at Vienna
1667–1748	John Bernoulli	1694–1778	Voltaire
1667–1754	Abraham DeMoivre	1712–1786	Frederick the Great
1685–1731	Brook Taylor	1737–1794	Edward Gibbon
1690–1764	Christian Goldbach	1751	Diderot's *Encyclopédie*
1707–1783	Leonhard Euler	1769	James Watt's steam engine
1717–1783	Jean le Rond d'Alembert	1769–1821	Napoleon Bonaparte
1718–1799	Maria Agnesi	1770–1827	Ludwig van Beethoven
1728–1777	Johann Lambert	1776–1783	American Revolution
1736–1813	Joseph Louis Lagrange	1789	Washington President
1749–1827	Pierre Simon Laplace	1789	French Revolution
1752–1833	Adrien-Marie Legendre	1798	Eli Whitney's cotton gin
1777–1855	Carl Friedrich Gauss	1798	Ecole Normale founded
1781–1848	Bernhard Bolzano	1798	Bonaparte in Egypt